Confocal Microscopy for Biologists

Confocal Microscopy for Biologists

Alan R. Hibbs

BIOCON
Melbourne, Australia

Kluwer Academic / Plenum Publishers
New York, Boston, Dordrecht, London, Moscow

Library of Congress Cataloging-in-Publication Data

Hibbs, Alan R.
Confocal microscopy for biologists/Alan R. Hibbs.
p. cm.
Includes bibliographical references and index.
ISBN 0-306-48468-4
1. Confocal microscopy. I. Title.

QH224.H53 2004
570.'28'2—dc22

2004044172

ISBN: 0-306-48468-4 (hardback)
ISBN: 0-306-48565-6 (eBook)

233 Spring Street, New York, New York 10013

http://www.wkap.nl/

10 9 8 7 6 5 4 3 2 1

A C.I.P. record for this book is available from the Library of Congress

Printed in the United States of America

In memory of my Father

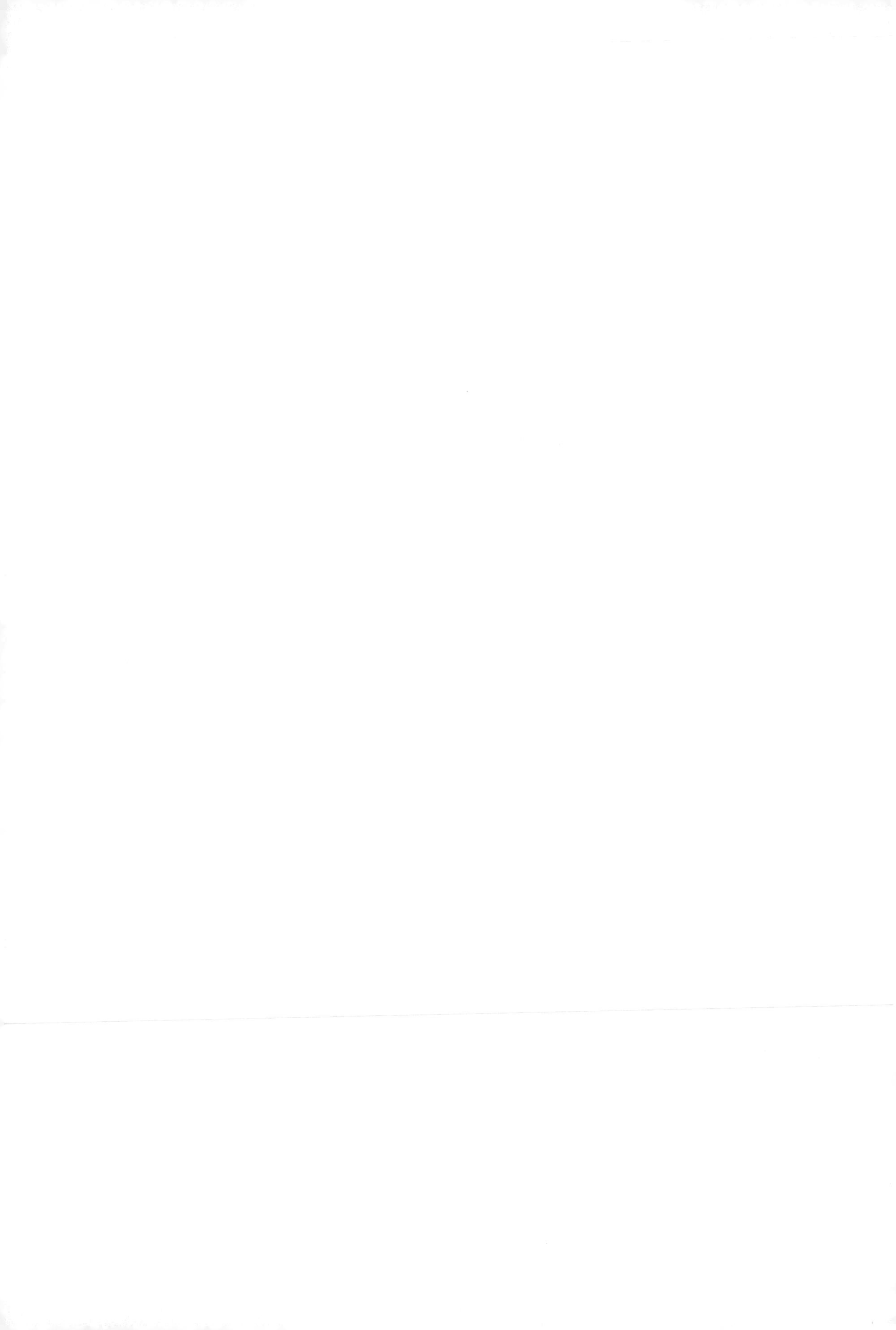

Preface

There has been a great upsurge in interest in light microscopy in recent years due to the advent of a number of significant advances in microscopy, one of the most important of which is confocal microscopy. Confocal microscopy has now become an important research tool, with a large number of new fluorescent dyes becoming available in the past few years, for probing your pet structure or molecule within fixed or living cell or tissue samples. Many of the people interested in using confocal microscopy to further their research do not have a background in microscopy or even cell biology and so not only do they find considerable difficulty in obtaining satisfactory results with a confocal microscope, but they may be mislead by how data is being presented.

This book is intended to teach you the basic concepts of microscopy, fluorescence, digital imaging and the principles of confocal microscopy so that you may take full advantage of the excellent confocal microscopes now available.

This book is also an excellent reference source for information related to confocal microscopy for both beginners and the more advanced users. For example, do you need to know the optimal pinhole size for a 63x 1.4 NA lens? Do you need to know the fluorescence emission spectrum of Alexa 568? Access to the wealth of practical information in this book is made easier by using both the detailed index and the extensive glossary.

The main emphasis in this book is on the laser scanning confocal microscopes used in biology. However, a variety of related imaging technologies (multi-photon microscopy, Nipkow disk confocal microscopy, CCD cameras etc) are also discussed.

This book has been written using a wide variety of sources of information, including personal discussions with many people involved in confocal microscopy, a number of important books on confocal microscopy and articles in a variety of scientific journals. Specific contributions (including contributed images) from individuals or organizations are acknowledged where appropriate, but I would like to take this opportunity to acknowledge the enormous contribution from a wide range of people that have made the writing of this book possible. In Chapter 15 "Further Reading" I have listed a number of books and articles – all of which have contributed in a variety of ways to the writing of this book.

I would like to particularly thank Ian Harper for openly sharing material on teaching confocal microscopy and for helpful criticisms and corrections to drafts of the text. I would also like to thank Stephen Cody for help and ideas during the initial preparation of this book. Many other people have contributed in many ways to the success of this book; in particular I would like to thank Martin Harris, Damian Myers, Mark Lam, Guy Cox, Ian Gibbins, Mark Prescott, Malcolm Gourlay, Brian Jones, Nick Klonis, Leann Tilley and Jens Rietdorf who contributed greatly by reviewing various sections of this book.

I would like to extend a special thankyou to Klaus Lingelbach and members of his laboratory for accommodating my "writing distractions" while both working on this book and contributing to the research program in the Department of Biology at the Philipps University in Marburg, Germany. I would also like to extend a special thanks to Franz Grolig for the many interesting discussions and fun time together teaching confocal microscopy in Germany. I would also like to thank Peter Lipp for many interesting discussions on confocal microscopy. I would like to sincerely thank the Alexander von Humboldt Foundation for their support while my family and I were resident in Germany in 2001.

I would also like to thank a number of people from the various confocal microscope manufacturers that have helped greatly in providing technical information about their instruments. I would particularly like to thank Andrew Dixon, Chris Power, Anna Smallcombe, Graham Brown, Nick Manison, Marina Pekelis, Chris Johnson, Richard Kerr, Jeffrey Larson, Werner Knebel, Carola Thoni, Jan Schroeder, Christine Ludwig, George Weiss, Gavin Symonds, Solveig Hehl, Tetsuhiro Minamikawa, Kazushi Hyakumura, Terry McCann, George Kumar, Fedja Bobanovic, Bernhard Zimmermann, Ralf Engelmann, Alex Watkins, Lee Gale, Baggi Somasundaram and Mizuho Shimizu.

I would like to extend a special thankyou to Jim Pawley and the staff and students of the annual "3D Microscopy of Living Cells" course held at the University of British Columbia in Vancouver. A variety of people involved in this course have contributed greatly to my understanding and appreciation of confocal microscopy and related technologies over the past few years. A number of individual faculty members of this course have also been involved in reviewing various sections of this book. I would particularly like to thank Steve Potter, Glen MacDonald, Mark Cannell, Aurelie Snyder, Anda Cornea, Steve Adams, PC Cheng, Bai-Ling Lin, Irina Majoul, Rainer Duden, Iain Johnson, Stephan Hell, Dan Axelrod, Andres Kriete, Felix Magadant, Michael Weis, Ernst Keller and Elaine Humphrey. I would also like to thank the many students and staff that have been involved in this course over the past few years.

I would like to extend a special thankyou to Jack Martin and Jane Moseley from St Vincent's Institute of Medical Research in Melbourne for encouraging me to first embark on giving training courses in confocal microscopy.

I would like to thank Kathleen Lyons and her assistants at Kluwer/Plenum Publishers for their important contribution to the production of this book.

I would like to thank my sister, Denise Hibbs, for her valuable help in final proof reading of the document.

Finally, I would like to thank my wife, Kirstin and my two sons, Ben and Cameron, for their support and understanding while spending so many hours writing when I should have been out enjoying life with them!

At the end of the day the only way to become adept at confocal microscopy is to gain "hands on" experience, so get in there and try out your novel ideas!

Alan R. Hibbs

Melbourne, Australia
March, 2004

ahibbs@biocon.com.au

http://www.biocon.com.au/

Contents

Chapter 1

What is Confocal Microscopy?

Confocal microscopy is a major advance upon normal light microscopy, allowing one to visualise not only deep into cells and tissues, but also to create three dimensional images and to follow specific cellular reactions over extended periods of time.

Although confocal microscopes have been in existence for over 20 years, they have only become widely available since the introduction of small reliable lasers, and powerful and relatively cheap computers. In the late 1980's the first commercially available confocal microscopes became available from Bio-Rad, Leica and Zeiss. The confocal microscope is now a highly sophisticated instrument with superb optics and imaging capabilities. Obtaining an image using a confocal microscope is not particularly difficult, but taking full advantage of the real versatility of these instruments is quite another matter – hence the need for this book.

Advantages of a Confocal Microscope

Optical sectioning ability - *can image cells/tissues internally*

3D reconstruction - *subcellular location of labelling*

Excellent resolution - *close to the theoretical limit of 0.1 to 0.2 μm*

Specific wavelengths of light used - *greatly improves multiple labelling*

Very high sensitivity - *capable of collecting single molecule fluorescence*

Digital images - *easy manipulation and merging of images*

Computer controlled - *complex settings can be programmed and recalled*

This chapter gives a brief overview of the principles of confocal microscopy, while later chapters go into considerable detail on many aspects of using confocal microscopy in biomedical research. Although this book is primarily concerned with laser scanning confocal microscopy (LSCM), other important developments in light microscopy, such as the Nipkow disk high-speed confocal microscope, multi-photon microscopy, deconvolution and the use of CCD cameras are also discussed in some detail where appropriate.

OPENING NEW AREAS FOR LIGHT MICROSCOPY

Light microscopy has a long history of development that culminated in the production of high quality optics in the latter part of the 1800's. Microscopy was a very important part of the biological and medical sciences throughout the first half of the 1900's, but by the 1950's developments in light microscopy gave way to a great deal of interest in using electron microscopy in biological research due to the much higher resolution of these instruments. Light microscopy continued to be used in the biological sciences, particularly as a tool for medical diagnosis, but was considered an established, but relatively un-exciting technique in the biological research community.

Types of Confocal Microscopes

Laser scanning confocal microscopes – *scan a finely focussed laser spot across the object to create an image, using a "pinhole" to remove out of focus light.*

Nipkow disk confocal microscopes – *use a special "Nipkow" disk to scan several thousand "pinholes" across the image, these "pinholes" remove out of focus light and also allow for very fast scan rates.*

Slit scanning confocal microscope – *use a slit in place of a pinhole to remove out of focus light. Capable of relatively fast scanning that can be viewed directly by eye. No longer generally available.*

Multi-photon microscopes – *scan a far-red pulsed laser across the sample to generate fluorescence from dyes that are normally excited by much shorter (often UV) wavelengths. Out of focus light is removed by the fact that the laser intensity is sufficient for multi-photon excitation of the fluorophore only at the focal plane.*

The invention of the confocal microscope (first patented by Marvin Minsky in 1957, see page 349 in Chapter 15 "Further Reading", Minsky, M (1957) US Patent #3013467 and Minsky, M. (1988) "Memoir of inventing the confocal laser-scanning microscope") did not elicit a great deal of interest in the microscopy community at the time, and was probably unheard of by most users of light microscopes in the biological sciences. The invention of a number of other important technologies (such as the laser and small reliable computers) allowed the true power of confocal microscopy to became available to the biological research community. The first commercial instruments became available to the biological research community in the late 1980's. The early instruments created considerable excitement in the biological research community, particularly amongst scientists who had previously made very little use of microscopy in their research. The reason for this excitement was only partly due to the improved imaging capabilities of the confocal microscope compared to conventional light microscopy. Perhaps more important for the upsurge in interest in these instruments was the potential of using these instruments to undertake "molecular" studies on single cells and tissue samples. Conventional epi-fluorescence microscopy was being used extensively at that time to locate molecules of interest within fixed biological samples using fluorescently labelled antibodies. A lot of studies were also being done on ion-physiology using video microscopy to study the dynamics of cellular ion fluxes (particularly calcium). The introduction of the laser scanning confocal microscope to cell and molecular biologists meant that finally there was an instrument available that had great versatility in fluorescence imaging and could readily image live cells – an area of study where other forms of light microscopy have great difficulty in producing high quality images.

The confocal microscopes produced in the late 1980s and early 1990s were instruments with excellent optical characteristics, with high resolution, reasonable sensitivity and good optical slicing ability. However, these instruments were at times frustratingly slow for performing quite basic image processing. Even though the early instruments had software for 3D reconstruction – they took over an hour to render a modest stack of images into a 3D rendition!

In the last 10 years, mainly due to the widespread interest in confocal microscopy, there has been a great upsurge in interest in light microscopy. This has resulted in considerable improvement in the design of the objective lenses used in light microscopy (particularly lenses for imaging into live biological specimens), the development of a large number of fluorescent probes, and many significant improvements in computers such that a specialist "imaging" computer is no longer required for most applications.

The full potential of confocal microscopy is yet to come – the trend being to use these instruments to do single cell biochemistry, and even single molecule biochemistry. This includes not only the ability to follow intracellular reactions and fluxes, but also the ability to establish the relationship between specific macromolecules within sub compartments of the cell – the mainstay of biochemistry over the past 40 years, but which was, until now, performed on cell extracts rather than whole living cells or tissues.

MULTI-SKILLED ENDEAVOUR

Confocal microscopy requires a wide range of skills, including the ability to operate complex computer software, an understanding of light microscopy, a detailed knowledge of the biology of the system under investigation, and an understanding of the physics of light. Images can be readily obtained from the confocal microscope with only very limited knowledge, but a broad range of skills is required to take full advantage of the instrument. It can also be "dangerous" to base important aspects of your research on an exciting, but easily misleading technology. For example, if you lack experience in cell biology you may perform experiments that are fundamentally flawed and thus your interpretations will be less valid than they should be. Likewise, if your background is in cell biology and you have very little understanding of the physics of the confocal microscope and light microscopy or the nature of digital images, your attempt at quantifying the label in your images may not only produce meaningless data - but perhaps, more importantly, you may be drawn to the wrong conclusion!

A good knowledge of fluorescence, biochemistry and immunology is important for designing fluorescence-labelling experiments using confocal microscopy. Furthermore, an understanding of lasers and the nature of light

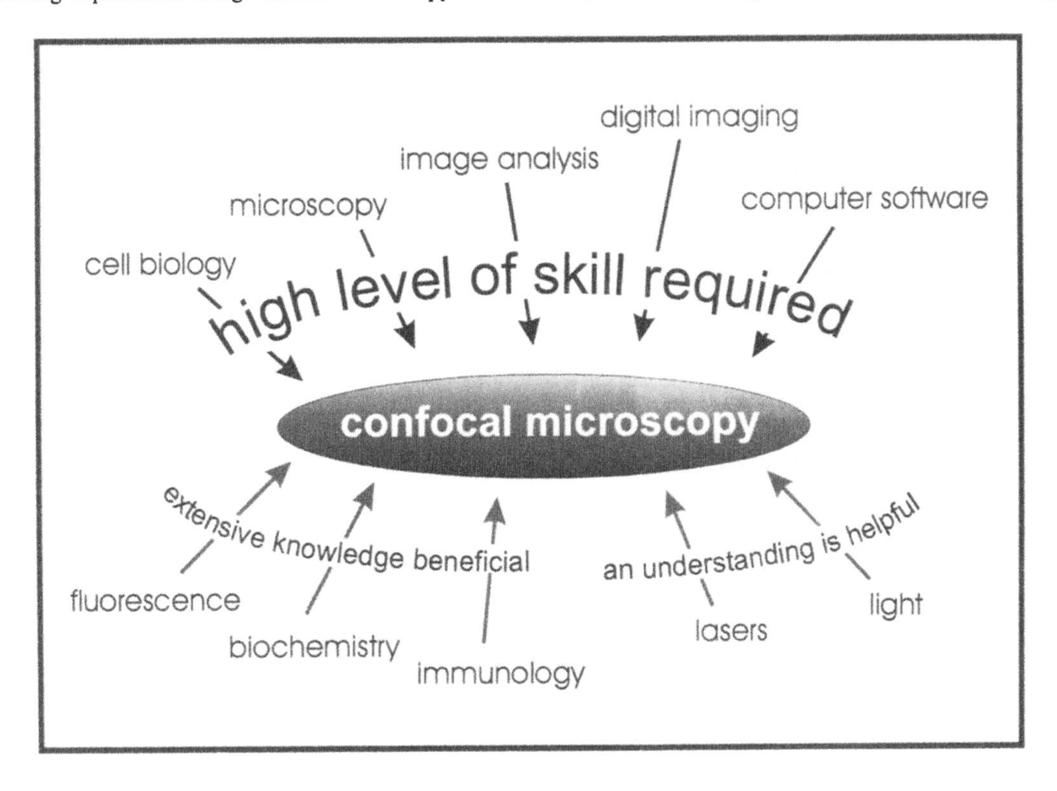

will also be beneficial, although not essential for taking advantage of the full capabilities of the confocal microscope.

As you gain a better understanding of the various technologies that are involved in confocal microscopy you will not only produce more reliable data, but you will find that the confocal microscope is capable of gathering a great deal more information from your sample than just a "pretty picture" - mind you the images you can get can be quite stunning!

LASER SCANNING CONFOCAL MICROSCOPY

There are many design aspects to a confocal microscope that make these microscopes much more versatile than a conventional fluorescence microscope. Although the confocal microscope is often thought of as an instrument that can create 3D images of live cells, the large number of operational features available on these instruments actually allow many creative methods of examining not just the structural details, but also the dynamics of physiological and developmental processes in living cells and tissues.

The principle of confocal microscopy is that the out of focus light is removed from the image by the use of a suitably positioned "pinhole". This not only creates images of exceptional resolution, but also allows one to collect optical slices of the object, and to use these slices to create a 3D representation of the sample. A simplified diagram of the layout of a typical laser scanning confocal microscope is shown in Figure 1-1. The Nipkow disk confocal

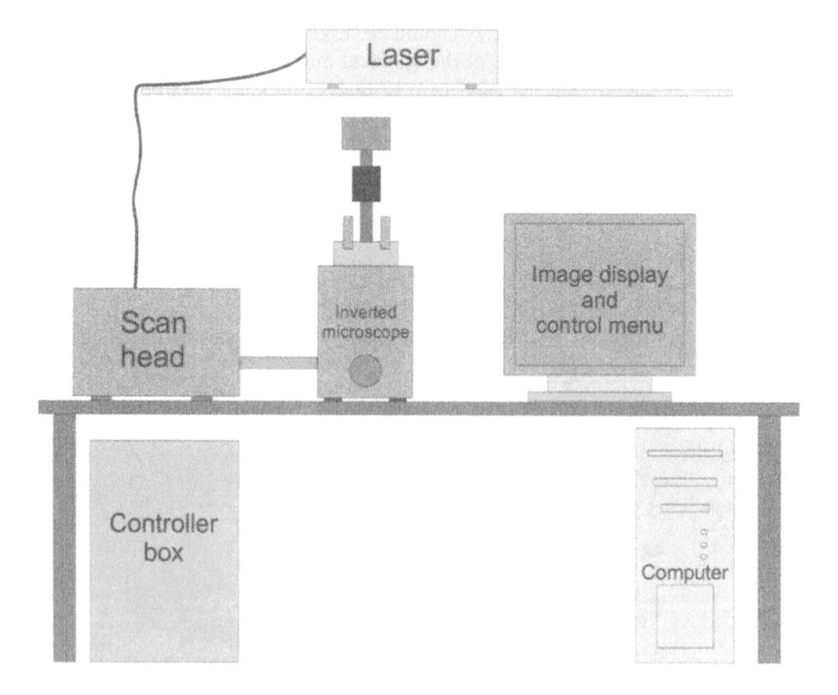

Figure 1-1. Confocal Microscope Layout.
Although the principle of confocal microscopy involves the use of a "pinhole" to remove out of focus light, there are many other aspects to this microscope that add greatly to its versatility. This includes the ability to use laser light of very specific wavelengths and the ability to collect and manipulate images in digital format. The basic layout for a confocal microscope includes a conventional light microscope (in this case an inverted microscope), often with epi-fluorescence optics installed, one or more lasers for irradiating the sample, a scan head for scanning the laser beam across the sample, a controller box containing much of the optics and some of the electronics needed for acquiring images, and finally a computer for both controlling the microscope and collecting the images.

Modes of Operation

Single channel fluorescence imaging - *very high sensitivity*

Dual/triple channel fluorescence imaging - *exactly co-aligned images*

Transmission imaging - *true colour images can be obtained*

Backscattered light (reflectance) imaging - *the highest resolution*

microscope and the multi-photon microscope do have many aspects of their design in common with the laser scanning confocal microscope, but due to both their many unique design characteristics, and different capabilities, they are discussed in detail in subsequent sections of this chapter. A conventional microscope (in this case an inverted microscope, but many systems incorporate an upright microscope instead) is at the heart of the instrument. Just as in conventional light microscopy, the objective lens of the microscope forms the image in confocal microscopy, and hence the quality of the final image, and particularly the resolution, is dependent on the quality of the objective used (for a more detailed discussion of the various attributes of an objective lens that determine the resolution and imaging quality of the lens see Chapter 2 "Understanding Microscopy", page 31).

Other components of a confocal microscope include the laser(s), a scan head, an optics (or controller) box and a computer. The scan head is responsible for scanning the laser beam across the sample, and the optics box (in some confocal microscopes the optics components are all mounted in the scan head) is used to separate light of different wavelengths (colours). A critical part of the confocal microscope is the "pinhole" (located in the scan head) whereby the "out of focus" rays of light are rejected. Photomultiplier tubes (PMTs) are used to detect the light in laser scanning confocal and multi-photon microscopes, whereas CCD cameras are used in Nipkow disk confocal microscopes. Digitisation of the amplified signal results in a "grey-scale" image, which can be readily coloured by suitable computer algorithms.

A computer, and the controller box containing additional computer cards, is required for capturing the image and controlling the microscope. Although older confocal microscopes, and some current models, have many manually adjusted controls on the scan head itself, most recently developed instruments have all of the controls (optical filter changes, pinhole adjustment, etc) fully automated, and their control built into the software.

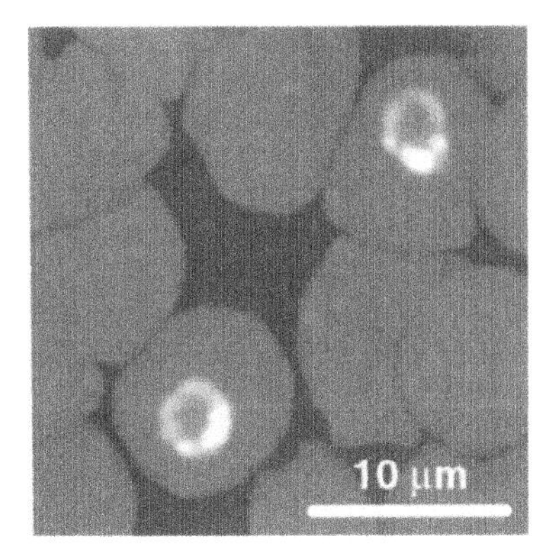

Figure 1-2. Confocal Microscope Image.
Ring stage malarial parasites within human red blood cells were fixed with acetone:methanol (1:1) and stained with ethidium bromide. Images were collected using a Zeiss 63x 1.4 NA oil immersion objective lens using a Bio-Rad MRC-600 confocal microscope – an excellent instrument that first brought confocal microscopy into "routine" use in the biological sciences in the late 1980's and early 1990's.

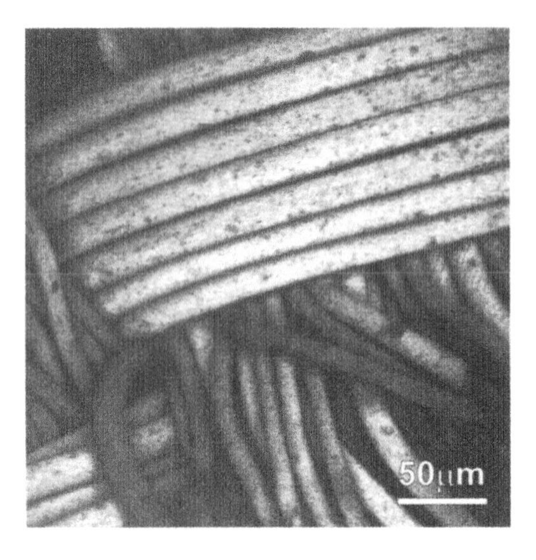

Figure 1-3. Extended Focus Confocal Microscope Image. A series of optical slices of highly fluorescent cloth were collected and are then displayed as an "extended focus" image. The images were collected using a Bio-Rad MRC-1024 confocal microscope using a Nikon 40x 0.6 NA long working distance "air" lens with coverslip thickness correction collar, and are described in more detail on page 138 in Chapter 4 "Image Collection".

From the brief description above one can see that there are many aspects to the confocal microscope besides the "pinhole" that eliminates out of focus light. This "coming together" of a number of technologies means that the confocal microscope has become a very versatile instrument in recent years, but also means that there is a considerable diversity of skills required to take full advantage of the instrument. Confocal microscopy is now a well-established technology in the field of fluorescence microscopy, but there are many new developments on the horizon for greatly extending this technology.

The image formed by a laser scanning confocal microscope shows exceptional clarity when displayed as a single optical slice (Figure 1-2), and wonderful 3D information when shown as a 3D projection (Figure 1-3). The principles of laser scanning confocal microscopy are discussed in some detail below, but information on the lasers, optical filters, dichroic mirrors etc are explained in Chapter 3 "Confocal Microscopy Hardware", page 65. Specific information on individual confocal microscopes of each of the major manufacturers is provided in "Appendix 1: Confocal Microscopes" (pages 355 - 441).

Epi-fluorescence Illumination

Most confocal microscopes use the epi-fluorescence illumination design (Figure 1-4) for fluorescence imaging. Epi-fluorescence imaging is also normally used in conventional wide-field fluorescence imaging using mercury or xenon arc lamps. In epi-illumination light from a powerful arc lamp or laser is directed to the sample by means of a "beam splitter" (a mirror that will reflect the illuminating light, but allow the returning fluorescent light to be transmitted through to the eye-piece of the microscope or to a suitable light detector). In the epi-fluorescence method of illumination the objective lens of the microscope functions as both the condenser and the objective and so the problem of careful alignment of the two lens systems is eliminated.

Epi-fluorescence illumination when used in a conventional fluorescence microscope has the important safety feature of not allowing the high-energy light (often short wavelength UV light) to directly enter the eyepieces of the microscope. In a similar manner, epi-illumination in the confocal microscope has the advantage that most of the irradiating laser light passes through the sample and is not directed towards the highly sensitive light detectors. As the amount of irradiating light is very much greater than the fluorescent light that is being detected, removal of most of the excitation light by passing the light through the sample results in a much improved detection sensitivity.

How is the Image Formed?

The image in a confocal microscope is created by scanning a diffraction-limited point of excitation light across the sample in a raster formation (Figure 1-5). Two scanning mirrors are used, one to scan at high-speed in the "x-x" direction (appearing as horizontal scanning across the computer screen) and the other to scan somewhat slower in the "y" direction (appearing as the vertical on the computer screen). The irradiating laser light is used to excite a suitable fluorophore, applied to or naturally occurring within the sample. A small amount of the fluorescent light emitted from the sample passes back through the objective and is separated from light of a second fluorophore or unwanted reflected excitation light by the use of suitable partially reflecting dichroic mirrors.

The "in focus" light, after passing through the confocal microscope "pinhole", is detected by a very sensitive light detector called a photomultiplier tube (PMT). The analogue signal from the photomultiplier tube is converted to a digital form that contains both information on the position of the laser in the image and the amount of light coming from the sample.

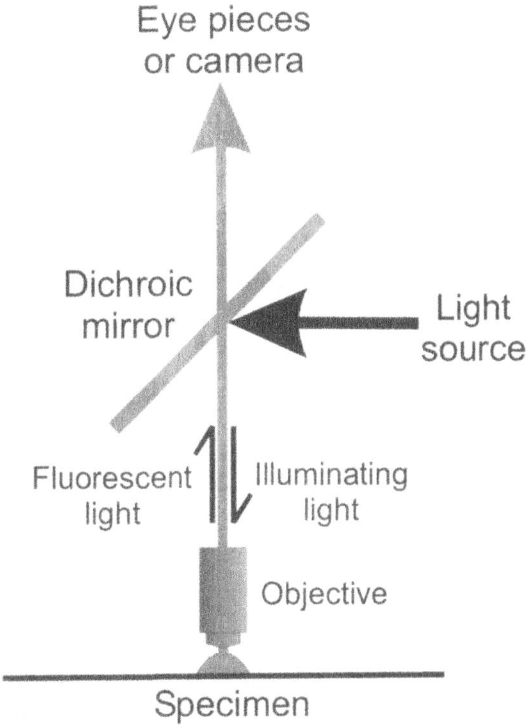

Figure 1-4. Epi-illumination.
In epi-illumination the illuminating light passes through the microscope objective to the specimen, and the returning fluorescent light passes back through the same objective. A dichroic mirror, that selectively reflects specific wavelengths of light, is used to separate the fluorescent from the illuminating light. The fluorescent light is directed to the microscope eyepieces or light detectors. Conventional fluorescence microscopy and fluorescence confocal microscopy use the epi-fluorescence configuration, both in upright and inverted microscope configurations.

The image is displayed on a computer screen as a shaded grey image, normally with 256 levels of grey (8-bit) (Figure 1-2), even if the image is collected as a 12-bit image with 4096 levels of grey. The image can be suitably coloured, and the intensity levels adjusted, later, for presentation or publication. For example, in Figure 1-2 the contrast in the image has been carefully adjusted to demonstrate the strong staining of the parasite nuclear DNA (white), surrounded by the cytoplasmic ring structure that contains RNA that is staining to a lesser extent (light grey). The host red cell cytoplasm is shown lightly staining (dark grey) with ethidium bromide (non-specific low level cytoplasmic staining). The background exhibits almost no staining and is thus shown in very dark grey. This image is only possible by confocal microscopy as the concentration of ethidium bromide stain used creates far too high a background to see any parasites by conventional epi-fluorescence microscopy. In this example, the excellent image clarity is due to both the thin optical slice imaged (explained in detail shortly), and the ability to readily manipulate the peak and background levels displayed when imaging by confocal microscopy. This image has been carefully adjusted both during collection and afterwards by digital image processing to create a clear delimitation of the red blood cells, the parasites and the parasite DNA. This demonstrates the "power" of confocal microscopy in creating images of exceptional clarity, but also highlights the inherent dangers associated with image manipulation.

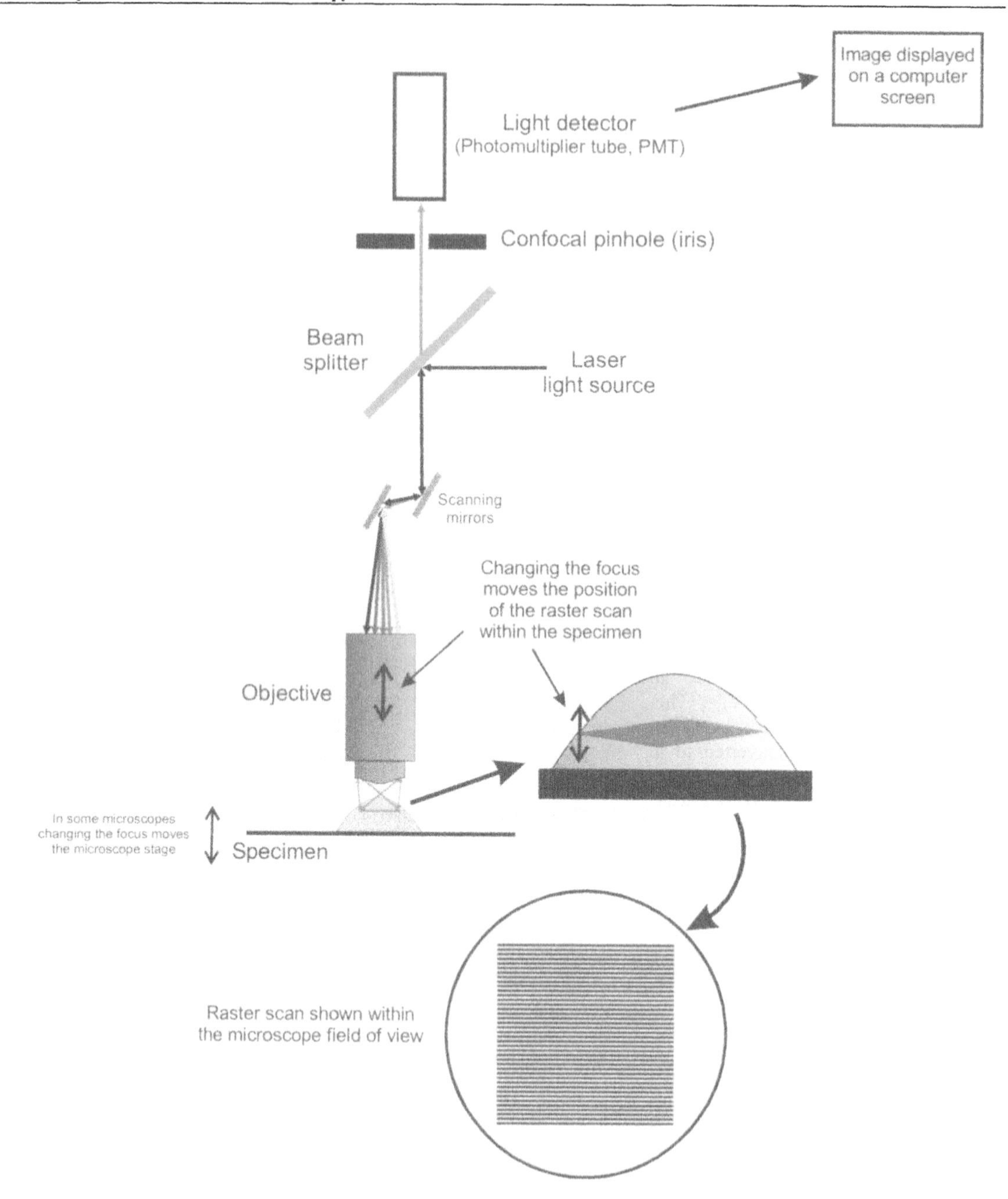

Figure 1-5. Image Formation in the Confocal Microscope.
In the laser scanning confocal microscope the image is formed by scanning a finely focussed laser beam across the sample in a raster pattern. One of the scanning mirrors scans the laser beam at a relatively high speed in the "x" direction (the horizontal scan on the computer screen). The other scanning mirror scans much slower in the "y" direction (the vertical scan display on the computer screen). The speed at which these mirrors scan determines the time taken to collect a single image. A typical collection time for a single image is ¼ to 2 seconds per scan, depending on the pixel size of the image collected. The focal position of the objective determines the level within the sample from which the image is derived (i.e. only the focal plane is visible). The image collected from a single focal plane is called an "optical slice".

How is Out of Focus Light Removed?

One of the great difficulties with conventional epi-fluorescence microscopy is that the out of focus light in the image can greatly detract from the quality of the image. Single cells or very thin tissue samples can normally be imaged by conventional fluorescence microscopy with relative ease, although as we have seen above even thin samples may be much better visualised by confocal microscopy. However, the confocal microscope really comes into force when you need to image multicellular samples, such as complex tissues. Background fluorescence from the surrounding media or from the surface on which the cells are attached can also be readily eliminated by confocal microscopy.

The thin optical slice of confocal microscopy is obtained simply by eliminating the light rays that originate from out of focus positions, or more correctly, other focal planes, within the sample (the normal "blur" seen by conventional microscopy of thick specimens). Light from a single focal plane is collected by focusing the fluorescent light through a small aperture or confocal pinhole (see Figure 1-6). Light rays from other focal planes are not correctly aligned with the confocal iris, and are thus eliminated from the image. This ability to remove out of focus light has the effect of producing an image that consists of information from only a very thin focal plane (a thin "z" section), and is often referred to as an "optical slice". A series of these "z" sections can be combined in a number of ways, outlined below, that allow one to visualise the 3D structure of the original sample.

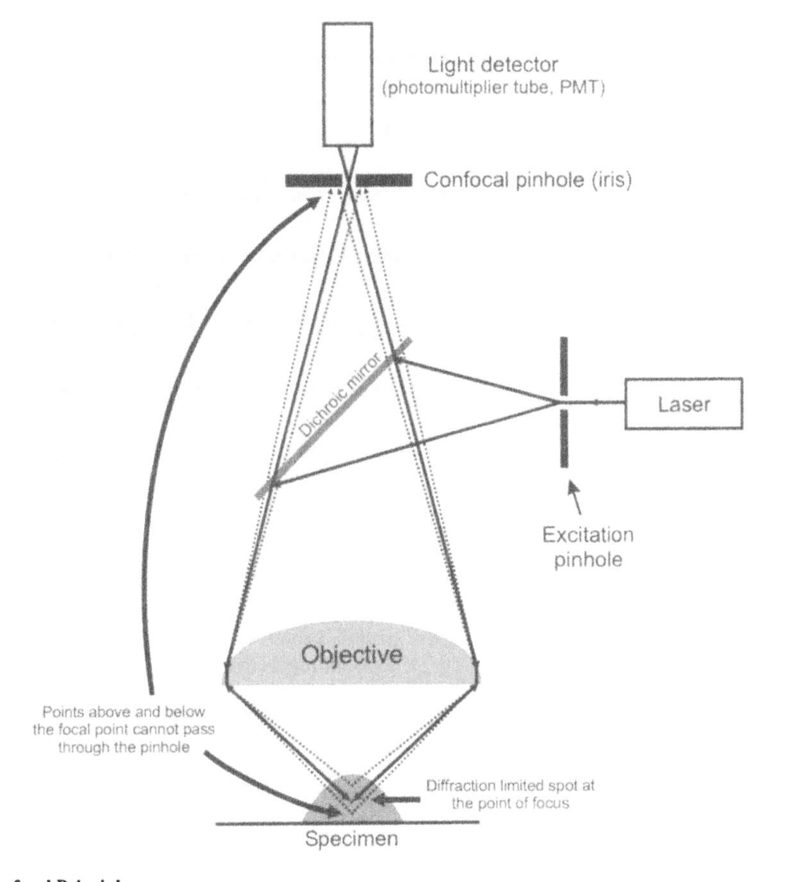

Figure 1-6. Confocal Principle.
Out of focus light rays (……..) are eliminated from the image by the use of a confocal pinhole or iris, through which only focussed light rays (_____) can pass. Or more correctly, only light from a single focal plane is allowed to pass through the pinhole. Moving the position of the objective lens or specimen (changing the focus) changes the position of the focal plane within the specimen. In this way a series of optical slices can be obtained from a confocal microscope.

Optical Slicing

At the heart of confocal microscopy is the ability to take thin "optical slices" through the cell or tissue of interest as outlined above. A single optical slice through an acridine orange stained live malarial parasite is shown in Figure 1-7), and the imaging of a single focal plane is shown diagrammatically in Figure 1-8. The ability to create optical slices not only greatly improves the quality of the image but also provides a great deal of information on the 3D structure of the object. Take care when examining a single optical slice, as collecting an image from a single focal plane may significantly alter the "look" of the image compared to conventional light microscopy - explained in more detail in Chapter 4 "Image Collection", page 101.

The advantage of eliminating out-of-focus light is clearly demonstrated in Figure 1-7, where a single thin optical slice has been obtained from within the centre of a live malaria infected red blood cell stained with acridine orange.

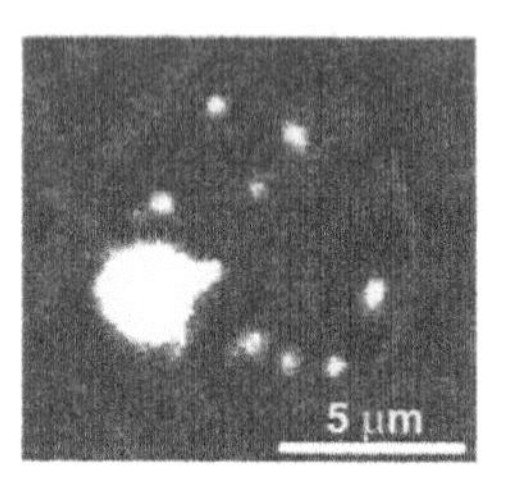

Figure 1-7. Single Optical Slice.
Live malarial parasites were stained with acridine orange and imaged live in a small incubation chamber using time-lapse imaging on an MRC-600 confocal microscope with a Zeiss 63x 1.4 NA oil immersion objective lens. This image shows a very thin (approximately 0.5 µm) optical slice through the centre of a malaria-infected red blood cell, which has also cut through the parasite (the large fluorescent structure in the bottom left) and several cytoplasmic vesicles (small dots distributed within the red cell cytoplasm). This image is reproduced from Hibbs and Saul (1994) with permission from Elsevier (see page 349 in Chapter 15 "Further Reading" for further details).

The parasite is visible as a very brightly staining structure in the bottom left hand corner (acridine orange bound to DNA). Small brightly staining "vesicles" (these mildly acidic vesicles are faintly stained with acridine orange) are visible in the cytoplasm of the infected red blood cell. A faint outline of the red blood cell membrane can also be discerned. The level of staining in the parasite is more than 1000x that in the small vesicles. The optical slicing nature of the confocal microscope allows one to obtain excellent images of very lightly staining structures in close proximity to very brightly staining structures. In conventional epi-fluorescence microscopy the light from the brightly staining parasite would "flare out" and obscure the small, lightly stained, vesicles. If the epi-fluorescence light source used in conventional fluorescence microscopy is attenuated to lower the "flair out" then the small very faintly fluorescing vesicles would no longer be visible. The instrument gain (sometimes denoted PMT voltage) and background (offset) levels have been suitably adjusted on the confocal microscope to result in the small cytoplasmic vesicles showing a peak intensity very similar to the parasite itself (white). Therefore, this image can only be used as an indicator of the number of vesicles present and cannot be used to quantify the relative level of labelling between the parasite and cytoplasmic vesicles or the size of the vesicles. The "speckled" appearance of the image in Figure 1-7, compared to the image in Figure 1-2, is mainly due to the statistical nature of light (Poisson noise). If the light gathered from the sample is very low (as is the case with these slightly fluorescent vesicles and red blood cell membrane) the statistical nature of light comes to dominate the image. This is discussed in more detail in on page 36 in Chapter 2 "Understanding Microscopy".

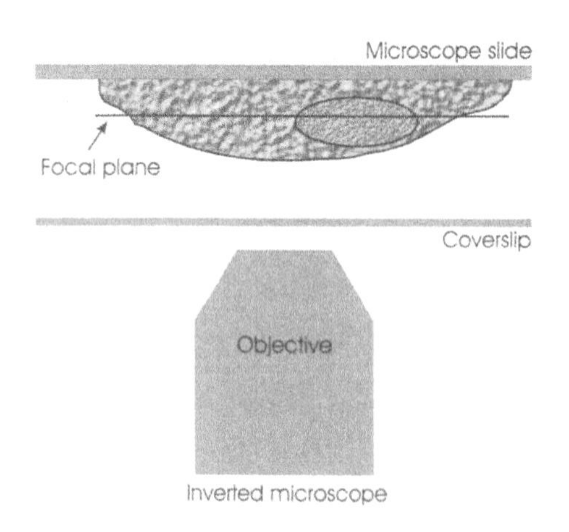

Figure 1-8. Location of Thin Optical Slice.
The confocal microscope creates an image of the focal plane by using a "pinhole" to eliminate all out-of-focus light, thus creating a thin "optical slice" of the object.

Interestingly, this "optical slicing" capability of a confocal microscope can be demonstrated by looking down through the eyepieces of a conventional epi-fluorescence microscope and manually moving the fine focus up and down. The in focus "optical sections" can be readily distinguished from the background "blurred image". This is because the human vision system is quite capable of momentarily "rejecting" the out of focus light, by both vision acuity and the ability of the brain to perform pattern recognition to differentiate "in focus" detail from out-of-focus "blur". Photographing such an image (whether by conventional film or CCD camera) will produce a "blurry" image, which is in contrast to the high clarity image collected by confocal microscopy.

NIPKOW DISK CONFOCAL MICROSCOPY

The Nipkow spinning disk confocal microscopes use a special disk (the Nipkow disk) that contains an array of fine "pinholes" (Figure 1-9) to eliminate out-of-focus light. While the disk is spinning at high-speed the light is focussed through the fine moving "pinholes", resulting in the sample being irradiated with an array of fine points of light. The resultant fluorescence from each of the points is focussed back up through the pinhole array and the "in-focus" light detected by a CCD camera (or even viewed directly if you wish).

There are two basic confocal microscope designs currently used in the biological sciences that utilize the Nipkow disk; the CARV confocal microscope, produced by Atto Bioscience, which uses a single Nipkow disk containing many thousands of "pinholes", and the Yokogawa Electric Corporation scan head (assembled into full confocal microscope systems by companies such as PerkinElmer and VisiTech International), which contains both a "pinhole" array Nipkow disk and a micro-lens array disk that are spun in tandem. The advantage of the Yokogawa micro-lens array scan head is that a more intense level of excitation light (for each moving spot) can be used to elicit fluorescence from the sample. Nipkow disk confocal microscopes are used extensively for live cell imaging, but are also used in the material sciences, particularly in the silicon chip manufacturing industry (for example, the MX50-CF spinning Nipkow disk confocal microscope produced by Olympus).

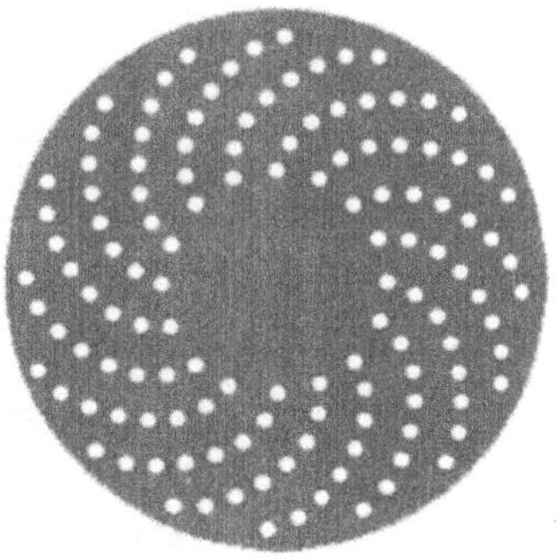

Figure 1-9. Nipkow Disk.
The Nipkow disk used in the spinning disk confocal microscopes contains an array of fine "pinholes" through which the irradiating and fluorescent light is focussed. The disk is spun at relatively high speed to create a high-speed "real time" confocal image.

The Nipkow spinning disk confocal microscopes are capable of very high speed scanning. The actual speed of scanning is determined by the sensitivity (and hence expense!) of the CCD camera that is attached, and of course, the fluorescence intensity of the sample itself. Routinely 10 to 20 frames per second are achieved (for a 512 x 512 pixel image, which would take approximately 0.5 sec on a conventional laser scanning confocal microscope). Much higher speeds of up to 360 frames per second, are possible, when using a very high quality cooled CCD camera. These instruments are particularly suited to live cell imaging, and are capable of capturing very fast ion fluxes and other dynamic processes within living cells.

The Nipkow disk confocal microscopes also have the added advantage of causing less fading of the fluorophore. Although the physical reason for the lower level of photo damage is not fully understood, it appears that the fact that the sample is irradiated with a large number of relatively low intensity spots of light is important. Although the overall laser power used may be similar to a single spot laser scanning confocal microscope, the intensity at each focussed spot is much lower, and hence there is significantly less destruction of the fluorophore (and less damage to cellular integrity). For this reason Nipkow disk confocal microscopy is well suited to long term studies of living cells, particularly studies of embryo development, cell division, cell movement etc.

Nipkow disk confocal microscopes are particularly suited to live cell imaging

The design of the Nipkow disk confocal microscope is described in some detail on pages 91 - 93 in Chapter 3 "Confocal Microscopy Hardware" and the microscopes available are described on pages 431 - 441 in "Appendix 1: Confocal Microscopes".

MULTI-PHOTON MICROSCOPY

Multi-photon microscopy is an exciting method of fluorescence microscopy that allows one to image significantly deeper into tissue samples compared to confocal microscopy. The multi-photon microscope consists of a light microscope, scan head, optics box and computer, but unlike the laser scanning confocal microscope (single photon excitation), the laser used for irradiation of the sample is a high-energy pulsed infrared laser that is capable of generating high levels of multi-photon fluorescence within the sample. The multi-photon laser is used to excite fluorophores that would normally require shorter wavelength blue or UV light to create fluorescence. The very high light intensity at the focal point at the peak of each pulse of light is such that 2, or even 3-photon (hence the term multi-photon) excitation can occur to a significant extent (Figure 1-10). See Chapter 8 "What is Fluorescence?" (page 187) for a discussion of multi-photon fluorescence in greater detail.

The multi-photon laser is often used to excite "UV" dyes such as DAPI and Hoechst in multi-labelling experiments, but visible light dyes that absorb in the blue region of the light spectrum, such as fluorescein, can also be readily excited using the multi-photon pulsed infrared laser. Appropriate tuning of the pulsed infrared laser wavelength will allow one to excite most "UV" dyes, or even to excite "UV" and visible dyes with 3- and 2-photon excitation respectively, simultaneously. Unfortunately the non-linear nature of multi-photon fluorescence means that the best wavelength for excitation cannot be readily determined by simply looking up the light absorbance curve of the fluorophore of interest in an established database, as one can for single-photon (conventional) excitation. You can't even be sure that the fluorophore will be excited at all by the pulsed infrared laser! The best way to determine whether you can use a multi-photon microscope with a particular dye, or dye combination, is to either try a range of

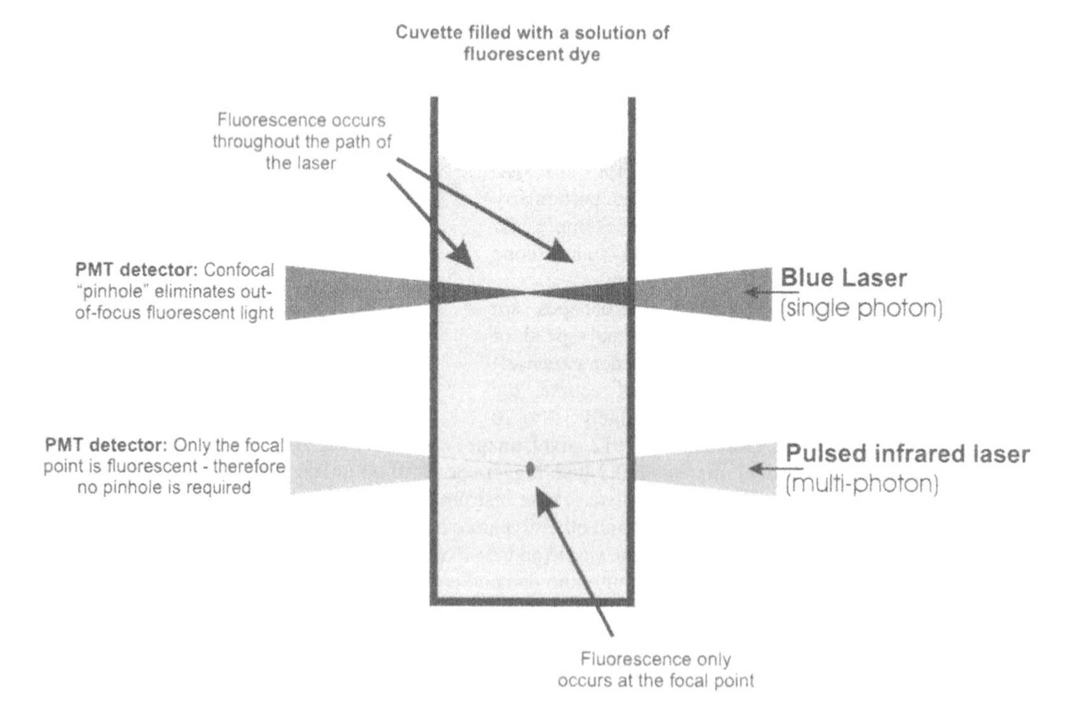

Figure 1-10. Multi-photon Induced Fluorescence.
In a multi-photon microscope fluorescence is generated by 2- or even 3-photon excitation of the fluorophore at the point of focus only, by using a high energy pulsed infrared laser. The infrared light has the ability of penetrating into biological tissue significantly further than visible wavelength light. However, the pulsed infrared laser is a very expensive component, and the non-linear nature of multi-photon excitation does mean that the behaviour of fluorophores will not be easily predicted, necessitating a trial and error approach to the best dye and wavelength combinations to use for particular experiments.

infrared wavelengths yourself, or to look on the internet for information on multi-photon excitation. There is also a multi-photon Listserver that may be particularly useful for providing information on specific technical problems (http: //groups.yahoo.com/group/mplsm-users/, see page 327 for more details on subscribing).

Multi-photon microscopy can be used to excite "UV" dyes – and to image deeper into tissue samples

Multi-photon microscopy is used extensively in areas of research where imaging depth is very important. Image acquisition in relatively thick tissue samples, such as brain slices, benefits greatly from using a multi-photon microscope. However, one must keep in mind that even though the multi-photon microscope can image significantly deeper into tissue than a conventional confocal microscope (approximately 200 μm, compared to 100 - 150 μm when using a confocal microscope) the depth of imaging is still relatively small when compared with the size of the tissue sample. For example, you are not going to be able to determine the size of a tumour lesion by multi-photon microscopy, unless the tumour is very small indeed.

Multi-photon microscopes may also be equipped with normal non-pulsed lasers and detectors for conventional confocal microscopy, allowing them to become versatile imaging instruments for a variety of applications. However, multi-photon capability does add a very significant further expense to the cost of the instrument (the pulsed infrared laser is a very expensive item).

IMAGING CAPABILITIES

There are a variety of image collection modes commonly available on a confocal and multi-photon microscopes. The simplest is to excite a single fluorescent dye with one laser line, and to collect images by directing the fluorescent light into a single detection channel using a broad emission optical filter. Other, more complex, modes include dual, triple or even four channel labelling, transmission imaging and backscatter light imaging. These imaging techniques, and multiple detection channels, can be combined to provide multi-coloured confocal images, or grey-scale transmission or reflectance images with coloured fluorescence image overlays. However, the different imaging modes in a confocal microscope are all collected as separate grey-scale images (even if they are displayed as coloured images or as a combined multi-coloured image on the collection screen). The primary images that are saved to disk are grey-scale images, not the multi-coloured combined image. These grey-scale images can be readily combined and coloured after collection.

Fluorescence Imaging

The simplest labelling method using a confocal microscope is to label the cells with a single fluorescent dye and then image using the single channel mode (Figure 1-11). The advantage of single channel labelling and detection is that there is no need to worry about "bleed-through" from, or into, another channel, and so the broadest possible emission spectrum can be collected into a single channel (resulting in an apparent increase in the sensitivity of the instrument).

A low level of "background" fluorescence is often used, particularly in single labelling applications, to enable you to locate the fluorescence of interest within the structure of the cell. The left hand panel in Figure 1-11 shows a low level of labelling of the parasite (particularly visible in the lower left cell) that can allow one to locate the fine "dots" of fluorescence, which are in this case associated with the red blood cell cytoplasm rather than the parasite itself. In the right hand panel of Figure 1-11 the parasite is not visible (and nor is the outline of the red blood cell), making the exact location of the punctate fluorescence in this image difficult to determine.

In spite of the simple nature of single channel labelling, combining the single label with a second, or even third label, or a transmission image, to accurately locate the fluorescence of interest within the cell is often important. Combining various imaging modes is a way of determining whether the label is actually located within cells, and is not simply associated with background fluorescence on the microscope slide. Single optical slices from a confocal microscope can be quite difficult to interpret for the novice as they may show very limited amounts of fluorescence within the single slice – some structural context within which to place the protein or label of interest is important.

Combining the fluorescence image with a transmission image (Figure 1-12) is a very effective way of providing this structural context for the label of interest. In Figure 1-12 the fine punctate fluorescence, due to fluorescently labelled transferrin which is associated with surface receptors and some endocytic vesicles, would be very difficult to interpret without the structural information on the location of the cells provided by the transmission image.

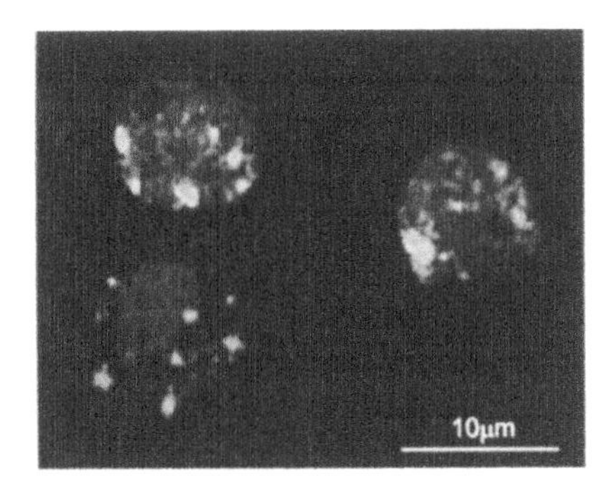

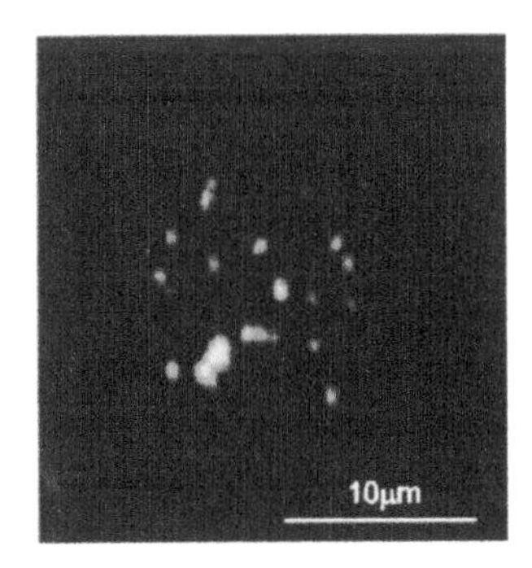

Figure 1-11. Single Labelling.
Immunolabelled malarial parasites (fixed using acetone:methanol, 1:1) showing punctate fluorescence within the malaria infected red blood cell. The cells were labelled using a fluorescein (FITC) labelled secondary antibody and primary antibodies against the malarial proteins QF120 (left panel) and FEST (right panel). Non-specific "background" fluorescence from the parasite (left panel) can be useful for locating the fluorescence of interest (the punctate green fluorescence). In the absence of suitable "background" staining (right panel) one can have considerable difficulty in determining the physical location of the labelling in relation to the parasite or red blood cells. The colour of these images has been adjusted to be similar in colour to the real colour emission of the FITC fluorophore used ("Green"), although the original images were collected as grey-scale images only. These images were collected using a Bio-Rad MRC-600 confocal microscope with a Zeiss 63x 1.4 NA oil immersion objective lens. See Kun et al 1997 (see page 349 of Chapter 15 "Further Reading") for further information on the immunolabelling protocol used to obtain these images.

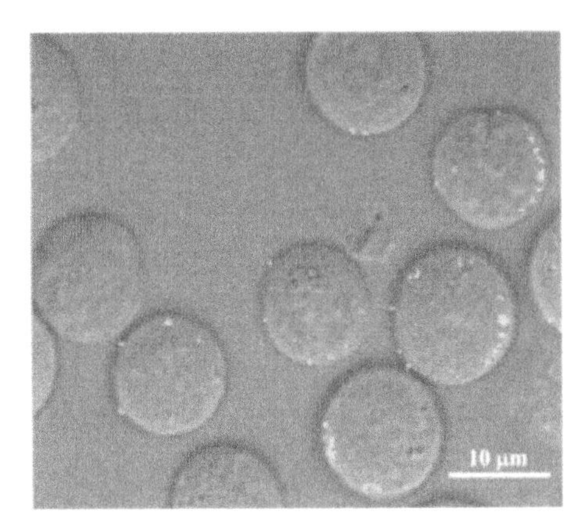

Figure 1-12. Fluorescence and Transmission Combined.
Fluorescently labelled transferrin (yellow/green) and a grey-scale transmission image are merged to demonstrate the location of the fluorescence relative to the position of the cells. Most of the transferrin in this experiment is on the external surface of the cells (directly labelled FITC-transferrin was incubated with transformed human lymphocyte (MOLT-4) cells). Cells were imaged live while gently resting on the coverslip base of a microscope incubation chamber on a Zeiss inverted microscope. The images were collected with a Bio-Rad MRC-600 confocal microscope using a Zeiss 63 x 1.4 NA oil immersion objective lens. Further information on the experimental conditions used to obtain this image can be found in Moss et al (1994) – see page 349 in Chapter 15 "Further Reading".

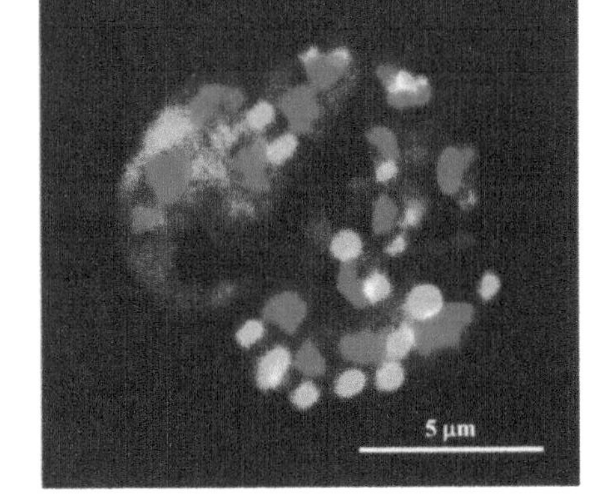

Figure 1-13. Dual Labelling.
Malarial parasites labelled with a fluorescently labelled antibody against a merozoite protein, RAP1 (green) and propidium iodide (red), which labels predominantly DNA. Two separately, but simultaneously, collected grey-scale images were combined using Confocal Assistant. The cells were fixed using acetone:methanol (1:1) prior to immunolabelling. Images were collected using a Bio-Rad MRC-600 confocal microscope using a Zeiss 63x 1.3 NA oil immersion objective lens.

The transmission image can also be used effectively to determine the "focus level" within the specimen. As has been discussed above, a confocal image results in a very thin "optical slice" through the sample. However, the position within the sample is quite difficult to determine, particularly when only a single optical slice has been collected. The transmission image, which is not confocal (i.e. it is just like a "normal" bright-field microscope image that contains both in focus and out of focus detail), can be used to locate the central region of the cell. This is carried out by focusing the rim of the cell carefully using the transmission channel – this will result in a confocal image that is located within the central part of the cell. Care should be taken if you use this technique to locate the central region of the cells when using DIC (Differential Interference Contrast) imaging because in this case the "transmission" image shows significant "shadowing", creating a pseudo 3D image that has some optical slicing characteristics. Excellent transmission images can usually be obtained when imaging live cells, but fixed cells, often used in immunolabelling, may not show up well by transmission imaging. This is mainly due to the refractive index of the mounting media being very close to that of the fixed cellular material.

The merged fluorescent and transmission image shown in Figure 1-12 also illustrates how interpretation of these images should be done with considerable care. In this merged image the two cells towards the right hand side appear to have several "dots" of fluorescence located just inside the rim of the cell. This could at first be interpreted as indicating that this punctate fluorescence is associated with endocytic vesicles within the cellular cytoplasm but still in close proximity to the plasma membrane. However, closer examination of this image (perhaps not as readily seen in the printed version) reveals that the focal plane is slightly "above" the cell rim, which results in the optical slice being obtained from an area of the cell not quite at the expected periphery of the cell. This method of using a transmission image to provide structural information about the cell or tissue sample is described in more detail in Chapter 4 "Image Collection" (page 101).

In many experiments, determining a fluorescence location within a subcellular compartment using transmission imaging, as described above, may not be sufficient. Locating the protein or probe of interest in relation to other cellular parameters or proteins is often important. Combining two or more fluorescence images is often a convenient way to provide considerably more detail on not only the location, but also the molecular or organelle associations of the protein or probe of interest. The second label in immunofluorescence is often a DNA stain as shown in Figure 1-13. In this case the individual malarial parasites (merozoites) within the infected red blood cell are shown in red (stained with propidium iodide) and a rhoptry-associated antigen (RAP1) is shown in green (immunolabelled with a an FITC labelled secondary antibody).

Nipkow spinning disk confocal microscopes are particularly suited to time lapse imaging

Despite the superficial similarity between the two labels (both showing punctate fluorescence within the infected red blood cell) the dual labelled image in Figure 1-13 clearly shows that the RAP1 protein is associated with structures outside the area of the merozoite nucleus. Careful analysis of a 3D reconstruction from a series of optical slices demonstrates that the RAP1 protein is always associated with one end of the merozoite nucleus. However, this is not as evident in the dual labelled single optical slice shown in Figure 1-13. Thus examining the sample in 3D may help to establish the relationship between various fluorescently labelled structures.

When the location of two different fluorophores is clearly separated within the image, as in Figure 1-13, and where the physical difference between the two structures (the green fluorescence is more "rounded" compared to the red fluorescence) makes the separation even more distinct, analysis of the results obtained from dual labelling is a very simple process. However, when the two labels are highlighting structures that overlap, or are of very similar appearance, you will need to have confidence that the two fluorescent channels are showing the two fluorophores correctly and that there isn't any misleading fluorescence bleed-through between the channels. See Chapter 4 "Image Collection" (page 101) and Chapter 10 "Confocal Microscopy Techniques" (page 239) for further discussion on the difficulties encountered in multi-labelling applications.

Time-lapse Imaging

Confocal and multi-photon microscopes are particularly well suited to time-lapse imaging. These instruments are capable of a wide range of experiments, from very high speed imaging of ion fluxes to the imaging of developing embryos over an extended period of many hours or even days. The laser scanning confocal and multi-photon microscopes collect images at relatively low speed (approximately 1 frame per second for a 512 x 512 pixel image), but this scan rate can be increased substantially by a variety of techniques, such as reducing the image collection size, or even by collecting only a single line across the cell or cells of interest. These techniques of increasing the effective scan rate of a laser scanning and multi-photon confocal microscope are discussed in some detail in Chapter 4 "Image Collection" (page 101).

The software available for operating the confocal microscope usually includes excellent time delay collection modules that can be used to collect a series of images from fractions of a second to minutes or even hours apart. One of the serious limitations of collecting extended time series by confocal microscopy is that very small movements in the cell, particularly in the z-direction (i.e. moving in or out of the focal plane), or movement of the focal point of the confocal microscope due to temperature fluctuations or mechanical vibrations, will result in movement of the cell structure being imaged. This is not normally a problem for time-lapse collection over a period of a few minutes, but longer time periods may require considerable care to avoid focus changes.

One of the remarkable things about time-lapse imaging of cellular events is that previously unforseen events may become quite obvious, with patterns, rhythmic responses and transient associations becoming readily apparent. Although presenting time-lapse imaging in a seminar as a short movie is very effective, portraying time-lapse results in a conventional scientific journal is much more difficult.

As has been previously mentioned, the Nipkow disk based confocal microscopes are particularly suited to time-lapse imaging due to the low level of photo-damage and the very fast imaging speeds that these instruments can attain. However, high speed line scanning, using a conventional laser "spot" scanning confocal microscope can be used for following ion fluxes in living cells.

3Dimensional Information

There is a wealth of 3D information in a single optical slice (a normal single image taken using a confocal microscope), even before rendering a stack of such slices into a 3D image. Furthermore, collecting a series of optical slices for 3D reconstruction may not always be possible, perhaps because of a high degree of photobleaching or maybe because of movement of the cells under study. In this case a single optical slice, or a small number of optical slices may be all that is available. However, from this limited amount of 3D information it is possible, with sufficient experience, to derive a substantial amount of 3D information about the sample.

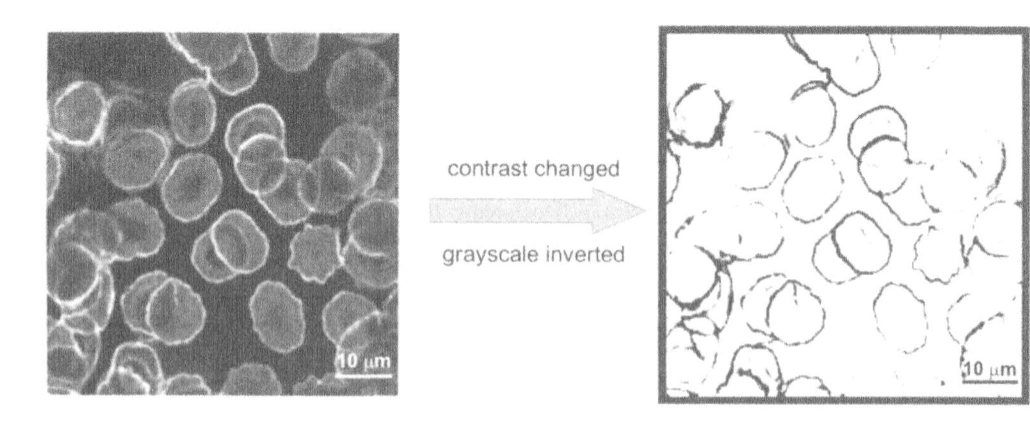

Figure 1-14. Surface Labelled Red Blood Cells.
A single optical slice through the centre of surface labelled red blood cells (fixed with acetone:methanol, 1:1, and labelled with an band-3 antibody) results in a ring of fluorescence. The image on the left has been collected to strongly show the "ring" of surface fluorescence, with also a low level of internal fluorescence visible (background fluorescence using this antibody). The image on the right has been manipulated to emphasize the "rings" of surface label only (by changing the contrast and inverting the image). This image was collected using a Bio-Rad MRC-600 confocal microscope using a Zeiss 63x 1.4 NA oil immersion objective lens.

A single image from a confocal microscope contains information on the subcellular location of the fluorescence (internal, cell surface, or external to the cells), even when displayed as a single optical slice. In Figure 1-14 human red blood cells have been surface labelled with an antibody against Band 3, a red blood cell membrane protein. The ring of fluorescence seen in Figure 1-14 is typical of surface fluorescence, with the gain (or photomultiplier tube voltage) and black level adjusted to show a low level of background labelling on the inside of these red blood cells and a bright ring of fluorescence located at the plasma membrane (Figure 1-14, left hand panel). The image can be further manipulated after collection (using suitable image processing software) to further emphasize the surface labelling (right hand panel of Figure 1-14).

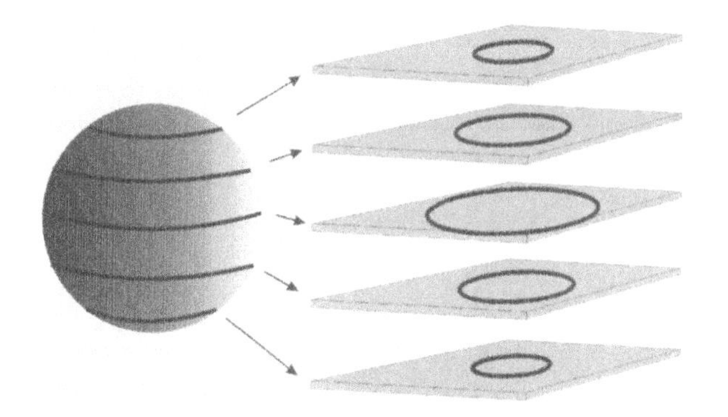

Figure 1-15. Optical Sectioning.
A series of thin optical slices contains a wealth of information on the 3D structure of the object being imaged. In this diagram slicing the "ball" structure results in a series of "rings" of varying size if the object is surface-labelled. These slices can later be rendered into a 3D image in a number of ways, including that of "rocking" the image to give the impression of a three dimensional structure.

A series of optical slices (shown diagrammatically in Figure 1-15) contains a great deal more information on the subcellular location of the fluorescence, including the shape and relationship between labelled intracellular structures. The position of the optical slices within the cell can often be determined by the size and shape of the "ring" shaped fluorescence emanating from the surface of the cell (the smaller rings in Figure 1-15 are toward the top of the cell, whereas the larger rings are evident when the optical slice is towards the centre of the cell). If you do take a series of optical slices down through the cell or tissue sample you will know from the slice number what position you are within the sample – well almost, as optical effects can result in considerable discrepancy between the physical movement of the focus adjustment and the optical position within the specimen!

In Figure 1-16 human MOLT-4 cells have been surface labelled with fluorescent transferrin at low temperature (on ice) and the cells imaged immediately by confocal microscopy before the transferrin was internalised (left hand panel), and after warming to 37°C to allow the transferrin to be internalised (right hand panel). Surface fluorescence can be seen to display punctate fluorescence as a rim of labelling (the transferrin has formed aggregates on the cell surface), which is particularly noticeable in those optical slices taken through the central region of the cell. Optical slices taken towards the top and bottom of the cell may at first appear to be labelled internal structures, but are, in fact, "dots" of fluorescence located on the lower and upper surface of the cell. Internal labelling can be readily identified using a series of optical slices (Figure 1-16, right hand panel) by the fact that the punctate fluorescence will remain as internal "dots" rather than form a rim of punctate fluorescence when the optical slice is located in the central part of the cell.

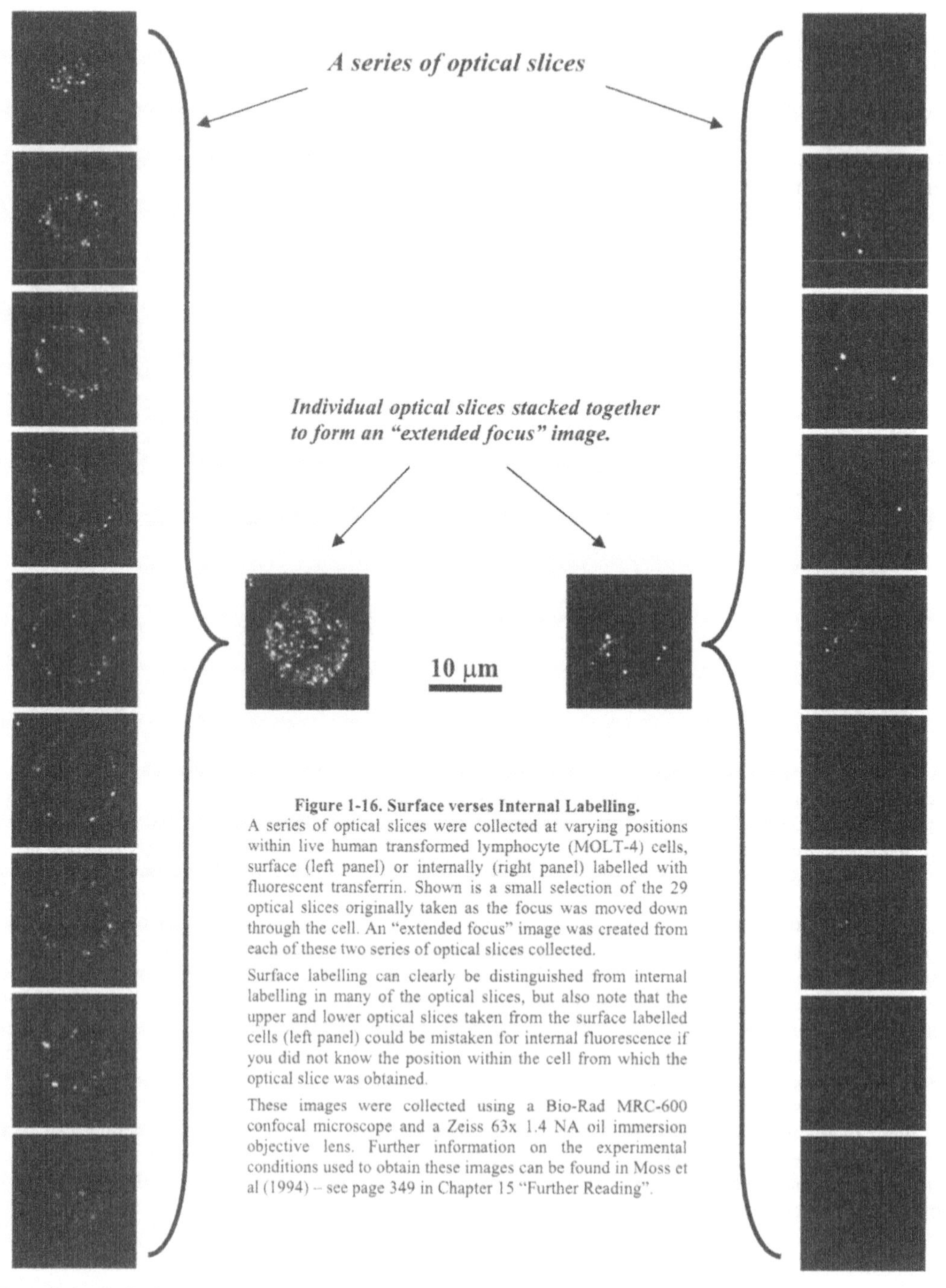

Figure 1-16. Surface verses Internal Labelling.
A series of optical slices were collected at varying positions within live human transformed lymphocyte (MOLT-4) cells, surface (left panel) or internally (right panel) labelled with fluorescent transferrin. Shown is a small selection of the 29 optical slices originally taken as the focus was moved down through the cell. An "extended focus" image was created from each of these two series of optical slices collected.

Surface labelling can clearly be distinguished from internal labelling in many of the optical slices, but also note that the upper and lower optical slices taken from the surface labelled cells (left panel) could be mistaken for internal fluorescence if you did not know the position within the cell from which the optical slice was obtained.

These images were collected using a Bio-Rad MRC-600 confocal microscope and a Zeiss 63x 1.4 NA oil immersion objective lens. Further information on the experimental conditions used to obtain these images can be found in Moss et al (1994) – see page 349 in Chapter 15 "Further Reading".

The surface and internal punctate fluorescence can be projected into an "extended focus" image, but care should be taken in interpreting this type of image as the extensive surface fluorescence shown in the optical slices in Figure 1-16 (left hand panel) could be mistaken for "internal" fluorescence in the "extended focus" image.

These images were collected from live MOLT-4 cells attached, using poly-L-lysine, to the coverslip of a microscope imaging-chamber (to eliminate movement of the cells). Great care should be exercised when collecting a series of optical slices of live cells not to unduly bleach the fluorophore. If quantitation of the fluorescence is not being attempted then an improvement in image display can be achieved by adjusting the contrast of the optical slices to eliminate any "gradient" of fluorescence created by fading of the fluorophore as subsequent optical slices are collected.

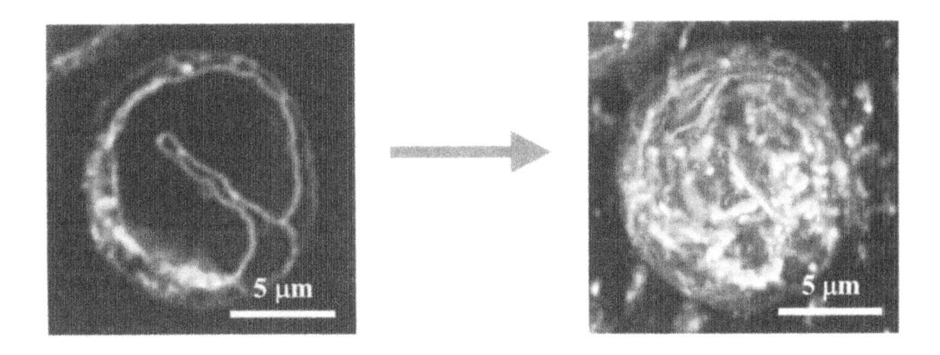

Figure 1-17. "Extended Focus" Image.
Human transformed lymphocyte (MOLT-4) cells, labelled with NBD-ceramide, were imaged with a single excitation line (488 nm blue laser line) on a Bio-Rad MRC-600 laser scanning confocal microscope using a 63x 1.4 NA oil immersion objective lens. The resulting fluorescence was collected in a single channel as a series of optical slices (one such optical slice is displayed in the left panel). The images have then been displayed as a series of optical slices "stacked" on top of each other – often called an "extended focus" image (right panel).

Simple 3D reconstruction of a series of optical slices can be readily carried out using software provided by the microscope companies. 3D information can be presented in a variety of forms, which is explained in some detail in a later Chapter. However, full manipulation in 3D requires separate software. This includes, in the more sophisticated software, the ability to do mathematical calculations on volume and distance on the 3D volume. A variety of 3D rendering programs is briefly described in Chapter 6 "Imaging Software" (page 166).

A very simple method of representing the 3D structure of the specimen is to make an extended focus image by simply adding the optical slices together, as mentioned above. This type of 3D representation has already been shown in Figure 1-16, but is shown again in Figure 1-17 to demonstrate that an "extended focus" image may not always reveal more information. In this case the image has become much too complex and "filled in" to readily interpret the location of the fluorescence. In the left hand panel of Figure 1-17 the deeply indented nuclear membrane, the plasma membrane and cytoplasmic structures can be readily seen, whereas in the "extended focus" image in the right hand panel the cellular structures are obscured by the extensive labelling of membranous and cytoplasmic structures.

A more dynamic way to display 3D information is to use the computer to produce a series of projections from various rotational viewpoints and then animate the rotation series so that the object appears to spin or rotate in space, giving the illusion of a 3D image. High-level 3D rendering software also gives you the opportunity to render in various degrees of transparency, and even to cut the object at any angle for displaying internal structures. Another method of 3D rendering is to create a red and green offset image that gives a distinct impression of 3D when visualised with the appropriate coloured glasses.

The most spectacular 3D rendering method is to create stereo pairs as shown in Figure 1-18. In the top panel a dual labelled image of human foreskin dendritic cells (green) are shown intertwined with the individual cells of the foreskin itself (the nuclei of which are shown in red). This image can be viewed with "3D" glasses designed to be used with stereo pair images, or, with some practice you can see this image in spectacular 3D simply by "crossing" your eyes (see figure legend for details on how this is done). You can readily follow the long projections from the green dendritic cells as they dive down through the skin cells – this is very difficult to see without rendering in 3D, and would be almost impossible to deduce from the data contained in individual optical slices.

The lower panel in Figure 1-18 shows an equally spectacular 3D rendition generated from a series of optical slices taken of cells (shown in blue) growing as a clump attached to Cytodex (dextran) beads (shown in green). Unfortunately 3D rendering techniques are not particularly suitable for scientific publication (unless we could all learn to "see" the stereo pairs shown in Figure 1-18 in 3D by "crossing" our eyes!). The generation of 3D images is certainly useful for elucidating structural associations that are not readily visible in the 2D image, and which could then be described in some detail in a publication.

Figure 1-18. Stereo Pair Images (facing page).

Top Panel

Paraformaldehyde fixed human foreskin tissue was labelled with an FITC labelled antibody specific for dendritic cells (green), and cell nuclei were labelled with propidium iodide (red). A series of optical slices were collected on a Bio-Rad Radiance confocal microscope using a Zeiss 63x 1.4 NA oil immersion objective lens and are displayed here as a maximum projection stereo pair off-set by 15°. This image was created from a series of optical slices collected using an immunolabelled sample kindly provided by Scott McCoombe, University of Melbourne, Australia. These images are quite spectacular if viewed as a stereo pair (see below).

Bottom Panel

HT29 cells were grown attached to collagen type I coated Cytodex (dextran) beads and labelled with Hoechst (blue, nucleus) and wheat germ agglutinin labelled with FITC (cell membrane and Cytodex beads, green). This image is a maximum intensity stereo pair projection (off-set by 15°) from a z-series collected on a Zeiss LSM 510 META confocal microscope using a 25x 0.8 NA water immersion objective lens with coverslip thickness correction collar. This image is used with permission from James B. Pawley and Group 4 (Kerstin Höner zu Bentrup, David W. Dorward, Mikkel Klausen and Kristien J.M. Zaal) of the 2003 course on "3D Microscopy of Living Cells" held in Vancouver, Canada.

How to View Stereo Pair Images

Viewing these images as a stereo pair can be done using stereo glasses, or with some practice by simply focusing each eye on a separate image of the pair. These images are specifically designed to be viewed with "cross-eyed" vision. To try this look directly at the images from a distance of about 30 cm and look cross-eyed at one pair of images. You should be able to see a 3rd image in the middle of the two. If you concentrate on this 3rd image and slowly bring the image in to focus you will be amazed to find that the 3D image is located away from the page (set back or forward) in quite startling detail.

As an aid to viewing these stereo images black dots have been placed above each of the panels – try crossing your eyes and bringing these two dots to merge as a single black dot in the middle of the pair of images. Once the black dot has merged as one (the black dot above each of the images will still be visible in your peripheral vision) the central image will be visible in stereo.

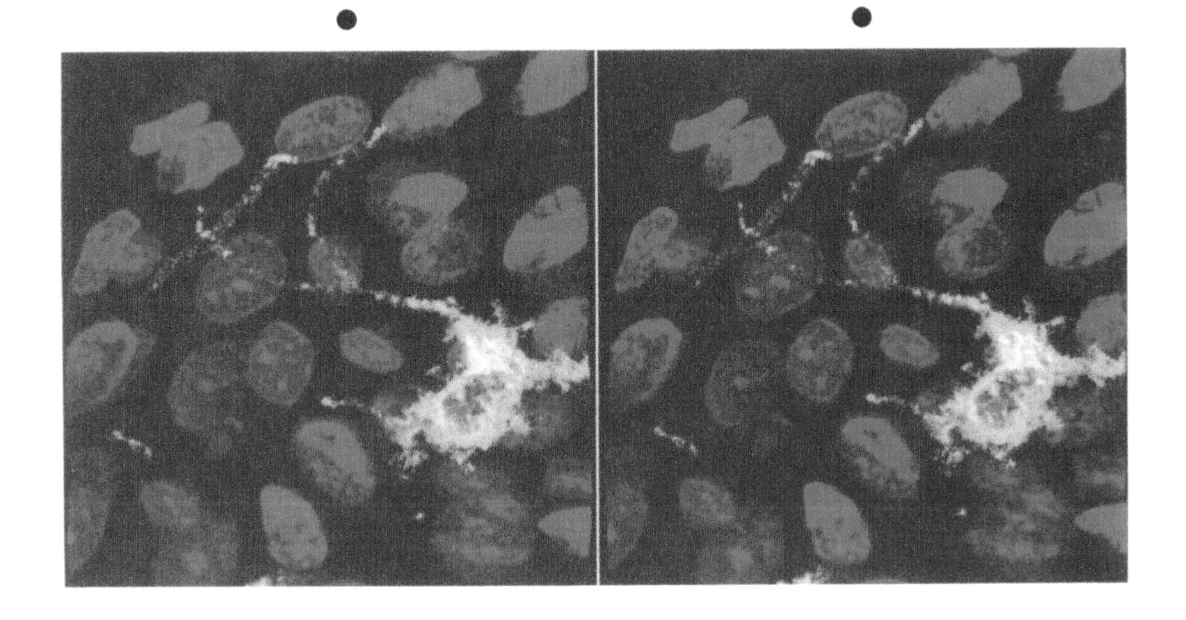

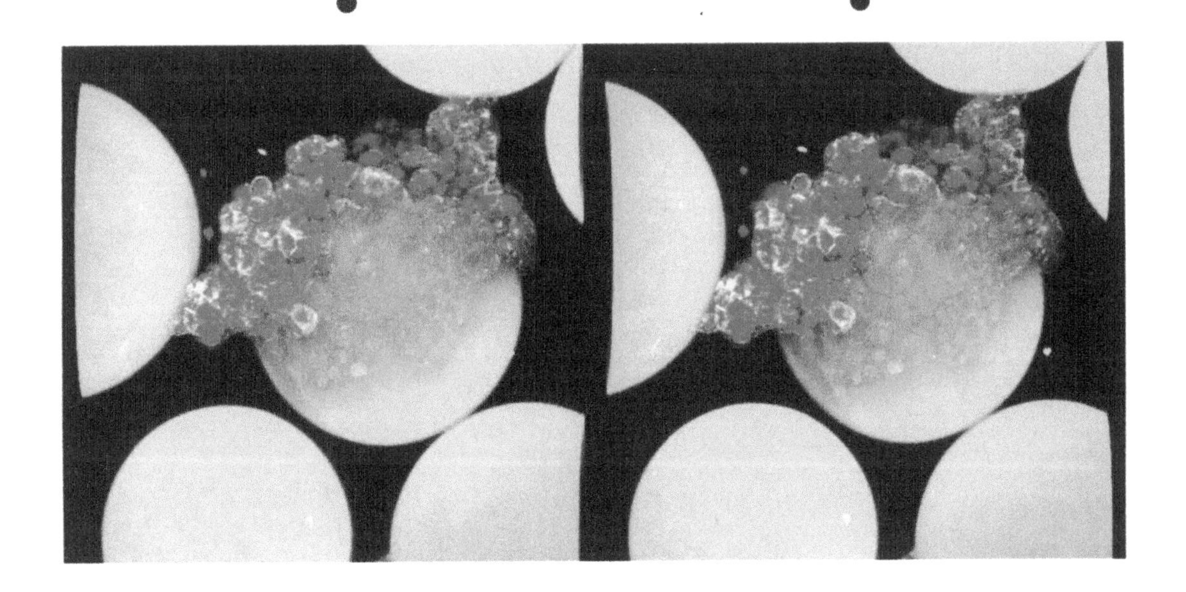

JUST FANCY IMAGES - OR A NOVEL APPROACH TO BIOLOGY?

Confocal microscopy does produce some quite remarkable images. Not only can these images be aesthetically quite beautiful, but they can also provide a wealth of structural information. However, the real power of confocal microscopy is the ability to not only probe the molecular structure of cells, but to follow molecular changes in dynamic living cells and tissues. One should always keep in mind, when imaging biological samples, that the aim of the investigation is to obtain "biological information". This may seem obvious, but quite often you will find yourself attempting to obtain the "perfect" image while trying to satisfy aesthetic criteria, when in fact you have already obtained the relevant biological information!

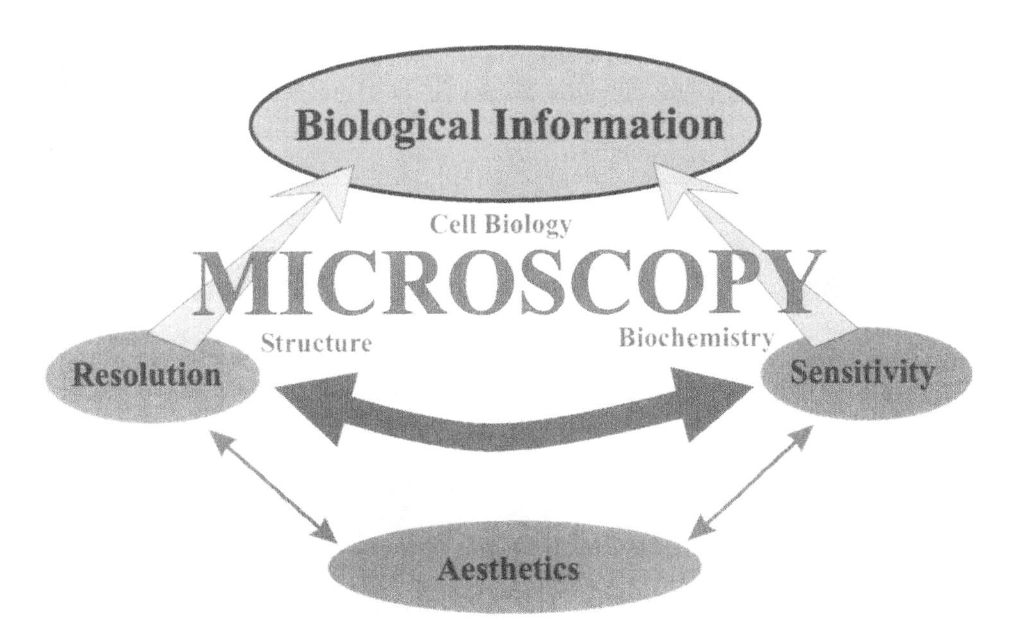

Live Cell and Tissue Imaging

Microscopy has traditionally been involved mainly with fixed and embedded biological samples. The exciting development with the ready availability of confocal and multi-photon microscopy for biological research is the ability to produce excellent images of the structure of live biological samples, and, as will be discussed below, to take microscopy beyond the pursuit of structure – to the development of techniques for following biochemical reactions and molecular associations in living cells.

Biochemistry of Single Cells

A remarkable development in the science of biochemistry in the past few years is the ability to determine the molecular composition and biochemical dynamics of biological samples. This includes the ability to follow a bewildering array of inorganic ions, including such transient ions as Nitric Oxide. The pH of intracellular compartments can be readily determined, and the membrane potential of intracellular organelles can be readily estimated. The mobility of fluorescently tagged macromolecules can be estimated using the interesting technique of Fluorescence Recovery after Photobleaching (FRAP).

Perhaps the most remarkable development of the "biochemistry of single cells" is the technique of Fluorescence Resonance Energy Transfer (FRET) for determining the molecular associations of the biological constituents of single cells. This technique has been made especially powerful with the discovery and subsequent development of derivatives of the Green Fluorescent Protein (GFP) that is allowing us to unravel the molecular associations of complex protein interactions during cell growth, and even the complex multi-cellular events associated with embryo development.

Applications of Confocal Microscopy

Confocal microscopy is now being used in a very large number of fields of biology. In fact, the only real limit is your imagination to come up with a suitable fluorescent molecule to highlight the reaction or structure of interest.

Below is a short list of the diverse applications where confocal microscopy has become very important. This list is not intended to be comprehensive, but simply to give you some idea of the diverse areas where confocal microscopy is now being utilised.

Some of these areas of study have their origin in early studies in fluorescence microscopy in the 1970's and 1980's (particularly studies of calcium fluxes within live cells), but perhaps the more interesting developments are fluorescent dyes that give dynamic information on the metabolic state of the cell, which in the past has usually been studied biochemically by physically disrupting the cell and isolating organelles or particular molecular complexes. The ability to label several cellular constituents simultaneously is creating a very powerful approach for probing the interrelationship between different cellular structures and metabolic fluxes within living cells.

A few applications

- **Immunolabelling** - a traditional area of fluorescence microscopy that has been greatly extended by confocal microscopy.
- **Organelle identification** - mitochondria, lysosomes, nuclei and endoplasmic reticulum etc., can all be readily identified using suitable fluorescent probes.
- **Protein trafficking** - Fluorescent Proteins (GFP, EFP, YFP etc) can be used to track intracellular protein movement by forming a fusion protein with the protein of interest.
- **Locating genes on chromosomes** - hybridisation and fluorescence PCR can be used to locate individual genes to specific locations on individual chromosomes.
- **Analysis of molecular mobility** - recovery from fluorescence bleaching can be used to measure the mobility of macromolecules within living cells.
- **Multiple labelling** - greatly enhanced by confocal microscopy by the perfect co-alignment of simultaneously collected images.
- **Live cell imaging** - very high sensitivity of the confocal microscope, and the ability to image within a cell or tissue sample make live cell work most exciting.
- **Transmission imaging** - unstained live cells and tissue samples can be readily imaged by bright-field, Phase or DIC.
- **Subcellular functions** - *pH* gradients, membrane potentials, free radical formation etc, can be readily measured by using specific fluorescent dyes.
- **Ion concentrations** - accurate measurement of intracellular ion concentrations (Ca^{2+}, Na^{+}, Mg^{2+}, Zn^{2+}, K^{+}, Cl^{-}, H^{+} etc), including following rapid intracellular ion concentration changes.

WHY USE A CONFOCAL OR MULTI-PHOTON MICROSCOPE?

There are a great number of advantages of using confocal microscopy compared to conventional epi-fluorescence microscopy, but the two that stand out the most are, (i) the ability to eliminate out-of-focus light, and (ii) the greatly increased versatility when attempting multiple labelling or gathering 3D information. The confocal microscope also has its limitations – which you should be aware of if you wish to make full use of the instrument's capabilities.

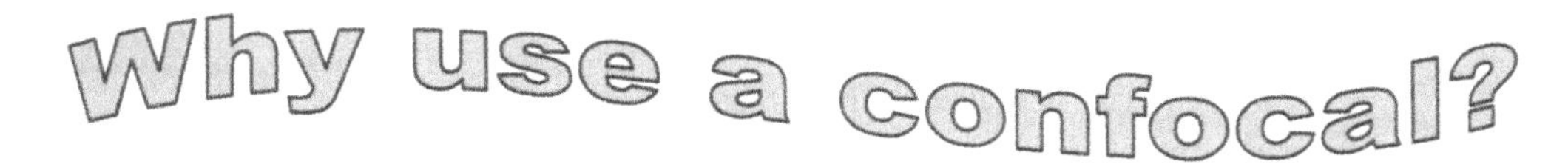

Benefits

- *Better resolution*
- *Increased sensitivity*
- *Optical slice capability*
- *Multiple-labelling*
- *Digital images*
- *Quantitation*

Drawbacks

- *Resolution limitations*
- *Mishandled data*
- *Expensive*
- *Complex to operate*
- *Reporter molecule required*
- *High intensity laser light*

Advantages

Better Resolution

The limit of resolution of a light microscope (i.e. the ability to distinguish two closely adjacent objects) is determined by the numerical aperture of the objective and the wavelength of light. For visible light, using the best objective available, the limit of resolution in the "X-Y" direction is approximately 0.1 to 0.2 µm (Figure 1-19). This limit is approached using a very high quality light microscope that has been carefully and correctly set up for bright-field imaging using a high Numerical Aperture (NA) objective lens. Conventional epi-fluorescence microscopy, even when set up correctly rarely has the same level of resolution. A confocal microscope extends the resolution of the light microscope very close to the theoretical limit of 0.1 µm. The "Z" resolution of the confocal microscope, at approximately 0.5 µm, is somewhat lower than the "X-Y" resolution.

Resolution Limit

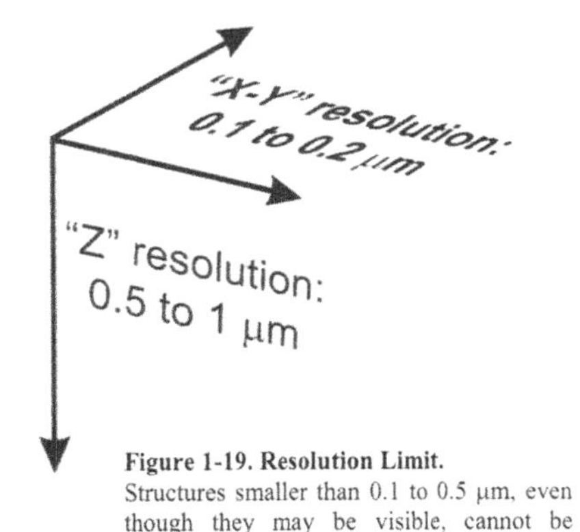

Figure 1-19. Resolution Limit. Structures smaller than 0.1 to 0.5 µm, even though they may be visible, cannot be resolved by conventional light microscopy confocal microscopy, or multi-photon microscopy.

High Sensitivity

High resolution without a high level of sensitivity would be of little use in biology as there would often not be any visible object to "resolve". Hence sensitivity is most important in fluorescent microscopy of biological samples. A laser scanning confocal microscope has high level of sensitivity compared to conventional epi-fluorescence microscopy. However, the sensitivity of the light detectors (photomultiplier tubes) is only part of the storey. In addition to having high sensitivity light detectors, the confocal microscope has the advantage of being able to vary the scan speed, and accumulate and average images. This versatility of image collection greatly increases the sensitivity of the instrument, although with the disadvantage that cellular movement may distort the image. This distortion is less marked if line averaging / accumulation or slow scanning is used rather than whole image accumulation. The confocal microscope also has the advantage of being able to readily attenuate the amount of laser light irradiating the sample. This means that a low laser level can be used to orientate the image, and then a high, but relatively short exposure can be used to collect a high quality image.

3D Imaging

3D Information

The ability to take optical slices through biological material *in situ* (i.e. without the need to fix and physically slice the cells) has meant that a wealth of 3D information is much more readily available. As discussed above, 3D information is available directly from the single optical slices obtained from the confocal microscope. However, reconstructing a 3D representation of a series of slices often results in a stunning image that may reveal structures and architectural relationships not obvious by conventional microscopy.

The information obtained with single cells is usually in accordance with that expected by more traditional light or electron microscopy. However, when analysing tissue samples the information obtained using the confocal microscope can often be "in conflict" with that obtained by more traditional means. The information obtained using a confocal microscope is probably closer to the real structure of the tissue, but stark differences between live cell imaging and more traditional histological examinations may create difficulties in acceptance and publication. It appears that we have come to accept the distorted world of fixed and sliced tissue as "reality"!

Multiple Labels

Perfectly Aligned Dual and Triple Label Images

Dual and triple fluorescence images collected by the confocal microscope are perfectly aligned (as long as the optics of the instrument have been correctly aligned). The images are collected as separate grey-scale images (although a merged colour image is often displayed during collection), and can be readily merged at a later time. Furthermore, perfectly aligned transmission images can also be simultaneously collected. The transmission image can be merged later with the fluorescent image, giving a perfectly aligned multiple image that can be used to locate the fluorescence within the cell or tissue structure. When using the krypton-argon-ion laser (3 colour laser) obtaining real colour transmission images is also possible.

Digital Images can be Readily Manipulated

Digital Images

The advantage of digital images is that they are stored as a computer file that can be readily accessed later to form merged or mosaic images. Furthermore, colour can be readily added, not just for the aesthetics of presentation, but also to clearly distinguish the level of label of interest. Images that show a number of cells can be easily cropped to show one particular cell, and structures of interest can be readily enhanced by simply altering the brightness and/or contrast of the image.

Quantitation of Label

Quantitative

Quantitative information can be readily obtained with great accuracy from the digital image. However, the reliability of the quantitative data not only depends on the accuracy of collection, but also on the nature of the labelling. When attempting to quantify an image it is essential to be confident that the original experiment is itself quantitative. For example, excessive labelling with a suitable dye or antibody will result in saturation of binding sites and will give quite misleading results when attempts are made to quantify the data.

8-bit confocal microscope images are collected with only 256 levels of grey (or 4096 levels in a 12-bit image). These levels are artificially set by the detector gain and offset controls on the microscope and so great care should be taken to only compare images collected with the same settings when attempting image quantitation.

- ADVANTAGES -

Limitations

Objects smaller than 0.1 to 0.2 μm cannot be resolved

Resolution Limited by Wavelength of Light

Although a confocal microscope pushes the limit of resolution to the theoretical limit of light microscopy, a confocal microscope does not resolve better than about 0.1 μm. This limit in resolution is very important in biology as many subcellular structures are at, or below, this size. Objects smaller than this resolution limit can be visible if a suitable dye is used (such as phalloidin to stain sub resolution microtubules), but if they are not resolved then two closely associated structures will appear as one. The limit to the resolution of a light microscope may result in misleading interpretations of the data when one is using two different dyes to establish whether the structures of interest are separated within the cell. Colocalisation of the two dyes may simply mean that the structures are not being resolved by the microscope.

- LIMITATIONS -

Digital Images can be very Easily Mishandled

Digital images are terrific for creating dual and triple merged images and mosaic displays. However, they can be readily cropped or altered to give the distinct impression of labelling in structures that are in fact not labelled. Honesty in image manipulation is very important. Care should be taken when colouring images that the colour facilitates the interpretation of the experiment, rather than adding further complications to already complex data.

Expensive Instrument to Buy and Maintain

Confocal microscopes, and particularly multi-photon microscopes, are very expensive instruments and so are often shared by a diverse range of people within a department or institute. This can create problems of correct care and maintenance of the instrument. Poorly aligned mirrors in a confocal microscope scan head and the laser will both degrade the quality of the image and seriously impact on the sensitivity of the instrument. The krypton-argon-ion laser is particularly expensive to maintain and has a relatively short useful life span.

Difficult to Operate

Considerable training of personnel will be required to gain maximum benefit from a confocal microscope. Obtaining an image using fixed highly fluorescent cells is easy, but working with live cells and low levels of fluorescence can be quite challenging. If you are working with mainly fixed cells or tissue slices you may benefit from the simplicity of using a high quality CCD camera on a conventional fluorescence microscope. Fluorescence imaging using a CCD camera is much more limited compared to confocal microscopy, but significantly easier to use.

Reporter Molecule is Required for Detection

You need a fluorescent tag – which may be very bulky, influence ion concentrations, and be toxic to live cells!

An externally added reporter molecule is almost always required to take full advantage of the capabilities of a confocal microscope. This molecule may contain a bulky fluorescence group in addition to the ion chelator, drug or antibody etc specific for the particular ion or molecule being studied. The reporter molecule may also alter the location or concentration of the ligand of interest simply by binding to the molecule or ion under study. Care in the design and execution of experiments will minimise the problems of bulky fluorescent molecules. The fluorescent molecule may also be toxic to live cells, particularly when irradiated. Take care to use the lowest concentration of the fluorescent dye as possible, and to collect images with the lowest possible laser intensity as soon as the cells have taken up sufficient dye.

Damaging High Intensity Laser

The high intensity of the laser when focussed to a fine spot within the sample can result in photo damage to both the dye being used, and to other cellular constituents. Using suitable antifade reagents can control photo bleaching of the dye. Antifade reagents work well in fixed cell preparations, but are not as effective and are often toxic to live cells.

The laser can cause damage!

Chapter 2

Understanding Microscopy

To take full advantage of the exciting developments in confocal and multi-photon microscopy, it is important to have some understanding of light microscopy. It is not only useful in selecting the correct objective lens and microscope for the task at hand, but this knowledge will also allow one to interpret with greater confidence the images obtained. This chapter gives a brief introduction to light microscopy, with a strong emphasis on the practical aspects that you need to know about to take full advantage of the light microscope when working with a confocal or multi-photon microscope. The specific instrumentation required for confocal microscopy is dealt with in detail in the following chapter (Chapter 3 "Confocal Microscopy Hardware", page 65).

For a more detailed understanding of microscopy please refer to the specialist books listed in Chapter 15 "Further Reading" (page 347). In particular, the book "*Fundamentals of Light Microscopy and Electronic Imaging*" by Douglas B. Murphy (Wiley-Liss, 2001) is an excellent introduction to light microscopy for biologists. Douglas Murphy's book covers microscopy (including a chapter on confocal microscopy) and electronic imaging in considerable detail, but with concise and careful layout that makes his publication ideal for both the novice and more advanced user. Another excellent introduction to a practical understanding of microscopy, image formation and setting up a light microscope correctly is the book "*Light Microscopy: an Illustrated Guide*" by Ron Oldfield (Wolfe, 1994).

ESSENTIAL OPTICS

Light has the interesting characteristic of behaving as both a wave and a particle (which is known as a photon). These dual aspects of light are sometimes difficult to comprehend as existing together, due to the macro-world in which we live not having equivalent phenomena to which we can compare them. Some knowledge of the properties of light is most helpful in understanding both the limitations and the possibilities that relate to the use of light microscopy in biology.

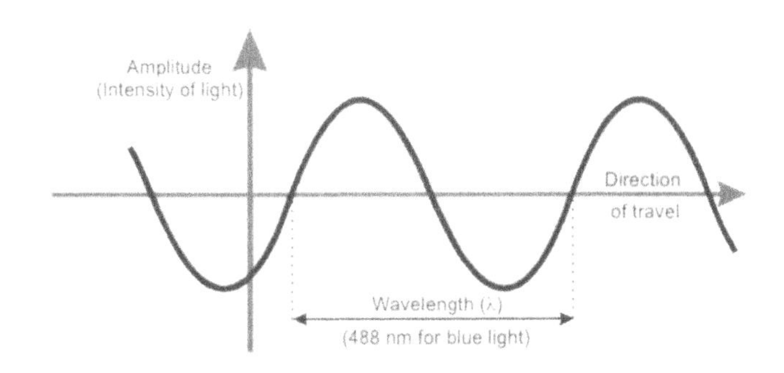

Figure 2-1. Wavelength of Light.
The wavelength (λ) of visible light ranges from approximately 400 nm (blue) to around 650 nm (red). Light has an amplitude (intensity) that is shown in a wave diagram as the height of the wave above or below the central axis.

Wave Nature of Light

Visible light is a small part of the electromagnetic spectrum that ranges from very long wave radio waves to extremely short gamma rays. Visible light has a wavelength (denoted λ, see Figure 2-1) that ranges from around 400 (blue) to 650 nm (red). The wavelength of the light will decrease as the light traverses material of a higher refractive index (such as glass). However, the frequency (cycles per second) of the light does not alter as the speed of the light also decreases on traversing the glass. The measurement given for the wavelength of specific colours of light is that derived in a vacuum, although the wavelength in air is almost identical for most practical purposes.

The wavelength of the electromagnetic radiation determines the ultimate size of structures visible by microscopy. Blue light corresponds to a maximum resolution of approximately 0.25 μm (the exact value depends on the wavelength of blue light used, the numerical aperture of the lens and the way in which the sample has been prepared). In contrast, electrons, which have a much shorter wavelength than visible light, are capable of imaging down to the molecular size of 0.1 nm. A high quality research light microscope is capable of a resolution close to the theoretical limit of 0.5 to 0.2 μm for visible light microscopy, but unfortunately using normal light microscopy one cannot exceed this limit. There are unusual forms of light microscopy that can exceed this limit – but they are still at the early development stage. Although eukaryotic cells are considerably larger than the wavelength of light, intracellular organelles and prokaryotic cells are similar to, or smaller than, the wavelength of light used for imaging. These important biological structures are thus at the limit of resolution of a light microscope.

Interacting Light Waves

The formation of an image in a light microscope involves the interaction of the light with the object, particularly the bending of light rays as they pass very small structures (diffraction), and of course the bending of the light rays by the various lenses in the microscope itself (refraction). The way in which light interacts, both with light itself and with other matter, is the key to both the formation of an image and for understanding the resolution limit of a light microscope.

Light waves that are in the same physical location, including light rays that are intersecting, although the two rays are on a different trajectory, can interact to produce what is known as interference (Figure 2-2). This can result in constructive interference, where the intensity of light (amplitude) is increased, or destructive interference, where the intensity of light decreases. The maximum possible increase is when the two waves are exactly aligned, of the same wavelength and in phase with each other, with the resultant wave being the sum of the amplitude of the

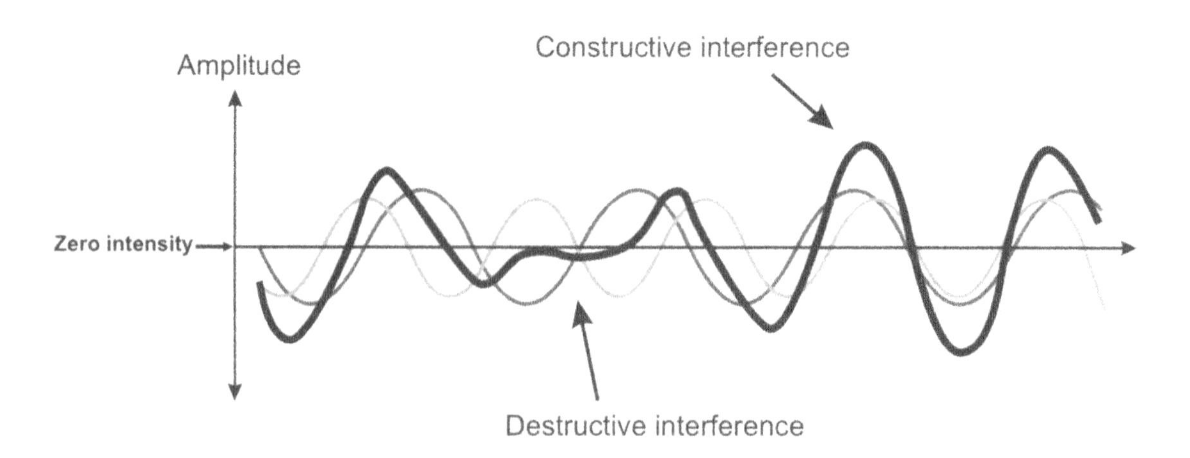

Figure 2-2. Interacting Light Waves.
Light waves interact in a way that is termed destructive (loss of light) or constructive (increased light) interference. The interaction can completely cancel out all incident light as long as the wavelength, direction of travel etc. of both light waves are identical but in opposing amplitude. A more realistic interaction between light waves results in partial destructive and partial constructive interference to give the final waveform as shown in darker outline in this diagram.

individual waves. Likewise, a complete cancellation would only occur when the interacting waves are of the same intensity, the same wavelength and propagating in the same direction, but ½ wavelength out of phase. Of course, in reality, these conditions are rarely met. Interacting waves rarely completely cancel each other, nor do they combine to create a wave of twice the magnitude. What usually happens is that the waves are partially interacting, creating a complex wave, which is a sum of both of the waves – including both constructive and destructive interference. Light

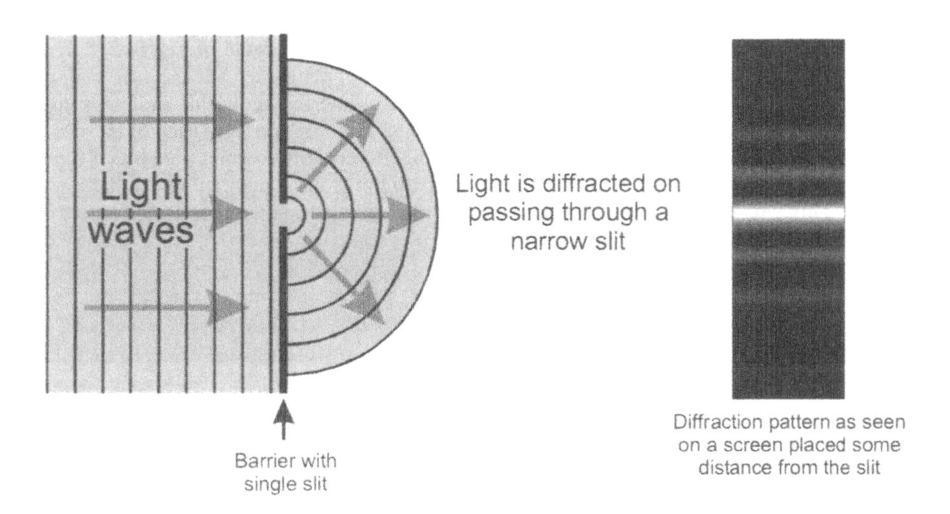

Figure 2-3. Diffraction of Light.
As light passes through a narrow slit the waves are diffracted, resulting in the light spreading out from the slit. The light waves that are diffracted interact in a way that is often described as interacting "wavelets" to form a pattern of light and dark areas. A screen placed some distance from the slit will show a characteristic banding-pattern as shown diagrammatically on the right of the above diagram.

waves can pass through each other, interacting to create varying localised intensities, and continue onwards unaffected by the encounter with the other wave. The process of interference is responsible for creating the real intermediate image of an object in a light microscope.

The importance of constructive and destructive interference in light microscopy is discussed in more detail shortly, but first we need to look further at another characteristic of light called diffraction. On passing close to the edge of an object light appears to "bend" around the object, in the same way as waves on a lake will bend around a pier and proceed to shore. This "bending" is referred to as diffraction. Diffraction is particularly marked when the light interacts with structures that are close to the wavelength of the light used for illumination. In the macro-world we are all so familiar with, the wavelength of light is very short compared to the everyday objects we see the light interacting with. A shadow is cast by a lamp post simply because the wavelength of light is considerably smaller than the diameter of the post. Sound waves or radio waves, which are much longer ("cm" in length) than the wavelengths for visible light ("nm" in length), will "bend" around the post.

In the diagram shown in Figure 2-3 the light is passing through a narrow slit that is smaller than the wavelength of the light. A high level of diffraction results in significant "bending" of the emerging light waves. The amount of "bending" or diffraction is determined by the relative size of the slit compared to the wavelength of light. A very wide slit will result in almost no discernible bending of the light rays, but making the slit narrower will result in a much greater spread of light (much greater diffraction). However, the slit does not have to be narrower than the wavelength of the light for a significant level of diffraction to occur.

One may expect that the diffracted light would create a "fan" or "cone" of even illumination emanating from the slit. This "cone" of illumination does exist, but is highly uneven, due to the complex interactions between the diffracted light rays. The diffracted light emanating from the slit will illuminate a screen placed some distance from the slit in a characteristic banding pattern, with a strong band of light in the centre and surrounded by bands of light with decreasing intensity (see the right hand panel in Figure 2-3).

In the above example light is shown to interact even when passing through a single fine slit. If you increase the complexity by adding a second slit there is a greatly increased level of interaction (Figure 2-4). In this case two small slits (somewhat smaller than the wavelength of light being used) are placed approximately one wavelength apart. The light waves that emerge from the slit are highly diffracted due to the relatively small size of the slits, fanning out in a wide arc from each of the slits. The two diffracted wave fronts now interact to create a complex pattern of constructive and destructive interference. Placing a screen at some distance from the double slit will result in a characteristic banding-pattern as shown diagrammatically in the right hand panel of Figure 2-4. The band with the highest intensity is now no longer the light directly opposite the slit opening as was the case with a single slit, but the region between the two slits (a region of maximum constructive interference). This banding pattern is due to constructive and destructive interference between the two wave fronts that emanate from each of the slits.

Diffraction is an important characteristic of light that allows one to construct a microscope that can image very small objects, but diffraction is also the phenomenon that eventually limits the size of an object that can be resolved by light microscopy. This will be discussed in some detail shortly, but first another important aspect of the wave nature of light, the process of refraction, will be introduced.

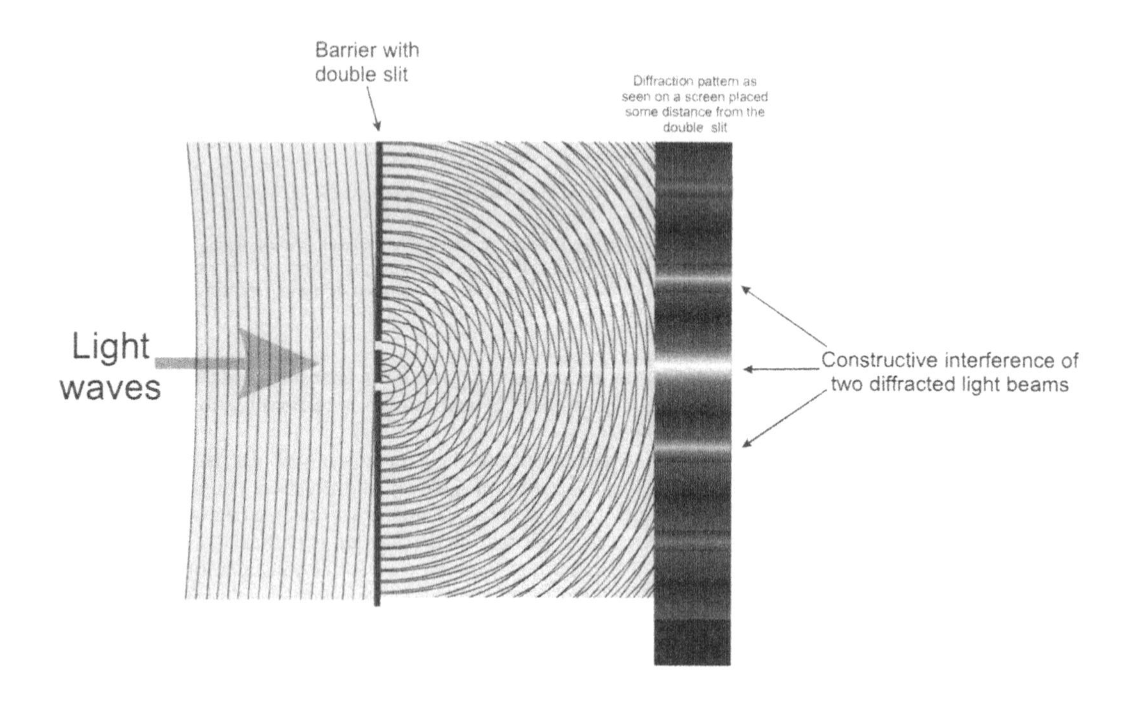

Figure 2-4. Interference Pattern using Two Slits.
Light waves will create a curved wave front by diffraction on passing through two slits that are smaller than the wavelength of irradiating light. If two slits are placed approximately one wavelength apart the waves will interact to create a banding pattern of destructive and constructive interference. This pattern of interference can be seen in the above "rings" of waves. However, this banding pattern is not readily visible when using light, unless the light is shown on a display screen as shown on the right hand side of this figure.

Refraction and the Bending of Light Rays

When light passes from one media to another with differing refractive indices (for example, when going from air into glass) the light will appear to "bend" on entering the second media (Figure 2-5). This phenomenon is known as refraction, and is the basis for focusing light rays using the curved surface of a lens. The amount by which the light rays are bent is determined by the angle at which the light rays approach the higher refractive index material and the refractive index difference between the two materials. Light rays that approach the material at 90° to the surface (zero angle "*i*", or "normal" in Figure 2-5) will slow down on entering the higher refractive index material, but will not be bent. Light waves that approach the surface at a large angle (a large angle "*i*" in Figure 2-5) will also slow down, and there will also be a change in the direction of propagation of the light ray. When the light ray exits the material of higher refractive index the light will again be bent, but this time away from the "normal" by exactly the same amount that the light was bent towards the normal on entering the glass, resulting in a light ray that is now travelling in the same direction as the original, but off-set by an amount determined by the angle "r" and the distance travelled through the glass.

The refractive index of materials that are able to transmit light is defined as the ratio between the velocity of light in the two materials (see Figure 2-5). However, at a practical level the refractive index is usually determined by comparing the light transmitting properties of the material with liquids of known refractive index. The common light transmitting materials used in microscopy are, air, water, glass and immersion oils. The index of refraction of air is very close to 1.000 (no refraction). In contrast the index of refraction of the glass used to make the lenses in the microscope is as high as 1.515. The media used for live cell microscopy has a refractive index of very close to that of water (1.333), which is significantly different to that of air and glass. Note that different types of glass can have

Material	Refractive index (η)
air	1.000
water	1.333
alcohol	1.360
glycerol	1.470
lead glass	1.70 - 1.80
soda lime glass	1.50 - 1.54
coverslip glass	1.515
immersion oil	1.515

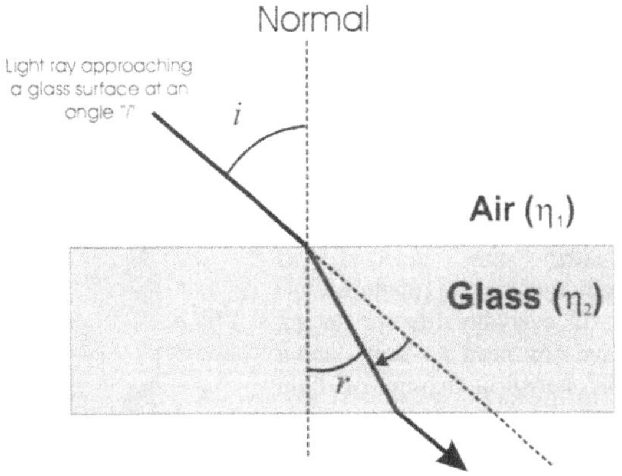

$$\text{Refractive Index } (\eta) = \frac{\text{Velocity of light in a vacuum (air)}}{\text{Velocity of light in material}}$$

Normal

Light ray approaching a glass surface at an angle "*i*"

i

Air (η_1)

Glass (η_2)

r

$$\eta_1 \sin i = \eta_2 \sin r$$

(Snell's law)

Figure 2-5. Refractive Index.
Light waves passing from a material with a lower refractive index (for example, air, with a refractive index of 1.000) to a material with a higher refractive index (for example, glass, with a refractive index of 1.515) will slow down. Light rays that enter the higher refractive index material at an angle "*i*" will be bent towards the "normal" (as shown in the diagram on the right). The light will be returned to the original direction (but the path will not be the same) once it emerges on the opposite side of the glass. The refractive index of various transparent materials relevant to microscopy is shown in the table on the left.

widely differing refractive indices (for example lead glass compared to soda-lime glass, Figure 2-5). This difference in refractive index between the lens, the sample and sometimes the intervening space, is an important consideration in microscopy, particularly when imaging live cells, and is discussed in more detail in the following section on "Basic Microscopy".

Light passing at an angle from one media to another with different refractive indices is always bent – this requires careful consideration of the path that the light will take, not only through the coverslip and immersion media and the lens, but also through the sample itself. Traditional light microscopy of fixed and embedded samples is designed to minimise the refractive index mismatches and thus refraction changes as the light passes through both the specimen and coverslip. In this case the mounting media is designed to have a very similar refractive index to the glass coverslip, the immersion media and the glass of the lens. There is a serious refractive index mismatch when using a dry objective (in which case the thickness of the coverslip is critical, which means that the coverslip must be correct for the design of the lens – discussed in more detail shortly), and when imaging live cells in water-based media.

When imaging live cells there is less scope for managing the refractive index of the sample compared to fixed cells. The cells, being close to the refractive index of water, are seriously refractive index mismatched with the lens. Using a water immersion objective will alleviate much of this mismatch, but even this is not a perfect system as the cells themselves are not an exact refractive index match to the water based cell incubation media. In fact, the cells have a highly variable refractive index. The nucleus, subcellular organelles and cell membranes all have somewhat different refractive indices compared to the cytoplasm, all of which are somewhat different to the surrounding aqueous media.

Counting Photons

Although many of the characteristics of light can be readily explained by describing light as an electromagnetic wave, light is also made up of discrete energy packets called photons. This means that sometimes light needs to be treated as distinct particles with statistical characteristics. Light travels as a wave, but it is created and detected as individual quanta (photons).

In everyday life we do not have any need to worry about the statistical nature of light because there are so many photons (for example, when reading this page) that individual photons are of no concern to us. However, when dealing with the very low light levels which are often encountered in fluorescence and particularly confocal and multi-photon microscopy, the statistical nature of light becomes an important factor in the creation of high quality images.

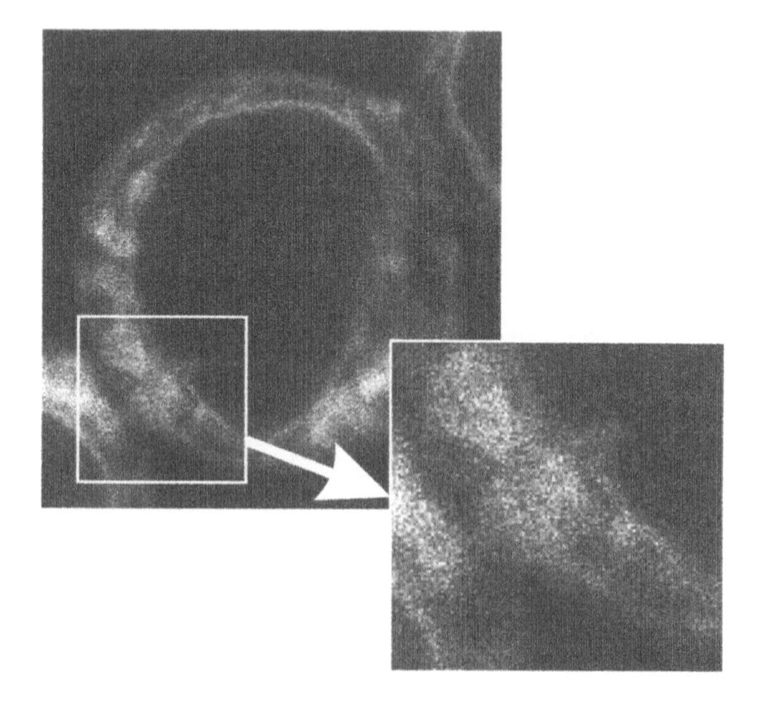

Figure 2-6. Poisson Noise.
Images collected from samples that exhibit low levels of fluorescence will appear "speckled" due to the statistical nature of light. In this example, HL-60 cells have been labelled with a relatively low level of DiI and then the image of the living cells collected using a single scan with a Leica SP confocal microscope using 488 nm blue laser light.

The statistical nature of light

will already be familiar to anyone who has tried imaging a relatively poorly labelled sample, as annoying "speckle" in the image (Figure 2-6). In the example shown in Figure 2-6 live transformed human lymphocytes (HL-60 cells) have been labelled with a relatively low amount of the membrane dye DiI. To avoid photobleaching the sample, or harming the integrity of the cells, a relatively low laser intensity has been used to capture this image using a single frame scan. This "speckle" is associated with the cellular structures labelled, including the central part of the nucleus, but is not associated with the "background" that lies out-side the confines of the cell. The observed "speckle" in the image is due to relatively sudden intensity changes between individual pixels in the image – due to statistical variation in the number of photons that reach the detector (Poisson statistics). "Speckle" in the background, particularly when no sample is actually present, is due to "noise" in the electronics of the instrument. Most confocal microscopes have a relatively low level of instrument noise, which only becomes obvious if the PMT voltage or gain level is set to maximum. In Figure 2-6 the gain level is not sufficient for the "background" noise to become evident.

When fluorescent light intensity levels are low in confocal microscopy there may be as few as 1-10 photons detected per pixel, even though the grey levels shown may range between 0 and 255. Individual photons arrive at the detector at random time points, creating a degree of variation in the number of photons allocated to each pixel in the image. During the dwell time for collection of light for a particular pixel, there may be, for example, only 10 photons detected, but due to the random distribution of the incoming photons there will be a variability of $\sqrt{10}$ (i.e. there will be on average 10 photons +/- 3 photons). This means that the next pixel, with, for example, the exact same amount of fluorescent material as the first, could vary by as much as 6 photons compared to the previous pixel simply on the basis of random fluctuations in the arrival of the photons. This statistical characteristic of light is sometimes referred to as "shot noise", but is more correctly called "Poisson noise".

Line averaging, frame averaging, slowing the scan rate, or opening the pinhole will result in more photons reaching the detector for each pixel in the image. Increasing the number of photons collected will result in an increase in signal that is greater than the increase in the "noise" – thus decreasing the ratio of signal to noise and lowering the effect of Poisson noise. This is why excellent images can be generated from very poorly stained fixed biological tissues by averaging 20 or more individual scans. Although frame averaging is often not practical on live cells due to movement, a considerably improved image may often be collected by slowing the scan rate or allowing a small number of "line-averages" during the scan.

Opening the confocal pinhole will also greatly increase the amount of light reaching the detectors, greatly reducing the level of Poisson noise in the image. However, opening the confocal pinhole will decrease the resolution of the microscope, particularly in the z-direction. A small increase in the size of the confocal pinhole is often a useful compromise for collecting better quality images.

Image Collection by "Photon Counting"

"Photon counting", where electronic circuitry is used to "count" each photon detected by the photomultiplier tube, can be carried out on some confocal microscopes. Photon counting is particularly good at generating excellent images from very low intensity fluorescence from fixed biological material (many scans may need to be accumulated, for enough photons to be "counted" per pixel, to generate an acceptable image brightness).

However, the image collection mode more commonly used in confocal and multi-photon microscopy, is for a voltage, proportional to the PMT output current, to be digitised (this will be a combination of signal from photons that have entered the photomultiplier tube, including combinations of several photons, and spurious electronic signals generated by the detector itself), and is then displayed as a greyscale level for each individual pixel.

BASIC MICROSCOPY

A basic understanding of microscopy, while not essential for obtaining an image on a confocal microscope, does mean you will be able to both push the microscope to its limits, and have much greater appreciation of the physical constraints placed on light microscopy. There are many books that attempt to describe the formation of an image and the importance of various optical components in a compound microscope (see Chapter 15 "Further Reading" for recommended books, page 347). I have found, however, that the diagrams and accompanying explanations in these texts are often not easy to understand. A much better way to understand the optics of the light microscope are web based tutorials like those found on the Olympus (www.olympusmicro.com/primer/) and the Nikon (www.microscopyu.com/articles/confocal/index.html) web sites. These tutorials have a number of excellent diagrams and clearly written text, but more importantly the authors have put together a number of interactive Java applets that let you, for example, change the NA of a lens and watch the change in the cone of light that traverses the optical path of the microscope.

Image Formation using a Simple Lens

The formation of the image in a light microscope is best understood by first briefly looking at image formation in a simple lens. A simple single lens can be used to create a magnified image of an object (Figure 2-7). In this example the object has been placed further away from the lens than the focal length of the lens (f). Light emanating from the object (including light diffracted by the object) is focussed by the lens to create a magnified image at a distance further away from the lens than the focal position f_i on the opposite side of the lens to the object. The diffracted light rays interact at the position of the image plane to create what we perceive as an image. At the back focal plane (f_i) of the lens, an image is not formed, but a characteristic diffraction interference pattern will be found. The resolving power of a simple lens is limited by a number of optical aberrations that seriously degrade the image. The objective of a microscope may contain a large number of individual lens elements to correct for many of these optical aberrations.

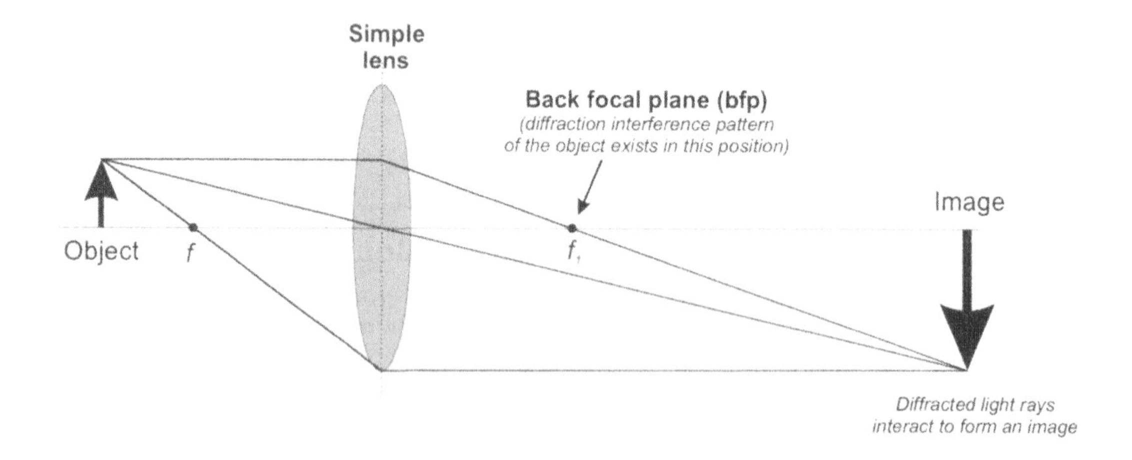

Figure 2-7. Simple Lens.
Image formation using a simple lens is shown in this diagram, where the object (a right-way-up arrow on the left of the diagram) is placed outside the focal point (f) of the lens. A real, but inverted image will be produced on the other side of the lens at a distance from the lens greater than the focal distance f_i. This type of image formation can be readily demonstrated using a simple hand held lens by looking through the lens at an object further away than the focal distance on the other side of the lens and placing your eye outside the focal distance of the lens. Note that this is different to the image formed by placing a simple magnifying lens close to the object (in which case the object will be closer than the focal distance of the lens), and a large "virtual" right-way-up image will be visible by eye.

Compound Microscope

A light microscope uses the same principle as the simple lens to create an image, but it is known as a compound microscope because it uses a combination of an objective lens and an eyepiece lens to form the final image. The most critical component of a microscope is the objective, which forms the intermediate image and is the determining element limiting resolution in the compound microscope. The eyepieces further magnify the image formed by the objective. In laser scanning confocal and multi-photon microscopy, the light from the objective is directed to the scan head and associated optics through a scan lens without passing through the eyepieces of the microscope.

The theory of image formation in a light microscope was proposed by Abbe in the late 1800's, and explained how diffraction at the object plane produces light that is later focussed to form the image. Smaller features diffract light at larger angles, and will only be imaged if the objective can accept these highly scattered light rays (a high NA objective, see page 42).

The most widely accepted method of illumination and microscope adjustment for light microscopy is called Köhler illumination (see box on next page). This method of illumination recognizes a series of conjugate focal planes (Figure 2-8). One set of conjugate focal planes consist of the image forming light rays (conjugate field-planes shown on the left in Figure 2-8) and the other set consist of the illuminating light rays (conjugate aperture-planes shown on the right in Figure 2-8).

The illuminating rays project an image of the light source (sometimes a lamp filament, but often a ground glass screen) onto the front focal plane of the condenser (position 2′ in Figure 2-8). This is where the variable iris of the condenser is located, which should be adjusted so that the image of the lamp fully fills this diaphragm. The illuminating light rays then pass through the specimen. As there is no image of the light source formed at the position of the specimen, the sample is uniformly illuminated, even when a lamp filament is used as a light source. An image of the lamp source is then formed again at the back focal plane (BFP) of the objective (position 3′ in Figure 2-8).

The other set of conjugate planes, the image forming rays (the conjugate field-planes), are focussed at the position of the specimen (position "2" in Figure 2-8), and also form an image at the position of the field iris (position "1" in Figure 2-8) and the intermediate image plane (position "3" in Figure 2-8) and of course on the retina of the eye.

The correct adjustment of the substage condenser, which is described in some detail in the following section, is important for obtaining maximum resolution and good contrast when imaging by bright-field conventional microscopy or transmission imaging on a confocal microscope. As has been mentioned previously, conventional epi-fluorescence microscopy, confocal microscopy and backscatter imaging utilise the objective lens as the condenser, resulting in a perfectly matched condenser lens without the need for further adjustment. A multi-photon microscope often has external detectors that collect the fluorescent light emanating from the sample before the light enters the combined objective/condenser lens.

Removing "dust" from the image

If the microscope has been correctly set up for Köhler illumination (see box on following page) then there will be several conjugate image planes (Figure 2-8), which are all in focus at the same time. "Dirt" superimposed on an image of the sample will be due to dust or grime on a lens component close to one of the conjugate imaging planes. The lens element on the back of the eyepiece is often the main culprit in forming such images.

Dust or oil smears in other parts of the optics of the microscope will not be "seen" as particles etc, but may detract greatly from the image. A common problem that seriously affects the image, but is not visible as distinct "particles" is oil (and associated dust) on the front lens element of the objective or the front lens of the condenser. Oil that has not been cleaned properly from an oil immersion lens may attract dust particles that will remain attached to the lens on addition of further oil.

If "dirt" particles are "in-focus" from other optical positions within the microscope, possibly the microscope has not been set up correctly for Köhler illumination.

Köhler Illumination

Bright and even illumination, with high resolution and good contrast for transmitted light imaging is best achieved by setting up the microscope for Köhler illumination. Although epi-fluorescence imaging (including fluorescence imaging in a confocal and multi-photon microscope) does not require setting up the lamp and substage condenser for Köhler illumination, correct set up for bright-field microscopy is required for laser scanning transmission imaging.

1. Check that the microscope lamp is focussed on the front aperture of the condenser (close the condenser aperture and observe the image of the light source on the leaves of the aperture).
2. Focus a specimen (a high contrast stained slide is best). Do not change the focus controls of the microscope further during the following procedures.
3. Partly close the field diaphragm.
4. Focus the image of the field diaphragm by adjusting the condenser focus knob.
5. Centre the image of the field diaphragm within the field of view by using the adjustment screws on the condenser.
6. Open the field diaphragm until its margins are just outside the field of view.
7. Remove an eyepiece and look down the microscope tube and view the bfp of the objective (or use a Bertrand or "phase" lens, which is sometimes provided on a convenient optical filter wheel on research microscopes). An image of the condenser iris should be visible.
8. Adjust the aperture of the condenser iris to about 3/4 fill the bfp of the objective. Closing the condenser iris further will increase contrast, but with a loss in resolution (the full NA of the objective is not being utilised). The 3/4 position is a compromise between resolution and contrast.
9. Control the intensity of the illumination by inserting Neutral Density optical filters or by adjusting the voltage setting on the lamp – or the intensity of the laser and detector gain on the confocal microscope. Do not use the field diaphragm or condenser iris to adjust the light intensity.

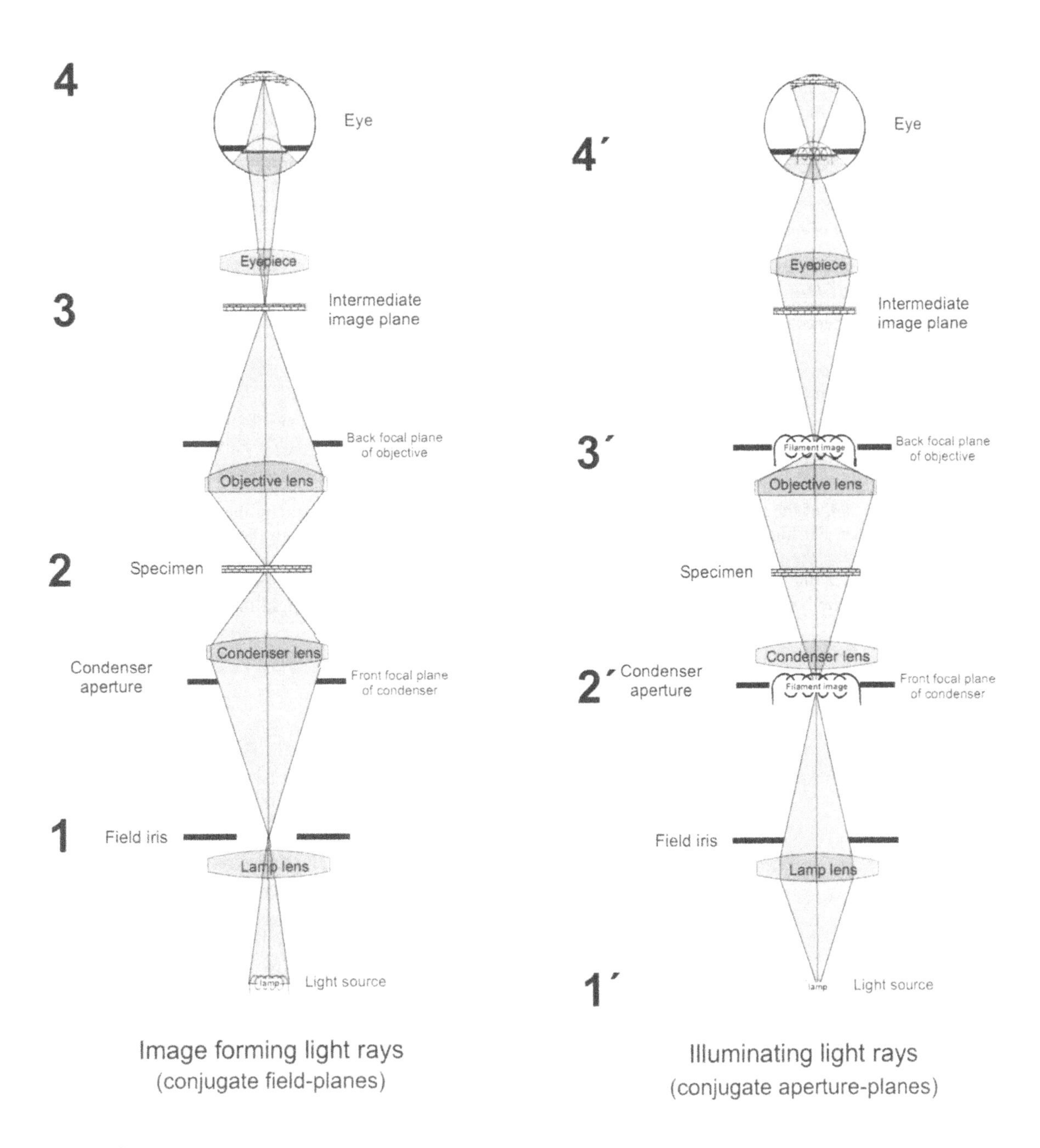

Figure 2-8. Image Formation and Köhler Illumination.
Image formation in a compound microscope adjusted for Köhler illumination is based on having a series of conjugate focal planes. Four conjugate field-planes or image-planes are shown in the diagram on the left (1 - 4) and four conjugate aperture-planes are shown in the diagram on the right (1′- 4′). Köhler illumination for bright-field microscopy and transmission imaging on a confocal microscope is obtained by correctly adjusting the focus and the opening on the condenser iris (see Köhler illumination box on the previous page). However, in epi-fluorescence imaging (including conventional and confocal or multi-photon fluorescence microscopy) and backscatter imaging the objective lens also becomes the condenser. The substage condenser is not utilised for illuminating the specimen in these imaging modes.

Numerical Aperture (NA)

The Numerical Aperture is the angle of light intake to a lens (Figure 2-9), and is the defining feature in determining the resolution of the lens. A relatively low NA of 0.45, as shown in the left hand diagram of Figure 2-9 has a half angle, α, of 27°. This is in contrast to the higher NA of 0.75 shown in the right hand diagram of Figure 2-9, with a half intake angle of 48°. The intake angle of the lens is important in determining its light gathering properties and the resolution limit. This means that a higher NA lens will not only result in a brighter image in low light conditions (such as fluorescence microscopy), but also that the wider angle of light collected will mean that highly diffracted light rays emanating from the sample will be collected by the lens. These highly diffracted rays determine the limit of resolution of the lens, and so the higher lens NA, the smaller the object that can be resolved by the microscope.

Importance of the Condenser

> **Don't be confused:**
>
> "Field Aperture" is used to adjust the area of the sample illuminated.
>
> "Condenser Iris" is used to control the angle of the cone of illuminating light.

In epi-fluorescence microscopy, which includes confocal and multi-photon fluorescence microscopy, the sample is illuminated by light that is directed towards the specimen through the objective lens. In other words, the objective lens acts as both the condenser, for providing the illumination, and the objective for forming the image. When imaging by bright-field microscopy (including transmission imaging using a confocal microscope) the NA of the illumination provided by the condenser is critical for obtaining the maximum resolution from the objective. For high resolution imaging matching the NA of the condenser with the NA of the objective is important. Changing the condenser every time you change the objective is not practical, therefore a condenser of high NA (as high as the highest NA objective to be used is preferable) is normally used. Closing the iris on the condenser lowers the NA of the condenser, by diminishing the angle of the cone of light that is used to illuminate

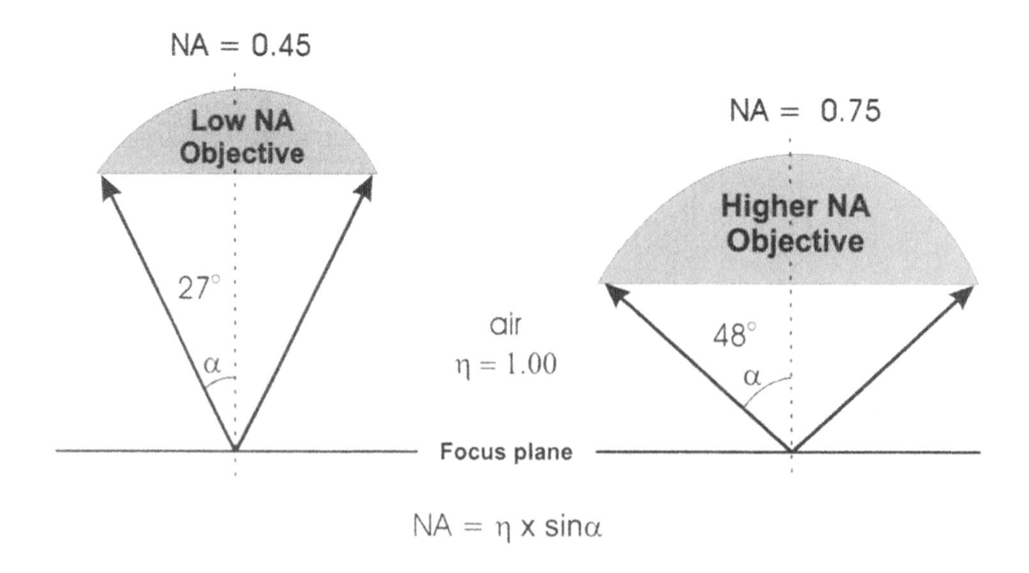

Figure 2-9. Numerical Aperture (NA).
The numerical aperture of a lens is an important factor in determining the resolution of the lens. A relatively low NA lens (0.45 in the example on the left above) has a relatively small light intake angle (the half angle α, having a value of 27° in this example), compared to the higher NA lens on the right (NA=0.75) with half intake angle of 48°.

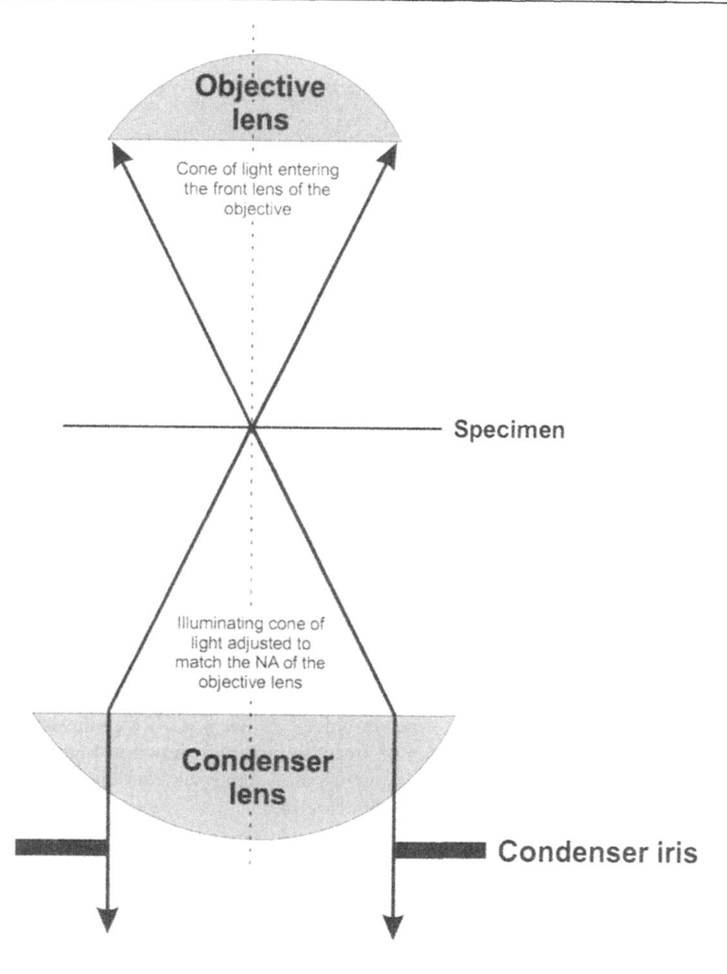

Figure 2-10. Importance of the Condenser.
The condenser is a critical component in the formation of a high-resolution image. In conventional bright-field and confocal microscope transmission imaging the substage condenser should be correctly adjusted for Köhler illumination (see box on page 40), where the cone of illuminating light is made to match the NA of the objective.

the sample (see Figure 2-10). Using a high NA condenser on a relatively low NA objective will result in "overfilling" of the front lens of the objective (i.e. the cone of light will be too wide for the NA intake angle of the objective). This will result in unwanted reflections that cause flare in the image, seriously degrading its contrast. The correct setting for the diaphragm of the condenser is somewhat below optimal for maximum resolution (a rule of thumb is to set the condenser iris to 3/4 of the diameter of the field of view when setting up Köhler illumination, see box on page 40 above). This will result in some loss in resolution, but a greatly increased level of contrast – such that the "effective" resolution (i.e. the structures that can be seen in the image) is in fact higher. Altering the contrast by changing the condenser iris only works in conventional bright-field microscopy or when using transmission imaging on a confocal or multi-photon microscope, and is not relevant when using conventional epi-fluorescence, confocal or multi-photon fluorescence imaging.

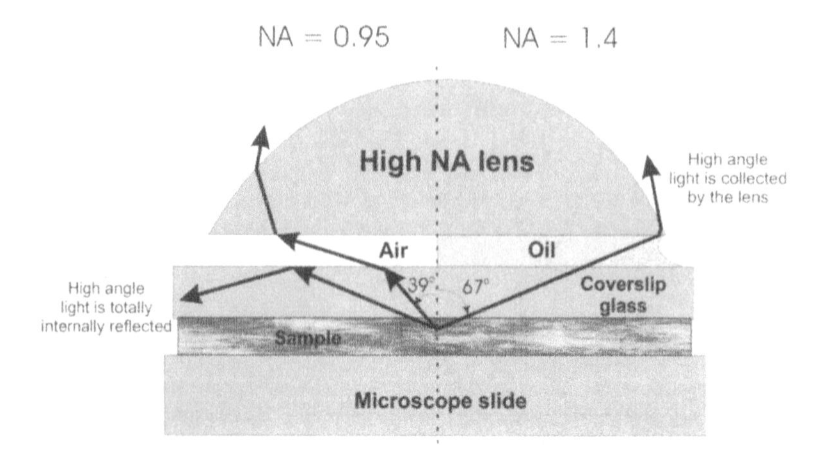

Figure 2-11. Immersion Lens.
Having a high refractive index immersion media (in this case oil) allows light rays of a much higher angle to be collected by the lens. This is the reason why an oil immersion lens has the highest NA (1.4) available for regular use on a light microscope.

Immersion Media

The NA of a "dry" or "air" objective (i.e. an objective that does not use any immersion media) is limited to approximately 0.95 due to total internal reflection of higher angle light rays (Figure 2-11). Obtaining higher resolution with a light microscope necessitates the use of a higher refractive index immersion media between the coverslip and the front lens of the objective, and an objective lens specifically designed for the immersion media used. The immersion media can be water, glycerol, or in the diagram shown in Figure 2-11, oil. Immersion oil has an index of refraction of 1.515, very close to glass, which allows much higher angle light rays to enter the objective (up to 67°, compared to only 39° maximum for an air objective).

In Figure 2-11 the refractive index of the immersion oil is closely matched to the glass of both the coverslip and the lens. In traditional fixed and embedded samples prepared for microscopy, the mounting media used for the sample imaged with an oil immersion objective should also have an index of refraction close to glass. This means that a properly embedded microscope sample will have continuity in the index of refraction of the media between the front lens of the objective and the focal plane under observation within the sample. The close matching of refractive index is important for obtaining the highest possible resolution from a light microscope, but also has the other important implication of greatly simplifying the light path between the sample and the lens, resulting in high-resolution images irrespective of the focal position within the sample.

Using an oil immersion lens to image live cells in an aqueous media results in a serious refractive index mismatch between the lens/oil/coverslip continuity and the growth media. Cells that are attached to the coverslip can be imaged at high resolution, but as the image plane moves further into the sample, the optical difficulties created by the refractive index mismatch between the lens and the aqueous media become more pronounced. In particular, spherical aberration (where the peripheral light rays are not focussed to the same point as more direct light rays, see page 48) seriously degrades the quality of the image. Water immersion lenses are a partial solution to the problem of refractive index mismatch. With a water immersion lens the immersion media closely matches the refractive index of the aqueous culture media used to hold the cells. However, even with a water-immersion objective lens, tissue or cells themselves will create significant optical aberrations (see "Optical Aberrations Caused by the Specimen", page 49 below).

Coverslip Thickness

In light microscopy, the coverslip is an integral part of the optics of the microscope. The correct thickness of the coverslip is important for high-resolution imaging, and is critical when using a relatively high NA air objective. Most objective lenses are designed for use with a 170 μm thick coverslip (a number 1.5 coverslip, although there is some variation between coverslips and so, for particularly critical applications, you should measure the thickness of the coverslip with a micrometer). Some objective lenses are equipped with an adjustable collar for correcting for different thicknesses of a coverslip, while others are used without a coverslip, such as the "dipping" objectives often used in physiological experiments.

A significantly thinner coverslip (or even no coverslip) can be used when imaging fixed and embedded samples with an oil immersion lens, as the glass of the lens, the immersion media, the glass of the coverslip, and the sample, are all designed to have a closely matching refractive index. This means that the high angle light rays, as shown in Figure 2-11, project in essentially a straight line from the focus plane to the lens. However, if material of different refractive index, such as air, is placed between the coverslip and the objective lens, then the paths of higher angle light rays are significantly displaced as they traverse the boundary between the glass and the air. This will result in the focal position within the sample not only being physically different to that formed when a matching refractive interface is used, but changing the focal position of the lens will alter the distance that light will travel through the air gap compared to the glass coverslip – which will result in a further change in the physical location of the focal plane. The effect is more severe for light rays that are emanating from the sample at a high angle, and so not only does the effective focal position within the sample not directly reflect the movement of the objective, but the light rays towards the outer region of the lens will be focussed to a different point to those closer to the central axis of the lens (this is known as spherical aberration, and is discussed in some detail shortly). A refractive index mismatch between the sample and the lens (for example, when using an oil immersion lens to image live cells) will also create problems of mismatched focal shift and spherical aberration as described for the air lens above).

Incorrect coverslip thickness – or using no coverslip at all – is often a cause of somewhat out-of-focus images when using conventional light microscopy. However, in a confocal microscope the out-of-focus light is rejected by the confocal pinhole, and so the effect of coverslip thickness mismatches is a serious loss of signal, rather than an out-of-focus image.

IMAGING MODES

The imaging modes available on a light microscope can be divided into two broad categories – imaging using light that is transmitted through the sample, and imaging using light that emanates back from the sample in the opposite direction to the illuminating light (backscatter and epi-fluorescence imaging). In this section a brief overview of the optics of the various imaging modes is given, but the more practical aspects of collecting images and setting the various optical filters etc are dealt with in other chapters (particularly Chapter 4 "Image Collection", page 101).

Bright-field Imaging

This is the traditional mainstay of light microscopy that everyone is very familiar with. It is the mode of imaging that has been described in considerable detail above, and has been used for over a 100 years to study fixed and stained biological samples. However, when you attempt to image unstained live cells with bright-field imaging you will find difficulty in creating a reasonable image of the cells, and some cells may be almost invisible. Increasing the contrast in the image is possible by closing the condenser iris, but this is at the expense of some loss of resolution! A more acceptable solution is to use one of the variations on bright-field imaging described below, where a variety of optical interactions with the specimen are exploited to create a high contrast image of living cells.

Transmission Imaging

In a laser scanning confocal microscope a bright-field image is often referred to as a transmission image. The name is derived from the fact that the illuminating light passes through the sample to form the image – in contrast to the epi-fluorescence mode of imaging where the light is directed back up through the objective lens. Transmission imaging is often used on unstained live cell material very effectively when using a confocal microscope because enhancing the contrast of the image is relatively easy by simply adjusting the gain and black level controls on the transmission-imaging channel. In this way a reasonable image of live cells can be obtained. However, if you want to create a really stunning live cell transmission image to combine with your fluorescence image try employing DIC optics as described shortly.

The transmission image obtained using a laser scanning confocal microscope is not confocal. The reason for this is that the illuminating laser light passes through the sample and substage condenser, where either a photomultiplier tube or a photodiode is used to detect the light. The light has not passed through the confocal pinhole and thus out-of-focus light is not rejected, creating a conventional style transmission image even when collected by the confocal microscope. Laser scanning transmission images are excellent for providing a physical context for fluorescent labelling, but they can only be used for 2D imaging and cannot be used to create a 3D image.

Remove the polariser when using the laser on a confocal microscope for DIC imaging

Phase Contrast Imaging

Phase contrast imaging is an excellent bright-field imaging mode that allows one to visualise living cells that have not been stained and thus have very little natural contrast. This method of imaging of live cells is familiar to most cell biologists as the method of choice for imaging cultured cells for the purpose of determining how many cells are in a culture.

Phase contrast imaging is a technique requiring specialised objectives and condenser settings that allow the high angle diffracted rays to pass through the objective, but the lower angle rays that have passed directly through the sample are rejected. This creates an image of high contrast of cell membranes, organelles and microstructures that have diffracted the illuminating light. The relevant optical components for phase contrast imaging can be readily installed on all light microscopes using relatively cheap components. Phase contrast imaging is rarely performed when collecting transmission images using a laser scanning confocal microscope, as the quality of the image is not sufficient for high-resolution imaging. A more effective, but expensive, method of imaging unstained cells by transmission imaging when using a laser scanning confocal microscope is DIC imaging as described below.

DIC (Nomarski) Imaging

One of a number of variations of Differential Interference Contrast (DIC) imaging is often installed on high-end light microscopes. One particularly common variation is called Nomarski optics. DIC imaging can be used to very effectively image unstained live cells and other biological material with very little natural contrast. The optics involved in DIC imaging are quite expensive and so rarely is DIC optics installed on cheaper microscopes which are used for cell culturing etc. The image produced by DIC optics exploits very small differences in phase as the light passes through the sample (mainly due to small changes in refractive index of the sample). DIC imaging requires several specialised optical components in addition to the conventional objective and eyepiece lenses of the light microscope.

The illuminating light used for DIC imaging is polarised, using a polariser in the case of conventional light microscopy, but taking advantage of the polarised light of the laser when using a laser-scanning microscope. In fact the polariser should be removed from the light path when using the laser for DIC imaging, otherwise the light intensity will be greatly reduced. The laser must be correctly aligned for DIC imaging. This should have been done when the confocal microscope was originally installed, and will not need re-alignment unless the laser is replaced. However, if you are having difficulty in creating high quality DIC images using the confocal microscope, even though top quality images are available by conventional optics, the laser is possibly out of alignment.

The light used for DIC imaging also passes through a special optical device called a Wollaston prism. One Wollaston prism is located below the condenser and the other prism is located directly behind the objective, sometimes in a slot directly behind where the objective is mounted on the microscope turret, and sometimes on a optical filter wheel. The Wollaston prism should be carefully adjusted while imaging to create a suitable DIC image with an acceptable "shadowing" effect. Adjustment of the Wollaston prism will alter the degree of "shadowing" and the apparent angle of the "shadow". The Wollaston prism can be detrimental to the quality of the fluorescence image when using high-resolution imaging. To test whether your Wollaston prism is causing problems choose a sample with simple and clearly defined fluorescent structures (fluorescently stained red blood cell membranes, relatively large fluorescent latex beads or sub-resolution latex beads are all excellent) and collect fluorescent images using the laser scanning confocal microscope at very high zoom settings. Collect images with and without the Wollaston prism in place – if you have a problem you will see an offset shadow image of the object that will seriously detract from the quality of a high-resolution image. If you do the test with a relatively complex image (for example, fluorescently labelled microtubules) you will notice a significant loss of quality in the image, but you may have difficulty in determining whether this is due to the Wollaston prism or other factors.

Warning!

The Wollaston prism, used in DIC imaging may affect the quality of the fluorescence image

The DIC image of live cells which have very low natural contrast is quite impressive. The cell outline, and intracellular detail, is shown as excellent grey-scale shadowing with a pseudo 3D effect. The 3D effect of DIC imaging is not true 3D as the shadowing effect is produced by differences in refractive index rather than differences in shape or height.

DIC imaging when implemented on a laser scanning confocal microscope uses the transmission channel to collect excellent pseudo 3D images. The DIC image is obtained without the light passing through the confocal pinhole. However, the creation of an image by DIC optics does result in some optical sectioning capability. This is not as pronounced as that obtained with the confocal pinhole, but a reasonably acceptable image can be obtained using DIC optics even when the focal position of the microscope is located, for example, at one end of a cell. Without the presence of the DIC optics this position of imaging would produce a very out-of-focus transmission image. Although the DIC image is to some extent an optical slice, DIC images do not give the immediate impression of being an optical slice – that is, a DIC image looks like a 3D image no matter what part of the cell is imaged.

> **DIC and Phase Contrast images can be generated from bright-field (transmission) images using QPm software from Iatia**

Another excellent way of creating DIC images using a confocal microscope is to use QPm, innovative software developed by Iatia (see Iatia Ltd., page 337). QPm is capable of creating excellent phase and DIC images using conventional bright-field images or transmission images collected on a confocal microscope using conventional bright-field optics. Iatia's QPm software is capable of constructing pure phase maps (independent of intensity information) from which the full range of phase contrast images can be generated (such as DIC, Phase contrast, Hoffman Modulation etc). The phase map is generated from a series of three z-stepped bright-field images without the need for any additional optics. This technology generates exceptionally high quality phase and DIC images for locating fluorescence label when imaging with a confocal or epi-fluorescence microscope equipped with a z-stepper. Furthermore, this collection of the Phase information is not affected by plastic culture dish bases or covers. The only drawback of this technology is that the creation of the phase or DIC image does require the collection of three z-sections and some processing time – which means this technology cannot be readily used as a device for initially locating the cells, but can be used very effectively to provide topographical information on the location of the fluorescence obtained by confocal microscopy.

Backscatter Imaging

The laser scanning confocal microscope is capable of creating excellent images using backscattered or reflected light. Confocal backscatter imaging is particularly useful for creating excellent 3D images of the surface of materials (including the surface of a tissue or cell sample). However, most biological material is relatively transparent at visible wavelengths, although a number of structures within the cells do scatter the light. It is possible to create 3D images of cellular material using optical slices collected internally by the collection of this internally generated backscattered light. Backscatter imaging, including example images, is discussed in more detail on page 107 in Chapter 4 "Image Collection".

Fluorescence Imaging

The main imaging mode used in confocal microscopy is that of epi-fluorescence microscopy. Conventional epi-fluorescence microscopy (using a Hg or Xe lamp) is also often installed on a confocal microscope so that you can more easily move around the sample to locate the cells that you may wish to look at in more detail laser scanning confocal microscope. If epi-fluorescence illumination is installed on your microscope directing the light back up through the eyepieces instead of the scan head is a simple matter of moving the correct sliders etc to place the correct dichroic mirrors and optical filters in the light path. The details of fluorescence imaging on a confocal and multi-photon microscope are described on page 102 in Chapter 4 "Image Collection".

OPTICAL ABERRATIONS

There are a number of optical aberrations that result in distortion of the image. High quality research grade microscopes and their accompanying objective lenses are produced with considerable skill to alleviate most of the common optical aberrations. However, some objective lenses are designed for specific applications and may therefore not be fully corrected for a variety of aberrations. More importantly, some optical aberrations can be

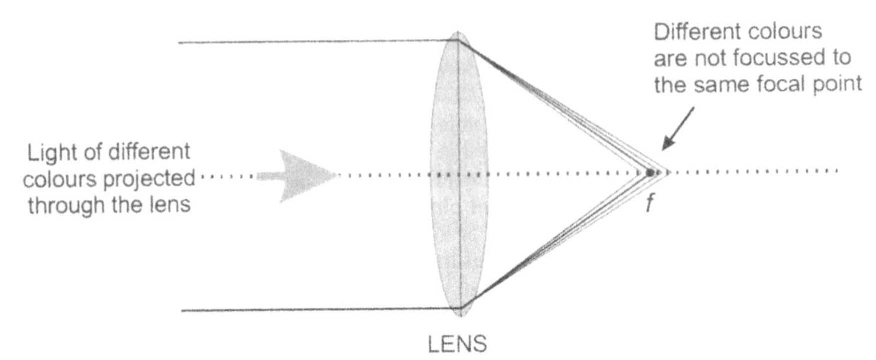

Figure 2-12. Chromatic Aberration.
Chromatic aberration is simply when light of different wavelengths is focussed to a different point by a lens. Chromatic aberration is an inherent problem with a lens that works by refraction (a mirror lens does not suffer from chromatic aberration), and is carefully corrected in most objective lenses. Chromatic aberration can be also caused by refractive index changes within the sample itself, resulting in problems in multi-colour labelling applications even when using highly corrected lenses.

inadvertently introduced into the microscope optics by the manner in which the object is presented (see discussion on the use of a coverslip above). Optical aberrations are also inherently generated by the very nature of the sample being imaged. This is particularly relevant when imaging biological samples that are relatively transparent to visible light, thus allowing a confocal or multi-photon microscope to image relatively deeply into biological tissue. However, the biological tissue will interact with both the light used for illumination and the returning fluorescence or backscattered light, creating many optical aberrations. Of particular concern is spherical aberration, described in some detail below.

Chromatic Aberration

Chromatic aberration is when light rays of different wavelength (colour) are not focussed to the same point by the lens. All lens systems that utilise refraction to focus the light result in problems with chromatic aberration (mirror lenses, as used in a reflecting telescope do not suffer from chromatic aberration). Most objective lenses are highly corrected for chromatic aberration. However, this correction is for specific wavelengths only, and different lenses are corrected for specific colours. Achromat lenses are corrected for two colours, red and blue (656 and 486 nm respectively), and apochromatic lenses are corrected for three colours (red, green and blue). The nomenclature used to describe the objectives with various optical corrections is shown on page 58.

Spherical Aberration

Spherical aberration is due to the light rays on the periphery of the lens focusing to a different point (usually closer to the lens) than the light rays that pass more directly through the lens (see Figure 2-13).

Spherical aberration is an important factor in optical microscopy. Not only does the lens system in the microscope need to be corrected to minimise spherical aberration if the formation of a high resolution and high contrast image that is in focus is to be achieved, but also the optical interaction between the microscope and the sample may result in the introduction of serious spherical aberration problems. Spherical aberration will be manifested as a poorly focussed image in conventional light microscopy, but when using a confocal or multi-photon microscope, spherical aberration will create a range of serious optical problems, even though these may not at first appear obvious to the inexperienced user. The most important effect is an apparent loss of fluorescence intensity (the instrument appears to lack sensitivity), but spherical aberration can also result in serious distortion of the image, particularly in the z-direction. This is particularly severe when creating 3D reconstructions of a series of optical slices, and even a more serious problem when attempting to make 3D measurements from such an image stack. The problems of spherical aberration created by the sample itself are discussed in more detail below.

Coma

Coma aberration is when off-axis light rays are not focussed to a point. The effect is to create "streaking" of objects towards the periphery of the viewing field.

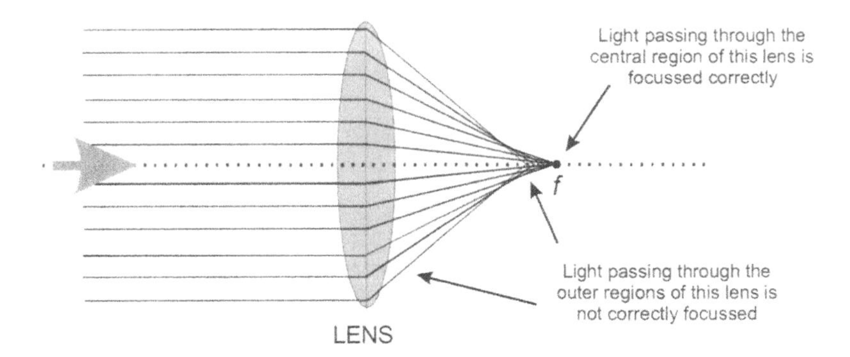

Figure 2-13. Spherical Aberration.
Spherical aberration is where the light rays that pass through the periphery of the lens do not focus to the same point as those light rays that pass through the central region of the lens.

Astigmatism

Astigmatism is formed when the focal position of the lens is different in two different planes at right angles to each other. Vertical and horizontal light rays come to focus in different image planes, creating an oval shaped image of a point source of light.

Curvature of Field

Field curvature is due to the image plane being a concave spherical surface, rather than flat. This is manifested in a conventional light microscope as the peripheral region of the field of view not being focussed to the same position within the specimen compared to the central region. The same is true of a confocal microscope, except that the image will more likely appear to have a concentration of structure in either the central or peripheral areas of the field of view (the out-of-focus light, being rejected by the confocal pinhole will mean that regions of the image that are not in the focal plane will not be seen).

High quality "Plan" lenses are highly corrected for field curvature, but at the expense of having a large number of lens elements within the objective. This greatly increases the expense of the lens, and can result in less light being transmitted by the lens.

Optical Aberrations Caused by the Specimen

The specimen itself is part of the optical system of a light microscope. Traditionally specimens for light microscopy have been fixed, very thinly sectioned and then embedded in a mounting media with a similar refractive index to the sample. This type of mounting has the effect of minimising any optical differences within the sample. However, confocal and multi-photon microscopy is often used for imaging live cells where the first problem is that the media (usually water based growth media) is of a different refractive index to the lens and the coverslip. This difference in refractive index is often overcome when imaging relatively deeply into live tissue by using a water immersion lens. Unfortunately, not all optical aberrations can be corrected by using a water immersion lens, as the cells themselves interact optically with the light used to form the image. In particular, various subcellular compartments, such as the nucleus and organelles are often a somewhat different refractive index to the surrounding cytoplasm or the incubation media. This type of refractive index mismatch results in a significant level of spherical aberration, often in localised regions of the sample. For example, when imaging cells, a cellular nucleus in front of the image plane will seriously distort the image. In conventional light microscopy cellular structures will affect the focusing of the sample, but in confocal and multi-photon microscopy the effect is much more complex. Optical distortion may result in part of the sample below the nucleus ending up in a different image plane than the one expected from the depth within the sample. This will often be readily recognized distortion when taking a z-section or creating 3D reconstructions, but may not be as readily recognized, but no less serious, when taking a simple 2D single optical slice.

Resolution in a Light Microscope

The ability to distinguish two adjacent objects is considered to be the measure of the resolution limit of a light microscope. Unfortunately there is a fundamental limit to the resolution that can be achieved (approximately 0.5 to 0.2 μm, depending on the wavelength of light used).

The interaction of light with very small objects (smaller than the wavelength of the irradiating light) results in the formation of a characteristic diffraction pattern (Figure 2-14), which is known as an Airy disk. The first thing to note is that this diffraction pattern is larger than the object that is being imaged (in this example a small bead is shown diagrammatically at the bottom of the figure). The second thing to note is that the image is not at all like the small bead used as the object, but is made up of a series of circular rings of light (shown as a light intensity graph in the top part of Figure 2-14). The diffraction pattern typically shows an intense central "spot" with a series of fainter, but distinct bands surrounding the central disk.

The radius of the diffraction limited Airy disk R is given by

$$R = \frac{1.22\,\lambda}{NA_{objective} + NA_{Condenser}}$$

Where λ is the wavelength of light, and NA is the numerical aperture of the lens (see page 42 for a discussion on the NA of a lens).

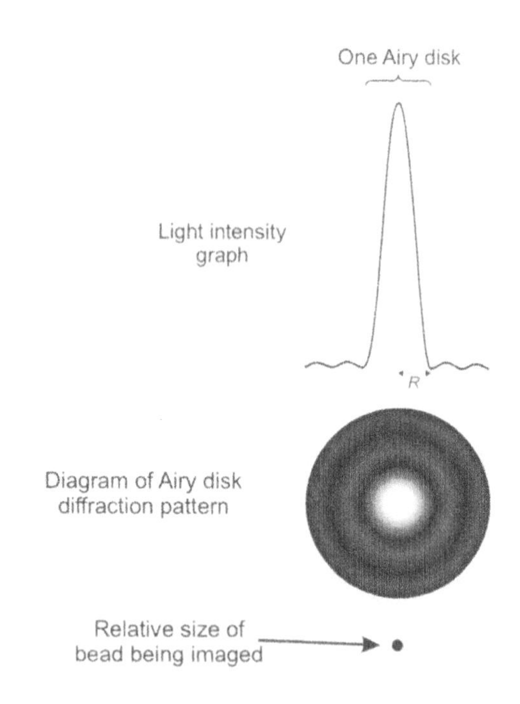

Figure 2-14. Airy Disk Diagram.
The light microscope has a limit to the size of an object that can be resolved due to the diffraction-limited formation of an "Airy disk". This diffraction pattern can be readily seen when imaging a small bead with a diameter at or below the wavelength of light used for imaging. However, this diffraction pattern is also present in a regular microscope image which can be looked on as an infinite array of points being imaged. The radius of the Airy disk R is defined as the distance between the first dark ring of the diffraction pattern and the mid-point of intensity of the central bright disk. One Airy disk (often used as a defined pinhole setting on a confocal microscope) is the distance between the first dark ring of the diffraction pattern on either side of the central disk (i.e. the size of the central spot).

In epi-fluorescence illumination, including confocal and multi-photon fluorescence imaging, where the objective also becomes the condenser, the diffraction limited disk radius R is given by

$$R = \frac{1.22\,\lambda}{2NA} \quad \text{or} \quad \frac{0.61\,\lambda}{NA}$$

For blue light of 488 nm using an oil immersion lens of 1.4 NA the value of R becomes

$$R = \frac{1.22 \times 488}{2 \times 1.4} = 213\text{ nm} \quad (0.213\ \mu\text{m})$$

This value is the limit of resolution when using blue light under ideal conditions and with a perfect lens. In reality, the resolution achieved by a confocal microscope is close to this limit, but does not exceed this value. The idealised Airy disk diffraction pattern shown here will only be visible using high quality optics on a correctly set up light microscope. Optical aberrations, particularly spherical aberration (see page 48), may result in an Airy disk diffraction pattern where the rings of light are no longer readily discernable.

Two small beads can only be resolved if the two Airy disks are sufficiently separated such that there is a drop in intensity between the two diffraction limited spots (Figure 2-15). The Raleigh criterion defines the limit of resolution in a light microscope as being when the two Airy disks are separated by a minimum distance such that the first dark ring of one diffraction pattern coincides with the centre of the bright disk of the other Airy disk (shown diagrammatically in panel "B" of Figure 2-15). The outer rings of the diffraction disk can overlap, but if the central disks overlap, then the two Airy disks (and hence the two beads) cannot be resolved. Once the two beads are close

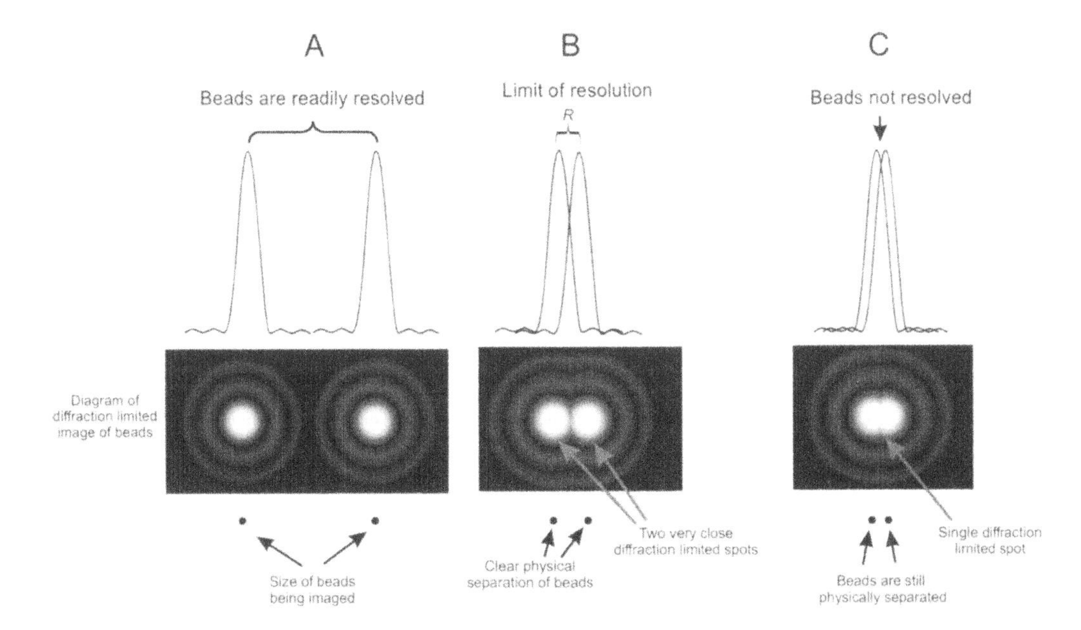

Figure 2-15. Resolution Limit in Light Microscopy.
The light microscope has a finite limit to the resolution that can be obtained due to the size of the diffraction limited Airy disk. The two small beads shown diagrammatically in "A" have clearly separated diffraction limited Airy disks. The closer spaced beads in "B" are at the limit of resolution of the microscope (separated by the Raleigh criterion *R*, which is equal to the radius of the Airy disk), and the closely spaced beads in "C" are not resolved by the microscope.

enough, but they don't need to be physically touching, the microscope will create only one Airy disk (this is shown diagrammatically in panel "C" of Figure 2-15). At this point the objects are closer together than the resolution limit of the microscope. Many biological structures, for example transport vesicles and microtubules are considerably smaller than the diffraction limited resolution of a light microscope.

The Airy disk also exists in three dimensions, in which case this diffraction pattern is known as the Point Spread Function (PSF). In three dimensions the diffraction rings spread out in all three directions, resulting in an oval shaped central spot (the z-resolution is less than the x-y resolution) with radiating rings of light.

When the individual bead or cellular structure itself is below the resolution limit of the microscope the object can still be imaged. Light rays passing the small bead will be diffracted, or in the case of fluorescence the bead itself will be emitting light. Although a very small bead can be imaged, the size of the image (Airy disk) will not necessarily reflect the size or shape of the object itself (Figure 2-16). In this figure a series of small beads are shown diagrammatically, with their corresponding image (as a diffraction limited Airy disk). Beads larger than the wavelength of light used for imaging (for example, an 8 μm diameter red blood cell) will be readily resolved by the microscope. This means that the size of a red blood cell can be accurately measured by imaging in a light

PSF = Point Spread Function
(3D diffraction limited Airy disk)

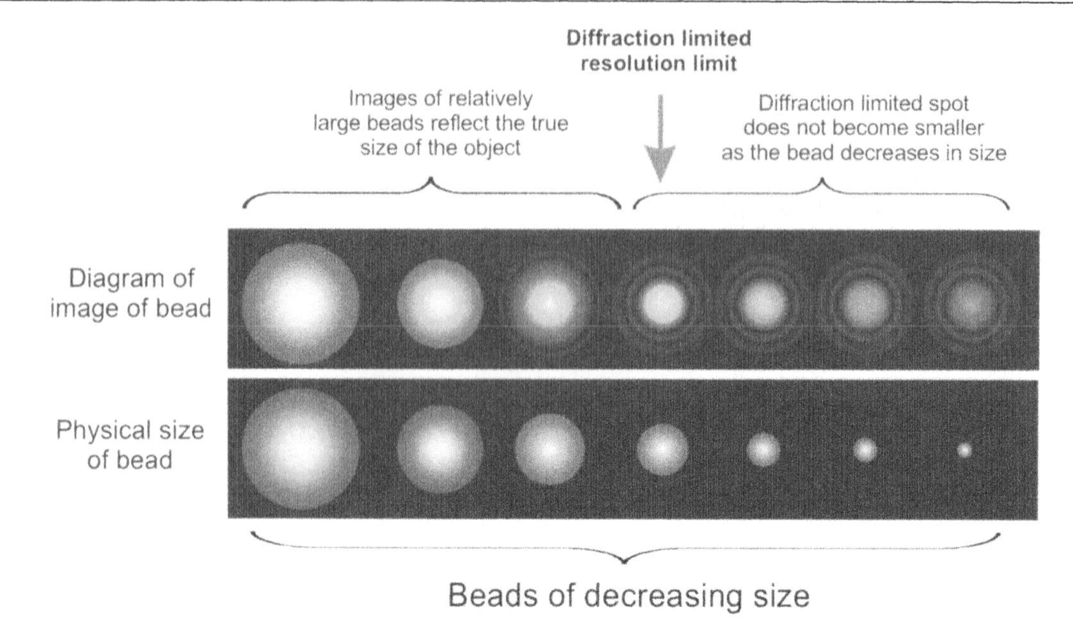

Figure 2-16. Small Objects May Appear Larger than they Actually are.
Due to the formation of a diffraction limited Airy disk in the light microscope a very small object may appear to be considerably larger than in reality. This phenomenon is demonstrated diagrammatically above where a bead of decreasing physical size is observed to decrease in size until the diffraction limited resolution of the microscope is reached (under ideal conditions the best resolution using blue light is approximately 0.20 μm). As the physical size of the bead further decreases the images of the diffraction limited Airy disk retains the same size, although there will be a decrease in intensity.

microscope. As the object size decreases (see the decreasing, but still relatively large beads on the left hand side of the lower panel of Figure 2-16) the size of the image of the object decreases (left hand side of upper panel). Once the resolution limit of the microscope has been reached, objects that are smaller than this limit will appear to be approximately the same size (see the series of progressively smaller beads depicted in the right hand side of the lower panel in Figure 2-16), that is, they will appear as the size of the diffraction limited Airy disk (right hand side of upper panel). As discussed above, the wavelength of light and the NA of the lens determine the size of the Airy disk. Objects will never appear smaller than approximately 0.20 μm in a light microscope – the smallest Airy disk size when using visible light with a high NA objective lens.

The resolution limit of light microscopy has important implications in biology as many of the small structures (organelles, microtubules etc) within cells are close to or below the resolution limit of light microscopy. This means the objects may be imaged if suitable dye or refractive changes are evident, but the object as depicted in the image will appear significantly larger than the real structure. This means that two or more transport vesicles, which are in the order of 0.05 μm diameter, that are physically close to each other, may appear as one vesicle. Furthermore, the size of this vesicle will appear the same as if there was only one vesicle present. This has serious implications for dual labelling, where apparent co-localisation may simply be due to the two fluorescent tags being present in close, but physically separate vesicles. Any attempt at determining the "size" of such small structures within the cell will result in a serious overestimation of the physical dimension of these small structures. In the case of transport vesicles, the degree of size "over estimation" may be as high as 10 fold (i.e. the 0.05 μm diameter vesicle may appear as approximately 0.5 μm diameter in the image as a diffraction limited spot).

Another important implication of the diffraction limit in light microscopy is that fibrous structures such as actin filaments or microtubules will appear significantly thicker than the actual tubular structure being imaged. Although you may not be over concerned that the image of the microtubules is much thicker than in reality, there is a problem if you are interested in the arrangement of the filaments within the cell as several filaments lying in close proximity to each other may result in what appears to be a single filament within the image.

COMPONENTS OF A MICROSCOPE

The basic components of a conventional light microscope are quite familiar to most people, consisting of a microscope body to which are attached the objectives, the microscope stage, condenser and eyepieces etc. However, there are a lot of extra items that are added to a research microscope that can make their operation quite a daunting task if you are not a regular user of a microscope. It can be frustrating to find that every microscope is subtly different, in even trivial matters of where a particular control is located – particularly when you are trying to quickly switch from say, confocal imaging, to bright-field imaging, or to conventional epi-fluorescence imaging.

This section gives a brief overview of the main components that you will encounter when adjusting a light

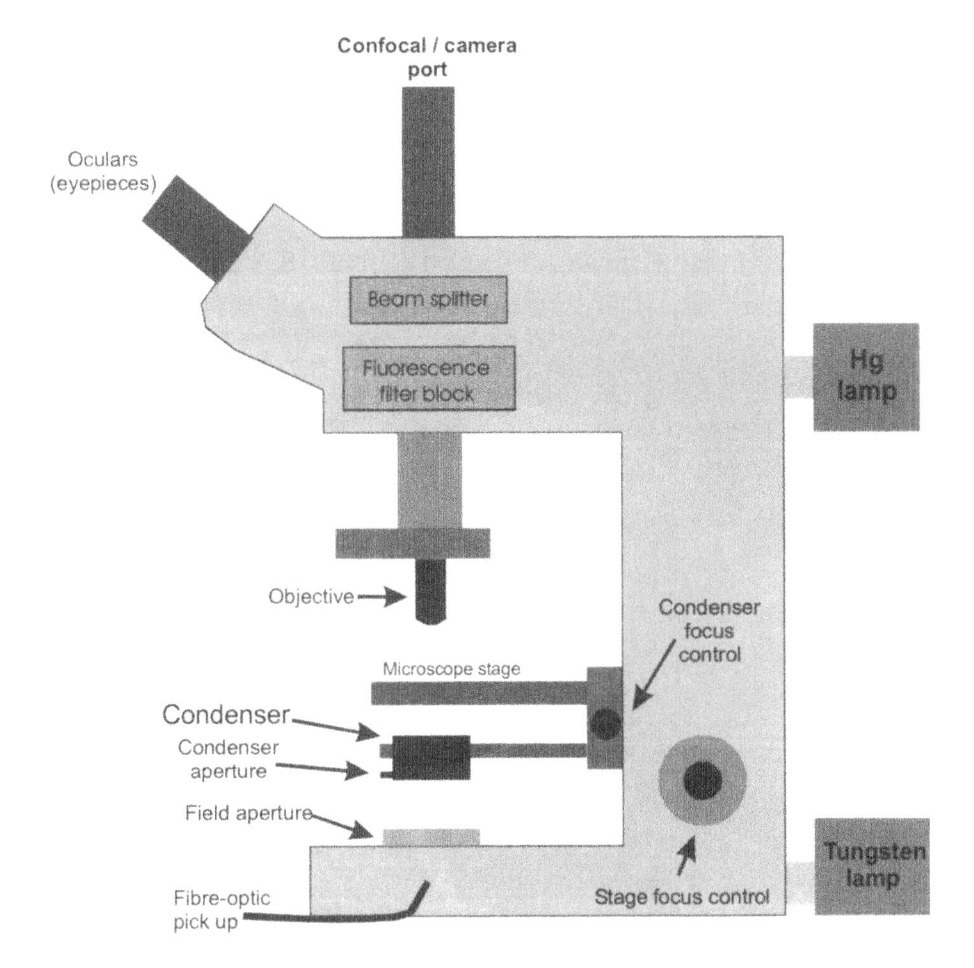

Figure 2-17. Upright Microscope.
The traditional upright microscope design is used in most laboratories as the instrument of choice for examining fixed and stained cell and tissue samples. The upright design is also used in electrophysiology experiments, where the live cell or tissue sample is imaged from above using a "dipping" objective. Imaging live cell and tissue samples on an upright microscope, without a "dipping" objective, is quite difficult, as you will require a special sealed imaging chamber that can be inverted on the microscope stage. The confocal microscope scan head is attached to an upright microscope via the camera port on the top of the microscope. The fibreoptic pick up (or in some cases the detector itself) for the confocal microscope transmission detector is located under the condenser at the base of the microscope.

microscope. However, if you are not confident in the use of a microscope don't hesitate to ask more experienced people – having someone point out the correct adjustment of particular components on the microscope you are using is a lot easier than spending your time trying to work out how to correctly operate the microscope from the more generalised diagrams etc in this and other publications. There are a number of books on light microscopy (listed in Chapter 15 "Further Reading", page 347) that go into considerably more detail on the various components of a light microscope than this chapter. The web sites of the various light microscope manufactures are another good source of information on various microscopes.

Microscope Layout

A light microscope can be of either the upright (Figure 2-17) or inverted (Figure 2-18) variety. Both forms of the microscope consist of the same basic parts (discussed shortly), but due to the different position of the components you will need to take some time familiarising yourself with the basic layout of, say, an inverted microscope, if you

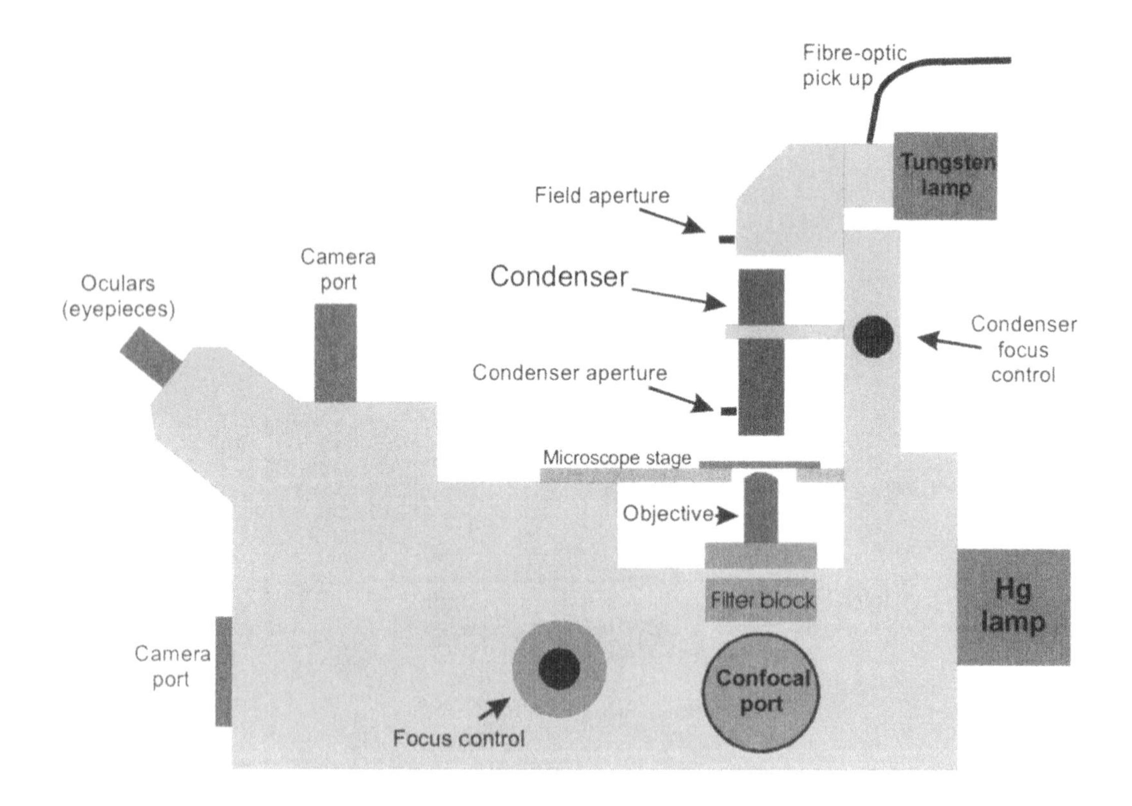

Figure 2-18. Inverted Microscope.
The inverted microscope is usually purchased when live cell or tissue samples are to be imaged. However, the modern infinity-corrected objective design does mean that the resolution and sensitivity of the inverted microscope is essentially the same as the upright microscope. For this reason an inverted microscope can be used to obtain excellent high-resolution images of fixed and stained cell and tissue samples. The confocal microscope scan head is attached directly underneath the objective. The fibreoptic pick up (or detector) for transmission imaging is located above the condenser on top of the microscope.

are a regular user of an upright microscope. The upright microscope is the traditional format used for imaging conventional stained and immunolabelled microscope slides. Due to familiarity most people prefer the upright microscope for routine work, but the inverted microscope now has essentially the same optical efficiency and resolution as an upright microscope and so there is no need to use an upright microscope to obtain the highest quality images.

An inverted microscope (Figure 2-18) is very different in layout compared to an upright microscope. However, the optical components involved are basically the same, with all of the adjustments and changes that can be made on an upright microscope. Previously, the inverted microscope resulted in a lower quality image compared to an upright microscope, mainly due to the large number of optical components required within the inverted design. In more recent years most microscope manufacturers have now produced microscopes with infinity corrected objective lenses, which results in fewer optical components needed in the inverted microscope, and thus an improvement in the quality of the image. Furthermore, the confocal microscope scan head attaches close underneath the objective in an inverted microscope and so the quality of image obtained is identical to that obtained when using an upright microscope, where the scan head is attached directly above the objective on the camera or video port of the microscope.

An inverted microscope can be used for most work for which you would normally use an upright microscope. If you will be requiring an instrument for live cell imaging (which is best done on an inverted microscope) you will find conventional immunolabelling and imaging of stained microscope slides can be imaged to the same standard on the inverted compared to the upright microscope.

There is one important area of work where you will not be able to use an inverted microscope – which is when doing electrophysiology studies where you cannot image the cells or tissue sample from below. In this case you will need to use a "dipping" objective on an upright microscope.

The following discussion on various components of a light microscope refers to the diagrams of the upright and inverted light microscopes shown in Figure 2-17 and Figure 2-18 respectively.

Care of the Microscope

A light microscope is a delicate and accurately machined piece of equipment, but if looked after with care it will continue to produce superb images for many more years than the lifetime of the attached electronics and computer equipment! Damage to the objective lens on a light microscope is particularly detrimental to the quality of the image, and is often the most expensive part of the instrument.

Cover the Microscope! Microscopes are provided with a simple cover to prevent dust accumulating on the sensitive optical components of the microscope – make sure you replace the cover when you have finished your work. Dust is a serious problem in microscopy, not only because the particles themselves may detract from the quality of the image, but also because particles of dust contain abrasive and sometimes reactive salts that can create havoc with sensitive optical components. This can be particularly serious when you attempt to clean off the dust using a lens tissue – the dust particles can create scratches in the lens! This is particularly important when cleaning a lens or optical filter element that has a dielectric coating as an anti-reflection coating or as used in interference filters.

Optical Components

The objective lens and the eyepieces are the main optical components of a light microscope required to create and magnify the image. However, there are a number of other optical components required for various imaging methods in light microscopy. The following section briefly describes the most important optical components associated with the light microscope that are used in confocal and multi-photon microscopy.

Objectives

The objective is the most critical part of the microscope as it forms the initial image, determines the resolution of the microscope, and provides most of the magnification. Take great care of the objectives, as they are the most expensive, and probably the most easily damaged part of the microscope. In general the price of the objective is a good indicator of the quality of the objective, but not always – there are some specialised objectives that are particularly expensive, such as extra-long working distance objectives, that are of no advantage unless you have a special requirement for the specialised feature.

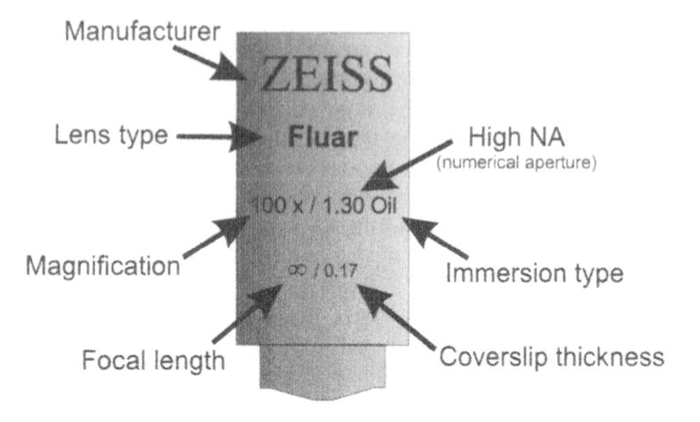

Figure 2-19. High Resolution Oil Immersion Lens.
The barrel of an objective contains a number of important labels that indicate the essential characteristics of the lens. Most of the markings are universal across all manufacturers, but be aware that the actual name of the lens (in this case "Fluar") may vary between manufacturers.

FLUAR	Fluorite lens which transmits visible and UV light.
100 x	nominal magnification of the lens.
1.30	numerical aperture (NA) determines the resolution obtained and the amount of light gathered. The higher the NA, the better the resolution.
Oil	immersion media to be used
∞	infinity corrected objective
0.17	coverslip thickness of "1½" (170μm).

There are a number of important parameters marked on the side of an objective (Figure 2-19 and Figure 2-20), a number of which are discussed below. If you are unsure of the markings on your own objectives please consult the information provided by the particular lens manufacturer (usually available on the Internet).

The name of the objective ("Fluar", in Figure 2-19 and "N-PLAN", in Figure 2-20) indicates the special characteristics of the objective, but unfortunately the naming system is somewhat different between each of the manufacturers. "PLAN" or "PL" normally refers to the flatness of field, and "Fluar" or related words refer to the lens construction material, fluorite. Fluar lenses are particularly transparent to UV light and so are often the lenses of choice for Hg lamp UV excitation. UV confocal microscopy uses short wavelength UV light for excitation, but most confocal microscopy is concerned with using visible light for both excitation and emission. Although multi-photon microscopy can be used to excite "UV" dyes, neither the light used to excite the dye (pulsed infrared light) nor the fluorescence emission (blue, green or red) are in the UV range of the light spectrum.

A "PLAN" or "PL" lens has been carefully corrected to alleviate unevenness in the field of view (i.e. the lens has a flat field of view). The "flatness" of the field of view is most important in confocal microscopy, as the "out of focus" area of the field of view will not be seen under the confocal microscope. A lens that is not perfectly flat (particularly noticeable in a low magnification lens) can be improved when using the confocal microscope by "zooming" in and using only the central part of the lens.

In Figure 2-20 the "N" refers to the lens being achromatic, which means that this lens has been carefully colour corrected for two colours (red and blue). An "APO", or apochromatic objective refers to a lens that has been corrected for three different colours (red, green and blue).

When a less colour-corrected lens is used objects may appear to have coloured "halos" associated with "edges" within the image. In confocal microscopy colour corrected lenses are important when performing dual or triple labelling. In the single label mode there is a less stringent requirement for colour correction (but remember that the excitation and emission wavelengths are still of different colour).

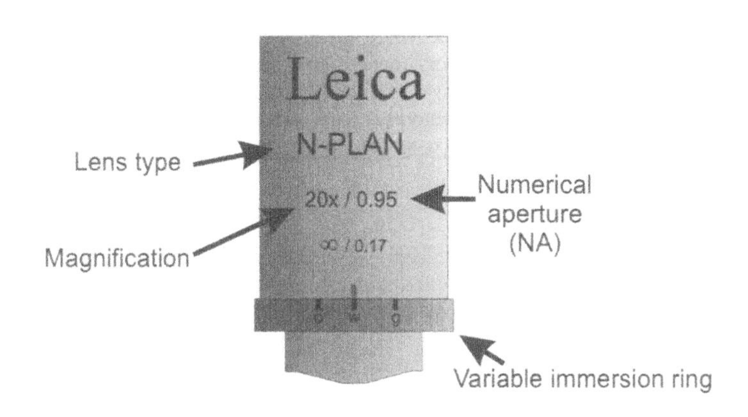

Figure 2-20. Multi-immersion Lens.
High quality multi-immersion lens for oil, water or glycerol.

PLAN	total field of view is in focus (i.e. a flat field of view).
N	achromatic, highly corrected for chromatic aberration (two colours, red and blue, focussed to the same point).
20x	nominal magnification of the lens.
0.95	numerical aperture (NA).
∞	infinity corrected objective
0.17	coverslip thickness of "1½" (170μm).
O, W, G	variable ring settings for using oil, water or glycerol as the immersion media.

The NA (Numerical Aperture) of the lens has been explained in some detail above, the important matter to remember is that the higher the NA, the higher the resolving power and the greater light gathering capability of the lens.

The nominal magnification of the objective is marked as "100x" in Figure 2-19 or "20x" in Figure 2-20. This magnification is an approximate value. The exact value varies for each individual lens, and must be estimated empirically (using a calibrated grid) if the exact value is needed. However, for most applications the manufacturer advised magnification is adequate. Furthermore, if your confocal microscope was set up correctly by the manufacturer when the instrument was installed the correct magnification factor should be already entered into the software used to control your confocal microscope.

Microscope objectives come in a wide range of different optical designs, with only some of the important features marked on the barrel of the objective, as discussed above. Removing the objective from the microscope (particularly on an inverted microscope) is a good way to familiarise yourself with the markings that are shown on the barrel of the objective.

This chapter does not attempt to fully describe the many objective designs that are available. If you find markings on the objective that you are not familiar with and they are not discussed here, then one of the best ways to find out more is to consult the web site or published information of the manufacturer of the lens. A number of lens manufacturers produce extensive brochures that outline not only the characteristics of their lenses, but may also have valuable information on optics and lens design in general. A knowledgeable company technical person, or the administrator of an imaging facility will often have a detailed knowledge of the various lenses that are available.

A microscope objective may be designed to be used as a dry "air" objective (for relatively low magnification), or for use with a variety of different immersion media (see box below). The lens shown in Figure 2-19 is an oil immersion objective, whereas the lens in Figure 2-20 is a "multi-immersion" objective. This multi-immersion objective has a variable ring on the body of the lens that can be rotated to accommodate water, glycerol or oil as the immersion media. Make sure that the immersion correction collar is at the correct setting, and that an immersion objective is always used as an immersion objective and not as a "dry" objective.

Objectives below 40x are usually air objectives. The 40x objective can be air, water, glycerol, or oil immersion, depending on the design of the particular lens. Higher magnification objectives are usually oil immersion objectives, although there have been a number of water objectives of high magnification produced in recent years.

Take great care when using a dry or water immersion objective on a microscope that has oil immersion objectives installed. You may find that quite often all of the objectives are smeared with oil! You may even find that the previous user has nicely cleaned the "oil" immersion objective, but that the other objectives have been left with a coating of oil!

Objective Names

Achromat: Colour corrected for red and blue (656 and 486 nm). Spherical correction for yellow-green (540 nm).

Fluorite: Lens elements made of fluorite or related compounds with very low colour dispersion and excellent transparency through to UV. Maximum possible NA is 1.3. Often used in fluorescence microscopy.

Apochromat: Highly colour corrected for red, green, blue and dark blue. Also spherical aberration corrected for green and blue.

Plan-Apochromat: Highly colour corrected as above for Apochromat, but with additional correction to give a flat field of view. Highest NA of the regular type of immersion objective lens (NA 1.4). Widely used in "top-end" research grade microscopes for high-resolution imaging.

Eyepieces (Oculars)

The eyepieces of a light microscope are an integral component for conventional light microscopy. In some microscopes the colour and spherical aberration correction is incorporated into the eyepieces, making the use of the correct eyepieces with the objective for which they are designed most important. However, when using a confocal or multi-photon microscope the eyepieces are not used for the laser scanning imaging system. The eyepieces will then only be used to locate the sample, before switching over to laser scanning.

Immersion Objectives

"Dry" or "Air" lenses: used with a coverslip, but no immersion media. Coverslip thickness is critical, with some lenses having a coverslip thickness correction collar.

Water immersion: used with water as the immersion media between the lens and the coverslip. Less optical aberrations when imaging into cell media or fresh tissue samples.

Glycerol immersion: relatively high refractive index immersion media, primarily for imaging fixed cell preparations.

Oil immersion: immersion media that matches the refractive index of the lens and fixed and embedded samples.

"Dipping" objective: specialised objective for live cell physiological studies where the lens is "dipped" directly into the culture media without an intervening coverslip (see Chapter 12 "Imaging Live Cells" (page 279).

Beam Splitter

A beam splitter diverts light from the specimen, to either the eyepieces, to the camera (or both), or to the confocal microscope scan head. Forgetting to change the position of the beam splitter when changing from bright-field to confocal imaging is a common frustration – even for more experienced people! The beam splitter can sometimes be set in an intermediate position in which a percentage of the light is directed to the eyepieces while the remainder is directed to the confocal microscope scan head. Having 100% of the light directed to the confocal microscope scan head is desirable when laser scanning to gain the maximum sensitivity from the confocal microscope.

Looking after the Objective

The condition of the objective lens is critical for obtaining high quality images on a confocal microscope. A damaged or "dirty" front lens element will result in a "blurred" image when using the microscope in bright-field or laser scanning transmission imaging. However, when imaging by fluorescence microscopy the damaged or "dirty" front lens element may result in a poor quality image (the resolution and sensitivity may appear to be not as good as you are accustomed to), this is particularly noticeable when imaging by confocal microscopy. Always have a top quality fluorescence sample handy that you can use on the microscope if you suspect that the images are not as good as they should be.

One of the common causes of poor quality images in confocal microscopy (or any light microscopy) is for the front lens of the objective to be smeared with immersion oil from a previous user or previous experiment of your own. This is particularly noticeable when using a water immersion or "dry" lens. In this case the smeared oil will not be miscible with the water, or in the case of air, still present, when imaging. Image deterioration due to oil smeared on the front lens can still be there even if you are imaging with an oil immersion objective. The poor quality image can also be due to the "old" oil being contaminated with dust etc that has become attached to the "sticky" surface of the oil, or perhaps the previous oil used is not compatible with the immersion oil you have now applied.

Inspecting the Objective: Visually inspecting the front lens element of an objective for any dust, oil, scratches or chipping is quite easy. The best way is to remove the objective from the microscope turret and to use a low powered "dry" objective (10x or 20x are fine) or ocular, hand held, to inspect the front surface of the lens that is giving trouble. If you are working in a multi-user facility consult with the person in charge before removing any of the objectives from the microscope turret! You can also use a small 10x hand magnifying glass.

Cleaning the Objective: Great care should be taken when cleaning an objective. Have a good supply of lens cleaning tissue near the microscope and carefully wipe the tip of the objective before and after using. This is important even if you are using a "dry" objective, because, as discussed above, oil or glycerol can often be accidentally smeared across the front lens of these objectives. Most of the time a simple gentle wiping of the front lens element of the objective is sufficient to clean the objective (but make sure you inspect the front lens element with a low powered objective or ocular as discussed above).The lens tissue should be wiped across the lens in one direction only and then discarded. A more effective method is to use a small amount of ethanol or methanol on the lens tissue as described below.

If a simple wiping action does not remove the contaminating oil or dust particles from the front lens element, then a solvent may be required to clean the lens. One of the best, and probably least harmful, solvents to use is ethanol. Ethanol should remove most grime found on the front lens element. Methanol can also be used in place of ethanol. Other, more "aggressive" solvents may readily remove the "dirt", but they can also create problems with the glue used to cement the lens elements into place. Under no circumstances should you ever soak or leave the objective in contact with any solvents.

A cotton wool bud (Q-tip) covered with several clean layers of lens cleaning tissue and then dipped in ethanol or methanol and used in a single swirling motion around the tip of the objective lens is highly effective in removing dirt or oil caught between the lens casing and the front lens element itself.

Optical Filter Block (or Optical Filter Wheel)

A conventional light microscope set up for epi-fluorescence imaging will have a wheel or slider that contains a variety of optical filters and dichroic mirrors (often referred to as filter blocks) that are used for fluorescence imaging using a mercury or xenon arc lamp. The microscope filter blocks are not used for imaging when using the laser scanning confocal microscope, and so they should be removed from the light path by choosing the correct setting on the microscope – otherwise you may have strange effects happening (such as blocking the excitation light!).

The confocal microscope scan head contains a separate selection of dichroic mirrors and optical filters (or, in the case of Leica and Zeiss, devices to split the light into the various colour components). These optical filters are quite separate from those used for epi-fluorescence microscopy, except in the case of the Nikon C1 confocal microscope, where conventional Nikon epi-fluorescence imaging filter blocks are used in a separate optics unit to separate the various colours emitted from the Nikon C1 scan head.

Condenser

The condenser is designed to focus a cone of light of uniform intensity onto the specimen (see page 43). Like an objective, the condenser has a numerical aperture and can be highly corrected for chromatic aberration (apochromatic). The condenser is mounted below the microscope stage (or above in an inverted microscope). The condenser can be readily moved in all directions to align this lens with the optical axis of the microscope for correct illumination. In the confocal microscope the fibreoptic pickup cable is located behind the condenser, picking up the laser light that has passed through the sample and the condenser.

In conventional bright-field light microscopy the correct adjustment of the condenser (see Köhler illumination, page 40) is very important for both resolution and contrast within the image. However, when using a laser scanning confocal microscope using the fluorescence-imaging mode the substage condenser is not used (in this case the "condenser" is in fact the same objective lens used to transmit the excitation light to the sample). When using the transmission-imaging mode in a laser scanning confocal microscope the condenser is critical in determining the effective NA of the objective, exactly the same way as when using bright-field microscopy.

Condenser Iris

The condenser Iris (also known as condenser diaphragm) determines the angle of the cone of light that is used for imaging the sample (see Figure 2-10 on page 43). In other words the NA of the condenser is adjusted using the condenser iris to match the NA of the objective lens being used. In Köhler illumination (see page 40) for transmission imaging the condenser iris is adjusted so that the illuminating light almost fills the back focal plane of the objective. Closing the condenser iris slightly (to ¾ of the total area) will result in a "better" image due to an increase in image contrast – although at the expense of a small loss in resolution. When using the transmission channel on the confocal microscope you have much better control over the level of contrast of the image by adjusting the gain and black level on the instrument and thus creating excellent transmission images is often possible when the illuminating light fully fills the back focal plane of the objective (adjust the condenser iris while scanning to obtain the best image).

When collecting transmission images using the confocal microscope the condenser should be correctly focussed for Köhler illumination, including placing immersion oil between the condenser front lens and the microscope slide for high resolution imaging. However, in confocal microscopy the transmission image is rarely used at the limit of the resolution of the microscope. The fluorescence image is normally the image for which you would like the best resolution – and as mentioned above the fluorescence image is not affected by the settings on the condenser as the fluorescence image is collected by epi-illumination.

Field Aperture

This is a variable aperture (or sometimes called a diaphragm), which is usually located in the base of an upright microscope and in the top of the overhead arm of in an inverted microscope. The field aperture is used to vary the area of the specimen that is illuminated. This is important for correctly setting up bright-field illumination (Köhler illumination) to give high-resolution images with good contrast. The setting of the field aperture is important in confocal microscopy when using transmission imaging, but the fluorescent light, as discussed above, being collected by epi-illumination, does not pass through the field aperture.

Focus Controls

The focus control will either move the microscope stage (for upright microscopes) or move the objectives themselves (an inverted microscope). The outer knob is the coarse focus control, and the inner knob is the fine focus control. The motorised focus control used in z-sectioning is often attached to the focus control knobs of the microscope itself. In most microscopes the z-stepper motor is attached to the fine focus control, but in the Zeiss microscope the z-stepper is best attached to the coarse control knob of the microscope. In the Zeiss microscope the fine focus knob does not allow for accurate positioning of optical sections.

Although the focus control of a research microscope provides very accurate positioning for z-sectioning, even more accurate and substantially faster z-control is available by using a piezo-electric stage z-control (a small device fitted on a conventional microscope stage) or a piezo-electric objective z-control. The advantage of the piezo electric z-axis control is that this type of z-stepper provides not only more accurate positioning, but at a very substantially higher speed. This can make true z-imaging possible, where the laser scan is in the x-z direction rather than the more conventional x-y direction.

Microscope Stage

The microscope stage on most research microscopes is quite large for accommodating any additional equipment, such as heated microscope imaging chambers that may be required. The large size of the stage is convenient for placing cell and tissue incubation chambers on the instrument, but can result in difficulty, particularly on an inverted microscope in being able to see what is marked on individual objectives. Some manufacturers provide a slip on sleeve that can be inverted for use on an inverted microscope; otherwise with most objectives on an inverted microscope you will be always trying to read things upside down! A small torch or portable light near the microscope can be very handy for illuminating the dark recesses underneath the microscope stage.

Most microscope stages are made of a relatively thick plate of aluminium, which acts as both a very good heat conductor and a heat sink. This is most important when you are imaging live cells, as the microscope stage may not be at the temperature at which you are attempting to maintain the cells (usually 37°C). If you are using a heated incubation chamber and objective heater to maintain your cells at the correct temperature, but the microscope stage is not being heated, try not to place the microscope incubation dish containing your cells directly on to the microscope stage. If you even momentarily place the incubation dish on the microscope stage the thin cell layer at the base of the dish will be very quickly equilibrated to the temperature of the microscope stage!

Microscope Attachment Ports

Research microscopes often have a variety of ports for the attachment of cameras, video cameras and the confocal microscope scan head. The confocal microscope scan head is normally mounted onto the microscope and correctly aligned by the technical representatives from the company that provides the instrument. However, the scan head can be moved to different attachment ports, or even to a different microscope. A special attachment collar and in some circumstances an additional lens will be required to use the laser scanning mechanism in the new position.

Confocal Microscope Scan Head Attachment Port

The scanning laser beam from the confocal microscope scan head is designed to enter the optical system of a microscope as close to the objective as possible. This eliminates any degradation of the image created by the additional optical elements within the microscope. The position of choice for mounting the confocal microscope scan head is usually directly above the objectives (the video camera port) on an upright microscope, and directly below the objectives (usually via a custom made port) on an inverted microscope. However, it is often possible to mount the confocal microscope scan head at other locations, for example onto the video outlet on an inverted microscope. On some instruments the confocal microscope scan head is mounted from the underneath of an inverted microscope by cutting a hole in an optics table directly under the microscope. This configuration has the advantage of creating more workspace around the microscope for additional instrumentation, and is the method of choice for the Bio-Rad multi-photon microscope installed with external detectors.

Camera Attachment Port

Many microscopes have a port available for directly attaching a 35 mm SLR camera. The automatic exposure meter of the camera can then be used for taking excellent photographs. However, today it is more common to attach a CCD camera to a light microscope. Although a relatively cheap and readily available CCD camera used for general photography can be readily attached to a light microscope (search the web for companies that sell suitable attachment collars) and produce excellent images of brightly lit samples, you will probably find that a camera that is suitable for fluorescence microscopy is considerably more expensive. Even though you have an excellent imaging device in a confocal microscope, it may be useful to be able to take quick and easy images with an attached CCD camera as well. A camera can be mounted on either a C-mount dedicated camera port, or on a tube-mount video camera port.

Fibreoptic Transmission Pickup

The confocal microscope can generate an excellent transmission image by collecting the light after it has passed through the sample and the condenser. The light detector (a photodiode) is sometimes located below the condenser (i.e. on the opposite side of the condenser to the sample), but often the light is relayed to an optics detection unit via fibreoptics (in which case the detector may be either a photodiode or a photomultiplier tube).

The transmitted light does not pass through the "confocal pinhole" within the scan head and so, although the quality of the image is very good, the image is not confocal. The phase or DIC optics of the microscope can be used to produce excellent phase and DIC images when collecting transmission images by laser scanning.

Light Sources

Lasers - see page 85 in Chapter 3 "Confocal Microscopy Hardware"

A number of light sources are available for light microscopy. Bright-field microscopy is normally carried out using a tungsten lamp with a variable voltage power supply to adjust the brightness, whereas fluorescence microscopy is performed using either a mercury arc lamp or a xenon arc lamp. These lamps provide a high level of light output across a wide range of wavelengths – from UV to infrared. The emission spectra of these various light sources is published on pages 30-31 of "*Fundamentals of Light Microscopy and Electronic Imaging*" by Douglas Murphy (see Chapter 15 "Further Reading", page 347). This section briefly explains the light sources

Vibration Free Imaging (using bicycle inner tubes!)

The microscope and attached scan head will need to be mounted on a vibration damping platform if high resolution images are going to be collected. Vibration will be evident in images from a confocal or multi-photon microscope as slightly "wavy" lines on the edges of objects (such as cell membranes), or as the blurring of images when collecting using an averaging filter algorithm.

Vibration damping can be accomplished with expensive gas-lift anti-vibration tables. However, very effective vibration damping can be achieved with very simple and much cheaper technology. If the microscope with attached scan head is placed on a heavy duty aluminium plate or stone platform then small size bicycle inner-tubes can be placed between the microscope support plate and a conventional table. Most building vibrations will not be transmitted through the air-cushion created by the bicycle inner-tubes, thus isolating the microscope from vibrations present in the building. Bicycle inner-tubes are not only highly effective in removing unwanted vibration artefacts, but are easy and cheap to install – and only need pumping with air every couple of years!

Although traditionally microscopy units are placed in the basement of buildings, there does not appear to be any appreciable difference in building vibrations that will affect light microscopy between the basement and the higher floors of the building. What usually causes the most annoying vibrations is equipment within the building - with the vibrations being efficiently transmitted relatively long distances in a modern concrete building.

commonly used for conventional light microscopy, whereas the lasers used in confocal and multi-photon microscopy are described in some detail in Chapter 3 "Confocal Microscopy Hardware" (page 65).

Mercury Arc Lamp

The mercury lamp is a very powerful source of light from UV to infrared and so is widely used in conventional wide-field epi-fluorescence microscopy. The mercury lamp produces a continuous emission across the visible spectrum, but does have specific emission lines that are significantly brighter (commonly used emission lines are 366, 405, 435, 546 and 578 nm). A large proportion of the emission is in the UV and infrared portion of the light spectrum. The UV and infrared wavelengths should be filtered if they are not specifically needed for the imaging method used as UV light will quickly destroy biological material and the infrared component will quickly overheat the sample.

Mercury lamps have a limited life span of about 200 hours (refer to manufacturers instructions), and should be changed at this number of hours rather than waiting for the lamp to "burn out". A worn mercury lamp can become unstable, resulting in poor quality fluorescence images, but more importantly mercury lamps can explode - releasing nasty mercury vapours into the room!

The mercury lamp must be correctly aligned and focussed for optimal illumination of the sample during fluorescence microscopy. Poor illumination in fluorescence microscopy is usually due to an out of date lamp or an incorrectly aligned lamp. Don't try and align or replace a mercury lamp if you have no experience – ask someone from the company that sold you the microscope, or other more experienced users than yourself.

Confocal microscopes are not always equipped with a mercury lamp as they are not required for creating confocal images. However, locating the sample and choosing the area to be imaged can be quite difficult using the laser scanning confocal microscope alone. Using a mercury lamp to locate the area of interest and adjust the microscope is very fast. Furthermore, your eyes are quite a remarkable light detecting instrument – if you take some time to look directly down the microscope using conventional wide-field epi-fluorescence microscopy you may find that fine detail and particularly movement, may be more easily observed with your eye than with a confocal microscope! Your eye is a dynamic and discriminating device and so simply taking a photograph of the image will not give anywhere near the same level detail! However, be careful not to bleach the sample with the mercury lamp!

Xenon Arc Lamp

A xenon arc lamp provides a high level of light output right across the visible spectrum. There are no major specific emission lines in the visible region of the light spectrum, making this lamp ideal where specific wavelengths of visible light are required (for example, in a spectrophotometer). There is significant emission in the UV and a much larger emission in the infrared. The xenon arc lamp can be used as an alternative to the mercury arc lamp for conventional epi-fluorescence microscopy.

Tungsten Lamp

This is the regular lamp used for normal bright-field microscopy. The lamp usually has an intensity adjustment for altering the intensity of the light for viewing, but which also alters the colour balance, which is important for correct colour balance when using traditional photographic film. With the introduction of digital cameras in microscopy the exact colour balance of the lamp is no longer important as adjustments can be readily performed after collection (or automatically during collection on some cameras).

Regular bright-field microscopy using the tungsten lamp is the simplest and least damaging method of locating the object to be later imaged by epi-fluorescence or confocal microscopy. The light intensity required to visualise the cells is significantly less than that used for fluorescence imaging. DIC optics allows one to see the outline and internal structure of unstained cells, which is particularly valuable for imaging live cell preparations.

Halogen Lamp

The quartz halogen lamp, which emits strongly right across the visible light spectrum, can be used in place of the tungsten lamp for bright-field imaging with suitable optical filters to remove the large infrared component of this light.

Chapter 3

Confocal Microscopy Hardware

The confocal and multi-photon microscope is comprised of a conventional light microscope to which are attached a number of important optical and electronic components. The additional components allow the laser light to be scanned across the sample, the returning fluorescent light to be split into various wavelengths and for the fluorescent, backscattered or transmitted light to be detected.

Confocal and multi-photon microscopes are traditionally thought of as microscopes that are capable of collecting images without the problem of out-of-focus "blur" detracting from the quality of the image, and are able to use a series of "optical slices" to create a 3D rendition of the object. The confocal "pinhole" has come to define confocal microscopy. The rejection of out-of-focus light is a very important aspect of confocal microscopy, but there is a lot more to a confocal microscope than the creation of optical slices. Confocal and multi-photon microscopes are highly versatile instruments that can collect multiple channel fluorescence, transmission (including phase and DIC) and backscatter images in perfect registration. These images can be combined, quantitatively analysed, stored, copied and easily published – in other words, a confocal microscope is a very powerful and versatile digital instrument that is a wonderful improvement on the traditional light microscope.

The various operating modes and the manner in which an image is created in a confocal microscope are covered in Chapter 1 "What is Confocal Microscopy?". This chapter is concerned with the various hardware components that come to make up the "confocal microscope". The general principles of each component are explained in some detail, but specific information on individual confocal microscopes from each of the manufacturers is described in more detail in "Appendix 1: Confocal Microscopes" (pages 355 - 441).

Technical Details on Specific Instruments

(Appendix 1: Confocal Microscopes)

Laser Scanning Instruments:

Bio-Rad	page 356
Carl Zeiss	page 382
Leica	page 402
Nikon	page 415
Olympus	page 420

Nipkow Disk Instruments:

Atto Bioscience	page 431
Yokogawa	page 435
PerkinElmer	page 438
VisiTech	page 441

Multi-Photon Instruments:

Bio-Rad	page 356
Carl Zeiss	page 382
Leica	page 402

INSTRUMENT DESIGN

There are two fundamentally different confocal microscopes available for biological research today – the laser "spot" scanning confocal microscope, in which, as the name suggests, the laser beam is scanned across the sample to create an image, and the Nipkow spinning disk confocal microscope, where a series of "pinholes" in a spinning disk result in a large number of points of light being scanned across the sample simultaneously. A variation on the "spot" scanning confocal microscope is the "slit" scanning instrument, where a narrow

Confocal Microscopy for Biologists, Alan R. Hibbs. Kluwer Academic / Plenum Publishers, New York, 2004.

band or slit of light is scanned across the sample. These instruments have the advantage of being significantly faster than the spot scanning systems, and the image can be seen "live" by looking down the microscope eyepieces. Slit scanning instruments have been produced by Meridian Instruments and Bio-Rad, but are no longer in production.

There is also a third design of confocal microscope – the stage scanning confocal microscope – that is used in the materials sciences, but rarely used in biology. The stage scanning confocal microscope, as the name implies, moves the sample instead of the focussed point of light to build up an image. These instruments were the forerunners of the present day laser scanning confocal microscopes, and produce excellent images as long as the sample is very rigid – hence their continuing popularity in the materials sciences (particularly microchip manufacturing). As the stage scanning confocal microscope is rarely used in biology, it is not discussed further in this chapter.

Although this chapter is divided into separate sections that describe laser spot scanning, Nipkow disk and multi-photon microscopes, a number of important hardware components common to all confocal microscopes will be described in detail under the heading "Laser Scanning Confocal Microscopes". For example, the separation of fluorescent light of different wavelengths is an important component of the multi-photon and Nipkow disk and laser scanning confocal microscope, but is dealt with in detail only in the section on laser scanning confocal microscopes.

LASER SCANNING CONFOCAL MICROSCOPES

Laser "spot" scanning confocal microscopes, more commonly referred to simply as laser scanning confocal microscopes, utilize a single diffraction limited laser point that is scanned across the sample in a raster pattern. This section is concerned with the various hardware components required for scanning the laser, separating various wavelengths of light and detecting the light.

Instrument Layout

The basic layout for the laser scanning confocal microscope is shown in Figure 3-1. This generalized diagram is based in the Bio-Rad Radiance confocal microscope, but the general principles are the same for each of the manufacturers. Figure 3-2 shows a photograph of the Zeiss META confocal microscope in which the light microscope and attached scan head are located on a vibration isolation table adjacent to the work bench used to house the dual display monitors. The same basic layout is also used for a laser scanning multi-photon microscope (shown in Figure 3-19), with some important differences, which include the type of laser and the position of the light detectors, both of which are highlighted in the section dealing with multi-photon hardware (pages 94 - 97). The Nipkow disk confocal microscope, although having a number of hardware components in common with the laser scanning confocal microscope, has a basic design that is quite different, and is described in some detail on pages 91 - 93.

At the heart of the laser scanning confocal microscope is a conventional light microscope, which may be either an upright or inverted microscope (an inverted microscope is shown in Figure 3-1). The light microscope itself may contain a range of optional optical components that make a research microscope a highly versatile, but somewhat complex instrument to use. The confocal microscope does make good use of some of the additional components of the light microscope (particularly DIC optics), but the central core of confocal microscopy involves fluorescence imaging, where most of the necessary optical components are housed in either the scan head or associated controller or optics box.

The laser scanning confocal microscope scan-head, which is discussed in detail shortly, is attached directly to a camera port, video outlet, or a custom designed outlet on the light microscope (Figure 3-1). On some Bio-Rad instruments the scan head is mounted directly underneath the microscope through a custom-made hole in the vibration isolation table. The scan head contains the scanning mirrors, associated optics for scanning the laser beam across the sample, and a number of optical components for separating both the different laser lines and the differently coloured fluorescent signals that are collected from the sample. The number of optical components incorporated into the scan head varies for each of the manufacturers. Some manufacturers have a minimal set of components located within the scan head, with many of the optical and electronic components being housed in an associated optics or controller box. Other manufacturers have incorporated all of the scanning and light separation hardware within the scan head, which results in a considerably larger scan head. Originally all confocal microscopes had the optical and electronic components either inside or directly attached to the scan head (this even included the laser). To varying degrees each of the manufacturers have attempted to remove some of the optical and electronic components to separate boxes connected by fibreoptics. The trend to locate a number of important components remote from the scan head itself is driven by a number of factors, including the ease of mounting and aligning a small scan head, and the removal of all possible sources of vibration (for example, the cooling fan on the laser) away from the microscope. There is also a manufacturing advantage in not requiring expensive design changes to the scan head when an improved optical or electronic component is added to the instrument.

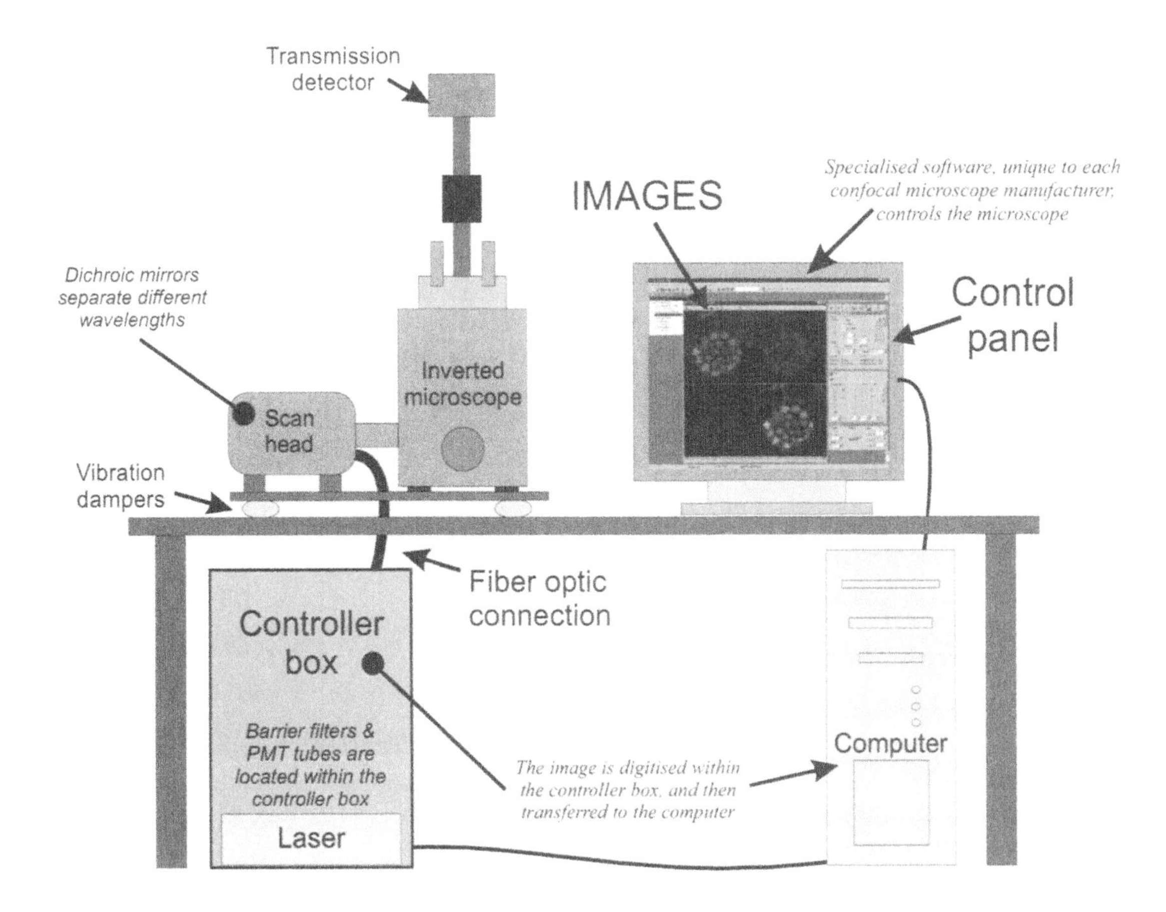

Figure 3-1. Simplified Layout of a Laser Scanning Confocal Microscope.
The laser scanning confocal microscope comes in a wide range of layouts. This diagram represents a simplified view of a typical laser scanning confocal microscope, based on the layout for the Bio-Rad Radiance series of confocal microscopes. The essential components are the conventional light microscope (represented here by an inverted microscope), a scan head, a controller box (sometimes known as the optics box, containing a varying number of optical components, depending on the model and manufacturer of the confocal microscope), a laser (or lasers) and a computer. The various components are connected via fibreoptic coupling. The scan head is controlled to varying levels (depending on the model and manufacturer) by an attached computer. Images are displayed "live" during collection on the computer screen using software unique to each confocal microscope manufacturer and then saved as image files.

The scan head and the attached microscope are often mounted on an anti-vibration platform (as shown in Figure 3-1) to remove unwanted building vibration when imaging at high resolution. The vibration isolation platform may consist of a small custom table that supports just the microscope and attached scan head, as in the case of the Leica and Zeiss confocal microscopes. Alternatively, you can construct your own simple vibration isolation table using bicycle inner tubes (see page 62 in Chapter 2 "Understanding Microscopy".

A conventional computer running MS Windows is connected to either the controller box, as shown in the example given in Figure 3-1, or directly to the confocal microscope scan head as on a number of other instruments.

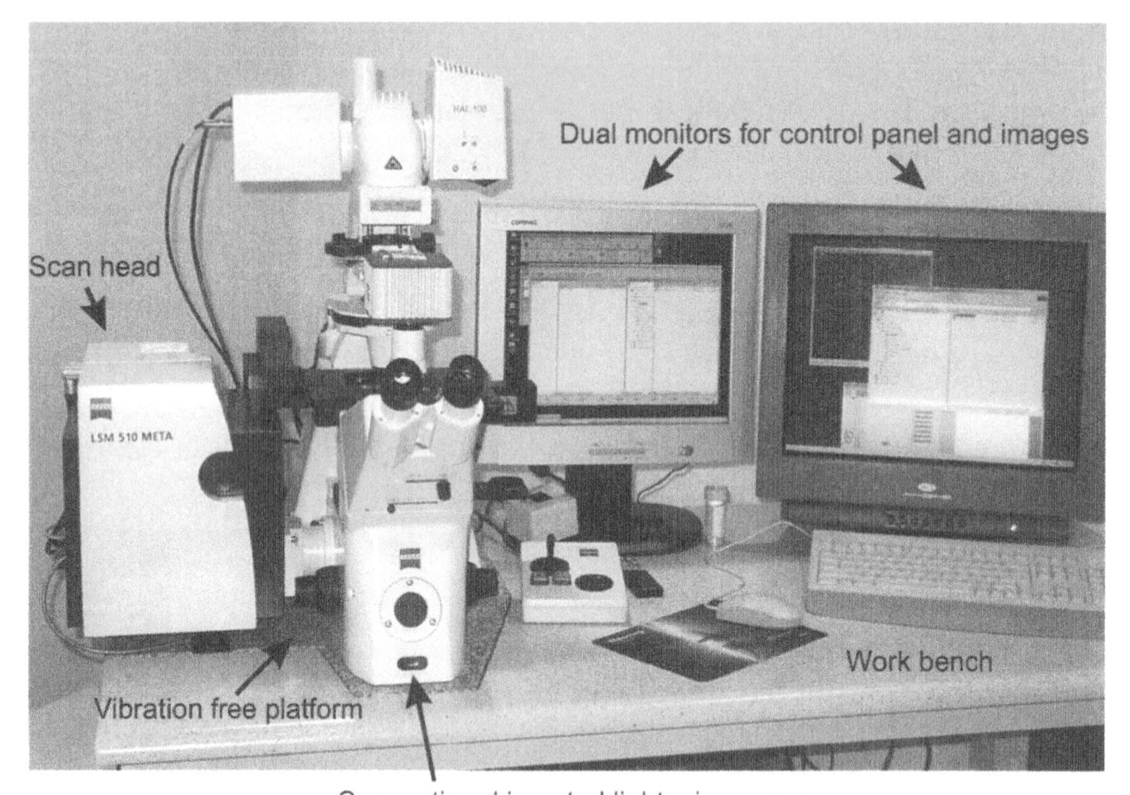

Figure 3-2. Zeiss 510 META Confocal Microscope.
The Zeiss META confocal microscope is mounted on a custom made work bench within which a vibration free platform is installed. The relatively large scan head on this confocal microscope contains all of the dichroic mirrors, optical filters, scan mirrors, pinholes and detectors for both "conventional" and spectral (META) detection channels. The laser (not shown) is mounted below the work bench and is connected to the scan head via a fibreoptic connection. The photograph was taken, with permission, of the Zeiss 510 META confocal microscope installed at the Baker Heart Research Institute, Melbourne, Australia.

The computer is provided with proprietary software to drive the scan head, and in more recent instruments to control the positioning of various optical filters, dichroic mirrors and for adjusting the confocal pinholes settings. Not all confocal microscopes are fully "computerised", but the trend has been to computer control more and more of the hardware. This has a number of important advantages, but does have the serious disadvantage of moving the operator further away from the optics of the instrument. Interestingly, the trend has been to produce software that more closely resembles the optical layout of the instrument, which ensures that the correct settings are more easily enabled.

Two laser scanning confocal microscopes, the Olympus Fluoview 300 and the Nikon C1, are still manufactured with manual control for most of the optical components. These instruments provide excellent resolution and optical sectioning capabilities at a considerable cost saving to the user.

Control of the instrument and display of your images requires a computer screen. Some manufacturers have opted for a large single screen to display both the control information and the images as they are collected, but others have incorporated the use of two computer monitors. The advantage of using two screens is that one can be used to display the control information, while at the same time you have a full screen to display the images at the resolution of the monitor – this is particularly important, as will be described in Chapter 4 "Image Collection" (page 101), when you wish to collect high resolution images in real time.

A number of different lasers are available for use in laser scanning confocal microscopy (discussed in detail on page 86). The laser(s) are housed either within the controller box (Bio-Rad Radiance confocal microscopes) or mounted separately (most other confocal microscope manufacturers) as shown in Figure 3-3. Mounting the laser directly on the confocal microscope scan head has been discontinued due to the possible transfer of vibrations from the cooling fan disturbing the image. Multi-photon microscopes normally have the pulsed infrared laser directly optically coupled (although not physically attached) to the scan head. The laser used, laser line selection and the intensity levels of the laser are software selectable on most instruments.

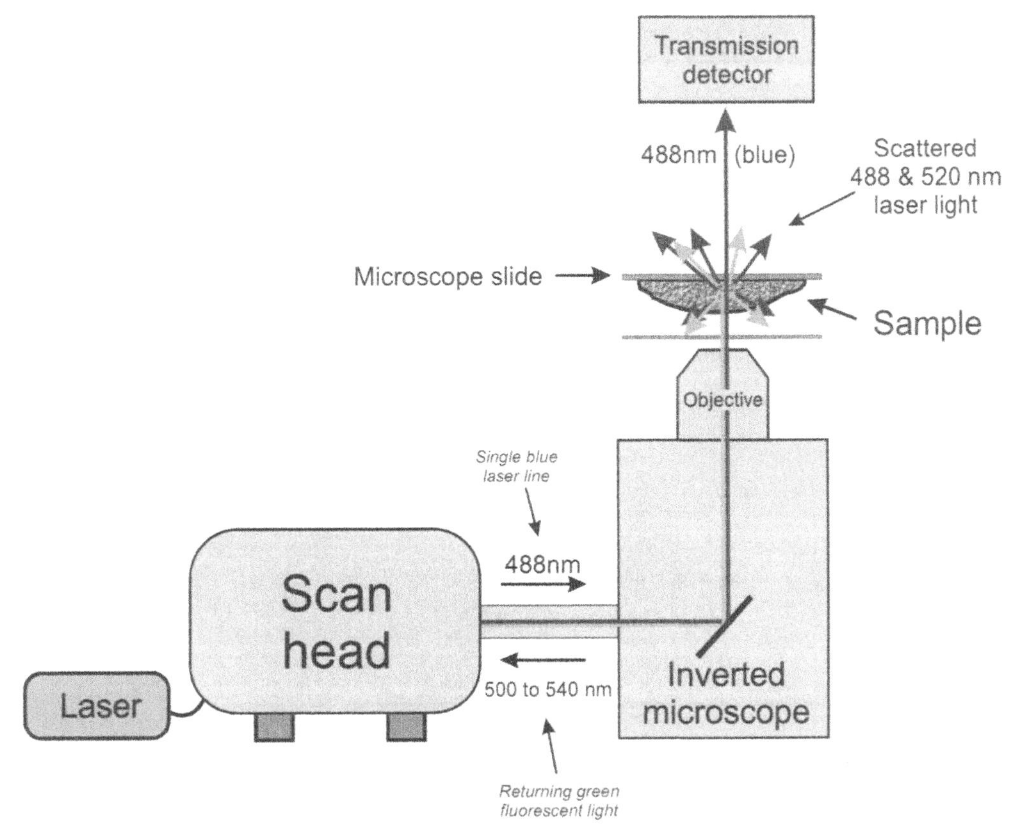

Figure 3-3. Irradiation of the Sample.
In this example 488 nm blue laser light is used to irradiate the sample, which is located on the stage of a conventional inverted light microscope. A small amount of backscattered 488 nm blue light and a small amount of fluorescent light of a range of wavelengths (predominantly 500-540 nm green light in this example) finds its way back through the scan head where the light is separated into various regions of the light spectrum and detected by the photomultiplier tubes. A relatively large proportion of the irradiating 488 nm blue light is transmitted through the sample (some of which will be diffracted by the sample) and is picked up by the condenser and passed on to a photodiode or photomultiplier tube detector. In this way simultaneous multi-channel fluorescence and transmission images can be collected.

Irradiation of the Sample

How many photons find their way to the detector?

The sample is irradiated using a specific laser line (see Figure 3-3, where the 488 nm blue light from the argon-ion laser is used). The laser light is focussed on the sample using an objective lens on the conventional light microscope. Under ideal conditions this laser spot is "diffraction limited", that is the spot is not infinitely small, as its size is determined by the diffraction of light (determined by the NA of the lens and the wavelength of the irradiating light, see Chapter 2 "Understanding Microscopy", page 31). The diffraction-limited spot is scanned in a raster pattern both across the sample (x-scanning, at relatively high speed), and down the sample (y-scanning, at a much lower speed) using scanning mirrors located within the scan head.

The focussed spot of laser light is scattered (by reflection and diffraction) in all directions after striking the sample (Figure 3-3). Some of the scattered light can be collected by the objective lens and used to create an image (known as backscatter or reflectance imaging), particularly of the surface topography of the sample, but also of internal structures, often after they have been labelled with highly reflective reagents such as silver-enhanced gold particles. A large proportion of the laser light passes directly though the sample, some of which can be collected by

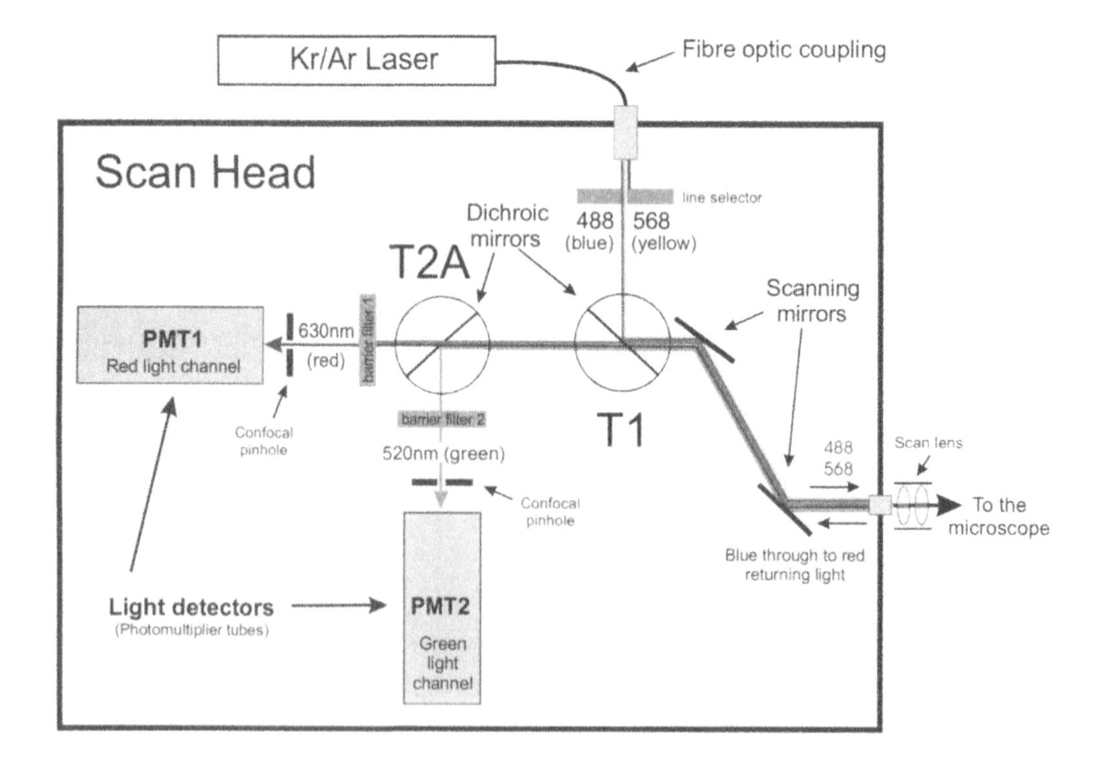

Figure 3-4. How Does the Scan Head Work?
The confocal microscope scan head contains a variety of optical components. However, the individual components that are included in the scan head vary between each of the confocal microscope manufacturers. Some manufacturers provide an additional "controller" or "optics" box in which are mounted various optical and electronic components, resulting in a significantly smaller scan head. The above idealised scan head is based on the Bio-Rad MRC series of confocal microscopes – which contains all of the necessary optical components, except for the laser (which is fibreoptic connected to the scan head). This scan head is ideal for demonstrating the important hardware components required for laser scanning confocal microscopy, but for details on specific instruments please refer to "Appendix 1: Confocal Microscopes" (pages 355 - 441).

T1 contains a triple dichroic mirror (see Figure 3-5 for spectral details), and T2A contains a 560DCLP mirror (see Figure 3-6).

Barrier filter 1 is a E600LP optical filter (see Figure 3-8), and barrier filter 2 is a HQ515/30 optical filter (see Figure 3-7).

the front lens of the condenser and transferred to a sensitive detector (the transmission detector). The transmitted light can be used to create very high quality transmitted light images that are usually depicted as a grey-scale image, but can be displayed as real-colour transmission images by collecting three separate transmission images using three differently coloured laser lines (blue, green and red laser lines).

A small proportion of the laser light used to irradiate the sample will result in the excitation of suitable fluorochromes that have been used to label the sample or of fluorophores already present within the sample (autofluorescence). The fluorescent light originating from the sample (for example, green fluorescent light from excited fluorescein molecules within the sample, denoted 520-540 nm returning fluorescent light in Figure 3-3) is emitted in all directions. A small proportion of this fluorescent light will be picked up by the front lens of the objective, and directed back into the scan head where the light is de-scanned by reflecting off the scanning mirrors. The fluorescent light is then separated from the irradiating blue laser light, and if required the resultant fluorescent light can be further separated into various regions of the light spectrum. Each channel of fluorescent light is then focussed through a confocal pinhole before being detected by a sensitive light detector called a photomultiplier tube. The output of the photomultiplier tube is digitised during a very short, but highly accurate, time period delegated to an individual image pixel (~1-3µs).

The image is collected as changes in signal (changes in light intensity), and is always collected as a grey-scale image - no matter what wavelengths or colours of light were collected originally. Colour is added to the image by using colour Look Up Tables (LUT) for display purposes. Practical information on collecting confocal images, including the use of colour in confocal microscopy, is discussed in detail in Chapter 4 "Image Collection" (page 101) and Chapter 5 "Digital Images in Microscopy" (page 145).

Scan Head and Associated Optics

The laser scanning confocal microscope scan head houses various optical components (Figure 3-4). However, as discussed above, the optical layout, and the optical components housed within the scan head vary widely between the various confocal microscope manufacturers. Most confocal microscope scan heads contain the scanning mirrors, scan lens, dichroic mirrors, pinholes and sometimes also the barrier filters, photomultiplier tubes and laser attenuation and line selection filters (as shown diagrammatically in Figure 3-4). The specific components included in the scan head from each of the major confocal microscope manufacturers are described in detail in "Appendix 1: Confocal Microscopes" (pages 355 - 441).

Originally the "scan head" from most manufacturers contained all of the optics required for delivery of the laser light (with the laser often physically attached to the scan head) - and the optics required for separation of fluorescent wavelengths and their detection. In more recent years there has been a trend to separate each of the major components (the laser, the optics and the scan mechanism) into separate, relatively independent devices that are now often connected by fibreoptics. Some manufacturers have taken this separation of components further than others. One particular manufacturer of confocal microscopes in the mid 1990's (Optiscan Imaging) produced a small scan head that contained only the scanning mirrors – with all other optical components placed in a remote optics box connected by a single fibreoptic connection. The Bio-Rad and Nikon confocal microscope scan heads are also relatively small and designed to be attached to a variety of microscopes from a number of microscope manufacturers. The other manufacturers (Leica, Zeiss and Olympus) have significantly larger scan heads that are directly attached to their own brand of microscope.

The following discussion is based on the idealised scan head shown in Figure 3-4. This scan head is particularly useful for this discussion as the scan head contains all of the optical components and has a relatively simple and easily understood configuration. Variations on this basic scan head design are discussed later in regard to unique features offered by various confocal microscope manufacturers.

Scanning Mirrors

Within the scan head are two scanning mirrors (Figure 3-4). One mirror scans the focussed laser light in the horizontal or x-direction at relatively high speed, and the other in the y-direction at considerably lower speed. The laser must be aligned central to the axis of movement of these mirrors; otherwise a scanning movement artefact may become evident in your images. Furthermore, the scanning mirrors need to both be in the same optical position for optimal scanning, which creates a number of technical difficulties when constructing a laser scanning confocal microscope. The alignment of the scanning mirrors are factory set, but adjustment of the alignment mirrors that direct the laser onto the scan mirrors may result in miss-alignment at the scanning mirrors.

The earlier Leica confocal microscope utilised a novel single mirror dual axis scanner that reduces optical losses. However, the more recent K-scan configuration has 3 scanning mirrors. The Bio-Rad scanning mirrors incorporate two concave mirrors between the scan mirrors to alleviate optical problems associated with the requirements to have both mirrors in the same optical position. Some confocal microscopes are designed with resonant scan mirrors for higher speed scanning.

The returning fluorescent or backscattered light is "de-scanned" by the scanning mirrors, resulting in a non-scanning beam being projected through the colour separation filters, the confocal pinhole and directed to the light detectors.

The speed of the scanning mirrors can be varied on most confocal microscopes (under software control). Changing the speed of the x-scan mirror will alter the frame capture rate, and consequently the dwell time for each pixel within the image, but it does not necessarily alter the number of pixels within the image. On the other hand, changing the y-scan rate is often used to increase the frame capture rate while decreasing the number of pixels in the image. This type of high-speed imaging is used as a simple way of speeding up the scan rate for locating the cells ("Fast" or "x2" and "Fastest" or "x4" on a Bio-Rad instrument, and "Fast x-y" on a Zeiss instrument).

The angle of the scan mirrors can also be altered on a number of instruments. Changing the mirror scan angle, in conjunction with the confocal microscope zoom function, can be used to image a relatively small, defined area of the sample, with a box size chosen to suit the dimensions of the object under observation. This is one of many handy features on a confocal microscope and allows you to position the area to be viewed at high-resolution without moving the x-y stage controls or physically rotating the sample.

Confocal Pinhole (Iris)

The confocal pinhole (shown in front of each photomultiplier tube in Figure 3-4) is an essential component of a confocal microscope, allowing this remarkable microscope to create optical sections within thick cell and tissue samples, that are otherwise extremely difficult to image. In earlier model confocal microscopes the confocal pinhole was of a fixed size, although often optimised for the resolution (NA) and magnification of a specific objective. However, with the movement of confocal microscopy into the biological sciences in the early 1990's a variable size pinhole (or at least a range of fixed sizes) became very important for allowing one to collect images under optimal iris settings using a range of objectives, and for opening the pinhole beyond the optimal diameter for the objective in use when it is necessary to allow more light to reach the light detectors. Opening the pinhole greatly increases the sensitivity of the instrument, although at the expense of loss in z-resolution (and loss in effective x-y resolution due to out-of-focus light detracting from the image). The Bio-Rad Radiance confocal microscope incorporates a Signal Enhancing Lens (SELS) that has the effect of greatly increasing the amount of out-of-focus light directed to the detectors without the need to insert a much larger confocal iris (see page 364 in "Appendix 1: Confocal Microscopes"" page 364). This lens is highly effective in live cell imaging where high sensitivity is more important than resolution.

The Nipkow disk based confocal microscopes do not have a variable size pinhole (discussed in detail below). The Nipkow disk has a pinhole size approximately optimised for a specific objective (often a 100x 1.4 NA oil immersion objective). The multi-photon microscope, also discussed in detail below, does not require a confocal pinhole to create optical sections. The inherent optical slicing capability of multi-photon microscopy means that all fluorescent light is from the focal plane, thus allowing the use of "external" detectors for increased sensitivity.

There are three major implementations of a physical "pinhole" in a confocal microscope. The majority of confocal microscope manufacturers provide a variable, but very fine micron sized pinhole (created by using two overlapping L-shaped blades), through which the light from the objective is focussed.

Why have a variable pinhole?

A pinhole size of 1 Airy disk: the pinhole size is often set to 1 Airy disk - the physical size of the Airy disk is dependant on the NA and the magnification of the lens.

An "open" pinhole lets more light through: opening the pinhole beyond 1 Airy disk allows a great deal more light through to the detectors. This additional out-of-focus light may allow one to visualise structures that are otherwise beyond the sensitivity of the microscope.

Closing the pinhole: increases the x-y resolution and particularly the z-sectioning capability of the microscope.

Due to the small size of this type of confocal pinhole correct alignment of the iris within the optical path of the scan head is critical. In contrast, the confocal pinhole in the Bio-Rad confocal microscope is placed in front of the light detectors (or in front of the fibreoptic pickup for directing the light to the detectors), and due to the use of infinity optics in the confocal microscope scan head the variable "pinhole" iris is millimetres in size. This large sized round shaped iris is more easily aligned and readily varied in diameter. There is a separate iris for each of the detection channels, each of which can be independently adjusted using the appropriate software.

An innovative alternative design to the iris nature of the confocal pinhole is the single-mode fibreoptic design of the Optiscan Imaging produced confocal microscope. In this case the single-mode optical fibre not only acts as the launch fibre for the laser, and the pick-up fibre for the fluorescent signal, but also as a highly effective confocal "pinhole". This design has the significant advantage of greatly simplifying the size and design of the scan head, but does have the disadvantage of requiring a change of the optic fibre if a different sized "pinhole" is required.

The software on most confocal microscopes will now automatically set the correct pinhole setting for 1 Airy disk diameter (usually by pressing a button on the "pinhole" menu). This quick method of selecting the optimal "pinhole" size is very convenient, but on most instruments this setting is dependent on you specifying which objective is in use (and that the correct NA and magnification of this objective is correctly set in the software). Only if you are using a fully automated microscope (in which the objective turret is changed under computer control) can you safely assume that the confocal microscope software "knows" which objective is being used (as long as each objective is inserted in the correct position in the objective turret).

The optimal confocal pinhole size under ideal conditions of confocal imaging is determined by the size of the Airy disk. However, the fluorophore may not create sufficient fluorescence for sufficient signal to be collected at the theoretical optimal "pinhole" size. In this case the "pinhole" will have to be opened further to collect sufficient light to create an acceptable image. A larger iris setting allows more of the fluorescent light to enter the light detectors, but has the effect of reducing the optical sectioning ability of the microscope (i.e. thicker optical sections are produced). A "fully open" iris (the diameter of an "open" pinhole varies between different confocal microscopes, from approximately 3 Airy disks in a Bio-Rad instrument to about 10 Airy disks in a Zeiss instrument) has minimal optical sectioning ability, producing an image similar to a conventional epi-fluorescence microscope image. A good compromise for the size of the "pinhole" is to use a setting of 2 x the Airy disk for most applications, at least when first attempting to collect an image. If very high resolution is required then calculating the theoretical "pinhole" size required may be worthwhile. Further closing of the "pinhole", beyond the 1 Airy disk limit, does result in a discernable increase in resolution – but produces a significant loss of light! For maximum resolution on a well stained sample, using a stable fluorophore, an almost fully closed "pinhole" may be appropriate, with collection of a large number of screen averages to generate an image with a good signal to noise ratio (i.e. a high quality image). The correct setting of the confocal pinhole during image collection is discussed in more detail in Chapter 4 "Image Collection" (page 101).

Optical Filter Separation of Fluorescent Light

An important aspect of laser scanning confocal microscopy is to be able to separate both the irradiating laser light from fluorescent light emitted by the sample, and to separate the fluorescent light into various regions of the light spectrum for multi-labelling applications.

This section is concerned with the use of dichroic mirrors and optical filters (see boxes below) for separating light of different wavelengths. A later section (page 81) describes the use of prisms and diffraction gratings to separate various wavelengths of fluorescent light. However, most instruments that employ highly sophisticated spectral separation techniques still make use of a number of dichroic mirrors and optical filters to assist in the separation of specific wavelengths of light. The exception to this is the Leica TCSP2-AOBS laser scanning confocal microscope that is designed without the use of any conventional mirrors or optical filters (see page 78). A variation on the dichroic mirror separation of light is to use a polarisation beam splitter (used in the Bio-Rad Radiance instruments), which exploits the highly polarised nature of the irradiating laser light compared to the fluorescent light emitted from the sample to separate the irradiating laser light from the returning fluorescent light. The Bio-Rad Rainbow confocal microscope employs the use of a series of long pass and short pass optical filters to separate out defined regions of the light spectrum, but the description of this instrument has been delegated to the section on spectral imaging (page 85) as this technique is highly suited to deriving spectral information from the sample.

A number of optical components that are required for separating different wavelengths of light are often located within the confocal microscope scan head, but in some instruments important optical components may also be located within an associated "controller" or "optics" box. The distribution of the various optical components depends on the manufacturer and the model of confocal microscope, and is discussed in more detail "Appendix 1: Confocal Microscopes" (pages 355 - 441). The following discussion is based on having the optical components located within the confocal microscope scan head. However, the principles involved are the same no matter where the optical components are physically located.

The terminology used to describe the characteristics of both dichroic mirrors and barrier filters can be somewhat confusing due to a number of different conventions in use (see the on page box 79 listing some of the more common optical filter terminology). Dichroic mirrors and optical filters are designed to absorb or reflect a particular range of wavelengths and to allow other wavelengths to pass through the mirror or optical filter. A dichroic mirror can reflect a very narrow range of wavelengths (Figure 3-5), or a relatively broad range of the light spectrum (Figure 3-6) from the surface of the mirror. These mirrors have a very high efficiency (reflecting a very high proportion of incident light), but it should be remembered that even with a high quality dichroic mirror there is still a small amount of light that may penetrate the optical filter. This can result in unwanted reflected light within a fluorescence image (this will be discussed more shortly), but can be used to advantage to create a reflected light image as an overlay on the fluorescence image if the reflected light is from physically distinct structures compared to the fluorescent light.

The idealised confocal microscope scan-head shown in Figure 3-4 uses a dichroic mirror (denoted T1 in Figure 3-4, with the spectral properties shown diagrammatically in Figure 3-5) as the primary means by which the irradiating laser light is directed to the sample. This particular dichroic mirror is designed to reflect three regions of coloured light (centred around 488, 568 and 647 nm light from the krypton-argon-ion laser) and for other regions of the light spectrum to pass through the mirror. In Figure 3-4 the blue 488 nm and yellow 568 nm lines from the laser are reflected to the scanning mirror and then on to the sample, but the red 647 nm line from the krypton-argon-ion laser is not being used and so has been blocked at the laser line selection filter. The irradiating light is scanned across the sample, as discussed in some detail above, and fluorescent light of a wide range of wavelengths emanating from the sample will pass back up through the scanning mirrors and directly through the dichroic mirror at T1 (only a narrow band of fluorescent light around 488 and 568 and 647 nm, the wavelength used for irradiation, will not be able to pass back through the mirror). Any backscattered blue or yellow light, still of the same wavelength as the irradiating laser lines, will be reflected by this triple dichroic mirror back towards the laser and thus directed away from the detection channels.

Normally, backscatter imaging using the original laser light requires a special partially reflective mirror in place of the double or triple dichroic mirror. There can, however, under some circumstances, as discussed above, be a small amount of reflected light (the same wavelength as the original irradiating light) that can penetrate this double or triple dichroic mirror. This reflected light can be used to advantage in some samples to create a backscattered light image, but care must be taken not to confuse this reflected light with fluorescent light emanating from the sample.

What is a dichroic mirror?

Dichroic mirrors are designed to reflect specific wavelengths and to transmit other wavelengths of light. Dichroic mirrors are the means by which the irradiating light can be separated from the fluorescent light, and also a convenient method by which the fluorescent light can be divided into different regions of the light spectrum.

The dichroic mirrors are surface coated with emulsions to create reflectors of extremely high specificity. However, these surface coatings can deteriorate. An occasional quick visual surface inspection may show up any problem before it affects the quality of the images collected.

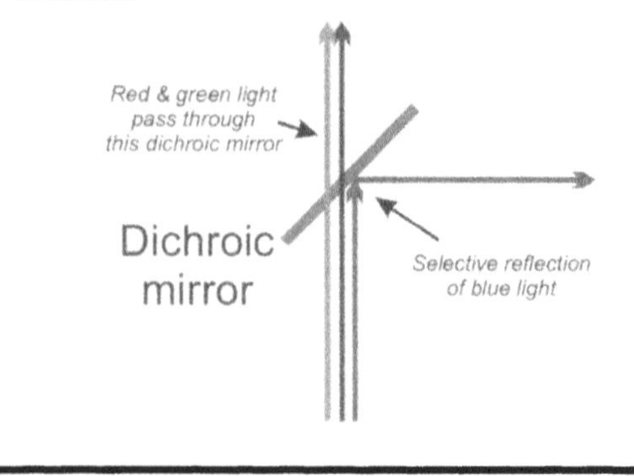

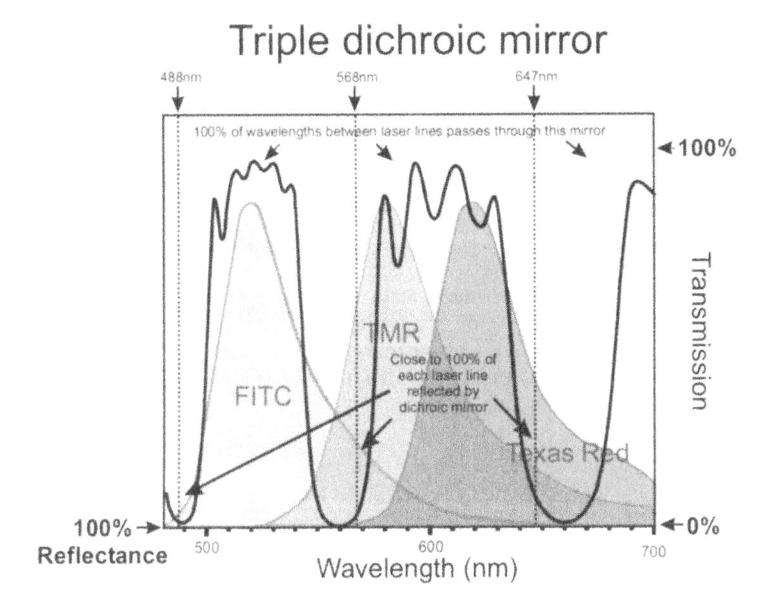

Figure 3-5. Triple Dichroic Mirror (T1). This dichroic mirror (position T1 in Figure 3-4) is used to separate the irradiating laser light (in this case 488 nm blue, 568 nm yellow and 647 nm red laser lines of the krypton-argon-ion laser) from fluorescent light emanating from the sample.

The emission spectra of the common fluorochromes, FITC (fluorescein), TMR (tetramethyl rhodamine) and Texas Red, and the laser emission lines 488nm (blue) 568nm (yellow) and 647nm (red) of the krypton-argon-ion laser are marked on the diagram as a reference guide to the optical characteristics of the mirror.

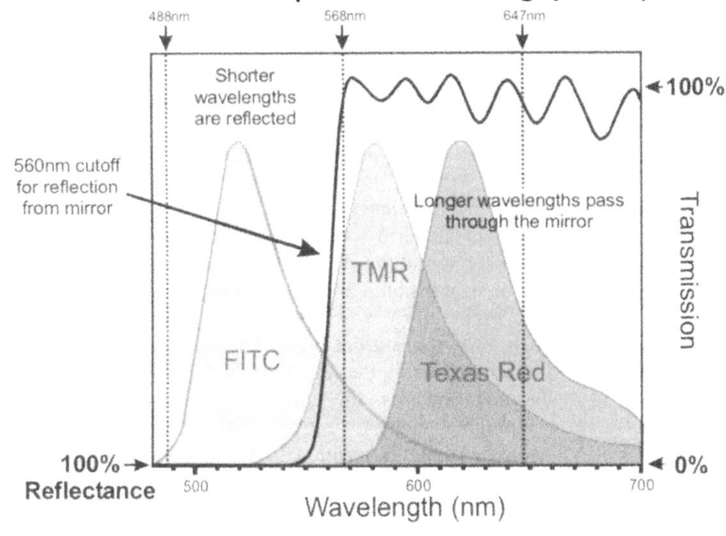

Figure 3-6. Long Pass Dichroic Mirror (T2A). This 560 DCLP (dichroic long pass) mirror is used in position T2A (see Figure 3-4) to separate the returning fluorescent light into shorter (green) and longer (red) wavelengths.

The emission spectra of the common fluorochromes, FITC (fluorescein), TMR (tetramethyl rhodamine) and Texas Red, and the laser emission lines 488nm (blue) 568nm (yellow) and 647nm (red) of the krypton-argon-ion laser are marked on the diagram as a reference guide to the optical characteristics of the mirror.

A second dichroic mirror located at position T2A in Figure 3-4 is designed to reflect short wavelength blue, green and yellow light and to transmit red light (Figure 3-6). The term 560DCLP (Figure 3-6) refers to the wavelength of light (560 nm) where 50% of the light is reflected and 50% allowed through the dichroic mirror (DC), and to the fact that all wavelengths longer (Long Pass, LP) than 560 nm (red light) pass through this mirror. This simple dichroic mirror is the primary means by which the returning fluorescent light is split into short (green) and long (red) wavelengths. Any blue or yellow reflected light that has penetrated the dichroic mirror at position T1 will be reflected into channel 2 (PMT2) by this second dichroic mirror. On the other hand, any red reflected light will penetrate this second dichroic mirror. This type of bleed-through can be removed by the use of suitable barrier filters (discussed below).

Barrier filters located in front of the confocal pinhole in the scan head in Figure 3-4 are used to further discriminate between different wavelengths of light after the initial splitting of the light between long and short wavelengths by the dichroic mirror T2A. Barrier filter 2 (Figure 3-4), in this case a HQ515/30 narrow band optical filter (the spectral characteristics of this filter are shown in Figure 3-7) is used to more closely define the wavelengths of light (a narrow band of 30 nm green light centred around 515 nm) that can enter the photodetector of channel 2 (PMT2). A narrow band optical filter is useful for minimising any bleed-through from the other channel, but there can be a considerable loss of light, depending on where in the emission spectrum of the fluorophore the relatively narrow "window" of light permeability of the filter is located. In contrast, barrier filter 1 shown in Figure 3-4 is a long pass filter (E600LP, edge filter long pass, Figure 3-8) that allows red light to enter detection channel 1 (PMT1). This optical filter is used to further define the wavelength cut off for entry to the channel 1 detector. The term "edge filter" refers to the sharp cut off characteristic of this particular optical filter. The dichroic mirror in position T2A has a cut off wavelength of 560 nm (Figure 3-6), whereas the long pass filter in filter position two has a cut off of 600 nm (Figure 3-8), thus further narrowing the region of the light spectrum that is directed towards the channel 1 detectors (PMT1). Although a dichroic mirror with a cut off of 600 nm could be placed in position T2A, it is often convenient to have access to a number of different dichroic mirror and optical filter combinations to define the region of the light spectrum that enters each of the detection channels to suit particular fluorophore combinations.

What is a barrier filter?

Barrier filters are simply pieces of coloured glass (or other suitable material) that absorb particular wavelengths of light, and allow the transmission of other wavelengths. For example, red glass absorbs all wavelengths of light except red - hence the filter appears red, as red is the only light that can get through the filter.

Barrier filters can be of broad specificity that allow either short wavelengths to pass through (Short Pass filters), or they may allow only long wavelengths through (Long Pass filters). Other optical filters may allow only a very narrow range of wavelengths through (Narrow Band Pass filters).

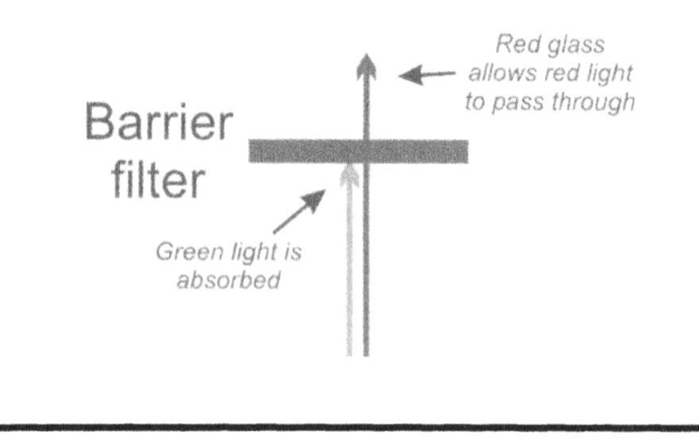

What is an Interference Filter?

Interference filters are constructed by depositing very thin dielectric layers onto a glass substrate. The layers are carefully coated to a depth of ½ or ¼ wavelengths, such that light of specific wavelengths will be transmitted or reflected from the filter by constructive or destructive interference. The dielectric layer is then coated with a silicon layer to protect the very thin layers from abrasion.

Interference filters are characterised by having very sharp cut-off wavelengths, and very narrow bands of transmission or reflectance. Interference filters can be readily made to transmit or reflect any area of the visible light spectrum, making them ideal for demanding applications like multi-channel imaging in a confocal microscope.

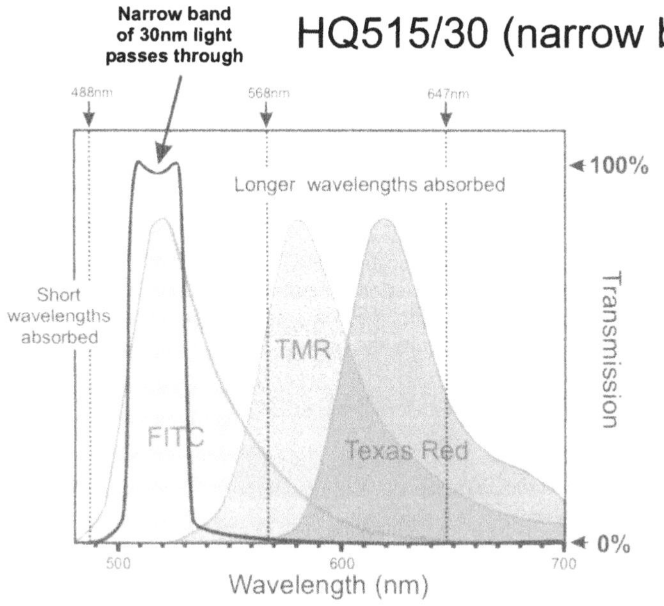

Narrow band of 30nm light passes through

HQ515/30 (narrow band) filter

488nm 568nm 647nm

Longer wavelengths absorbed

Short wavelengths absorbed

TMR

FITC

Texas Red

100%

0%

Transmission

500 600 700

Wavelength (nm)

Figure 3-7. Narrow Band Optical Filter.
HQ515/30 narrow band optical filter (barrier filter 2, as shown in Figure 3-4) allows a narrow band of light (in this case a 30 nm band of green light) to pass through. Light with a wavelength longer or shorter than this narrow band is absorbed or reflected by the filter. This type of optical filter is particularly useful in dual or triple labelling for minimising the detection of red fluorescence in the detection channel set up to detect green fluorescence.

The emission spectra of the common fluorochromes, FITC (fluorescein), TMR (tetramethyl rhodamine) and Texas Red, and the laser emission lines 488nm (blue) 568nm (yellow) and 647nm (red) of the krypton-argon-ion laser are marked on the diagram as a reference guide to the optical characteristics of the mirror.

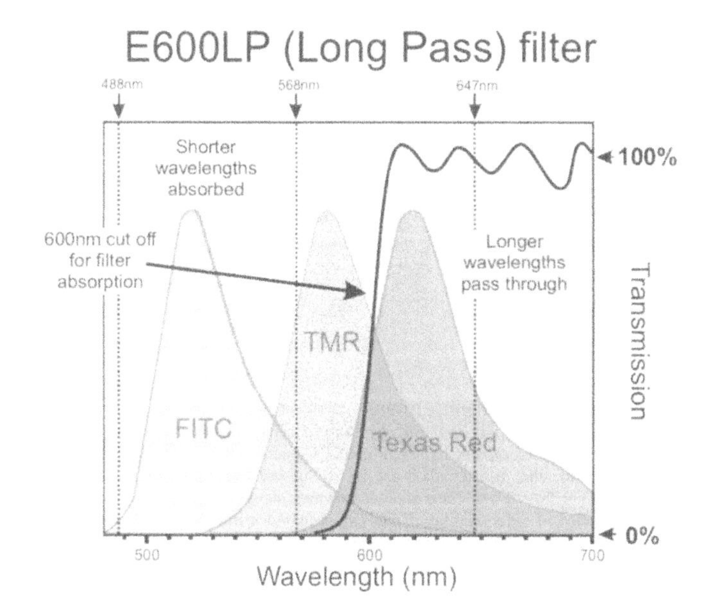

Figure 3-8. Long Pass Optical Filter.
E600LP, a long pass edge filter (barrier filter 1, as shown in Figure 3-4) that allows longer wavelength light (in this case longer than 600 nm) to pass through, but absorbs shorter wavelengths. This optical filter can be used to eliminate reflected blue laser light that may obscure low levels of fluorescent light emanating from the sample. This filter is sometimes referred to as an orange glass filter (OG).

The emission spectra of the common fluorochromes, FITC (fluorescein), TMR (tetramethyl rhodamine) and Texas Red, and the laser emission lines 488nm (blue) 568nm (yellow) and 647nm (red) of the krypton-argon-ion laser are marked on the diagram as a reference guide to the optical characteristics of the mirror.

Acoustic Optical Beam Splitter (AOBS)

The Acoustic Optical Beam Splitter (AOBS), is a highly efficient and versatile beam splitter (Figure 3-9) used by Leica as an optional device that replaces the primary dichroic mirror in the Leica SP2 AOBS confocal microscope (also see "Appendix 1: Confocal Microscopes–Leica Confocal Microscopes", page 402).

AOBS – replaces the primary dichroic mirror in the Leica SP2 AOBS confocal microscope for separating the irradiating laser light from fluorescent light emanating from the sample

The AOBS device allows not only a highly efficient separation of fluorescent light from the irradiating laser lines, but also very fast switching between alternative laser lines for multi-labelling applications.

The AOBS element is a tellurium oxide crystal that has a refractive index that can be varied by transducing radio frequency sound waves through the crystal. A single radio frequency acoustic signal can be used to separate a very narrow band of light (for example 488 nm blue light) from the broad emission spectrum emanating from the sample. Further narrow acoustic frequencies can be applied to separate other specific wavelengths for multi-labelling applications. This type of spectral separation is traditionally carried out by the primary dichroic mirror in a confocal microscope. The advantage of the AOBS system is both the speed of switching, and the narrow wavelength of light that can be separated. Essentially a single wavelength of light can be separated from the full light spectrum. This extremely narrow band selection means that less of the returning fluorescent light is rejected by the beam splitting AOBS device compared to a more traditional dichroic mirror system – thus increasing the sensitivity of the instrument.

The AOBS device, being capable of very high-speed line switching, can also be used to collect images using different excitation lines for each alternate scan line in the image. Laser line switching can also be achieved by electronic switching of Acoustic Optical Tuneable Filters (AOTF) used to attenuate the intensity of the laser light, but the AOBS element can both select out specific narrow bands of light and switch between multiple bands at very high speeds in the one optical device.

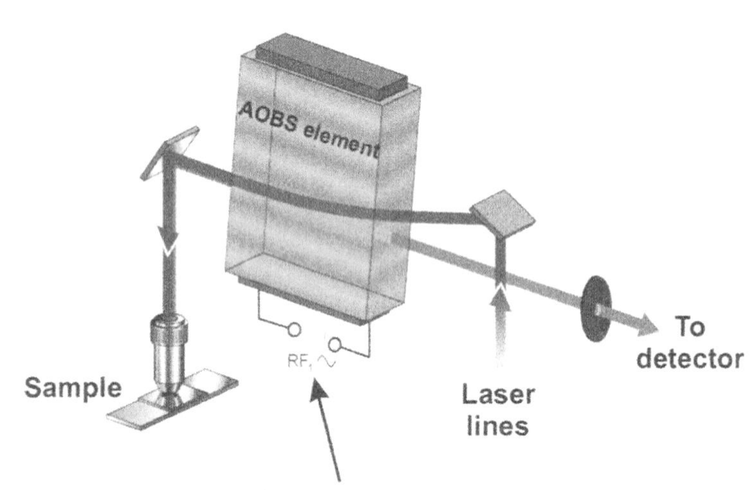

Figure 3-9. Acoustic Optical Beam Splitter (AOBS).
Using an AOBS is an extremely fast and efficient way of separating the laser line used for excitation from the returning fluorescent light in a confocal microscope. The AOBS element is used as a replacement for the primary dichroic mirror in the Leica SP2 AOBS confocal microscope. Applying specific radio frequency acoustic waves across the tellurium oxide crystal will result in the separation of individual laser lines (more than one line can be selected at once by applying multiple acoustic frequencies). Wavelengths not selected will pass directly through the crystal. This figure is adapted from a diagram kindly supplied by Werner Knebel, Leica Microsystems, Mannheim, Germany

Hardware Settings for Dual Labelling.

In Figure 3-10 a detailed diagram demonstrates the adjustments required in both the scan head and the controller box when the Bio-Rad Radiance confocal microscope is set up for dual labelling (in this case green light emitting FITC and red light emitting Texas red). This type of hardware setup is applicable to all confocal microscopes that rely on dichroic mirrors and optical filters to separate various regions of the light spectrum.

When dual labelling, the fluorescent light coming from the sample is split into long and short wavelength components by the dichroic mirror (560DCLP) located in the scan head. This dichroic mirror has a cut-off at 560 nm, wavelengths shorter than this (green light) are reflected into channel 1, and wavelengths longer than 560 nm (red light) are directed to channel 2. Light passes through a variable sized confocal iris for each of the channels before being transferred to the controller box via a large diameter fibreoptic cable. The fluorescent light that is transferred to the controller box passes though suitable barrier filters before being detected by the individual photomultiplier tubes (PMT tubes).

In Figure 3-10 the initial dichroic mirror has divided the wavelengths into green and red light, with the subsequent barrier filters used to further limit the wavelengths that will be detected by each of the two photomultiplier tubes. Channel one has a narrow band barrier filter installed (HQ515/30) that results in most wavelengths being absorbed, with a narrow band of 30 nm (centred on 515 nm) allowed to pass through. This narrow band of light corresponds to the main green emission peak of fluorescein (FITC).

The second channel in this example (Figure 3-10) makes use of a long pass barrier filter (E600LP) that will allow all wavelengths longer than 600 nm (long wavelength red light) to pass through. The filter will absorb all shorter wavelengths. The effect of this optical filter is to allow only the fluorescent light coming from Texas red to enter the photomultiplier tube used for channel 2, while blocking any fluorescent light coming from the vicinity of tetramethylrhodamine (TMR) from entering this channel. Changing the barrier filter to the E570LP would result in light from both tetramethylrhodamine and Texas red passing through to photomultiplier tube 2.

Careful choice of dichroic mirrors and barrier filters allows one to perform dual and even triple labelling with only two detectors (if only two detectors are available then triple labelling cannot be done simultaneously). However, a number of confocal microscopes have at least three fluorescence detection channels.

The spectral separation technologies used by Leica and Zeiss are discussed in more detail shortly. These instruments use other optical components either in addition to dichroic mirrors and optical filters or even as a complete replacement of these more traditional components.

Optical filter terminology

DF = Discriminating Filter

LP = Long Pass

SP = Short Pass

KP = Kurz Pass (short pass)

DRLP = Dichroic Long Pass

DCLP = Dichroic Long Pass

EFLP = Long Pass Edge Filter

NB = Narrow Band

HQ = High Quantum efficiency

OG = Orange Glass

RG = Red Glass

WB = Wide Band

ND = Neutral Density

An optical filter is described by a simple letter and number notation, for example 515LP, means that only longer wavelengths (Long Pass) are transmitted, and "515" denotes the wavelength at which 50% of the light can pass through the filter. A narrow band optical filter will be denoted with two numbers, for example 515/30, where "515" denotes the central wavelength in which a narrow band of 30 nm is transmitted by the filter.

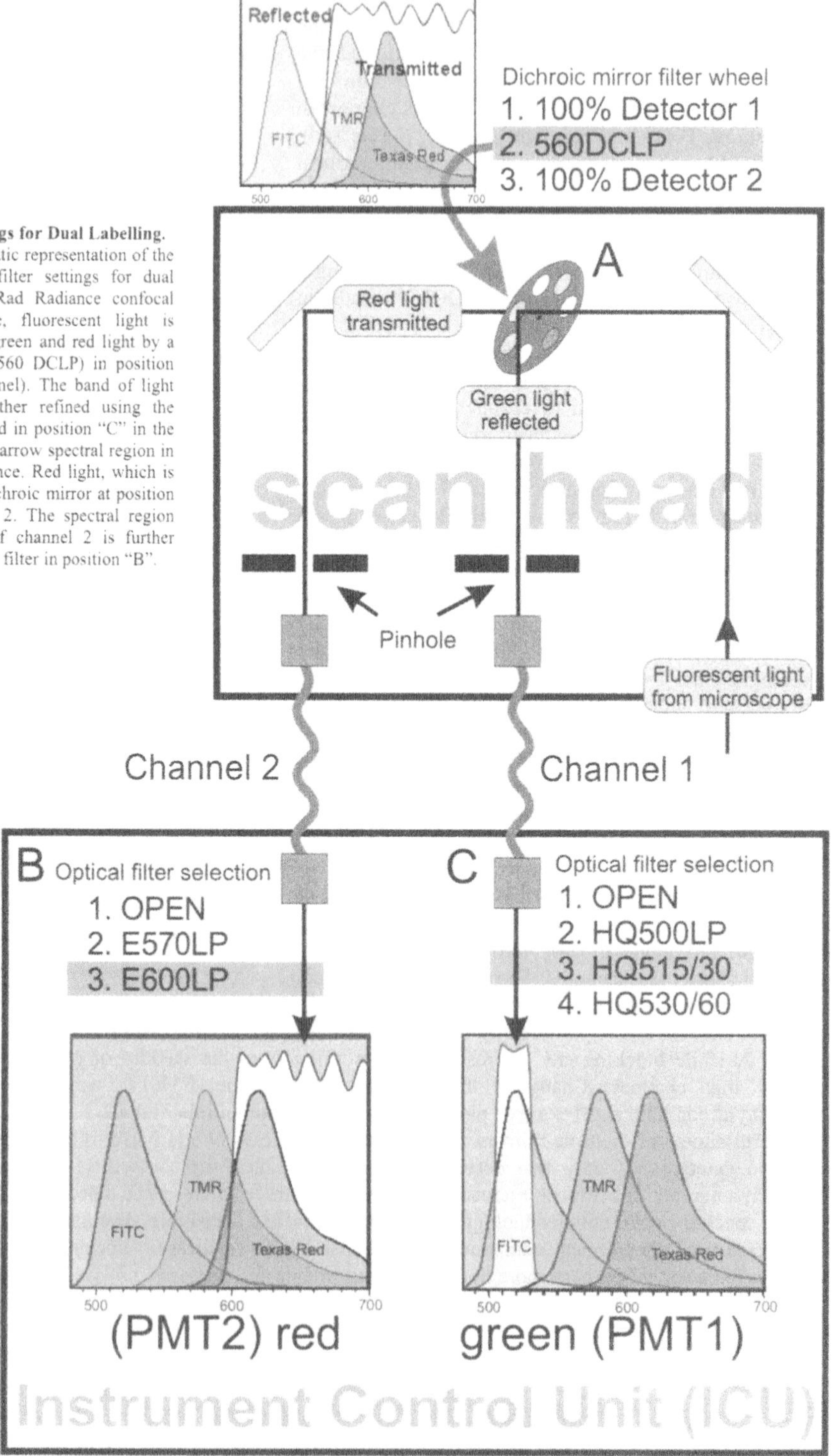

Figure 3-10. Hardware Settings for Dual Labelling. This figure shows a diagrammatic representation of the dichroic mirror and optical filter settings for dual labelling, based on the Bio-Rad Radiance confocal microscope. In this example, fluorescent light is separated into its component green and red light by a dichroic mirror (in this case 560 DCLP) in position "A" in the scan head (top panel). The band of light directed to channel 1 is further refined using the HQ515/30 barrier filter (located in position "C" in the Instrument Control Unit) to a narrow spectral region in the vicinity of FITC fluorescence. Red light, which is allowed to pass through the dichroic mirror at position "A", is directed into channel 2. The spectral region allowed to enter the PMT of channel 2 is further defined by the use of a E600LP filter in position "B".

Spectral Separation of Fluorescent Light

An alternative method of separating various wavelengths of fluorescent light in confocal microscopy is to use spectral separation. The Leica SP confocal microscope uses a prism (Figure 3-11) to separate the light into its component colours, which are then directed to individual detection channels. The Zeiss META confocal microscope uses a diffraction grating (Figure 3-12) to spread the light across a multi-PMT array, and the Olympus FV1000 confocal microscope uses a diffraction grating and a variable slit to direct the light to individual detection channels (Figure 3-13). Bio-Rad has implemented a spectral separation technique based on using optical interference filters with particularly sharp cut-off wavelengths (Figure 3-14). Although this method of spectral separation is based on using traditional optical filters, the Bio-Rad technique has been included in this section as it is more akin to the spectral separation technologies discussed here rather than that of the instruments based on using optical filter methods.

Spectral separation has a number of advantages over the more traditional method of using individual dichroic mirrors and optical filters. Of particular importance is the greatly increased versatitly of the instrument in both defining spectral regions that are directed to individual channels, and the ability to separate fluorophores with highly overlapping emission spectra using a process called linear unmixing. Although spectral separation does have significant advantages it also increases the complexity and consequently the purchase price of the instrument.

The prism used in the Leica SP confocal microscope separates the fluorescent light into its component colours, and then a computer controlled slit mechanism is used to collect a specific range of wavelengths in up to four individual detection channels (a diagrammatic representation of a two channel system is shown in Figure 3-11, and a more detailed description of the hardware involved is available on page 406 in "Appendix 1: Confocal Microscopes"). The front surface of each variable width slit is a mirror that reflects the light to the next variable slit mechanism. In this way there are no physical "gaps" in the light spectrum between each individual channel. Changing the position and the size of each slit is simply a matter of moving a "slider" on a graphical display on the computer screen (see page 411). Up to four spectral channels for fluorescence and backscatter imaging, as well as a separate transmission detection channel, are available on the Leica SP confocal microscope.

Spectral Separation in Confocal Microscopy

Leica SP
Optical prism to separate the light into its component colours, which are then directed to individual detectors.

Zeiss META
Diffraction grating to spectrally spread the light across a multi-PMT array detector.

Olympus FV1000
Individual diffraction grating and variable slit for each spectral detection channel.

Bio-Rad Rainbow
Dual optical filter wheels to separate defined regions of the light spectrum for each detection channel.

The Leica spectral separation technique can be used to direct defined regions of the light spectrum to individual detectors for multi-channel labelling. Alternatively, a series of images collected from relatively narrow spectral regions (lambda stack) can be obtained. Spectral information can be used to further characterise individual fluorophores or to separate fluorophores with highly overlapping emission spectra by linear unmixing.

A different method of spectral selection has been implemented on the Zeiss META confocal microscope (for a technical description of this instrument see page 388 in "Appendix 1: Confocal Microscopes"). In this instrument the fluorescent or backscattered light is separated into its constituent colours using a diffraction grating, and then the full spectrum spread over an array of light detectors (Figure 3-12). Each miniature photomultiplier tube, which forms part of the array detector (multi-PMT array), collects a defined region of the spectrum from each pixel in the image.

Individual detection channels can be created by combining adjacent PMT tubes in the multi-PMT array (shown as channel S1 (ChS1) and channel S2 (ChS2) in Figure 3-12, and channels S1 to S8 in the Zeiss computer software implementation shown on page 396). Alternatively, linear unmixing can be used to separate the complex spectrum collected from several closely overlapping fluorophores. This is achieved by using stored spectra of the individual fluorophores, spectra collected from individually labelled controls or spectra from specified (single labelled) regions of the sample. Fluorophores with highly overlapping spectra (for example, four "green" fluorescent molecules) can be separated in this manner (see page 400).

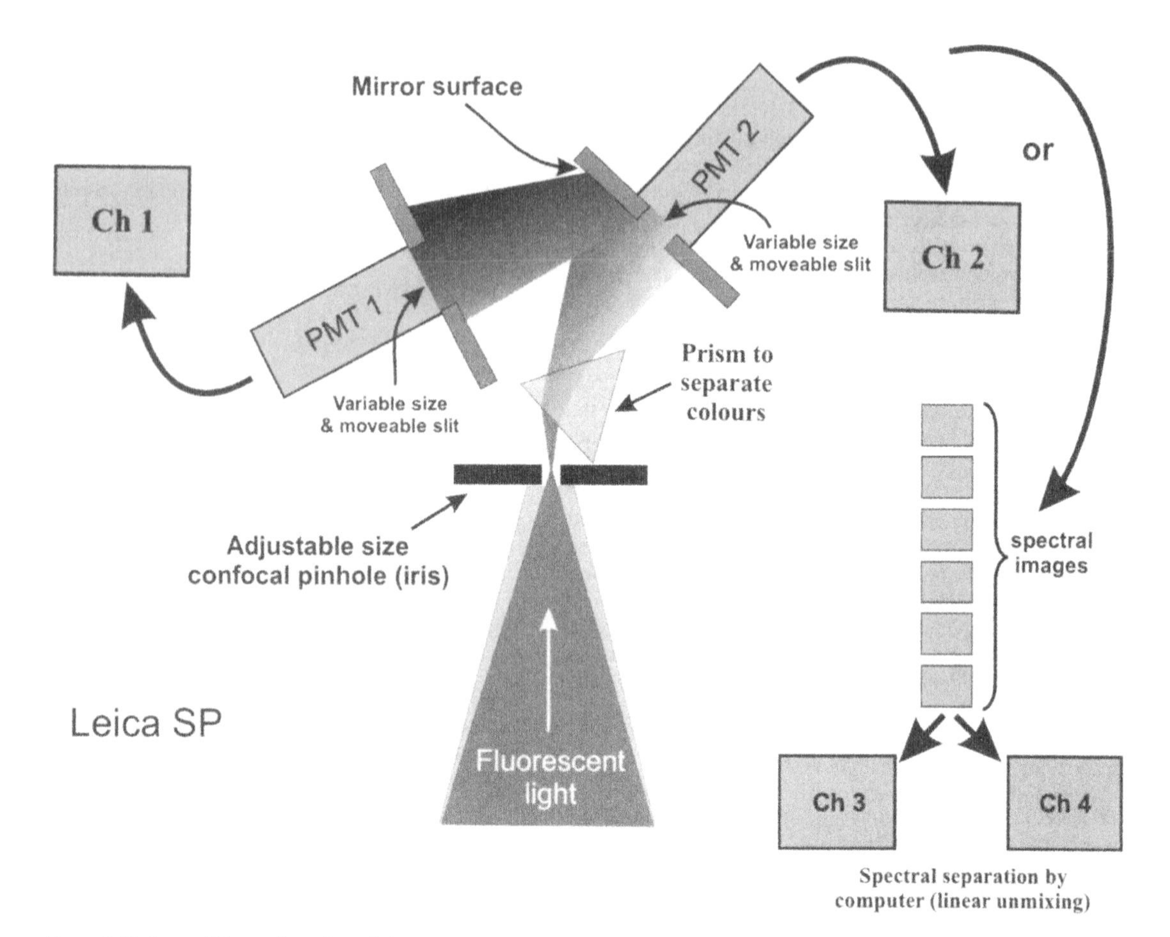

Figure 3-11. Spectral Separation using a Prism.
Spectral separation of fluorescent light using a prism is implemented on the Leica SP series of confocal microscopes. In these instruments fluorescent light is separated into its component colours using a prism, and individual channels (up to four) collect various "windows" of the light spectrum by the use of variable width slits. Each slit has a mirrored surface to direct light that does not enter the first detector to the next slit/detector. Light from up to four defined regions of the light spectrum is directed to individual photomultiplier tubes (PMT 1 and PMT2 in the two channel example shown above). The output from these detectors is then directed to channel 1 (Ch 1) or channel 2 (Ch 2). Alternatively, a series of spectral images can be collected using narrow slit widths, and spectral separation of individual fluorophores achieved by liner unmixing (in the above diagram shown as channel 3 (Ch 3) and channel 4 (Ch 4)). This figure is a simple diagrammatic representation of the spectral separation technique used in the Leica SP confocal microscope. For a more detailed description of the hardware involved please refer to the "Leica TCS SP2 Confocal Microscope" section on page 405 in "Appendix 1: Confocal Microscopes".

Successful separation of fluorophores by linear unmixing in the Zeiss META instrument does require that the fluorescent signal is of sufficient intensity across several miniature light detectors within the multi-PMT array for a non-ambiguous assignment of the component of the spectrum contributed by that particular fluorophore. As the fluorescent light is spread over several detectors (rather than a larger region of the light spectrum being directed to a single detector), there is a detection limit at which the META detector is no longer reliable. For this reason the Zeiss META confocal microscope is also supplied with up to three "traditional" detection channels that use dichroic mirrors and optical filters to separate various regions of the fluorescence spectrum.

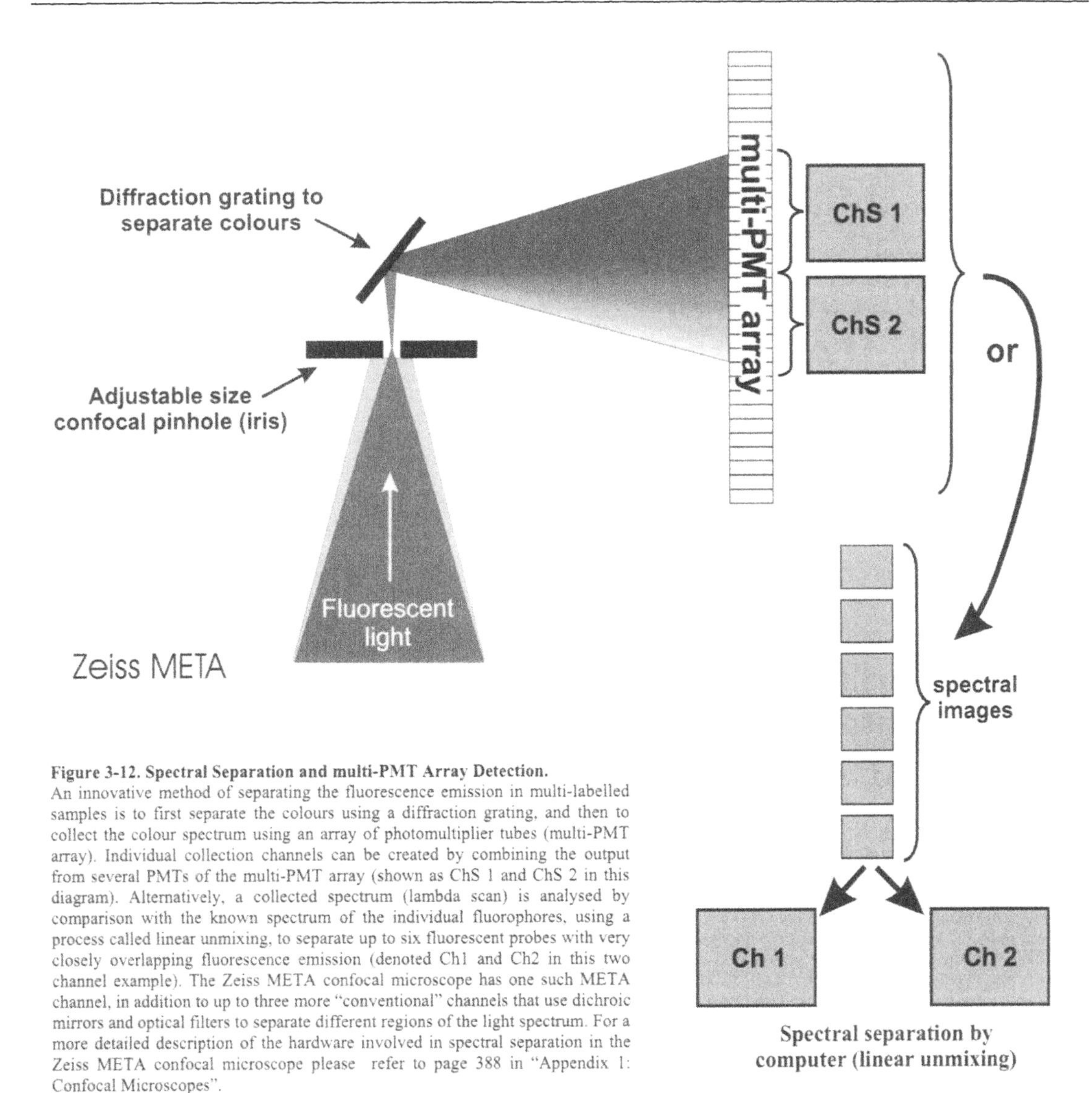

Figure 3-12. Spectral Separation and multi-PMT Array Detection.
An innovative method of separating the fluorescence emission in multi-labelled samples is to first separate the colours using a diffraction grating, and then to collect the colour spectrum using an array of photomultiplier tubes (multi-PMT array). Individual collection channels can be created by combining the output from several PMTs of the multi-PMT array (shown as ChS 1 and ChS 2 in this diagram). Alternatively, a collected spectrum (lambda scan) is analysed by comparison with the known spectrum of the individual fluorophores, using a process called linear unmixing, to separate up to six fluorescent probes with very closely overlapping fluorescence emission (denoted Ch1 and Ch2 in this two channel example). The Zeiss META confocal microscope has one such META channel, in addition to up to three more "conventional" channels that use dichroic mirrors and optical filters to separate different regions of the light spectrum. For a more detailed description of the hardware involved in spectral separation in the Zeiss META confocal microscope please refer to page 388 in "Appendix 1: Confocal Microscopes".

Spectral separation also available on the Fluoview 1000 confocal microscope from Olympus (Figure 3-13, see also page 422 in "Appendix 1: Confocal Microscopes"). This instrument has two spectral separation channels and up to two further conventional optical filter based detection channels. The spectral separation channels utilise dichroic mirrors for innitial separation of defined regions of the light spectrum. The light is then separated into its component colours using a diffraction grating, and the region of the light spectrum directed to each individual detector is adjusted using a moveable and variable size slit in front of the photomultiplier tube. Each of the two individual spectral separation channels available can be used as independent imaging channels (shown as Ch1 in Figure 3-13), or alternatively, a series of images can be collected using a narrow slit width (lambda stack) and linear unmixing used to separte fluorophores with highly overlapping fluorescence emission spectra (shown as Ch1 and Ch2 in Figure 3-13).

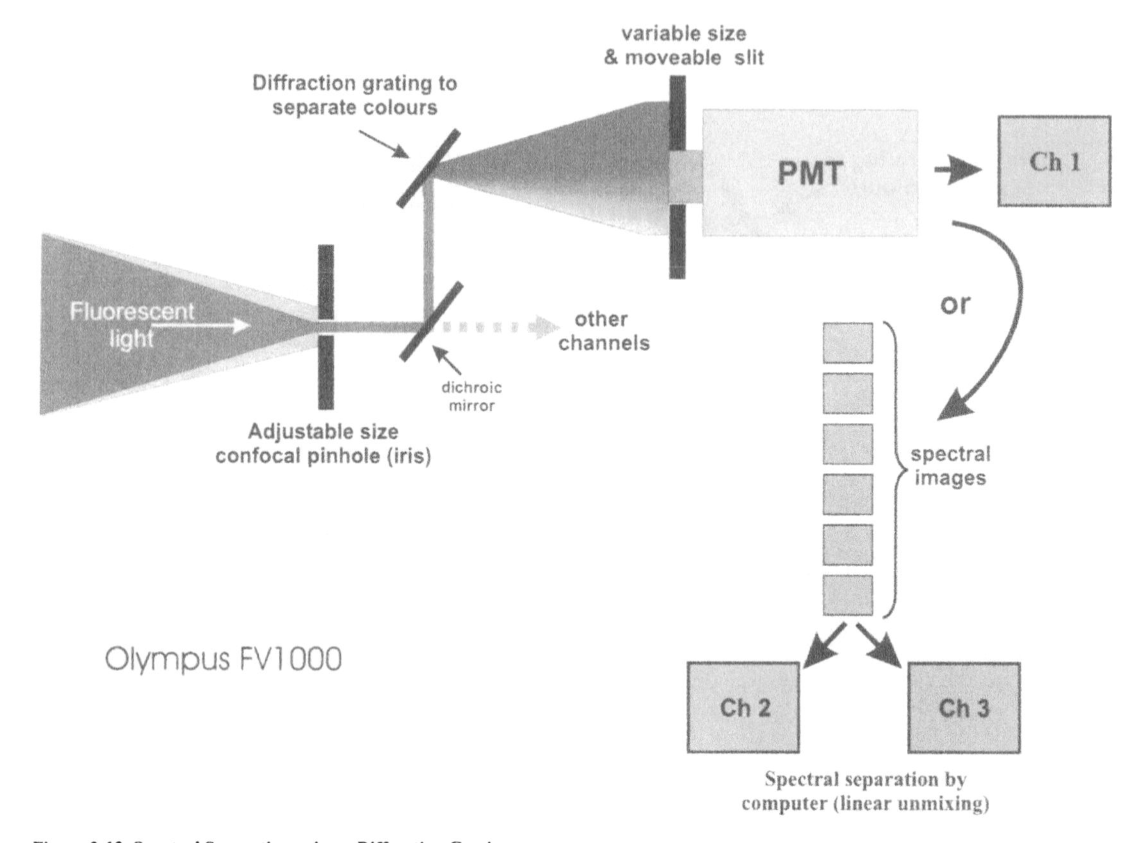

Figure 3-13. Spectral Separation using a Diffraction Grating.
The Olympus Fluoview 1000 confocal microscope has two spectral separation channels and up to two conventional optical filter based detection channels. The spectral separation channels utilise a dichroic mirror for initial separation of the returning fluorescent light. A diffraction grating is then used to separate the fluorescent light into its component colours. The region of the light spectrum directed to each spectral channel is determined by the size and position of the variable slit placed in front of the photomultiplier tube. The output from the photomultiplier tube is then directed to channel 1 (Ch 1, as shown above) or channel 2. Alternatively, a series of images can be collected using a narrow slit to collect defined spectral regions (down to a band pass size of 1 nm). The spectral emission collected for each image is directed to individual single PMT detection channels. Linear unmixing can then be used to spectrally separate fluorophores with overlapping emission spectra (Ch2 and Ch3 in this example). The Olympus Fluoview 1000 confocal microscope is described in more detail on page 423 of "Appendix 1: Confocal Microscopes".

Spectral separation has also been implemented on the Bio-Rad Radiance 2100 Rainbow confocal microscope using sharp cut-off interference filters. An optical filter wheel housing a series of long pass optical filters and a second filter wheel housing a set of short pass optical filters (Figure 3-14, and also see page 363 in "Appendix 1: Confocal Microscopes") are used to define the region of the light spectrum that will be directed to each individual channel (for example, Ch1, as shown in Figure 3-14). This technique can be used to "create" spectral "windows" for multi-channel labelling applications.

These spectral "windows" may be as wide as the full visible spectrum as defined by the optical filters available, or alternatively a very narrow band of light (down to 10 nm) can be collected. A spectral (lambda) scan can be collected by sequentially scanning a series of 10 nm steps.

Closely positioned, or even overlapping, spectral "windows" that show significant bleed-through between individual channels can still be used for multi-labelling applications by applying the process of linear unmixing. In this case spectral unmixing is applied to two or three simultaneously collected relatively broad spectral "windows" (in contrast to the Zeis META, where linear unmixing is applied to a lambda stack). Previously collected fluorescent images using the defined spectral "windows", or spectral information already available in the computer, is used to calculate the contribution to the fluorescence intensity of each channel from the individual fluorophores. The Bio-Rad spectral separation technique, in conjunction with linear unmixing, is a powerful way of separating fluorophores with overlapping emission spectra in live cell imaging applications due to the significantly greater sensitivity of the instrument when using spectral reassignment on individual channels collected from a relatively broad spectral region.

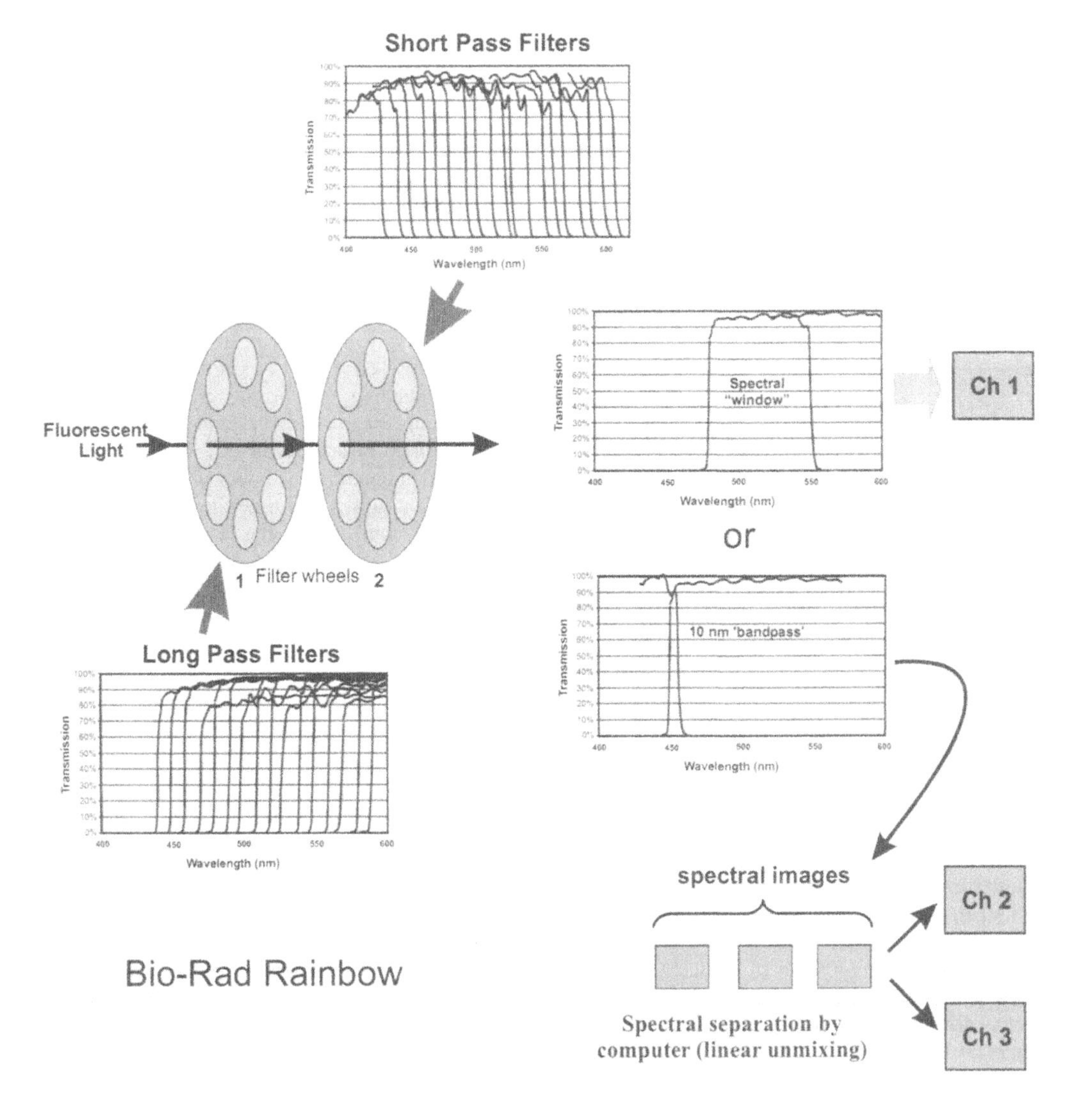

Figure 3-14. Optical Filter Wheel Spectral Separation.
The Bio-Rad Radiance 2100 Rainbow confocal microscope is capable of spectral separation of the emitted fluorescent light using a series of optical filters. A combination of long pass and short pass interference filters, with very sharp cut-off wavelengths, are used to collect the spectral emission in defined spectral regions (down to a band pass size of 10 nm). The spectral emission collected is directed to individual single PMT detection channels. The instrument can be used as a multi-channel confocal microscope (a single channel, Ch1, s shown in this diagram), or alternatively a series of narrow spectral bands can be used to create a spectral (lambda) stack of the sample. Linear unmixing (Ch2 and Ch3) has been implemented on the Rainbow confocal microscope using up to three channels (up to three spectrally separated images) to separate fluorophores with overlapping emission spectra (shown as a 10 nm band pass in the above diagram, but wider spectral regions can be used to greatly increase the sensitivity of the instrument). The Bio-Rad Rainbow confocal microscope is described in more detail on page 363 of "Appendix 1: Confocal Microscopes". This figure is derived from diagrams kindly provided by Chris Johnson, Bio-Rad, Australia.

Lasers used in Confocal Microscopy

A large range of lasers is available for use on the confocal microscope (Table 3-1). Traditionally the argon-ion and the krypton-argon-ion mixed gas lasers have been used for laser scanning confocal microscopy. However, in more recent years there has been a move away from using the multiple line krypton-argon-ion laser due to the high maintenance cost and low reliability of this type of laser. The trend is now to use more than one laser, and to combine the different laser beams using fibreoptics. This has the advantage that one can, for example, use the very stable argon-ion and helium-neon (HeNe) lasers in the one imaging experiment.

> **Argon-Ion Laser**
> **488 (blue), 514 (yellow)**
> Highly reliable laser, excellent blue (488 nm) line. The "workhorse" of single channel labelling.

Widely used Lasers

The argon-ion gassed laser is widely used in laser scanning confocal microscopy and flow cytometry. The Argon-ion laser is a very reliable laser that is significantly cheaper to run, although with somewhat more limited versatility, compared to the krypton-argon-ion laser. The argon-ion laser is suitable for dual labelling and simultaneous transmission imaging, but is not capable of triple labelling or real colour transmission imaging.

When dual labelling with the argon-ion laser lines there may be some emission signal bleed-through into the other detector due to the relatively close wavelengths of the two laser lines (488 and 514 nm). The argon-ion laser can be readily combined with other lasers by fibreoptic coupling.

The krypton-argon-ion laser provides great versatility due to the ready availability of three laser lines (488, 568 and 647 nm). These lines can be used for single, double or triple labelling, or for the creation of real colour transmission images. The krypton-argon-ion mixed gas laser requires refurbishment of the mirrors and re-gassing after about 1200 hours (1 to 2 years running, depending on how much time the laser is switched on). The red (647 nm) line of the krypton-argon-ion laser is particularly susceptible to serious diminishment or complete loss after the laser is used for some time. This indicates that the laser is ready for refurbishment, but the laser can continue to be used for many more hours if the 488 nm (blue) and 568 nm (yellow) lines are the only ones required. In other words, if the laser is only used for single and dual labelling there is no need to rush into getting the laser re-furbished when only the red line has diminished.

> **Kr/Ar Mixed Gas Laser**
> **488 (blue), 568 (yellow) & 647 (red)**
> Excellent for dual and triple labelling, but relatively expensive and has a short life span (particularly the red line).

The helium-neon laser is supplied as a single wavelength laser with either a green (543 nm), yellow (594 nm) or red (633 nm) emission line. When the HeNe laser is used in combination with the argon-ion laser one can perform triple labelling with very cost-effective and stable lasers. The argon-ion and HeNe lasers have a much longer useful life span compared to the Krypton-argon-ion laser, resulting in much lower maintenance costs. The only disadvantage with using HeNe lasers is that they may not have sufficient power for performing photobleaching FRAP or FRET experiments.

> **Helium-Neon Lasers (HeNe)**
> **543 (green), (594) yellow & 633 (red)**
> Highly reliable lasers that are often used as the 2nd or 3rd laser line in multiple labelling experiments.

The violet or blue-laser-diode is a relatively new laser that has become available for confocal microscopy. The wavelength of emission (405 nm violet light) is sufficiently within the absorption spectrum of a number of important "UV" dyes, such as DAPI and Hoechst DNA stains for this laser to be used as a replacement for the expensive and technically difficult UV confocal microscope. However, the emission wavelength is not optimal for the excitation of the important UV ratiometric dyes used in calcium imaging and intracellular pH estimations.

Specific UV lines of the argon-ion laser can be readily utilised for UV confocal microscopy. However, due to the high power level required to run this laser, water-cooling of the laser may be necessary. The difficulties of using a high powered UV laser and the problems associated with the optics of conventional light microscopes not being optimised (or even suitable) for UV light make the UV confocal microscope expensive and somewhat difficult to use.

> **Violet/Blue-Laser-Diode**
> **405 nm (blue/UV)**
> Possible replacement for the UV laser in confocal microscopy. Can be used to excite many "UV" dyes.

Table 3-1. Lasers used in Confocal Microscopy

A wide range of lasers is available for use in confocal and multi-photon microscopy. The following table lists some of the commonly available lasers and includes their most common emission wavelengths. For specialist applications other lasers are available, and further emission lines, not listed for the lasers in this table, may also be available.

LASER	Emission Lines	Suitability	NOTES
Gas Lasers: A range of gassed and mixed gassed lasers is available for confocal microscopy. Many of these lasers require forced air-cooling, or even water-cooling if higher power levels are used.			
Krypton-argon-ion (Kr/Ar)	488 nm blue 568 nm yellow 647 nm red	FITC TRITC Cy5	Excellent laser for double and triple labelling applications, but a relatively expensive and high maintenance cost laser. The red (647 nm) line is particularly susceptible to loss of power. Must be cooled by forced air circulation.
Argon-ion	457 nm blue 477 nm blue 488 nm blue 514 nm yellow	FITC	A highly reliable gassed laser that is particularly suitable for excitation of dyes such as the widely used fluorescein (FITC). Must be cooled by forced air circulation.
Krypton-ion	568 nm yellow 647 nm red	TRITC	Can be used as the second laser in dual labelling experiments as this laser is a more stable laser than the Krypton-argon-ion laser.
Helium-cadmium (HeCd)	325 nm UV 442 nm blue	CFP/YFP Lucifer yellow	A highly reliable laser with relatively short wavelength blue and UV lines. Used for CFP/YFP FRET pair imaging, lucifer yellow and many other dyes.
Helium-neon (HeNe) Gassed Lasers: The relatively low power of these gassed lasers is usually sufficient for confocal microscopy, although you may find that the laser may need to be used at a relatively high power setting. These lasers do have the advantage of being very stable, having a long life and do not require forced air-cooling.			
HeNe green	543 nm green	TRITC	Highly reliable laser that does not require a cooling fan. Can be used as a replacement for the 568 nm line of the krypton-argon-ion laser.
HeNe orange	594 nm orange	Lissamine Rh	Highly reliable laser that does not require a cooling fan. Can be used as an alternative to the second 568 nm line of the Krypton-argon-ion laser.
HeNe red	633 nm red	Far red dyes	Highly reliable laser that does not require a cooling fan. Often used as the 3rd line for triple labelling.
Solid State Lasers: These lasers have the advantage of being relatively low cost, easy to use and low maintenance. These lasers are also small in size and do not require forced air or water-cooling.			
Violet/Blue-laser-diode	405 nm violet	UV dyes	Solid-state laser that can be used to excite a number of UV dyes, including DAPI and Hoechst DNA stains.
Blue-laser-diode	488 nm blue	FITC/Alexa 488	Solid-state blue laser than can be used as a replacement for the argon-ion 488 nm blue laser line.
Green-laser-diode	532 nm	Red dyes	Solid-state green laser of higher power than the green HeNe laser.
Yellow-laser-diode	561 nm	Red dyes	Solid-state yellow laser as replacement for 568 nm Kr/Ar line.
Orange-laser-diode	594 nm orange	TRITC	Solid-state orange light emitting laser.
Red-laser-diode	637 nm	Far red dyes	Solid-state red laser that can be used as the 3rd line in triple labelling experiments.
Ultraviolet (UV) Lasers:			
Argon-ion	275- 528 nm UV	UV dyes	Water-cooled UV laser.

Laser Line Selection

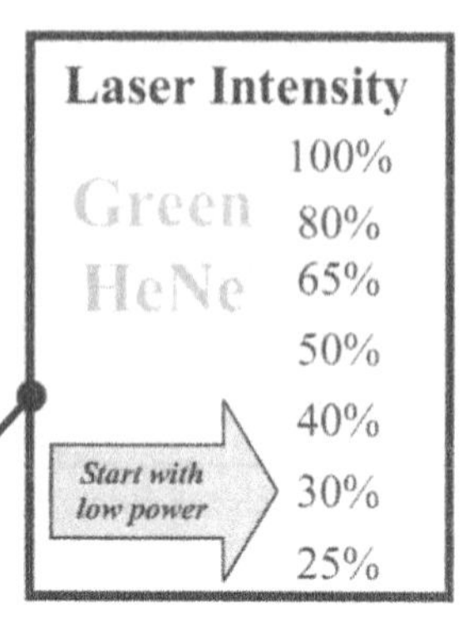

A single laser can have several laser lines, and in addition several lasers can be connected to the confocal microscope via fibreoptic couplings. Therefore you need an efficient means of selecting individual laser lines, or combinations of lines, for multiple labelling applications. Traditionally laser line selection was achieved by placing optical filters or mirrors in front of the laser. This tried and true technology is still used by many confocal microscope manufacturers, although these days the selection of the correct optical filter is usually under computer control. High-speed electronic switching between laser lines is now possible on a number of instruments.

The selection of individual laser lines is important, even for example, when imaging cells labelled with a single green emitting fluorophore such as fluorescein. Using the single 488 nm (blue) line on the argon-ion laser is better than running the risk of inadvertently exciting other molecules within the preparation with other laser lines that may be present in the laser emission. Selective use of individual laser lines also improves the specificity of dual labelling experiments. Using the 488 nm (blue) and 568 nm (yellow) laser lines from the krypton-argon-ion laser simultaneously for dual labelling may sometimes result in an unacceptably high bleed-through between detection channels. Increasing the specificity of the fluorescence may be achieved by irradiation of the sample with each of the lines individually and then subsequently merging the images to create a dual labelled image.

ND filters have discrete values for attenuation of laser light

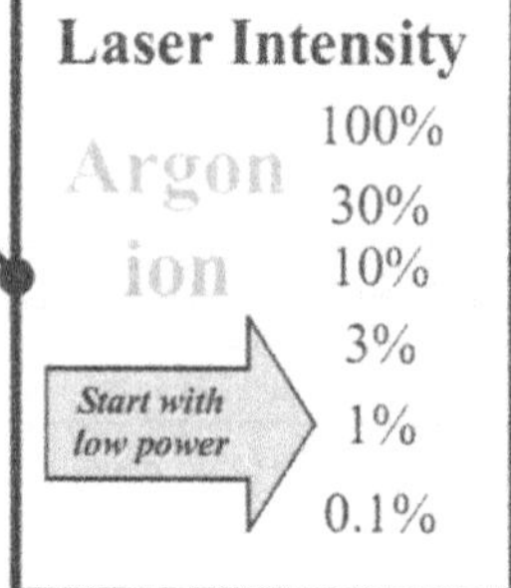

Acoustic Optical Tuneable Filters (AOTF, see Figure 3-15) are a very fast and efficient means of selecting the correct laser line in a confocal microscope. AOTF filters are now used as versatile, although rather expensive, replacements for traditional optical filters by a number of confocal microscope manufacturers. The acoustic frequency transmitted through the tellurium oxide crystal in the AOTF filter determines the exact wavelengths of light that are deflected and thus allowed to enter the optical system of the microscope. The real power of an AOTF filter lies in the fact that laser line selection can occur extremely fast by electronic switching. Under suitable software control the instrument can select individual laser lines fast enough to not only make individual scans with different laser lines, but to even collect individual scan lines (alternating line scanning) with different laser lines irradiating the sample.

Attenuation of Laser Light

Full control of the intensity of the laser light that will *reach the sample* is most important in confocal microscopy. One needs to be able to adjust the relative intensities of the individual lasers, particularly in multiple labelling experiments. Individual laser lines and different lasers have significantly different power outputs. Furthermore, laser intensity at the sample needs to be kept to a minimum to avoid unnecessary photo bleaching of the sample, particularly when imaging live cells. This may necessitate continual adjustment of the laser power as one switches between looking for suitable cells (with very low laser power) and collecting high quality images (with somewhat higher laser power).

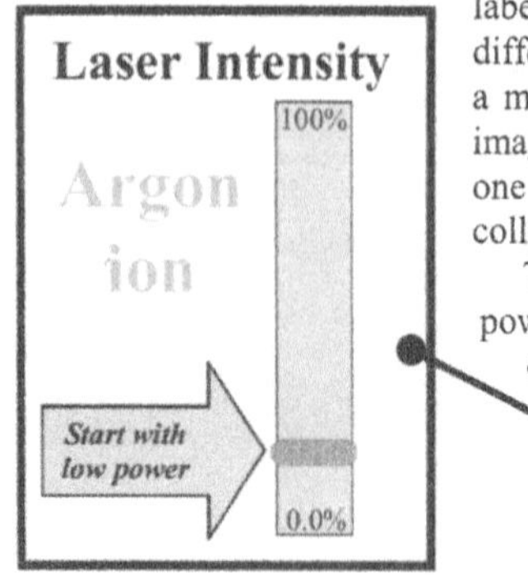

The laser light intensity level at the sample is controlled by adjusting both the power setting on the laser and by using suitable optical filters to attenuate the light output from the laser. The power setting for a number of lasers can be readily adjusted, either by a continuous power control setting as used by Leica (a physical knob on the desk top) and Zeiss (under software control), or a two-step software selectable adjustment provided by Bio-Rad. The power setting on the laser is often

AOTF filters allow one to continuously vary the intensity of the laser light that reaches the sample

adjusted to a relatively low level to extend the life of the laser. However, the laser power setting must be sufficient to create a stable laser intensity level. The continuously variable laser power setting is not easily reproduced by returning the knob or on-screen slider to the previous position, therefore, when a reproducible laser intensity is required it is better to keep the power setting constant and to vary the intensity of the laser beam by adjusting the optical filters in the light path.

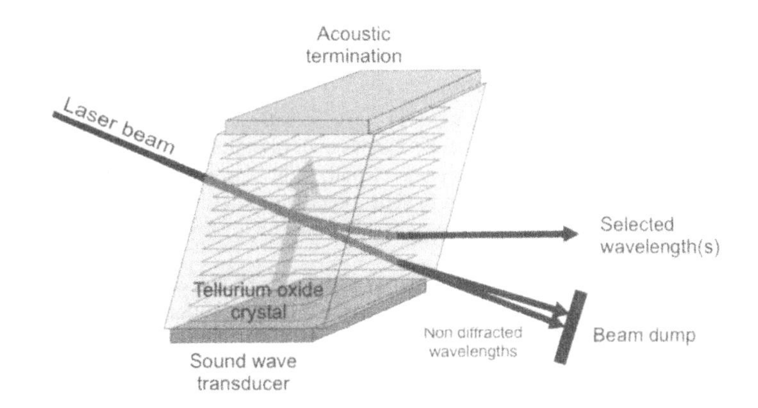

Figure 3-15. Acoustic Optical Tuneable Filter (AOTF).
Laser line and beam intensity can be selected very quickly using an Acoustic Optical Tuneable Filter (AOTF). In this device high-frequency sound waves, when transmitted through a crystal of tellurium oxide, result in changes in refractive index that are highly specific for the frequency of sound applied. Multiple laser lines can be deflected by applying multiple acoustic frequencies. The intensity of the emitted light depends on the power of the acoustic wave. Adjustment of the frequency and power of the acoustic wave allows one to separate out multiple wavelengths of the desired intensity. This figure is derived from a diagram kindly provided by Guy Cox, Australian Key Centre for Microscopy and Microanalysis, University of Sydney, Australia and Chris Johnson, Bio-Rad, Australia.

A series of Neutral Density filters are often used to attenuate the amount of laser light permitted to pass into the scan head of a confocal microscope. Individual optical filters are brought into place by selecting the appropriate percentage of laser light required on the software control panel (discrete percentage settings, rather than a continuous slider, usually indicate the use of Neutral Density filters). Use the lowest laser intensity possible (i.e. the highest Neutral Density filter available) so as to minimize the problems of fluorescence saturation, bleaching and photo toxicity.

An alternative method of attenuating the laser power that reaches the sample is to use an AOTF (Acoustic Optical Tuneable Filter, see Figure 3-15). As described above, an AOTF filter can be used to select individual laser lines, but in addition an AOTF filter can be used to control the total amount of laser light permitted to pass through by varying the power of the acoustic wave applied to the tellurium oxide crystal. This method of attenuating the amount of laser light that reaches the sample can normally be recognized by the presence of a "continuous", i.e. 0 to 100% adjustment slider. One advantage of using an AOTF filter for adjusting the amount of laser light reaching the sample is that a selected laser intensity can be used again for a later experiment (as long as the laser is left switched on and the laser power setting is not adjusted). The AOTF filter has other important advantages, such as flyback blanking (resulting in reduced photobleaching) and region of interest (ROI) scanning (see page 127 for further details).

Always try imaging using a very low intensity of laser light for positioning and focusing, and then increase the laser intensity for the final image collection. Great care is required to minimize the laser intensity and exposure time when imaging live cells, otherwise you will not only end up fading your fluorophore, but you may also create toxic free radicals that may alter the biochemical characteristics, if not the morphology, of your cells.

Some lasers have a relatively low power output compared to others. For example, the helium-neon laser has a significantly lower maximum power output compared to the argon-ion or krypton-argon-ion lasers. This means that attenuation filters are usually set much higher when using HeNe lasers. In fact, the HeNe laser is routinely set to 50-100% of full laser power for live cell imaging. On the other hand, a setting of 25% on the argon-ion or krypton-argon-ion laser would be quite destructive to both the fluorophore and the cells (this laser is often used at 1% or lower).

Light Detectors

The ideal light detector would be capable of collecting all photons that land on the detector, have a very low level of "noise", be equally sensitive across all wavelengths and have a linear response to a wide range of light intensity. Naturally, no available light detectors come close to this ideal situation. Photomultiplier tubes (PMTs) are used as highly sensitive light detectors in most laser scanning confocal microscopes, but CCD cameras are used in the Nipkow scanning disk confocal microscopes. The transmitted light in a laser scanning confocal microscope is routinely detected using a photodiode, but a photomultiplier tube can also be used for detecting the transmitted light.

Photomultiplier (PMT) Tubes

Photomultiplier tubes are used as very sensitive light detectors in the confocal microscope. Each imaging channel in the confocal microscope has its own photomultiplier tube. For example, when dual labelling a dichroic mirror is used to split the longer wavelength light (for example, red) from shorter wavelength light (for example, green), each of which is directed to separate photomultiplier tubes. The photomultiplier tubes are located within the scan head of most confocal microscopes, but may be located in a fibreoptic coupled optics or controller box in some confocal microscopes.

The effect of using a separate photomultiplier tube for each channel on a confocal microscope is first, that the number of separate channels that can be collected simultaneously is determined by the number of PMT tubes available, and second, that the total light hitting each photomultiplier tube is what determines the signal output of that PMT tube – the wavelength of light hitting the photomultiplier tube is irrelevant as far as the confocal microscope is concerned. This means that the signal display from each channel is quite correctly displayed as shades of grey, with the lovely colour of confocal microscope images being added later by the computer. Even the traditional green / red dual labelled images are false colour computer generated representations of the signal received by the photomultiplier tubes. The red and green coloration may or may not represent the true "colour" or wavelengths of light being collected from the sample.

The photomultiplier tubes in your confocal microscope may be individually matched for the region of the light spectrum that is usually directed to them. For example, the channel 1 photomultiplier tube may be more sensitive in the blue/green region of the light spectrum compared to PMT 2 or PMT 3. Bio-Rad uses prism enhanced photomultiplier tubes that significantly increase the amount of light that can be collected in each channel.

The Zeiss META confocal microscope (see Figure 3-12 and "Appendix 1: Confocal Microscopes-", page 382) has the unusual feature of having an array of at least 32 photomultiplier tubes. These photomultiplier tubes are designed to collect all of the light from a defined spectral region of the sample by using a diffraction grating to split the light into a spectrum of wavelengths that are then collected in 5 nm intervals by the photomultiplier array (see page 83). The identity of individual fluorophores can be derived from this spectrum by comparing the values obtained with those of a stored spectrum of the compound.

Photodiode Light Detectors

A light sensitive photodiode is sometimes used as the means of collecting a transmission image. Confocal microscopes have either a photodiode detector or a fibreoptic pickup port mounted below the sample (in an upright microscope) or above the stage (in an inverted microscope) that can either detect the laser light that passes through the sample directly, or transmit the light via fibreoptic cable to a photomultiplier tube. Most of the laser used to irradiate the sample does, in fact, pass right through the sample and so there is normally sufficient light to generate an excellent transmission image – as long as the sample itself is sufficiently thin and transparent to allow the transmitted light to pass through.

A number of confocal microscopes may require a mirror on the microscope itself to be moved, or at least a setting change to be made on the computer to allow a mirror to be placed in the light path for directing the transmitted light to the photodiode detector. Failure to obtain a signal when attempting to image in the transmission channel is usually due to forgetting to either set this mirror correctly, or for leaving another optical item or shutter incorrectly set on the microscope. The fully automated microscopes now becoming available make all necessary adjustments once you select "transmission" imaging on the confocal microscope control panel.

Instrument Control Panel

The Leica and Bio-Rad confocal microscopes (and previously Optiscan Imaging) are provided with a control panel with programmable physical knobs to operate many microscope functions (this feature is an optional "extra" on the Bio-Rad instruments). Altering the gain, focal position, black level etc., using a simple physical knob is a lot easier than finding the correct menu item or virtual "knob" on a computer screen using a mouse. These physical knobs are highly versatile due to the ease with which they can be programmed to perform a number of commonly used operations. Further details on these programmable knobs are provided in the Leica and Bio-Rad sections of "Appendix 1: Confocal Microscopes" (page 355).

NIPKOW DISK CONFOCAL MICROSCOPES

The Nipkow spinning disk is an innovative device that has been available for many years, being used in reflectance imaging confocal microscopes for at least as long as single "spot" laser scanning confocal microscopes have been available. However, excellent confocal microscopes utilising a Nipkow disk for imaging fluorescent samples in biology have been available only in more recent years. Confocal microscopes based on the Nipkow disk utilise an array of pinholes in the spinning Nipkow disk to reject out-of-focus light emanating from the sample, resulting in the formation of an in-focus optical slice. The image can be observed "live" by looking through an eyepiece lens attached to the scan head, or the image can be "captured" using a suitable CCD camera.

Nipkow spinning disk confocal microscopes are particularly suited to live cell imaging due to the relatively low light intensity at each laser spot, and the high speed image acquisition that these instruments are capable of.

Confocal microscopes that utilise a Nipkow disk are produced by a number of manufacturers. These instruments come in two basic designs, one has a single Nipkow disk containing a large number of pinholes (for example the instrument produced by Atto Bioscience), and the other, with an additional micro lens array disk, is produced by a number of manufacturers (PerkinElmer, VisiTech International) using a scan head designed and manufactured by Yokogawa (all of these instruments are described in detail in "Appendix 1: Confocal Microscopes", pages 355 - 441).

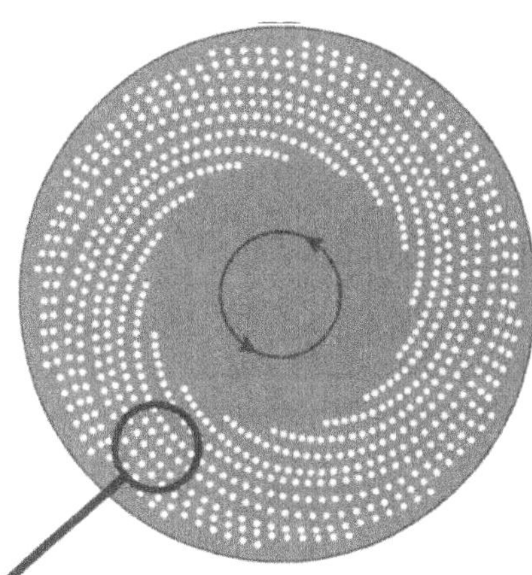

Figure 3-16. Spinning Nipkow Disk.
The Nipkow disk consists of an array of "pinholes" arranged in a spiral pattern. The disk in spun at relatively high speed, resulting in the simultaneous scanning of a large number of spots across the sample.

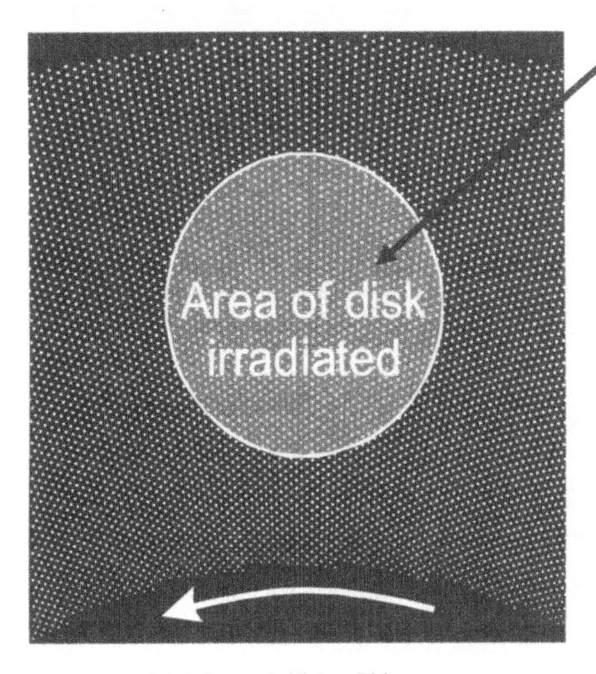

Figure 3-17. Pinhole Pattern in Nipkow Disk.
A large number of "pinholes" in the Nipkow disk are irradiated with a broad beam of light from either a laser or a mercury arc lamp. The sample can be directly viewed while the disk is spinning or a CCD camera used to collect the images. This diagram shows the general pinhole configuration for a Nipkow disk, rather than the specific layout for a particular Nipkow disk design.

Nipkow Disk

The design of the Nipkow spinning disk confocal microscope is based on using a disk containing an array of accurately placed "pinholes" (Figure 3-16). The disk is spun at high-speed while a relatively broadly spread light source is directed through the disk on to the sample (Figure 3-17). This results in the simultaneous scanning of a large number of focussed light spots across the sample. A high quality optical slice can be very quickly derived from the fluorescent light returning back up through the pinholes of the spinning disk. Out-of-focus light that is not focussed back up through these pinholes is blocked by the disk. The size of the pinholes in the Nipkow disk is designed to approximate the specified 1 Airy disk size for a specific objective (for example, a 100x 1.4 NA objective), and will be less than ideal for optimal z-sectioning when using other objectives.

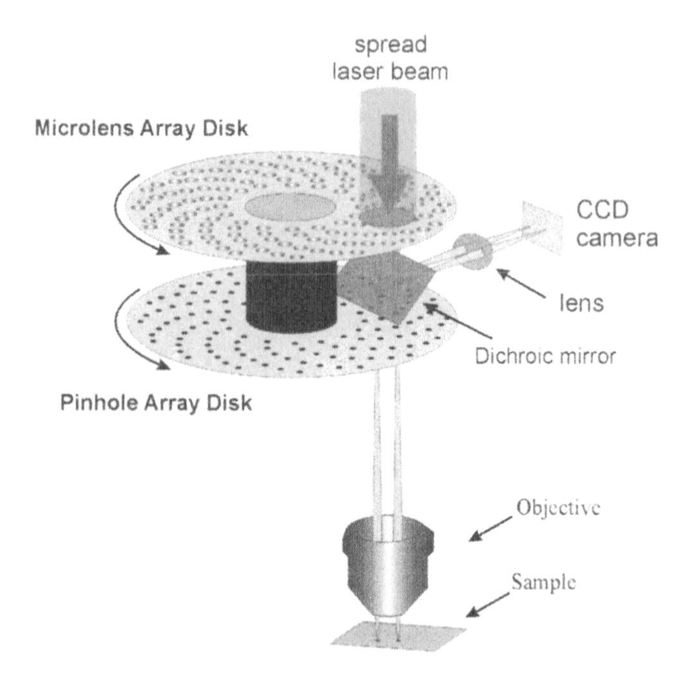

Figure 3-18. Nipkow Disk Micro-lens Array.
The Nipkow disk micro-lens array spinning tandem disk unit (CSU10) produced by Yokogawa is the basis of the high-speed spinning disk confocal microscope produced by PerkinElmer, VisiTech and others. In this microscope a spread laser beam is directed onto the upper surface of the micro lens array disk. The light that passes through each micro lens is focussed onto a "pinhole" in the pinhole array Nipkow disk. This results in the illumination of the sample with many thousands of fine dots – allowing for the creation of a confocal image that can be viewed by eye or a CCD camera. This diagram is modified from a diagram kindly provided by George L. Kumar, PerkinElmer, USA.

The Nipkow disk is particularly suited to live cell imaging, whether attempting to image high-speed calcium fluxes, or for following the development of living embryos over many hours. This is due to both the high-speed imaging capabilities of the instrument, and the fact that the large number of light points that are scanned simultaneously across the sample result in a relatively low level of light intensity at each spot. This results in significantly lower levels of photobleaching and phototoxicity. The Nipkow disk is spinning at high-speed, resulting in 340 to over 1000 frames per second. The actual image capture speed of the instrument is, however, highly dependent on the speed and sensitivity of the attached CCD camera and the amount of fluorescence emanating from the sample. Image capture rates of 10 to 20 frames per second (at 512 x 512 pixels) are routinely available, with speeds of 100 frames per second or higher possible when using high-sensitivity cooled CCD cameras.

Nipkow disk confocal microscopes can also be used very successfully for high-resolution imaging of fixed samples. However, there are a number of limitations when compared to a laser spot scanning instrument. The first noticeable limitation is that there is no provision for zooming, which means one cannot collect images with correct Nyquist sampling simply by increasing the zoom. When using a Nipkow disk instrument the magnification of the objective becomes important, in addition to the NA of the objective, in determining the resolution limit obtained. For example, if you use a 63x NA 1.4 objective on a laser spot scanning system you can readily satisfy Nyquist sampling by increasing the zoom on the instrument, whereas on a Nipkow disk based instrument it may be necessary to use a 100x 1.4 NA objective for high resolution imaging. The 100x objective will optically enlarge the image to more closely satisfy Nyquist sampling.

One further limitation of a Nipkow disk instrument is that the z-resolution of the instrument is not as good as that of a laser spot scanning instrument. One of the major factors in limiting the z-resolution is the spacing of the pinholes in the Nipkow disk. Closely spaced pinholes will allow significantly more light through the disk, but with the disadvantage that out-of-focus light may inadvertently cross the disk through several pinholes. The effect of allowing more out-of-focus light through the disk is to limit the z-sectioning ability of the instrument. The second micro-lens array disk in the Yokogawa scan head, discussed below, is designed to direct more light through the pinholes without the need to increase either the size or number of pinholes in the Nipkow disk.

Micro-Lens Array

A variation on the basic Nipkow spinning disk confocal microscope is the incorporation of a second spinning disk that contains an array of micro lenses (Figure 3-18), and is the basis of the compact confocal microscope scan head produced by Yokogawa (see page 435). This type of scan head has the advantage of focusing the light from each micro lens onto the corresponding pinhole of the second disk, increasing the light intensity of each individual scanning spot. The relatively small and robust scan head manufactured by Yokogawa contains a Nipkow disk, micro lens array disk and primary dichroic mirrors for splitting the irradiating light from the fluorescent light. This scan head is directly attached to a conventional light microscope (usually an inverted light microscope for live cell imaging). A CCD camera and a laser light source are attached to the scan head. A dual camera attachment device is available from some manufacturers. A three colour CCD camera can also be attached for full-colour imaging.

A number of manufacturers assemble a Nipkow spinning disk confocal microscope using the Yokogawa scan head. This includes PerkinElmer, VisiTech International and a number of distributors of VisiTech instruments (for example Solamere and McBain Instruments) that make significant contributions to various components used in these microscopes (see "Appendix 1: Confocal Microscopes", page 355).

Light Sources for Nipkow Disk Confocal Microscopy

A relatively broad beam of light is used to illuminate the Nipkow disk, which means that mercury or xenon arc lamps, as well as lasers, can be used for illumination. However, the Yokogawa Nipkow disk micro lens array system (CSU10 an CSU21) is normally operated using a laser light source, whereas the Atto Bioscience CARV instrument utilises the light from an arc lamp. The relatively small proportion of the total surface area of the Nipkow disk that allows light through the disk does mean that each laser spot is of relatively low intensity, even when using a high-intensity laser beam. The low light intensity at each scanning spot is increased significantly by addition of the micro-lens array disk. There are, however, as discussed above, important advantages for live cell imaging in having a relatively low light level at each focussed spot.

CCD Camera Detectors

The image in a spinning Nipkow disk confocal microscope can be viewed directly (by looking through a viewing port on the scan head) or the image can be captured using a sensitive CCD camera. There has been a dramatic improvement in the efficiency and sensitivity of CCD chips for light detection over the past few years. Although relatively cheap and reliable CCD cameras that can be attached to a light microscope are now available, imaging live cells using fluorescence microscopy does require a very sensitive (and consequently expensive) CCD camera. An attachment is manufactured by VisiTech International that allows two CCD cameras to be attached to the Yokogawa scan head for simultaneous capture of two fluorescence channels. Each CCD camera collects grey-scale images of a region of the light spectrum that is determined by the dichroic mirrors and optical filters present in the scan head. A three colour CCD camera can also be attached to the Nipkow disk scan head to collect full-colour images. A three colour camera does have the advantage of collecting spectral information across the full region of the light spectrum directed to the CCD camera, but the design of a three-colour CCD camera does mean that speed or resolution will be compromised. Collection speed in a confocal microscope using a spinning Nipkow disk is usually limited by the sensitivity and read-out speed of the attached CCD camera (a grey-scale camera, rather than a three colour camera is used for high speed imaging), rather than the frame re-fresh rate of the spinning Nipkow disk.

Nipkow disk confocal microscopes are particularly suited to live cell imaging

High-speed imaging - *for example, following vesicle movement or ion fluxes in living cells.*

Low light intensity – *excellent for limiting photo damage when time lapse imaging of cell division or embryo development.*

MULTI-PHOTON MICROSCOPES

The multi-photon microscope has many features in common with the confocal microscope, and is, in fact, often an additional imaging method added to an existing confocal microscope. The fundamental difference between a confocal and multi-photon microscope is the laser used to generate fluorescence within the sample, although there may be additional components, such as external detectors (discussed below) that are unique to multi-photon microscopy. Multi-photon excitation of a fluorophore (see Chapter 1 "What is Confocal Microscopy?", and Chapter 8 "What is Fluorescence?", page 187) is when two relatively long wavelength photons (usually red or infrared photons) arrive almost simultaneously at a fluorophore and result in the excitation of an electron to a higher energy level that reflects the sum of the energy of the two arriving photons. This means that fluorophores, such as fluorescein, which is conventionally excited by a single 488 nm blue photon can be also excited by the simultaneous arrival of two 970 to 780 nm far red photons. The fluorescence emission (green light of around 520 nm) will be the same in both conventional and multi-photon excitation.

The advantages of multi-photon microscopy are that the longer wavelength red and infrared light used to irradiate the sample is able to penetrate biological tissue to a much greater depth than blue or shorter wavelength light. Furthermore, multi-photon excitation can be used to excite dyes that are traditionally excited by short

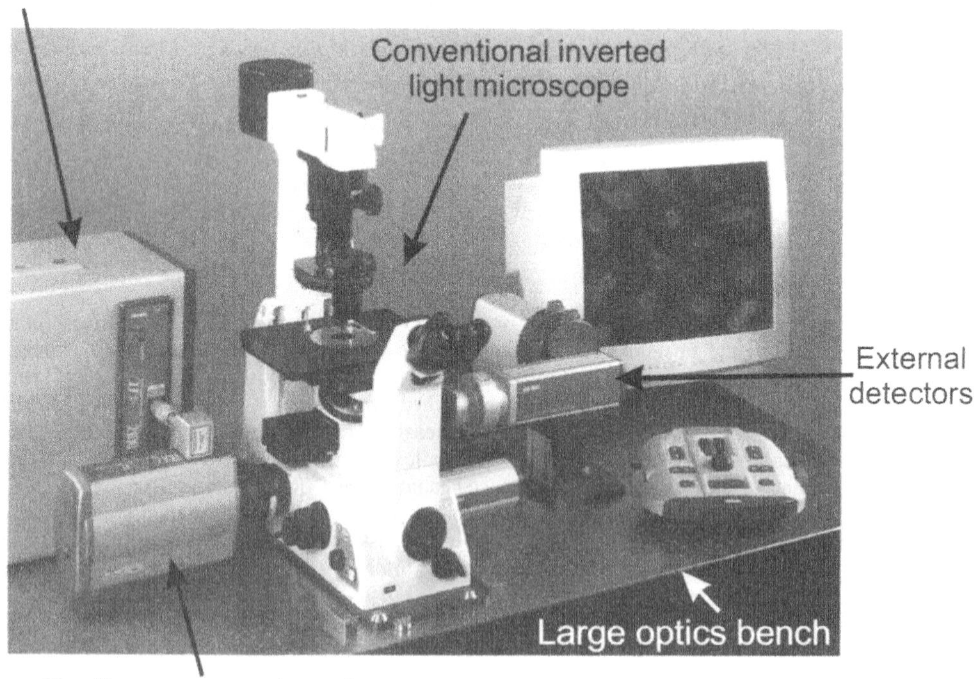

Figure 3-19. Multi-photon Microscope.
This photograph is of a Bio-Rad Radiance 2100 multi-photon microscope. The central part of the microscope is a conventional (in this case inverted) light microscope to which is attached the relatively small Radiance scan head. The Ti:sapphire pulsed infrared laser is attached directly to the scan head. On the opposite side of the light microscope to the scan head are the external detectors used for detecting fluorescent light derived from multi-photon excitation. This photograph was kindly provided by Bio-Rad Cell Sciences, UK.

wavelength UV light (for example, DAPI or the Hoechst DNA stains and calcium ratiometric emission dyes such as Indo1). Not only can these dyes be excited by the far less destructive infrared light, but the problems of microscopy using UV light are eliminated. In a multi-photon microscope "UV" dyes are excited with long wavelength infrared light and the fluorescence emission is in the visible light region of the spectrum and so therefore there is no need to have the microscope designed to handle UV light even though "UV" dyes have been used.

One of the difficulties in using multi-photon microscopy is that the excitation efficiency and best infrared wavelengths to use is not readily deduced from the single-photon absorption spectrum. This may mean that not only should a number of wavelengths be tried, but that a number of different dyes may need to be tested. Information from other users of multi-photon microscopes is a great way to find out which dyes are best suited to your experiments.

Microscope Design

The multi-photon microscope (Figure 3-19) utilises most of the scanning and detection mechanisms of the laser scanning confocal microscope, with the important exception of the unique pulsed infra-red laser used and the option of being able to attach "external" or "direct" detectors (see below). Due to the relatively large size of the pulsed infrared lasers and associated visible wavelength pump laser a multi-photon microscope is often assembled on a relatively large optics table. Although it is possible to fibreoptic couple a pulsed infrared laser, there are a number of technical difficulties associated with maintaining the correct spacing and profile of the infrared pulse such that direct coupling of the laser to the microscope is normally used.

Multi-photon microscopes are manufactured by Bio-Rad (page 356), Carl Zeiss (page 382) and Leica (page 402) using a variety of pulsed infrared lasers (see Table 3-2). Pulsed infrared lasers for multi-photon microscopy can also be attached to the Olympus laser scanning confocal microscope scan head.

Pulsed Infrared Lasers

The pulsed infrared lasers used in multi-photon microscopy (Table 3-2) are produced by two main laser manufacturers, Coherent and Spectra Physics. The pulsed infrared lasers used in multi-photon microscopy are termed ultrafast as they generate pulses in the picosecond (10^{-12}) to femtosecond (10^{-15}) range. The earlier pulsed infrared lasers were relatively difficult to operate as they required considerable skill in achieving mode-locking (the pulsed mode) and changing the output wavelength was not a trivial matter. Considerable improvement in the design of these lasers now means that the latest lasers are relatively compact, are readily mode-locked and the wavelength can be changed by simply moving a slider on a computer screen. However, these lasers are still expensive, which is a limiting factor in the wider application of multi-photon microscopy in biology.

The mode-locked titanium sapphire (Ti:sapphire) laser used in multi-photon microscopy is capable of very high peak outputs in the pico- to femto second range with a broad spectral emission from 700 to 100 nm. The Ti:sapphire laser is "pumped" using a CW (continuous wave) laser, originally a high power water cooled argon-ion laser, but in more recent years the trend has been to utilize frequency doubled solid state diode pumped lasers. The change to diode pump lasers has resulted in rugged and relatively compact laser systems for multi-photon microscopy.

Titanium sapphire pulsed infrared lasers may be of a fixed wavelength of infra red light (for example the 800 nm Vitesse pulsed infra red laser from Coherent), or they may be tuneable over a range of infra red to red wavelengths (in the range 680 to 1000 nm). Both Coherent and Spectra Physics produce a number of tuneable titanium sapphire lasers. The Tsunami laser from Spectra Physics and the Mira laser from Coherent require manual adjustment for modifying the emission wavelength. Unfortunately changing the wavelength on these lasers is not as simple as turning a single knob, but requires careful adjustment of the tuning knob, the power to the laser and to make sure the laser stays mode locked (in the pulsing state). The introduction of computer controlled wavelength tuning (for example, the Mai-Tai from Spectra Physics and the Chameleon from Coherent) has made wavelength selection as easy as moving a slider on a computer screen. This has now changed the way in which one can approach multi-photon microscopy. One can now simply try different wavelengths for efficiency of excitation of your fluorophore while imaging by multi-photon microscopy.

Compact single emission wavelength pulsed infra red lasers are also available (for example, the Vitesse laser with a fixed 800 nm pulsed infra red output from Coherent). For applications that can readily utilise the output of a single emission laser this may be a simpler solution to providing pulsed infra red light for multi-photon microscopy.

Table 3-2. Lasers used in Multi-Photon Microscopy

A range of ultrafast pulsed infrared lasers are available for use in multi-photon microscopy. These titanium sapphire (Ti:sapphire) lasers are usually "pumped" with a relatively high energy output visible light laser. Multi-photon excitation can be elicited using both femto-second (10^{-15}) and pico-second (10^{-12}) pulsed infrared lasers. This table lists the pulsed IR lasers commonly used in multi-photon microscopy. A number of other pulsed IR lasers are available, particularly for applications in physics and chemistry, but are rarely used for multi-photon imaging in biology.

LASER	Emission Lines	Suitability	NOTES
Spectra Physics			
Tsunami	690-1080 nm	Visible to UV dyes	Pulsed IR Ti:sapphire laser with wide tuneable range. Acousto-optic mode lock that is self starting and self sustaining, allowing for stable output of a pulsed infrared beam over many hours or even days. This Ti:sapphire laser has a relatively large footprint with manually adjusted wavelength tuning. Variable pulse width from 35 femto seconds to 100 pico seconds is possible. Separate pump laser required.
Mai Tai	750-850 nm 780-920 nm	Visible to UV dyes	Pulsed IR laser with limited but very simple computer controlled tuneable range, using the same acousto-optic mode locking and self starting method of the Tsunami laser. A solid state Millennia diode pump laser is incorporated into the laser box. The Mai Tai Ti:sapphire laser is a relatively small and compact laser.
Coherent			
Mira	690-1000 nm	Visible to UV dyes	Pulsed IR Ti:sapphire laser with a wide tuneable range. Mode locking achieved by the use of Kerr Lens mode locking, a solid state mode locking technique that is very stable but does require careful adjustment to initiate mode-lock. A diode-pumped green Verdi laser with high power output and good reliability is used to drive the Ti:sapphire laser. Tuneable across the full range of the Ti:sapphire crystal (700 to 1000 nm) without the necessity to change any optical components or with any sacrifice in power output. Capable of pico- or femtosecond pulse generation.
Chameleon	720 – 930 nm	Visible to UV dyes	Pulsed IR Ti:sapphire laser with a very wide computer controlled tuneable range. Mode locking achieved by Kerr Lens mode locking. Diode-pumped Verdi laser is incorporated into a single box with the Ti:sapphire laser.
Vitesse XT	700-760 nm 760-860 nm 780-920 nm	Visible to UV dyes	Pulsed IR laser with computer controlled limited tuneable range. This laser is no longer available, having been replaced by the Chameleon tuneable laser.
Vitesse	Fixed 800 nm	Visible to UV dyes	Pulsed IR laser with fixed wavelength.

Direct (External) Detectors

Multi-photon fluorescence only occurs at the focal point of the laser (see page 12 in Chapter 1 "What is Confocal Microscopy?") and so there is no need to use a confocal pinhole. In other words, all the fluorescence emanating from the sample is from the focal plane. Although multi-photon microscopes often operate by collecting the fluorescent light from the sample after it has passed through the scanning mirrors, confocal optics and through the confocal pinhole, the inherent ability to reject out of focus light in a multi-photon microscope means that the light detectors can be placed external to the scan head (Figure 3-20). When external detectors are implemented there is at least a 30% increase (and possibly as high as 100% increase) in the amount of light that can be collected.

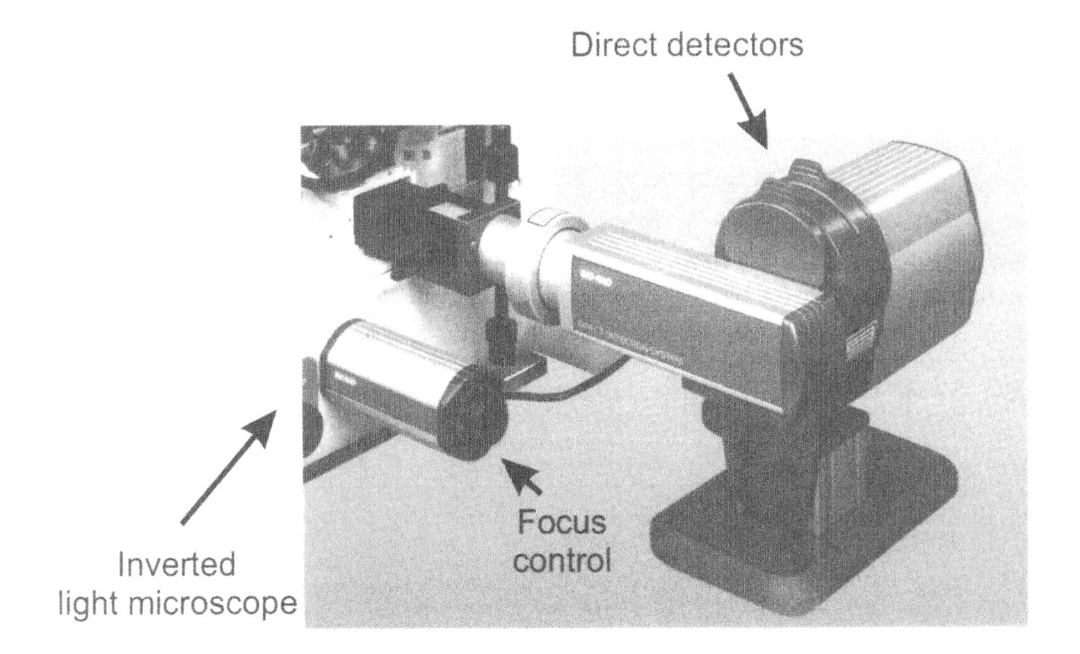

Figure 3-20. Direct Detectors on a Multi-photon Microscope.
A close up view of the external detectors used for detecting fluorescent light derived from multi-photon excitation in the Bio-Rad Radiance 2100 multi-photon microscope. Fluorescent light derived from multi-photon excitation can be detected using the photomultiplier tubes used for confocal microscopy, but as all fluorescent light is from the focal region of the microscope there is no need to de-scan the light or to use a confocal pinhole. Using external detectors in multi-photon microscopy results in at least 30% increase in sensitivity of the instrument. This photograph was kindly provided by Bio-Rad Cell Sciences, UK.

CONFOCAL MICROSCOPE MANUFACTURERS

There is a wide range of confocal and multi-photon microscopes available from a number of manufacturers. Details on the hardware of the individual instruments manufactured by the major confocal microscope manufacturers are described in "Appendix 1: Confocal Microscopes" (pages 356 - 441). A summary of major confocal microscope manufacturers is shown in Table 3-3.

Table 3-3. Confocal Microscope Manufacturers

This table briefly outlines the type of instrument produced by each of the major confocal microscope manufacturers. See "Appendix 1: Confocal Microscopes" (pages 355 - 441) for detailed technical information on the instruments they produce.

Laser Scanning Instruments

All laser scanning confocal microscopes are capable of multi-channel fluorescence, transmission and backscatter imaging. The following companies design and manufacture a number of confocal microscopes, and all manufacturers, with the exception of Bio-Rad, also design and manufacturer the associated light microscope.

Manufacturer	Description
Bio-Rad Page 356	**Laser scanning confocal and multi-photon microscopes.** Bio-Rad has been producing confocal microscopes since the very early days of using confocal microscopes in biology. Bio-Rad's latest instruments are based on using a relatively small scan head that can be easily changed from one microscope to another. They have achieved this by placing most of the optical components in an associated optics box that can be placed under the bench.
Carl Zeiss Page 382	**Laser scanning confocal and multi-photon microscopes.** The Zeiss "META" confocal microscope can be used as a relatively "conventional" confocal microscope for multi-channel labelling or the innovative META spectral detection system can be implemented as one of the available channels. The META separation technique can provide excellent separation of a number of fluorophores with very similar (overlapping) spectra.
Leica Page 402	**Laser scanning confocal and multi-photon microscopes.** The Leica SP confocal microscopes utilise a prism and computer controlled variable slit mechanism to separate the fluorescent into four different channels. Leica also manufacture confocal microscopes with detection channels using dichroic mirrors and optical filters.
Nikon Page 415	**Laser scanning confocal microscopes.** The Nikon C1 confocal microscope is designed with a relatively small scan head that is attached to an associated optics box by fibreoptic coupling. The optics box houses conventional optical filter cubes used in the Nikon epi-fluorescence microscope. These filter cubes are designed for specific multi-labelling applications and are readily swapped over for different experiments.
Olympus Page 420	**Laser scanning confocal microscopes.** Olympus manufacture the Fluoview confocal microscopes, which are based on using dichroic mirrors and optical filters to separate out the various wavelengths of light. Olympus manufacture both a computer controlled confocal microscope (FV500) and a model with manually operated dichroic mirror, optical filter and pinhole changes. Pulsed infrared lasers for multi-photon microscopy can be readily attached to the Olympus confocal microscope scan head.

Nipkow Disk Instruments

Manufacturer	Description
Atto Bioscience Page 431	**Nipkow spinning disk confocal microscope.** Atto Bioscience manufacture a Nipkow spinning disk based confocal microscope, which includes the scan head, CCD camera and associated software.
Yokogawa Page 435	**Nipkow / micro-lens array disk confocal microscope scan head.** Yokogawa manufacture a confocal microscope scan head that contains a Nipkow disk that incorporates a second disk containing a micro lens array. The scan head can be purchased directly from Yokogawa, or you can purchase an assembled confocal microscope that includes this scan head as well as CCD camera and associated microscope from a number of other companies.
PerkinElmer Page 438	**Nipkow spinning disk confocal microscope (Yokogawa scan head).** PerkinElmer provide a complete system, including the scan head, the CCD camera, microscope and associated software.
VisiTech Page 441	**Nipkow spinning disk confocal microscope (Yokogawa scan head).** VisiTech assemble a complete system, including the scan head, the CCD camera, microscope and software. VisiTech also manufacture a dual CCD attachment. A number of distributors of the VisiTech confocal microscope (Chromaphor Analysen-Technik, McBain Instruments, Quorum Technologies, Solamere and Visitron Systems) contribute various individual components to the instrument they sell.

WHICH CONFOCAL MICROSCOPE SHOULD I USE?

Presented with a bewildering array of different confocal microscopes the question is often asked as to which design is the best. The first thing to note is that most people using a confocal microscope don't actually have a "choice" of instruments; they are usually limited to those instruments that are readily available to them. This often means that the only confocal microscope that can be used on a regular basis is the one available in your department or institute. However, it is worth considering which confocal microscope is best suited for a particular task, of great value when a confocal microscope is first purchased, but also because it is always possible to arrange for time on another instrument for particularly critical experiments.

As has been discussed throughout this chapter, there are two fundamentally different confocal microscopes to consider – the Nipkow disk confocal microscope and the laser spot scanning confocal microscope. In addition, the special attributes of the multi-photon microscope also need to be considered.

The Nipkow disk confocal microscopes are particularly suited to high-speed and light-sensitive live cell imaging. Although the total level of light directed onto the sample may be similar to that used in a spot scanning confocal microscope, the Nipkow disk scanning confocal microscope has a much lower light intensity per individual scanning spot. The relatively low light intensity in each scanned spot does mean that damage to the fluorophore and cellular macromolecules is considerably reduced. Most fluorophore and cellular damage is due to free radical damage generated by the laser light, and so low light intensities may allow free radical scavengers naturally present in live cells to remove most free radicals before they can damage the cells. The Nipkow disk confocal microscope can collect images at greatly increased speed compared to a laser spot scanning confocal microscope. Image collection is routinely 10 to 15 frames per second, but can be as high 100s of frames per second if there is sufficient light emitted by the fluorophore and the attached CCD camera is of high enough sensitivity. Highly transient calcium "sparks" and "puffs" and other ionic fluxes can be readily followed in real time using a Nipkow disk confocal microscope.

Long term studies using live cells are also particularly well suited to Nipkow disk confocal microscopy. The extremely low level of cell damage and fluorophore fading caused by the laser light means this type of microscope can be used to follow metabolic and structural changes in cells over many hours or even days. Low levels of cellular damage are particularly important when imaging live embryos. The Nipkow disk scanning confocal microscopes are capable of long term studies involving the development of embryos – in which the embryo remains viable throughout the experiment.

If Nipkow disk confocal microscopes are so good at imaging live cells without damage then why not use them all the time? There are some disadvantages to using a Nipkow disk instruments that need to be considered. The main disadvantage is that the microscope is not capable of instrument zooming, which means that it is not possible to exploit the full resolving power of the objective used. The optical resolution cannot be matched to suit the Nyquist sampling requirements of the object. Furthermore, the Nipkow disk confocal microscopes, due to inherent design limitations of the instrument, cannot readily perform a number of confocal microscopy techniques – although innovative approaches, such as using the pinholes to bleach thousands of spots for FRAP studies, can be utilised.

For routine work requiring high quality imaging of fluorescently immunolabelled cells or tissue slices, for the imaging of GFP transfected cells, and for analysis using many different fluorescent dyes the laser spot scanning confocal microscopes are the best choice. All of the laser scanning confocal microscopes discussed in this book are capable of producing excellent high-resolution images with great sensitivity. In fact, the variation encountered in sample preparation is probably greater than the variation in sensitivity between most instruments. However, there are significant differences between the instruments that make choosing an instrument for a particular task somewhat difficult. The major difference between the confocal microscopes produced by each of the major manufacturers lies in the versatility of the instrument, rather than in any inherent optical or electronic performance criteria. Some laser scanning confocal microscopes have somewhat limited capabilities (for example the Nikon C1 confocal microscope). However, this does not mean that this instrument is inferior for your particular application. If your experiments do not require the sophisticated light separation techniques available on the more elaborate Zeiss and Leica manufactured instruments, then you may find the simplicity of the Nikon or Olympus instruments appealing.

The Multi-photon microscope has the unique characteristic of being particularly suited to live cell, and particularly live tissue imaging. The extra depth penetration of the infrared laser and the clarity of images obtained significantly deeper into tissue samples makes this instrument valuable for performing live cells studies. However, there are a number of inherent difficulties associated with predicting the efficiency of exciting particular fluorophores, and the greatly increased cost of purchasing the titanium sapphire pulsed infrared laser and associated optics, make this microscope a major investment for live cell imaging.

Chapter 4

Image Collection

The laser scanning confocal and multi-photon microscope creates an image by scanning a laser across the sample and then recording either the fluorescent or backscattered light emanating from this scanned area. The laser light that has passed through the sample can also be collected to simultaneously obtain a transmission image. An image in a confocal microscope is not usually visible through the microscope, becoming visible after suitable digitisation and subsequent display on a computer screen. However, a live image can be directly viewed when using a Nipkow spinning disk confocal microscope.

Collecting images on a confocal microscope is a very simple matter, but to get the most out of the instrument, understanding what the different scan modes mean, what zooming the image actually does etc., is not inherently difficult but can be quite daunting due to the complexity of the instruments. This chapter is intended as an introduction on how images are collected on the confocal microscope, including practical hints on how to collect high-resolution images and how to obtain the maximum biological information from your sample.

Although the principles involved in image collection are the same, there are important hardware and image collection software differences between the various confocal microscope manufacturers. "Appendix 1: Confocal Microscopes" (pages 355 - 441) has information on the hardware and software controls for each of the major confocal microscopes.

The manipulation of the digital images obtained from the confocal microscope is discussed in more detail in Chapter 5 "Digital Images in Microscopy" (page 145), and Chapter 6 "Imaging Software" (page 163).

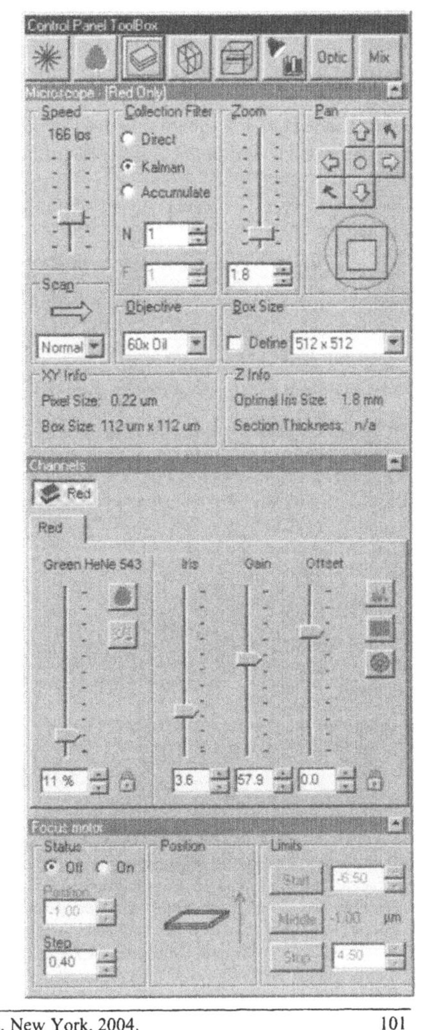

Figure 4-1. Confocal Microscope Control Panel.
The layout of the control panel is quite different for each of the major confocal microscope manufacturers, but the items displayed are essentially the same, as shown in this example from the Bi-Rad Lasersharp software. Please refer to "Appendix 1: Confocal Microscopes" (page 355 - 441) for information on the software interface used by each of the major confocal microscope manufacturers.

IMAGE COLLECTION MODES

The "work horse" of confocal microscopy in the biological sciences is fluorescence image collection. However, obtaining images by other methods, such as backscatter and transmission image collection is also important, often for providing structural information on the location of the fluorescence.

Fluorescence Image Collection

In the fluorescence imaging mode fluorescent light is separated into component colours by a variety of optical filters, prisms or diffraction grating and detected with very sensitive photomultiplier tubes, or in the case of the Nipkow disk confocal microscopes, a sensitive CCD camera (see Chapter 3 "Confocal Microscopy Hardware", page 65). The instruments are designed to direct light of specified regions of the light spectrum to the either a single detector or, in the case of the Zeiss META confocal microscope, to an array of detectors. In both cases the detectors are designed to detect the amount of light that falls on them irrespective of the wavelength, although the detector may be more sensitive to light in particular regions of the spectrum. The level of fluorescence in various parts of the image in channel one, for example, will simply be the amount of light detected from the region of the light spectrum directed to that channel. In other words, the image is an intensity range of a particular region of the light spectrum (Figure 4-2). The image is often conveniently coloured after collection (this can be done immediately on-screen as the scan proceeds, or later after the image has been saved to disk using a Colour Look Up Table, LUT) to depict the region of the light spectrum from which the light was collected in a specific colour. For example, when imaging FITC or Alexa Fluor

> All images collected on a confocal microscope are collected as grey-scale images, with colour being added after collection.

Channel 1

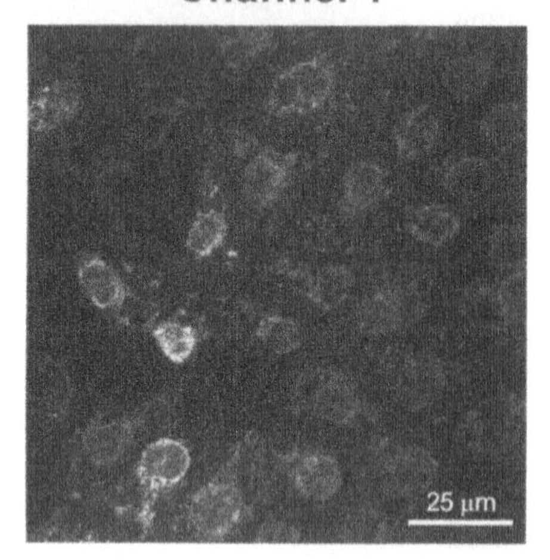

Light with a wavelength from 500 to 540 nm

Channel 2

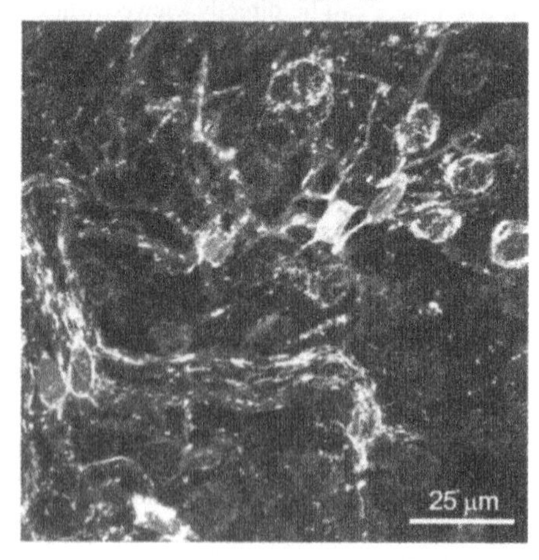

Light with a wavelength longer than 600nm

Original grey-scale images

Figure 4-2. Dual Channel Fluorescence Imaging.
Fluorescence images are collected as grey-scale images that are subsequently coloured for display. Each collection channel is displayed as a separate imaging display box on the computer screen. In this example, a live human heart muscle tissue biopsy sample was stained with MitoTracker Green (left panel) and MitoTracker Red (right panel) and imaged using a Bio-Rad MRC-1024 laser scanning confocal microscope using dual channel collection with standard green/red dual channel optical filter sets (for example those filters suitable for FITC/TRITC dual channel collection). A single optical slice was collected approximately 10 µm away from the coverslip, within the heart tissue, using a Nikon 60x 1.4 NA oil immersion objective lens. These grey-scale images are shown individually in colour and as a merged RGB (multi-coloured) images in Figure 4-3 (page105).

488 the image is often coloured green, but don't forget the instrument has simply detected light of a defined spectral region – and this may not necessarily be the same as the colour shown on the computer screen!

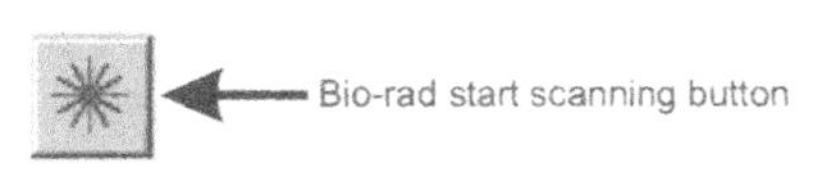

The controls for image collection are provided on the main control panel of the software for the particular instrument you are using. There is usually a single "button" used for start and stop image collection (the control panel for the Bio-Rad Radiance confocal microscope is shown in Figure 4-1), and the start/stop scanning "button" is shown to the left.

Most confocal microscopes also have a convenient "fast" scan mode for adjusting the focus and x-y position of the scan. This "fast" scan mode often involves the collection of a lower resolution scan that is not to be confused with altering the scan speed of the main collection mode. In the Bio-Rad instruments this "fast" scan is achieved by setting the scan mode to "Fast" (also called "x2") or "Fastest" (x4), but in the case of the Zeiss instruments this "fast" mode is called "Fast X-Y" scanning (see diagram to the right).

Zeiss "fast scan" button

Find

Fast XY

Start

STOP

Stop

Most of the settings for fluorescence imaging are now operated via the controlling software. The earlier model confocal microscopes required physical changes in settings on the scan head (including physically changing the optical filter blocks, or moving sliders in the scan head). The Olympus Fluoview confocal microscope is one of the few manually operated confocal microscopes still being manufactured, but you will find there are a number of other models from various manufacturers with manual controls that are still in use in many laboratories. The Nikon C1 confocal microscope also uses manual dichroic mirror and optical filter changing by utilizing the Nikon conventional epi-fluorescence microscopy optical filter blocks housed within a fibreoptic coupled optics box. Manual changing of optical filters and dichroic mirrors does not in any way detract from the optical performance of the instrument, but manual control can limit the versatility of the instrument – for example, when performing a dual labelling experiment alternate line collection is not possible on a manually adjusted instrument. Fast sequential switching is only possible using fast motorised or electronic switching of optical filters etc. Some confocal microscopes do not use traditional optical filters / dichroic mirrors, but make use of electronically controlled filtering devices (such as the Acoustic Optical Tuneable Filter, AOTF) or prisms and variable width slits to select the wavelengths of interest. Detailed information on the various optical components used to separate out light of different wavelengths is provided in Chapter 3 "Confocal Microscopy Hardware" (page 65), and the hardware design for each of the major confocal microscope manufacturers is discussed in detail in "Appendix 1: Confocal Microscopes" (pages 356 - 441).

Use the "fast" scan mode to adjust the focus and x-y position of the scan

The confocal microscope images displayed on the computer screen can be coloured during the scan by applying a coloured Look Up Table (LUT) to the grey-scale image, even during collection (Figure 4-3). A "real time" multi-coloured merged image (lower panel in Figure 4-3) of the collected images can also be displayed. In this case the merged image is displayed as a 24-bit red-green-blue (RGB) multi-colour image, with the image collected by detection channel one on the confocal microscope allocated to the green display channel and channel two allocated to the red display channel, with no image available for the blue channel in this example. Thus, red-green-blue 24-bit colour images are a very simple way of displaying up to three channels simultaneously in the same image.

The human eye has different levels of sensitivity to different regions of the light spectrum (our eyes are most sensitive to green, less to red, and show very poor sensitivity to blue). This variability in the eye's ability to detect different colours means that the images should be collected by displaying each channel as a grey-scale image. In this way the gain and black level (intensity) of each image can be more readily adjusted to the correct setting. If you use green and red to display the two channels while making adjustments you many find that the gain setting does not truly reflect the level of light in that channel, but in fact the ability of your eyes to detect that particular colour (this is even more important if you have any degree of colour sight impairment).

All of the confocal microscopes with computer controlled optical settings have a means of saving instrument and software settings used for collecting images in subsequent experiments. How this is done is different for each of the instruments, but the principle is the same – the settings for FITC imaging, for example, can be recalled and used when you require those settings, not necessarily for imaging FITC, but perhaps a fluorophore with a similar emission spectrum, such as MitoTracker Green. The Zeiss confocal microscope control software has the handy tool of being able to open any previously saved image from the microscope and to "copy" the settings from that image to your current image collection window. This is an excellent feature, particularly when you have spent some time adjusting different settings on the microscope and want to repeat the exact conditions at a later date. "Appendix 1: Confocal Microscopes" (pages 356 - 441outlines the basic system used by each of the confocal microscope manufacturers for saving and recalling a specific series of instrument settings.

Use a pre-set "method" to quickly obtain an image – then make your own adjustments

There are a large number of settings that can be altered on a confocal microscope to optimise the resolution, the sensitivity and for specifying the region of the light spectrum to be directed to a particular channel. These settings are discussed in some detail later in this chapter, but first the other two collection methods commonly used on a confocal microscope, backscatter and transmission image collection, will be discussed.

Hints for "finding" your sample

Due to the thin optical slice imaged by a confocal microscope the sample can often be difficult to locate, particularly when attempting to image relatively thin single cell layers.

Start with Conventional Light Microscopy:

- Find the sample using regular bright-field, phase contrast or DIC imaging (by looking down through the eyepieces). This has the added advantage of causing very little damage to the cells.
- Switch to conventional epi-fluorescence wide-field microscopy and choose an area of the sample for imaging by confocal microscopy. Take care - the wide-field epi-fluorescence mode can quickly fade your sample.

Switch to Confocal Microscopy:

- Choose a previously used "method" for the fluorophore in use, or adjust the optical filters or spectral selection sliders to the correct settings for the current fluorophore.
- Choose a relatively low laser setting (particularly important when imaging live cells).
- Open the confocal pinhole.
- Turn the PMT tube gain (or voltage) high.
- Start the "fast" scan mode.
- Carefully alter the fine focus control (manually) while scanning. The "image" may appear as a bright "band" across the screen if you move the focus control while the instrument is scanning. Try not to move the focus knob during the brief pause between scans.
- If you cannot find the sample, try opening the pinhole further and again carefully move the focus control on the microscope.
- If you still cannot find the sample try increasing the laser power (be careful – the laser will still bleach your sample, even if you there is no image on the screen!).

Use the transmission image to locate your cells: A simple way of locating single cell layers is to use transmission imaging on the confocal microscope to locate the position of the cells. The transmission image is not confocal (i.e. is not a single focal plane), which means it is often possible to image the cells even when the focus is not correctly adjusted.

Note: The Zeiss confocal microscope has a special "Find" for locating the sample (see the Zeiss scan button diagram on the previous page) – which works by taking a fast line-scan and automatically adjusting the gain and choosing the best PMT gain setting to display an initial image. However, other instrument settings are not adjusted automatically using the "Find" mode.

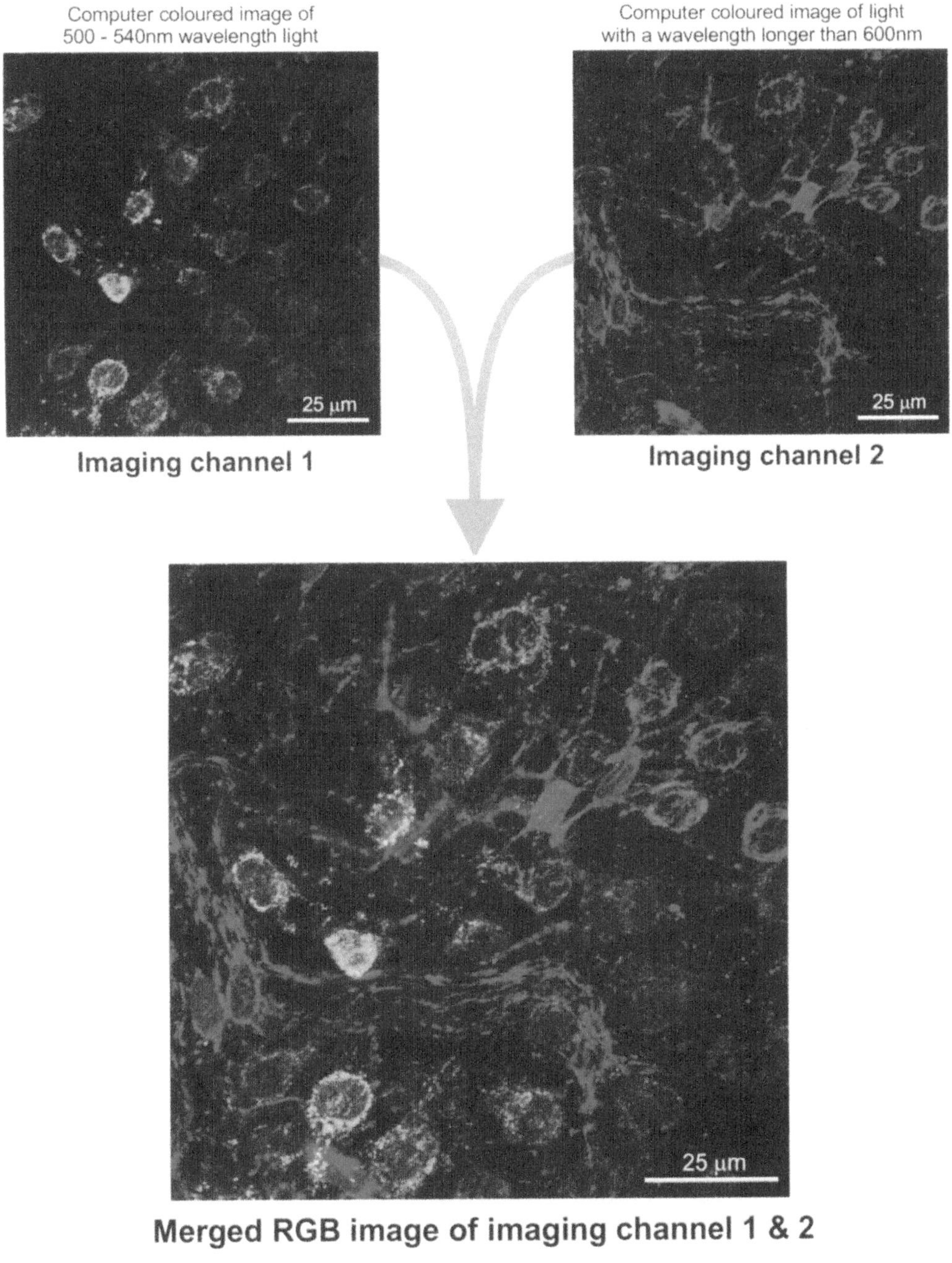

Computer coloured image of
500 - 540nm wavelength light

Computer coloured image of light
with a wavelength longer than 600nm

25 µm

25 µm

Imaging channel 1

Imaging channel 2

25 µm

Merged RGB image of imaging channel 1 & 2

Figure 4-3. Merged Dual Channel Fluorescence Image.
Live human heart muscle was stained with MitoTracker Green (channel 1) and MitoTracker Red (channel 2) and imaged with a 60x 1.4 NA oil immersion objective using a Bio-Rad MRC-1024 laser scanning confocal microscope with dual channel collection (as described in Figure 4-2, page 102). Confocal microscope images are collected as grey-scale images (see Figure 4-2) that are pseudo coloured by the computer (channel 1 and 2 above) during "live" display if required. The merged multi-colour image shown in the lower panel is generated during collection by directing the output from imaging channel 1 to the green channel and the output from imaging channel 2 to the red channel of an RGB (red, green, blue) image. A merged multi-colour RGB image can also be readily obtained from the individual grey-scale images after collection using image processing software (see page 249).

Backscatter (Reflectance) Image Collection

The confocal microscope is capable of producing excellent "backscatter" or reflected light images (Figure 4-4). Although this method of imaging is often referred to as reflectance imaging, a more correct term is "backscatter" imaging, to emphasise the point that the light being collected is that which is "scattered" by the sample rather than the direct reflection from the sample. In fact, the direct reflection from the sample often seriously degrades the quality of the image. If you find you have a bright spot in the image, and possibly "rings" of brightness then you may have a problem of the direct reflection from the specimen entering the photomultiplier tube. The simplest way to remove this unwanted reflected light from your image is to place a polariser in the beam path to remove most of the polarised laser light that is directly reflected from the sample. The light that is interacting with the sample and being "scattered" is effectively randomly polarised and so progresses through the polarising filter in sufficient quantity to create an image. If you do

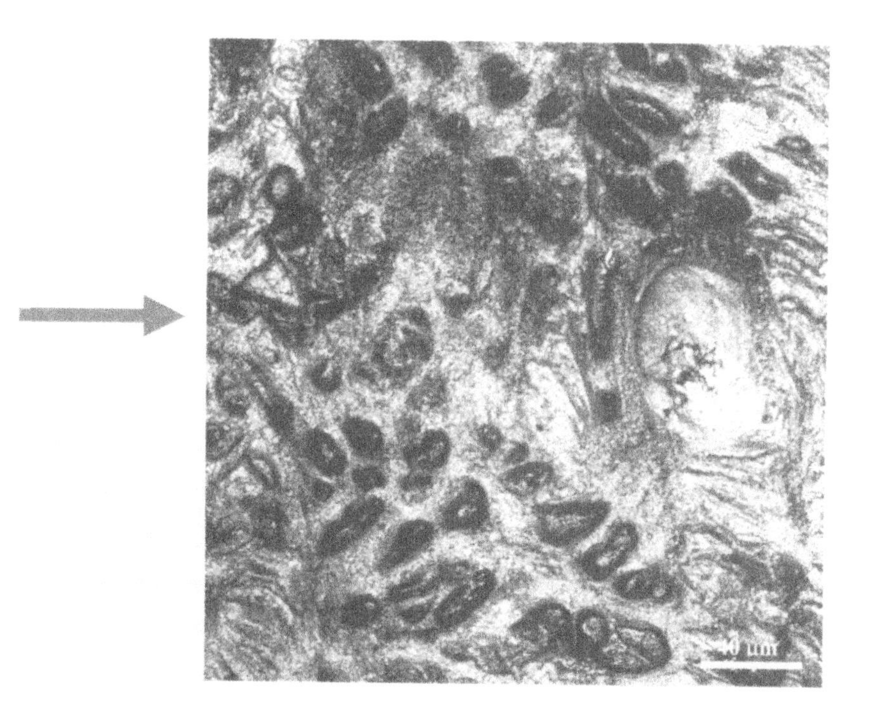

Figure 4-4. Backscatter Image of Pig Skin.
A series of optical slices were collected using backscattered light from a 488 nm blue laser line (9 slices out of a total of 40 collected are shown on the left) while scanning the laser across the surface of a small piece of excised pig skin. The pig skin was gently placed on the lower coverslip surface of an inverted microscope imaging chamber and imaged with a Nikon extra long working distance (ELWD) "dry" 40 x 0.6 NA objective lens with a coverslip thickness adjustment collar, using a Bio-Rad MRC-1024 confocal microscope. A simple "extended focus" image (above) shows excellent surface topography of the skin. A backscatter image of internal structures can be created with more translucent samples.

Backscatter (reflectance) images are high-resolution optical slices

not have a polariser that can be used, then an alternative is to use the confocal microscope zoom function to image a smaller portion of the field of view and then to use the "pan" function on the control panel to move the scan region to one side of the field of view.

The backscattered light image is often used to determine the surface topography of the sample (for which this imaging method is used extensively in the materials sciences), but in biological samples backscattered images of internal cellular structures can often be constructed. Some cellular structures, for example, the haemozoin found in relatively mature malarial parasites, are so highly reflective that they can often be imaged without the necessity to adjust the optical settings of the confocal microscope specifically for backscatter or reflectance imaging! However, this can be a problem when attempting to collect fluorescence images of these parasites as the strong backscatter signal from the haemozoin crystals may be mistaken for fluorescence.

To collect a backscattered light image the irradiating laser light (blue 488 nm light from the argon-ion laser in the example in Figure 4-4), that is scattered from the sample, needs to be detected by the photomultiplier tubes or CCD camera (see box on this page). Any surface with a change in refractive index, for example the surface of the coverslip will have a high level of backscatter, but luckily the backscatter-imaging mode is confocal – i.e. only the focal plane is visible, which allows high quality backscatter images to be obtained from structures less than a few microns away from the coverslip.

How to Collect a Backscatter Light Image

Use a pre-set "method" for backscatter or reflectance imaging if one is available.

If no method is available, adjust the optical filters, dichroic mirrors or spectral selection to allow the irradiating light to be directed back up through the scan head to the PMT tubes. This can often be accomplished by placing a partial dichroic mirror as the main device for separating the incoming laser light from the light emanating from the sample. A partial mirror will direct a portion of the laser light to the sample, and because the mirror is a partial reflector at all wavelengths, a portion of the reflected laser light will also pass through the mirror and be directed to the photomultiplier tubes.

Removing the intense reflection spot and diffraction rings:

Try using a polariser if available to remove the direct reflection spot.

Try scanning a smaller area by adjusting the confocal microscope zoom, and then "pan" the zoom area to one side of the field of view.

Transmission Image Collection

A mirror located above the condenser on an inverted microscope (and under the condenser on an upright microscope) directs light, via a fibreoptic cable, to the controller box where the photodiode or photomultiplier tube transmission detectors are housed. In most of the newer confocal microscopes the mirror is automatically brought into position when you set the instrument up to perform transmission imaging. However, on many older model confocal microscopes the mirror may have to be switched to the correct position to obtain a transmission image. Likewise, make sure you switch the mirror back again when attempting to locate an object using conventional bright-field microscopy by looking to the microscope eyepieces – otherwise there will be no light directed through the eyepieces. In some confocal microscopes the transmission detector itself may be located behind the condenser, instead of using a fibreoptic cable to transfer the signal to a remote detector.

Excellent grey-scale transmission images can be collected using a confocal microscope (Figure 4-5), but as the collection does not involve a "pinhole" the image created is not confocal - out of focus light is still part of the image, just as in conventional light microscopy. This means that the transmission image will go out of focus as the microscope focus control is moved. This characteristic can be readily exploited to "find" your cells. The transmission image will be visible as an out of focus "blur" long before you have brought the fluorescence in to focus.

> Transmission images are NOT confocal

Transmission images can be collected on a confocal microscope using phase contrast, DIC (Figure 4-5), Hoffman modulation and many other optical techniques. The relevant optical components are simply placed in the light path of the microscope as for conventional bright-field microscopy. However, due to the polarised nature of the irradiating laser light, optical techniques such as DIC imaging (see Chapter 2 “Understanding Microscopy”, page 31) that utilise polarised light are best performed without the polariser in the light path.

Take care when using DIC optics that the Wollaston prism is not unduly interfering with the quality of the fluorescent image. Some implementations of DIC will produce a slightly offset double image. At lower zoom settings there may be a slight deterioration in the fluorescent image, but at higher zoom settings the offset “shadow” of the main image will make the fluorescence image unacceptable.

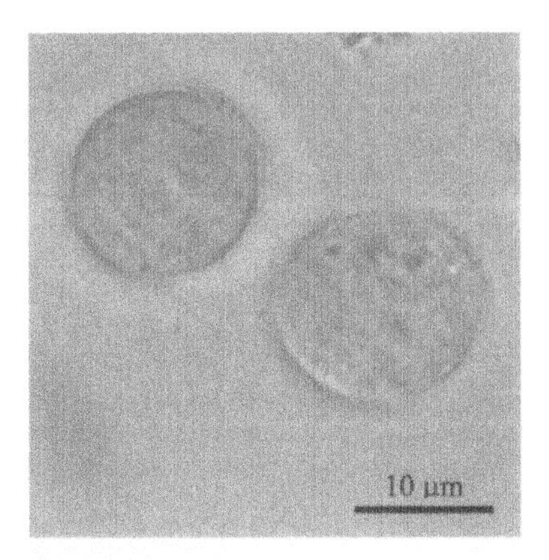

Transmission image

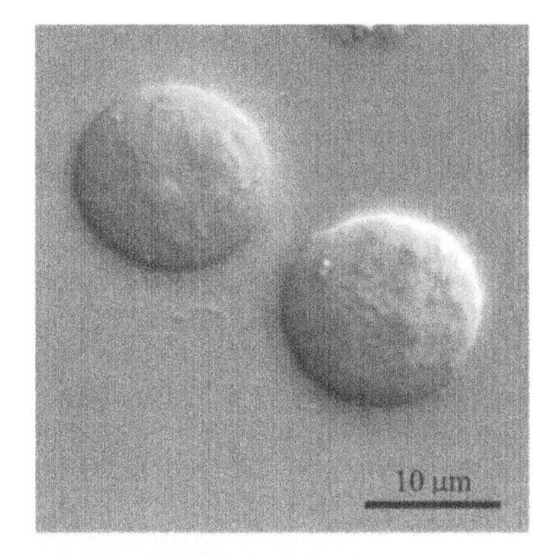

DIC image

Figure 4-5. Transmission and DIC Image.
The transmission image collected on the confocal microscope, although of high quality, is not confocal (i.e. the image goes in and out of focus as the focus control is altered). The left hand transmission image of live HL-60 cells was collected on a Leica SP confocal microscope with a 63x 1.32 NA oil immersion objective with normal bright-field (Köhler) illumination using the 488nm blue line on an argon-ion laser. The image on the right is of the same two cells, but collected using the confocal microscope transmission detector with DIC optics present. The pseudo 3D image formed when using DIC optics is a shadowing effect due to changes in refractive index rather than true 3D structure. DIC images can also be generated from transmission images using the QPm software from Iatia (see page 47).

The grey-scale transmission image can be coloured if desired, or more commonly kept as a grey-scale image and overlaid with a colour fluorescence image. The background grey-scale transmission image can be used very effectively to physically locate the position of the fluorescent label within the cells or to simply locate the position of individual cells. Background grey-scale transmission images with an overlay of coloured fluorescence can be readily created using the proprietary software that is used to operate your confocal microscope, other image manipulation software or the channels or layers in Photoshop (see Chapter 10 “Confocal Microscopy Techniques”, page 239).

Careful adjustment of the gain and offset levels can create a superb transmission image, even when using unstained cells that are very difficult to image by conventional transmission microscopy. This is due to the ability to greatly increase the contrast in samples that show only very small differences in absorbance of transmitted light. The transmission image created when imaging live single cell layers is significantly better than the transmission image obtained from fixed cells.

A grey-scale transmission image can be generated by any of the laser lines on the confocal microscope. However, if you wish to use the transmission channel to collect images of stained cells or tissue sections try different laser lines as some will be absorbed by the dyes used for staining the sample. Unfortunately thick tissue samples cannot be imaged satisfactorily by transmitted light on the confocal microscope. Transmitted light can still be used on a relatively thick tissue sample to locate the tissue for later fluorescence confocal imaging (as long as the tissue is thin enough and transparent enough to allow sufficient light to pass through to form an image). Excellent transmission images can be obtained from thinly sliced tissue samples.

Real Colour Transmission Image Collection

Confocal microscopes equipped with a krypton-argon-ion laser are capable of generating "real colour" transmission images. In this way conventionally stained tissue sections (for example, H&E stained tissue slices or DAB immunostaining) can be imaged using a confocal microscope.

The three laser lines (blue, yellow and red) of the krypton-argon-ion laser when correctly balanced constitute white light, which means that careful balancing of the light intensity and gain of each individual channel on the confocal microscope can produce a real colour transmission image.

The Bio-Rad MRC-1024 confocal microscope has three transmission detectors that can be independently controlled to create a very effective real colour transmission image "live" on screen in a single mixer panel (see "Appendix 1: Confocal Microscopes", page 356). Other confocal microscopes can also create a very acceptable real colour transmission image by sequentially collecting three individual grey-scale transmission images using each of the three laser lines. Individual grey-scale transmission images can be merged after collection using the confocal microscope imaging software, or other imaging software (for example, the powerful, but free image processing software NIH Image or Image J, or commercially available general image processing software such as Photoshop). The colour balance of the three individual channels can be readily adjusted to create a high-quality real colour transmission image.

When using the Bio-Rad MRC-1024 confocal microscope to collect the three transmission images simultaneously, set the laser power to high. Otherwise, the continual switching to "high laser" each time the scanner is activated (the default setting on the Bio-Rad instruments), will result in changes in colour balance as the 3 laser lines stabilise. Don't forget to switch the laser back to "low" after finishing work on transmission imaging, otherwise the laser lifetime will be seriously diminished.

Photon Counting

Although imaging in light microscopy is ultimately a question of counting photons, most instruments operate by measuring the electronic output from the photomultiplier tube – without any attempt at "counting" individual photons. The photon counting mode is a method implemented by operating the photomultiplier tube in a pulse-counting mode. The electronic circuitry associated with the photomultiplier tube is set to count only those pulses over a pre-determined threshold.

The photon counting mode is particularly useful for collecting images of very low fluorescence levels. In fact, samples with a high light level will result in saturation or "pile up" of photons and thus an underestimate of the number of photons reaching the photomultiplier tube. You may be surprised to learn that the number of photons per pixel for a single scan in a confocal microscope may be less than 10 for even a relatively "bright" image when using fluorescent samples, and hence too much signal is rarely a problem with fluorescence imaging.

Photon counting is usually carried out using the "accumulate" collection filter (see page 121). The collection is stopped when sufficient photons have been collected to result in a visible image on the display monitor, or alternatively, a very low light level image is "contrast stretched", after collection, to display the relatively low grey levels present in the image. Don't be alarmed by the apparent "lack" of an image when collecting using the photon counting mode as subsequent alterations in the brightness levels of the image may indicate that a good quality image, although at low intensity grey levels, has been collected.

LASER INTENSITY AT THE SAMPLE

The intensity of the laser light *at the sample* is what matters when imaging biological samples. If the laser intensity is too low there may be insufficient fluorescence to create an acceptable image, and if the laser intensity is too high the fluorophore may be destroyed by photobleaching, or in the case of live cells the cells themselves or at least the biochemical and physiological processes within the cells may be damaged or destroyed. A combination of altering the power setting on the laser and adjusting the Neutral Density or AOTF filters to attenuate the laser light is used to control the intensity of the laser at the sample.

Details on the various lasers used in confocal microscopy are given on page 86 to 88 in Chapter 3 "Confocal Microscopy Hardware", including various methods of attenuating the intensity of laser light that is directed at the sample. This chapter covers the more practical aspects of controlling the laser for obtaining high quality images.

Laser Power Level

The laser power level refers to the power output of the laser itself, and as discussed above, this is only one important factor in determining the laser intensity at the sample.

The laser power level should be kept relatively low most of the time - i.e. only use as much laser power as necessary to obtain an acceptable image and at a minimal level sufficient to achieve stable output from the laser (set the detector gain to high and laser power to lowest stable position, then attenuate the laser intensity with an Acoustic Optical Beams Splitter (AOBS), Neutral Density (ND) or Acoustic Optical Tuneable (AOTF) filters, discussed in more detail below).

Under normal imaging conditions keeping the laser power level low not only helps to maintain the level of fluorescence in the sample and the integrity of the cells, but also significantly extends the life of the laser itself (particularly important for extending the notoriously short life span of the krypton-argon-ion laser).

Most manufacturers provide a continuously variable power level adjustment for the laser. However, the Bio-Rad instruments have a dual level power setting for the krypton-argon and argon-ion lasers. The level should be set to "low power" at all times (this is accomplished by flicking a switch on the laser of the MRC models or by Software control on the Radiance instruments), unless attempting real colour transmission imaging (see page 110 above).

The power adjustment on a laser is not easily reproduced – i.e. if you change the position of the dial to a previously used level then the power output of the laser will be only very approximately what the output was before the adjustment (use ND or AOTF filer adjustments when accurate laser intensity adjustment is required).

Switching Between Bright-Field and Confocal Imaging

One of the seemingly simple procedures in confocal microscopy, switching between normal bright-field light microscopy and laser-scanning, can often be quite frustrating when you start using a microscope with which you are not familiar. Although the position of the major optical elements of a light microscope will be recognizable no matter which microscope you use, the method by which various optical components are adjusted and brought in or out of the light path varies greatly between not only manufacturers – but even between different models from the same manufacturer!

A number of top-end research microscopes produced by the major microscope manufacturers are now fully automated and can be directly controlled by the computer used to drive the confocal microscope scan head. For example, when using a fully automated Zeiss microscope with the LSM 510 META confocal microscope scan head it is possible to switch between bright-field microscopy and laser scanning by simply pressing the appropriate "button" on the computer screen.

If you don't have the luxury of having a computer controlled microscope then take careful note of the slider or wheel positions needed to direct the light to either the scan head or the eyepieces. Take particular care to make sure that the optical filter blocks used in conventional epi-fluorescence microscopy are moved to the no-filter position when using the scan head.

Laser Intensity Level

The laser intensity level is the amount of laser light that reaches the sample. This is what is important in confocal imaging and is particularly critical in live cell microscopy.

ND or AOTF filters (see Chapter 3 "Confocal Microscopy Hardware", page 65) are used to control the laser intensity (as distinct from laser power above). These filters, in combination with the power setting on the laser are what determine the intensity of laser light that irradiates the sample. These filters are commonly changed through settings on the confocal microscope control software, but some instruments may require you to physically move a slider holding the ND filters. AOTF filters provide continuous adjustment of laser intensity between 0% and 100% (where 100% is determined by the current laser power setting). ND filters, on the other hand, provide step-wise adjustment of the laser power.

Adjusting Laser Intensity

AOTF – **Acoustic Optical Tuneable Filter,** *provides continuous adjustment of the laser intensity at the sample.*

ND Filters – **Neutral Density Filters,** *allow the selection of several laser intensity levels.*

See page 88 in Chapter 3 "Confocal Microscopy Hardware"

Starting with a relatively low level of light intensity is important so as to minimise laser damage to the sample. The ND filter is normally set to 0.3% or 1% for the first test scan when using an argon-ion or krypton-argon-ion laser, with settings of 30% or more of laser light very quickly fading the fluorescence in most samples (even in fixed samples containing antifade reagents). The red and green HeNe lasers are normally run at a higher intensity setting than the argon-ion and krypton-argon-ion lasers simply because these lasers normally have a lower maximum power rating. The HeNe lasers may often be operated at 50 to 80% of their maximum power.

An exception to always using a low power setting on the laser is when you are conducting FRAP (Fluorescence Recovery After Photobleaching) experiments, where you may need a significantly higher laser power setting to perform the bleaching part of the experiment. The best way to attenuate the laser intensity for pre and post bleach imaging when performing FRAP experiments is to set the laser power to maximum and then attenuate the laser intensity at the sample using the AOBS, AOTF or ND filters. If the laser power setting is not adjusted during photobleaching the laser intensity at the sample can be readily set back to the original imaging level for collecting the post-bleach image by altering the position or setting of the appropriate optical filters. When you change the ND, AOBS or AOTF setting the power level of the laser does not change – allowing a reasonable level of reproducibility of the intensity of the laser at the sample.

Take care not to bleach your sample while attempting to locate the cells – *keep the laser intensity as low as possible*

One final point on minimising the laser intensity at the sample is to use a relatively low level of laser intensity to image your sample by transmission imaging to first locate your cells, before switching to a higher laser intensity for fluorescence imaging – this can often be done by displaying both the transmission and fluorescence images simultaneously on the collection screen.

INSTRUMENT SENSITIVITY CONTROL

The sensitivity of the laser scanning confocal microscope is primarily controlled by adjusting the voltage of the light detecting photomultiplier tubes. However, some instruments have an additional signal amplification circuit that is called the "gain" level, and just to make matters more confusing, some instruments only have a single "gain" level adjustment that may include both the photomultiplier tube voltage and an additional amplifier circuit.

What really matters in the collection of an image is the signal to noise ratio (i.e. how strong is the "signal" compared to the amount of "noise" in the detector), and that every photomultiplier tube has a particular voltage and region of the light spectrum at which this ratio is highest. This means that there is an ideal voltage setting and region of the light spectrum at which all confocal microscope images should be collected. At a practical level one may not be able to collect the image at the "ideal" voltage, or have the luxury of using a photomultiplier tube with maximum sensitivity in the region of the light spectrum being collected (although most manufacturers do provide detectors for each channel that have maximum sensitivity in the region of the light spectrum generally collected with that particular channel). In spite of the fact that there is a photomultiplier tube voltage that has the best signal to noise ratio, most people simply adjust the gain on the instrument to create a reasonably bright image that makes full use of the available grey-scale levels (Figure 4-7) without any regard to obtaining the best signal/noise ratio in the image.

Gain Level or Photomultiplier Tube (PMT) Voltage

Changing the "gain" or photomultiplier tube voltage on a confocal microscope is now normally accomplished by software controls, but on some older confocal microscopes the gain control may be a physical knob on the scan head itself. The Leica confocal microscope photomultiplier tube voltage is software controlled, but one of the programmable array of physical knobs that sit on the desk can be set to adjust the photomultiplier voltage. The photomultiplier tube voltage is sometimes denoted in Volts (100s of Volts) or as a percentage of the total Voltage allowed.

An increased gain or photomultiplier voltage has the effect of creating a brighter image, and likewise decreased gain results in a duller image. The gain level (Figure 4-6) should be set on the confocal microscope such that the brightest areas of the image, or Region of Interest (ROI) being imaged, are displayed close to their maximum possible value (white on a grey-scale display). To display large areas of the screen at the maximum value will result in parts of the image not showing the correct grey level (higher light levels than the maximum that can be recorded will be counted as simply being at the highest level, and thus image information will be less). To help set the correct gain level and avoid saturation or under-illumination of the image or ROI use the SETCOL (sometimes called GlowOver, GlowUnder, or Range Indicator) colouring Look Up Table (LUT) - see Figure 4-13 on page 131.

The low signal setting (only present in the Bio-Rad MRC confocal microscopes) results in further amplification of the signal obtained from the photomultiplier tube.

Black Level or Offset Control

The black level (also known as off-set control) is used in conjunction with the gain level (or PMT Voltage) to optimise your image collection. The black level control sets the detected light value that you wish to call zero (Figure 4-6). The black level control sets the lowest brightness level in your image to whatever value you wish, but this would normally be set to a low level (below 10 on an 8-bit image), giving an image with excellent contrast. Setting the black level to zero (the default setting) on the computer interface control panel will normally result in the "black" or background areas of the image containing a low level of grey (almost black). This low level of signal in the background areas of the image is to be expected due to statistical noise in the light detection system of the confocal microscope. A good starting point is to set the background to a level slightly above "0", in this way a very low signal will not be accidentally lost by having the black level set too low.

Falsely setting the black level to achieve a grey-scale setting of "0" (black) in the background when there is a significant level of "background" fluorescence may appear to result in a "better" image, but will also result in the loss of low level labelling information in the image. However, a low level of background information may play havoc with attempts to quantitatively compare different areas of the image (see Chapter 5 "Digital Images in Microscopy", page 145). This is particularly noticeable when attempting to quantify punctate fluorescence within a defined region (for example within a single cell). A good way to collect images with a relatively high background is to collect the image with a significant "background" and then remove this background later if you wish to quantify the level of fluorescence in the image.

For specialised applications the black level may sometimes be set abnormally high. A very high set black level will result in an image with a high background, which may sometimes be beneficial for displaying lightly stained cells for positioning highly fluorescent points of interest within the cell. However, the image should be preferably collected with maximal information and any adjustments for highlighting particular structures being made after collection.

Digitisation

For each voxel (3D pixel) of the sample scanned, a number is generated depending on the amount of signal coming from the photomultiplier tube. This signal represents the total signal from the photomultiplier tube (true "signal" plus "noise"). Only in the photon counting mode is the signal used for digitisation representative of individual photons (and then, only at low light levels).

Each pixel in the image is assigned a value between 0 (black) and 255 (white) for an 8-bit image, or 0 (black) and 4095 (white) for a 12-bit image. Images with many more grey levels (16-bit or even 32-bit images) can be created, but the increased files size and increased computing power required for collection and manipulation is not currently feasible for confocal microscopy. Some image processing software uses a reverse numbering system for grey values within the image ("0" being white and "255" being black). The software will usually have a simple procedure to reverse or invert this system to show the image in the correct shading to that when the image was first collected.

> 8-bit images = 256 grey levels
> 12-bit images = 4096 grey levels
> 16-bit images = 65,536 grey levels

In many confocal microscope applications, where low levels of fluorescence are being detected, the number of photons is considerably less than the number of grey-scales available - even in an 8-bit image. For example, as mentioned previously low light level images may contain as few as 10 photons per pixel. This means that for most applications an 8-bit image has sufficient range to display the full range of grey levels present within the sample. However, where extremes of fluorescence are detected 12-bit (4096) grey-scale image collection may be an advantage. The greater grey-scale range will be particularly beneficial for image quantitation, but can also be used to create an image with an acceptable number of grey-levels even when the image was collected at a very low light level. However, there is no need to display more than 256 levels of grey, as neither the screen nor our eyes can handle even this number of grey levels!

If there were a very large number of grey levels available, and the hardware could readily handle the large dynamic range, there would be less need to set the gain and black levels on the instrument as all light levels could be digitised and then suitably displayed later. This feature may become available on future confocal microscopes, but is currently not practical due to the very large files sizes involved.

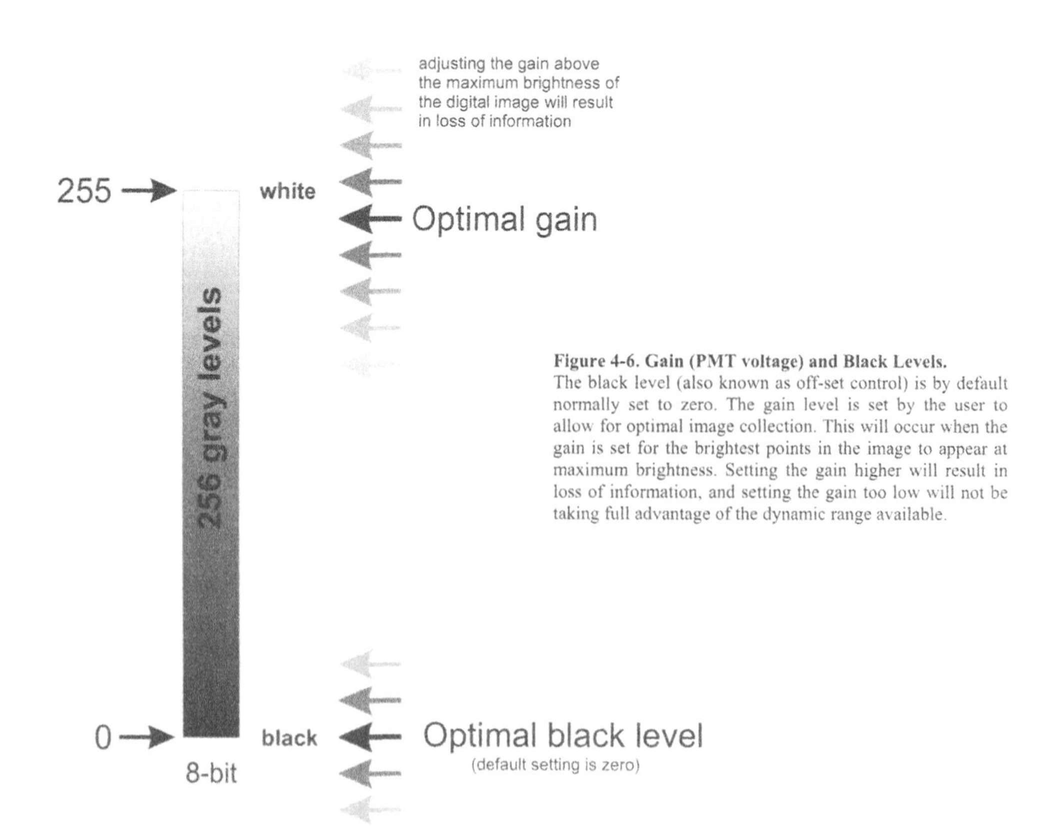

Figure 4-6. Gain (PMT voltage) and Black Levels.
The black level (also known as off-set control) is by default normally set to zero. The gain level is set by the user to allow for optimal image collection. This will occur when the gain is set for the brightest points in the image to appear at maximum brightness. Setting the gain higher will result in loss of information, and setting the gain too low will not be taking full advantage of the dynamic range available.

PINHOLE ADJUSTMENT

At the "heart" of the confocal microscope is the confocal pinhole, where out-of-focus light is removed, resulting in the formation of optical slices. This is the basis of the ability of the confocal microscope to image deep within the cell or tissue samples, and the key to the creation of 3D images of cells.

Alignment of laser light through the confocal pinhole was for many years a cause of much frustration both for the users and the manufacturers, but in more recent years the pinhole has been designed such that alignment has been done in the factory and should not require further alignment in the field, or that re-alignment is handled by manufacturer provided software and is now relatively routine. The Bio-Rad confocal microscope scan heads, in contrast to the other manufacturers, are designed around the principle of infinity optics, which allows for a much larger pinhole size. This greatly simplifies the alignment of the laser light through the pinhole, and a relatively large circular iris can be used to vary the size of the aperture. Although the alignment of the confocal pinhole is a lot less troublesome than in the past, some instruments still need frequent pinhole alignment for optimal performance – particularly when a new optical filter setting is used.

> **One Airy Disk "Button"**
>
> Make sure you have the correct lens selected "on screen" when you press the "1-Airy Disk" pinhole size selection "button"

The optimum size of the confocal pinhole depends on the NA and the magnification of the objective. Most confocal microscope manufacturers now have a software "button" that allows one to set the "optimal" pinhole size of one Airy disk, or in the case of the Zeiss 510 confocal microscope, to specify the optimal pinhole size for the z-step size selected (make sure you have the correct objective set in the software options). Although one Airy disk is often used, don't assume just because one Airy disk is considered "optimal" that this is the best setting for your experiment at hand. A good starting point for imaging is to "fully open" the pinhole to maximise the amount of light that can reach the detectors. Once the sample has been located the pinhole may then be closed further to improve the apparent resolution of the instrument by removing out-of-focus light. You will find that closing the pinhole further than one Airy unit (Figure 4-7, right hand image) will be one of many factors that may increase the apparent resolution of the microscope – but with a significant loss in the amount of light reaching the detector. Likewise, a further opening of the pinhole will result in some loss in resolution, but there will be a very large increase (approximately 10x) in the amount of light that can pass through to the detectors (Figure 4-7, left hand image). The image intensity displayed for "open", "one Airy unit" and "closed" pinhole in Figure 4-7 has been adjusted, by altering the gain setting on the instrument, to allow the three images to be shown at approximately equal intensity – the actual light level that reaches the detector is shown diagrammatically in the accompanying bar graph. The light intensity is approximately 10 fold higher for "1 Airy disk" compared to "closed" and a further 10 fold higher for an "open" pinhole.

A "fully opened" pinhole will eliminate most, but not all, of the optical sectioning capability of the confocal microscope. However, the exact dimensions of "fully open" varies for each of the confocal microscope manufacturers. In Figure 4-7 "fully open" means a physical opening of 600 µm, which is approximately 4.9 Airy disk units in the Leica SP confocal microscope used in these experiments. The pinhole opening of 4.9 Airy units has resulted in a relatively poorly resolved image in Figure 4-7 (left hand image) as these live cells have excellent 3D structure, with a lot of fluorescence in out-of-focus regions of the sample contributing to the background "haze" seen in the "open" pinhole image. In thin fixed samples this loss of z-axis resolution may be tolerated due to the relative lack of 3D structure in the sample (such samples can often be more easily, and with excellent results, imaged using a simple CCD camera attached to the microscope). Thus imaging fixed samples with a relatively open pinhole may greatly increase the level of sensitivity of the instrument without a large loss in resolution. Tissue samples, particularly relatively thick tissue biopsy material, cannot usually be imaged with an open pinhole as this will result in far too much

> **What is an Airy Disk?**
>
> An Airy disk is a diffraction pattern created by an object under the microscope. Rings of bright/dark pattern may be seen around small objects at high magnification. The amount of overlap of neighbouring Airy disks is one way to describe the limit of resolution of the microscope - when the two Airy disks merge the object can no longer be resolved (see page 50).

signal from out-of-focus regions of the sample. In the Bio-Rad Radiance confocal microscope a Signal Enhancing Lens (SELS) can be incorporated into the light path to direct more out-of-focus light to the detectors without the need to replace the existing confocal iris with a wider opening (see "Appendix 1: Confocal Microscopes - Bio-Rad Cell Science Division", page 364).

Having some understanding of the reason why a particular pinhole size is optimal for a particular lens is great, but you should also take a flexible approach to imaging and simply try collecting images with the pinhole at various size settings and observe what improvements, if any, are evident in the images you collect. Start with an open pinhole (greatly increasing the chances of finding the sample, as well as requiring significantly less laser intensity to create the initial image) and then close the pinhole down until sufficient detail is discernable within the image, or if present, select the "one Airy disk" setting on the instrument.

In dual or triple labelling experiments the images should be collected at the same pinhole setting for each channel. If a different pinhole setting is used for each channel then the "z" section represented by each optical slice will have a different "z" resolution for each of the channels. Different size "z" sections between channels will result in some confusion when the dual or triple images are merged as to whether you have co-localisation or just close proximity of the label. Ideally the pinhole size should be adjusted to take into account the wavelength of light used in each channel (one Airy disk = 0.6λ/NA, for fluorescence imaging – see Chapter 2 "Understanding Microscopy", page 31).

Optimal "pinhole" settings (Bio-Rad)

(based on 488 nm reflected blue light)

Objective	N.A.	"pinhole" size (mm)
10x	0.45	0.79
40x	1.3	1.09
60x	1.4	1.53
100x	1.4	2.55

The optimal "pinhole" size is based on the theoretical size of the Airy disk, calculated as shown below:.

$$\text{"pinhole" diameter (mm)} = \frac{73.2 \times \lambda \times Mag_{Obj}}{N.A.} \times 10^{-6}$$

λ = wavelength (nm)

N.A. = numerical aperture of the objective lens

Mag_{Obj} = magnification of the objective

Calculations for confocal microscopes from other manufacturers will show the same relationship between pinhole size, NA and magnification, but the size of the pinhole will be shown in "μm" instead of "mm".

Optimal "pinhole" size

- **One Airy disk sized pinhole:** the exact size of the pinhole depends on the NA and magnification of the objective. This setting is often the best compromise between resolution and sensitivity.
- **Very small pinhole:** A pinhole size below the size of one Airy disk severely limits light throughput – but does result in an apparent increase in x-y resolution due to further removal of out-of-focus light.
- **Large pinhole:** Results in almost a non-confocal image – but a large increase in light throughput (i.e. the sample will appear significantly brighter).

Note: when dual labelling collect both channels with the same effective confocal pinhole size (if possible, adjust the size for different wavelengths of light for critical applications).

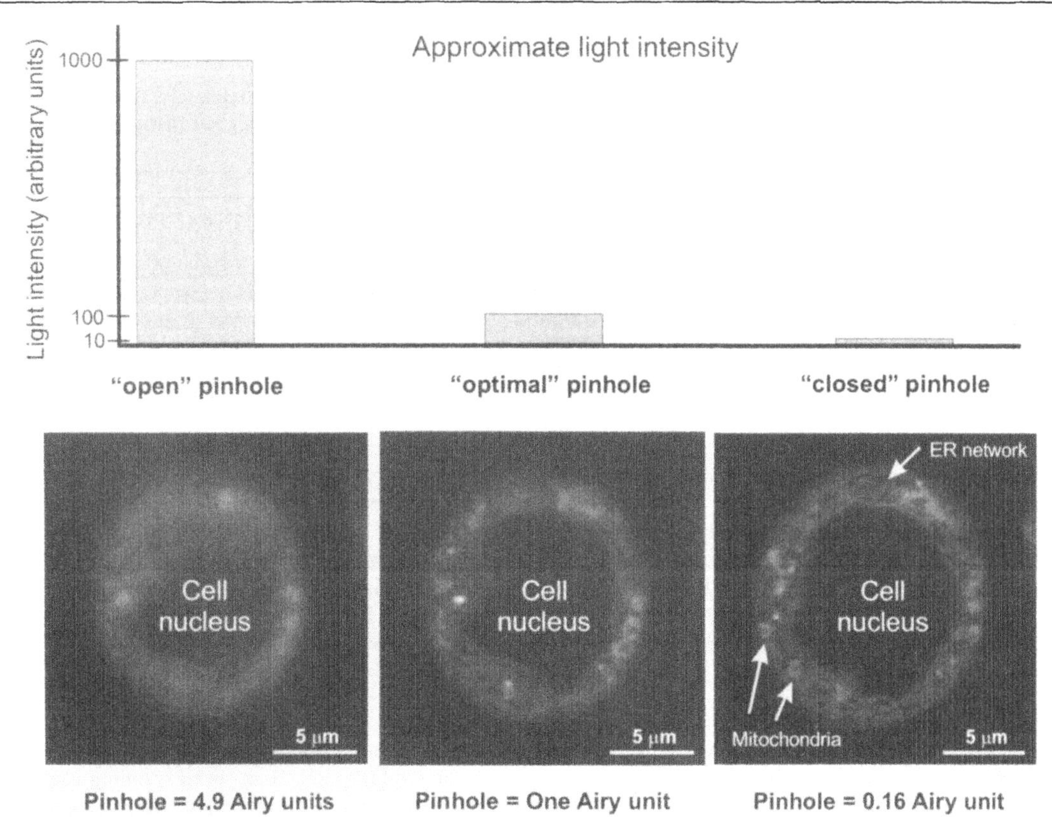

Figure 4-7. Pinhole Settings.
The confocal pinhole should be initially set to "open" (left hand image) to maximise the amount of light collected for "finding" your sample. This will minimise the laser intensity required, and thus lower the level of photo damage while imaging. The pinhole is often set to one Airy disk (centre image) after initially locating the cells with a more "open" pinhole setting. The physical size of the pinhole for one Airy disk will vary depending on the NA, magnification of the lens and which confocal microscope you are using. In this example a "fully open" pinhole is 600 µm, one Airy disk is 122 µm and a "closed" pinhole is 20 µm. Low levels of fluorescence may require keeping the pinhole open beyond one Airy disk to allow more light to reach the detectors at the expense of a loss in definition within the image. If you do have a high level of fluorescence you may be able to significantly increase the apparent resolution of the instrument by closing the pinhole further than one Airy disk (right hand image). Optical x-y resolution does not increase greatly on further closing the pinhole, but the apparent x-y resolution (i.e. the ability to distinguish features within the image) may be significantly increased due to the removal of out-of-focus light that greatly detracts from the quality of the image. The above images of HL-60 cultured suspension cells dyed with $DiOC_5$ were collected on a Leica SP laser scanning confocal microscope using a 63x 1.32 NA oil immersion lens with varying degrees of closure of the confocal pinhole. The images were collected from the same cell, but there has been some movement of cellular organelles between images.

The light intensity shown in the above images has been adjusted using the "gain" level on the instrument to give approximately the same light intensity for each image – the intensity of light reaching the detector is much greater in the "open" pinhole image compared to both the "one Airy disk" and "further closed" settings (shown diagrammatically in the above bar graph).

SCAN SPEED

The rate at which the finely focussed laser "spot" in a laser scanning confocal microscope is scanned across the sample is termed the scan speed of the instrument (usually between 100 and 500 Hz, but expressed in terms of microseconds per pixel in the case of Zeiss instruments). The rate at which a single image is collected in confocal microscopy (the frame acquisition rate) is also sometimes incorrectly termed the "scan speed", but the frame acquisition rate includes both the scan speed and other scan parameters such as line averaging.

Most laser scanning confocal microscopes can only scan at a relatively slow speed due to the scanning mechanism used to drive the scanning mirrors. Most confocal microscopes use galvanometer driven mirrors to scan the laser beam across the sample. However, a number of different scanning mechanisms have been implemented on various instruments over the years, including significantly faster resonant scanning mirrors.

Laser scanning confocal microscopes typically have a frame acquisition rate of approximately 1 frame / second (for a 512 x 512 pixel image with no line averaging). There are, however, specially designed instruments, such as the Leica TCS SP2 RS resonant mirror scanning confocal microscope that have significantly increased frame acquisition rates. Furthermore, the "high-speed" Nipkow disk based confocal microscopes manufactured by Yokogawa, PerkinElmer, Atto Bioscience and others, have a different design in which the scanning mirrors are "replaced" by a Nipkow disk that allows confocal imaging at very high speeds (over 100 frames / sec, but more normally up to 40 frames / second – depending on the CCD camera used). The high scan rates of the Nipkow disk based confocal microscopes are achieved by simultaneously scanning several hundred "spots" of light across the sample (see Chapter 3 "Confocal Microscopy Hardware", page 65).

Bio-Rad Scan Rates

SLOW - longer dwell time

NORMAL - single pass (500 Hz)

Fast (x2) - less pixels (faster y-scan)

Fastest (x4) - even less pixels (faster y-scan)

The line per second (lps) scan rate can also be altered by changing the slider in the top left corner of the control panel (see page 101).

Most manufacturers denote their scan speed by the frequency of the scanning mirrors (in Hz, or lines per second, lps), but be aware that Bio-Rad denotes the scan speed in an on-screen notation of "Normal", "Faster", "Slow" etc in addition to lines per second. These scan speed rates are documented in Hz in the Bio-Rad user manual (see boxed text above). However, the "FAST" scan rates on the Bio-Rad instruments are not a simple matter of both mirrors scanning faster, but are in fact an increased "y" scan rate and the formation of larger pixels.

Why have Different Scan Rates?

The faster the mirror scanning mechanism operates in a conventional laser scanning confocal microscope, the less time there is for the collection of the fluorescent light for each individual pixel. High-speed scanning is a great advantage when there is the need to follow fast physiological changes or to follow movement in living cells, but the disadvantage is that there may not be enough fluorescence signal to provide a sufficiently high signal above the inherent noise level of the instrument and the Poisson noise of low light levels. It is important to be able to slow down the speed of scanning to improve the quality of the images collected, although, as will be discussed shortly there are a variety of ways of increasing the quality of the signal collected, including choosing the appropriate scan rate, without directly changing the scan speed of the mirrors.

Leica Scan Rates

200 Hz – slow speed scanning

400 Hz – most common speed

800 Hz – increased scan speed

1000 Hz – further increased scan speed

The Leica scan rate may also be altered by implementing line or screen averaging.

A conventional laser scanning confocal microscope has an upper limit on the scan speed that is determined by the characteristics of the mechanism used to move the scanning mirrors. Unfortunately this is a relatively low speed in most laser scanning confocal microscopes. However, tips on how to collect images at significantly higher speed using a conventional laser scanning confocal microscope are discussed in some detail later in this chapter.

Is there an "Optimal" Scan Speed?

The laser scanning confocal microscope usually has an "optimal" scanning speed that gives a reasonable rate of scanning (about 1 frame per second for a 512 x 512 pixel image without averaging). In most laser scanning confocal microscopes a line scan speed of approximately 400 Hz is a good starting point for imaging. In the Bio-Rad confocal microscopes this speed is designated "Normal" scan rate (a scan rate of 500 Hz). Most instruments will default to this optimal value for scanning speed when the instrument is first turned on, and is normally the scan speed you will use for adjusting the image, although you may wish to use a faster scan rate for finding the cells of interest.

Decreased Scan Speed – for Improving your Image

Decreasing the scan rate (for example, down to 200 Hz on the Leica system, but denoted "SLOW" scan on the Bio-Rad system - at about 160 Hz) will result in a significantly higher level of fluorescence signal collected – usually resulting in a significantly higher quality image, i.e. an image that has less "noise". Noise in an image is seen as random points or speckle in the image. The image can sometimes be so "noisy" that it consists almost entirely of single pixels of high value against a low background (an image with a lot of "speckle").

Do you need to change the scan speed?
Greatly improved images can be obtained by line and screen averaging – without changing the speed of the scanning mirrors.

Decreasing the scan speed has the disadvantage of increasing both the fading of the fluorophore and tissue damage caused by the laser. This increased dwell time may need to be offset by lowering the laser intensity used. The scan speed, laser intensity and instrument gain all need to be adjusted relative to each other to determine the settings that give the "best" image (the highest resolution, the least "noise", and highest contrast) and result in the least amount of fluorophore fading.

Decreasing the scan speed results in considerable difficulty in moving the sample using the controls on the microscope stage. This is due to the relatively long "lag" between individual scans that appear on the computer screen. To facilitate moving around a sample try scanning first at a relatively high speed and then drop the speed down when a suitable area for imaging has been found. As mentioned previously, some instruments, such as the Zeiss laser scanning confocal microscope, have a "fast scan" mode for finding your cells of interest and then the image is collected using the "collection mode", which is at a significantly lower scan speed.

Use a slow scan rate or line averaging for imaging low fluorescence levels in live cells that may have some movement

Decreasing the scan rate to increase the pixel "dwell" time is not always necessary for obtaining an improved image. A more convenient way may be to collect either line averaged or screen averaged images as discussed in some detail below.

Slow Scanning for Live Cells

Slowing down the scan rate (or applying line averaging) can be used very effectively to collect high quality images from live cells (Figure 4-8). Relatively small movements within, and between, live cells can result in a very "blurry" image when screen averaging, but line averaging or slow scanning rates will result in a much higher quality image. This is due to the effective "dwell" time of the laser being increased. The collection time for each pixel in the image is only as long as the "dwell" time of the laser – whereas screen averaging may result in a considerable delay between collection of subsequent second and third passes of individual pixels.

Fast Scanning for Moving the Sample

High speed scanning rates, as mentioned above, can be most useful for initial positioning of the sample under the microscope. If the scan rate is increased to 800 or 1000 Hz the screen collection rate is significantly increased, lowering the amount of delay when you move the sample. Raising the scan rate will decrease the amount of signal for each pixel, resulting in a relatively "noisy" image, but seeing the location of the cell or tissue structure may be all that is required initially. The scan rate can then be slowed to collect a higher quality image of the area of interest. Furthermore, the high-speed scanning rate can be effectively used to locate the cells even with very low levels of fluorescence by using the transmission-imaging mode while scanning at high-speed. The very fast scan rates denoted "Fast" (x2) and "Fastest" (x4) on the Bio-Rad instruments and "Fast x-y" scanning on a Zeiss instrument are achieved by having a lower sampling density, resulting in less than 512 image pixels being displayed in a 512 x 512 box size. This becomes quite noticeable on the "Fastest" or "x4" scan rate, where the individual image pixels can be seen as small boxes within the image. The faster scan rate is achieved by increasing the scan speed of the frame (or Y) galvanometer, resulting in less "lines" in the digitised image (i.e. a 512 x 512 pixel image will have only 256 "lines", with each line duplicated to retain the original 512 x 512 image size). The pixels visible in the image are not removed by scan averaging.

"Low Signal ON" results in a greatly increased signal amplification when "SLOW" scanning on the Bio-Rad MRC series of confocal microscopes

Fast Scanning to Catch Movement

FAST scan is excellent when moving the sample while imaging

The laser scanning confocal microscope is relatively slow at capturing images. However, where speed of acquisition is important, such as following cellular or subcellular movement or studying fast ion changes within the cell, there are a number of techniques that can be used to greatly increase the scan speed. These techniques are discussed in some detail in Chapter 10 "Confocal Microscopy Techniques" (page 239). Briefly, these techniques include two main approaches. The first being "line scanning", which means that the sampling rate is as fast as a single-line scan (msec time frame), and is then displayed as an image with time in the vertical axis and the line scan at a fixed position within the tissue or cell in the horizontal axis. In this way very high speed scanning of a very narrow region of the cell can be obtained.

An alternative method of increasing the scan speed is to decrease the box size (i.e. decrease the number of pixels in the image), and then to use the confocal microscope zoom to increase the effective magnification of the region of interest within the sample to give approximately the same pixel resolution as that portion of the image in the original 512 x 512 or 1024 x 1024 image. In this way the speed of acquisition of images can be greatly increased, and is particularly suitable for following fast ionic (particularly Ca^{2+}) changes within the cell.

The Nipkow disk based confocal microscopes are particularly suited to high-speed scanning. As discussed above, the frame acquisition rate of a Nipkow disk confocal microscope is determined by the speed of the attached CCD camera. These instruments routinely collect images at greatly increased speed (10 to 20 frames per second) compared to a conventional single spot scanning confocal microscope (approximately 1 frame per second), and are capable of very high-speed scanning (over 300 frames per second).

Bi-directional Scanning

A laser scanning confocal microscope normally scans the laser and collects the emitted fluorescence during only one direction of the line scan. The laser is normally blocked on the return line scan (flyback blanking, see page 89) to avoid excessive bleaching of the sample. Some confocal microscopes are capable of doubling the frame scan rate by using bi-directional scanning. Bi-directional scanning can sometimes create problems of scan "offset" on each alternate line. This can be corrected on the Leica confocal microscopes by adjusting the "phase" control (usually programmed to one of the bench top knobs), or automatically, in the Zeiss confocal microscopes.

COLLECTION FILTERS

The default setting for image collection on a confocal microscope is continuous single screen scanning. This is the simplest manner in which to display an image "live" while making adjustments to the microscope. However, on a number of instruments these "live" images are not saved to a computer file and furthermore, single scan images may not be of sufficient quality. There are a number of other modes of image collection (referred to as collection filters) that may be used to significantly improved image quality. A number of image processing filters can also be applied to the image after collection (see Chapter 5 "Digital Images in Microscopy", page 145).

Continuous Single Frame Image Collection

Single frame image collection as mentioned above is the default image collection mode on most confocal microscopes and is a useful mode for finding suitable cells for analysis, for focusing the microscope and for finding the optical slice of interest within the sample. This mode is also used to establish the upper and lower limits for z-sectioning. Once a suitable part of the sample has been determined, a number of "filters" can be used to greatly improve the collected image.

Single Frame Capture Mode

Some instruments have a separate "single frame capture" mode for collecting images that you wish to save. On these instruments (for example the Leica confocal microscope) the image displayed by the continuous single frame scanning mode cannot be saved directly. The "single scan" button, usually on the right hand half of the Leica control panel will result in the scanned image being saved to a file. This "single scan" button will collect an image not only at the pre-set scan speed, but also with the number of line or frame averages that have been previously set.

Line Averaging

In the line-averaging mode each line is scanned a specified number of times and the average displayed. Line averaging will slow the apparent speed of image collection (the exact speed will be dependent on the number of line averages specified). Line averaging, just like screen averaging and Kalman, discussed below, will significantly reduce the "noise" level in the image, often resulting in a significantly "better" image. Line averaging also has the added advantage of not being as affected by movement as screen averaging (see discussion under "Slow Scanning for Live Cells" above). Line averaging, in a similar way to decreasing the scan speed, can be used to great advantage when imaging live cells to increase the amount of detail in the image, and as long as the movement is relatively slow the image will not appear distorted (see Figure 4-8). If some movement of the object occurs during scanning with line averaging the resultant image will be slightly distorted, but remain reasonably well "focussed". This is in contrast to screen averaging, discussed below, where even very small sample movements will seriously degrade the sharpness of the image (see image "D" in Figure 4-8). Line averaging is considered superior to frame averaging even for fixed and embedded samples because any movement or instrument vibration during image collection has less affect on the quality of the image.

Screen Averaging

Most confocal microscope manufacturers provide direct screen averaging (as distinct from Kalman averaging, explained below). The on screen image that you see is an average of all images collected so far (you can set the number of images to be collected). Averaging is very effective in removing unwanted "noise" from the image by greatly improving the signal to noise ratio. This is particularly noticeable when attempting to image very low light level samples. In fact a sample that will initially hardly show any fluorescence with a single scan may become an excellent image after a number of screen averages. The image may need to be "contrast stretched" for optimal viewing (see Figure 5-12 on page 158). Screen averaging is best done on fixed samples as very small movements of the object will result in a significantly "blurred" image (see Figure 4-8). Line averaging, discussed above, is a much better alternative when imaging objects with some movement – such as live cells.

Kalman Averaging

Some confocal microscope manufacturers provide an alternative screen averaging collection filter called Kalman averaging, where there is a bias towards the last collected image. Images collected in this mode are averaged as they are collected, up to a maximum collected of 255. There is often a dramatic improvement of a low light level image during the first 10 to 20 images collected. Beyond 30 images collected there is rarely any further image improvement. However, when the fluorescence level is particularly low a discernable improvement in image quality may be apparent even at more than 100 images averaged.

The weighted bias towards the last image collected when Kalman averaging is an effective way to follow the movement of small objects, such as subcellular particles, within the image. Kalman averaging will create a ghosting effect of the moving cell or particle.

Kalman averaging, like screen averaging, is not normally used to collect high-resolution images of moving cells, as this will result in a blurring of the image. If cell movement is a problem then a slower scan rate (or line averaging), as discussed above, may greatly improve the image and significantly increase the sensitivity without unduly distorting the image.

Collection to Peak

In this collection mode the images collected are summed until any part of the image becomes a value of 255 (white) in an 8-bit image or 4906 in a 12-bit image, at which stage collection stops and you can save the image. The images are not averaged, in contrast to screen and Kalman averaging discussed above, but are simply summed together. This type of collection is valuable for low light level samples, but does not lower background "noise" as efficiently as Kalman, line average, or screen average collection.

Collection to peak is particularly suited to low light level samples that have a relatively even spread of low level fluorescence. If a small part of the image (even a single pixel) is excessively bright, image collection will be terminated before the low level of fluorescence in the remainder of the image has become visible.

Exponential Collection

In this mode the confocal microscope will continue collecting and summing images until the scan is stopped, or the number of images specified is reached. This method of collection is useful for low light level images that have a part of the image that is relatively bright (perhaps an area of non-specific contaminating fluorophore). This type of image would not be suitable for "Collection to Peak ", described above, as the low level staining of interest would still not be visible when collection is stopped, when the part of the image having excessively high fluorescence reaches a peak level of 255 (or 4096 in a 12-bit image). Exponential collection is often the method of choice when using the Photon Counting mode.

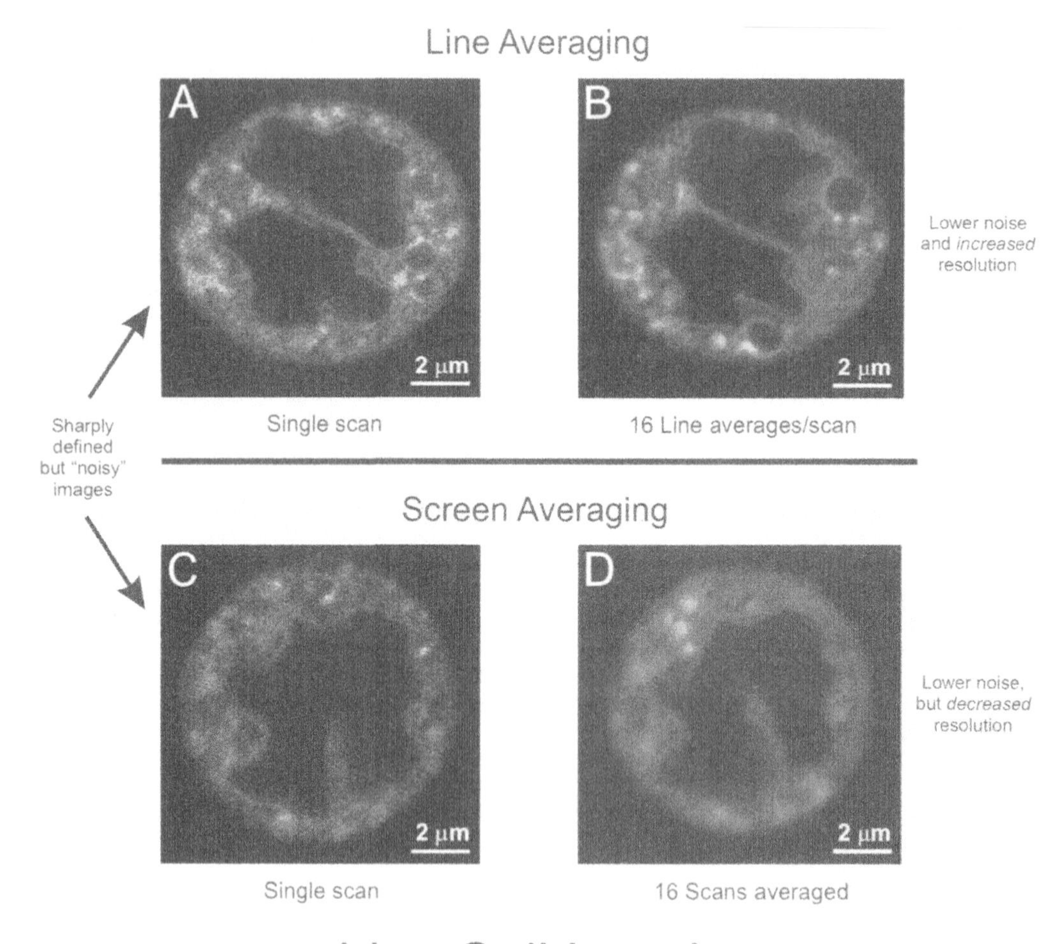

Figure 4-8. Line verses Screen Averaging for Live Cell Imaging.
Line and screen averaging greatly improve the signal to noise ratio, thus improving the "quality" of the image. This means that finer detail can be resolved (as shown in image "B" above, compared to the single scan in image "A"). However, when imaging live cells there is always a small amount of movement that will destroy the resolution of the image if screen averaging is used (compare image "B", a 16 line-averaged scan with image "D", an average of 16 frames). In live cell imaging line averaging is a better way to improve the quality (higher signal to noise) of the image. Live HL-60 cultured suspension cells were labelled with $DiIC_5$ and these images collected using a Leica SP laser scanning confocal microscope using a 63x 1.32 NA oil immersion objective.

IMAGE COLLECTION BOX SIZE

512 x 512 pixel image size –
a good starting size for most experiments

The image in a confocal microscope is collected into a set box size on the computer screen. This box size can be changed from 128 x 128 pixels (or even smaller if required), up to a maximum box size of 2048 x 2048 pixels (Figure 4-9) on most instruments. The smaller the box size the smaller the computer file generated, and the faster the image can be collected. Most image collection is carried out at a box size of 512 x 512 pixels (the default setting on most instruments). This allows one to have almost 4 complete images on the screen together at 1:1 or screen resolution (if your screen is set to a display resolution of 1024 x 768 pixels or higher), which is most useful in multiple labelling experiments.

When collecting into a box size of 1024 x 1024 pixels or larger the image is not fully displayed on the computer screen. Moving the scroll bars will be necessary to display the area of the image that does not fit on the screen. This is particularly noticeable when displaying multiple images from a dual or triple labelling experiment on the screen simultaneously – in which case a large portion of a 1024 x 1024 pixel image will not be shown on the screen. On most instruments the image box size can be displayed either directly at screen resolution (i.e. as a 1024 x 1024 image that may not quite all fit on the display screen) or as an "automatic" size which will resize the image to fit all of the image on the screen (but the image is now not displayed at screen resolution, but as a smaller image). Great care should be taken to note which method you are using, as accidentally, for example, using the "auto" sizing function to display two dual labelling images when collecting at 1024 x 1024 will result in the images being

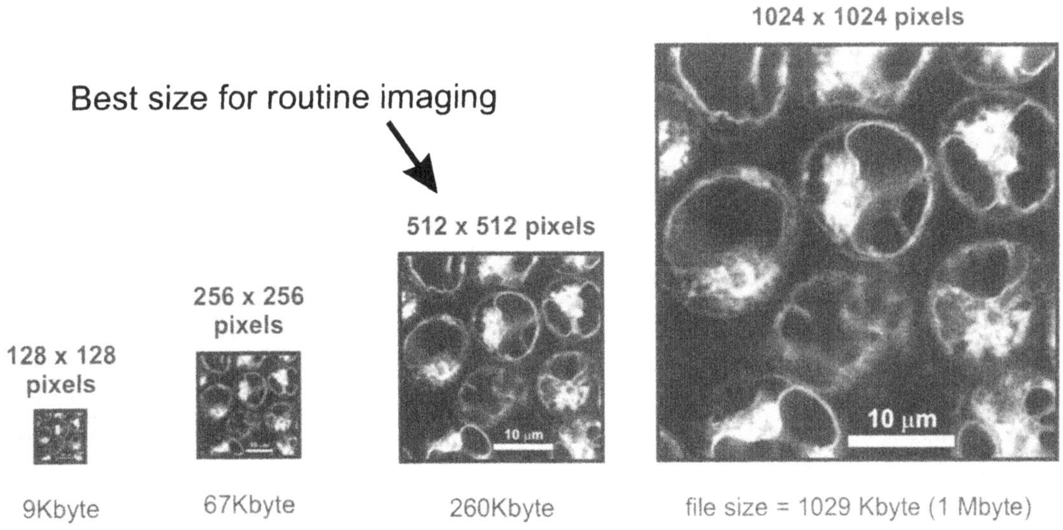

Figure 4-9. Image Collection Size (Box Size).
Images can be collected on a Bio-Rad MRC-600 confocal microscope using a Zeiss 63x 1.4 NA oil immersion objective lens with different image box sizes (the number of pixels in the image). The normal collection size is 512 x 512 pixels, with the maximum box size often being larger than 1024 x 1024 pixels. The smallest box size may be as small as one pixel. In many confocal microscopes the shape and orientation of the collection box can be altered to orientate the scan position with the object being studied. The above images are of a single optical slice through live human MOLT-4 cells labelled with NBD-ceramide, a lipid probe that concentrates in the Golgi apparatus, but also stains most intracellular membranes.

displayed at an image box size of slightly less than 512 x 512 pixels – which will result in some loss of on screen information as the image size is reduced simply by removing pixels. However, when the size of the image displayed on the screen is less than the set box size, the image is still being collected and saved at the set box size and is simply being displayed at a smaller box size to fit within the active window on the screen. Take care when optimising the image in critical applications that you are working at screen resolution.

> Use a large box size for image collection if you intend to enlarge a small area of the image later for presentation

When working at screen resolution on images with large box sizes, where only a part of the image is displayed on the screen, can be confusing. This can sometimes result in optimising the image composition (for example, the position and size of individual cells) for only the portion of the image visible on screen.

The 1024 x 1024 pixel box size creates a relatively large file (1 Mbyte for a single image), this is of particular concern when collecting a series of optical slices. Therefore care should be taken to only collect large box sizes if the increased number of pixels is needed. The advantage of a larger box size is that the image being collected with more pixels, can be later cropped or particular regions of the image can be further enlarged for presentation and publication.

Changing the image collection box size does not normally change the area of the sample that is being scanned. However, take care as some box size changes (particularly when changing the aspect ratio of the scan box) can result in a change in the zoom factor of the microscope. For example, changing the box size from 512 x 512 to 512 x 768 on some confocal microscopes will result in an apparent change in the magnification of the microscope – in addition to the expected increased scan area! If size comparisons are important make sure you take careful note of either the pixel size or the field of view for a given number of pixels.

An increased scan box size increases the number of pixels collected, and so will appear as a larger image on the screen. This effectively means that you have collected more pixels per unit area across the scanned area of the sample, increasing the resolving power of the microscope (Figure 4-9). The larger the box size the more the image can be later enlarged by suitable software without the pixels becoming discernable. However, as will be discussed below under "Optimal Image Collection" the resolution obtained in a confocal microscope is determined by a combination of the number of pixels and the confocal microscope zoom factor employed. Simply having more pixels in the image does not necessarily result in a higher resolution image of the original object.

The 512 x 512 pixel image size will display well as an almost full screen image (at screen resolution) on a large lecture theatre screen using a digital projector, without the individual pixels being discernable. The higher resolution of the printed page means that a 512 x 512 pixel image will be optimally displayed at a size that fills one column of a printed A4 sized page. A higher resolution image (for example 1024 x 1024 pixels) will not be advantageous for display or printing, unless a small selected area of the image is required to be displayed, or the image will be printed at a size large than approximately 8 x 8 cm.

As discussed previously, a smaller box size from the standard 512 x 512 pixels is often used to collect images at relatively high speed (up to 10 frames/sec in a conventional laser scanning confocal microscope). This is because the scan speed of the mirrors is the same when you make the box size smaller, but the area of the scan is considerably reduced. The greatly increased speed is useful when following physiological responses in live cells as discussed in some detail above.

Most laser scanning confocal microscopes allow one to choose a box size of various dimensions, even down to a size of only 1 pixel in any one direction (i.e. you can have a box size of 1 x 512 pixels, 512 x 1 pixel or even 1x 1 pixel). The scanned area can also be rotated on many instruments so that you can scan a narrow defined area of interest, thus limiting the area of the sample that is exposed to damaging laser light.

The collection box size on a Nipkow disk confocal microscope is determined by the pixel size of the CCD chip and the total zoom factor of the camera used. There is thus less flexibility in manipulating the collection box size in these instruments. Nipkow disk confocal microscopes are capable of digitally zooming the image *after* collection, but due to the CCD camera constraints on image capture these instruments are not capable of increased resolution by decreased scan area as described here for laser scanning instruments. The zoom factor in a Nipkow disk instrument is determined by the microscope objective used, whether binning is operating and the magnification of the optovar image magnifier) in use.

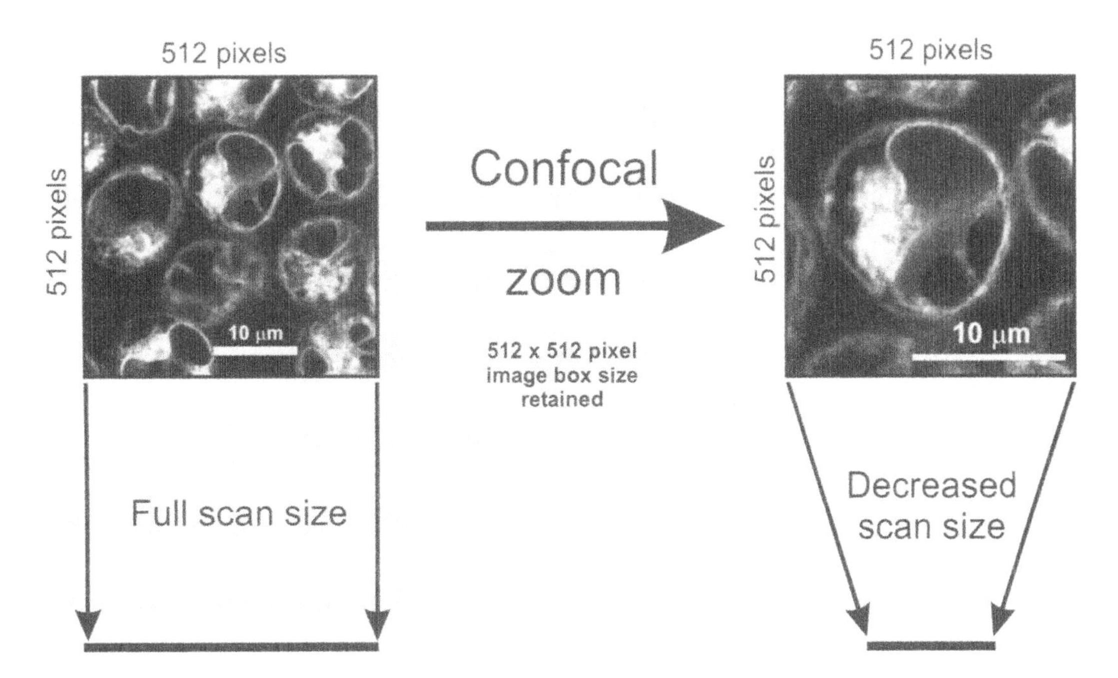

Figure 4-10. Scanning Zoom.
Decreasing the scan size, but collecting the image within the same box size, results in increased magnification and increased resolution. The ability to alter the zoom level on the confocal microscope allows one to readily match the number of pixels with the optical resolution of the lens being used (the Nyquist sampling criteria). These images show NBD-ceramide labelled live MOLT-4 cells as in Figure 4-9, page 123).

CONFOCAL MICROSCOPE ZOOM

One of the advantages of laser spot scanning confocal microscopy is the ability to zoom in on a particular area of interest within the sample. The ability to zoom not only increases the effective magnification of the microscope, but also allows one to image only the area of interest within the field of view (for example, a single cell). An important effect of using the zoom function is to allow one to match the resolving power of the objective lens with the number of pixels used to display the data (the Nyquist sampling criteria, discussed in more detail shortly).

There are two important types of zoom involved in confocal microscopy that should not be confused as they both have very different effects on the resolution of the image. Scanning zoom, performed by scanning a smaller area of the sample, increases the magnification and the resolving power of the microscope. On the other hand, digital zoom, carried out after image collection, increases the image display size, but does not result in an increase in the resolving power of the instrument.

> **Scanning zoom results in "true" magnification of the image**

Scanning Zoom

The first method of zooming, which involves scanning the laser over a smaller area of the image (Figure 4-10), creates a real magnification by simply displaying this decreased scan area within the same size box on the computer screen. The effect is that the 512 x 512 pixels of the box display are now obtained from a smaller area of the sample, increasing the resolution of the microscope.

The amount of zoom that results in a real increase in resolution is limited by the theoretical limit of resolution of light microscopy (0.1 to 0.2 µm, when using a high NA lens), and the "noise" level within the image. Zooming the image beyond about 6x results in a larger image, but no further increase in resolution. This is referred to as "empty magnification".

The confocal microscope zoom feature is often used on a slightly lower magnification lens (for example a 60 x), to increase the magnification by confocal microscope zooming, rather than using the higher magnification (100x) lens. This is due to the fact that the slightly lower magnification 60x lens may be of the same, or even higher NA compared to the 100x lens, thus having the same or even increased resolution and increased light gathering capacity. A further advantage of the lower magnification 60x lens is that when the zoom is set to the lowest value (usually 1, but can be 0.7) there is an increased field of view compared to the higher magnification 100x lens.

Make sure your image is displayed at screen resolution when high-resolution imaging

One important consideration when using the zoom feature on the confocal microscope is that the same amount of laser light is now scanned across a smaller area of the sample. This has the effect of increasing the light intensity within the area being imaged. A high zoom level can result in a very significant increase in laser light intensity at the sample and may thus result in fast fading of the fluorophore.

Panning While Zoomed

When using the zoom function the area of the sample under observation can be readily "panned" by controls on the computer. In this way fine adjustments in positioning the image can be made without touching the x-y controls on the microscope stage. This "panning" control does not have any effect when the confocal microscope zoom function is not being used (zoom factor of 1, or 0.7 on some instruments).

Zooming for Optimal Resolution

High-resolution imaging by confocal microscopy may necessitate adjusting the zoom factor to suit the resolving power of the microscope objective in use in order to satisfy the Nyquist criteria for correct sampling. The Nyquist Sampling Theorem states that, when a continuous, analogue image is digitised, the information content of the signal will be retained only if the diameter of the area represented by each pixel is at least 2.3x smaller than the optical resolution limit of the microscope (see on-line tool in box below). This means that a 100x 1.4 NA objective with a theoretical resolution limit of 0.2 µm will require a pixel size of 0.08 µm.

Zoom Settings for Optimal Resolution

Obj	10x	40x	60x
NA	0.45	1.3	1.4
Zoom	7.2x	4.9x	3.7x

Settings for a microscope with a tube magnification of 1x and a raster size (box size) of 512 x 512 pixels

Most confocal microscopes now display the size of the image pixels on the control panel display. However, if the size of the pixels is not displayed try establishing the size of the pixels by counting the number of pixels in a displayed scale bar. For example, a 5 µm scale bar should be 62 pixels long for a pixel size of 0.08 µm. Alternatively, one can adjust the zoom factor until there are approximately 4 pixels across the smallest object discernable within the image. A less precise, but very effective method of determining the correct zoom setting is to make adjustments while carefully examining the image. You will quickly notice that zooming beyond a particular setting does not increase the detail discernable in the image.

The inter-plane spacing between individual optical slices, using the same optical conditions should be about 3x larger than the x-y spacing (i.e. approximately 0.2 to 0.3 µm step size for the above example using a 100 x 1.4 NA objective).

The optimal zoom factor is dependent on both the NA and the magnification of the objective (see separate box). Note that the low magnification 10x objective requires a higher zoom factor compared to the 100x objective for displaying the full resolution limit of the objective.

Online tool to calculate sampling criteria:
www2.bitplane.com/sampling/index.cfm

Over Sampling and Under Sampling

Using a lower zoom setting (under sampling) has the advantage of increasing the field of view and the amount of fluorescent light gathered per pixel. This will not be optimal for resolving small objects within the image, but will result in reducing the rate of photobleaching and greatly increase the number of cells, for example, that can be imaged in the same display screen.

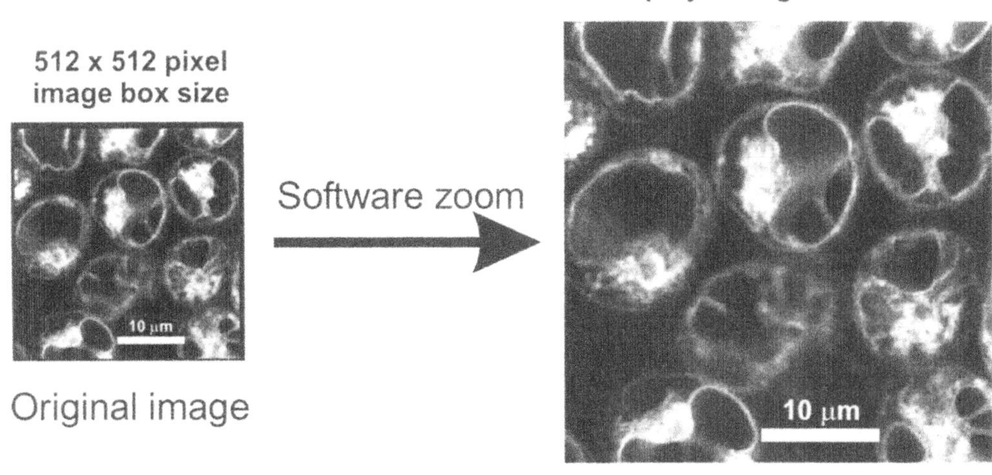

Figure 4-11. Digital Zoom.
Increasing the size of the image via software does not increase the resolution of the image, the same number of pixels are simply spread over a larger area by enlarging the size of the individual image pixels. NBD-ceramide labelled live MOLT-4 cell images were collected as described in Figure 4-9 (page 123).

Using a higher zoom setting than strictly necessary for high-resolution imaging (over sampling) will produce a larger image that may be easier to view and will provide additional data that could be suitable for deconvolution, but may result in significantly more fading of the fluorophore.

Digital Zooming after Image Collection

The other method of zooming is to increase the size of the image after collection (Figure 4-11). Software, or digital zooming, increases the size of the image, but does not increase the resolution of the image. Software zoom is often useful when preparing an image for presentation or publication. Either an increased size is required (for example when making a seminar presentation), or a cropped image of particular cells shown at an increased size (perhaps for a figure in a publication).

Software zooming is very effective as long as the image is not increased in size to such an extent that the image pixels become visible. Using some computer software the pixels will not become visible because the image will be "interpolated" (ragged pixels smoothed over) on changing the size, but this does not increase the resolution of the image. In fact, interpolation of an image "blurs" the image to remove the sharp edged original image pixels.

Collecting the image with a large box size (for example 1024 x 1024) will allow for a lot more zoom during later processing compared to an image collected at a smaller box size (for example 128 x 128). As discussed above a box size of 512 x 512 is found to be a reasonable compromise between resolution and files size.

Software zooming simply increases image size – with no increase in resolution

REGION OF INTEREST (ROI) SCANNING

Confocal microscopes equipped with acoustic optical tuneable filters (AOTF) are capable of region of interest (ROI) scanning (see page 89 for a description of how an AOTF filter works). Regions of interest can be defined by drawing an area on the imaging screen. The area can be a defined rectangle or circle, or even a randomly drawn shape. Several areas can be drawn, often at any angle or shape and imaged simultaneously. A small defined ROI will be scanned significantly faster than a full sized image. Furthermore, only a small region of the specimen will be subjected to high intensity irradiating light.

When scanning a defined ROI the scan speed and distance scanned by the x-mirror is not altered, whereas the y-scan is limited by the ROI defined. For example, a small ROI marked in the middle of the imaging area will result in a scan speed that is equivalent to a large horizontal imaging box size that has the same height (y-dimension) as the ROI. Several small ROI aligned horizontally across the screen will scan at relatively high-speed as the y-scan mirror only has to traverse the size of the largest ROI defined. However, if several ROI are defined in a vertical orientation then the y-scan mirror will have to complete a scan from the top of the highest ROI to the bottom of the lowest ROI (which may involve scanning the whole of the defined imaging box size).

IMAGE COLOUR

The colours seen in images obtained from a confocal microscope are entirely false, that is the grey-scale image obtained from the microscope is coloured by the computer for presentation only. Even the lovely multi-coloured images in publications are in fact grey-scale images that have been simply coloured for publication. Colour in digital images is discussed in more detail in Chapter 5 "Digital Images in Microscopy" (page 145). This section is concerned with how colour is utilised in confocal microscopy to clarify the presentation of complex data.

The ability to colour the images obtained from a confocal microscope means that colour can be used to great advantage to represent a particular fluorophore or to emphasize a particularly interesting degree of labelling within the sample. Conventionally green is used for green fluorescent dyes (for example fluorescein, Alexa Fluar 488 etc.), red for red emitting fluorescent dyes (for example, tetramethyl rhodamine, Alexa Fluar 568 etc) and blue for far-red fluorescent dyes. However, the computer can colour the images any colour you like – so don't assume that a "green" image is a "green" fluorophore.

A green colour gradient, for example dark green, light green, yellow and white is very useful for clearly showing labelling in subcellular structures. The same image coloured only green will often be very difficult to see clearly (particularly for colour impaired individuals). This becomes quite critical when presenting images in a seminar (green alone does not show up well on a lecture theatre screen) or for publication (again, green alone does not print well). The grey-scale images obtained from a confocal microscope can be coloured either during, or after collection, by using a colour LUT (Look Up Table). The image can be later converted to a 24-bit full colour image if necessary.

Dual labelling is routinely displayed as one chromophore shown in red and the other in green. This colouring scheme is conventionally used to represent red and green fluorescent dyes, but the actual individual dye emission spectrum may be considerably removed from the pure green and red used to make up the dual labelled image. Red and green colouring is chosen simply because the computer displays all colours on the screen as a mix of the primary colours (red, green/yellow, and blue). Using red and green as the colours for dual labelling greatly simplifies the manipulation and particularly the merging of colour images. However, displaying all three channels in grey-scale during image collection for critical applications is recommended so that the intensity setting of the individual channels is not influenced by the sensitivity of your eyes to different colours (this is particularly critical for people with impaired colour vision). The "yellow" colour observed in dual labelled samples is simply a combination of red and green as shown on the computer screen, and does not represent "yellow" light being emitted by the sample.

During multi-channel image collection on a confocal microscope a composite image using different colours to represent the different collection channels is often displayed alongside the grey-scale or single colour individual channel images. In this way an initial assessment of the degree of labelling overlap within the two or more probes within the sample can be made. Unfortunately this multi-coloured merged image is not normally saved when saving the newly acquired images. Merged multi-coloured images can be created later using either the original confocal microscope software or a variety of other image processing software (see Chapter 6 "Imaging Software", page 163), including the widely available Photoshop program.

In Figure 4-12 a colour gradient LUT (GEOG) has been used to good affect to demonstrate the uptake of fluorescent dye by live malarial parasites. In the "GEOG" colour LUT low levels of fluorescence are depicted in dark blue, with progressively "warmer" colours (green, yellow, red then white) being used to great affect to emphasize areas of the cell with increased fluorescence. After 40 minutes incubation in the presence of the fluorescent dye the food vacuole of the parasite is shown as a strongly staining spot (coloured red by the colour gradient). The position of the colour in the gradient can be readily adjusted, for example the red dot at 40 minutes could be coloured with the white part of the gradient if further emphasis of the level of dye uptake is required.

Using colour gradients can greatly enhance the message you are trying to convey with your images. However, use colour sparingly in a presentation. For example, too many different colour gradients in a presentation can greatly detract from your overall message.

The advantage of using a Look Up Table rather than colouring the image directly (by creating a 24-bit full colour image) is that the colour can be easily changed, and as we will see later, colour in confocal microscopy can be used to display your data in a much more informative and imaginative way than simply providing a "single" colour image for presentation. Colour in digital images is discussed in more detail in Chapter 5 "Digital Images in Microscopy" (page 145).

Using Colour to Optimise Image Collection

A very useful colour LUT for optimising image collection is the special "SETCOL" colour look up table, also known as GlowOver or GlowUnder by some manufacturers (Figure 4-13). This LUT colours your image mainly in shades of grey, but the very low intensity levels (black and near black) are coloured green, and the very high intensity levels (white and near white) are coloured red. There are some variations between the various manufacturers, but the principle remains the same, in that high and low pixel values are coloured in a colour distinct from the bulk of the image pixels. This type of colour LUT can be used to very effectively keep your image collection within the limits of the digitised image.

Use SETCOL (or GlowOver) to optimise image collection

When the image is collected with the SETCOL active you should adjust the gain on the confocal microscope such that the image shows almost no red (just a few pixels tipping red), and almost no green (again, just a few pixels tipping green). The bulk of the image will be displayed as a grey-scale LUT, which has the advantage of being the most suitable "colour" for our eyes to detect relevant details. Using SETCOL you can be confident that your image has been collected not only between the two limits required by digitisation, but that you are taking full advantage of the full range of grey levels available.

After using SETCOL for a while you will find that your judgment of correct grey levels will be sufficient for most applications without using SETCOL. However, in particularly difficult or critical imaging situations SETCOL will be a valuable aid to collecting optimised images.

For optimal image collection - display images in grey-scale while collecting

The colours in images obtained from a confocal microscope are computer generated

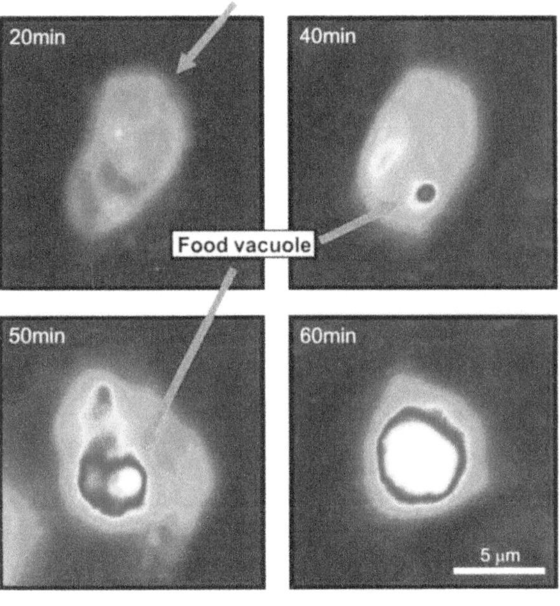

"GEOG" Colour Look Up Table (LUT)

Increasing fluorescence intensity

Figure 4-12. Colour in Images.
All colours in images collected on a confocal microscope are computer generated. The original image collected from the microscope is a grey-scale image, with the colour being applied by using a colour Look Up Table (LUT). In this figure a colour gradient (GEOG), which provides a gradient of colour from cool blue colours to hot red and then white colours, is used to emphasize the areas of the cell that is showing increased level of label with time. This image is of a live human malarial parasite labelled with trace amounts of residual fluorochrome that are released from fluorescently labelled latex beads (microspheres). The fluorescence intensity of the parasite was followed over a period of 60 minutes. These images were collected on a Bio-Rad MRC-600 confocal microscope using a Zeiss 63x 1.4 NA oil immersion objective lens. This image is reproduced with further annotation from Hibbs *et al* (1997) – see page 349 in Chapter 15 "Further Reading" with permission from Wissenschaftliche Verlagsgesellschaft, Stuttgart.

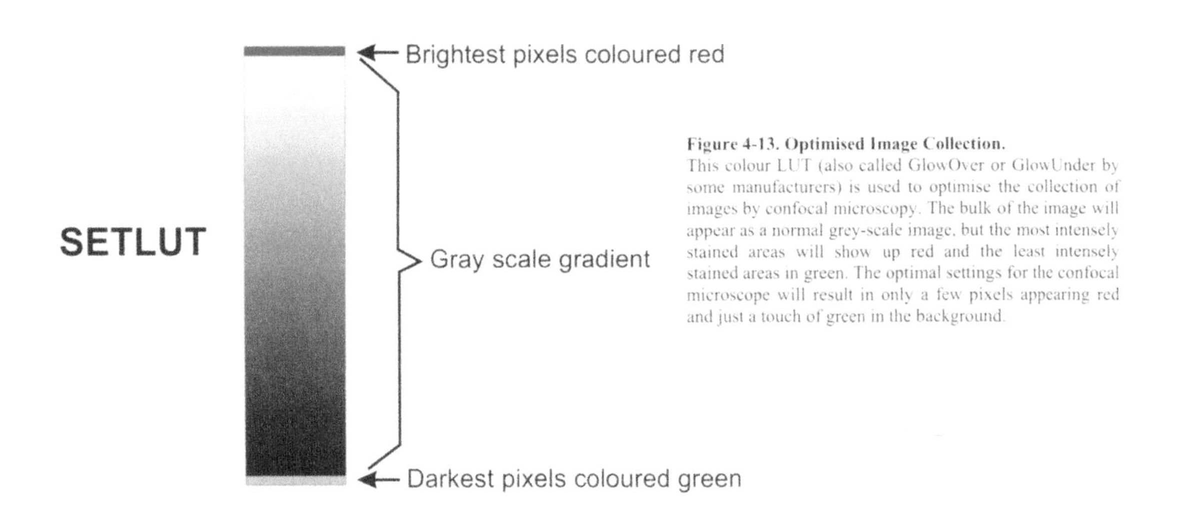

Figure 4-13. Optimised Image Collection.
This colour LUT (also called GlowOver or GlowUnder by some manufacturers) is used to optimise the collection of images by confocal microscopy. The bulk of the image will appear as a normal grey-scale image, but the most intensely stained areas will show up red and the least intensely stained areas in green. The optimal settings for the confocal microscope will result in only a few pixels appearing red and just a touch of green in the background.

OPTIMAL IMAGE COLLECTION

Optimal image collection is a "balancing act" between many conflicting settings on the confocal microscope (see Figure 4-14). The initial aim is to locate the region of the sample or cells of interest using as low a laser power as possible (with a high gain setting and a relatively "open" pinhole) -- but don't stop there! If you want to obtain maximum information from high-resolution images you will need to spend some time adjusting the pinhole, gain and zoom on the confocal microscope.

The Zeiss 510 confocal microscope has a "Find" button that quickly optimises the instrument gain setting for your sample by using fast line scanning. This button gives a good first approximation of the correct gain setting, but there are a number of other important variables that may need to be adjusted. Take care to make sure the laser power and intensity levels are set low and the pinhole is opened before attempting to use the "Find" button, otherwise you may inadvertently photo-bleach your sample. Other instruments don't have an automated object "Find" function and so all the starting parameters that you set on the instrument are entirely of your own choosing. Correct adjustment of the pinhole, zoom, laser intensity, gain and black level, even after you have obtained an image will be necessary for collecting high- resolution images that take full advantage of the instrument.

Nyquist Sampling – make sure you collect sufficient pixels across a structure of interest to both separate the structure from neighbouring structures, and to have confidence that the object is "real" and not simply a result of optical or electronic noise.

In confocal microscopy generating an image that satisfies Nyquist sampling criteria is important (i.e. the objects of interest within the image should be displayed with at least 2, but preferably 4 pixels across the structure). Your object of interest may be simply the nuclei of the cells (in which case only relatively low resolution will be required). However, if your interest is in the harder to image subcellular organelles or structures then you will need to aim at creating images that are at the limit of the resolution of the objective lens you are using. Satisfying the Nyquist sampling criteria in high-resolution imaging will require considerable zoom adjustment, and care not to inadvertently photo bleach the sample. The exact amount of zoom can be determined empirically by simply zooming up and down and observing how much structure is visible or not, alternatively you can use the information available on the lens, such as the NA, to determine the optical resolution of the lens in use and then to zoom appropriately to match the known resolution with the size of the pixels created by the combination of zoom function and lens characteristics (see page 126 above for more details on optimising the zoom).

Once the cells of interest have been located and a reasonable zoom level has been attained it is important to now close down the pinhole – often resulting in a somewhat "noisy" or speckled image. If you are imaging fixed cell preparations you can produce a respectable image by turning up the laser power until you have sufficient fluorescence to produce a relatively low noise image. However, a better way to produce a better quality high-resolution image is to leave the laser power relatively low and to either perform line averaging or screen averaging (or Kalman averaging) – or a combination of both. Fixed samples on a stable microscope can be imaged for several minutes using screen and line averaging to produce very high-resolution images with good contrast and low noise.

If you are imaging live cells you rarely have sufficient fluorescence in a single scan to produce a high quality image, but you will need to take considerable care when line or screen averaging not to produce a blurred image, caused by cell or subcellular movement during collection of the images. As has been discussed previously, live cells image better using line scan averaging rather than screen scan averaging.

Once a respectable image has been collected and saved the brightness and gamma of the scanned image can be changed later, depending on the type of display or presentation required.

WARNING - for critical imaging always display images at screen resolution (i.e. 1:1, or 100%)

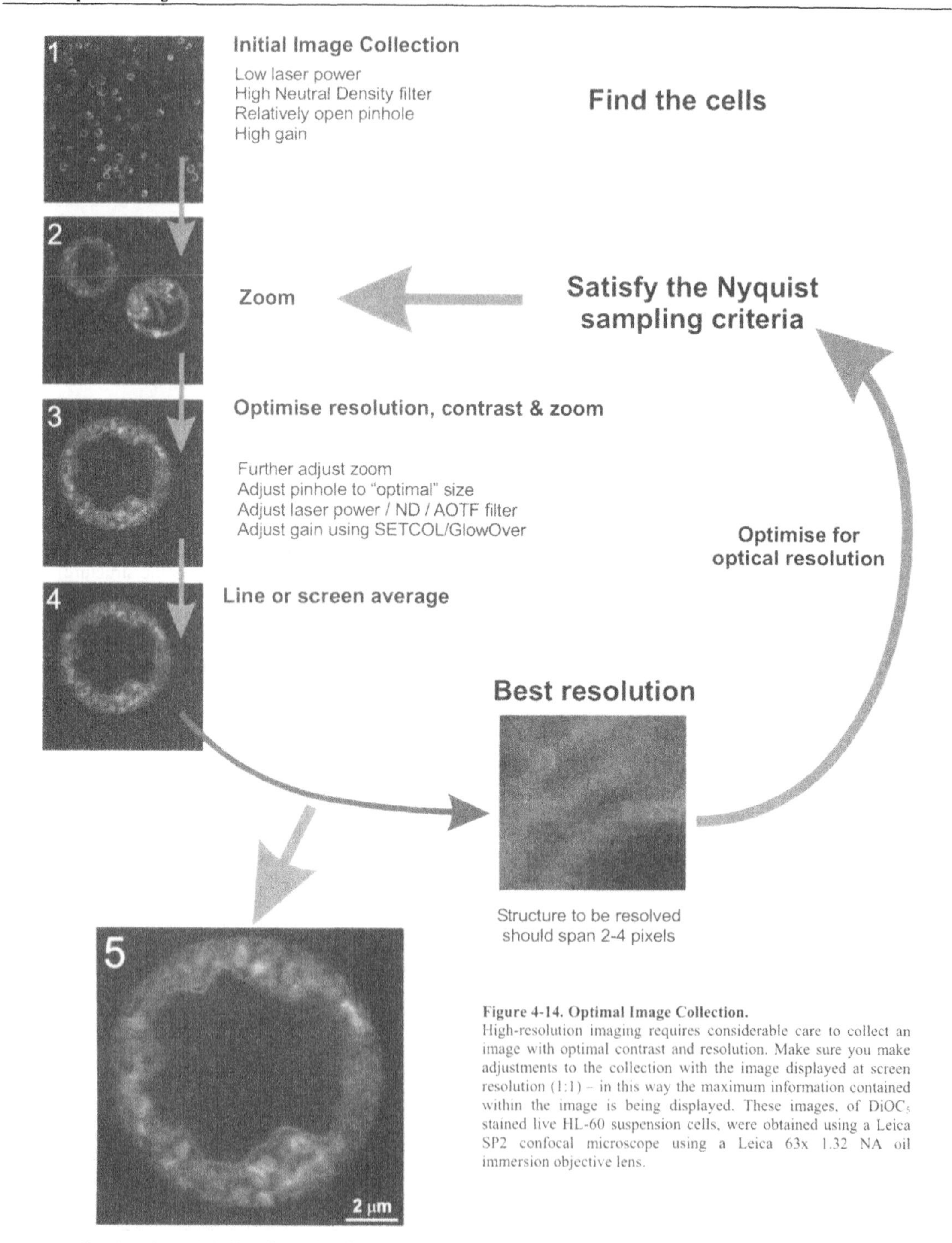

Figure 4-14. Optimal Image Collection.
High-resolution imaging requires considerable care to collect an image with optimal contrast and resolution. Make sure you make adjustments to the collection with the image displayed at screen resolution (1:1) – in this way the maximum information contained within the image is being displayed. These images, of $DiOC_5$ stained live HL-60 suspension cells, were obtained using a Leica SP2 confocal microscope using a Leica 63x 1.32 NA oil immersion objective lens.

Understanding "Noise" in an Image

On a number of occasions throughout this book the terms "noise", "image quality" or "better image" have been used in discussing the collection of images from a confocal microscope. This section briefly discusses what is meant by "noise" in an image, and what is meant by a high-quality image. Most biologists have no trouble recognising a "good" image when they see one, but what is often much more difficult is to determine the source of the problem "noise" and what can be done to improve the quality of the image. In many instruments the "noise" level within the instrument can be readily defined, but in confocal microscopy there are a number of interacting and often conflicting requirements such that a clear definition of what constitutes a "high-quality" image is not universally applicable. An excellent starting point for looking at noise in confocal microscope images are the two papers by Zucker and Price in *Cytometry* (44:273-294 and 44:295-308, 2001, see page 350 in Chapter 15 "Further Reading", for more details).

Traditionally the efficiency of electronic instrumentation is determined by the signal to noise ratio (S/N), often given as the formula,

$$S/N\ ratio = \frac{\text{Signal - Background}}{\text{Standard Deviation of Background}}$$

where the standard deviation of the background is a measure of the "spread" of the background signal in the image. A high "noise" level in the background will be manifested as a high level of "speckle" in the background, which is due to individual pixels having a significantly higher value than surrounding pixels. However, this formula is not particularly useful in confocal microscopy as the background is often extremely low (i.e. the electronic "noise" of the instrument is very low). A more realistic formula includes the spread of the signal within the formula, often expressed as the coefficient of variation (*CV*),

$$CV = \frac{\sigma\ \text{(standard deviation)}}{\mu\ \text{(mean intensity)}}$$

where "standard deviation" refers to the spread of the intensity levels within the image (determined by using the Histogram function). This formula is a good measure of the level of "noise" in the instrument, but can only be used on an image with uniform fluorescence intensity across the field of view (i.e. a fluorophore solution, or a piece of fluorescent plastic), or with at least a region of even fluorescent intensity (for example, fluorescent latex beads). A biological sample will have a high level of variation within the sample, irrespective of the level of "noise". However, a low "noise" image will appear as relatively "smooth" with little or no "speckle".

One way to greatly improve the S/N ratio in a confocal microscope is to simply open the pinhole – but this will also result in a significant loss of resolution and a large increase in out-of-focus light within the image plane. Confocal microscopy is always a balance between resolution and signal intensity. Matters are further complicated when imaging live cells as movement of the cells and subcellular organelles, as well as fading of the fluorophore and damage to the cells all must be taken into account when determining the settings on the confocal microscope. Confocal and multi-photon microscopy is always a compromise between these different competing requirements. There is never a perfect setting for the instrument; you will always need to take into account the nature of the sample and the type of information that is required.

Collecting high-quality images is always a compromise between a number of conflicting requirements:

High-resolution – using a small pinhole size will increase the optical resolution of the microscope, but will restrict the number of photons reaching the detector.

Low "noise" – slow scanning or frame averaging will significantly lower the "noise" level, but may not be possible on moving cells.

Fast imaging – high-speed cellular events may require high-speed imaging with a significant increase in the level of image "noise".

Sources of "noise"

There are a variety of sources of "noise" in a confocal microscope. The different types of "noise" can be divided into noise created by the instrument itself (electronic and detector noise) and noise due to the statistical nature of light (Shot noise). Unwanted light in the image plane is also sometimes referred to as noise. This can include a wide range of problems, including bleed-through between channels and out-of-focus light that penetrates the confocal pinhole (particularly when the pinhole is not closed to at least one Airy disk). However, the only "noise" in this category that will be considered here is stray light that may affect the background light intensity – even in the absence of a sample.

> **Photomultiplier "sweat spot"**
> Every photomultiplier tube has a particular voltage at which the signal is highest relative to the noise (the highest S/N ratio)

Light detectors: The highly sensitive light detector used in the laser scanning confocal and multi-photon microscope, the photomultiplier tube, does produce some "noise" that is due to the release of electrons within the device in the absence of light. This type of noise is rarely encountered in confocal microscopy unless the photomultiplier tube is used at maximum gain. Try imaging at maximum gain *without* any laser light to determine the detector and electronic noise in your system. Photomultiplier tubes do have a specific voltage at which the noise level is at the lowest level. This should be the preferred voltage for operating the instrument, but for versatility and ease of use most confocal microscopes allow for user selection of a large range in voltage.

Electronic circuitry: The electronic circuitry used for signal amplification in a confocal microscope does not contribute greatly to the noise level in an image. Try imaging at maximum gain *without* any laser light to determine the detector and electronic noise in your system. Try turning out the room lighting to eliminate any possibility of stray light from the room lighting contributing to the detector noise level.

Stray light: Light that is not from the focal plane in a confocal microscope is considered "stray" light. Opening the pinhole will increase the amount of "stray" light entering the photomultiplier tube, but there are other potential sources of stray light that may affect the image. Room lighting rarely interferes with a confocal microscope image, but it can cause problems with a multi-photon microscope when a confocal pinhole is not used. Light reflecting internally within the optics of the microscope and scan head may also result in a significant level of light being detected. Stray light is not strictly "noise", but is signal due to light that is not derived from the image plane of the sample. Try imaging at maximum gain and relatively high laser power with *no sample present*. Compare the level of signal before and after turning off, or blocking the output from the laser. This will give you a measure of the level of electronic (or dark noise) compared to "noise" from stray light in the instrument.

Poisson statistics: Even when a steady light source is used to illuminate the sample each individual photon has a certain probability of arrival. The degree of uncertainty in the arrival of each individual photon is known as Shot noise. The noise level or error in arrival of a photon is related to the square root of the number of photons. This means that in low light conditions there will be a large variation between individual pixel intensities. For example, the arrival of 4 photons will have an error of 2 photons (a 50% error rate), whereas 100 photons will only have an error rate of 10 (10%). Poisson statistics is the main cause of poor image quality in low light conditions in confocal microscopy. Increasing the level of fluorescent light intensity, averaging over several scans or slowing the scan rate will all reduce the level of Shot noise. Opening the pinhole will also lower the Shot noise (increasing the amount of light that reaches the detector), but with loss in z-sectioning ability and apparent x-y resolution.

3D IMAGE COLLECTION

The first step in 3D image collection is to collect a "stack" of optical sections (Figure 4-15). This is carried out by setting up a series collection using the software used to operate the confocal microscope. The stepper motor on the confocal microscope will then be used to collect a series of images at defined steps through the sample. The stepper motors are highly accurate devices, but there may be some difficulty in returning to the same place of focus after progressing through the stack. This lack of return to the origin is a particular problem with some microscopes where the stepper motor operates through the fine focus control. A z-stepper that uses a piezo-electric stepping device on the microscope stage or to move the objective is usually more accurate. If you do find the lack of return to the origin is a problem then make sure you start the automatic collection from the position you have left the microscope focus at when setting up the 3D "start" and "stop" positions. In this way the series of slices will start at the correct position, even if the stepper motor does not return the focus to this position correctly after collection.

A series of z-section optical slices hold a wealth of information on the location of structures and the physical relationship of structures within the sample. Take careful note of the 2D images collected and become adept at extracting valuable 3D data from these 2D slices. The optical slices that you have collected can be rendered into 3D structures in a variety of ways. 3D rendering is a very involved and fast changing area of study in which powerful and relatively cheap computers are allowing one to perform relatively sophisticated analysis that was once the domain of specialist image processing work stations. If you are interested in learning more about 3D reconstruction and image manipulation you should consult "The Image Processing Handbook" by John C. Russ (see page 353 in Chapter 15 "Further Reading").

3D reconstruction can be performed on the collected "stack" of images using either the software provided by the confocal microscope manufacturers, or a variety of other specialist 3D rendering software. Most confocal microscope manufacturers provide a simple set of 3D reconstruction tools that are quite effective – and in fact may be all that you will need. However, for more sophisticated reconstruction and analysis there are much better dedicated 3D rendering programs available (see Chapter 6 "Imaging Software", page 163).

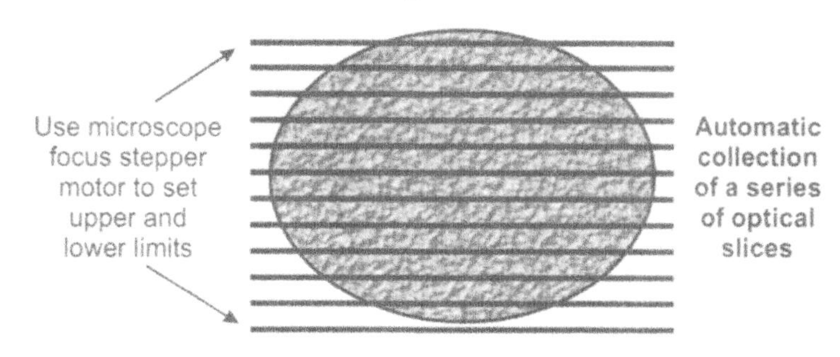

Figure 4-15. 3D Image Collection.
The microscope focus stepper motor is used to set the upper and lower limits for optical slicing. Once you have set the appropriate scan speed and collection filters you can then collect the z-series automatically. The correct number of optical slices for full resolution in the z-direction is determined by the z-resolving power of the lens.

Extended Focus Image

A relatively simple, but often effective, way of displaying 3D information is to produce an "extended focus" image (Figure 4-16). This type of image is normally presented as a stack of maximum intensity in each slice. This has the advantage of producing a "solid looking" almost 3D rendition of the object. This is particularly useful if you are interested in the surface profile of the object, or the labelling is sufficiently localised to allow one to "see" further into the sample. A number of different renditions of the "extended focus" image can be produced, including minimum projection and different degrees of transparency. This will allow a much better view of internal fluorescence.

Rocking Motion Image

A further extension of the "extended focus" image is to create a series of projections from different angles (usually between 15° and −15° from the centre line) and then to use 3D rendering software to create a "movie" of the object "rocking" from side to side. This type of image presentation creates an excellent 3D impression – often allowing one to "see" further into the object than is possible using a simple "extended focus" image. The software provided by the confocal microscope manufacturers will produce excellent "rocking motion" 3D images, but if you need to vary the transparency or remove a portion of the object to "see" inside you will need to use a more sophisticated image analysis and 3D rendering software provided by specialist software developers.

A full 360° rendition of the image stack is also possible (taking considerably more time), and can then be displayed as an interactive rotating image by using suitable software. With some software it is possible to also slice through the image stack while positioning the 3D reconstruction at any angle. This level of sophistication is usually only available in specialist image processing software packages.

24 Individual Optical Slices

1 2 3 4 5 6 7 8 9 10 11 12 13 14 15 16 17 18 19 20 21 22 23 24

"Extended focus Image"

"barring" caused by insufficient optical slices

A

50μm

24 optical slices combined
(each slice collected 1 μm apart).

Figure 4-16. Extended Focus Image.
A series of optical slices can be combined in a number of ways to represent the 3D information present. In this example the optical slices displayed on the left have been combined into an "extended focus" image above by simply adding the maximum intensity from each slice and then dividing by the number of images (a single maximum intensity projection). In image "A" above the physical spacing between each individual optical slice (1μm) is too far apart, resulting in a "barring" pattern on the cloth fibres. In contrast, image "B" below has closer spaced optical slices (0.3μm), resulting in "smooth" cloth fibres in the extended focus image'. A piece of fluorescent cloth was imaged using a Nikon 40x NA 0.6 extra long working distance (ELWD) "air" objective on a Bio-Rad MRC-1024 confocal microscope.

"Extended focus" image using closer spacing of optical slices

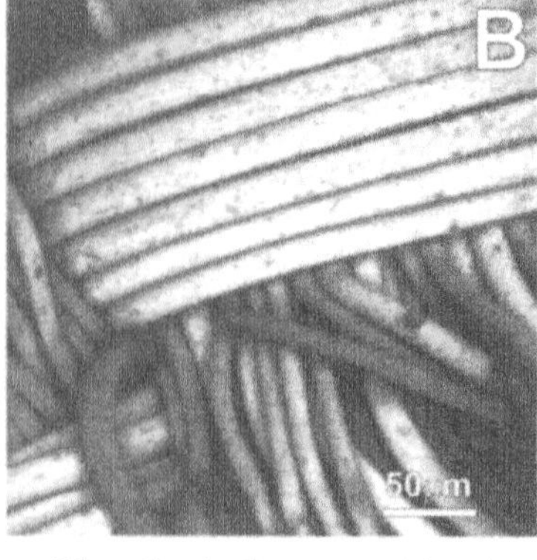

70 optical slices combined
(each optical slice collected at a z-spacing of 0.3μm)

Red/Green Stereo Pair

A stereo 3D image of your object can be created by displaying two projections of the image from different angles in two different colours (red and green). If you look at such an image using red/green glasses you will often find a reasonable 3D representation of your object. The red/green stereo pair is particularly good for some images but not others – you will need to experiment with your images to see which ones work and which do not. The disadvantage of the red/green stereo pair is that the image is not at all good for viewing without the red/green glasses – and such glasses are not always available!

Offset Stereo Pair

Excellent 3D renditions of your images can be obtained by creating two projections of your stack of optical slices from two different angles. These are then displayed side by side and viewed either with "stereo" glasses or by focusing one eye on each of the images (often in a "cross-eyed" mode). If you are able to see these images without using stereo glasses then you will find this a most satisfying and quite a spectacular way of creating 3D renditions of confocal images. In a darkened lecture hall these types of 3D renditions are particularly spectacular as the 3D image will be not only very clear with spectacular 3D, but the image will appear to be suspended in mid-air in front of the screen!

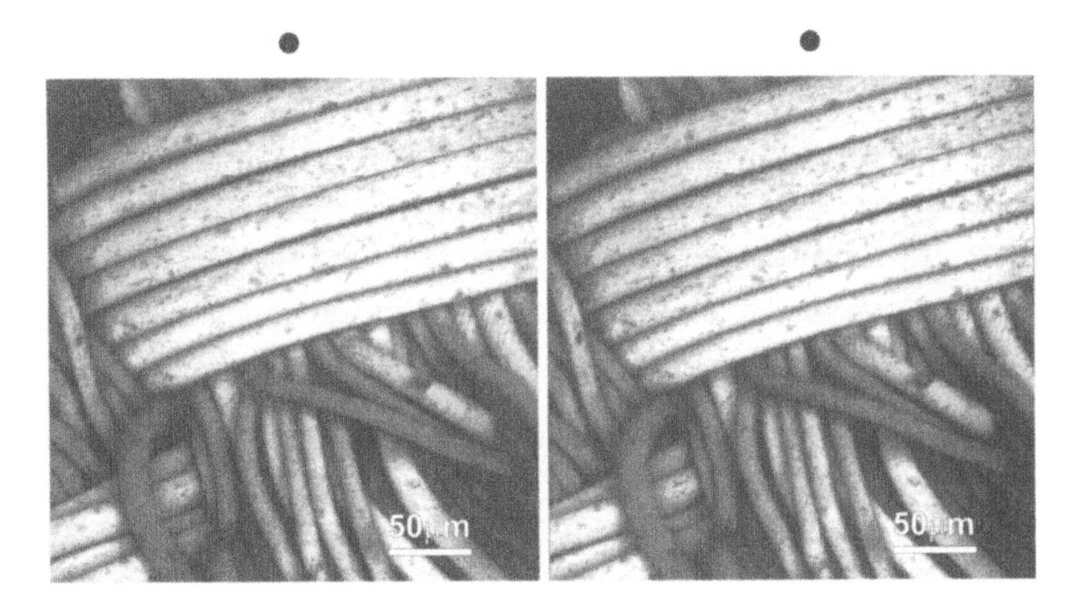

Figure 4-17. Cloth Stereo Pair Image.
This stereo pair image of cloth can be viewed using stereo glasses, or with some practice a stereo view can be achieved by simply focusing each eye on a separate image of the pair. To try this look directly at the images at a distance of about 30 cm and look cross eyed at the images. Try bringing the black dots above each image into a single dot in the middle of the page. You should be able to see a 3rd image in the middle of the two. If you concentrate on this 3rd image and slowly bring the central image in to focus you will be amazed to find that the 3D image is located away from the page (set back or forward) in quite startling detail (see Figure 1-18 on page 22 for further details).

A piece of fluorescent cloth was imaged using a Nikon 40x NA 0.6 extra long working distance (ELWD) "air" objective on a Bio-Rad MRC-1024 confocal microscope. 70 optical slices were collected and then displayed as an offset stereo pair using the Bio-Rad LaserSharp software.

TIME RESOLVED IMAGE COLLECTION

Time resolved image collection (Figure 4-18) can be carried out by using the available settings on the software provided by the confocal microscope manufacturers. You can set the microscope to collect images automatically several times a second, every few minutes or only every hour or less. Time resolved imaging is particularly useful for following biological processes in live cells. However, care must be taken with time resolved image collection that the continuing exposure to the laser doesn't unduly damage the cells or fade the fluorochrome.

Fast time resolved imaging on a conventional laser scanning confocal microscope will be limited by the speed of the scanning mirrors. Faster scanning can be achieved by imaging a smaller box size (less pixels) or increasing the scan speed – but ultimately the speed at which you can realistically image is about 8 frames per second. To image faster than this would require a different approach such as line scanning – or use of a different type of confocal microscope, such as the Nipkow disk confocal microscope.

The Nipkow disk confocal microscope is capable of greatly increased scanning speeds compared to a laser scanning confocal microscope. The Nipkow disk scanning head incorporated into the PerkinElmer and VisiTech instruments, manufactured by Yokogawa, has a micro-lens associated with each "pinhole" in the Nipkow disk (see page 91 in Chapter 3 "Confocal Microscopy Hardware" and the Nipkow disk section of "Appendix 1: Confocal Microscopes" (page 431), for details on the design of these instruments). Nipkow disk based confocal microscopes are capable of much higher speeds (routinely 20 frames per second, but can be higher than 100 frames per second – depending on the CCD camera attached) compared to a laser spot scanning instrument. These instruments are particularly suited to time lapse imaging as not only can they image at high frame rates, but they also operate on a very low level of light – allowing one to image live cells without damage over many hours, or even days if you can keep the cells alive on the microscope for this long!

If extended time collection is carried out you will need to make sure that the cells do not move during the experiment – unless, of course, cell movement is what you are interested in.

Focus movement during long time-lapse imaging is a serious problem that is not easy to solve. One of the major causes of focus drift is differential heating and cooling of microscope components and chambers. The only way to get close to eliminating focus drift is to use a heated enclosure that fully covers the microscope and associated components (see Chapter 12 "Imaging Live

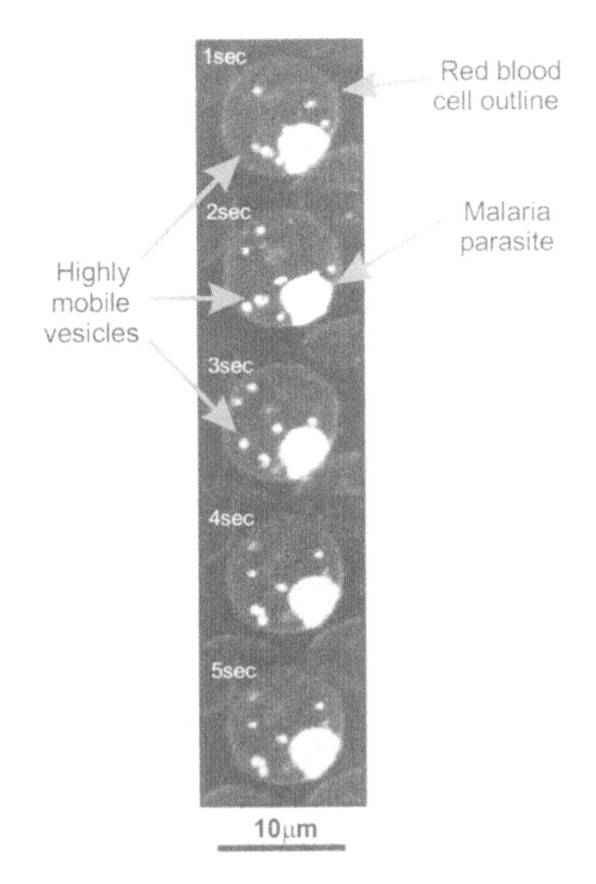

Figure 4-18. Time Resolved Image Collection.
A series of images were collected 4 times per second for several minutes on a Bio-Rad MRC-600 confocal microscope using a Zeiss 63x 1.4 NA oil immersion objective lens with an image collection box size of 128 x 128 pixels. The above display is a small collection (at 1 second intervals) of the total number of images collected (several hundred). Small acridine orange stained vesicles within the malaria-infected red blood cells are shown to be highly mobile in these live cells, whereas the parasite is relatively immobile. Further information on the experimental conditions used to obtain these images can be found in Hibbs and Saul (1994) – see page 349 in Chapter 15 "Further Reading".

Cells", page 279). If you are taking images at relatively long time intervals try "refocusing" the sample before each new image is collected. Transmission imaging can be used to determine what part of the image is the correct focal plane.

For high speed imaging, or imaging for extended time periods, consider using a Nipkow disk confocal microscope.

Time resolved imaging is also used for collecting a series of images when performing a FRAP (Fluorescence Recovery After Photobleaching) experiment (see page 252). When doing FRAP studies the time-lapse program on the computer is used to collect a series of images after the initial "burn" using the laser. Some confocal microscopes have specialist software for performing FRAP experiments (making initial image collection easier) that specifies the position and power of the laser "burn" and specifies the time interval for the series of images collected during the fluorescence recovery period.

A series of images collected over time can be displayed as a series of individual images (Figure 4-18), but movement can be much more readily conveyed by converting the images into a "movie" display. A movie can be created simply by moving through the series quickly using the software that is used to operate your confocal microscope. You can also use simple programs such as Confocal Assistant, NIH image or Image J to display a series. Alternatively you can convert the series to the "AVI" "movie" format that can be readily displayed on most computers. A number of confocal microscopes now store the images as single TIF files (with the complete series being saved in a specified directory). The free software NIH image, and the Windows equivalent, Scion Image or Image J, are capable of creating "AVI" movies from such a series of TIF images. An "AVI" movie has the added advantage that it can be readily displayed within a PowerPoint presentation.

IS THERE A "STANDARD" SAMPLE

To be able to test the settings on a confocal microscope using a standard sample would be very convenient. A standard sample, with a guaranteed level of fluorescence, that was permanent and widely available would be wonderful. There are a variety of ways of creating a "standard" sample for confocal microscopy, unfortunately none of them are fully satisfactory.

The following "test" samples are quite useful:

1. **Fluorescent solution:** A dilute solution of common fluorophores (such as fluorescein or tetramethylrhodamine) can be used to measure the level of fluorescence in each channel in the confocal microscope. A solution of fluorophore is added to a small microscope chamber, preferably sealed (to eliminate oxygen and evaporation), and the sample imaged on the confocal microscope. Any fading of the fluorophore is offset by diffusion of fresh fluorophore from the surrounding media into the area of the sample that is being imaged. Unfortunately the solution cannot be stored for long, but making up a new solution is quite simple. Imaging into a water-based solution of a fluorophore may result in a significant drop in fluorescence with depth due to problems of spherical aberration. This is particularly noticeable when using an oil immersion lens. Therefore you should image at the same depth (or take a depth profile) when using a fluorophore solution as a "standard" sample.

2. **Fluorescent plastic:** Using fluorescent plastic (used for making signs in advertising) is a very convenient way of creating a "standard" slide. Red and green fluorescent plastic is readily available - simply cut a piece into the shape of a 1" x 3" microscope slide and image the plastic using the confocal microscope. The plastic does suffer from fading if the laser level is too high, or if you image for too long in the same place. However, if you keep the laser power level relatively low, the problem of fading can be almost eliminated. The level of fluorescence of the plastic is quite stable when stored at room temperature in the dark. There is a significant variation in the amount of fluorescence depending on the depth within the plastic you are imaging. This variation is strongly dependent on the lens being used (dry, water immersion, oil immersion etc), but due to the relatively high refractive index of the plastic you will find the change of fluorescence intensity with depth is quite different from imaging into a water-based solution as discussed above.

3. **Latex beads and Quantum dots:** Fluorescently labelled latex beads and highly fluorescent Quantum dots are available from a number of companies, and can be readily used as "standard" samples to determine the performance of various aspects of the confocal microscope. Molecular Probes sells a set of pre-made slides containing a variety of coloured latex beads that can be used for testing the alignment of the different channels

in a confocal microscope. A wide variety of latex beads is available from a number of manufacturers for use in flow cytometry (as standards in cell sorting). However, latex bead "standards" are not as reliable when used in confocal microscopy compared to flow cytometry as the confocal microscope images a thin optical slice through the bead (in flow cytometry the fluorescence intensity from the whole bead is measured). Relatively large multi-coloured beads (up to 15 µm in size) are used for testing the alignment of the various optical components of the microscope, and can be embedded in agarose to form a 3D sample. Small latex beads (even as small as 0.03 µm, which is significantly smaller than the resolution of a high NA lens) or subresolution Quantum dots can be used to determine the point spread function (PSF) of your microscope.

4. **Pollen grains:** Pollen grains are highly fluorescent and widely available and often have lots of fine detail that can be imaged by backscattered light and by fluorescence confocal microscopy. A microscope slide containing pollen grains for confocal microscopy can be readily prepared using the "sticky tape" mounting method outlined for moss spores below.

5. **Moss spores:** A convenient and readily prepared sample makes use of fluorescent moss spores. Spores from the giant moss, *Dawsonia superbe*, which grows in moist fern glades in S.E. Australia (there are bound to be many suitable spores from other species around the world), are found to be an excellent test of resolution of the confocal microscope. These spores can be readily mounted direct in immersion oil by dusting the spores onto clear "magic" sticky tape and then mounting the tape under a coverslip using a drop of immersion oil. The sticky side of the tape, which holds the spores, should face the coverslip. The spores will be held firm by the glue on the tape. The spores contain very fine projections from the surface that are quite difficult to image by confocal microscopy - but are a good test of your microscope. The spores are fluorescent in both the red and green channel of the typical FITC/TRITC dual labelling settings. The fluorophore is highly resistant to fading and is stable for years. The spores can also be readily imaged by backscatter (or reflectance) imaging, in which case the fine surface spikes can be more easily resolved by the higher resolution obtainable by backscatter imaging.

6. **Autofluorescence of Hair:** An excellent way of providing a low fluorescent "standard" sample is to image your own hair! Hair is autofluorescent across a broad range of the light spectrum. The level of fluorescence is relatively low, but this makes this sample ideal for testing the sensitivity of the instrument. Human hair is placed on a microscope slide and covered with a drop of water or glycerol and covered with a coverslip. The low level of autofluorescence in the hair is quite stable and does not unduly photo bleach. Fine structure within the hair can be resolved if the instrument is sufficiently sensitive.

7. **Your own cell preparation:** A convenient source of "standard" material to test the confocal microscope is to use your own cell preparation. Immunolabelling is particularly good as test slides as the cells are fixed and the fluorescence is stable for many months if stored at -20°C. The only drawback is that each antibody and cell preparation is different. This means that a "standard" immunolabelled slide is not commercially available - but you'll soon get to know your own cells and antibody. Just prepare a few extra slides and use them over several months to test the functioning of the confocal microscope.

LOOKING AFTER YOUR IMAGES

Finally, after having collected lots of valuable images, having a good filing system is important. This is particularly important when the instrument is used by a number of different people or groups. Don't forget that all the original data is in electronic format, which requires careful file naming to find that "wonderful" image some months or years later. The other aspect of electronic data is that the date can be very easily destroyed! Careful backup is essential.

The software used to collect your images on the confocal microscope will often provide a convenient database management system for your images. However, take care, because you may not always have access to the software used to operate the particular confocal microscope you are currently using to collect your images – and the software from the manufacturer of the confocal microscope may change over the years, possibly limiting access to your "old" files. The following hints for file naming and backup are intended to safeguard you against changes in software and the high probability of forgetting what that "wonderful" image was called!

Filing your Images

Retain all original confocal microscope images collected. These should be collected and burned to CD as soon as possible to save against accidental loss or overwriting. All changes to the images should be saved under a new file name that is different from that of the original image.

The confocal microscope hard disk has a limited amount of space for the temporary storage of images (no matter how big the hard disk seems at the beginning!). It's a good idea to remove all files from the confocal microscope hard disk at the end of each session of work on the confocal microscope, or at least at a regular weekly or monthly backup session. There is no guarantee that the files will be safe on a multi-user system. The initial login requirement (with password) may simply be to direct your collected images to your own subdirectory on the hard disk. This login is not always a security system for your files – in most systems anyone can access your files once they have logged in under their own password!

Hints for looking after your files

- Don't leave files on the confocal microscope hard disk!
- Use a simple unique file name
- Keep a log book of all files collected
- Retain all files collected in original format
- Collect high quality images for publication!

Choosing a Unique Naming System

Even though you can have a long and descriptive file name for your image, including the date and perhaps a number as the initial part of the file name, is often convenient. The date can also be included in the folder name where the images for a particular date are collected. I often find a numbering system preceding the date means that the files can be correctly ordered alphabetically, even when different months are involved.

A logbook of all files collected is the best way of remembering what the image is some days or weeks later. Be very careful if you wish to use the annotation or "notes" function of the confocal microscope software as this data may be lost if the file is saved under another format – and furthermore, may become difficult to access when using other imaging software, or even later versions of the confocal microscope software!

Some older versions of the confocal microscope software used to operate the instrument may only allow the archaic 8-digit file name. This does put considerable constraint on naming files, but I have found a simple numbering system based on the date works extremely well. If this is combined with a simple identifier for the directory name containing all the images from a single operator for each day then there is never any future problem with finding files.

Finally, collect high quality images for publication as you do your experiments – and make special note of what these files are called. This may seem obvious, but often in the rush of finishing off an experiment you will find not enough care is taken to make sure the image is of publication quality, often necessitating a repeat of the experiment to collect suitable images – or even worse, because you can't find the image you want!

Managing your Images

Later versions of the "Windows" operating system now have a reasonable image management system. In Windows ME and higher you can select display "Thumbnail" from the "View" option on the folder menu bar. This option will then give you a small thumbnail view of each image in the directory. If you want to see a slightly larger view of the image then simply select the file / thumbnail of interest and a larger view will be shown in the left side of the window. If you are not using the folder "thumbnail" view option you can still see the "thumbnail" of each image by selecting the file. If this larger view is not visible perhaps your window size is too small – simply enlarge the window to see the image. The only drawback of this "thumbnail" viewing method is that when a file contains a "stack" of images only the first image in the stack is shown – which, unfortunately is often blank!

A further enhancement of the file management system of Windows ME and higher is a special folder called "My Images", which is located within the folder "My Documents" (which should be located on your desktop). Any image placed within this folder will be displayed automatically as a thumbnail, and the larger thumbnail visible in the bottom left corner of the window when you select a file can be zoomed and rotated (by 90°). The image can be seen in even more detail by clicking on the "Image Preview" icon on the small menu bar above the image in the corner of the window. This simple image management system is a great way to keep track of lots of single image files.

Unfortunately, as discussed in the previous paragraph you don't have the opportunity to display different images from a multi-image file.

Data Backup

The choice media at present for backup of important confocal microscopy images is the CD. CDs, although becoming somewhat "small" for archiving large collections of images (particularly time series), are a widely available device that can be used in both a PC and Mac computer. They are very cheap and once "burned" can be read on any computer with a CD-ROM drive (which is almost every computer made since the early 1990's!). Eventually larger capacity DVD disks will probably be the choice of media for backup. Tape backup, which was popular some years ago, and is still used in some large facilities, while having a very large capacity does suffer from slow access to the data and less reliable long term storage when compared with the CD.

If you work in a large department or Institute you may find that the facility provides a method of regular backup of all images collected on the confocal microscope. While such a facility backup system is admirable, my advice is do not rely on such a system for your valuable images – burn your own CDs after each session on the confocal microscope.

A number of confocal microscope imaging facilities now use a RAID disk system to archive confocal microscope images. A RAID system consists of a large array of hard disks with a built in redundancy that allows one to swap a malfunctioning disk without any loss of data. Although a RAID system is an excellent way of storing large amounts of data in a very secure environment, you can't take the RAID disks with you – so burning a CD is still the most secure way of ensuring you still have access to the data in the future.

Zip disks (and their larger cousin Jaz disks) are a convenient means of transferring large files between computers, but are expensive for long term storage of large amounts of data.

Although you may have access to a network with a large capacity, don't use this as your only storage place for image files. Often the network administrator will not be impressed by large amounts of data being retained on the network, but perhaps more importantly you are reliant on the operation of the network to access your files. If the network is down or there has been a hard disk failure on the network, then you still have access to your image files if you have stored them on CD.

Chapter 5

Digital Images in Microscopy

The images obtained from confocal microscopes are in a digital format. This greatly simplifies manipulation of the images for merging, colouring, cropping, printing etc. An understanding of what a digital image is will prove most valuable in not only getting the best out of your images, but also for interpreting your own results and for understanding the images obtained by others.

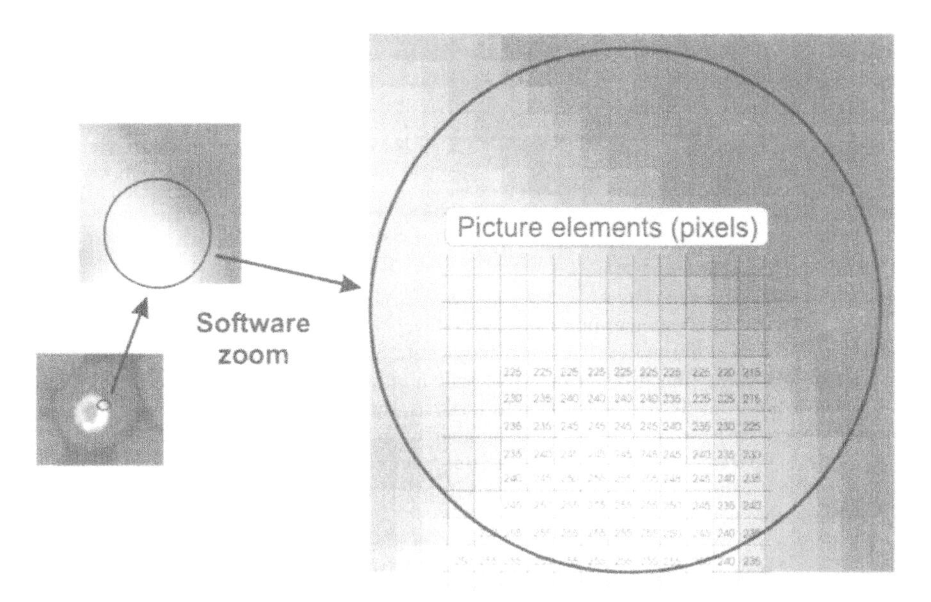

8-bit image = 256 grey-levels (0 = black to 255 = white)

Figure 5-1. Picture Elements (Pixels).
Digital images are displayed as squares of intensity values that are known as pixels (Picture Elements). An 8-bit grey-scale image has a maximum of 256 shades of grey for each pixel. A 12-bit image has up to 4096 shades of grey, and although this number of grey-levels cannot be displayed directly on a computer screen or printed page, the increased bit-depth can be used to optimise the display or for quantitative analysis. Digital images are best displayed at a size that is sufficiently small such that the individual pixels are not visible. The optimal size for displaying a digital image on a computer screen is at screen resolution – but the best size for printing is much more complex, depending on the resolution of the printer and the method of printing. The above image is of a malarial parasite stained with ethidium bromide, and imaged using a Bi-Rad MRC-600 laser scanning confocal microscope.

IMAGE PIXELS

The basic element in a computer image is the pixel, or picture element (Figure 5-1). The initial image obtained from the confocal microscope simply indicates the intensity of light at each point in the sample. The light level collected from each point is encoded as a number, and depicted as a grey-scale point on a computer screen (Figure 5-2), which simply means that the level of light in each pixel is depicted as a shade of grey on the computer screen. These initial image files contain either 256 shades of grey (8-bit images) or 4096 shades of grey (12-bit images) or even 65,000 shades of grey (16-bit images). Each pixel within the image is denoted by a number between 0 and 255 (or 4095 for 12-bit images), with the number zero representing black and the number 255 representing white at the two extreme ends of the range (Figure 5-2). In some software this convention is reversed - the "0" level pixels are then shown in white and the highest level (255 or 4095) are shown in black. Image processing software can readily "invert" these images to produce the more conventional zero = black images.

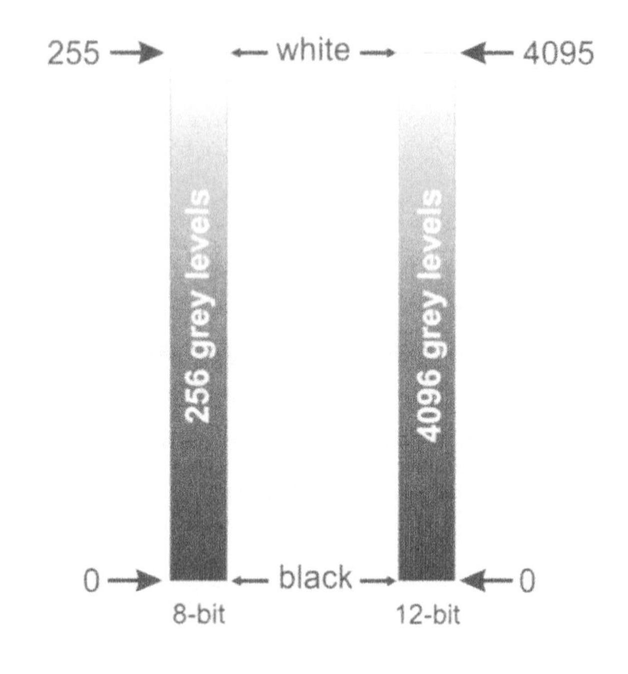

Figure 5-2. 8-bit and 12-bit Grey-Scale.
An 8-bit grey-scale image contains 256 grey levels for each pixel (0 to 255) and a 12-bit grey-scale image contains 4096 grey levels (0 to 4095). In confocal microscopy black is usually designated as "0". However, this may be reversed in some imaging software.

Neither a computer screen, nor a printer is capable of displaying even 256 levels of grey, never mind the much greater number of grey-levels in a 12-bit image. Furthermore, our eyes cannot distinguish such a large number of grey levels even if they could be printed or displayed. Our eyes are capable of detecting a very large light range (much larger than that shown on a computer screen), but they cannot distinguish between two very close grey levels. However, image analysis software (discussed in more detail in Chapter 6 "Imaging Software", page 163) can analyse the full number of grey levels available in the image. This is most important for both quantitative analysis of the images or later manipulation of the contrast and brightness controls. A large number of grey-levels can be more easily visualised if different grey levels are shown as different gradations of colours (discussed in more detail below).

One of the advantages of collecting grey-scale images with a high number of grey levels is that although the initial collection may result in an image that at first appears to be very dark and poorly exposed (all of the pixels being at the low end of the 12 or 16-bit grey-scale) you still end up with a perfectly respectable image once the display levels are adjusted. For example, if you were to collect an 8-bit image all in the lowest two grey levels ("0" and "1"), you would have a very poor image no matter how much you tried to "brighten" up the image by altering the contrast and brightness. However, if you collect the same image using 12-bit acquisition you would have many grey levels in the low end of the image ("32" grey levels are equivalent to "2" grey levels in the 8-bit image). These grey-levels can be adjusted so that they are spread out across the full display level of the screen. You will end up with 32 grey levels in a spread of 256 grey

The computer display is only 8-bit (256 grey levels) or less – no matter how many grey levels are in the original image

levels if you are using an 8-bit (24-bit colour) display card, quite an acceptable number of grey levels for creating a respectable image.

Images collected on a confocal microscope consist of a defined number of pixels, often referred to as the image collection box size (see page 123 in Chapter 4 "Image Collection"). For example, the image may consist of 512 x 512 pixels (a total of 262,144 pixels) or perhaps 1024 x 1024 pixels (a total of 1,048,576 pixels). Every one of these pixels then has 256 or 4095 grey levels, depending on the bit-depth of the image as discussed above.

A confocal microscope image is saved as a computer file, which consists of information on microscope settings, image information such as image size, as well as the image itself. Image files may also contain multiple images, such as images from multiple channel acquisition or from several time points. Information is contained within the file "header" (text information at the beginning of the file) for the software to determine the number of images in the file, their dimensions, perhaps image colour information, time or z-section intervals and a number of microscope settings.

The true advantage of digital images is that they are stored as a series of discrete numbers, thus allowing one to make many enhancements and changes to the image by using software that alters individual pixels. The altered image can then be saved as a new file, although it is important to keep an archived copy of the original image file, as one can easily and inadvertently compromise the image quality by the wrong choice of file formats or image processing manipulations. File formats with lossy compression (e.g. JPEG) will slightly alter every intensity value each time the file is saved. Many image manipulations will also alter the image values, and particularly when done repeatedly, may seriously degrade the quality of the image.

COMPUTER SCREEN RESOLUTION

The computer screen display also uses pixels. However, these screen pixels are not to be confused with the image sample points described above. An image has a defined number of sample points (image pixels) such as 512 x 512, but these do not have a defined size. Their size is simply determined by the resolution of the display monitor

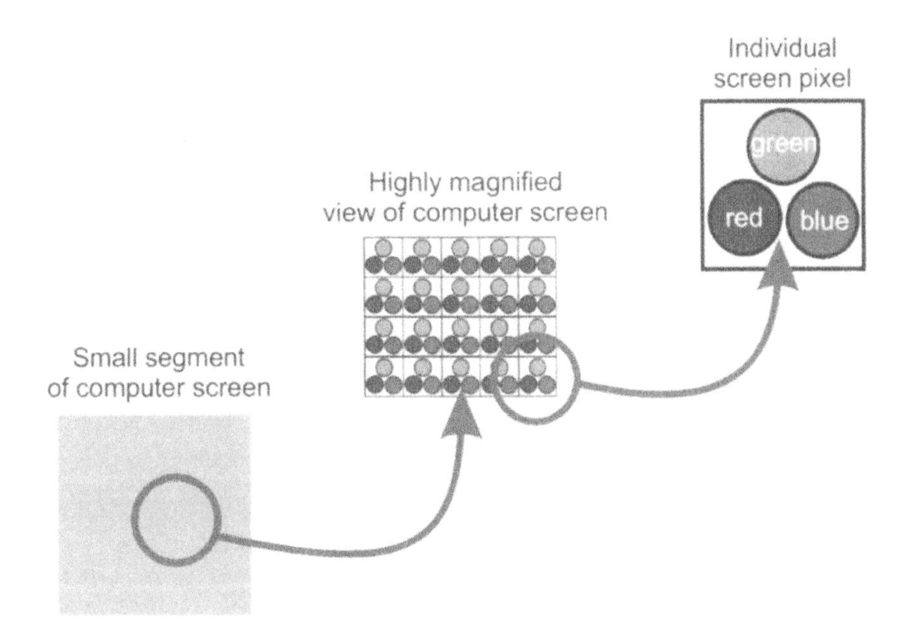

Figure 5-3. Computer Screen Dots.
The computer screen is made up of individual coloured dots (red, green and blue). The minimum size of a display pixel consists of just one dot of each colour. However, a lower display resolution on a high quality monitor may result in many sets of the triplet colours making up one screen pixel. An equal balance of the 3 colours creates grey. These coloured "dots" may appear as "bars" on some screens.

and the digital zoom setting used to display the image. Although the size of each point in an image is defined by the digital zoom and screen display settings, the appearance of the image is also determined by the physical resolution of the monitor itself.

The size of the computer screen pixels is determined both by the resolution of the screen (the better the screen the higher the resolution) and the display resolution software setting. How an image is displayed on a computer screen, and particularly how changing the display resolution of the computer monitor alters the size of the displayed image is discussed in some detail in the following sections.

Screen Dots (dpi)

As discussed above, digital images consist of a defined number of pixels, and are displayed on a computer monitor that has a defined display resolution – also defined in pixels! Underlying all of these pixels is the physical display limit, which is determined by the particular computer screen used.

The computer screen displays images by using coloured dots (Figure 5-3). These coloured dots consist of only three colours (red, green and blue), and changes in intensity between these three colours determines all of the different colours that can be generated on a computer screen (this is explained in more detail on page 154). The number of dots per inch (dpi) specifies the physical resolution limit of the monitor. The closer these dots are placed to each other the finer the resolution of the screen (for example a 0.25 dot pitch display, with 101 dots per inch, is a higher quality monitor compared to a 0.28 dot pitch display, with 90 dots per inch). The number of dots per inch is

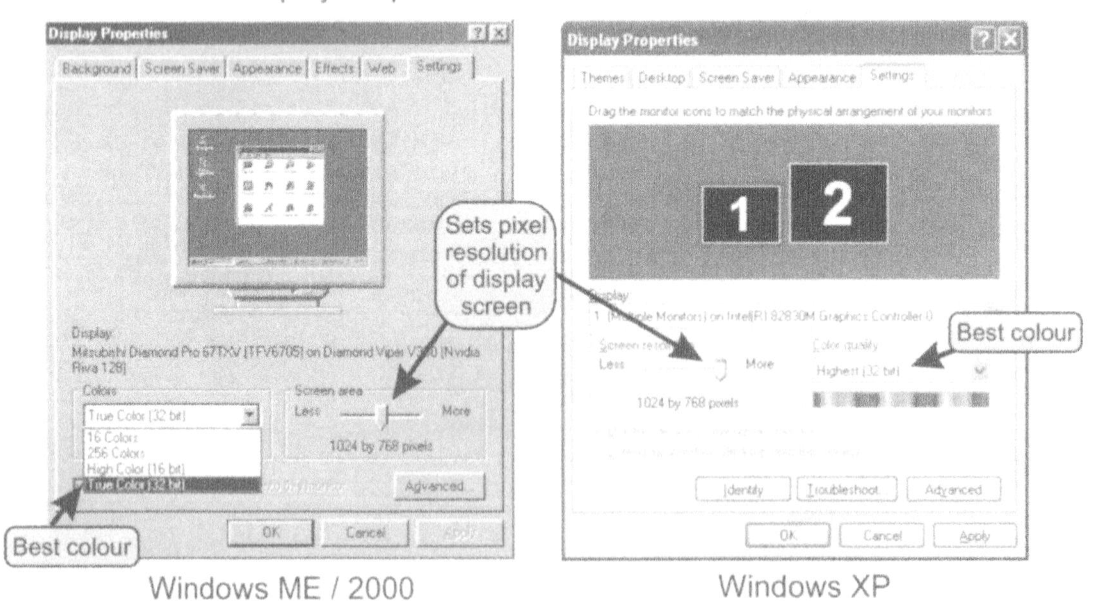

Figure 5-4. Changing Display Resolution.
The screen display resolution is readily changed using the "display settings" from the Windows "control panel". The optimal display resolution for a high quality screen is 1024 x 768 pixels or higher. The colour setting should also be set to "True Colour (32-bit)" for the best colour rendition. Many computer monitors can now display at a higher resolution than 1024 x 768, but this is rarely necessary for displaying images obtained from a confocal microscope. "True Colour (32-bit)" (above left) is in fact 24-bit colour (see page 156) with an extra 8-bit channel that is sometimes used for improved display of animated images (alpha channel transparency). The equivalent setting on a Macintosh computer is 24-bit colour display.

one of the main factors that determine the clarity of the screen display. The Sony Trinitron tube has a somewhat different design, resulting in tiny "bars" of colour – but otherwise the principle of using red green and blue colours is the same. The liquid crystal display (LCD) used in laptop computers and "thin" desktop computer monitors also has a screen resolution that is defined by the number of dots per inch, although the LCD display is a very different technology and looks more like "bars" of colour, rather than dots.

The number of dots per inch on the computer screen determines the upper physical limit of the display resolution, but the screen resolution display (i.e. the 1024 x 768 or 800 x 600 pixels etc., used for display) can also be set below that of the physical limit of the screen itself.

The computer screen also has a range of intensity levels that can be displayed between "black" (no light) and "white" (all three colours on full). Although you can theoretically have a large number of grey levels generated by the display card, display monitors can only display, at the most, 8-bit (256 levels of grey). The phosphors on the computer monitor may struggle to distinguish even this number of grey-levels. A more realistic estimate of the number of shades of grey that a computer monitor can display is perhaps only 100 grey-levels. A LCD display, as used in laptops and many newer model desktop monitors has a further reduced number of possible grey levels.

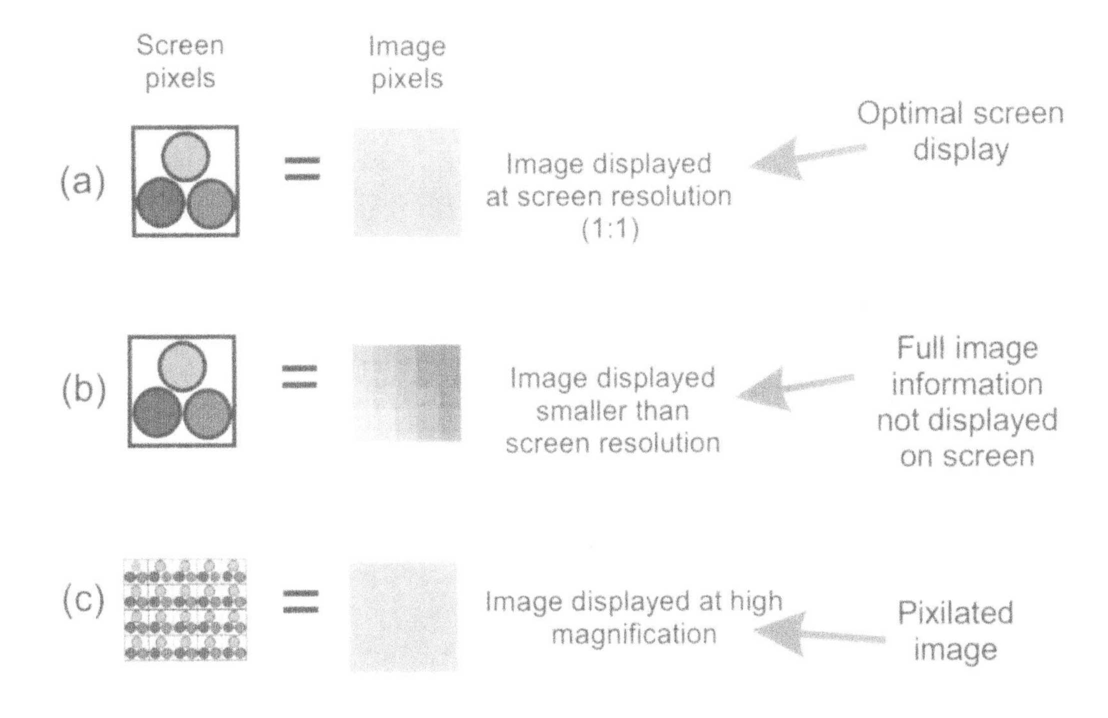

Figure 5-5. Image verses Screen Pixels.
Digital images are displayed using the screen pixels of the computer monitor. The image can be displayed (a) at a resolution that is equal to the screen resolution, in which case the screen and image pixels are the same. The image can also be displayed at a lower (b), or higher (c), resolution than the pixel resolution of the screen.

Display Resolution

The total number of pixels used for display determines the display resolution of the screen. For example, a high-resolution display will normally have 1024 x 768 (or more) display pixels, whereas a low-resolution display may have only 640 x 480 display pixels. The upper limit of your display resolution will be determined by both the physical limit of your monitor (in dpi) and the memory available on the graphics card installed.

The display resolution of the computer screen can be readily adjusted using the "display settings" from the Windows "control panel" (Figure 5-4). The maximum display resolution available is determined by the amount of memory available on the video card in your computer, but only up to the physical limit of your computer monitor as discussed above. However, going beyond a 1024 x 768 pixel screen display is rarely required for images obtained using a confocal microscope. Increasing the display resolution, number of colours and the refresh rate will impose additional computational demands on your computer's video graphics card, which may slow the computer down, or result in erratic image display.

The colour setting for the display should be set as high as possible ("True Colour 32-bit" is best, although High

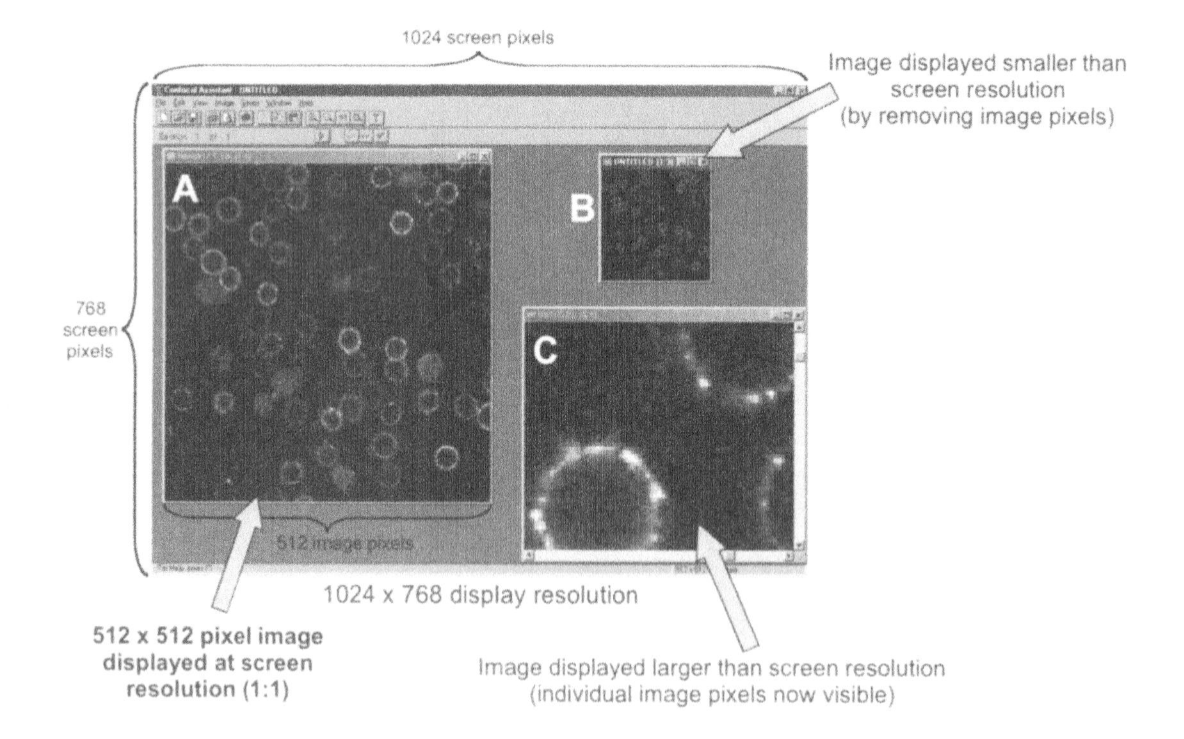

Figure 5-6. Changing the Image Size.
The optimal size for displaying a confocal microscope image is at the screen resolution of the computer monitor being used (image A), which is known as a 1:1 display. Displaying the image smaller than the screen resolution of the monitor (image B) results in some image data not being displayed. Over-magnifying the image results in a pixilated image (image C). Some image manipulation programs may automatically interpolate (add more pixels) to create a non-pixilated image when the image is increased in size – but this does not mean increased resolution, just a larger but somewhat "blurred" image.

Colour 16-bit is quite acceptable if your computer cannot handle the higher setting). If the colour resolution is set lower some colours may not appear of "photographic" quality on the computer screen. The effect is most noticeable when there are large areas of a single colour that is not being adequately depicted using the lower colour settings, in which case the quality of the display may appear quite poor. "True Colour 32-bit" is the same as 24-bit colour (8-bit per red, green and blue channel), with an extra byte (8-bit) that is sometimes used for improved display (alpha channel transparency). When displaying images from a confocal microscope the extra "byte" is not used for display, but is important for the fast and efficient computer handling of the image.

The screen display pixels are the same as the image pixels only when the image is displayed at screen resolution (denoted 1:1 on the image display menu, or referred to in some programs, for example Photoshop, as "actual pixels" or 100% display size). However, the image is often displayed (or printed) at a different display resolution to that of the screen itself (Figure 5-6). Displaying the image smaller than the screen display resolution is often convenient for comparing different images. However, one should keep in mind that the smaller image doesn't simply have smaller image pixels, but in fact some of the image pixels are simply not being shown on the screen! Some confocal microscopes have an "auto" display size as an option for image display – be careful, as "auto" sizing may result in the image not being displayed at screen resolution, resulting in less than the full image information being displayed, even though the "full size" image will be saved. As you are adjusting the instrument zoom, photomultiplier tube voltage, pinhole size etc based on the image you see on the screen, collecting images on "auto" may result in less than optimal adjustment of settings for collection. "Auto" adjustment may also result in other strange effects as the image intensity and contrast levels may also be adjusted automatically for display (but not in the saved image!).

Increasing the image size beyond the display resolution (by digitally zooming the image) will result in a pixilated image (image "c" in Figure 5-5). This is because the increased image size is created by increasing the size of the individual image pixels (replication of display pixels to make the image pixels larger) being displayed, until

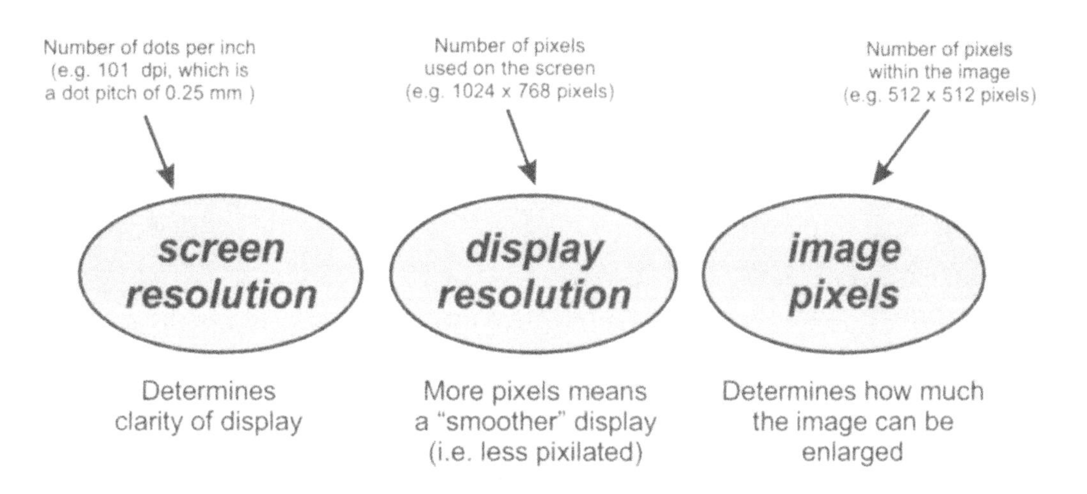

Figure 5-7. Resolution on a Computer Screen.
Screen resolution (dpi) determines the clarity of the display. The display resolution (for example 1024 x 768 pixels) determines the smoothness of the image. The number of pixels in the image (512 x 512 pixels for most confocal microscope images) determines how much the image can be enlarged before becoming pixilated.

eventually they are distinctly visible on the screen. Digital zooming does not always result in individual image pixels becoming visible as some software interpolates the image to create what appears to be a non-pixilated image, but in fact is due to the creation of a range of display pixel intensities for each image pixel. Although an interpolated image appears "smooth" when zoomed, there is no additional information present in the image – i.e. the image is bigger, but structures are not better resolved.

The optimal screen display resolution for viewing confocal microscope images is normally 1024 x 768 pixels (adjusted by using the "Display Properties" menu, Figure 5-4). At 1024 x 768 pixel screen resolution a "standard" 512 x 512 pixel image is displayed fully on the screen – but be careful with 1024 x 1024 or large size images as they will either not be fully on the screen at once (scrolling will be necessary) or a less than optimal display size will be used to allow the full image to be seen on the screen. A screen display resolution higher than 1024 x 768 pixels will result in a more complete display of a relatively large image, but be careful as the individual display pixel may no longer be readily discernible. In other words, the display resolution has gone beyond the resolution limit of your eyes. A lower screen resolution than 1024 x 768 pixels results in a full size 512 x 512 pixel confocal microscope image not being fully displayed on the screen when shown at screen display resolution (1:1 or 100% image display).

The optimal size for displaying images obtained from a confocal microscope is at the display resolution of your computer screen. The physical size of the image on the screen will be determined by the screen resolution setting of the monitor (Figure 5-8). Displaying the image at screen resolution will mean that maximum information in the original image is displayed. Images shown at display resolution have the entire original image data displayed, as each image pixel is represented by a screen display pixel (Figure 5-5a). When the confocal microscope image is displayed smaller than the screen resolution then some individual pixels will need to be combined into one screen pixel (Figure 5-5b). Alternatively, if the confocal microscope image is displayed larger than the screen resolution then many screen pixels will represent a single image pixel (Figure 5-5c). If the increase in image size is large, then the image will appear pixilated (Figure 5-6). However, image manipulation programs, can be requested to re-sample the image (interpolate the pixels) to create an array of "new" pixels so that the image does not appear as pixilated (discussed in more detail in Chapter 6 "Imaging Software", page 163).

In summary, there are three levels of image resolution that you need to be aware of (Figure 5-7): the screen resolution, the display resolution, and the number of pixels in the image. The optimal size for viewing confocal microscope images is at the display resolution, where an image pixel is equal to a screen display pixel.

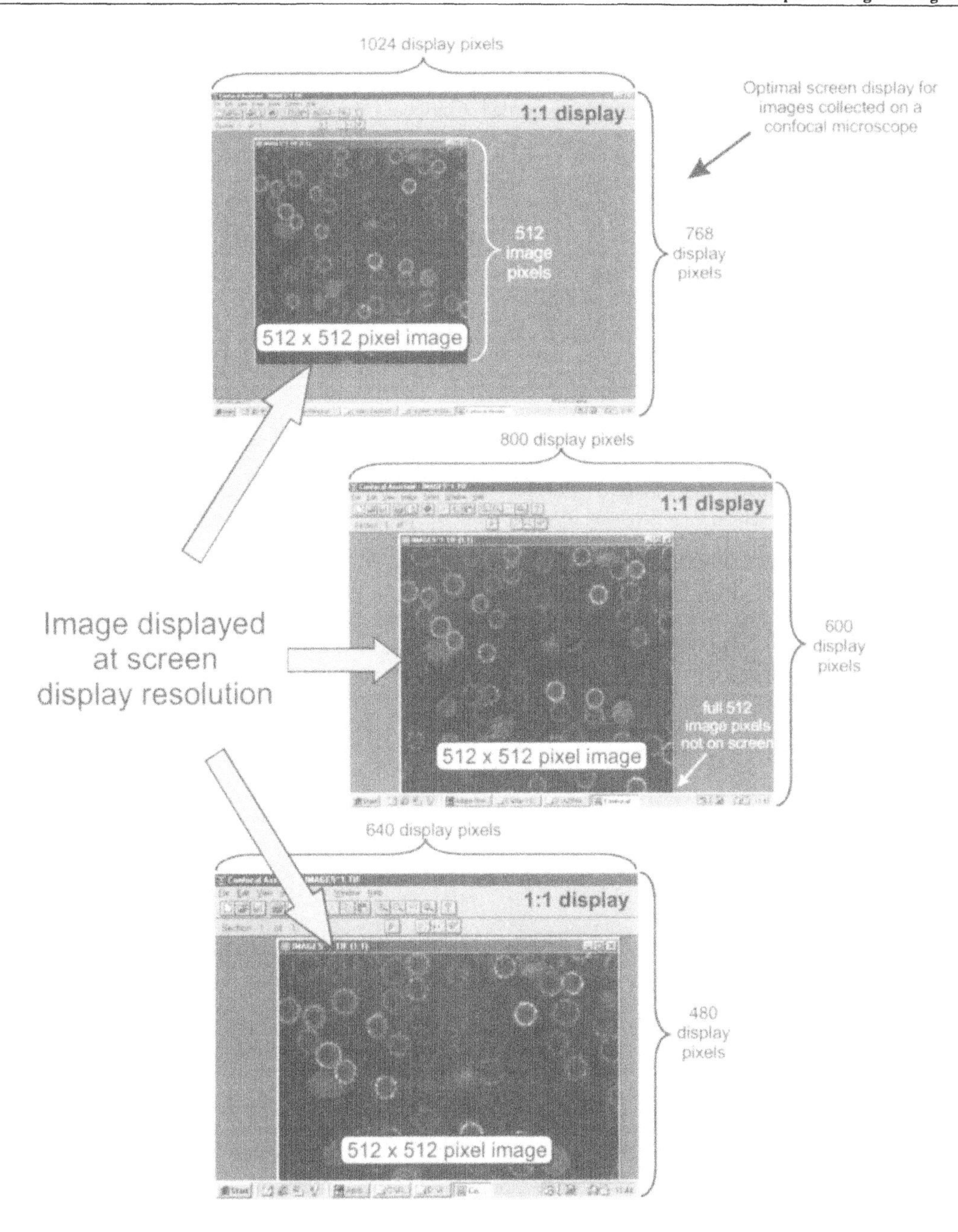

Figure 5-8. Image Display at Different Screen Resolutions.
Changing the computer monitor display resolution results in a 512 x 512 pixel image being displayed at different sizes. This occurs when the image pixels exactly equal the screen display pixels (1:1 or 100% display). Altering the computer monitor screen display resolution alters the size of the display pixels and thus the size of the image displayed. Changing the display resolution from 1024 x 768 pixels to 640 x 480 pixels has effectively resulted in a digital zoom of the image (i.e. the display pixels, and hence the image pixels are displayed larger on the screen). Enlarging the display pixels by lowering the display resolution will result in the display pixels, and hence the image pixels, becoming visible on the computer screen.

COLOUR IN DIGITAL IMAGES

Digital images obtained from a confocal microscope consist entirely of intensities for each individual pixel, and no colour information from the original sample is recorded. Confocal microscope images are therefore collected as grey-scale images and either subsequently coloured after collection, or displayed as coloured images during collection by applying a colour Look Up Table (LUT) to each signal channel. Using colour in confocal microscopy is covered in some detail on page 128 (in Chapter 4 "Image Collection"). This section deals with how we perceive colour and how colour is generated using digital image display devices such as computer monitors and printers.

Our means of colour perception greatly influences what colour we actually "see", a very important consideration when struggling to print an image that looks wonderful on the computer screen.

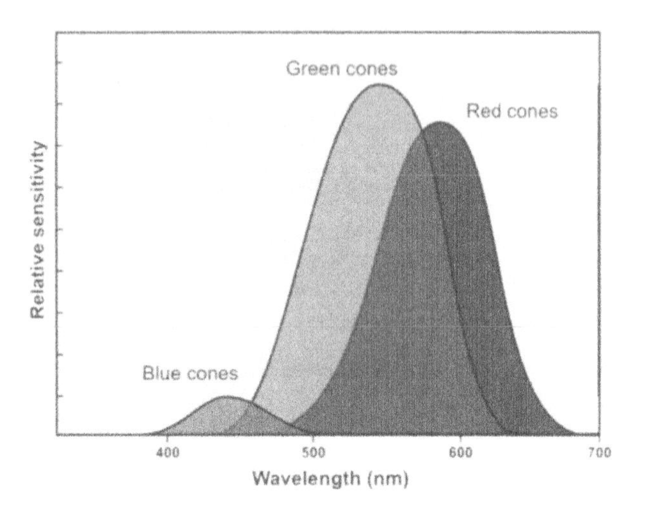

Figure 5-9. Human Perception of Colour.
We have 3 types of colour receptors called cones. One type is most sensitive to red, the other green and the third blue light. The millions of colours that we can perceive are made up of the differences in excitation of these three receptors.

Human Perception of Colour

The human eye has three colour receptors (cones), one that is mainly sensitive to red light, one to green / yellow light and the third to blue light (Figure 5-9). The huge range of colour that we can perceive is entirely made up of variations in the level of detection of the light by these three chromophores.

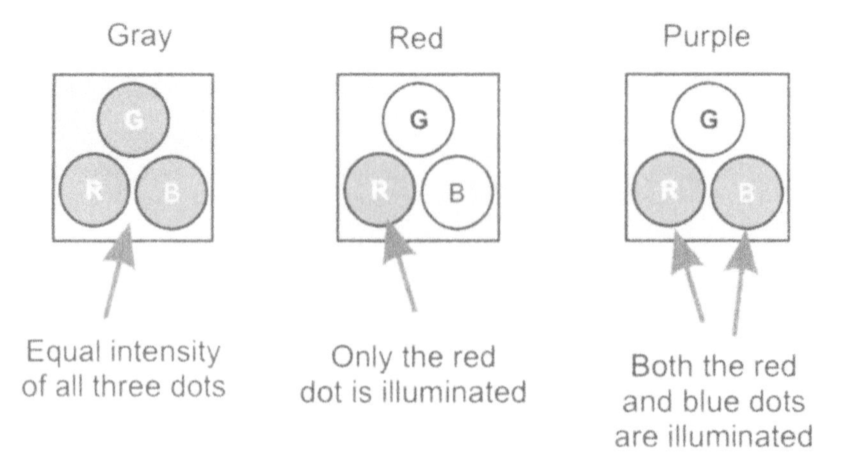

Figure 5-10. Colour Generation on a Computer Screen.
Individual screen pixels consist of a red, green and blue dot. Equal intensity of these three colours results in grey (left panel). Illumination of individual red (central panel), green or blue dots results in the display of these primary colours. Simultaneous display of a red and green dot (right panel) results in the secondary colour purple. Altering the relative intensity of these three individual colours results in the generation of the millions of different colours that we can perceive on a computer screen.

Partially colour-blind people (approximately 1/10 males are red/green colour impaired) have either a missing chromophore, or the chromophore is not of the correct characteristics and greatly overlaps in its absorption with one of the other chromophores.

An interesting characteristic of human colour vision is that colours are considered in the context of both their surroundings and the known purpose of the object being examined. For example, white paper looks white in both sunlight and indoor light. This is despite the fact that the balance of wavelengths of light (colour) bouncing off the paper in sunlight compared to indoor lighting is quite different. The paper in reality should look a different colour, but we interpret the light reflected from the paper as white anyway!

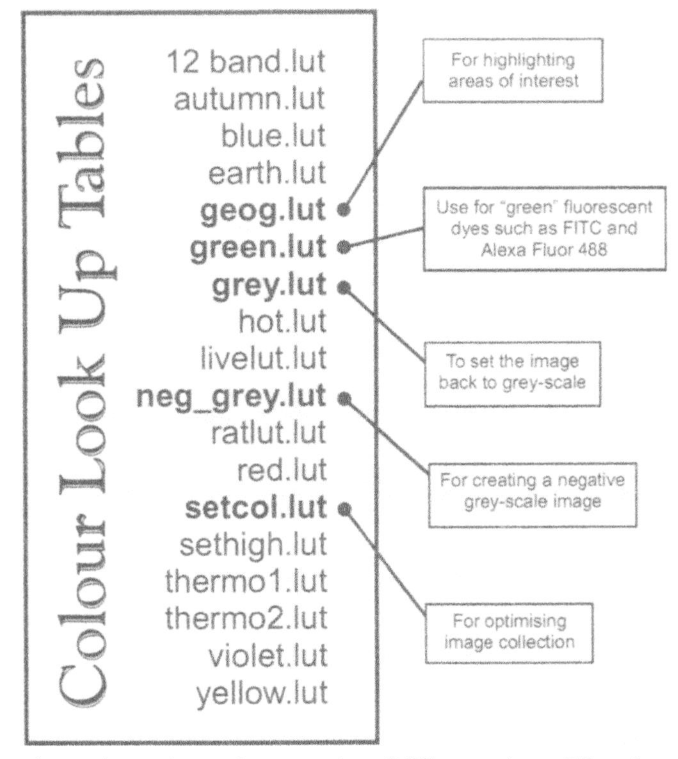

Computer Colour Generation

The computer generates all of its wonderful colours in a similar way in which humans perceive colour. That is, the computer uses only red, green (or yellow), and blue light to generate all the colours you see on the computer screen (Figure 5-10). Simply altering the proportions of these three colours give us the perception of different colours. When these colours are equally balanced at a high level of brightness the monitor will appear white. Lowering the level of brightness of the three colours without altering the balance between the colours creates varying shades of "grey".

The colour dots on a computer screen display are very close together, but can be readily seen with a low magnification hand lens.

Colouring your Image

Grey-scale images generated by the confocal microscope can be coloured in two fundamentally different ways (Figure 5-11). Immediately after collection (or even during collection) a colour LUT can be applied (see box above) to highlight particular areas of the image based on varying intensities of the individual pixels. A LUT may be applied to all the pixels in a grey-scale image so that they possess a colour range indicative of the emission spectrum of the dye being used. Later, when the image is used for producing prints for publication the LUT coloured image may be converted to RGB (red, green, blue) 24-bit colour.

Colour Look Up Tables (LUTS)

A grey-scale image can be readily coloured with a number of different LUTS provided with the software used to control your confocal microscope, or available in stand-alone image processing software. You can also readily alter the LUTs provided to suit your own requirements. The LUT operates by specifying exactly what colour a pixel of a particular grey-scale will have in the image. LUT colours are usually designed to form a gradient of colour that will highlight the fluorescence labelling. For example, the green LUT is a gradient of colour from dark green, light green, yellow/green, yellow and finally white. This colour gradient is much more effective for displaying green fluorescence than simply using green alone.

The autumn LUT is used in a similar way to the green LUT above, but this time for colouring images red. The autumn LUT is a gradient of colour from dark red to white and is much better at displaying red images than simply using red alone.

Several other LUTs are used for highlighting the area of interest within the image with specific colours. For example the "GEOG" LUT is a series of brightly coloured bands ranging from "cold" (blue) colour to "hot" colours (yellow and white). Suitable adjustment of the brightness and contrast of the image when using the "GEOG" LUT can be used to highlight an area of the image of interest with a colour that denotes the level of labelling in that compartment. In this way the comparison of the level of labelling between two different cells or subcellular compartments is visually very accurate.

An 8-bit colour LUT is sometimes referred to as "Indexed Colour". Applying an 8-bit LUT to a single channel grey-scale image is essentially the same operation as performed for a 24-bit RGB image, but with only one channel involved. Some software provides the means to apply an 8-bit Indexed Colour LUT to an image with 2 or 3 channels. Although this creates a multi-coloured image only 1/3 the size of a 24-bit colour image, the quality of the image will be greatly degraded and the brightness and contrast of each channel will be irreversibly altered. This type of colour image should only be used where size must be minimized, such as for web graphics.

Colour 24-bit (RGB) Images

Image manipulation software other than the proprietary software that controls your confocal microscope does not recognize the LUTs associated with your microscope. Therefore to use other software, the LUT coloured image may need to be converted to an RGB (red, green, blue) format image. The RGB format is a 24-bit full colour image where each pixel within the image now specifies the intensity of each of the red, green and blue components of the screen display pixels. An RGB image can be readily exported in a suitable format (for example TIF, see file formats on page 162) to other image manipulation software. A full colour RGB image is 3x the size of the original LUT coloured grey-scale image. This still holds true for single

> **An RGB image is now 24-bit,**
> a much larger file size
> (3x the size of an 8-bit image)

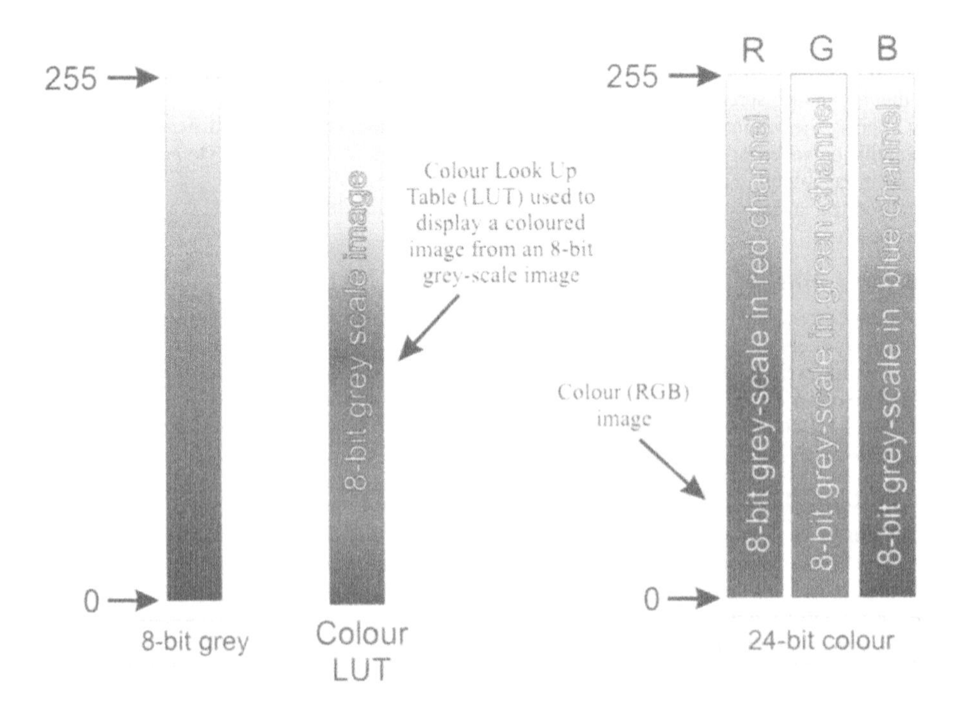

Figure 5-11. LUT Colour and RGB Colour Images.
Colour can be applied to a grey-scale image by using a Colour Look Up Table (LUT), in which case the original file remains a grey-scale image. A LUT coloured image can be converted to an RGB image (24-bit colour image), which is 3x the size of the original grey-scale image.

> **A LUT (Look Up Table) coloured image remains an 8-bit image - a relatively small file size**

colour images, where for example, the red and blue channels are set to zero – these "zero" channels still take up their full allocation of space in the 24-bit RGB colour image. The original 8-bit LUT coloured image only needs to be converted to a 24-bit colour image if you have more than one channel being used (i.e. you have a merged multi-labelled image from the confocal microscope). A single channel 8-bit, or Indexed Colour image is much smaller and gives exactly the same result whether shown on screen or printed.

Colour in Merged Images

Images displayed as merged images by your confocal microscope software are in fact not truly merged, but are retained as separate grey-scale images and are simply displayed on the screen overlaid in different colours (RGB, 24-bit colour display). Overlay colours have the advantage of allowing one to readily alter the intensity of either of the colour channels independently. Red and Green are the two colour channels routinely used for dual labelling, with blue being used for a third fluorescence channel or sometimes a transmission image. The merged multi-coloured image displayed "live" during collection on a confocal microscope can sometimes be saved as a 24-bit RGB colour image, or the original 8-bit grey-scale confocal microscope images that are saved can be used to create a 24-bit RGB colour image later using appropriate image manipulation software.

> **The Colour "Yellow"**
>
> Did you know that the colour yellow can be created in two fundamentally different ways -
>
> 1. *a mixture of red and green (i.e. two separate wavelengths of light together),*
> 2. *a single wavelength of light intermediate between red and green.*
>
> With careful adjustment of the intensity of the individual wavelengths of light involved both of these "yellows" can be made to appear exactly the same!

Combining red and green creates yellow (but see interesting aside in box on this page). The exact colour of the overlap will vary, depending on the ratio of the intensity of the two different colours (red and green). Great care should be taken when using the "yellow" colour as evidence of co-localization in high resolution images as small changes in the red/green intensity balance may result in parts of the image not appearing to co-localize, when in fact they do - but one channel is being "swamped" by the other channel.

Printer inks

Unfortunately the simplicity of forming a red/green merged image using the confocal microscope software, is somewhat lost when printing the image. Large commercial printers often use the complementary colours to Red, Green and Blue (Yellow, Cyan and Magenta) for printing on an opaque surface such as paper. Printing red and green on paper then requires various ink combinations, which may result in unpredictable results, described in more detail below.

Printing Colour

Unfortunately, as mentioned above, most large scale commercial printers generate colour images using a different set of colours compared to a computer monitor. These colours are cyan, magenta and yellow (CMYK, where K is black, which is often included in colour printing). These colours are the inverse of red, green and blue. For example Cyan (a pale blue colour) is the full light spectrum with the red part subtracted, Magenta (a pink colour) is the full light spectrum with the yellow/green removed.

There are important differences between the generation of colour on a computer and that of colour printing. Images displayed on the computer screen are visible because the monitor is emitting light of specific wavelengths (colours) that stimulate our colour receptors, whereas the printed page is reflecting light of great complexity that is also stimulating our colour receptors. In fact, it's a wonder we can ever see the same colour on a computer screen as we can see on the printed page!

This fundamental difference in how colour is generated on a computer monitor compared to that on printed paper is one of the reasons that printing an image that faithfully represents what appears on the screen is very difficult.

Printing is also further complicated by the fact that the printer inks are only an approximation of the ideal spectrum for each individual ink colour. Minor spectral contaminants may become obvious in specific ink combinations. This is not normally noticeable on a photographic type image, but may be witnessed as "rings" of colour on a printed confocal microscope image due to gradations of colour.

This difficulty in obtaining a print that faithfully represents that which is seen on the computer screen can be largely overcome by experience in manipulating the colour and intensity of the image that is to be printed. In high quality printing houses the computer screen colours can be matched with the printed output, but we are usually stuck with trial and error!

The various printers used to print colour images are discussed further on page 183 (in Chapter 7 "Presentation and Publication").

IMAGE MANIPULATION

One of the great advantages of obtaining images in a digital format is the ease with which the image can be manipulated for later presentation and publication. In this section various methods of attempting to "enhance" the image are discussed. These filters are computer algorithms that perform mathematical operations on the numbers that represent the intensity values within the image. These filters usually operate on blocks of values so that each pixel is modified based on the intensity values of the neighbouring pixels. Most filters are applied to the digital data simply by choosing the appropriate "filter" from the menu.

The following filters are usually present in your confocal microscope software. A variation of these basic tools is also available in most imaging software, including the widely used Photoshop program.

Contrast Stretching

Images should always be collected within the grey-scale range available (256 shades of grey when using 8-bit image collection, and 4096 grey levels when using 12-bit image collection). Use the SETCOL Look Up Table (see page 129). With this colour Look Up Table (LUT) the highest value pixels (255 or 4095) are shown in red, and the lowest value pixels (0) are shown in green. Adjusting the gain and black level to result in only a very small number of red or green pixels being present can be used to collect an optimised image. However, collecting a fully optimised

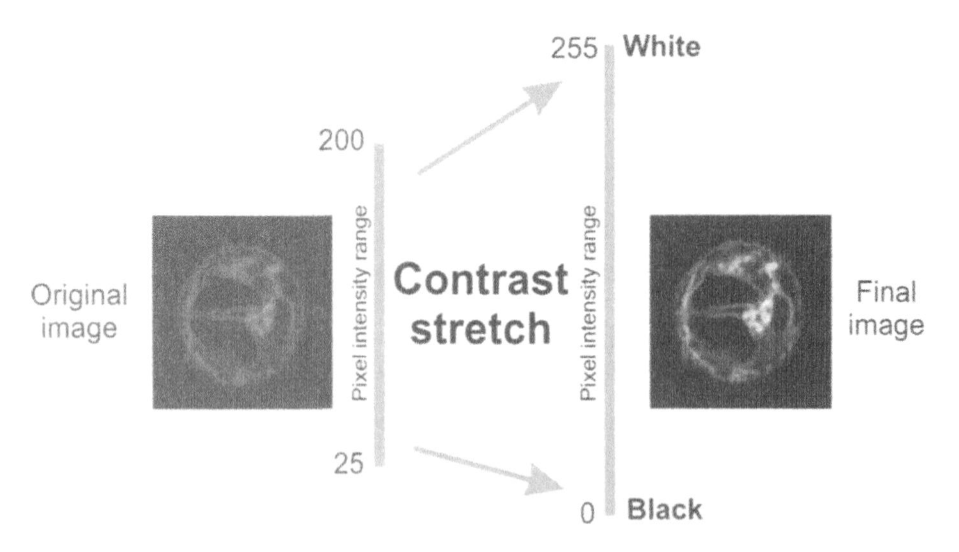

Figure 5-12. Contrast Stretch.
Images that have been collected with an intensity range less than the 256 grey levels possible with an 8-bit image can be contrast stretched to fully utilise the grey-scale range available. In the above example the original image has a range between 25 and 200, which can be stretched out to the full range of 0 to 255. This results in a higher contrast image as shown on the right.

image is not always possible. Making full use of the 256 shades of grey available may also be carried out by contrast stretching (Figure 5-12) after collection. Contrast stretching simply spreads the original grey levels (in this example ranging from 25 to 200) out to the full range of 0 (black) to 255 (white). In this way a higher contrast image that takes advantage of the full dynamic range available is created.

Using contrast stretching on a 12-bit image is an important method of making full use of the greatly increased dynamic range available in a 12-bit image.

Warning!
Contrast stretch cannot be "undone" once the file has been saved.

Contrast stretching will not appear to "brighten" the image if a small part of the image (as small as 1 pixel!) is already at full brightness. This can sometimes occur when a small fleck of highly concentrated label is present in the image (for example a spot of precipitated fluorochrome). The high intensity pixels may need to be removed prior to contrast stretching. Alternatively one can simply "stretch out" the brightness levels in the image using software brightness controls.

Take care not to use contrast stretch on images for which you may later wish to compare levels of fluorescence. The individual slices of a "z" series collected for 3D reconstruction can also be contrast stretched. However, this may seriously alter the balance of labelling within the 3D image.

Image Enhancing Filters

A number of computer algorithms are available in most imaging software for "enhancing" your image. These "enhancement" filters work by analysing each individual pixel and comparing that pixel with the surrounding pixels – but "enhancement" alterations cannot be reversed! These "enhancement" filters do not add information to the image. In fact, in spite of the term "enhancement" they actually detract from the information available in the original image. However, careful use of these digital filters can be valuable for preparing an image for presentation and publication.

Image Enhancement Filters
Median Filter
Edge Enhancement
Local Area Contrast
Sharpening Filter
Smoothing Filter
......and more....

The **Smoothing Filter** is particularly useful for removing "grainy" (high noise) areas of images that were collected from samples with below optimum fluorescence intensity. The Smoothing Filter can be followed by one of the edge enhancement or sharpening filters to "refocus" the image.

The **Median Filter** can be used to remove "noise" from an image. The Median Filter removes single pixel "spikes" of noise by looking for pixels that are much brighter than their neighbours and setting them to the median value of their neighbourhood. However, care needs to be taken when using the median filter or inappropriate banding patterns may appear within the image. Some "purists" argue that you should never use a median filter (or most other filters for that matter) because you should present your data as close to the original as possible. However, the careful use of filters may be important in allowing your images to be more clearly presented, particularly in a seminar and in publications. A final warning – if you intend to perform "quantitative" analysis on your images this should be done on the original image before using any "enhancement" filters.

The edge enhancement or **sharpening filters** come in a variety of forms (Edge Enhancement, Local Area Contrast, Sharpening). Each filter contains a different algorithm for increasing sharpness in the image. Finding the most useful filter or combination of filters is often a matter of trial and error. However, the following discussion on edge enhancement filters may provide some help.

The **Local Area Contrast Filter** is particularly useful for increasing the intensity of individual "dots" of fluorescence within the image. This type of filter should be used with care in images with relatively large areas of fluorescence as applying this algorithm may create a "grainy" image. This type of filter has been used to good effect in Figure 7-1 (page 177) in Chapter 7 "Presentation and Publication", where very poorly staining vesicles within the malaria infected red blood cell cytoplasm have been increased in intensity to clearly demonstrate their position within the cell (including time lapse movie images) without being concerned about the relative intensity of the vesicles or with comparing their intensity of labelling with other cellular structures (such as the intensely stained nucleus of the parasite in this example).

Edge Enhancement Filters are best used on images with relatively defined structures or edges within the image. This filter may result in an image that looks significantly more "in focus". The **Sharpening Filter**, similar to the edge enhancement filter, can be used to "re-focus" an image. However, the term "focus" in relation to confocal microscopy images has quite a different meaning to "focusing" a conventional bright-field microscopy image. In a conventional light microscope image a "blurred" or "out-of-focus" image is normally due to the light not being correctly focussed – and can be evident on quite low magnification images. In confocal microscopy the "blur" in the image is due to either the diffraction limit of the microscope, or due to movement of the object. The image will not be "out-of-focus" in the traditional sense as the pinhole has removed "out-of-focus" light. This means that with edge enhancement filters, although they will make the image appear more "in-focus", you need to remember that the original "blur" was not "out-of-focus".

Sharpening filters can also have the deleterious effect of "sharpening" or making more obvious any "noise" in the image. The edge enhancement and sharpening filters usually have the ability to set the pixel matrix size so that noise fluctuations within the image (seen as "speckle") are not themselves "enhanced".

QUANTITATIVE ANALYSIS OF IMAGES PIXELS

One of the advantages of digital images is that the intensity levels within the image can be readily quantified. Most imaging software, including the software provided by the confocal microscope manufacturers and general image analysis and manipulation software such as Photoshop, allow some means of measuring the intensity of the pixels within the image. However, sophisticated quantitative image analysis, particularly if you want to analyse 3D data sets, requires dedicated (and expensive) software.

Quantitation of digital images is carried out by displaying the pixel intensity values within the image. This can be carried out by using a line drawn across the image, by selecting the whole image, or by drawing a shape around the area of interest and using the software to create a histogram of the frequency of the pixel intensity values.

Don't forget that although the intensity of labelling within an image can be readily and accurately measured, this does not necessarily mean that the experiment is quantitative. In fact, it's quite difficult to quantify fluorescence imaging data as the nature of the original experiment often means that only qualitative data is available. A general

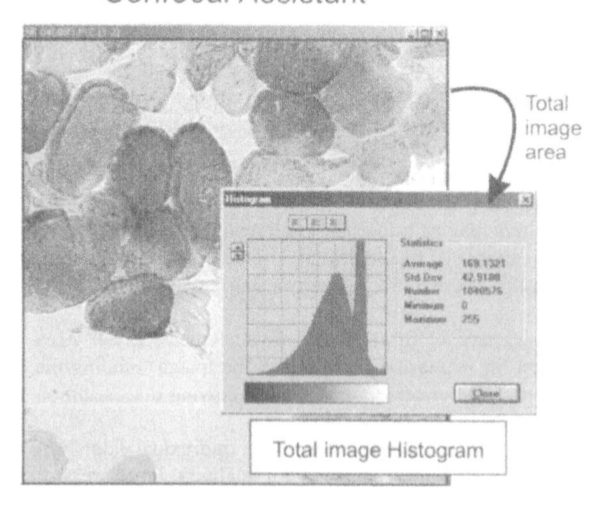

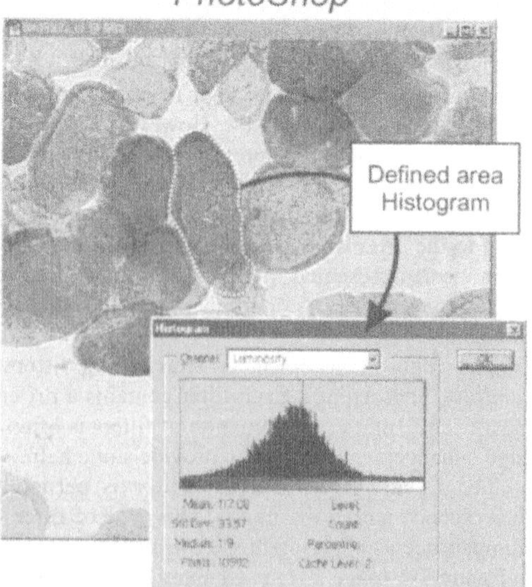

Figure 5-13. Histogram Analysis of Images.
Image analysis programs can be used to measure the exact value of any of the pixels within the confocal microscope image and to display these intensities as a histogram (described in more detail in Figure 5-14). In this example information on the total pixel intensities within the displayed image (on the left), for a specified area of the image (right panel) is shown as an image Histogram. The average, standard deviation, maximum and minimum pixel intensities are also shown for each image. These images are "real-colour" transmission images of immunostained human muscle fibres. The dark punctate staining is due to a high concentration of the antibody (stained with DAB) showing respiratory active mitochondria – with almost no staining in the background between muscle fibres.

rule for establishing experiments that can be readily quantitated is to use the lowest level of fluorescent label as possible, and to label all samples in a single batch together. The difficulties of quantitation of confocal microscope images, and particularly the difficulty of collecting quantitative data, are discussed in more detail in Chapter 10 "Confocal Microscopy Techniques".

NIH Image or Object Image (for Mac) and Scion Image (for PCs) and Image J (cross-platform) are free programs all based on NIH Image that are capable of most basic, and some quite advanced, image processing

Image Histograms

A histogram of the pixel intensities within the image is a very useful way of analysing the spread of labelling intensity within an image. At the simplest level a histogram of the whole image can be generated (Figure 5-13, left hand panel), or the histogram can be generated from a defined region of the image (Figure 5-13, right hand panel). A histogram can also be created from a line or in fact any defined area within the image. Most image processing programs, including the widely available NIH image, Photoshop and Confocal Assistant can be used to generate an image histogram.

An image histogram (Figure 5-14) consists of a graphical display of all the pixels in the defined area of the image being analysed. The pixel intensity (x-axis) is plotted against the number of pixels with that intensity (y-axis). The scale of the display can be altered to highlight pixel intensities that have very few pixels, or if you wish, to display the pixel intensities that have very high numbers. In the example in Figure 5-14 the histogram of only one channel (red) of a 24-bit full colour image is shown. The histogram analysis also includes statistical information on the average, standard deviation, maximum and minimum pixel intensities within the total image or defined region.

A special use of the image histogram is to generate a histogram of a very evenly stained sample (fluorescent plastic or a solution of fluorophore are ideal) and to use the resultant histogram to analyse the spread or "noise" within the image.

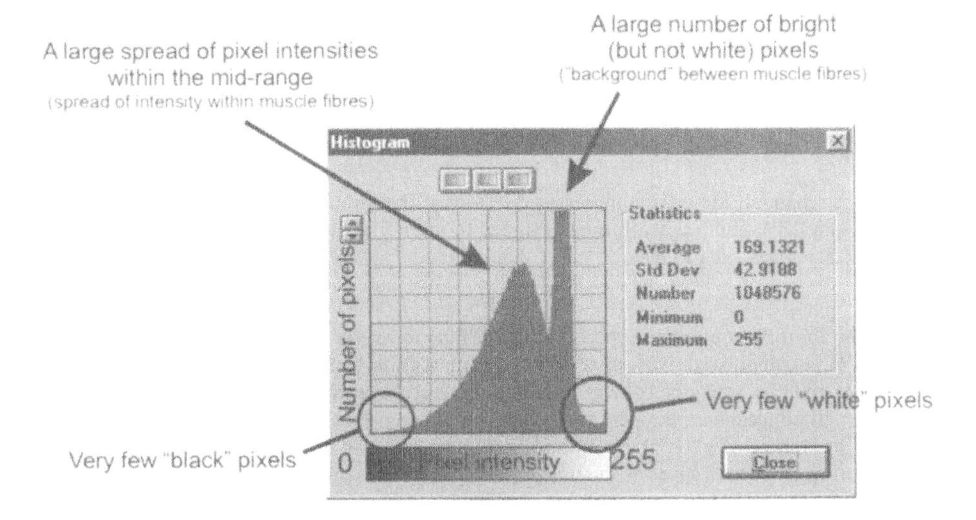

Figure 5-14. Image Histogram.
The image histogram is a valuable tool for analysing the distribution of pixel intensities within the image. The pixel intensity is plotted on the x-axis, and the number of pixels for each intensity level is plotted on the y-axis. In this example only the "red" channel of an RGB real-colour 24-bit image as shown in Figure 5-13 is displayed. This histogram clearly shows that there are a very large number of high intensity pixels (the very high peak on the right of the histogram display), but there are very few "white" or full brightness pixels (extreme right of pixel display). The main bulk of the pixels are spread in a broad peak within the middle of the intensity display. There are very few "black" pixels at the far left of the histogram display.

FILE FORMATS

Images collected on a confocal microscope are saved in a number of different file formats depending on the confocal microscope manufacturer concerned. Sometimes the file format is unique to a particular manufacturer (for example, the Bio-Rad "PIC" format), or more conveniently the file is of a type, such as TIF, that can be widely read by a variety of image analysis programs.

Most image file formats give you a choice of file compression (except the Bio-Rad PIC format, which is never compressed). Data compression is a great space saver (sometimes up to 1/10th the original file size is created). However, the down side of data compression is that in many types of file compression some information in the image is lost. This may not be noticeable at first, but repeated compression / decompression cycles (as in reading and saving using a format that compresses the files) may result in significant changes that will become very noticeable.

Always save the original confocal microscope file in a safe place, and then save all subsequent changes to the image to another file name. In this way the original uncompressed data can be retrieved at a later date.

The following file formats are widely used:

- **PIC:** The Bio-Rad PIC format is not compatible with any other PIC format (or any other format in fact!), but can be loaded as a "raw" image (bypassing the header bytes at the beginning of the file) by most image manipulation programs. A number of programs can load Bio-Rad PIC files directly, or by using "plugins". All images obtained from a Bio-Rad confocal microscope are collected in this proprietary Bio-Rad format, and can be translated to another format by many image manipulation programs, including Confocal Assistant, by simply saving the file in the new format. Bio-Rad PIC files are not compressed.

- **TIF:** Tagged-Image File Format is used to exchange files between applications and computer platforms. TIF is a flexible bitmap image format supported by virtually all paint, image-editing, and page-layout applications. The TIF format can be either compressed or uncompressed – but be warned, the TIF "standard" is notoriously non-standard! In particular, most multi-image TIF files obtained from confocal microscopes cannot be read by common image processing programs. They will need to be saved as single image TIF files before they can be accessed.

- **BMP:** Standard Windows Bit Map image format on DOS and Windows-compatible computers. You can also specify RLE compression.

- **GIF:** Graphics Interchange Format, a compressed file format that is often used for presentation of images on web pages. GIF is an LZW-compressed format designed to minimize file size and electronic transfer time. This format is particularly good for producing "clear" backgrounds, where the cells appear to "float" over other images or text.

- **JPG (JPEG):** The Joint Photographic Experts Group (JPEG) format is commonly used to display photographs on web pages. JPG is compressed, with the degree of compression being selectable when you attempt to save the file. Some information (but not colour) is lost from the image during compression. However, this will probably not be at all noticeable, unless the file is repeatedly compressed and uncompressed. This file format is commonly used for storage of photographs from digital cameras, but be careful when using compressed JPG file format for storing confocal images as the image is altered by the compression algorithm (even when there is no discernible difference). You may encounter difficulties when attempting quantitative analysis of JPG compressed images. JPEG compression can be avoided by setting image quality to "highest" when saving.

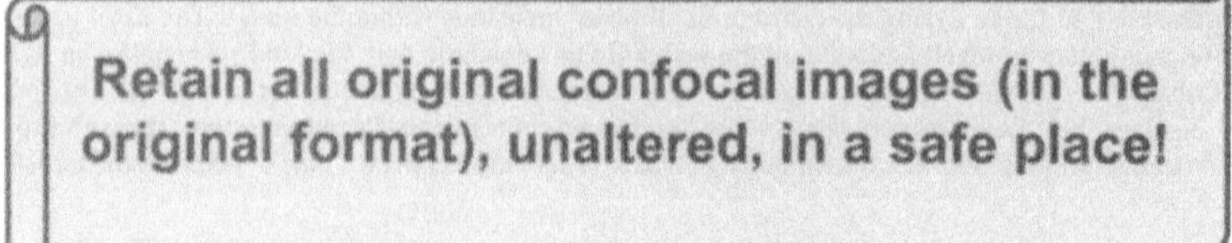

Chapter 6

Imaging Software

A wide variety of imaging software is now available. An understanding of how imaging software operates, and the advantages and limitations of various software packages is important for getting the most out of your images - and to lower the level of frustration in using the wrong software for the job at hand!

Imaging software can be divided up into three different areas, image collection, image analysis and diagram or

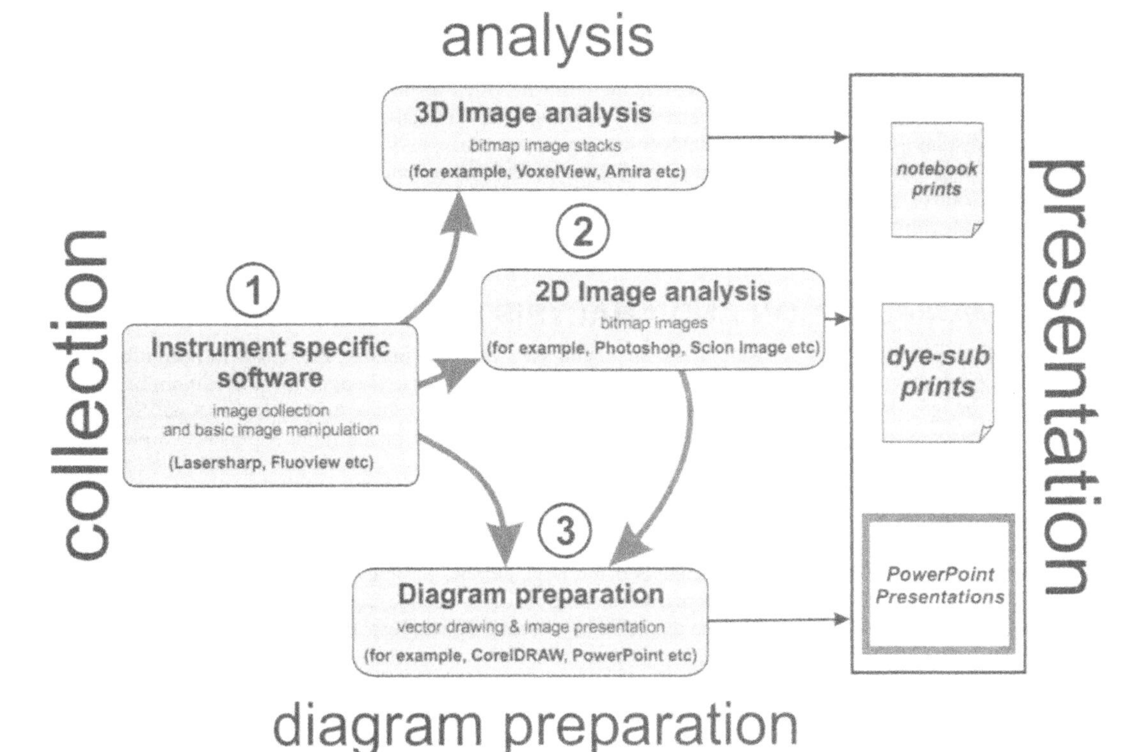

Figure 6-1. Imaging Software.
Confocal microscope image collection software (1) is provided by the microscope manufacturers and is specific to a particular model of microscope. You will also need one or more software applications at a workstation away from the confocal microscope to view, modify, merge, and perform quantification of the images (2) and further software for preparing images for publication (3).

image presentation (Figure 6-1).

The collection of the image from the confocal microscope is performed by proprietary software provided by the microscope manufacturer. This software also performs the task of controlling the operation of the microscope (changing laser intensity, choosing laser line, altering the pinhole size etc).

The next step, after the image has been obtained from the microscope, is called image analysis. The software used to control the microscope will perform a number of basic image analysis functions quite well. Further image analysis is performed using a variety of software. There is no single piece of software that provides all the image analysis you will need. As is explained below, different programs have fundamentally different methods of handling the image and so are suited to particular tasks. For example, Photoshop is particularly suited to printing of images, whereas VoxBlast is designed for 3D analysis of images and CorelDRAW excels in preparing figures for publication.

IMAGE COLLECTION SOFTWARE

As mentioned above, the software used for image collection is specific to each of the manufacturers of confocal microscopes. Although the controls required for manipulating the settings on the confocal microscopes are often similar or even the same, the software tools for accomplishing the task are quite different between each of the manufacturers. For this reason the software used to control the microscopes is not described in any detail in this section. In "Appendix 1: Confocal Microscopes" (see page 355) there is an introduction to the software used to control the microscope and collect images from each of the major confocal microscope manufacturers. However, for a detailed description of the software you will need to consult the manufacturer's manual or on-screen help provided with the software.

Image collection software provided by the various confocal microscope manufacturers is primarily designed to control the various optical parameters on the microscope (pinhole size, PMT gain, scan speed etc) and to display the scanned images on a computer screen. However, the software often also provides excellent, although somewhat basic, 2D and 3D image manipulation and analysis. Although there is dedicated software for sophisticated image analysis and display, the algorithms available on the software driving your confocal microscope are often a convenient starting point for image analysis. Having the confocal microscope software installed on other computers is also useful (if this is part of the agreement with the confocal microscope manufacturer) so that you can open files, access microscope information and perform simple 2D and 3D analysis without the need for additional software that may involve translating the images to a different format.

IMAGE MANIPULATION AND ANALYSIS SOFTWARE

In addition to the proprietary software that comes with your confocal microscope, you will need a basic image manipulation program (such as Confocal Assistant or Photoshop) to prepare images for publication, or simply for viewing your images on a separate computer from the one dedicated to the confocal microscope. Several other programs are capable of producing excellent 3D presentations and some of the more sophisticated programs can perform image analysis on the 3D image stack.

A number of excellent programs, such as Confocal Assistant, NIH Image, Scion Image etc are provided free of charge and can be downloaded from the relevant web sites. These programs are an excellent starting point for basic manipulation of images obtained from a confocal microscope. The real power of NIH Image and related programs such as Scion Image, Image J and Object Image is that a large number of plug-ins are available that greatly extend the basic feature set available in these programs. These plug-ins are available from a number of sites on the internet, or you can program your own and make them available to others by uploading them to the internet sites.

Imaging Software from Confocal Microscope Manufacturers

Some confocal microscope manufacturers provide free "stripped down" versions of their confocal microscope software to facilitate viewing and basic image manipulation of images obtained from the confocal microscope. The programs are marked as either free (by downloading from the appropriate web site), shareware (where basic functions are often free, with a small fee required for prolonged use or to unlock special features), or commercial applications (from relatively modestly priced basic programs to very expensive sophisticated applications).

Confocal Assistant: [Windows, free] see page 168 Basic image manipulation (including 3D) and ability to translate between many file types (but originally designed for viewing and basic manipulation of Bio-Rad proprietary "PIC" files). Simple dual and triple channel merging, mosaic presentation and 3D representation are available. Available from Bio-Rad web site, (ftp://ftp.genetics.bio-rad.com/Public/confocal/) or the Purdue flow cytometry web site, www.cyto.purdue.edu/flowcyt/software/cas402.htm

Zeiss Image Examiner: [Windows, free] Confocal microscope image database software from Carl Zeiss for opening and viewing images collected using a Zeiss confocal microscope. www.zeiss.de/C12567BE0045ACF1?Open

Leica Lite: [Windows, free] Confocal microscope image database and basic image manipulation software from Leica for viewing images collected on Leica confocal microscopes.

LaserSharp Analysis: [Windows, free] Confocal microscope image database and basic image manipulation component of the Bio-Rad LaserSharp confocal microscope software can be installed on computers remote from the confocal microscope for viewing images collected on a Bio-Rad confocal microscope and for performing basic image manipulation.

Basic Image Manipulation Software

A wide range of basic image manipulation software is available free of charge on the internet. This type of software is primarily designed for manipulation of photographs, but can be a valuable aid in the manipulation and viewing of images from a confocal microscope. A number of these programs perform many of the same functions as Photoshop, and may be an excellent choice if funding is limited or you do not require all the tools available in Photoshop.

Irfanview: [Windows, shareware] Basic image manipulation software. www.irfanview.com

GIMP: [Windows, Mac, Linux, shareware] Free basic image manipulation software for Mac and PC. www.gimp.org/

GraphicConverter: [Mac, shareware] Basic image manipulation software. www.lemkesoft.de

Microview: [Windows, free] Free 3D volume viewer and analysis software for Mac and PC. http://microview.sourceforge.net/

Xnview: [Windows, free] Free image viewer and converter. www.xnview.com/

Quicktime: [Windows, free] **Quicktime Pro:** [Windows, commercial Program that displays and exports many types of movies and images. An upgrade to Quicktime Pro [Commercial software] allows additional image import and export functions, and basic movie editing applicable to any image series such as time-lapse, z-series and rotations. www.quicktime.com/

Software for Presentation and Preparation of Figures

Commercially available image manipulation programs primarily designed for photography and presentation are very useful for producing high-quality figures and slides for presentation and publication.

Photoshop: [Mac, Windows, commercial] see page 173 Image manipulation program produced by Adobe. Sophisticated image manipulation and publication quality printing. A program that takes particular care to keep you informed of the exact pixel dimensions and display resolution of your image. Colour balance and manipulation are particularly strong features.

www.adobe.com/products/photoshop/main.html#

Plugins are available from Bio-Rad for reading Bio-Rad "PIC" files (ftp://ftp.genetics.bio-rad.com/Public/confocal/).

CorelDRAW: [Mac, Windows, commercial] see page 175 Image presentation program from Corel for the preparation of figures and diagrams. An excellent program for preparing complex diagrams, but take care when re-sizing images as the resolution at which the image is being displayed may not be obvious. CorelDRAW diagrams can be transferred to PowerPoint for slide presentations. Corel also produces an image manipulation program called Corel PHOTO-PAINT. www.corel.com/

PowerPoint: [Mac, Windows, commercial] see page 176 A program that is part of the Microsoft Office Suite, designed for preparing seminar presentations. Images can be readily changed in size to fit into a presentation – but take care with critical images not to accidentally down-size the pixel resolution of the image. Animation and the insertion of short "movies" are great features of this program. www.microsoft.com/office/powerpoint/default.asp

Sophisticated 2D and 3D Image Manipulation Software

A range of 2D and 3D image manipulation and analysis programs is available for working with images from a confocal microscope. Many of these programs are not specifically designed for use with confocal microscope images, but do contain a wealth of algorithms that can be used to great effect with confocal microscope images. Some programs are capable of extensive manipulation and analysis in 3D.

Free Software

NIH Image: [Mac, free] A program for image acquisition, image processing and data analysis – with lots of "plugins" for various image manipulation tasks written in the NIH macro language. Includes 3D presentation and making a "movie". http://rsb.info.nih.gov/nih-image/

Scion Image: [Windows, Mac, free] Image acquisition and analysis program from Scion Corp. The program is based on NIH image, with added features for the control of Scion frame-grabber boards. Free Windows and Mac versions available from Scion Corp, www.scioncorp.com/

Image J: [Mac, Windows, Linux, free] NIH Image ported to Java by the author of NIH Image. Extensive Java plugins available, and custom plugins may be written in Java or the macro programming language, http://rsb.info.nih.gov/ij/.

Object Image: [Mac, free] Extended feature set based on NIH Image. Additional features include non-destructive overlays, 3D viewing and drawing (http://simon.bio.uva.nl/1-Introduction.html).

Commercially Available Software

Amira: [Windows, commercial] A high-end graphically presented 3D image manipulation program. Includes various methods of 3D data display and analysis. www.amiravis.com/

AutoDeblur: [Windows, commercial] Deconvolution (including its use in confocal microscopy) software produced by AutoQuant. www.aqi.com/

AutoVisualize: [Windows, commercial] 3D rendering, image processing and analysis software from AutoQuant. www.aqi.com/

IDL: [Windows, Mac, Unix & Linux] (Interactive Data Language) Image acquisition and analysis program with a flexible programming language that readily allows for custom features. www.rsinc.com/idl/

Image Pro Plus: [Windows, commercial] Image manipulation program from Media Cybernetics that can be used to capture and manipulate microscopy images (www.mediacy.com/ippage.htm/). A version that has a number of confocal microscope specific algorithms, called Laser Pix, is available from Bio-Rad. http://cellscience.bio-rad.com/products/software/LaserPix/default.htm

Imaris: [Windows, commercial] A high-end 3D image manipulation and analysis program produced by Bitplane. www.bitplane.ch/ or www.bitplane.com/

IPLab: [Windows, Mac, commercial] Image acquisition and analysis program with many modules for specialized functions. www.scanalytics.com

LaserPix: [Windows, commercial] 2D and 3D image manipulation and data analysis program provided as an additional component from Bio-Rad. This program is based on Image Pro Plus, with some additional features specific for confocal microscopy.
http://microscopy.bio-rad.com/products/software/LaserPix/default.htm

MetaMorph: [Windows, commercial] Image acquisition and analysis program with many modules for specialised functions from Universal Imaging Corporation. www.image1.com/products/metamorph/

Openlab: [Windows, commercial] Image acquisition and analysis program with many modules for specialized functions from Improvision. www.improvision.com/

Slidebook: [Windows & Mac, commercial] Image processing software from Intelligent Imaging Innovations (III). www.intelligent-imaging.com/. Also includes software for high-speed FRAP analysis using a spinning Nipkow disk confocal microscope.

SoftWoRx: [Windows, commercial] Deconvolution software from DeltaVision that is sold separately or bundled as part of the DeltaVision deconvolution system. www.api.com/products/bio/deltavision.html

V++: [Windows] Digital Optics web site provides information on V++ image processing and analysis software for Windows. www.digitaloptics.co.nz/

Volocity: [Windows, commercial] 3D visualization and measurement program from Improvision. www.improvision.com/.

VoxBlast: [Windows, commercial] A high-end 3D image manipulation and analysis program.
www.vaytek.com/VoxBlast.html

Voxel View: [Windows, commercial] A high-end 3D image manipulation and analysis program previously produced by Vital Images, but no longer available.

Software for Archiving Images

Archiving images obtained from a confocal microscope is fraught with many difficulties. Each of the major confocal microscope manufacturers provide software for accessing both the images and the saved instrument settings. Stand-alone archiving software, that can handle images from a wide variety of instruments is of great value, particularly in a multi-user facility.

SIDB: [Windows, free] Scientific Image Database for archiving and annotating images from a variety of instruments is available free of charge (including the source code). This software was developed at the University of Utrecht, The Netherlands. http://sidb.sourceforge.net

Confocal Assistant (free software)

A simple way to display confocal microscope images

Confocal Assistant (Figure 6-2) is a small but very useful program provided free of charge from Bio-Rad (www.bio-rad.com). The program is primarily used for looking at confocal microscope image files (the Bio-Rad PIC format is often not recognised by other software), and for translating the Bio-Rad PIC file format to a variety of other formats (for example TIF, GIF, JPG, BMP etc).

If your Windows environment is set up to recognize the PIC extension as being associated with Confocal Assistant, then this program becomes a very quick and simple way of viewing confocal microscope images from Bio-Rad, and is also convenient for viewing confocal microscope images from other manufacturers.

A number of simple, but very useful, image analysis tools available with Confocal Assistant are explained in some detail in the following pages.

The movie display button Figure 6-2 is useful for automatically scrolling through a time series that has been collected. The speed of display can be readily controlled, making movie display of time lapse experiments from collecting intervals of seconds, minutes or even hours a simple matter of clicking on this button.

Confocal Assistant
Windows (not available for Mac)

- Displaying confocal microscope images
- Changing file type (*PIC, TIF, BMP, JPG* etc).
- Simple image merging
- 3D-reconstruction
- Time lapse display
- Creating mosaics
- Simple printing

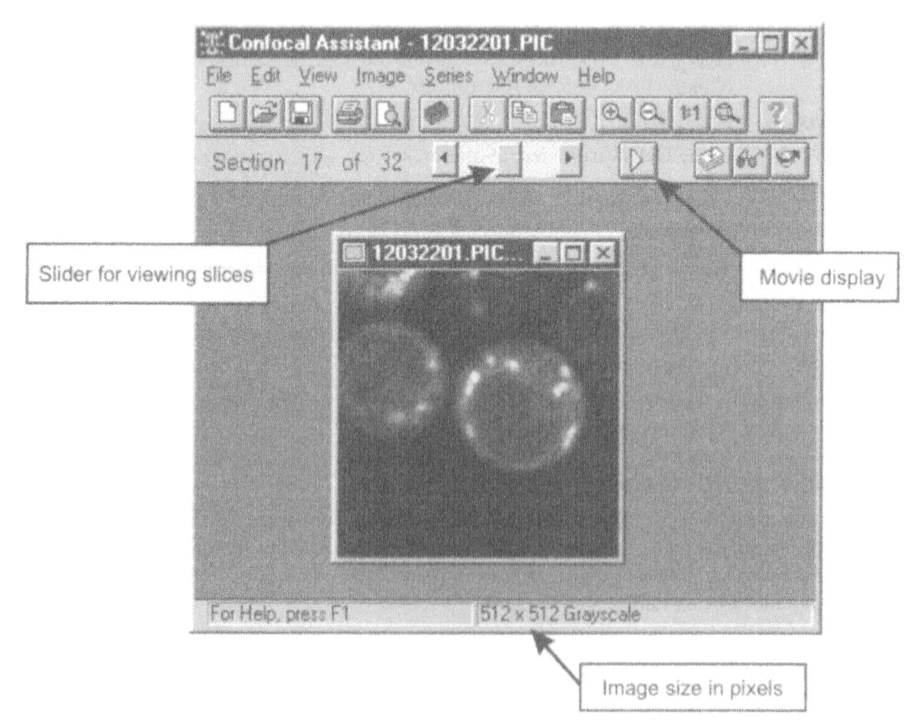

Figure 6-2. Confocal Assistant – Main Menu.
This small, but very useful program, is available without charge from Bio-Rad. Confocal Assistant is often used simply to view your image files as the program can directly read the Bio-Rad PIC format and the single image TIF format as used by Leica. There are also a number of useful image analysis tools available with Confocal Assistant, as well as the ability to save images in various commonly used image file formats.

Confocal Assistant continued ………

Converting PIC images to other file formats

Converting Bio-Rad PIC images to other file formats using Confocal Assistant is simply a matter of saving the file in the format of choice (BMP, GIF, TIF etc) by using the "save as" option from the file menu (Figure 6-3).

The **TIF** file format is a commonly used format (preferably non-compressed) for allowing the Bio-Rad confocal microscope images to be read by other software.

You should refrain from using lossy data compression on your images while you are still making changes. Only at the end when the image is ready for printing or sending to others should the file be saved in a format (such as **JPG**) which supports data compression. For sending large files across the Internet as an email attachment compression is essential (as much as a 10-fold reduction in size is possible).

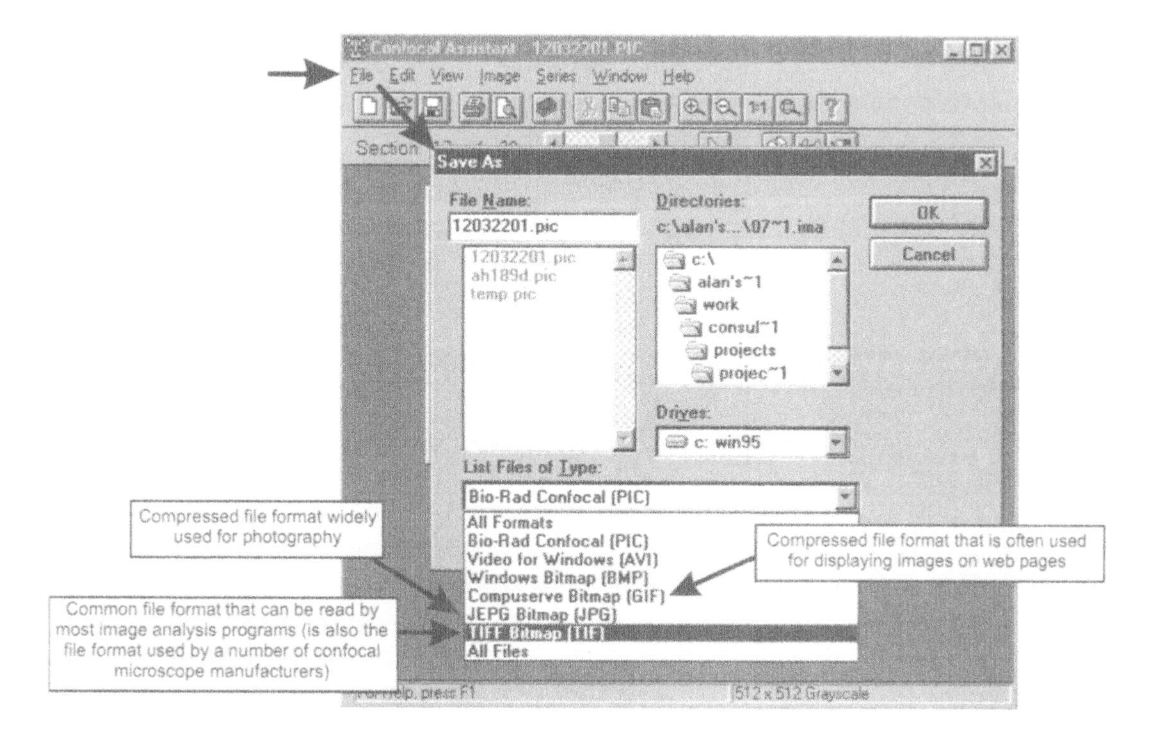

Figure 6-3. Confocal Assistant - Converting Bio-Rad "PIC" Files to Other Formats.
Images opened with Confocal Assistant can be readily converted to other file formats simply by selecting another format available from the menu when you save the file using the "save as" menu option.

Note: *It is good practice to never alter the original file collected from the confocal microscope. All alterations should be saved under a new file name.*

Confocal Assistant continued ………

The menu and tool bar

A series of options is available from the Confocal Assistant menu and tool bar (Figure 6-4). Although, the images in Confocal Assistant, can be readily displayed at different sizes by simply clicking the "+" or "-" icon, the actual image file size is not altered. To alter the image file size (i.e. the number of pixels in the image) the image will need to be resampled (see the "image" menu, discussed in Figure 6-5).

The series projection simply stacks a series of optical sections one on top of the other, resulting in what is sometimes referred to as an extended focus image. This type of display is very effective when the image has a relatively small number of highly fluorescent objects within the stack (for example, when imaging mitochondria in a single cell).

The "merge images" button allows one to select the image file for either the red, green or blue channel of the merged image (but see note below!). This simple merge function is particularly good for dual or triple labelling, but cannot be used to create combined colour and grey-scale transmission merged images.

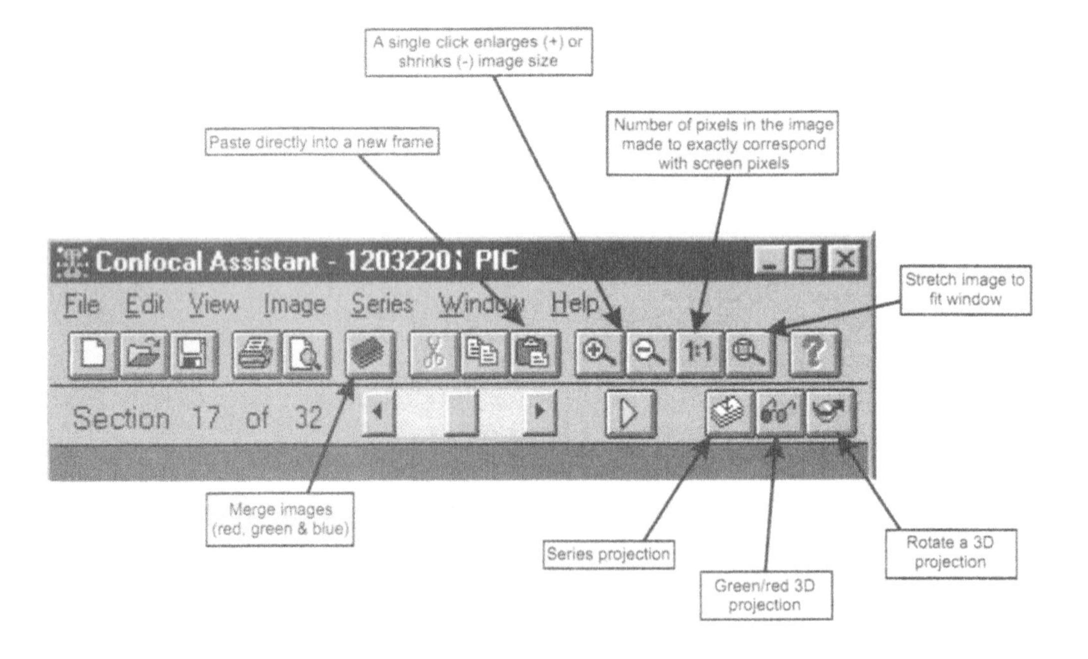

Figure 6-4. Confocal Assistant - Menu Bar and Tool Bar.
A number of simple algorithms are available for displaying the image, merging images and creating projections from an image series.

A problem with Confocal Assistant: *When merging images in Confocal Assistant make sure that the image files are located in the same subdirectory as the Confocal Assistant program - otherwise you may find you get the error "file not found". This problem is due to the current version of Confocal Assistant not recognizing the Win95 long file format in directory headings.*

Confocal Assistant continued ....

The "Image" menu

The image displayed in Confocal Assistant can be readily manipulated in a number of useful ways by using the "image" menu (Figure 6-5). Of particular relevance to microscope images is the ability to "contrast stretch" the image.

A limited Histogram function is available for analysing the pixel intensity of the whole image. Unfortunately Confocal Assistant does not have provision for Histogram analysis of defined sections (individual cells) within the image.

The colour Look Up Tables (LUT) provided with the LaserSharp software can be used in Confocal Assistant. The LUT colour image can then be converted to true RGB colour and saved under a new file format (for example the TIF format). The coloured image can then be readily manipulated in other image processing programs such as Photoshop.

Dual and Triple labelled images (in the same file) can be merged with the "Side-by-Side Merge" function. To merge images in separate files you will need to use the "merge" button on the menu bar (see previous page).

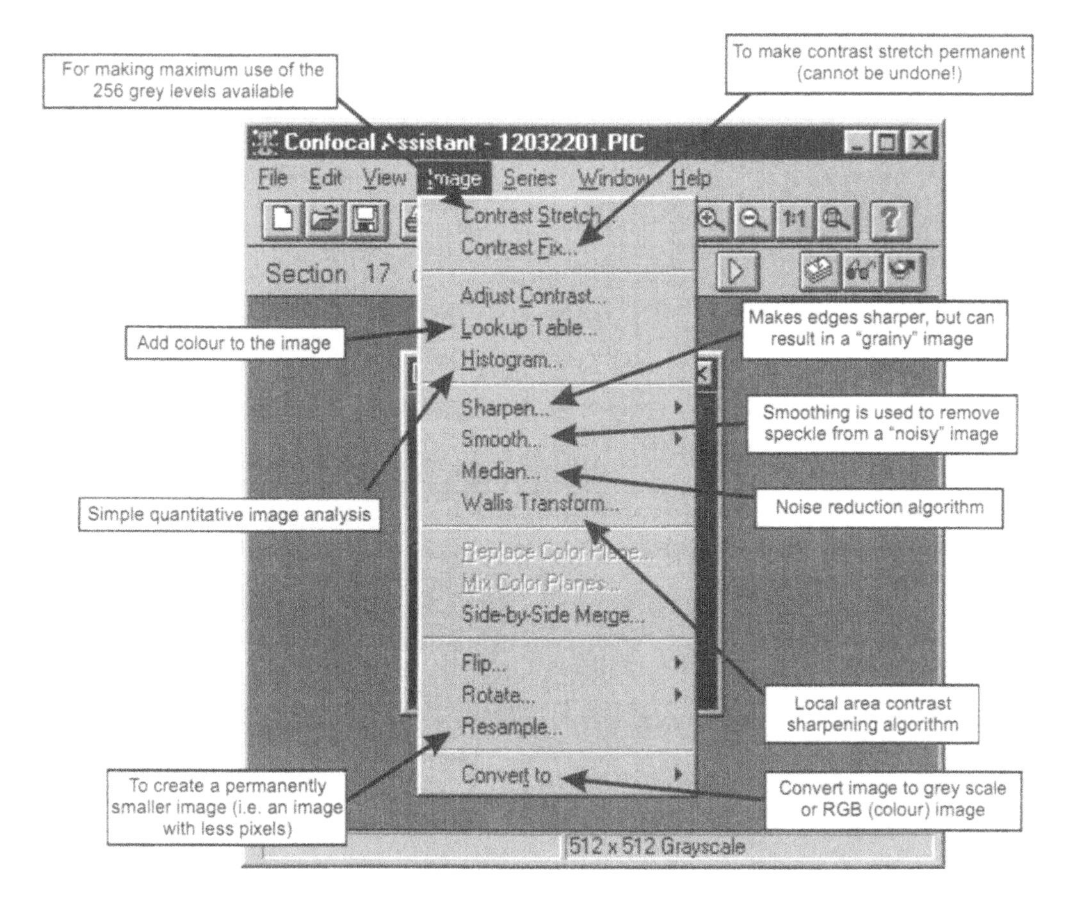

Figure 6-5. Confocal Assistant - Image Menu.
A series of reasonably basic, but very useful image analysis tools are available in Confocal Assistant. All these tools can be used on the Bio-Rad confocal microscope PIC file format, or on other formats (such as TIF) that can be used by other programs.

Confocal Assistant continued ....

The "Series" menu

A series of images collected on a confocal microscope as a stack of optical slices, or as a single optical slice collected repeatedly over time (a time series) can be readily displayed using Confocal Assistant.

The stack of optical slices can be manipulated in a fairly rudimentary way by Confocal Assistant (Figure 6-6). The slices can be readily displayed and each individual slice enhanced with the tools available in the "Image" menu. However, the 3D capabilities of Confocal Assistant are somewhat limited. For more detailed 3D analysis a more sophisticated program will be necessary.

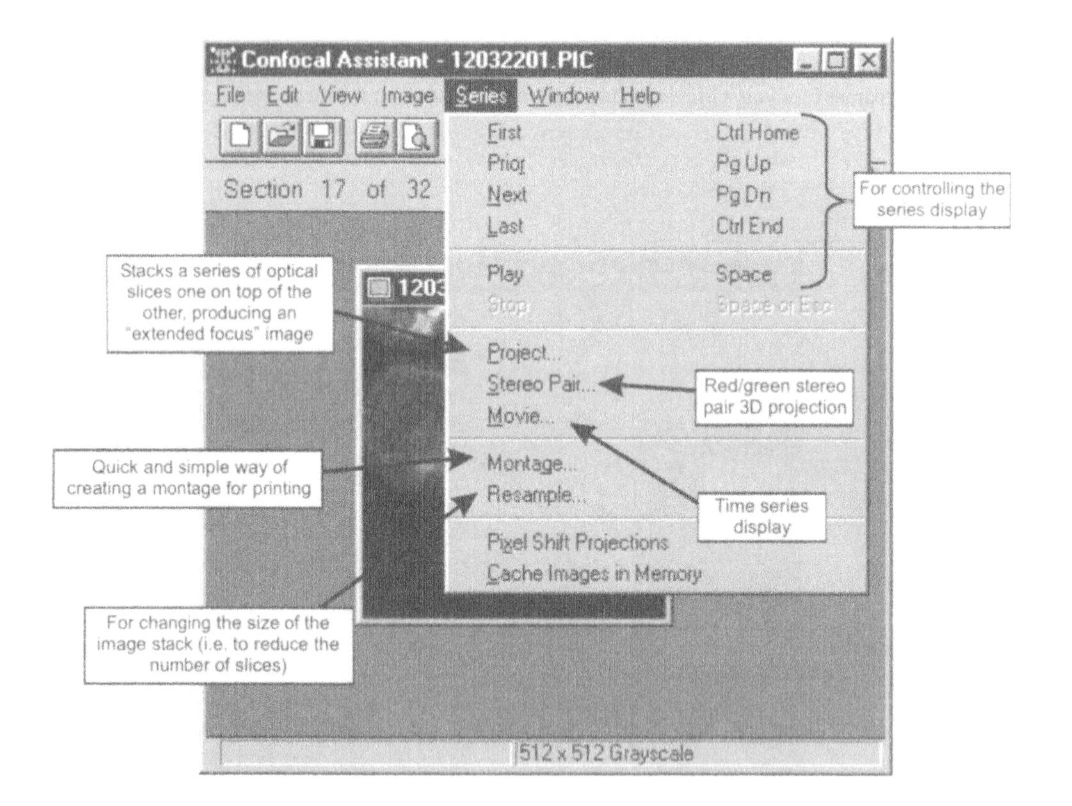

Figure 6-6. Confocal Assistant - Series Menu.
A series of confocal microscope images, created either as a stack of optical slices or as a time series, can be displayed in a number of simple but useful ways by using the options available under the Series menu.

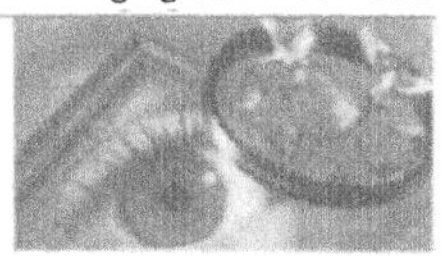

Photoshop

Bitmap image-editing program

Photoshop (Figure 6-7) is an excellent program for image manipulation for the purpose of publication, presentation and some data quantitation. This program is designed as a general program for image editing in photography, but can be readily used to manipulate images obtained from the confocal microscope. Photoshop can directly read single image TIF files generated by most confocal microscopes, but Photoshop cannot read the Bio-Rad PIC or multi-image TIF files. There are, however, Photoshop "plugins" that can be used to read the Bio-Rad "PIC" format files (ftp://ftp.genetics.bio-rad.com/Public/confocal/), or you can use Confocal Assistant to translate the Bio-Rad "PIC into a single image "TIF" format. Alternatively you can import the image as a "raw" image.

Photoshop

- Contrast / intensity changes
- Colour changes
- Merging images
- Creating mosaics
- Quantitation of images
- Publication quality prints

Photoshop is a complex program that does require some time to become proficient in its use. However, mastering basic manipulation of an image in Photoshop is not difficult, which allows one to produce Photoshop or TIF files that are suitable for publication when printed on a high quality dye-sublimation printer.

The number of pixels in the image is always displayed with the image in Photoshop. The image can be viewed on screen at the display resolution (i.e. 1:1 display of image and screen display pixels), or in a smaller or larger size. Photoshop always indicates the size of the image relative to the 1:1 display. This is most important when fine detail is being examined, otherwise image information can be inadvertently not displayed. This careful handling of the pixel resolution of the image is not possible in CorelDRAW (although CorelDRAW does have great flexibility that makes the program particularly suited to creating presentations and figures, see page 175).

Photoshop has a proprietary format called "PSD" that can handle several layers of image or text information within the file. However, most image processing programs will not recognize this format, in which case this format will have to be converted to another format such as "TIF" (with loss of independent layers). To convert a Photoshop file to a TIF format, first "Flatten" the image (remove all the layers) using the "Flatten image" command from the "Layer" menu bar button, and then save in the new file format.

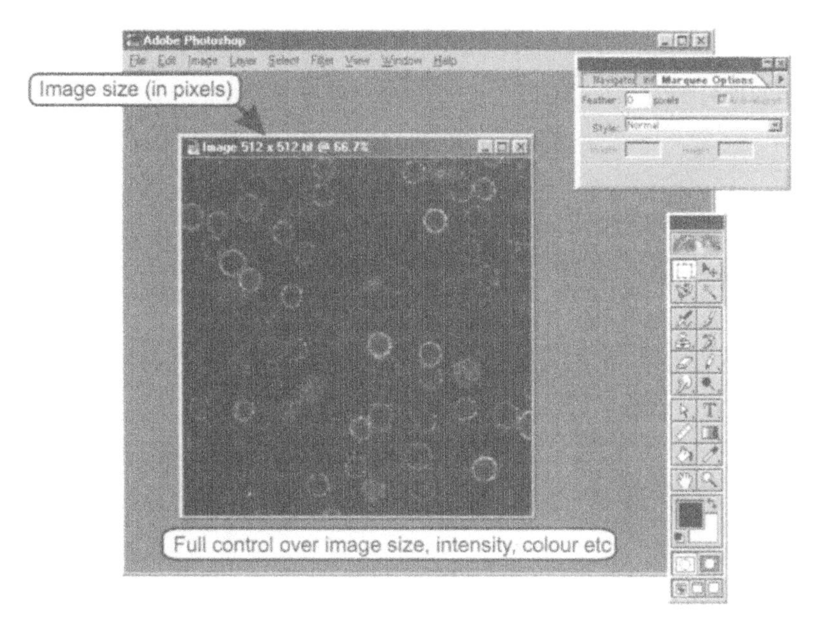

Figure 6-7. Photoshop – Image Display Screen.
Photoshop is specifically designed for manipulating digital images. Photoshop can read single image TIF files from a number of confocal microscopes, but cannot read Bio-Rad "PIC" files directly (unless you use the "raw" mode of image import). However, Confocal Assistant can conveniently be used to convert the file to a TIF file that can readily be read by Photoshop.

Dual and Triple labelled images can be merged using "layers" (see page 249) or "channels" (page 251) in Photoshop, although using the simple merge functions of other image processing programs designed for microscopy is probably easier. Photoshop can be used to merge dual or triple images that have been coloured in a colour gradient (Confocal Assistant only merges into separate red, green and blue channels), or to merge grey-scale transmission images and colour fluorescence images.

> **"Quick Photoshop for Research" by Jerry Sedgewick (see Chapter 15 "Further Reading", page 347) is a great guide for using Photoshop for basic image manipulation.**

The real power of Photoshop is the ability to manipulate the colour, contrast, brightness and hue of the total image or just small parts of the image. If necessary, even the individual pixels within the image can be edited. The powerful editing features are mainly accessed via the tool bar (Figure 6-8). Considerable care should be taken when manipulating an image, that you are still truthfully representing the image in your final display (see Chapter 7 "Presentation and Publication", page 177). Complex mosaics of images can be created using Photoshop. This can be particularly useful for creating interesting images for slide presentations where different views and magnifications need to be presented in the same slide.

Photoshop is particularly suited to high quality printing. The size of the image in pixels is always known, which means one can match the pixel output of the printer for maximum resolution prints. Furthermore, some idea of the "look" of the final printed image can be obtained by displaying the image on screen as a "CMYK" image that mimics the printer inks used for printing. This "print" display is not perfect as you are still viewing the image on a computer screen (using RGB colouring).

Photoshop can also be used as a tool for quantifying the pixel intensity within the image. The lasso tool is used to define the area of the image for which a histogram can be created. The histogram gives a graphical view of the range of pixel intensities, average, median and total number of pixels in the selected area.

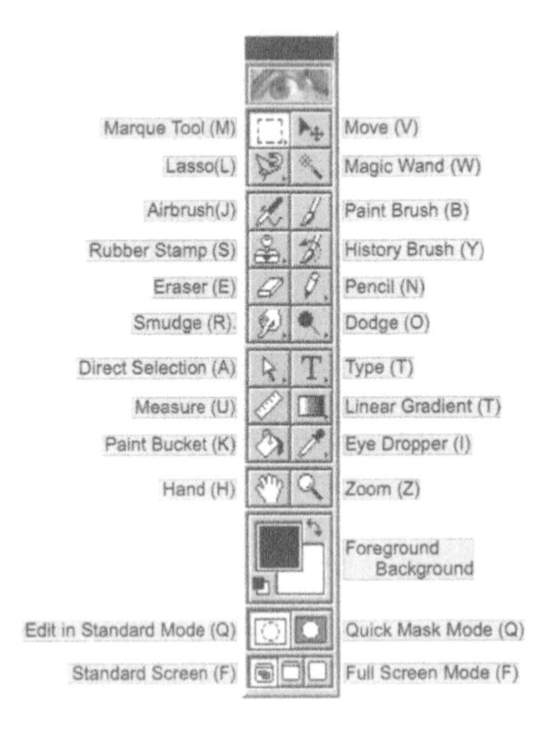

Figure 6-8. Photoshop Tool Bar.
Most Photoshop options are available from the Tool Bar. Double clicking on each of the tools will bring up a toolbox for adjusting the tool. Holding the mouse button down will display further tool options.

Some Handy Tools

Marque Tool - is used to define an area of the image (rectangle, circle, line, trimmer or crop) that can be deleted, moved or changed.

Lasso Tool - is used to select an area for deleting, moving or changing. With the magnetic lasso ( ) Photoshop will attempt to exactly follow an "edge" within the image (such as a cell outline).

Stamp Tool - is used to "stamp" a pattern ( ) within the image, or to "rubber stamp" ( ) a replica of a part of the image. The "rubber stamp" can be used to "remove" unwanted background by replicating background that is acceptable. Press "Alt + mouse button" to define the area to be replicated. Then replicate by dragging the "rubber stamp" tool across the image.

Magic Wand Tool - is used to select areas of the image with similar pixel colours or intensities. This can be used to delete areas or change the colour or intensity of select areas of the image.

History Brush - is used to "undo" changes on select parts of the image.

Dodge, Burn & Sponge - are terms from photography. The selected tool is activated by moving it across the image while depressing the mouse button. Dodge ( ) lightens, Burn ( ) darkens, and Sponge ( ) lightens the background.

Eyedropper Tool - is used to select colour from specific points in the image for application to other parts using the paintbrush or pencil tools.

CorelDRAW

A vector drawing program that can import bitmap images

CorelDRAW
- Creating diagrams
- Slides for presentation
- Computer presentations

CorelDRAW (Figure 6-9) is an excellent program for creating slides and diagrams, that incorporate images, for publication or presentation. The way that the program handles images is very different to that of Photoshop. Although full image information can be obtained, the normal means of manipulating bitmap images imported into CorelDRAW is to simply resize the image without any regard to the number of pixels in the image. This does have the advantage that it's very easy to change the image size to fit a diagram or mosaic being constructed. However, the disadvantage is that you may not be aware that you have over-sized the image (detracting from the quality of the image), or that you are under-sizing the image (and thus not fully presenting all the information available in the image). However, CorelDRAW does retain the original image pixels when you "copy" an image for transferring to another program. For example, if you have resized an image in CorelDRAW by dragging the corners of the selected image, you will find that when copying this "smaller" image into Photoshop that the correct number of pixels are still retained with the image. This is in contrast to Microsoft Word and PowerPoint (discussed on the following page), where resizing the image will result in an incorrect number of pixels in the image when exported to Photoshop.

Commercial printing houses can rarely handle CorelDRAW files directly. Translating the CorelDRAW file to a bitmap file for printing will usually result in seriously compromised image quality. Publication quality figures can be produced by printing to Acrobat Distiller from CorelDRAW (producing a PDF file), but make sure you set the parameters to high-resolution printing.

Confocal microscope images are best contrast stretched, merged, enhanced etc using Confocal Assistant, Photoshop, or other image editing software and then imported into CorelDRAW for producing slides or figures for publication.

Slides can be readily produced from CorelDRAW files by either using a slide maker (see Chapter 7 "Presentation and Publication", page 177, for further information), or by simply photographing the full screen view (press F9 when in CorelDRAW) on a high quality monitor. Alternatively CorelDraw figures can be selected and then pasted into PowerPoint for computer presentations.

Corel does produce a bitmap image-editing program called Corel PHOTO-PAINT, which can carry out most of the basic functions of image editing, but for full control of the bitmap image Photoshop is considered to be more versatile.

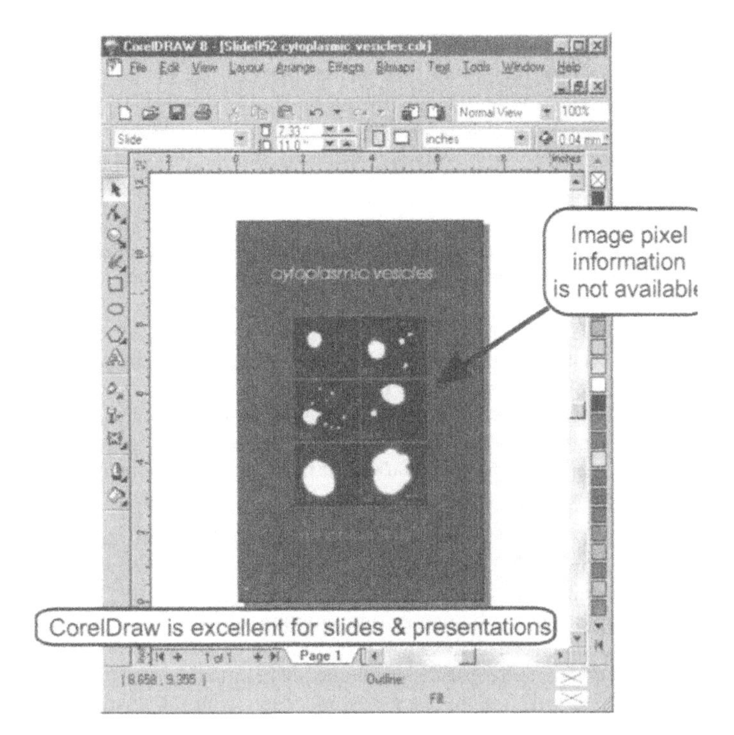

Figure 6-9. CorelDRAW.
The CorelDraw program is designed for producing diagrams (including figures containing images) for slides, posters PowerPoint presentations and publications. Images can be readily incorporated into the diagram as bitmap images (by importing a file or dragging a file into an open document). However, image manipulation (contrast, colour management, number of pixels etc) is better done using an image editing program such as Photoshop or Corel PhotoPaint – the final image then being imported into Corel DRAW. The figure shown in this diagram is reproduced from Hibbs and Saul (1994) with permission from Elsevier (see page 349 in Chapter 15 "Further Reading" for further details).

PowerPoint

PowerPoint is familiar to most people as the seminar presentation software bundled with the Microsoft Office Suite. This software is excellent for the presentation of images in a seminar, but keep in mind that this program is not specifically designed for the presentation of digital images from a microscope. Although PowerPoint can very easily be used to display and arrange digital images, care must be taken not to accidentally downsize the number of pixels in your image, or to display the image incorrectly for your screen resolution. For optimal presentation of images in a seminar try to design the seminar using the screen resolution you will have when using the computer projector (usually a maximum of 1024 x 768 pixels). If you work on a computer screen with a higher resolution you may find that your seminar preparation is not as good as you would have expected.

Do not use PowerPoint to "store" images

PowerPoint is a versatile program for presenting images in a seminar, but be very careful if you use this program to store or display images on your computer as the quality of the image can very easily become degraded by moving the image in and out of PowerPoint (this is explained in some detail below). Microsoft Word also has very strange ways of handling image size.

Problems associated with how PowerPoint handles images:

1. **The ease with which images can be resized in PowerPoint can create problems:** Images can be readily resized in PowerPoint by simply clicking on the image and dragging the corner anchors. This ease of image manipulation is great for displaying multiple images on a screen for seminar presentation, but keep in mind that every time you change the size the program interpolates the pixels of the image to create an image on the screen. This means that if you display the image smaller then the pixel resolution there will be pixels missing in the image. Likewise, if you display the image larger than the display resolution of your computer there will be pixels added! These changes to the image are often not of any concern when giving a seminar, but as described in more detail below, resizing the image can create havoc with the pixels in the image.

2. **Screen display resolution problems with PowerPoint:** When displaying a digital image in PowerPoint care must be taken if you require a 1:1 display of the pixels of the image and the display resolution of the projector or computer screen being used. To achieve a 1:1 screen display, correct adjustment of the image relative to the "page display" is most important. For example, if you have a 1024 x 768 screen display and wish to display an image of this pixel dimension as a full screen display using a projector or computer screen, the size of the image in PowerPoint will need to be adjusted (by moving the corner highlights) to fully fill the "page display" area on the screen. When you then switch to full screen display the 1024 x 768 image will be displayed at screen resolution. If you were to adjust the image to "100%" size using the menu options, then you would find that the full screen display will be larger than a 1:1 pixel display (100% in this case refers to the image size, i.e. 1:1, on the "page display" screen)

3. **Accidental changes in image pixels using cut-and-paste:** When using the cut-and-paste function in PowerPoint takes great care as to the displayed size of the image *before* you cut-and-paste. If you have resized the image to fit several images on the display screen of PowerPoint, and then copy the image and paste into a bitmap-editing program such as Photoshop you will notice that the number of pixels in the image reflects the small size that you had displayed the image in PowerPoint! Mind you, even displaying the image at "100%" using the menu options in PowerPoint does not fully solve the problem – in this case you will find there are now *more* pixels in the image! The only way around this dilemma is to store the original images in separate files, preferably using the universally readable TIF format (well, almost universal!).

4. **Extracting images from a PowerPoint display with correct resolution:** To maintain the correct resolution of the original image the image should be selected and then save as an image (right click on mouse and select "save picture as"). Do NOT cut the image from a PowerPoint display and paste into another application.

Chapter 7

Presentation and Publication

Digital imaging allows one to easily change not only the appearance of the image but also the underlying data used by the computer to form the image. Digital imaging is a very powerful technique for presenting data in an easily understood form, but can also result in misleading images that do not reflect the original data.

This chapter is concerned with how to produce publication quality images, and how to ensure that these images are representative of the original data.

HONESTY IN IMAGING RESEARCH

Presenting imaging research is fraught with difficulties when it comes to honestly representing your research. Everyone wants to show their data in the best possible light, but misleading presentation of data can result in the data appearing to strongly support a particular conclusion when in reality the truth is different.

In the following example of choosing specific images to represent the data, you will see that truthful presentation is not as straight forward as it may at first appear.

In the experiment presented in Figure 7-1, live malarial parasites have been labelled with acridine orange and then visualised by confocal microscopy. The parasite appears as a brightly staining structure that increases in size over a period of 40 hours. Within the infected red blood cell cytoplasm can be seen a number of small vesicles.

In Figure 7-1 a single cell is presented at each time point, a common way of presenting data in microscopy. The question arises; are these cells representative of the experiment as a whole? The answer, of course, is they are as representative as I choose them to be!

There are two points about the images in Figure 7-1 that could be very misleading. The first point to note is - how do you know that I didn't simply choose the images to fit the idea that the number of vesicles per cell peaks in the middle of the life cycle of the parasite? The other point is - do you think these vesicles are brightly stained?

The answer to the first question is, yes, I did choose the images to fit the idea that the number of vesicles per infected

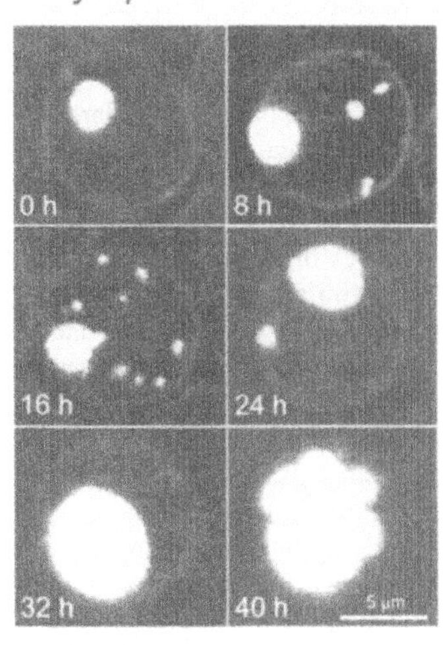

Figure 7-1. Honest Presentation of Images.
This image shows acridine orange staining vesicles within the cytoplasm of live malaria infected red blood cells. The distinct impression here is that the number of vesicles increases with the age of the parasite (from 0 to 40 hours) - but is this a true representation of the data? These images were collected using a Bio-Rad MRC-600 confocal microscope with a Zeiss 63x 1.4 NA oil immersion objective. This figure is reproduced with further annotation from Hibbs and Saul (1994) with permission from Elsevier (see page 349 in Chapter 15 "Further Reading" for further details).

cell peaks at 16 hours. How do you know I'm not cheating? Often in imaging research it is necessary to have considerable faith in the person presenting the work, but if possible one should try and back up imaging data with supporting data that is quantitative.

In this particular example 100 cells were chosen at random and the number of vesicles per infected red blood cell counted. The results are presented in Figure 7-2, which clearly demonstrates that the greatest numbers of vesicles are present between 16 and 24 hours. This quantitative data strongly supports the images in Figure 7-1. In fact the images in Figure 7-1 were chosen on the basis of the quantitative data of Figure 7-2.

If you look closely at the data presented in Figure 7-2, where each dot on the graph represents a single infected red blood cell, it would be possible to show different trends by simply presenting just a few images of your choosing. This is due to the large spread in the data set - not an uncommon event in biology!

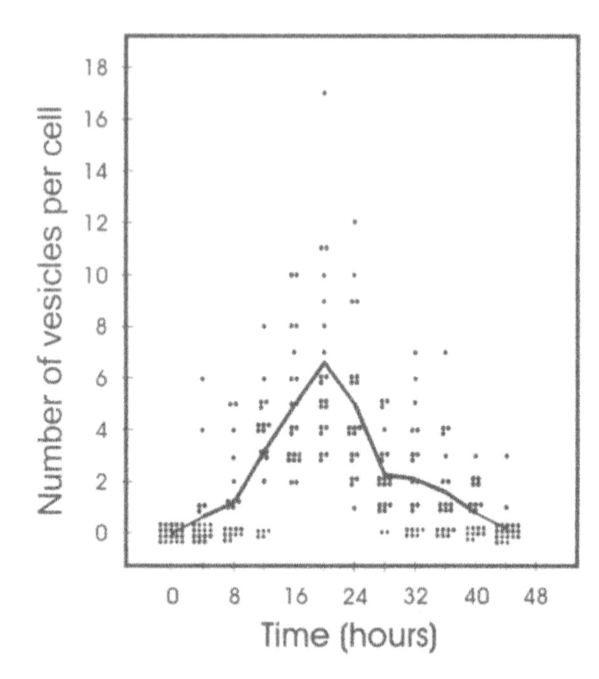

Figure 7-2. Quantitative Analysis of Cytoplasmic Vesicles.
The number of vesicles in the malaria infected red blood cell cytoplasm of 100 randomly selected cells was scored and the results are presented graphically in this figure. Each point on the graph represents a single infected red blood cell. This quantitative data clearly indicates that the number of vesicles per cell peaks at about 20 hours. The data depicted in this graph was used as the basis for selecting the cells shown in Figure 7-1. This diagram is reproduced from Hibbs and Saul (1994) with permission from Elsevier (see page 349 in Chapter 15 "Further Reading" for further details).

The answer to the second question, "are the vesicles brightly stained", is a simple "no". You may well think - but the images in Figure 7-1 show that the vesicles are very brightly stained. Even if you were to look at the original images as they were being collected you would see that they are brightly stained!

In reality these vesicles are at least 1000 x less bright than the parasite! So the real result is that they are in fact very poorly stained. This point is discussed in the body of the text where these images were published, but if you just looked at the images without reading the text of the article you would be misled. These images were never designed to convey a true representation of the level of label in the parasite compared to the vesicles; they were presented to simply show the number of vesicles at various stages of the parasite life cycle.

Another important point worth noting is that one cannot quantitatively collect (never mind display or print) two levels of fluorescence that differ by 1000 fold. To collect fluorescence with such a large difference requires manipulation of the gain and blackness levels on the confocal microscope such that the centre of the parasite, although registering 255 on the grey-scale of the image, is in reality a much higher value. To collect an image that could quantitatively display such large differences in fluorescence would require an instrument that could collect at least 12-bit (4096 grey levels) or even 16-bit (65,000 grey levels) grey-scale images, as against the 8-bit (256 grey levels) used to collect the images in the above discussion. However, even if you were to collect a 16-bit grey-scale image with 65,000 grey levels, you would not be able to display this number of grey levels on either your computer screen or by using a printer! The number of effective grey levels on the best display monitor is probably 200 grey levels (less than 8-bit) – that's if your eyes can see that many grey levels anyway!

The conclusion to draw from this discussion is that a simple set of images as shown in Figure 7-1 can present a number of difficulties when trying to honestly display the data. You need to be honest in how you present your own images - but don't assume that others are, without supporting evidence!

Some Points to Remember when Presenting Imaging Results

- **Single images may not be representative of the true data.**

  The example above, using acridine orange to stain malarial parasites, clearly demonstrates that an "obliging" cell can often be found that contains the labelling pattern you are seeking! Images of single cells must be accompanied by suitable supportive quantitative data, which may be as simple as also presenting a low magnification image that includes a number of cells, or a more elaborate statistical analysis of the frequency of the phenomena observed.

- **The intensities within the image may not be representative of the level of labelling in the cell.**

  The confocal microscope is a very sensitive detector of light. Very low levels of labelling can be made to appear quite bright and, as discussed above, very large differences in labelling within the one cell (or within the one image) may not be truly represented due to the limitations of the dynamic range available in a digital image.

- **Comparing the level of label between images means that the experiment would have to be done in a technically rigorous manner for the comparison to be meaningful.**

  Truly quantitative analysis in imaging research is very difficult. Most fluorescent labelling protocols are aimed at maximizing the level of label, rather than ensuring that the level of labelling obtained is a quantitative representation of the sample. If comparisons are to be made between images, one must ensure that the experiment is done in a manner that allows quantitative data to be obtained, and that the settings on the confocal microscope were exactly the same (or at least any differences in settings carefully recorded).

- **Setting of Grey-Scale Levels for Optimising the Dynamic Range.**

  The grey-scale range in a digital image can be readily altered, not only to take full advantage of the dynamic range available, but also to emphasize particular intensity levels within the image. For example, a non-linear "gamma" setting can be used for display to emphasize structures stained at low intensity.

- **Colour in an image may be very useful for understanding the image, but may also be misleading.**

  The application of colour Look Up Tables (LUT) can result in very small differences in levels of intensity within the image being greatly exaggerated. This phenomenon can be used to great effect, but can also lead to misleading representation of the data. For example, the "GEOG" LUT, which consists of several tight bands of different colour (from dark blue through red to yellow to white), with the transition between colours only being the result of a single grey-scale unit. Therefore a structure could become a different colour (and thus appear to be much more highly stained), even though there may be only a single grey-scale unit difference between it and the surrounding cell.

Dishonesty in imaging research is simply misleading yourself!

- **Changes to the image during the printing and publication process may significantly alter the interpretation of the labelling in the image.**

  Great care is needed during printing and publication to ensure that the colours and intensities are truly representative of the original experiment. A good example of where misleading results can occur is when depicting dual labelling as different colours (usually red and green). The printing characteristics of printer inks are significantly different than the red, green and blue dots on the computer screen. This difference between screen display and printing can significantly alter the apparent balance between the two chromophores. Even if the image is not intended as a quantitative analysis of the level of labelling, the imbalance in the depiction of the two colours can lead to confusion as to whether particular structures are labelled by both chromophores.

- **Dual labelling needs to be interpreted with caution.**

  Clearly separated structures can be unambiguously determined to be physically separate. However, dual labelling is often required to distinguish between two very closely spaced cellular compartments. It is possible to visualise cellular structures that are smaller than the theoretical limit of resolution of a light microscope (for example microtubules) if they are fluorescently stained. However, two different molecular labels that appear to exactly co-localize by confocal microscopy (appearing "yellow" in the merged image) may in fact be attached to physically separated structures that are smaller than the resolution of the microscope (and thus appear as one fluorescently labelled structure).

- **The published images become "the data".**

  Unlike biochemical data, published findings in imaging experiments often rely on the presentation of a very small number of cells. Unfortunately, once images have been published they often become "the data", and so it is very important to ensure that published images are "representative" of the data obtained in the original experiment.

- **Quantitation of confocal images is easy; quantitation of imaging data is very difficult!**

  The pixel intensity anywhere in the image can be readily determined with suitable software. Furthermore, one can create graphs depicting the level of labelling within individual cells or subcellular compartments. However, due to the difficulty of conducting the experiment in a manner such that quantitative levels of label are truly meaningful, caution should be exercised in presenting data as "quantitative".

PRODUCING PUBLICATION QUALITY IMAGES

The real key to producing high quality images for publication is to produce top quality images from the confocal microscope in the first place. Attempting to produce a good quality image from poorly collected data is time consuming and difficult. In the heat of the moment one can easily "overlook" the fact that collecting images suitable for publication will require particular attention to the layout and structure of the cells, the gain setting, and particularly how representative the image is of the overall results. Try to expend extra time during the experiment in collecting top quality images required for publication.

A few guidelines for publication quality images are as follows:

> **Collect top quality images in the first place** - rather than spending time later trying to "improve" the image!

- *Collect top quality confocal microscope images – making use of the full dynamic range (grey-scale) available,*
- *Make sure you collect images representative of the labelling performed,*
- *Collect the images in grey-scale mode,*
- *Use colour to make the results clear - but be careful not to confuse the data,*
- *Provide a top quality print of your image to the publisher when submitting an electronic version.*

Collecting Top Quality Original Data

There are two problems to be dealt with when collecting high quality images that are to be used for publication. The first is to ensure that the image itself is collected correctly, with the correct filter settings, laser line and most importantly, within the minimum and maximum grey levels that can be contained in a confocal microscope image. The Second problem is to ensure that the image is representative of the data on hand. If you cannot collect images from a number of cells or tissue slices, make sure that the images that are collected are representative of the experimental results obtained.

Closing the confocal "pinhole" further, and collecting images at a slower scan rate, and/or with screen or line averaging can often improve the quality of your images. Although the "on-screen" difference may appear marginal, after processing, printing and publishing the difference may be quite marked.

The commonly used 512 x 512 pixel image size is quite adequate for most computer presentations, slides and published images, but if the image is to be greatly enlarged (or a small section of the image will be later enlarged), it is advisable to collect a larger 1024 x 1024 pixel sized image. The optimal size for confocal microscope images is to match the image collection size with the pixel resolution and final display size of the screen or printer used.

Collect Images in "Grey-Scale"

Images should be collected in the "grey-scale" mode on the confocal microscope. The human eye is very sensitive to small changes in grey-scale, much more so than similar changes in colour intensity. Furthermore, the human eye has different sensitivities to different colours, which will create some problems when images in multi-labelling experiments are collected in red and green. The difference in sensitivity between colours is particularly pronounced in "colour blind" individuals. The best solution is to collect images in grey-scale (the mid-range of "SETCOL" and "GlowOver" are also grey-scale) and then apply colour later.

"Enhancing" your Image for Publication

A variety of enhancement tools are available when using imaging software (for details see Chapter 5 "Digital Images in Microscopy" (page 145). These "enhancement" tools should be used to "improve" the quality of presentation of your image. This is similar to the concept of careful selection of photographic paper and film exposure time in conventional photography. A poorly presented image can be distracting, particularly in an auditorium where there is little time to "ponder" over the image.

Suitable software image processing tools should be used to "sharpen" the image. An image that appears "out of focus" (even if this is the reality of a very high magnification image!) is very distracting. Distracting "dirt" can be removed from the image by careful "cropping", without being dishonest in the presentation of your data. Make sure you retain a copy of the original image collected from the confocal microscope so that if you are not happy with changes made you can go back and start all over again – besides the fact that the original image should be retained as it is the "raw" data of your experiments.

MAKING CONVENTIONAL SLIDES

Colour slides of very high quality, with colour that is very close to that seen on the computer screen can be easily produced. However, the colour balance characteristics of the film being used should be taken into account for the production of high quality slides.

The optimal image size for producing colour slides is to display the image on the computer screen at 1:1 resolution (i.e. each screen pixel represents an image pixel). For a 512 x 512 pixel confocal microscope image this will mean a 3/4 filled screen when the display is set to 1024 x 768 (the normal high resolution setting for a computer monitor). The slide produced will have an image that is 3/4 the size of the projected screen at optimal image resolution.

If you wish to produce a full screen (or full slide) view of the image then you will need to collect a 1024 x 1024 resolution image. Displaying this image somewhat smaller than optimal resolution to fully fill a computer screen will not unduly detract from the quality of the image.

Slides from a Computer Screen

High quality colour slides can be produced by simply photographing a computer screen. A 17" high quality "flat screen" monitor works best. The trick to producing good slides from the computer screen is to take great care with the focus (don't use auto focus), and to use an exposure time that is greater than the screen refresh rate. I find an exposure time of between 1/4 and 5 seconds the best (the screen refresh rate is usually between 60 and 80 times per second). Colour slide film produces very accurate colour reproduction of a computer screen. However, as mentioned above the colour balance in different films is different. Trial and error will quickly establish the colour balance of the particular film you are using. Colour balance may need to be adjusted in the image to produce an acceptable colour balance in a slide.

The camera light meter can be readily used to determine the correct exposure. However, you will need to keep in mind the rules of conventional photography - i.e. a lot of dark background may unduly influence the light meter in the camera. The solution to this problem is to compensate by underexposure if necessary. You may need to take several different exposures of your image until you become familiar with the characteristics of the film and camera combination you are using.

The quality of the image is improved by using a 80 mm lens or higher on your camera, rather than the standard 50 mm lens. The camera may need to be moved further away from the screen, but it will provide a better result in terms of straightness of lines within the image. Borders around an image are particularly prone to "pin cushioning" (frame lines not being straight) and so are best avoided if possible.

Colour Slide Makers

Image quality is better using a slide maker than on your computer screen

The best quality colour slides are made by using a colour slide maker. Although the quality from photographing the screen is reasonably acceptable, the quality of the result from a slide maker usually exceeds the quality of the image you see on the computer screen.

A slide maker is a device that uses a camera to take a photograph of a computer screen. However, there is an important difference in the way the colour in the image is generated, compared to a colour monitor, that makes the photographs taken with a slide maker superior to photographing a normal colour screen. Within the slide maker there is a grey-scale screen with a very small dot per inch (dpi) rating. The colour in the image is produced by displaying the red, green and blue components of the image at separate times on the screen. Using appropriate colour filters the individual grey-scale dots on the screen act as individual colour dots (in a colour computer monitor the colour is produced by 3 physically separated screen dots for each screen pixel).

The colour balance of different films is compensated by the exposure timing of the slide maker (commonly used film will have pre-set values in the slide maker). However, this compensation is not always perfect, and so test photos of critical images are advisable.

USING A COMPUTER PROJECTOR

The convenience of presenting your images (including moving images) and diagrams directly to your audience by using a computer projector will mean that this technology is now the main means of presenting digital images to an audience. Computer projection for digital display has improved greatly over the past few years. However, present computer projectors suffer from a relatively small dynamic range (few grey levels) and so care should be taken to ensure that critical images will display well in a large auditorium when using a computer projector. It is most important to match the display resolution of the computer projector with that of the computer used (often your own laptop computer). Do not use a resolution that is higher than that of the computer projector. Exceeding the resolution of the computer projector on the large screen will result in a very poor quality image!

PowerPoint Presentations

PowerPoint is a program that is particularly well suited to presenting your work in a seminar. PowerPoint is very popular because it allows you to provide text, diagrams and images (including movies) all in the same presentation. The program, and in particular the difficulties associated with handling images correctly when using PowerPoint, is discussed in detail on page 176 in Chapter 6 "Imaging Software".

Particular care should be taken when transferring images in and out of a PowerPoint presentation. Valuable pixels can be lost from the image on transferring images to another program. For example, when an image that is displayed in PowerPoint at 25% is "copied" to the clipboard and then "posted" into a new file in Photoshop, the image will no longer be the same size as the original! In fact, 2/3 of the pixels of the original image will be lost (which is not even the ¾ loss to be expected if one assumed a 25% display size was a 25% smaller image!). To make matters worse, an image displayed in PowerPoint at "100%" has more pixels added to the image on "pasting" into Photoshop!

PowerPoint allows re-sizing of your images by simply clicking on the imported image and dragging the corner (don't forget to have "constrain proportions" on, or you will end up with a distorted image!). However, this attribute can be easily abused – if you are displaying critical images try and display them at the resolution of your computer monitor. This is not as simple as you would like when using PowerPoint. If you display the image at 100% (using the menu) you will find that the image is in fact oversized on the final display screen. 100% display in PowerPoint means 100% relative to the editing screen! To display the image at the screen resolution you will need to know the screen resolution of your monitor (usually a maximum of 1024 x 768 pixels on a laptop computer or in the projected image of a computer projector, but sometimes higher on desktop computers). You will also need to know the pixel size (regularly 512 x 512 pixels, but often 1024 x 1024 pixels) of your confocal image. To display a 512 x 512 image at full screen resolution on a 1024 x 768 pixel display you will need to re-size the image to take up half the horizontal size of the editing page in PowerPoint (i.e. ½ 1024 = 512 pixels). When you then go to full screen display the image will be shown at the same resolution as that of the computer screen (i.e. a 512 x 512 image will be displayed at half the full width of a computer screen set to a display resolution of 1024 x 768 pixels).

Displaying your image at the pixel resolution of the display device you are using will result in the best quality image. Displaying the image larger than the pixel resolution of the monitor will result in a "fuzzy" image, and displaying the image smaller can also detract from the quality of the image. It is a good idea to get hold of a good quality digital photograph and try out different sizes etc to get a feel for the effect of changing the size of the image in PowerPoint. Photoshop is a program that allows you to keep track of the size of your image in pixels, but doesn't have the convenience of PowerPoint for seminar presentations.

SUBMISSION OF FILES FOR PUBLICATION

Although you may have the "perfect" image in digital form and expect an excellent quality of presentation in a journal to which you have submitted your article – remember that what is "perfect" for an image from a confocal microscope is still somewhat subjective. Therefore you should always try to submit a high quality print (preferably a dye sublimation print) along with your digital image file. In this way the publisher can try to match the printing with how you would like the image to appear. Photographs of people and scenery are often much easier to produce in a very acceptable high quality manner without referring to a printed version simply because the publisher will have some knowledge of what the photograph should look like. In confocal microscopy the person doing the printing for a publication normally has no idea as to what the image should look like when printed. A good high quality print submitted with the electronic copy of the image file will ensure that the final version very closely matches what you see on your computer screen. The display colour and intensity from different computer screens varies greatly and so you cannot even rely on the person doing the printing having the same view of the image on their monitor as you have on your own computer.

Microscopy journals, and often cell biology journals, have very high standards when it comes to publishing images, but other journals (particularly biochemical / molecular biology) may not routinely publish images and so with these journals it is most important to not only provide the best quality image for their printing process, but to also give clear instructions and examples of how the image is to be printed.

Printers use CMYK
(cyan, magenta, yellow and black)

Computer screens use RGB
(red, green and blue)

PRINTING FOR PUBLICATION

You may be required to prepare a printed version of your image for publication even when the journal concerned will use an electronic version of your digital image for producing the final print within your article. This is because, as discussed above, it can be very difficult for the person doing the printing to "know" what the final image should look like without a high quality example.

Colour printing has been discussed previously on page 157 in Chapter 5 "Digital Images in Microscopy". In summary here, it is important to recognize that the process of colour printing is very different to the manner in which colour images are presented on a computer screen. Printer inks consist of three basic colours, cyan, magenta and yellow (CMY) with black (K) often added as a separate ink to produce truly "black" black. These printer inks are the inverse of the colours used to produce colour images on a computer screen (for example cyan, a light bluish colour is the full visible light spectrum without red). Furthermore, the "colour" white is produced on a computer screen by all three colours (red, green and blue) together, whereas the "colour" white in printing is produced by the lack of ink on the printed page (the paper becomes the "white" part of the image).

The above fundamental differences in the way that printers produce colour images compared to computer screens means that there is considerable difficulty in producing a print that faithfully represents the image on the screen. In professional art studios the computer screen is adjusted to match the printer as closely as possible. However, we normally have to resort to trial and error to produce the correct colours.

One of the disappointments of printed images is that they are often of less vibrancy compared to the computer screen (slide images are usually as vibrant as the computer screen). One way of overcoming this problem is to "falsely" increase the intensity and contrast of the image. In other words, for acceptable presentation in a seminar it may be necessary to make the image on your computer screen appear somewhat too "bright" and with "garish" colouring.

Unfortunately, pure red and green do not print well on colour printers. When printing dual or triple colour merged images, better quality images can often be obtained by over emphasizing the intensity of the red and green. Alternatively a less "pure" green and red may be used to great effect. A more "yellowish" green, for example, prints particularly well.

If the individual image pixels are visible in a high quality print, it may be possible to "improve" the presentation by "dithering" the image. Dithering, a process by which "jagged" pixel edges are smoothed over, is available as an option in most print menus.

Inkjet Printers

High quality inkjet printers using glossy "photographic" quality paper can produce very high quality colour prints. These prints are now almost as good as the dye sublimation prints (see below). However, they are relatively slow to produce, and any slight defect in the ink supply during the print process can ruin a perfectly good image.

Some photographic quality inkjet printers are provided with special inks that result in true grading and mixing of the colours. These inks are less intense, but several dots are added to build up the intensity of colour required. Colour inkjet printing of confocal microscope images does work out quite expensive as the ink cartridges are expensive and a lot of ink is consumed in a "photographic" quality print. Large size printing for posters is usually done on a poster size inkjet printer. One of the problems with inkjet printers is the stability of the inks used. The light stability of the inks has improved a lot over the past few years, but if you need to store inkjet colour prints for extended periods they should be kept away from light.

Black and White Laser Printer

Very high-resolution black and white laser printers are available, but unfortunately they are not suitable for publication quality prints because of their lack of grey-scale printing. The method by which a laser printer produces grey shading is to print less black dots - there is no such thing as "grey" dots. These images can be quite acceptable for informal presentation, but are rarely suitable for publication. Higher quality laser printers have a greater number of "dots per inch".

Colour Laser Printers

Laser colour printers are now becoming much more affordable, and the quality of printing and the range of colour is very good, but not usually of publication quality. They are of sufficiently high quality to be used as "hand outs" or poster presentations. One reason that the a colour laser print is not of sufficient quality for publication is that, like the black and white laser printer, different shades of colour are created by depositing a different number of "dots" of colour – the inks are not deposited as shades of colour.

Dye Sublimation Printers (Colour and Grey-Scale Printing)

The best printer for "photographic" quality images for publication is a dye sublimation printer. This printer produces vibrant colours that are truly mixed together on the page. The process of sublimation (heating and mixing the colour dots) produces a glossy image that is of very high quality colour in which there is excellent shading between different colours within the image. Very high quality "grey-scale" images can also be produced on a dye sublimation printer by using the black ribbon only. Unlike the laser printer above, dye sublimation printers produce true shades of grey.

Dye sublimation printers are the best printers for "photographic" quality

The only downside of dye sublimation printers is that text and diagrams are not as "crisp" as that obtained with the laser or inkjet printers. It is often a good idea to produce a colour or grey-scale dye sublimation print without any text or borders for publication. The text and borders can be added later using Letraset, although many journals prefer to add their own text and borders.

Dye sublimation printers can have difficulty in printing particular colour gradients. This is not usually a problem when printing normal complex photographs, but the colour shading generated by confocal microscope colour LUTs can result in "banding" of colours. If this is noticeable in your printed images (and is not evident on the computer screen) then try a different make of dye sublimation printer. Evidently this problem is caused by minor contaminants within the inks of particular manufacturers. Dye sublimation printers are expensive, and the paper and ink ribbon is an added on-going expense. However, these printers are widely available at printing bureaus where you can have single prints done for a reasonable cost.

Printing on "Photographic Paper"

There are many photographic shops that now provide colour printing of digital images onto photographic film. The quality of these prints is extremely high – in fact, the prints cannot be distinguished from prints produced from colour film. The good thing about having prints done on photographic paper is that the dyes used in the colour process have been around for many years and are known to be very stable. The price of printing on photographic film is similar to having individual prints done from colour film negatives.

Table 7-1. Types of Printers

Various printers are available, with most of them falling into the following categories based on the method of printing used.

Printer type	Characteristics	Quality of print
B&W Laser:	The image is created by printing fine dots of black on conventional "photocopy" paper. The number of "dots per inch" determines print quality. Simply having fewer dots within a given area produces "grey" shading effects – the ink is never grey.	Very high quality text, reasonable image quality (useful for notebooks) but lack true grey-scale printing.
Colour Laser:	The image is produced in a similar way to B&W laser printers, but the ink is applied in four different colours (CMYK). Similar to B&W laser printers, there is no shading of colours – just more or less "dots" within a given area.	Reasonably good for presentations, but lacks any true colour gradient in the image.
Ink Jet:	The ink is deposited on the paper as small droplets (from three, CMY, or four, CMYK, ink cartridges). Printing on "photographic quality" glossy paper produces excellent images. Some photographic quality printers use "diluted" ink colours that can be deposited to varying degrees to create true colour shading.	Very high quality "photographic" quality possible when using special glossy paper. Oversize printing results in image pixels being readily visible.
Dye Sublimation:	Three colour ink (CMY) is deposited from an inking "ribbon" and heated to produce a slight "mixing" of the colour pixels. These printers produce excellent photographic quality prints with true colour mixing and shading.	Photographic publication quality images. A slight oversize printing is often tolerated because the image pixel boundaries are somewhat "blurred" by the heating process of mixing the colours. Also produce excellent grey-scale images.
Photographic:	The process of producing digital images on photographic paper is not strictly a printing process, but has been included in this list as an excellent alternative to the different printers available.	Excellent quality reproduction on regular colour photographic paper. Not only is the image quality excellent, but the dyes used in the photographic paper are known to be particularly stable.

Chapter 8

What is Fluorescence?

Fluorescence is the property whereby a molecule emits light at a specific wavelength when irradiated by light of a shorter wavelength. An exception is the case of multi-photon microscopy where light of a long wavelength (red light) is able to excite a fluorophore such that a short wavelength photon (blue or green light) is emitted. Using fluorescence to create an image in microscopy has been in use since the 1940's, but there has been a greatly increased interest in using fluorescent molecules in microscopy since the advent of the laser scanning confocal microscope in the late 1980's.

This chapter gives a brief introduction to fluorescence, particularly in regard to its application to fluorescence microscopy. An excellent book that covers most aspects of fluorescence (but not fluorescence imaging) is the text by Joseph R. Lakowicz ("Principles of Fluorescence Spectroscopy", Plenum Press, NY, 1999). The following chapter (Chapter 9 "Fluorescent Probes") contains further details on a large range of fluorescent compounds available for use in fluorescence microscopy.

WAVELENGTHS OF LIGHT

Visible light (see Figure 8-1) is a small part of the electromagnetic radiation spectrum that consists of a very broad range of wavelengths or energy levels from radio waves to gamma radiation. Relatively long wavelength (red) light consists of low energy photons. As the wavelength becomes shorter, the colour changes through red/yellow,

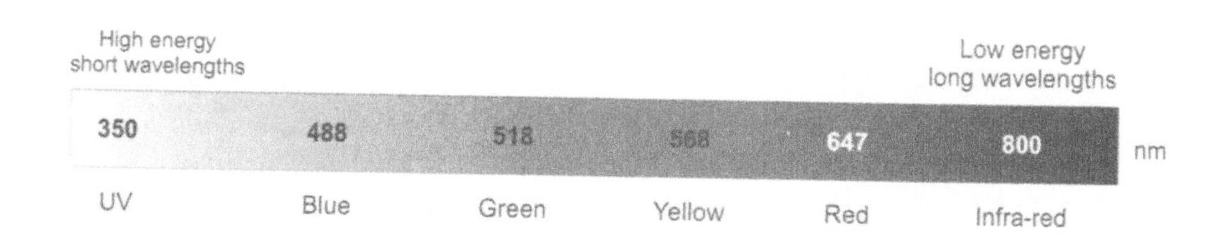

Figure 8-1. Visible Light Spectrum.
The visible light spectrum consists of short wavelength (high-energy) photons at the Ultraviolet (UV) end, and long wavelength (low energy) photons at the infrared end of the spectrum. In general, biological material absorbs longer wavelengths less, thus allowing infrared light to penetrate deeper into the tissue. Shorter wavelength UV light is particularly destructive to biological material – and UV lasers are quite hazardous as the rays are not only dangerous, but also invisible.

yellow, green, blue and purple before becoming invisible again at the UV end of the spectrum. The higher energy blue and invisible UV photons have a relatively short wavelength and correspondingly higher energy.

Confocal microscopy is typically limited to using visible light due to the reliability and availability of relatively cheap and small lasers that have strong lines of emission in the visible region of the light spectrum. There are confocal microscopes available that use UV lasers, but these instruments are significantly more expensive, are difficult to align correctly for multi-labelling applications, and the UV laser is significantly more hazardous to operate than visible wavelength lasers.

An exciting addition to confocal microscopy is the multi-photon microscope that makes use of red or far-red pulsed light to excite visible and UV light excited fluorochromes. The multi-photon microscope has the advantage that the relatively long wavelength infrared light more easily penetrates thick tissue samples compared to visible and UV wavelengths (see page 315 in Chapter 12 "Imaging Live Cells"). However, the multi-photon microscope, due to the non-linear optics involved, is not as predictable in its use as a conventional or 1-photon fluorescence microscope. In more recent years this has become less of a problem as lists of suitable dyes and the best wavelength for excitation by the multi-photon pulsed laser have become available (see Chapter 13, page 325).

Lasers are also used extensively in flow cytometry (the sorting and counting of single cells in suspension by laser activated fluorescence of molecular tags). The 488 nm blue line of the argon-ion laser is one of the main lasers used in flow cytometry, therefore a large number of fluorophores used in flow cytometry can be readily utilised in confocal microscopy.

A more detailed discussion of the lasers used in confocal (Table 3-1, page 87) and multi-photon microscopy (Table 3-2, page 96) is included in Chapter 3 "Confocal Microscopy Hardware".

EXCITATION OF A FLUOROPHORE

Conventional fluorescence microscopy and confocal fluorescence microscopy both utilise "conventional" or single-photon excitation to create fluorescence. This will always result in the emitted light being of a longer wavelength compared to the light used for the excitation. This is in contrast to multi-photon excitation, where two or more photons of long wavelength red light are used to elicit the emission of a short wavelength blue or green photon.

Single Photon Excitation

The most common form of eliciting fluorescence from a molecule is by single-photon excitation. In this case, when a population of fluorophore molecules is excited by an appropriate wavelength of light some of the molecules will absorb a photon of sufficient energy to boost an electron from the ground-state energy level (S_0) to an excited-state energy level (S_1) (Figure 8-2). Within the S1 energy level there are a number of vibrational levels (depicted as a number of lines in the S_1 level of the Jabłoński diagram). The excited electron will rapidly relax to the lowest vibrational level (still within S_1), with the subsequent drop back down to the ground-state energy level (S_0) resulting in

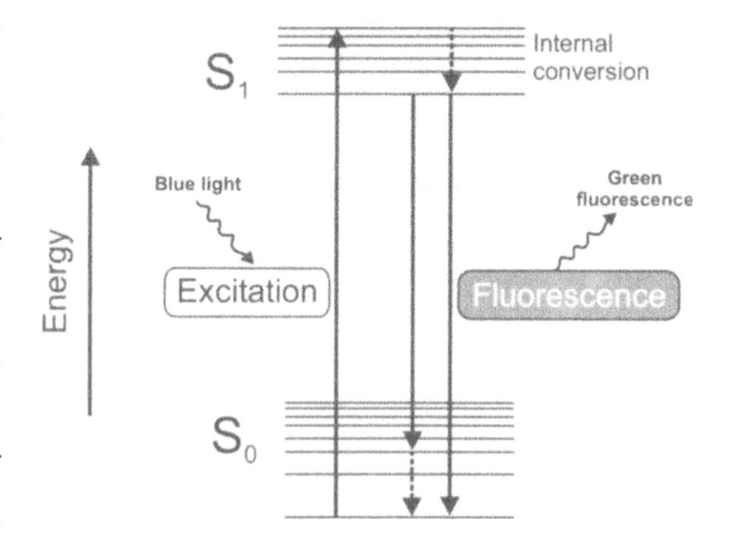

Figure 8-2. Single Photon Excitation.
A simple Jabłoński energy diagram depicting single-photon excitation, where an electron is promoted to a higher energy orbit by the absorption of a single photon (excitation). This energy is released as a photon of diminished energy or longer wavelength light (fluorescence). In this example relatively short wavelength blue light (488 nm, for example) is used to boost an electron from the S_0 to the S_1 energy level. Internal conversion (and production of heat) results in the electron dropping several vibrational levels (still within S_1), and finally a drop back to the S_0 or ground state – which results in the emission of a longer wavelength green photon.

the emission of a photon of light (fluorescence). The photon of light emitted is of a longer wavelength (lower energy) than the light used to excite the electron, due to the loss of energy (released as heat) as the electron drops several vibrational levels while still in the excited (S_1) state. The process of electron excitation and subsequent emission of a photon of fluorescence is extremely fast (usually with a half-life of $<10^{-8}$ seconds).

In single photon excitation, the fluorescence excitation spectrum of the fluorophore usually exactly matches its absorption spectrum. This means that one can readily predict the best wavelength for the highest fluorescence quantum yield by studying the spectrum – the peak wavelength of absorption will result in the highest level of fluorescence. This is in contrast to multi-photon excitation described in more detail below.

Other forms of light emission from the excited state, for example phosphorescence, have a much longer half-life of emission. Not all electrons that drop back to the ground state result in the emission of a photon of light, sometimes heat or chemical reactions result from this change in energy (discussed in more detail later under the heading "Photobleaching").

Multi-photon Excitation

Although conventional single-photon excitation of a fluorophore always results in the emission of a photon of a longer wavelength, multi-photon excitation results in the emission of a photon of higher energy (shorter wavelength) than the photon used to excite the molecule.

In multi-photon excitation (Figure 8-3) two relatively long wavelength photons (usually infrared light) are required to arrive at the one molecule almost simultaneously – resulting in the two lower energy photons boosting the energy level of the outer shell electron into a higher energy level (S_0 to S_1 energy levels) that would normally be associated with the absorption of a shorter wavelength single photon.

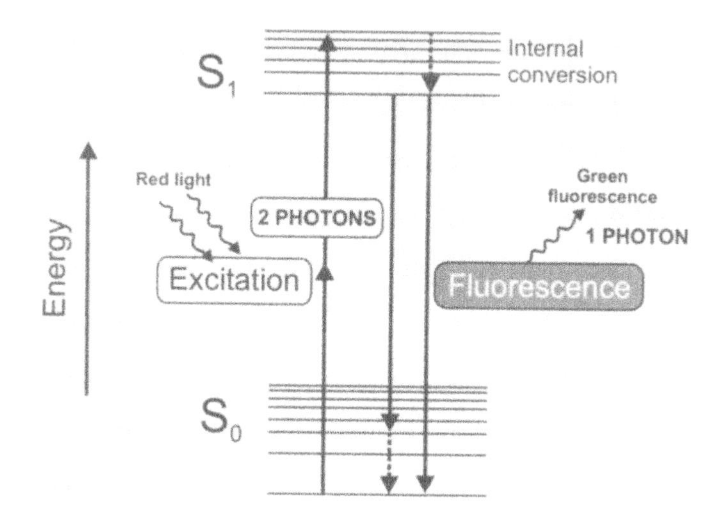

Figure 8-3. Multi-photon Excitation.
Jabłoński energy diagram showing how Multi-photon excitation requires the almost simultaneous arrival of two photons (usually of long wavelength infrared light) – which then results, after some internal conversion energy loss, in the emission of a shorter wavelength blue or green photon.

To enable multi-photon excitation to occur to such an extent that it becomes useful for fluorescence microscopy a very high light intensity is required. A special type of picosecond or femtosecond pulsed laser with very high intensity light at the peak of each pulse is required. Multi-photon excitation usually refers to either two-photon excitation, in which two photons are used to excite the fluorophore, or three-photon excitation where three photons are involved in exciting the fluorophore.

In single photon excitation the amount of emitted fluorescent light is proportional to the intensity of excitation light – however, in the case of two-photon excitation the emission intensity depends on the square of the incident light intensity. For three-photon excitation the emission intensity depends on the third power of the incident intensity. This difference between single-photon and multi-photon excitation means that almost all of the multi-photon excitation occurs at the focal point, where the intensity of the light is sufficient for a significant amount of multi-photon excitation to occur. This means that in multi-photon microscopy there is no need to use a pinhole to remove out-of-focus light as only light at the focal point of the lens has sufficient intensity to create fluorescence.

FLUORESCENCE EMISSION SPECTRA

Fluorescent molecules have characteristic excitation spectra (which is the same as the absorption spectra for single-photon excitation) and fluorescence emission spectra. The excitation and emission spectrum of a typical fluorophore is depicted in Figure 8-4, where the absorption maximum is in the region of blue light and the emission maximum is of longer wavelength green light. This type of spectrum is typical of a very common fluorescent molecule called fluorescein. Many different modifications and conjugates of fluorescein are used extensively in confocal microscopy.

The amount of light absorbed by the molecule per unit of concentration is called the extinction coefficient. The efficiency with which a fluorescent molecule converts absorbed light into emitted fluorescent light is known as the quantum yield. The quantum yield is determined by the molecular characteristics of the molecule, its local environment (conjugating the fluorophore to other molecules, such as a protein changes the quantum yield), and sometimes the wavelength of light used for excitation.

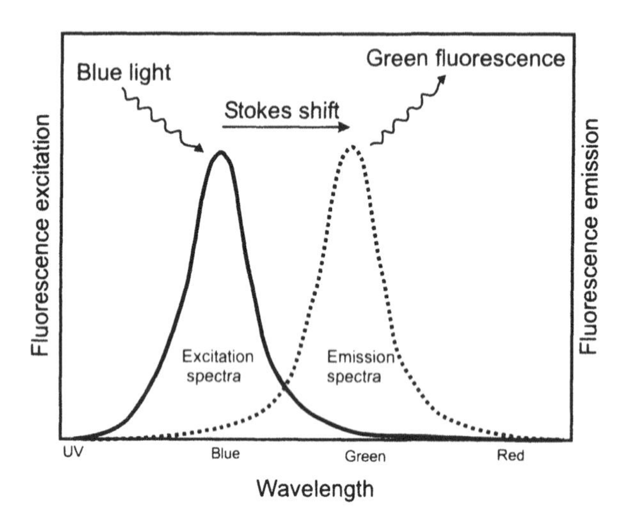

Figure 8-4. Excitation and Emission Spectra.
In this idealised example, blue light (short wavelength, relatively high energy) is used to excite the molecule, and green light (longer wavelength, and relatively lower energy) is emitted as fluorescence. The shift in wavelengths of emission compared to excitation is called the Stokes shift. This type of spectrum is typical of the common fluorophore, fluorescein.

The shift in wavelength between the absorption and emission spectra is called the "Stokes shift". A large Stokes shift makes collection of the light emitted from the fluorescent molecule relatively easy, as the emitted light is further separated from the wavelength of light used to excite it.

The most efficient light for exciting a fluorophore is the wavelength where maximum absorption by the fluorophore occurs (a high extinction coefficient). Exciting a fluorophore with a wavelength of light that is well away from the maximal absorption wavelength is possible (Figure 8-6), however, the amount of fluorescent light emitted will be significantly less.

In Figure 8-6 a shorter wavelength (purple light) is used to excite a fluorophore that has a maximum absorption in the blue region of the spectrum. This results in the amount of green fluorescence obtained being greatly reduced compared the example in Figure 8-4, where blue light is used to excite the molecule. Changing the wavelength of light (colour) used to excite a dye does not alter the wavelength (colour) of emission. For example, fluorescein emits green fluorescence irrespective of the wavelength of light used for excitation.

The fluorescence process is cyclical, being repeated until the fluorophore is destroyed by photobleaching (described in more detail below). However, not all molecules in a sample will be excited as the number of excited molecules is based on the probability of the molecule absorbing a suitable photon (which is proportional to the number of photons used to irradiate the sample). The fluorophore will also need to be in the ground state to accept such a photon. Although the fluorochrome very quickly drops back to the ground-state, there will be a significant

Fluorescein emits green light – irrespective of the wavelength of light used for excitation

number of molecules in the sample that are in the process of returning to the ground-state and thus not yet available for re-excitation.

Multi-photon fluorescence excitation spectra are much more difficult to obtain compared to single-photon excitation spectra. Furthermore, the excitation spectra or optima for multi-photon excitation does not exactly match the multi-photon absorption spectrum for the compound.

Environmental Influence on Fluorescence Spectra

The molecular environment of the fluorophore can significantly alter the absorbance spectrum of the dye. It is important to study the relevant spectrum of the dye of interest in the correct environment. For example, conjugation of the dye to protein will often alter the absorbance characteristics of a dye. Most spectra obtained from Molecular Probes or the Bio-Rad Web sites are for fluorophores conjugated to IgG.

The local environment of a fluorophore also has a significant effect on not only the absorbance spectrum, but also more importantly on the fluorescence emission spectrum of the dye.

Environmental effects on specific fluorochromes are exploited for determining the concentrations of specific ions or the pH of subcellular structures using specific fluorophores (see page 216 in Chapter 9 "Fluorescent Probes").

Density of packing of the fluorochrome is another environmental effect that can be usefully exploited to gather important cellular information. For example, the close packing of acridine orange molecules inside an acidic vesicle results in a change in the wavelength of the fluorescence emission (towards the red end of the spectrum). This phenomenon can be used as a measure of the pH of the vesicle under study. Another dye, called JC1, has a very marked shift in the wavelength of emission (from green to red) on close molecular packing. This dye specifically accumulates in the matrix of mitochondria with an active membrane potential, and thus the shift from green (unpacked) to red (packed) fluorescence is a measure of the respiratory activity of the mitochondria.

Although the above environmental effects have been successfully exploited to study cellular process, it should be kept in mind that other fluorescent dyes that are being used as indicators of cellular processes may also undergo excitation and emission changes due to their local molecular environment. These changes in emission due to the local environment may result in experiments that are difficult to interpret, or even generate misleading results.

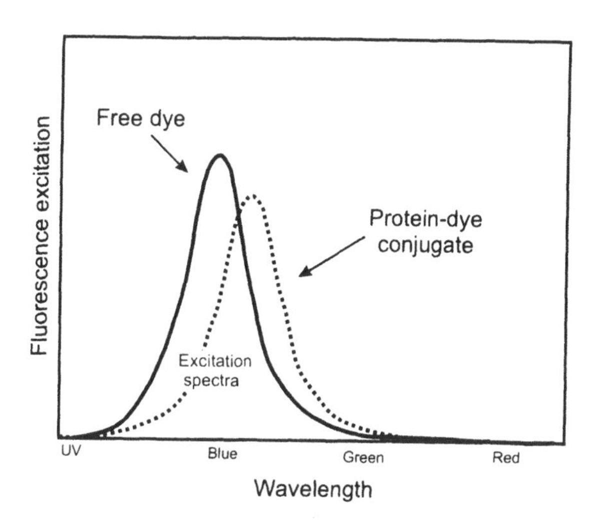

Figure 8-5. Alteration of Absorption and Excitation Spectra by Conjugation of Dye to Protein.
The characteristic spectrum of individual fluorescent dyes can significantly alter when they are conjugated to protein or other macromolecules. The resultant fluorescence spectrum is also often altered (shifted to longer wavelengths). The fluorescence spectra shown on the Bio-Rad web site are often of the fluorophore conjugated to IgG antibody molecules.

BRIGHTNESS OF A FLUOROPHORE

Different fluorophores vary greatly in the intensity of light that they emit. The level of fluorescent light emitted is proportional to both the extinction coefficient (the amount of light absorbed) and the quantum efficiency (the amount of fluorescent light emitted for a given amount of light absorbed) of the fluorophore.

Light can be highly destructive to biological material and so minimising both the amount of light and the time of exposure is important. A fluorophore that emits a low level of fluorescence will require a much higher level of irradiation or a much longer time of exposure to the light source. It is sensible to use fluorophores with the highest

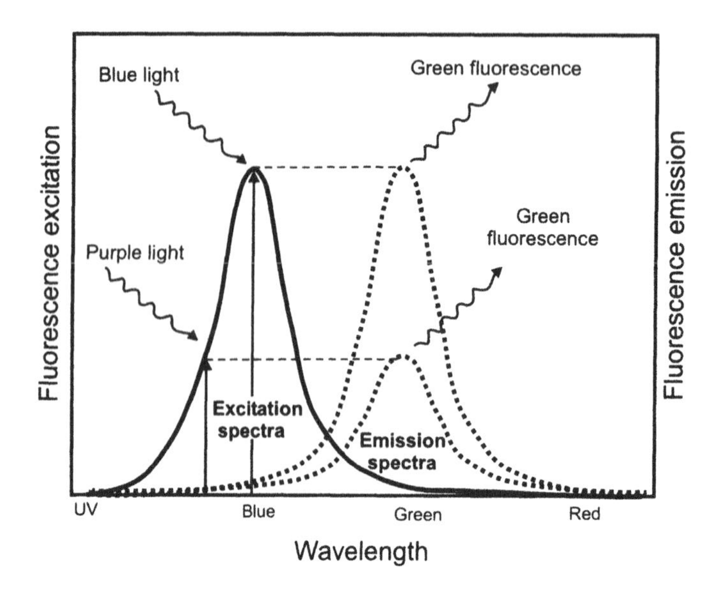

Figure 8-6. Reduced Quantum Yield using Sub-optimal Excitation Wavelengths. Excitation of a fluorophore at a shorter wavelength will still result in the same emission spectrum, but with diminished intensity. In other words, a green fluorescent molecule will always emit green light irrespective of the colour of the light used to excite it.

fluorescence output so that a relatively low level of light and a minimum of exposure time can be used for excitation. It therefore also makes sense, where possible, to excite a fluorophore with a wavelength of light that corresponds as closely as possible to the apex of maximum absorption.

A highly efficient fluorophore (high quantum yield) is particularly important when the level of labelling is low, for example when using a low affinity antibody. The amount of light used is important even when using fixed material as the fluorophore itself is often destroyed by high levels of irradiation. The efficiency of the fluorophore becomes critical when using live material, as high levels of light may cause considerable harm to living cells or tissue samples being studied.

WHICH EXCITATION WAVELENGTH IS BEST?

Most dyes have a range of wavelengths that can be readily utilised for excitation. The dye is most efficiently excited at the peak excitation wavelength (the figure listed in fluorescent dye Tables). However, many fluorescent dyes can be readily excited at either higher or lower wavelengths than the peak excitation wavelength quoted – with some loss of fluorescence yield.

In multi-labelling applications the choice of the correct wavelength of excitation may be critical (Figure 8-7) for minimising bleed-through (the detection of fluorescence emission intended for one channel being detected in the other channel). In the example shown in Figure 8-7, a fluorescein-antibody conjugate is very efficiently excited by the 488 nm blue line of the krypton-argon-ion laser. In contrast, a tetramethylrhodamine-antibody conjugate is only very poorly excited by the 488 nm blue line. However, the 568 nm yellow line of the krypton-argon-ion laser is efficient at exciting the TMR conjugate. This difference in ability of different fluorophores to be excited by the different laser lines can be readily

A note on interpreting fluorescence spectra:

The absorbance and fluorescence emission spectra shown in publications (including this book) have been "normalized" to show each of the spectra being compared at the same peak height. This does not mean that in your preparation the peak height will always be the same, in fact, the fluorescence intensity of individual fluorophores will vary greatly. Take care when estimating, for example, the degree of "bleed-through" expected with a particular fluorophore combination – would greatly increasing the height of the graph of one fluorophore help or hinder you in eliminating the "bleed-through" problem?

exploited in dual labelling experiments. For many dual-labelling experiments involving FITC and TMR conjugates both laser lines (488 and 568 nm) are used together. However, if bleed-through is a problem, sequentially exciting the two chromophores sequentially with the 488 nm and then the 568 nm laser line is possible. This will result in less bleed-through between the two collection channels. In some confocal microscopes each scan line in the image can be collected using each irradiating laser wavelength separately. This is accomplished by high speed switching between the alternate laser lines, and only collecting the relevant emission wavelengths on each scan of the laser.

When a probe is excited away from the peak absorption region the efficiency, and thus the brightness of the probe, will be diminished, but, as discussed above, the wavelength of emission is not affected. Some dyes, such as propidium iodide have a very broad excitation spectrum and so can be readily excited at wavelengths well away from the peak excitation wavelength. This makes dual labelling experiments possible, even with only one excitation wavelength (for example, FITC and propidium iodide can both be excited using 488 nm blue light, with FITC emitting green light and propidium iodide emitting red light). Other dyes, such as the new Alexa Fluor probes have quite narrow excitation spectra and so the use of an incorrect dye/laser combination may mean an almost total lack of fluorescence.

The blue (488 nm) and the yellow (568 nm) lines on the krypton-argon-ion laser, or the blue line on the argon-ion laser (488 nm) in combination with the green line on the Green HeNe laser (543 nm) can be readily used to excite most dyes. These lines are of sufficient power, and being of great reliability they are often the "workhorses" of confocal microscopy. Triple labelling may require the use of the red (647 nm) line of the krypton-argon-ion laser (which is notoriously unstable), or the red line of a Red HeNe (633 nm) laser.

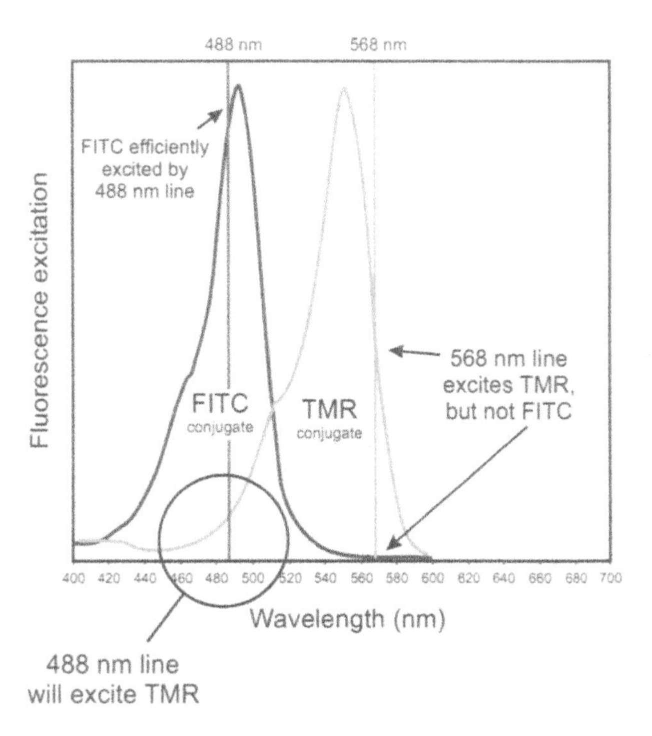

Figure 8-7. Which Excitation Wavelength Works Best?
Fluorescein-antibody conjugate (FITC-conjugate) is very efficiently excited by the 488 nm line of the krypton-argon-ion laser – but does not excite the Tetramethylrhodamine-antibody conjugate (TMR-conjugate). On the other hand, the TMR-conjugate is reasonably efficiently excited by the 568 nm (yellow) line of the laser, and also, although to a much lesser extent, by the 488 nm line.

This difference in efficiency of excitation by different laser lines can be readily exploited in dual labelling experiments. If bleed-through in dual labelling is a problem, then the FITC and TMR fluorophores can be excited individually by the 488 and 568 nm laser lines and the separate images combined after collection. In some instruments the individual laser lines can be switched so fast that each line in the image can be collected sequentially with each of the laser lines.

MOLECULAR STRUCTURE OF COMMON FLUOROPHORES

Many organic molecules show fluorescence when irradiated with light of suitable wavelengths. Typically such molecules contain a series of ring structures exhibiting a variety of double bonds. Initially fluorescence studies (including fluorescence microscopy) utilised many readily available organic molecules that are naturally fluorescent, such as the naturally occurring anti-malarial drug quinine (Figure 8-8). Quinine was the molecule observed in one of the earliest reports of fluorescence (Herschel J.F.W. (1845), Philosophical Transactions of the Royal Society of London 135:143-145), even though the phenomena that Herschel described could not be understood by any known scientific explanation available at that time.

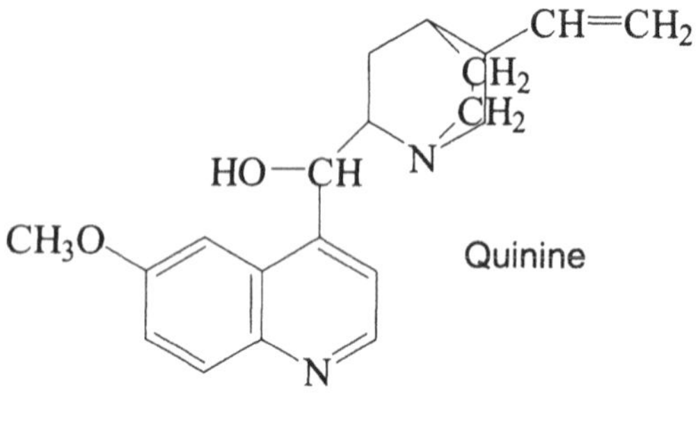

CH=CH$_2$ CH$_2$ CH$_2$ N HO—CH CH$_3$O N Quinine

Figure 8-8. Quinine Structure.
This naturally occurring anti-malarial molecule is fluorescent when irradiated with UV light. The carbon ring structures are a common motive to many fluorescent molecules.

Quinine, the bitter tasting component of tonic water, emits blue light when irradiated with UV light (this can be seen using sunlight), and the blue emission is enhanced by adding a less polar solvent such as alcohol – and is completely lost (quenched) by the addition of common salt (NaCl). In the later 1940's the first spectrofluorometers were developed as an assay for anti-malarial drugs, including quinine.

The UV induced fluorescence of quinine, and many of the other earlier dyes used in fluorescence studies, is a valuable tool in conventional epi-fluorescence microscopy – but are not as readily utilised by most modern confocal microscopes.

Although confocal microscopes that can excite UV dyes are available, and multi-photon microscopes are readily capable of exciting most UV dyes, confocal microscopes are generally designed to use visible wavelength excited fluorescent dyes. Consequently there has been a great deal of effort in recent years to develop a variety of dyes that are excited not simply within the visible light spectrum, but most efficiently at wavelengths emitted by the common lasers used in confocal microscopy (particularly the 488 nm blue line from the argon-ion laser).

Fluorescein, the Green Fluorescent Dye

The traditional workhorse of fluorescence microscopy is fluorescein (commonly known as FITC, Figure 8-9). This molecule is widely used for a variety of fluorescent labelling protocols and techniques. Various derivatives can be used as cellular markers, or when conjugated to antibodies FITC has been widely used as the fluorescent dye of choice for immunolabelling.

FITC was extensively used as a fluorophore long before laser scanning confocal microscopy became available, as the mercury lamp used in conventional epi-fluorescence microscopy has a very strong emission in the blue region of the visible light spectrum (where FITC shows maximum absorption).

The FITC absorption maximum is close to the main emission line (488 nm blue light) of the argon-ion and krypton-argon-ion lasers, and so in more recent years FITC has become a very important molecule for confocal microscopy. FITC has retained its popularity over the years despite the development of a range of "better" dyes as not only is the absorption maximum suitable for confocal microscopy, but the dye has a high quantum efficiency (the amount of fluorescent light emitted per unit of irradiated light is quite high). This means that in practical terms a relatively low light level can be used and still give sufficiently bright fluorescence to create good images. The big disadvantage of FITC is that this fluorophore fades rather easily if an anti-fade reagent is not used.

The level of fluorescence exhibited by FITC is greatly influenced by the local environment of the probe. For example, subtle changes in pH in different cellular compartments can result in changes in brightness of the fluorescence of FITC and its derivatives. It is particularly important to be aware of this phenomena when using FITC as a probe of live cells, but should also be kept in mind when performing immunolabelling on fixed material as the pH of the incubation media and particularly the mountant can have a large effect on the intensity of fluorescence observed.

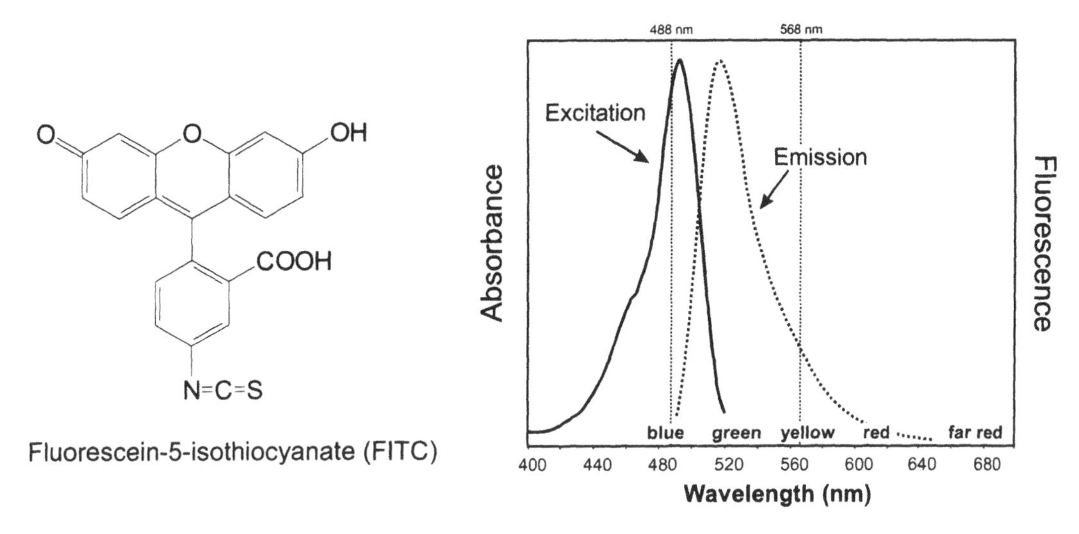

Figure 8-9. Fluorescein-5-isothiocyanate (FITC) Spectra.
Excitation and emission spectra of FITC when conjugated to IgG. FITC has traditionally been the fluorescent dye of choice for immunolabelling when conjugated to suitable antibodies. Blue (488 nm) laser light will very efficiently excite fluorescein, resulting in green fluorescence (with a peak emission at approximately 520 nm). The 568 nm (yellow) laser line does not create any fluorescence from FITC under normal irradiation conditions. This diagram was designed using spectra obtained from the Bio-Rad fluorescence Internet database site (http: //fluorescence.bio-rad.com/).

Tetramethylrhodamine (TMR), Red Fluorescent Dye

Tetramethylrhodamine-5-isothiocyanate, commonly known as TRITC (Figure 8-10), was traditionally often used as the second dye in dual labelling experiments with FITC. This dye has an absorption maximum in the yellow region of the visible light spectrum and emits fluorescent light in the red region, sufficiently removed from the green emission of fluorescein to make TMR suitable for dual labelling experiments. This dye may have sufficient absorbance at 488 nm (blue light) to permit dual labelling when only one laser line is available (which may create significant "bleed-through" problems, see the section "Which Excitation Wavelength is Best" above). TRITC can be reasonably efficiently excited by the 514 nm second line available on the argon-ion laser or the 568 nm yellow line on the krypton-argon-ion laser.

This fluorophore was also extensively used with epi-fluorescence microscopy prior to the advent of laser scanning confocal microscopy. A number of derivatives have been developed, for example Lissamine rhodamine, with somewhat different excitation and emission spectra.

For a variety of reasons these red emitting fluorophores often have significantly lower levels of fluorescence compared to FITC, in which case the less efficiently excited red fluorophore should be used with the higher affinity antibody when performing a dual immunolabelling experiment. However, these red fluorophores also often show less photobleaching compared to FITC.

Alexa Fluor Dyes

More recently a number of dyes have been produced, primarily by Molecular Probes, which may eventually supersede FITC. These include the Alexa Fluor dyes, which come in a variety of emission and excitation wavelengths and have a number of advantages over FITC. In particular they have an even higher quantum yield, and somewhat surprisingly they are hardly affected by photobleaching even in the absence of antifade reagents. They are also less affected by the local environment, for example changes in pH, compared to more traditional dyes such as FITC. The Alexa Fluor dyes are discussed in more detail on page 208 in the following Chapter 9 "Fluorescent Probes".

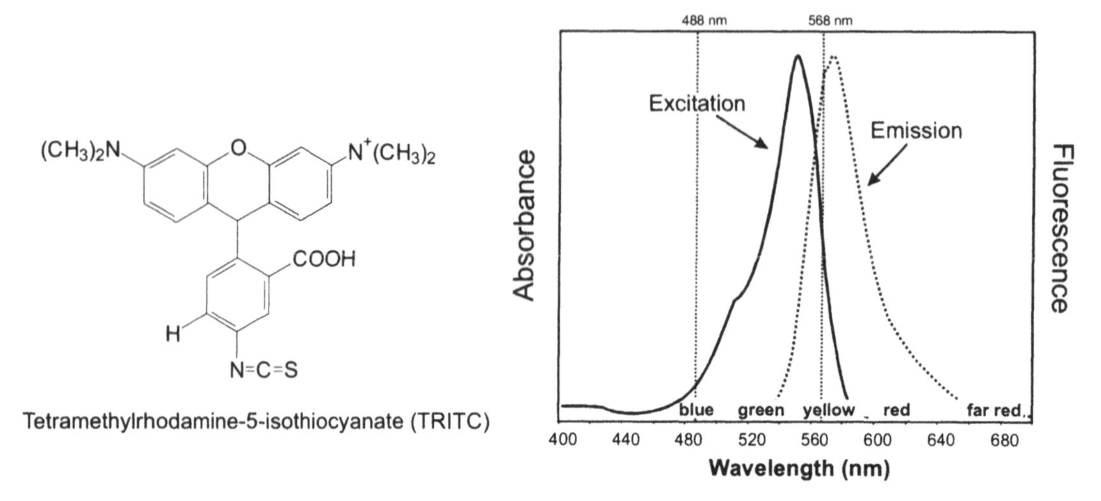

Figure 8-10. Tetramethylrhodamine-5-isothiocyanate (TRITC) Spectra.
TRITC is often the molecule of choice for the second label in dual immunolabelling experiments with FITC. TRITC is readily excited by the 568 nm (yellow) line of the krypton-argon-ion laser, but is also to a minor extent excited by the 488 nm (blue) line – which creates "bleed-through" into the other channel when dual labelling. This diagram was designed using spectra obtained from the Bio-Rad fluorescence Internet database site (http: //fluorescence.bio-rad.com/).

PHOTOBLEACHING

Unfortunately not all the light absorbed by a fluorophore is utilised in the creation of fluorescence. Some of the absorbed energy can be dissipated as heat as mentioned above, which may be of concern when using live cells where localised heating could result in loss of cellular function. Although the amount of heat created is very low, this heat is generated in a very localised environment where the laser is focussed within the sample such that it may not be at all obvious that local heating has occurred, except when gross damage to the cells or tissues has taken place. The infrared laser used in multi-photon microscopy creates considerably more localised heating than the visible light lasers used in confocal microscopy.

There is also the possibility that the excited electron will not return to the original ground state, but that the energy will be utilised to create chemical reactions (Figure 8-11), which is known as photobleaching - i.e. the fluorophore appears to become rather dull after irradiating.

Chemical reactions driven by the electron energy created during photon excitation is not only disruptive to biological systems, but is also highly destructive to the fluorophore itself. Free radicals thus created will readily react with any suitable molecule in close proximity. Free radical formation can result in the laser light becoming highly toxic to the cells, even though in the absence of the fluorophore the cells may tolerate a relatively high level of laser irradiation.

Another form of light emission is called phosphorescence. Phosphorescence has a much longer half-life of emission (with a decay time of seconds, or even minutes) compared to fluorescence and is usually only observed in the absence of oxygen.

Controlling Photobleaching in Fixed Samples

The utilisation of the electron energy for chemical reactions is characteristic of the fluorophore and thus cannot easily be controlled without chemically modifying the fluorescent molecule itself. However a number of reagents are available to help minimise the damage done by free radicals, which are the chief products generated by the high-energy electron. These reagents effectively "mop up" the free radicals formed by the irradiating light.

There are many commercially available antifade reagents, however one of the most effective is *n*-propyl gallate which can be readily prepared in the laboratory (see enclosed box on next page). This reagent is added to a simple glycerol mounting media, and can be stored at -20° for many months.

Samples mounted in *n*-propyl gallate will be stable for several weeks if stored at -20° (the glycerol mountant means that the slides do not freeze at this temperature, but bacteria are not able to grow and destroy the mount). For long-term storage, sealing the mount by running nail polish or VALAP around the edges of the coverslip, is essential.

Controlling Photobleaching in Live Cells

Most of the reagents used for helping to control the level of photobleaching in the sample are highly toxic to live cells. However, one relatively benign method is to use sodium ascorbate (Vitamin C) as an antifade reagent. The only problem is that a very high concentration of Vitamin C (2 mg/ml) is required to be effective. I have used this on live cells without any detrimental effect on protein trafficking. However, one should be aware that such high levels of Vitamin C may affect cellular functions in many subtle ways that may not be obvious.

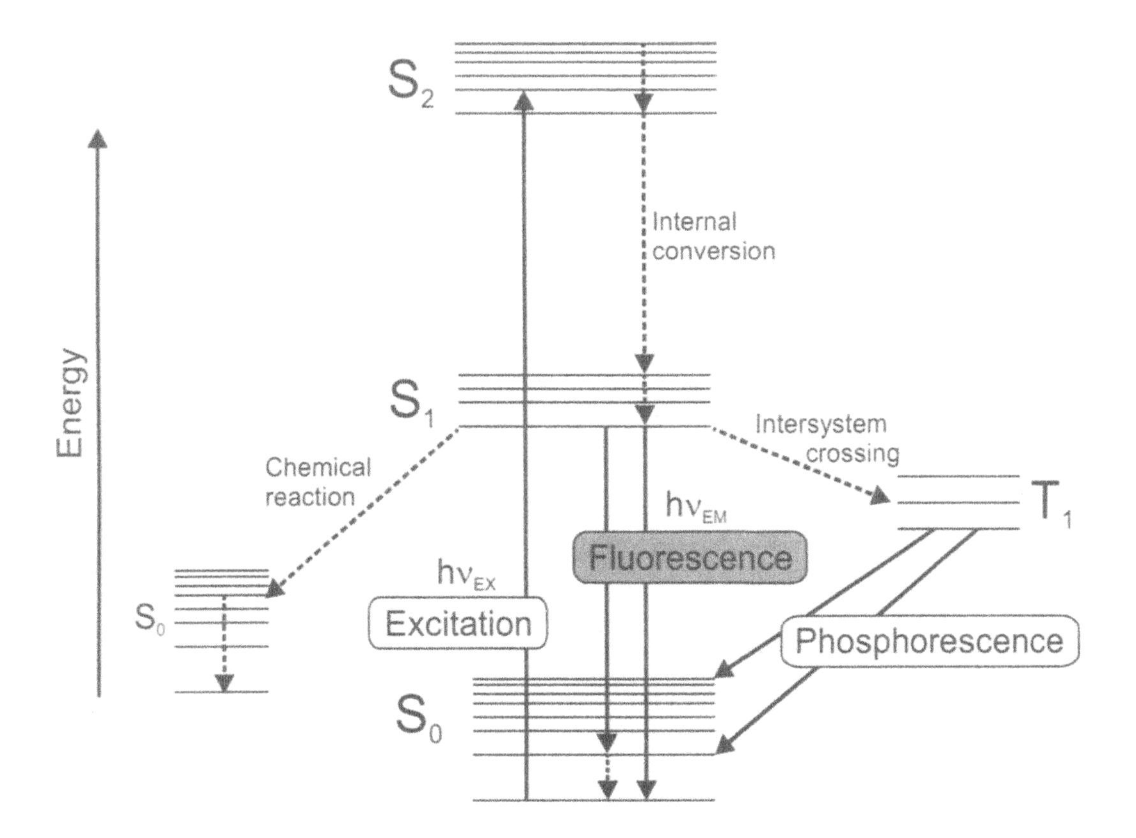

Figure 8-11. Chemical Reactions and Phosphorescence from Excited Electrons.
A more complex Jabłoński diagram, demonstrating how some electrons dropping back to a lower energy level do not result in fluorescence - they can result in phosphorescence or chemical reactions. One of the most common chemical reactions is the generation of free radicals, which are highly reactive chemical intermediates that will react with any organic molecules in the immediate vicinity. These reactions will not only destroy the fluorescent dye molecule, but will also quite possibly disrupt other important cellular macromolecules.

Different fluorophores have varying degrees of susceptibility to photobleaching and so often a careful choice of fluorescent molecule may minimise problems without resorting to toxic antifade reagents. For example, when immunolabelling, the new Alexa Fluor fluorophores from Molecular Probes show very little photobleaching when used in simple phosphate buffered saline solutions.

Don't forget that using a probe such as the Alexa Fluor dyes that are resistant to photobleaching does not alter the level of highly reactive free radicals generated by other molecules that may interact with the irradiating light within the cells or tissue sample. Antifade reagent will protect both the fluorophore and the cells or tissue under study.

Antifade Reagents

***p*-phenylenediamine** (PPD)
Johnson and Nogueira Araujo (1981) J Immunol Methods 43:349-350
A very effective antifade reagent but suffers from low photo-stability (unstable in room lighting!), and is toxic to live cells. Use 1mg/ml in 90% glycerol, pH 8.0.

***n*-propyl gallate (NPG) - highly recommended**
Giloh and Sedat (1982) Science 217: 1252-1255.
Non-toxic photo-stable antifade reagent. Dissolve 50 mg of *n*-propyl gallate (Sigma, not Aldrich) in 100 μl 2M Tris base (warm tube under hot tap water), and then add 900 μl glycerol. Mix well and store at -20°C in small aliquots.

DABCO 1,4-diazobicyclo[2,2,2]-octane.
A stable and readily available antifade reagent. Used at a concentration of 0.1%. Possible use for *in vivo* studies.

*** Ascorbic acid (Vitamin C)**
Reasonable antifade reagent that can be used for *in vivo* studies. However, the concentration required to be effective (2mg/ml in PBS or cell growth media) may be disruptive to live cells.

*** Trolux,** 6-hydroxy-2,5,7,8-tetramethylchoman-2 carboxylic acid.
A Vitamin E derivative that acts as an anti-oxidant, and can be used as an antifade reagent on live cells.

Commercially Available Preparations:

Vectashield
Slow Fade
FluoroGuard
Moliwal

To prolong the storage life of the labelled slide seal around the edges of the coverslip with nail polish for fixed cells, or VALAP for live cells (see page 283). The mounts can then be stored at -20°C (the high glycerol concentration means that the mount does not freeze) for several weeks or even months.

* = Compatible with live cell preparations

AUTOFLUORESCENCE

Fluorescence coming from organic molecules that are present in the sample being examined is often a source of problems for microscopists. This autofluorescence may obscure the fluorescence coming from the probe of interest. However, sometimes autofluorescence can be put to good use to produce superb images of cellular structures without the need for adding a fluorescent label.

The chitin of insects is highly autofluorescent, and can be utilised to produce excellent low magnification images of insect structure. In plants chlorophyll is also highly fluorescent when using 488 nm blue light and so can be utilised to successfully image plant structure by confocal microscopy. Autofluorescence can often be used to provide further structural information in fluorescently labelled samples.

Autofluorescence in Living Cells and Tissues

All cells contain an enormous array of organic molecules, some of which are naturally fluorescent. Reduced pyridine nucleotides (NADH and NADPH) are strongly fluorescent when irradiated with UV light. This fluorescence is concentrated in the mitochondria and so can be used as a marker of mitochondria in cells. However, its presence may cause considerable difficulty by obscuring the fluorescence associated with the molecule of interest. NADH autofluorescence can be substantially reduced by using suitable drugs to uncouple the mitochondria. Naturally this causes considerable disruption to the biochemical pathways within the cell.

In confocal microscopy NADH autofluorescence is not normally of concern because the wavelength of excitation is usually in the visible region of the light spectrum and so does not excite NADH. However, considerable interest is being generated in using multi-photon microscopy to image NADH in living cells. With this microscope, studying the bioenergetics of mitochondria *in situ* should be possible.

Many other organic molecules, including some cytochromes may emit fluorescence. The longer the wavelength of irradiation (lower energy photons) the less is the problem of autofluorescence. Even using the 568 nm (yellow) line of the krypton-argon-ion laser may significantly lower unacceptable levels of autofluorescence.

Stress also appears to increase the level of autofluorescence. In fact some cells when seriously compromised in health become very strongly autofluorescent. This type of autofluorescence is often characterised by being spread across a broad region of the light spectrum (fluorescence will appear in all channels on the confocal microscope), thus allowing one to distinguish autofluorescence from fluorescence originating from the probe of interest.

Fluorescence Created by Fixation

The act of fixing a cell or tissue sample for microscopy can result in unacceptable levels of autofluorescence. Aldehyde fixatives in particular create autofluorescence on cross-linking with cellular proteins. For this reason glutaraldehyde is usually unacceptable as a fixative for studies involving fluorescence. Formaldehyde or para-formaldehyde is a more acceptable choice of fixative. Possibly, the most suitable fixative from this point of view is either pure acetone or a combination of acetone and methanol. Acetone alone results in no autofluorescence, whereas the addition of methanol does result in a low level of autofluorescence. There are means of lowering the level of autofluorescence created by aldehyde fixatives in a sample by washing with a solution of sodium borohydride, but it's probably a better idea to avoid excessive autofluorescence by the correct choice of fixative in the first place. The problems of autofluorescence in fixed cell preparations are discussed in detail on page 275 in Chapter 11 "Fluorescence Immunolabelling".

Using Autofluorescence to Visualise Cell Structure

Not all autofluorescence, or fluorescence created by fixation should be looked on as detrimental to the experiment on hand. Autofluorescence can often be used to visualise cellular structure for either localising your specific probe, or for assessing the structure and integrity of the cells under study.

Autofluorescence

Aromatic amino acid residues
Present in most proteins. Excitation (200-340 nm) and emission (360-455 nm) does not interfere with visible light confocal microscopy. However, may be a problem with UV and multi-photon fluorescence microscopy.

Reduced pyridine nucleotides
NADH and NADPH are highly fluorescent (UV and multi-photon excitation).

Flavins and flavin nucleotides
Riboflavin, FMN and FAD are also fluorescent (UV to short wave blue excitation and 2-photon excitation).

Zinc-protoporphyrin

Chitin
Strong fluorescence using blue (488 nm) light.

Chlorophyll
Strong autofluorescence in both the green and red wavelengths when using visible light lasers.

Lipofuscin
Increased lipofuscin pigment with age of cells. UV and 490 nm excitation. Recognizable as pigment granules within the cytoplasm.

Dead cells
Many dead or dying cells become autofluorescent, exhibiting a broad range of excitation and emission (from UV through to red).

CHOOSING THE CORRECT DYE

There is no such thing as the perfect fluorescent dye, but if you take into account the following parameters when choosing, then there should be little difficulty in obtaining a dye eminently suited to your experiments. The number of new dyes being developed each year is quite remarkable. Furthermore, a number of these "new generation" dyes are much closer to the ideal fluorophore compared to many of the more traditional dyes.

The Ideal Fluorophore:

- **Appropriate wavelength of excitation**

  Choose an excitation wavelength that is close to the peak of absorption for the fluorophore you wish to use, or alternatively change fluorophores to suit the particular wavelengths (laser lines) you have available for excitation.

- **High quantum yield**

  Choose a brightly fluorescing fluorophore, and use the most highly fluorescent fluorophore for labelling the lowest abundant molecule in dual labelling applications.

- **Narrow emission spectrum**

  A broad emission spectrum is fine if you are only using one fluorescent label, but when using multiple fluorophores a narrow emission spectrum greatly reduces "bleed-through" into the other channels.

- **Suitable wavelength of emission for the optical filter sets available**

  Confocal microscopes often use removable optical filter sets to select for specific wavelengths of light. You may be restricted to particular fluorophores, depending on the dichroic mirror and optical filter combinations you have available.

- **Minimal susceptibility to photobleaching**.

  Some fluorophores have greatly reduced susceptibility to photobleaching, allowing longer collection times with reduced laser power, resulting in better quality images and less cell damage. Many of the traditional fluorophores such as fluorescein are particularly prone to photobleaching.

- **Minimal disruption to cellular metabolic processes**

  When labelling live cells it is most important that the fluorophore does not unduly disrupt the metabolic processes within the cell. One should always keep in mind that all fluorophores disrupt cellular processes to some extent – and many fluorophores that are sensitive to pH and ionic concentrations may themselves change, or buffer the pH, or change the ionic concentrations of ions in the cellular cytoplasm.

- **High specificity of label**

  A high specificity of the label is desirable. This is particularly important, for example, when using high specificity antibodies in immunolabelling – a relatively small amount of non-specific labelling of cellular structures with the fluorescently labelled primary or secondary antibody will greatly detract from the high specificity of the antibody used. Dyes used to highlight subcellular structures (such as organelles) often have a relatively low level of specificity, but can still be utilised due to knowledge of the expected morphology or location of the structure being labelled.

Chapter 9

Fluorescent Probes

A wide variety of molecules can be made into fluorescent probes. Sometimes these probes started out as molecular dyes that were known to bind to specific subcellular structures or molecules of interest. In more recent years, however, there have been a large number of non-fluorescent probes of cellular interest that have been made into fluorescent molecules by the addition of suitable molecular groups.

This chapter provides an overview of some of the interesting fluorescent probes that are available. Some probes are discussed in considerable detail, but others have been left out completely!

WHAT ARE FLUOROPHORES FOR?

Fluorescent probes are used in an increasingly wide range of biological applications. With the ready availability of very sensitive CCD cameras and confocal microscopes a lot of effort has gone into synthesizing fluorochromes with defined excitation and emission spectra as specific markers of cellular functions.

However, not all probes are created with perfect characteristics. Many probes, although used primarily for a particular purpose (such as FITC being used in immunolabelling), they can be influenced by other cellular functions or the environment. For example, FITC fluorescence is also influenced by the pH of the local environment of the probe. This could be of critical concern when attempts are made at immunolabelling live semi-permeabilised cells. In this case the specificity of the label would not be affected, but the amount of fluorescence observed may not directly reflect the amount of antigen present.

What are fluorochromes for?

Locating biomolecules
- immunolabelling
- receptor identification
- lipid analysis
- cytoskeletal proteins
- ion channel probes
- DNA probes

Cellular structure
- cell morphology
- organelle identification

Cellular ions
- pH measurements
- membrane potential
- reactive oxygen species
- inorganic ions (Ca^{2+}, Cl^-, Zn^{2+} etc)

Molecular movement
- endocytosis & exocytosis
- lipid transport
- membrane fluidity
- protein trafficking

Cell integrity
- live/dead cells
- apoptosis
- membrane integrity

Cellular functions
- signal transduction
- enzyme activity

Molecular genetics
- gene mapping
- chromosome analysis

and many more...

The fluorescence excitation and emission spectra depicted in this chapter were obtained from either the Bio-Rad fluorescence database Internet site (http://fluorescence.bio-rad.com), or the Molecular Probes Internet site (www.probes.com).

GETTING INFORMATION ON FLUOROCHROMES

There is a very large range of fluorescent dyes available for confocal microscopy, but information on various probes is scattered amongst a large number of different disciplines in the literature, making it somewhat difficult to track down relevant information.

Internet

It's a good idea to make use of the Internet for gathering information on suitable fluorescent probes (see Chapter 13 "The Internet", page 325). There are a number of confocal microscopy sites that have information on different fluorescent probes, but it is also important to look at related sites in flow cytometry, microscopy, multi-photon microscopy and fluorescence. A number of fluorophores used in confocal microscopy have been around for many years and studied extensively in more established areas of science.

Molecular Probes Catalogue and Internet Site

Molecular Probes Inc (Eugene OR, USA, www.probes.com) manufactures a very large range of fluorescent dyes for the research community. The Molecular Probes catalogue (available on their Internet site, as a printed book and on a CD-ROM) contains a great deal of information on thousands of fluorescent probes that are available, this includes valuable information on spectral characteristics, molecular weight and structure and chemical characteristics. Furthermore, there are many pages of excellent introductory information on both fluorescence and using fluorescence in microscopy. Most probes have relevant publications listed so that you can very easily delve into the fluorescence imaging literature to find information you need.

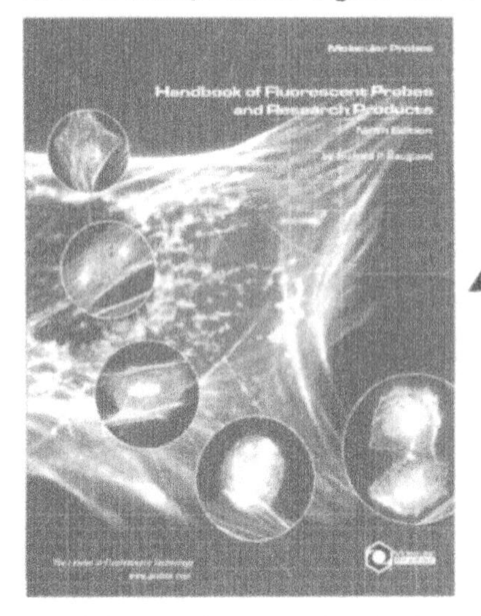

Reproduced with permission - copyright by Molecular Probes

Also available on CD-ROM and the Internet

The printed edition of the Molecular Probes handbook is well worth browsing through when you have a chance as you may pick up lots of clues that are relevant to your own work. The on-line edition contains all of the information of the printed edition, but is more difficult to browse (but much easier to search).

Molecular Probes also has a technical services department (Tech@probes.mhs.compuserve.com) which provides excellent technical advice, although in more recent years this service has suffered from overload and hence often a poor response. Molecular Probes does have access to a large database of literature related to fluorescent probes and so often their technical staff can get you on the right track with the literature, if not with technical advice itself.

Bio-Rad Fluorescence Spectra Internet Site.

Bio-Rad has established an Internet site (www.fluorescence.bio-rad.com) where one can overlay the excitation and emission spectra of fluorochromes of your own choosing (Figure 9-1). Up to 4 fluorochromes can be plotted on the same graph for ease of comparison of fluorescence spectra. Many of the common fluorochromes are included in their database, and no doubt many more will be included in the future. In fact, if they don't have your fluorophore of interest you should contact them and suggest that it is included.

In addition to the emission spectra of the fluorophores of interest, one can also overlay laser emission lines and the absorbance spectra of a large number of commonly used optical filters. This site will become most important for confocal microscopy as more fluorochromes are added to their database.

It should be noted that the fluorescence emission spectrum of many of the fluorochromes (for example FITC, Texas Red) is shown when the fluorochrome is conjugated to antibody. This is important, as the protein conjugated form of the fluorochrome will often show a significant spectral shift compared to the non-conjugated form.

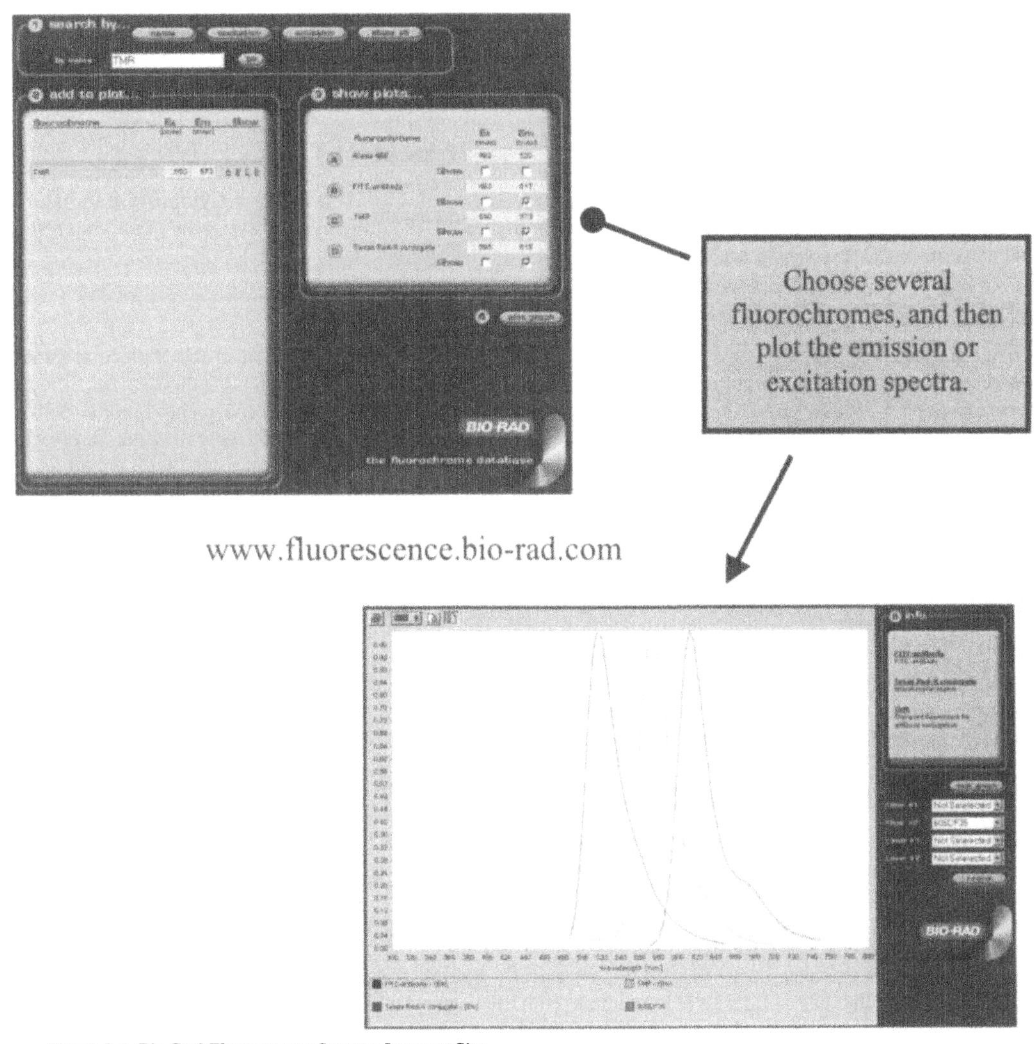

Figure 9-1. Bio-Rad Fluorescence Spectra Internet Site.
The emission and excitation spectra of a wide range of fluorochromes can be plotted, along with individual laser lines and optical filter characteristics, using the Bio-Rad fluorochrome database (reproduced with permission from www.fluorescence.bio-rad.com).

Confocal Microscopy Listserver Archive Site

It is worth browsing through the archive of the Confocal microscopy Listserver for information on fluorescent probes (http: //listserv.acsu.buffalo.edu/archives/confocal.html, see also Chapter 13 "The Internet", page 325, for more details). There are often questions that arise on how to label particular cellular processes or structures, or how to overcome technical difficulties with specific probes. If you want specific information on a fluorescent probe then why not send out a question on the list yourself (see page 326, for details on how to subscribe and send messages to the Confocal Listserver). Questions on specific probes often elicit a wealth of hidden hints and ideas in regards to the use and characteristics of both commonly used and unusual probes. Make sure you know the name of the probe and any related compounds so as not to create confusion in the answers you may receive.

Information from Flow Cytometry can be Very Useful

Flow cytometry (the counting and sorting of single cells in suspension by laser activated fluorescence of molecular tags) makes extensive use of fluorochromes for labelling cells. The flow cytometry literature and people working on FACS instruments (instruments that count or sort cells by fluorescence) are often valuable sources of advice and technical hints for fluorescence microscopy. Flow cytometry is not involved in imaging cells, but the FACS instruments often use the same lasers as those used in confocal microscopy. In particular the 488 nm blue line of the argon-ion laser is used extensively in Flow cytometry. The Purdue University Flow Cytometry Laboratories have an extensive site (http: //www.cyto.purdue.edu/index.htm) dedicated to flow cytometry, with some interesting additional hints on imaging in cytometry (http: //www.cyto.purdue.edu/flowcyt/confocal/confocal.htm).

WHAT FLUORESCENT PROBES ARE AVAILABLE?

The range of fluorescent probes now available is truly astounding. A very brief look at the Molecular Probes catalogue or web site will soon convince you that there is a bewildering array of probes on the market. Some of these probes have been around since fluorescence microscopy first became established as a science (in the 1940's), but others have only been very recently developed using extensive recent knowledge of the behaviour fluorescent molecules. Many of the earlier dyes used in fluorescence microscopy required UV excitation, but with the advent of the confocal microscope and the extensive use of visible light lasers in fluorescence microscopy and fluorescence cell sorting there have been a great number of visible light excited dyes developed.

In more recent years the chemistry of fluorescence has come to be understood in considerable detail. This has allowed the development of dyes with specific excitation and emission wavelengths (for example the new Alexa Fluor dyes from Molecular Probes). An understanding of the interaction of light with organic molecules has also resulted in these newer chromophores having significantly improved characteristics for use in cell biology. For example, many of the newer fluorescent probes have greatly increased quantum yield (brighter fluorophores) and are less susceptible to photo-degradation (photobleaching) compared to their more traditional counterparts.

This section is intended to give a brief overview of some of the fluorescent probes that are available, with some detail provided on specific probes as examples of the type of compounds in use today.

REACTIVE FLUOROPHORES

Reactive fluorophores can be attached to *lipids, carbohydrates, proteins etc.*

Many fluorescent probes are used directly as a stain for specific cellular macromolecules, ions, membrane potential, pH etc. A large range of fluorophores that can be used to label other molecules, such as antibodies, lipids etc, is also available. These probes can often be purchased as the reactive chemical (for example FITC, Texas red sulphonyl chloride, etc, see Table 9-1), purchased as a kit that contains all the necessary reagents for chemically linking and purifying the conjugate, or purchased already conjugated to molecules of interest (for example, BODIPY®-ceramide, FITC-IgG, etc).

Many of these fluorescent molecules are small in comparison to the molecule to which they are attached (for example FITC, with a molecular weight of approximately 300 Daltons, compared to an antibody of molecular weight 150,000 Daltons to which the fluorophore may be attached), but sometimes the attached fluorophore is as large as the molecule of interest (for example FITC attached to the drug Brefeldin A). However, when conjugating a fluorophore it is important to have the correct ratio of dye molecules to the molecule of interest – otherwise you will end up with less than optimal fluorescence emission. Too few dye molecules and the fluorescence will be lower than optimal, but too many dye molecules and the fluorescence will be quenched – resulting in lower fluorescence! It is also important to purify (normally by gel filtration) the conjugated probe from free fluorescent dye, otherwise you will end up with excellent labelling – but due to the free dye, and not your molecule of interest. This is explained in some detail on page 264 in Chapter 11 "Fluorescence Immunolabelling". You can quite easily attach the free dye molecule to your own macromolecule of interest, but commercial preparations, if available, are often of higher quantum yield and superior purity.

Most fluorophores also change their emission and excitation spectra depending on the molecular environment surrounding the probe. These effects are commonly used to determine local pH, ion concentrations and membrane potentials, but you should keep in mind that all probes are to some extent influenced by their local environment.

Table 9-1. Reactive Fluorophores

Reactive fluorophores are fluorescent groups that can be chemically attached to various molecules (antibodies, proteins, lipids, carbohydrates, etc) to allow them to be visualized by fluorescence microscopy. These fluorophores can be purchased already conjugated to a variety of macromolecules, or as the reactive form for conjugation to your own macromolecule/protein of interest.

PROBE	Abs λ (nm)	Em λ (nm)	NOTES
UV excited dyes (blue emission)			
Hydroxy coumarin	325	386	Succinimidyl ester
Amino coumarin	350	445	Succinimidyl ester
Methoxy coumarin	360	410	Succinimidyl ester
Cascade Blue	375;400	423	Hydrazide
Blue excited dyes (green emission)			
Fluorescein (FITC, reactive isothiocyanate)	495	519	Widely used fluorophore for labelling a variety of macromolecules, particularly antibodies. Good quantum-yield, but prone to photobleaching and is pH sensitive. Relatively cheap and readily available from a number of companies.
NBD	466	539	Green fluorescent molecule that is often used to label lipids. Gives good quantum-yield, but is prone to photo bleaching.
BODIPY®-FL	505	513	"FL' denotes Fluorescein equivalent. More photo-stable than NBD.
Alexa Fluor 488	495	519	Equivalent to FITC, but less prone to photobleaching. Available from Molecular Probes.
Cy2	495	510	Fluorescein equivalent cyanine dye is less prone to photobleaching compared to FITC and gives good quantum yield. Reported to not be excited by multi-photon excitation.
Blue excited dyes (red emission)			
R-Phycoerythrin (PE)	480;565	578	High molecular weight (240 kD) high quantum efficiency fluorescent protein. Often used in flow cytometry, often as Cy conjugates – see below.
PE-Cy5 conjugates	480;565;650	670	Cy dyes are often conjugated to PE to create a FRET pair where blue (488nm) excitation can result in far-red (670nm) emission.
PE-Cy7 conjugates	480;565;743	767	The Cy7 equivalent of the above Cy-PE conjugate, again resulting in longer wavelength red emission when using shorter wavelength blue or yellow excitation.
Red 613	480;565	613	PE-Texas Red. is a red emitting, blue excited dye.
Yellow excited dyes (orange/red emission)			
Tetramethylrhodamine (TRITC)	555	580	Lower quantum yield compared to fluorescein, but less prone to photobleaching. Readily excited by the 546 Hg lamp emission, and by the 543 nm green HeNe laser line & 568 nm Kr/Ar laser line. Emission in the red (580nm), well separated from FITC emission (519).
Lissamine rhodamine B	570	590	Readily excited by the 568 nm line of the Kr/Ar laser.
Rhodamine red-X	570	590	The reactive compound is more stable than Lissamine rhodamine, and the conjugate is less subject to hydrolysis at the pH used for conjugation.
Cy3	512;552	565,615	Rhodamine red equivalent. Resistant to photobleaching.
Alexa Fluor 532	531	554	Equivalent to rhodamine 6G, high quantum yield, and highly resistant to photobleaching.
Alexa Fluor 546	556	573	Equivalent to Cy3 or TRITC, high quantum yield, and highly resistant to photobleaching.
Alexa Fluor 568	579	603	Equivalent to Lissamine rhodamine. High quantum yield, and highly resistant to photobleaching. Readily excited by the 568 nm line of the Kr/Ar laser.
Red excited dyes (far red emission)			
Texas Red	595	615	Longer wavelength emitter, and so can be used as a 3rd label in triple labelling experiments. High quantum yield and lower background compared to FITC, TRITC or Lissamine rhodamine B. Difficult to detect the long wavelength red emission by eye, but readily detected by the confocal microscope.
Alexa Fluor 594	591	618	Equivalent to Texas red. High quantum yield, and highly resistant to photobleaching.
Cy5	625-650	670	Far-red dye for 3rd label. High quantum yield, and highly resistant to photobleaching.
TruRed	490,675	695	PerCP-Cy5.5 conjugate
APC-Cy7 conjugates	650;755	767	PharRed

Traditional Fluorescein (FITC), TMR and Texas Red Dyes

The green emitting fluorescein, red emitting tetramethylrhodamine and far-red emitting Texas red dyes are the fluorescent probes traditionally used for fluorescence immunolabelling (Table 9-2). Even though these probes can been largely replaced by more highly fluorescent and less fading prone alternatives, such as the Alexa Fluor and Cy dyes, they are worth examining in some detail (Figure 9-2) as they are still very widely used. One reason they are still used, besides being the dyes that are most widely known, is that they are often cheaper and more readily available, and if sensitivity and susceptibility to fading is not an important consideration they can be excellent fluorescent probes.

Fluorescein (or FITC, fluorescein isothiocyanate, see note below), has been very extensively used for immunofluorescence labelling for many years. FITC (the reactive compound) reacts with primary amines (e.g. lysine residues in proteins). Fluorescein (the name of the attached fluorophore) is readily excited by the 488 nm (blue) line of the krypton-argon-ion and argon-ion lasers, and also by the emission spectrum of the Hg lamp used in conventional epi-fluorescence microscopy.

The quantum yield (efficiency of excitation) is very high, although in more recent years a number of dyes with significantly higher quantum yields have been developed.

The fluorescence intensity of fluorescein is influenced by environmental factors such as pH. This can be exploited for investigating the pH of intracellular compartments, but creates difficulties when attempting to compare the levels of fluorescence as an indicator of the amount of probe present. This is not normally an issue when using the probe on fixed cell or tissue samples, but it is important in this case to maintain the correct pH for maximum emission (normally somewhat alkaline, e.g. pH 8.5).

The relatively broad fluorescence emission spectrum of fluorescein does

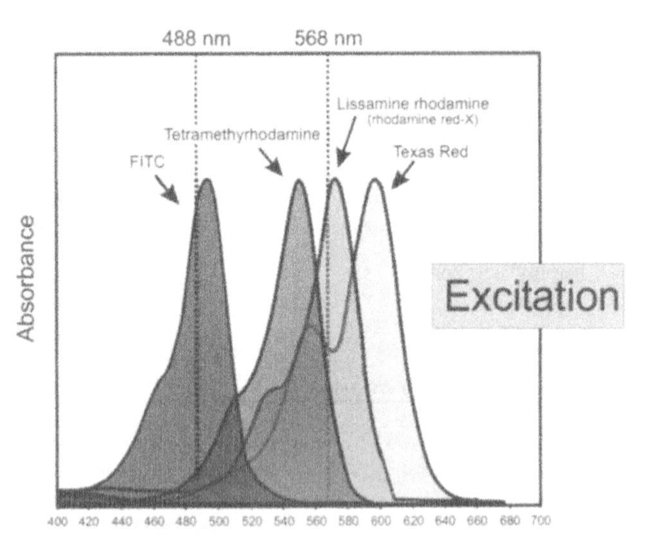

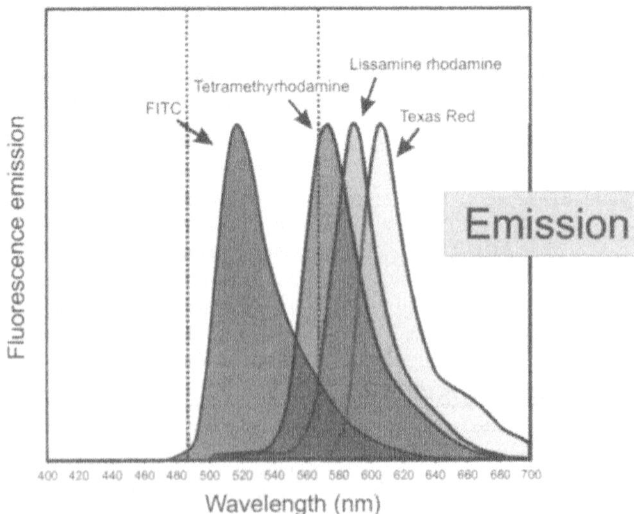

Figure 9-2. Excitation and Emission Spectra of FITC, TMR, LR and Texas Red. These spectra were obtained from fluorophore when conjugated to IgG. The vertical dotted lines are located at the 488 nm (blue) and 568 nm (yellow) emission lines of the krypton-argon-ion laser.

create some problems in dual labelling experiments, but choosing Lissamine rhodamine as the second fluorophore, rather than the widely used tetramethylrhodamine (TRITC) will greatly improve the separation of the two signals. Fluorescein can be readily excited by multi-photon lasers and so can be used in multi-photon applications. Fluorescein is also one of the few fluorescent dyes approved for clinical use in humans, making it an important dye for medical imaging.

Confused with names?

FITC: fluorescein isothiocyanate (the reactive form of fluorescein)
TMR: tetramethylrhodamine
TRITC: tetramethylrhodamine isothiocyanate (the reactive form of TMR)
TAMRA: the carboxylic acid form of TMR
LR: Lissamine rhodamine

Tetramethylrhodamine is widely used for the second fluorochrome in dual labelling experiments. It is efficiently excited by the 546 nm spectral line from the Hg arc lamp used in conventional epi-fluorescence microscopy. However, it is not very efficiently excited with the 568 nm (yellow) line of the krypton-argon-ion laser. Tetramethylrhodamine is available attached to a range of secondary antibodies, and can also be purchased as the reactive (TRITC) form that can be used to label your own antibodies.

Lissamine rhodamine is significantly more efficiently excited with the 568 nm laser line compared to tetramethylrhodamine. Furthermore the Lissamine rhodamine emission spectrum is further removed from the FITC emission spectrum, thus reducing the amount of "bleed-through" in dual labelling experiments. For these reasons Lissamine rhodamine should be used in preference to tetramethylrhodamine in dual labelling experiments.

The fluorescence emission spectra of these longer wavelength rhodamine green derivatives are not affected by changes in pH between 4 and 10, in contrast to the pH dependency of fluorescein emission mentioned above. This may be an important consideration if you are using live cells, but as mentioned above for fluorescein, when using fixed cell or tissue preparations pH dependency is not as important.

Texas red is another red emitting fluorescent dye that is still further into the red than Lissamine rhodamine. Texas red is a good choice for dual labelling experiments, as there is almost no spectral overlap with fluorescein. The 568nm yellow line of the krypton-argon-ion laser, although not ideally suited, reasonably efficiently excites Texas red.

Each of these fluorochromes can be purchased as a reactive form (unfortunately the Texas red reactive form is rather unstable) that can be readily conjugated to antibodies or proteins of interest. However, the wavelength of fluorescence emission of these probes is slightly longer, and the fluorescence emission is significantly quenched, when they are conjugated to proteins.

These probes are all subject to photobleaching, although using suitable antifade reagents such as *n*-propyl gallate significantly reduces the level of fading.

Table 9-2. Traditional Fluorescein and Rhodamine Dyes

These fluorescent dyes are still used extensively for single and multiple labelling applications, even though there are now better dyes available. If fading or sensitivity is not particularly an issue for your application then these dyes may be a cheaper alternative.

Fluorescent Dye	Abs λ (nm)	Em λ (nm)	Usual Excitation
FITC (fluorescein isothiocyanate)	495	519	488 nm Ar and Kr/Ar laser line
TMR (tetramethylrhodamine)	531	580	Hg lamp, red filter set
LR (Lissamine rhodamine)	570	590	568 nm Kr/Ar laser line
Texas red	595	615	568 nm Kr/Ar laser line

Alexa Fluor Dyes

Molecular Probes have developed a new series of fluorochromes called Alexa Fluor dyes (Table 9-3), mainly for use in immunolabelling. These sulphonated rhodamine derivatives have significantly improved characteristics compared to the more traditional fluorochromes such as fluorescein or tetramethylrhodamine.

The Alexa Fluor dyes have been designed to be maximally excited by the commonly available laser lines used in confocal microscopy (Figure 9-3). For example, Alexa Fluor 488 is specifically designed for excitation by the blue 488 nm line of the argon-ion or krypton-argon-ion lasers, and Alexa Fluor 568 is specifically designed to be excited by the yellow 568 nm line of the krypton-argon-ion laser. Other members of the Alexa Fluor family (Alexa Fluor 546 and 594) are designed to be optimally excited by the spectral emission lines of the Hg lamp used in conventional epi-fluorescence microscopy.

It should be noted that the naming convention used for the Alexa Fluor dyes is for the number associated with the dye name to be the recommended laser line (e.g. Alexa Fluor 568 is designated to be excited by the 568 nm line of the krypton-argon-ion laser) and is not the wavelength of peak excitation (the peak excitation for Alexa Fluor 568 is in fact 579nm)

The new Alexa Fluor dyes have narrower emission spectra compared to some of the more traditional fluorescent dyes, which decreases the level of crossover between different detection channels when attempting dual or triple labelling.

The Alexa Fluor dyes have a significantly higher quantum yield than traditional FITC and green rhodamine derived fluorophores, which results in a higher level of fluorescence for the same concentration of dye. They are almost unaffected by photobleaching when used with normal laser intensities (without the need for antifade reagents). In fact, these dyes show the least fading

Alexa Fluor

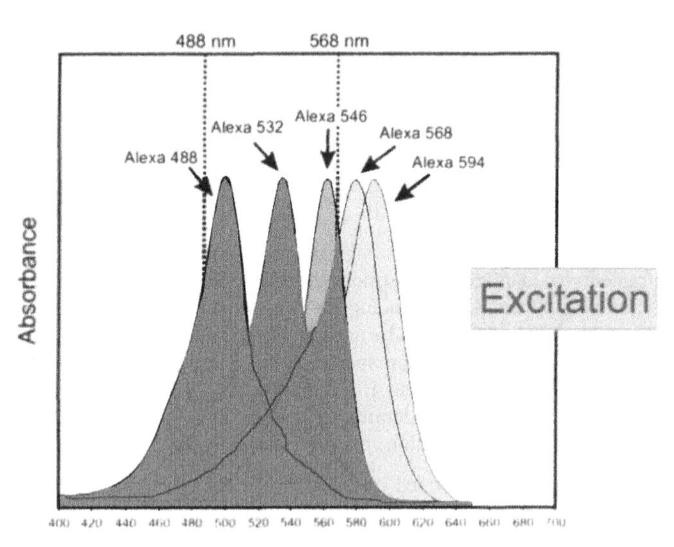

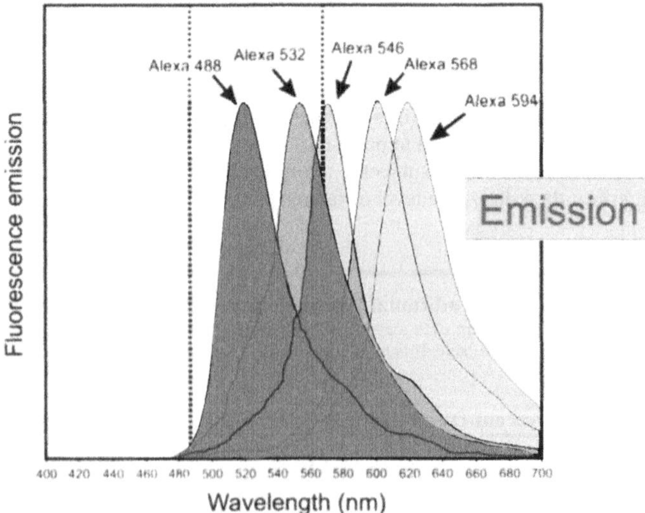

Figure 9-3. Alexa Fluor Fluorophores from Molecular Probes. Excitation and emission spectra are shown for Alexa Fluor dyes conjugated to IgG. The vertical dotted lines are located at the 488 nm (blue) and 568 nm (yellow) emission of the krypton-argon-ion laser.

when applied in a simple buffer solution, making them great for live cell studies as well as in fixed specimen analysis.

The Alexa Fluor dyes are not sensitive to pH changes over a broad range. The Alexa Fluor dyes are available as conjugates to a variety of secondary antibodies for immunofluorescence experiments, and as chemical derivatives that can be reacted with proteins or antibodies of interest.

Molecular Probes also provide kits (at a price!), which contain all reagents and reaction tubes (including G25 purification column) for successful conjugation to proteins. These kits make the process of conjugating the probe to your protein of interest very easy indeed. See page 264 in Chapter 11 "Fluorescence Immunolabelling" for a more detailed discussion of fluorescently labelling your own antibodies or proteins.

Considering the advantages of the Alexa Fluor dyes, you may well ask why use any other dyes? The answer is partly that you may not always need the advantages that the Alexa Fluor dyes provide (perhaps quantum efficiency and photo bleaching is not an important matter as your labelling is very good), or perhaps it is as simple as the fact that you have FITC in stock ready to use! If the level of fluorescence is relatively low or you are doing multiple labelled samples you should certainly consider the Alexa Fluor dyes, whereas in other situations you may find the traditional FITC and TRITC dyes quite adequate for your needs.

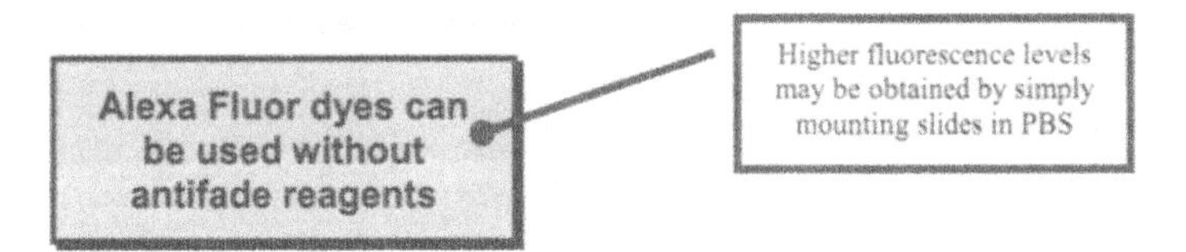

Table 9-3. Alexa Fluor Dyes

Alexa Fluor dyes are normally purchased conjugated to proteins, lipids or carbohydrates of interest. However, a number of the probes can be purchased either as their reactive form, or as a kit, which includes everything you need for conjugating the probe to your own antibody or macromolecule.

Fluorescent Dye	Abs λ (nm)	Em λ (nm)	Usual Excitation	Equivalent Dyes
Alexa Fluor 350	347	442	UV excitation	
Alexa Fluor 430	434	540	Blue laser excitation	
Alexa Fluor 488	495	519	488 nm argon and Kr/Ar laser line	Fluorescein
Alexa Fluor 532	531	554	Frequency-doubled Nd-YAG laser	Rhodamine 6G
Alexa Fluor 546	556	573	Hg lamp, red filter set	Cy3, TMR
Alexa Fluor 555	555	570	Hg lamp & 543 nm green HeNe	Cy3
Alexa Fluor 568	579	604	568 nm Kr/Ar laser line	Lissamine rhodamine B
Alexa Fluor 594	591	618	647 nm Kr/Ar laser line	Texas Red
Alexa Fluor 647	650	670	647 nm Kr/Ar laser line	Cy5
Alexa Fluor 660	650	690	647 nm & 633 nm laser lines	Cy5
Alexa Fluor 680	679	702	Xenon-arc & far-red diode laser	Cy5.5
Alexa Fluor 700	696	719	Xenon-arc & far-red diode laser	Cy5.5
Alexa Fluor 750	752	779	Xenon-arc or dye pumped lasers	Cy7

Cyanine Dyes (Cy2, Cy3 and Cy5)

The water-soluble cyanine dyes (Table 9-4) from Jackson Immuno Research Inc. (also sold by Amersham Pharmacia) are superior to the traditional fluorescein and green rhodamine derivatives for immunolabelling experiments. They are more photo stable, are significantly brighter and have lower background labelling than their traditional counterparts, FITC and tetramethylrhodamine. There are a series of Cy dyes available that can be used as replacements for the commonly used fluorescein and tetramethylrhodamine etc. However, a lot of people become familiar with the Cy dyes, not because of their superiority to the more common fluorophores, but because of the existence of Cy5, the far-red dye that is an excellent choice for triple labelling applications.

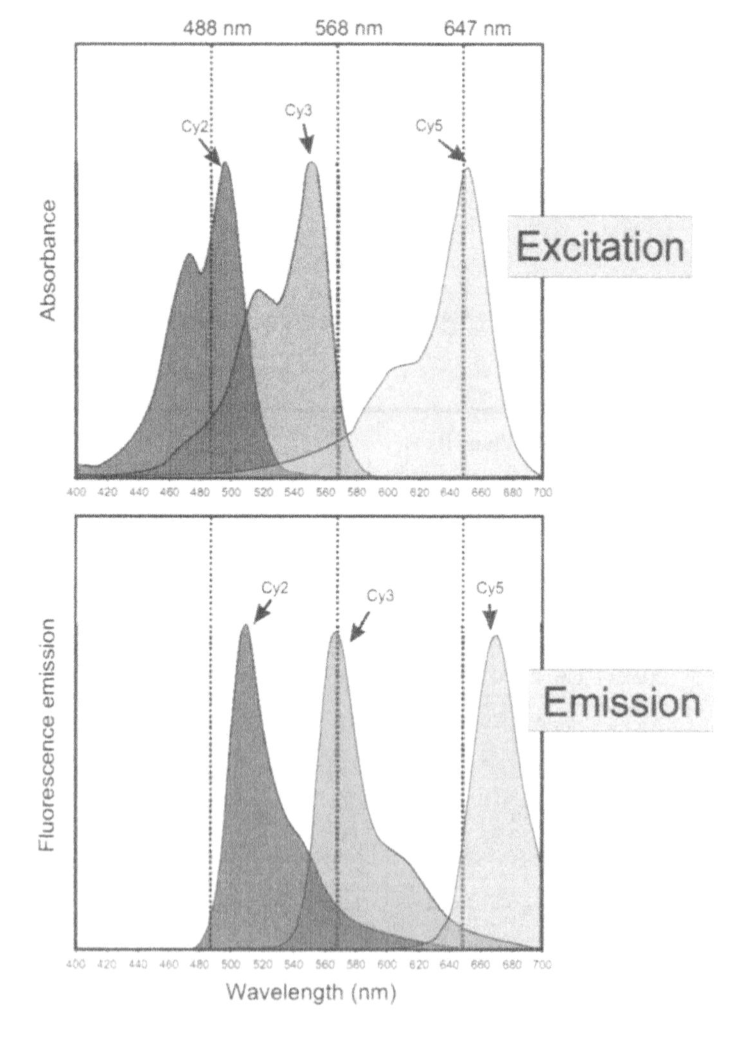

Figure 9-4. Excitation and Emission Spectra of Cy2, Cy3 and Cy5.
The Cy series of dyes from Jackson Immuno Research. Cy5 is an excellent choice for the 3rd label in triple labelling experiments. The vertical dotted lines are located at the 488 nm (blue), 568 nm (yellow) and 647 nm (red) emission of the krypton-argon-ion laser. These spectra were obtained from fluorophores conjugated to IgG.

Cy2, has excitation and emission spectra (Figure 9-4) similar to fluorescein (FITC). This dye can be used as a replacement for fluorescein in most applications due to the equivalent spectral characteristics, but is in most cases superior to fluorescein as it is less prone to fading and has a higher fluorescence yield. Interestingly, it has been reported that Cy2 is not readily excited by multi-photon lasers.

Cy3 has emission and excitation spectra similar to tetramethyl-rhodamine.

Cy5, the far-red fluorochrome, has an emission spectrum significantly further into the red than any of the more common fluorochromes, being optimally excited near 650 nm (red) and fluoresces maximally near 670 nm (far-red). Cy5 is efficiently excited by the 647 nm (red) line of the krypton-argon-ion laser or the 633 nm line of the helium-neon laser. The "far-red" emission of this dye makes this fluorochrome ideal for the 3rd label in triple labelling experiments by confocal microscopy or fluorescence-activated cell sorting (FACS) analysis. The wide separation of the Cy5 emission from that of shorter-wavelength-emitting fluorophores makes this fluorophore particularly valuable for multiple labelling experiments.

A significant advantage of using Cy5 over other fluorophores is the lower autofluorescence of biological specimens in the region of the red light used to excite this fluorophore. However, because of its emission maximum at 670 nm, Cy5 cannot be seen well by the eye, and therefore cannot be used with conventional epi-fluorescence microscopes, unless you are using a high sensitivity CCD camera to not only capture the images, but also to focus the microscope.

The Cy dyes can be purchased conjugated to many primary and most of the common useful secondary antibodies, streptavidin and egg-white avidin.

Cy5 fluorescence cannot be detected by the eye – but is readily detected by the confocal microscope

Cy5 - the far-red dye, ideal for triple labelling

Table 9-4. Cyanine (Cy) dyes

The Cy fluorescent dyes give good fluorescent yield and are relatively resistant to photobleaching. They can be used as replacements for the commonly used FITC and rhodamine dyes used in multiple labelling applications. The far-red Cy5 dye is particularly useful for triple-labelling experiments as the emission spectrum is distinctly further into the red than Cy3 or its rhodamine equivalents.

Dye	Abs λ (nm)	Em λ (nm)	Usual Excitation	Equivalent Dyes
Cy2	492	510	488 nm argon & Kr/Ar laser line	Fluorescein
Cy3	550	570	514 nm argon (50% efficacy) 543 nm green HeNe & 568 Kr/Ar	Tetramethylrhodamine Lissamine Rhodamine
Cy3.5	581	596	514 nm argon (50% efficacy) 543 nm green HeNe & 568 Kr/Ar	Lissamine Rhodamine
Cy5	650	670	568 nm Kr/Ar laser line	Far-red emission dye
Cy5.5	675	694	568 nm Kr/Ar laser line	Far-red emission dye
Cy7	743	767	647 nm Kr/Ar laser line	Infrared emission dye

NUCLEIC ACID PROBES

A wide variety of nucleic acid probes are available for fluorescence microscopy (see Table 9-5). For many years the UV excited DNA stains such as DAPI (4,6-diamindino-2-phenylindole) and several Hoechst stains were the mainstay of fluorescence microscopy. However, with the wider availability of laser scanning confocal microscopes these DNA stains could only rarely be used (they are not excited by the common laser lines available, although they can be excited by an expensive UV or multi-photon laser). This means that a lot of work has been done using visible light excited stains such as propidium iodide. Although propidium iodide is readily available and easy to use, it does suffer from a significant level of background staining and a very broad emission spectrum. The newer DNA stains, such as the SYTO and SYTOX dyes, are more specific for DNA and have a high quantum yield (high level of fluorescence) and are readily excited by visible light lasers.

A variety of nucleic acid fluorescent probes are able to enter live cells (such as DAPI, SYTO dyes, dihydroethidium etc) and can thus be used as vital stains to analyse living cells and tissue slices. Other DNA stains, such as propidium iodide, only enter cells with a damaged cellular membrane (dead cells) and thus can be used as a marker for dead cells within the culture / tissue slice. Combinations of such dyes are the basis for the "cell viability" kits that are available.

Acridine Orange (AO)

Acridine dyes are basic cationic dyes that were first isolated in 1870 as a dyeing agent. Acridine derivatives were found to be useful as antibiotic reagents in the early part of this century, and in the 1930's the quinacrine derivatives were found to be effective against malaria. In the 1920's acridine was found to be useful for staining cell nuclei. In 1940 acridine orange (Figure 9-5) was found to be an excellent fluorophore for fluorescence microscopy.

Acridine orange (3,6-dimethylaminoacridine) is a weak base that is soluble in water and when bound to compact chromosomal DNA fluoresces green, and when bound to RNA or non-intact DNA of dead cells fluoresces red.

Acridine orange readily diffuses across cell membranes at neutral pH, but becomes membrane-impermeable at acid pH. This results in acridine orange accumulating in acidic vesicles, particularly lysosomes. As the concentration in the acidic vesicle increases, the fluorescence emission is shifted more to the red end of the visible light spectrum. This change in fluorescence is due to tight packing of the acridine orange molecules.

Acridine orange has relatively broad excitation and emission spectra. In aqueous solution acridine orange exhibits two forms, the green fluorescent form with an emission peak at 530 nm, and a red fluorescent form with an peak emission at 640 nm. Due to the broad absorption spectrum, acridine orange can be excited by a variety of light sources, but this does create problems with bleed-through when attempting dual labelling with other fluorochromes.

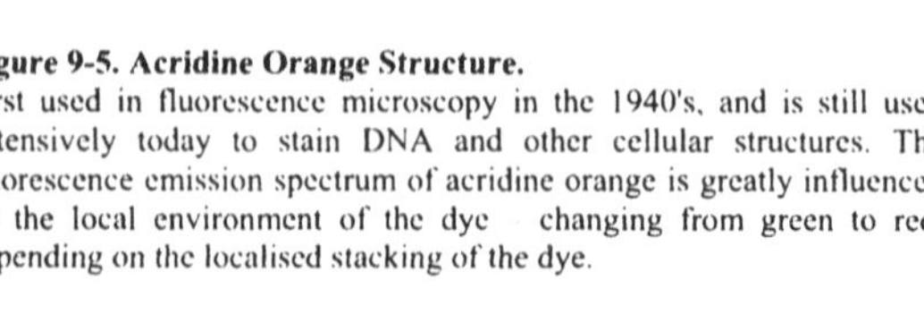

Figure 9-5. Acridine Orange Structure.
First used in fluorescence microscopy in the 1940's, and is still used extensively today to stain DNA and other cellular structures. The fluorescence emission spectrum of acridine orange is greatly influenced by the local environment of the dye – changing from green to red, depending on the localised stacking of the dye.

Acridine orange has been used for a variety of applications in cell biology. With its ability to distinguish the DNA of intact live cells compared to dead cells it was the forerunner of the current dyes available for distinguishing dead and live cells. Of significance also is the large range of molecules that will bind acridine orange to varying extents – and in the process result in an emission spectra between green and red – such that most cells show an extensive and fine labelling of cellular structure with a wide range of emission colours. This is most noticeable when using a Hg arc lamp to excite the fluorophore (blue light) and using a broad emission optical filter allowing you to see emission wavelengths from green through to red. This spectral property of acridine orange could be more extensively exploited today, especially using the spectral capabilities of the newer confocal microscopes, to elicit subtle molecular information from the intracellular milieu.

Although having many features that made it useful as a fluorescent dye, acridine orange has now been superseded in many situations by other more reliable and robust dyes. However, acridine orange still has a useful place as a general cellular fluorescent stain. Many of the current generation of DNA stains have been developed to overcome the shortcomings of DNA stains such as acridine orange, and are discussed in more detail below.

Propidium Iodide (PI)

This red fluorophore (Figure 9-6) intercalates double-stranded nucleic acids, thus making it an excellent dye for staining DNA. The amount of fluorescence increases greatly on binding to DNA. However, propidium iodide also stains double stranded RNA, located where the mRNA is bound to the tRNA within the ribosomes in the cell cytoplasm. This cytoplasmic staining is significantly less than the nuclear DNA, and if it is considered a problem it

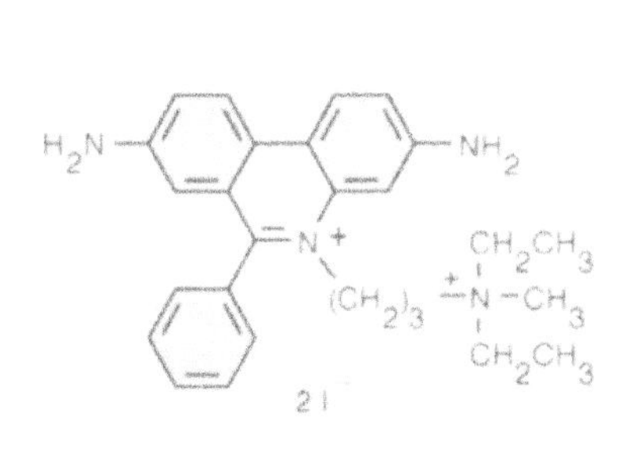

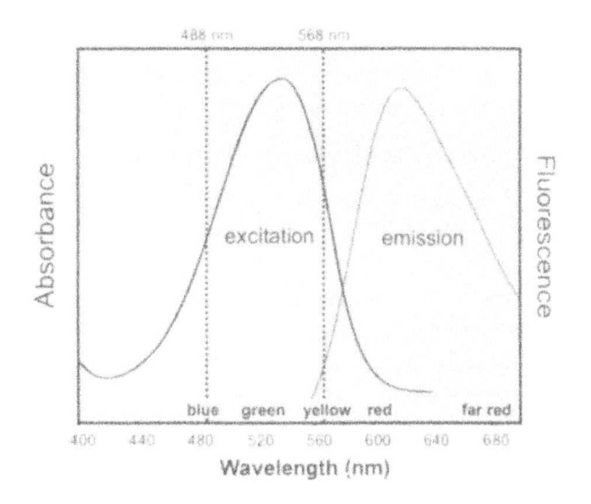

Figure 9-6. Propidium Iodide Structure and Spectra.
This red emitting fluorophore has very broad excitation and emission spectra. This means that the dye is suitable for dual labelling applications when only a single 488 nm blue laser line is available, but the down side is that there is considerable overlap in both excitation and emission with other fluorophores.

can be removed by digestion of the RNA with RNAase. If you are interested primarily in locating the nucleus within the cell then propidium iodide will be quite adequate even without removing the cytoplasmic RNA. However, if the cytoplasmic RNA is removed it is possible to do triple labelling with only a 2 channel confocal microscope. This would entail using the two channels to identify cytoplasmic located molecules, and then identifying the nucleus as being physically distinct, even though it would be imaged within the red channel.

Propidium iodide is readily excited by a wide range of light sources, such as the 488 nm (blue) line of the krypton-argon-ion laser, the 488 and 514 line of the argon-ion laser and the blue/yellow spectral lines of the Hg lamp used in conventional epi-fluorescence microscopy. This is due to the very broad excitation spectrum of this fluorophore (Figure 9-6, right panel).

Propidium iodide is frequently used as a secondary fluorescent stain to highlight cell nuclei in immunolabelling. Suitable optical filters can readily distinguish the propidium iodide red emission spectrum, from the emission spectrum of the common green fluorophore FITC.

Local environmental effects, such as the mounting media, can affect the emission spectrum of propidium iodide. This is not a problem with cell nuclei with well-condensed DNA, which bind propidium iodide with the characteristic PI-DNA red fluorescence. However, less condensed DNA does not bind propidium iodide as tightly and the surrounding media may influence the emission spectrum. For example, glycerol, which is widely used as a mounting media in fluorescence microscopy, may result in a shift in the emission spectrum such that the typical dual labelling optical filter sets are no longer appropriate. Instead of resorting to changing the filter set, it is a simple matter of mounting the sample with propidium iodide in buffer lacking glycerol (for example, phosphate buffered saline) to return the emission to the expected wavelengths. A number of common antifade reagents also influence the emission spectrum of propidium iodide. However, propidium iodide is relatively photo stable, thus allowing it to be used without antifade in PBS solution when used as a second label with the new Alexa Fluor dyes from Molecular Probes.

A large number of newer generation DNA binding dyes (for example YOYO, SYTO, TOTO etc) that have overcome some of the limitations have been developed. However, propidium iodide is cheap and easy to us, and still quite effective.

Ethidium Bromide (EthBr)

Ethidium bromide is structurally very similar to propidium iodide and is also often used as a DNA dye. Ethidium bromide readily stains double stranded RNA located in the cell cytoplasm, although this can be removed, if required, by the use of RNAase. However, ethidium bromide does produce more background staining than propidium iodide, as it appears to bind some protein constituents within the cell. This "background" staining can be used to advantage for displaying other parts of the cells at low levels of fluorescence. For example, the red blood cells shown in Figure 1-2 (page 5), show faint ethidium bromide staining, whereas the DNA of the malarial parasites located within two of the red blood cells is brightly stained (the cytoplasmic RNA of the parasite is also stained at an intermediate level).

UV DNA Dyes (DAPI, Hoechst 33258 etc)

These UV excited DNA binding stains are widely used in fluorescence microscopy. However, due to being excited by UV light, they are not particularly suited to conventional visible light confocal microscopy. They are readily excited by the UV laser used in UV confocal microscopy, but these microscopes are not widely available due to both the expense of the laser and the difficulties encountered with the optics of UV confocal microscopy. These stains are often readily excited using the pulsed infrared laser of multi-photon microscopy. The longer wavelength (700 nm) infrared light used in multi-photon microscopy is less detrimental to the cells (as long as the power level at the sample is not too high), and often penetrates the tissue to a greater depth. 3-photon excitation, using 1050 nm is also possible (allowing one to perform multi-labelling experiments).

DAPI is a vital stain at relatively low concentrations (having no discernable effect on cell viability), but irradiation of DAPI stained cells for only a few minutes with UV light (350 nm) will have a detrimental effect on cell viability.

The development of solid-state, relatively short wavelength lasers (for example, the 405 nm violet laser – see page 86) may result in an upsurge in interest in these "UV" excited DNA dyes. The violet laser is of sufficiently long wavelength to be essentially treated as a visible light laser in the optics of the microscope and scan head, but at the same time of short enough wavelength to excite these "UV" dyes.

SYTO Dyes (cell permeant)

A large range of SYTO cell permeant dyes, that can be excited by light ranging from UV through to far red, is available. The SYTO dyes can cross cell membranes of living cells, and so are often used as the "live cell" indicator in live/dead cell detection kits. These dyes are more specific for DNA compared to the more traditional dyes such as propidium iodide, and have a narrower wavelength of emission – more readily allowing for multi-labelling applications. Kits are available from Molecular Probes that contain small samples of a number of SYTO dyes for testing. It is a great idea to first purchase such a kit and to empirically determine the dye best suited to your labelling requirements.

Unsure of which DNA stain to use?

"Test Kits" are available from Molecular Probes with a range of SYTO or SYTOX DNA stains.

SYTOX Dyes (do not enter live cells).

The SYTOX dyes have been developed as highly specific DNA stains that are only taken up by dead (membrane compromised) cells. Like the SYTO dyes above, a set of these dyes, with varying wavelengths of excitation, can be purchased from Molecular Probes as a sampling kit. These dyes are used as the "dead cell" indicator in the "cell viability" kits that can be purchased from Molecular Probes. The relatively narrow excitation and emission spectra, and the range of excitation and emission wavelengths available with the various SYTOX dyes does mean that it is often possible to combine one of these DNA stains with many other fluorescent dyes in multi-labelling applications.

Table 9-5. Nucleic Acid Probes

There are a large number of DNA specific fluorescent probes available, and each of these dyes has individual characteristics (binding affinity, specificity for DNA or other macromolecules and the wavelength of excitation and emission).

PROBE	Abs λ (nm)	Em λ (nm)	NOTES
Traditional UV dyes: These dyes cannot be used with commonly available lasers (Krypton, Kr/Ar, HeNe etc) on a confocal microscope, but can be used with a UV laser, in conventional Hg lamp UV fluorescence microscopy and also with the new solid state violet (405nm) lasers. They can also be imaged with the pulsed infrared laser used in multi-photon microscopy.			
DAPI (4,6-diamidino-2-phenylindole)	358	461 blue	Semi-permeant AT-selective DNA stain. Also used to detect Mycoplasma. Readily excited by the pulsed IR laser of multi-photon microscopy. See page 214.
Hoechst 33258 (bisbenzimide)	352	461 blue	Permeant live cell AT-selective DNA stains. Minor groove-binding dsDNA selective (Hoechst 33342 has similar characteristics). Readily excited by the pulsed IR laser of multi-photon microscopy. See page 214.
Live cell labelling (cell-permeant dyes) These dyes penetrate live cells and so can be used as a nuclear marker in live tissue slices and cells.			
Acridine orange	500 (DNA) 460 (RNA)	526 (DNA) 650 (RNA) 650 (acidic)	Is readily taken up by live cells, and has a very broad range of fluorescence emission wavelengths, depending on what cellular components the dye is associated with (DNA, RNA, acidic vesicles, proteins etc). This broad specificity can be used to great effect as a marker of cellular structure in living cells. See page 212.
Dihydroethidium	518	605 red	A cell-permeant form of ethidium bromide, blue fluorescent until oxidized to ethidium within live cells (becoming a red fluorescent dye).
7-AAD	546	647 red	Poorly taken up by live cells, GC-selective DNA stain.
SYTO	range	blue to red	A large number of SYTO dyes are available with various wavelengths of excitation and emission (blue, green, orange and red). Sampler kits are available. These are low affinity DNA stains. See page 214.
DRAQ5	647, 568	670 red	Live cell DNA stain that can be excited by a range of wavelengths (optimally at 647nm) and emits in the red to far-red region of the spectrum.
Dead / fixed cell preparations (live cell membrane - impermeant dyes) Living cells do not, in general, take up these dyes. This means they are often used as markers for "dead" cells – i.e. cells that have lost their membrane integrity. They are also often used as a nuclear or chromosome marker in fixed cell preparations.			
Propidium iodide	536	617 red	This widely used red fluorophore is cheap, readily available, and strongly stains cellular DNA. The dye is not taken up by live cells, and so is often used as a marker for dead cells. dsRNA is also stained to some extent, but can be removed if necessary by treating the sample with RNAase. See page 213.
Ethidium bromide	518	605 red	Very similar in structure and staining characteristics to propidium iodide. The increased background labelling can be used to advantage to show the structure or position of the cells in the sample. See page 214.
Acridine homodimer	431	498 green	AT selective stain for DNA in fixed / dead cells.
Cyanine dimers	range	blue to red	The "TOTO" family of dyes. YOYO, BOBO, POPO, JOJO, LOLO etc. A large range of high affinity DNA stains is available with different binding characteristics and wavelengths of emission. The best way to determine which dye is best for your application is to purchase a sampler kit containing a selection of dyes.
Cyanine monomers	range	blue to red	TO-PRO, LO-PRO, BO-PRO, TO-PRO cyanine monomers ideal as counter stains.
SYTOX	445 504 547	470 blue 523 green 570 orange	These highly impermeable cyanine DNA stains are ideal as a dye for dead cells. They are also sold as one of the components of a Live/Dead cell detection kit. These are high affinity DNA stains. See page 214.
YOYO-1 iodide	491	509 green	Ultra sensitive DNA stain, dead cell stain, giving green fluorescence. One member of the cyanine dimers listed above.
APOPTRAK	647,568	670	Derivative of DRAQ5 that can differentiate between apoptotic & non-apoptotic cells.

FLUORESCENT ION INDICATORS

A large number of fluorescent probes that change their spectral response on binding specific ligands have been developed. The original work in this area concentrated on measurement of Ca^{2+} fluxes in live cells (using video microscopy, rather than confocal microscopy). However, in more recent years there has been a proliferation of reagents for estimating the subcellular concentration and flux of a variety of ions.

Fluorescent Ion Indicators

Ca^{2+}, Mg^{2+}, Zn^{2+}
Na^{+}, K^{+},
Cl^{-}, Br^{-}, I^{-}, thiocyanate (SCN^{-})
Cu^{+}, Ni^{2+}, Co^{2+}, Fe^{2+}
Al^{3+}, Ga^{3+}
Cd^{2+}, Hg^{2+}, Pb^{2+} (using Ca^{2+},indicators)
Caesium (Cs^{+})
Inorganic phosphate, Cyanide, Selenium
Thiols, sulphides
Nitrite (NO_2^{-})
Eu^{3+}, Tb^{3+} (Lanthanides)
pH, $\Delta\psi$ (membrane potential)
……… and many more

How do Ion Indicators Work?

Fluorescent probes that are specific for individual ions are molecules that bind the ion of interest with high specificity, and on binding there can be a change in either the brightness of fluorescence (an increase, or a decrease - termed quenching) or a change in the wavelength of emission.

The original ion indicator fluorophores were fluorochromes that were found to change fluorescence on binding the ion of interest. However, in more recent years considerable effort has gone into developing dyes that bind specific ions of interest. These new dyes are often developed from existing molecules that are known to bind the ion of interest (for example, specific Cu^{+} binding peptides). These non-fluorescent molecules are made into ion sensitive fluorophores by the chemical attachment of suitable fluorophores to the ion-binding moiety.

The high sensitivity of these probes ensures that there is considerable interest in developing probes for ion detection in many areas of biology. Not only can the metabolic fluxes within the living cell be studied, but also the environmental levels of contamination in living organisms can be analysed.

It should always be kept in mind that many ion specific probes may be influenced by related ions in the cell. Some fluorescent probes have a higher specificity than others, and even those probes that will react to various ions can often be used when you have confidence that either the ion of interest is the only ion concentration changing, or that the other ions are significantly lower in concentration in your system and will not unduly influence your results.

In this section, Ca^{2+} binding probes, pH probes and membrane potential probes will be discussed in some depth (they are the most extensively used ion indicator probes), with only brief mention being made of the large number of other ion indicator probes available. The principles involved with calcium, pH and membrane potential dyes can be readily applied to other ion indicators.

One final word of warning – most fluorescent probes are influenced by their molecular environment. This could mean that the fluorescence of the dye with specificity for your particular ion of interest may be influenced by other molecular interactions. This point is most important to remember when using fluorescent probes that don't yet have an established literature on their specificity and characteristics – and when using fluorescent probes in unusual cells or environmental conditions.

Do you have confidence that your ion indicator probe is specific for your ion of interest?

Esterase Derivatives

Living cells do not readily take up electrically charged fluorescent probes (they cannot cross the hydrophobic milieu of cellular membranes). This is not a problem with hydrophobic lipid soluble dyes such as acridine orange, NBD-ceramide (a fluorescent lipid) or $DiOC_6$ (a lipid soluble dye). However, many ion indicators are electrically charged and thus do not readily enter the cell.

The salt or dextran forms of ion fluorescent indicator probes must be microinjected or scrape loaded into cells. However, the AM (acetoxymethyl) ester derivatives can be readily loaded into live cells by simply incubating the dye in the presence of the cells. Once the AM ester derivative is taken into the cells intracellular esterase activity cleaves the AM ester moiety, resulting in the fluorescent indicator dye being retained within the cell, as it can no longer cross the plasma membrane. The loading conditions will determine the final destination of the probe inside the cells (whether in the cytoplasm or subcellular compartments).

One added advantage of using the AM ester derivative is that the AM form of the dye is often not fluorescent, only becoming fluorescent after cleavage of the ester group within the cell. This has the beneficial effect of excess dye in the media not being visible by fluorescence microscopy, allowing you to load the dye and then image the cells without the need for washing away excess external dye. Esterase activity is present in all living cells.

High concentrations of AM esters may have deleterious effects on the metabolic state or physiology of the cell due to the release of acetate, protons and formaldehyde – the by-products of ester cleavage. The efficiency of ester cleavage of the AM moiety can vary greatly between cell types.

Loading dyes into living cells:

If an AM ester is not available, there are many alternative ways in which fluorescent dyes can be introduced into live cells:

For example,

- Electroporation
- Microinjection
- Transfection
- Lipid soluble dyes
- Hypotonic shock
- SLO permeabilisation

See page 315 in Chapter 12 "Imaging Live Cells"

Ca^{2+} Indicators

Intracellular calcium levels modulate many cellular functions, often playing a key role in the cell response to external agents. It is important to not only measure the Ca^{2+} ion concentrations within individual cells or cellular compartments, but it is also essential to be able to measure the dynamics of Ca^{2+} fluxes within the cells. High transient fluxes of Ca^{2+} concentration are widely used to control a number of cellular responses. Individual fluorescent probes are particularly suited to measuring either Ca^{2+} concentrations or Ca^{2+} cellular fluxes.

Many fluorescent probes used to measure intracellular Ca^{2+} concentrations are based on the non-fluorescent Ca^{2+} chelators EGTA and BAPTA.

Very high speed calcium flux detection by confocal microscopy requires specialist confocal microscopy techniques such as line scanning, (which is discussed on page 256 in Chapter 10 "Confocal Microscopy Techniques") or alternatively the use of high speed confocal microscopes such as the Nipkow disk based confocal microscopes.

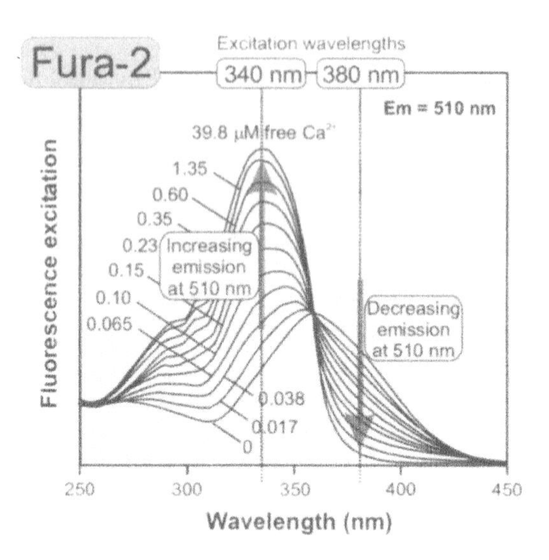

Figure 9-7. Excitation Spectra of Fura-2.
Fura-2 in solutions containing 0 to 39.8 μM free Ca^{2+}. Emission is at 510 nm. Ratiometric imaging is carried out by exciting Fura-2 with 340 and 380 nm light, and following the changes in fluorescence emission at 510 nm. This diagram is modified from Randi R. Silver (1998), Methods in Cell biology, pp 237-251.

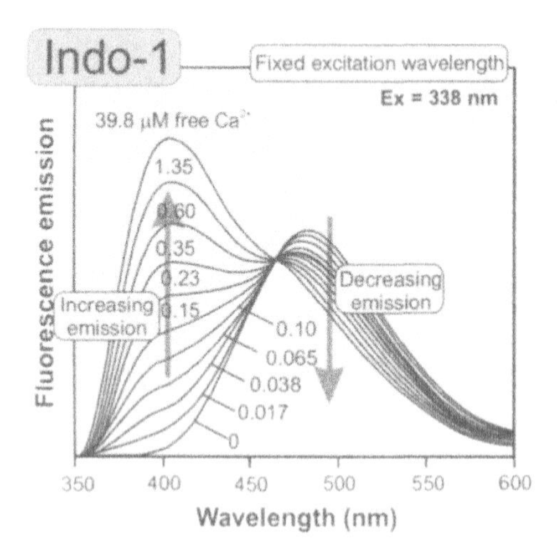

Figure 9-8. Indo-1 Emission.
Indo 1 in solutions containing 0 to 39.8 μM free Ca^{2+}. Excitation is at 338 nm (or 351 nm line from the argon-ion laser). Ratiometric imaging is performed by exciting the fluorophore with 338 nm light and following the changes in emission at 400 and 475 nm. This diagram is modified from Randi R. Silver (1998), Methods in Cell biology, pp 237-251.

Ideally one would want to measure the intracellular changes in Ca^{2+} concentration, and not simply the change in fluorescence intensity. Simple changes in fluorescence intensity may be due to changes in dye concentration within the cell and may not reflect the true Ca^{2+} subcellular concentration.

Ratiometric Ca^{2+} indicators (such as Fura-2 and Indo-1) can be used to establish the subcellular Ca^{2+} concentration. These dyes exhibit a shift in the excitation (Fura-2, Figure 9-7), or emission spectra (Indo-1, Figure 9-8) that is dependent on Ca^{2+} concentration. Exciting Fura-2 at the optimal wavelengths of 340 and 380 nm (Figure 9-10), and collecting the fluorescence emission at 510 nm can be used to accurately calculate the Ca^{2+} concentration independent of dye concentration. Indo-1 ratio imaging can be accomplished by exciting with a single wavelength (338 nm) and collecting the fluorescence emission at two wavelengths (400 and 475 nm), Figure 9-8.

The problem with the above ratiometric calcium dyes is that they both require UV excitation. UV lasers are used in some confocal microscopes, but due to the expense and associated problems of optical aberrations and limited depth of optical sectioning these microscopes have not become widely available.

There are many visible light calcium indicators available (see Table 9-6). However, they are not ratiometric dyes. They either show an increase or a decrease in fluorescence on binding calcium, but they don't show the wavelength shift of Indo-1 and Fura-2. For example, Fluo-3 (Figure 9-9) shows a very marked calcium dependent increase in fluorescence emission at 525 nm when excited by the 488 nm (blue) line of the argon-ion or krypton-argon-ion laser (Fluo-4 is a very similar dye, but with increased quantum yield). This probe can be used as a measure of calcium levels, but suffers from the problem of one not knowing if the increased fluorescence is simply due to an increase in dye concentration, rather than a true change in Ca^{2+} levels. The development of true visible wavelength ratiometric dyes for calcium imaging would be of great benefit to studies of cellular physiology.

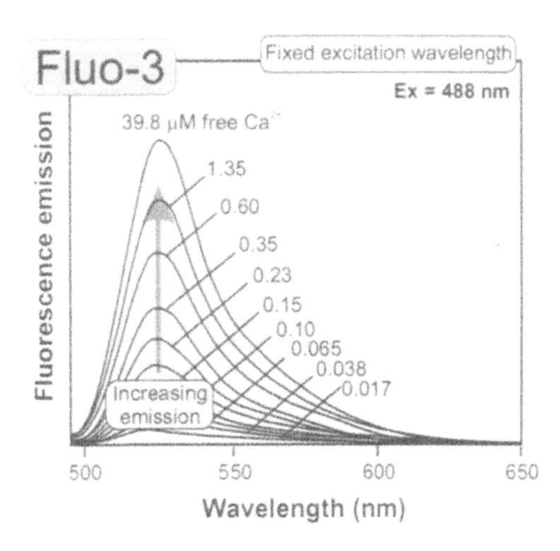

Figure 9-9. Fluo-3 Emission Spectra.
Fluo-3 spectra (very similar to Fluo-4) in solutions containing 0 to 39.8 μM free Ca^{2+}. Excitation is at 488 nm. The fluorescence emission increases in the presence of increasing concentrations of calcium. This diagram is modified from Randi R. Silver (1998), Methods in Cell biology, pp 237-251.

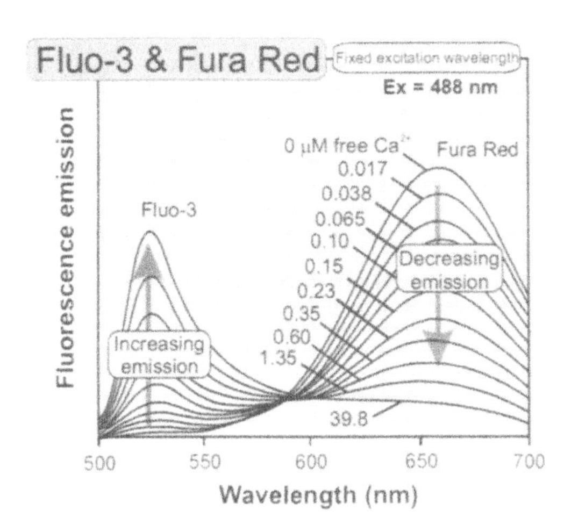

Figure 9-10. Fluo-3 and Fura Red Emission Spectra.
Fluo-3 in solutions containing 0 to 39.8 μM free Ca . Excitation is at 488 nm. A combination of fluo-3 and fura red can be used as a "pseudo" ratiometric dye for measuring intracellular calcium concentrations. This diagram is modified from Randi R. Silver (1998), Methods in Cell biology, pp 237-251.

To overcome the difficulty of estimating Ca^{2+} concentration a mixture of two different visible light dyes can be used. Fura Red is a visible light Ca^{2+} binding fluorescent dye that shows a decrease in fluorescence at 650 nm on binding Ca^{2+}. If this dye is used together with Fluo-3 (Figure 9-10) it is possible to create a ratiometric estimate of Ca^{2+} concentrations. The two dyes are excited at a single wavelength (488 nm, blue line of the argon-ion or krypton-argon-ion laser) and then the emission is measured at both 525 nm (green), which increases on binding Ca^{2+}, and 650 nm (red) wavelengths, which decreases on binding Ca^{2+}. The ratiometric image can be created by using the standard FITC/Texas-red image collection filters on the confocal microscope.

Ratiometric imaging with two dyes is much better than using a single dye only for estimating the true Ca^{2+} concentration within the cell. However, there is still the possibility that the two dyes may diffuse differently within the cells, resulting in differences in the concentration of the two dyes within subcellular compartments.

Table 9-6. Calcium Indicators

This table lists a number of the many calcium indicator fluorochromes now available. Individual Ca^{2+} indicators have varying dissociation constants and so are sensitive to differing levels of the Ca^{2+} ion. You will need to decide the Ca^{2+} level and rate of change you would expect in your system before you can choose the best available Ca^{2+} indicator. This table has been compiled from a variety of sources, including information in the Molecular Probes Handbook and from their web site (http://www.probes.com).

PROBE	Abs λ (nm)	Em λ (nm)	NOTES
UV Ratiometric Dyes: The following dyes are ratiometric (the level of Ca^{2+} can be readily determined independently of the concentration of the dye present), but being UV excited they are difficult to use in confocal microscopy.			
Fura 2	340 380	510 increases 510 decreases with increased Ca^{2+}	Ratiometric UV dye excited at two different wavelengths (340 nm & 380 nm) with emission monitored at 510 nm. The ratio of the emission from both channels gives a direct measure of the level of Ca^{2+}, independent of dye concentration.
Bis-Fura 2	Essentially the same as Fura 2		The attachment of two Fura molecules to a single BAPTA molecule has resulted in a Ca^{2+} probe with increased fluorescence, but with similar characteristics to Fura 2.
Indo-1	338	400 increases 475 decreases with increased Ca^{2+}	Ratiometric UV dye which is excited with a single wavelength (338 nm) and the changes in fluorescence monitored at both 400 nm & 475 nm. The ratio of the emission from both channels gives a direct measure of the Ca^{2+} concentration. See Figure 9-8.
BTC	400 464	533 increases with increased Ca^{2+}	Coumarin benzothiazole–based Ca^{2+} indicator, longer wavelength UV/blue light excited Ca^{2+} indicator dye that can be used as a ratiometric probe. May be possible to be used with the 488 nm argon-ion line and the 405 nm solid-state violet laser line.
Quin-1	348	485 / 398	An early generation fluorescent ratiometric calcium indictor dye. Not sufficiently bright or photo stable for live cell imaging.
Visible Dyes: A large number of dyes excited with visible light are available for following Ca^{2+} levels in living cells. Non of these probes are ratiometric, but a pseudo ratiometric imaging method has been developed using dye combinations.			
Calcium Green	506	533 increases with increased Ca^{2+}	Similar spectral properties compared to Fluo-3, but with significantly increased quantum yield. Particularly suitable for multi-photon excitation.
Calcium Orange	549	576 increases with increased Ca^{2+}	Longer wavelength calcium indicator, compatible with TMR filter sets.
Calcium Crimson	590	615 increases with increased Ca^{2+}	Longer wavelength calcium indicator.
Fluo-3	506	526 increases with increased Ca^{2+}	Readily excited with the 488 nm blue line of the Kr/Ar and argon-ion lasers. The level of fluorescence increases at 526 nm with increased Ca^{2+} concentration. See Figure 9-9.
Fluo-4	494	516	Similar to fluo-3, but with higher quantum yield. Also includes analogues fluo-5N, fluo-5F, fluo-4FF and mag-fluo-4 with lower Ca^{2+} binding affinity. See Figure 9-9.
Fura Red	≈470	≈650 decreases with increased Ca^{2+}	The level of fluorescence detected in the red (650 nm) channel decreases with increased Ca^{2+} concentration. There is also a small shift in wavelength of absorbance and emission depending on whether Ca^{2+} is bound or not. See Figure 9-10.
Fluo-3 / Fura Red in combination	488	526 increases 650 decreases with increased Ca^{2+}	The combination of these two dyes, Fluo-3 (which increases in fluorescence) and Fura Red (which decreases in fluorescence) with increased Ca^{2+} concentration can be used to create a pseudo ratiometric dye. See Figure 9-10.
Mag-fluo-4	494	516	Low affinity Ca^{2+} / Mg^{2+} binding probe used for measuring high intracellular Ca^{2+} concentrations. Related molecules are fluo-5N, fluo-5F and fluo-4FF.
Mag-indo-1	340	405/485	Ca^{2+} and Mg^{2+} fluorescent indicator probes.
Magnesium Green	506	531 green	Low affinity Ca^{2+} indicator for measuring relatively high Ca^{2+} fluxes in cells.
Rhod-2	552	581 yellow	Long wavelength calcium indicators that may be useful in cells with a high level of autofluorescence at the shorter wavelengths. Related probes are rhod-FF and rhod-FN.
X-Rhod-1	580	602 red	Long wavelength calcium indicator similar to rhod-2 above. Related to X-rhod-5N.
Oregon Green 488 BAPTA	492	517 green	Calcium indicator readily excited with the 488 nm blue line of the argon-ion laser with a high quantum yield. Almost identical absorption/excitation spectra to fluorescein.
Calcein	494 blue	517 green	Low affinity Ca^{2+} chelator and indicator of intact cells (see Table 9-10, cell tracers).

pH Indicators

A large number of fluorescent molecules demonstrate changes in excitation or emission spectra depending on the pH of the surrounding media. The changes in fluorescence intensity observed with many commonly used probes may create difficulties in interpreting the concentration of dye within subcellular compartments. However, the sensitivity of many dyes to pH changes can be used to directly measure the pH of the cells. One word of caution – the "true" pH of an intracellular compartment can only be measured by careful use of a ratiometric dye (similar in principle to the ratiometric Ca^{2+} probes described above), although a good indication of the acidity or otherwise of various subcellular compartments can be obtained with non-ratiometric dyes.

One of the earliest indicators used to estimate the pH of the surrounding media was fluorescein. However, this molecule is not retained well within cells, and is prone to photobleaching, making an accurate estimation of intracellular pH very difficult. Carboxy-fluorescein diacetate (CFDA) was the first pH indicator that was designed to be loaded as an AM ester precursor, and was further modified to form a more suitable pH indicator dye (better cell retention and a p*K*a closer to the intracellular pH of living cells) called 2', 7'-bis-(2-carboxyethyl)-5(6)-carboxyfluorescein (BCECF). Changes in intracellular pH from 6.5 to 7.5 result in a shift in BCECF absorption maxima from 503 nm to 482 nm accompanied by an increase in fluorescence emission at 520 nm, thereby allowing BCECF to be used as a ratiometric dye (although this is rarely done as it is far from ideal as a ratiometric dye).

Table 9-7. pH Indicator Dyes

Many fluorescent dyes show changes in fluorescence depending on the pH of the surrounding media. This pH sensitivity often creates problems when attempting quantitative or semi-quantitative fluorescence microscopy – but it has been exploited to create a series of dyes that are used specifically to determine intracellular pH.

PROBE	Ex λ (nm)	Em λ (nm)	Notes
FDA	490/440	520	Fluorescein diacetate, one of the earliest dyes used as a fluorescence pH sensor, but suffers from a number of disadvantages, including high susceptibility to photobleaching and the lack of retention in live cells.
CFDA	490/440	520	Carboxy fluorescein diacetate, produced as an AM ester for better access and subsequent retention within living cells.
BCECF	490/450	520	2', 7'-bis-(2-carboxyethyl)-5(6)-carboxyfluorescein, a further modified form of fluorescein that results in a dye that is much better as an intracellular pH indicator. Can be used as a ratiometric excitation dye – but is far from ideal and so often used as a simple indicator dye. p*K*a optimal for measurement of close to neutral pH levels.
SNARF	488	580/640	Seminaphthorhodafluors: A greatly improved fluorescence indicator dye that can be used as a ratiometric dye by excitation at 488 nm. Used for measurement of pH around neutral pH (between pH 7 and 8). Can be used as a ratiometric emission dye by exciting with 488 or 514 nm.
SNAFL	488	580/640	Seminaphthofluoresceins: High quantum yield pH indicator that can be used as a dual emission (580/640 nm) or dual excitation (520/580) ratiometric dye.
HPTS (pyranine)	450/405	510	Water soluble pH indicator, but as no AM form is available it will need to be introduced into cells by other means.
Oregon green	490/440	520	A fluorinated analogue of fluorescein that can be used as a pH indicator in the somewhat acidic pH range (pH 4.2 to 5.7).
LysoSensor	blue to green	Range	A range of LysoSensor fluorescence sensors for acidic compartments (pH 4.5 to 6.0), with varying wavelengths of excitation/emission. Some can be used as ratiometric emission and excitation dyes, but others are non-ratiometric.
LysoTracker	blue to red	Range	A range of fluorescent probes called LysoTracker are available primarily as markers of acidic compartments such as the lysosome. They are designed to accumulate in acidic compartments, but are not intended as indicators of the pH of the compartment (in contrast to the LysoSensor probes above). Non of these dyes are ratiometric.

More recently developed pH indicators are designed for detecting a specified range of pH. For example, the SNARF and SNAFL pH indicators are used to establish the intracellular pH in the physiological range, whereas the new LysoSensor probes are used to estimate the pH within acidic organelles. These indicators are visible wavelength ratiometric dyes that can be used for pH measurements using visible light laser scanning confocal microscopy. The intracellular fluorescence change in pH indicators is often calibrated using the K^+/H^+ ionophore nigericin, which equilibrates the intracellular and extracellular pH level in the presence of extracellular K^+.

A number of fluorescent pH indicators are also available as large molecular weight Dextran conjugates that have a number of advantages as intracellular pH indicators. The dextran conjugates may be less prone to photobleaching, are better retained within the cell, and are less likely to become compartmentalised within the cell. However, these large molecular probes will need to be loaded into the cells by either physical injection or endocytosis (often the route of choice for measuring the pH of intracellular compartments such as the lysosome). Fluorescent dextran conjugates may be particularly suited to measurement of intracellular pH during embryonic development due to the greater cell retention and resistance to photobleaching.

Membrane Potential Probes

Estimating the membrane potential of various subcellular compartments is a highly valued method of determining the metabolic state of the cell. The various mitochondrial stains described in the section on membrane probes can be used to varying degrees of success in determining the membrane potential of the mitochondrial inner membrane. Determination of membrane potential, although a very powerful tool in studying cellular metabolism, is a somewhat difficult technique as the dye and its interaction with the laser light used for irradiation can cause significant changes to the membrane potential. Some membrane potential dyes are retained in the mitochondria after the inner membrane potential has been dissipated. This characteristic can be used to advantage to establish which cells, or mitochondria, had an active membrane potential when the dye was added. Other membrane potential dyes are dependent on the maintenance of the membrane potential for them to be retained within the mitochondria.

Not all membrane potential dyes are for estimating the membrane potential of the mitochondrial inner membrane. Several membrane dyes, for example the DiI and DiO fluorescent dyes, can be used as indicators of the membrane potential of any membrane in which they have become embedded (including the plasma membrane of the cell). Other probes, such as the oxonol (including bis-oxonol) family of fluorescent probes, accumulate in depolarised cells. Increased depolarisation results in further uptake of these dyes into the cells.

Other Ion Indicators

The range of different ions important in cellular metabolism and of potential pollutants that can be measured by fluorescence microscopy is probably only limited by your imagination in coming up with a molecule that binds the ion of interest, and then to in attaching a suitable fluorophore such that the binding or release of the ion of interest will influence either the intensity or the wavelength of emission of the attached fluorophore.

A number of well established fluorescent indicator probes are widely used for such ions as sodium (the SBFI and sodium green indicator dyes), potassium (the PBFI indicator, an analogue of the sodium indicator SBFI that selectively binds K^+ ions), chloride (6-methoxyquinolinium derivatives such as SPQ) and magnesium (variants of the BAPTA chelator used for calcium ion concentration measurements such as magnesium green).

The highly transient nitric oxide ion (NO) can also be followed using fluorescent probes. These probes include 4,5-diaminofluorescein diacetate (DAF-2 diacetate, itself non-fluorescent), which reacts with NO present within the cell to form a fluorescent heterocycle (a non-reversible reaction), becoming trapped within the cellular cytoplasm.

Fluorescent indicators for a number of metal ions are also most useful in biology for determining the level of heavy metal contamination. Fluorescent probes are available, with varying degrees of specificity for the heavy metal ions Zn^{2+}, Cu^{2+}/Cu^+, Fe^{3+}/Fe^{2+}, Ni^{2+}, Hg^{2+}, Pb^{2+} and Cd^{2+}. The difficulty of determining the level of specific heavy metals by fluorescence microscopy is that relatively abundant Ca^{2+}, Mg^{2+}, Na^+ and K^+ metal ions may significantly interfere with the level of fluorescence observed. Greatly improved specificity for heavy metal ions may come from establishing fluorescent indicators derived from peptides that have a very high specificity for particularly heavy metals. It may be possible using peptides to derive fluorescent compounds that can readily estimate, for example, the low concentration of Cu^{2+} within the cellular cytoplasm even though many other metal ions, including a high abundance of Ca^{2+}, are present.

ORGANELLE PROBES

A wide variety of remarkable organelle specific fluorescent probes (Table 9-8) has been developed over the past few years. Initially, many of the original organelle specific probes suffered from a marked lack of specificity! However, the newer generation probes appear to be much more specific for the specified organelle. Furthermore, a number of the more recently developed probes can be fixed *in situ* after loading and then the sample probed with, for example, antibodies against specific proteins within the organelle. This can be useful for confirming that the fluorescent probe is associated with the organelle of interest, or for establishing whether particular proteins are present within the organelle.

These organelle specific probes are loaded directly into live cells. Some of the probes also measure physiological parameters within the organelle. For example, many of the mitochondria specific probes are also indicators of the membrane potential across the mitochondrial inner membrane.

The organelle markers are remarkably specific for the organelle for which they are intended, but it should always be kept in mind that they may also be associated with other subcellular structures that are not related to the organelle for which they are intended markers. This problem of specificity is particularly important if you are using these probes in unusual cells or complex tissues.

Specific dyes are available for:
Mitochondria
Lysosomes
Golgi
Endoplasmic reticulum

The following organelle specific probes are often derivatives of existing fluorescent probes that showed some promise as organelle specific probes. The further development of these organelle probes, and the design of novel probes, has been a collaborative effort between many research scientists and Molecular Probes.

Mitochondria Specific Probes

Mitochondria can be identified in living cells by using a variety of fluorescent probes (Table 9-8). Many of the dyes with high specificity for mitochondria accumulate in actively respiring mitochondria, whereas other stains rely on the unique lipid environment of the mitochondria for their specificity.

The mitochondrial specific probes used initially (the methyl esters rhodamine 123 and tetramethylrhodamine) are readily taken up by actively respiring mitochondria, but are not taken up by mitochondria that lack a membrane potential – and quickly leach out of the cell during fixation. The newly developed dyes, MitoTracker Red, MitoTracker Yellow and MitoTracker Green are aldehyde fixable mitochondrial specific dyes with varying dependence on a high membrane potential. With these dyes, a distinction can be made between functional and non-functional (in terms of oxidative metabolism) mitochondria. Retention of the dye during aldehyde fixation allows one to directly compare the staining pattern with antibodies directed against proteins specifically associated with the mitochondria.

The MitoTracker fluorescent probes come with a variety of different excitation and emission characteristics, thus allowing one to perform dual labelling experiments with other mitochondrial markers. They are, however, not identical in their mitochondrial staining characteristics. For example, MitoTracker Red is readily taken up by actively respiring mitochondria that have a strong membrane potential. However, once the mitochondrial membrane potential is dissipated, either through the use of uncouplers, inhibitors or during cell death, the MitoTracker Red dye does not readily diffuse out of the mitochondria. On the other hand, MitoTracker Green specifically associates with mitochondria irrespective of their membrane potential. These two fluorescent probes can be used together to monitor the respiratory capacity of the mitochondria within cells or tissue samples. They are not particularly suited to following changes in membrane potential in living cells due to the relatively high retention rate of MitoTracker Red within mitochondria with a dissipated membrane potential.

Other probes that can be useful for identifying mitochondria, whether they are active in oxidative phosphorylation or not are nonyl-acridine orange (which stains to the mitochondria specific lipid, cardiolipin), and some DNA stains (particularly some Hoechst and SYTO DNA stains) that bind strongly to mitochondrial DNA.

Some mitochondrial probes indicate a high membrane potential – **which means they only stain active mitochondria**

Table 9-8. Organelle Specific Probes

Organelle specific probes exploit a number of physical and functional characteristics of the organelle to create a fluorescent probe that, hopefully, only stains the organelle of interest. These dyes have been extensively tested only on specific cell types. If your work involves unusual cell types, or mutant phenotypes, independent confirmation of the specificity of the probe may be necessary.

PROBE	Abs λ (nm)	Em λ (nm)	NOTES
Mitochondrial probes			
The following dyes stain mitochondria with varying degrees of specificity. Some of the dyes depend on an active membrane potential, whereas others exploit the unique molecular environment of the mitochondria irrespective of the membrane potential.			
Rhodamine 123	507	529	Original membrane potential dye. This dye is readily taken up by actively respiring mitochondria (having a high membrane potential), but is lost once the membrane potential is reduced. The dye is lost on fixation. Sensitive to photobleaching.
Tetramethylrhodamine	549	574	An early membrane potential dye with similar characteristics to Rhodamine 123.
MitoTracker Green	490	516	Accumulates in mitochondria regardless of the membrane potential. Accumulation appears to be enhanced in mitochondria with a high membrane potential, but the molecule is retained within the mitochondria even after the membrane potential has been dissipated. The dye is not fluorescent until taken up by the mitochondria. This dye is aldehyde fixable, allowing dual labelling with immunolabelled permeabilised and fixed cells.
MitoTracker Red	578	599	A longer wavelength mitochondria specific dye that selectively stains only actively respiring mitochondria. This dye is aldehyde fixable.
MitoTracker Orange	551	576	Similar in staining characteristics to MitoTracker Red, but with a shorter wavelength (yellow) emission spectrum.
JC1 & JC9	498 monomer 593 aggregate	525 monomer 595 aggregate	These mitochondrial membrane potential sensitive dyes shows two characteristic emission spectra green fluorescence when located in mitochondria with low membrane potential, and red fluorescence when concentrated in mitochondria with a high membrane potential. Dissipation of the membrane potential results in rapid loss of the dye from the mitochondria and subsequently the mitochondria become green.
nonyl-acridine orange	495	519	A mitochondria specific probe (binding to cardiolipin) that is not dependent on the membrane potential.
Mitochondrial antibodies	conjugated to various fluorophores		Antibodies against known mitochondrial protein can be used to identify mitochondria in fixed and permeabilised cells only. Antibodies against a number of proteins involved in electron transport are available commercially.
Golgi specific probes			
The Golgi apparatus can be specifically stained using a variety of fluorescently labelled lipids. Care should be taken to establish that the probe stains the Golgi apparatus in your particular cells (preferably using a Golgi specific antibody). These lipids are metabolised within the Golgi, resulting in subsequent redistribution of the fluorescent moiety to other cellular membranes.			
BODIPY® FL C5-ceramide As free fluorophore, or complexed with BSA	505	511 green 620 red	This green fluorescent lipid specifically concentrates in the Golgi apparatus, where the close stacking of the dye results in longer wavelength (red) fluorescence. The dye can be used to locate the Golgi apparatus and to study lipid dynamics in living cells.
BODIPY® TR C5-ceramide As free fluorophore, or complexed with BSA	589	617 red	This red fluorescent lipid specifically concentrates in the Golgi apparatus. This dye is particularly useful when doing dual labelling experiments, as there is no green emission.
NBD C6-ceramide As free fluorophore, or complexed with BSA	466	536 green ≈620 red	This probe is suitable for labelling the Golgi in both live and fixed cells. However, the NBD fluorescent group is more prone to photobleaching compared to BODIPY®.
See Table 9-9 for information on using the above lipid probes as membrane markers and cell tracers.			
Golgi specific antibodies	conjugated to various fluorophores		Antibodies against known Golgi specific proteins have extremely high specificity for the Golgi apparatus, but unfortunately organelle specific antibodies can only be used on fixed and permeabilised cells.

Organelle Specific Probes continued.........

PROBE	Abs λ (nm)	Em λ (nm)	NOTES
Endoplasmic reticulum specific probes			
There has been difficulty in specifically labelling the endoplasmic reticulum, but it appears that with the new ER-Tracker Blue-White fluorophore from Molecular Probes there may at last be a stain that doesn't also include mitochondria.			
$DiOC_6$	484	501 green	A hydrophobic membrane dye that can be used as a general membrane stain (see Table 9-9), or as a reasonably specific stain for the ER. However, mitochondria are also stained to a significant extent.
$DiIC_6$	549	565 red	A longer wavelength (red) emission fluorophore with similar characteristic to $DiOC_6$ discussed above. See Table 9-9 for information on using as a membrane stain.
$DiOC_7$	482	504	Similar to the C_6 probe above, but specific for the ER of plants.
Rhodamine B hexyl esters	556	578	An ER stain that also accumulates in mitochondria of living cells. Is relatively non-toxic at low concentrations.
Fluorescent Brefeldin A	503	510	Brefeldin A, an inhibitor of protein transport, when fluorescently labelled appears to bind specifically to the endoplasmic reticulum of living cells.
ER-Tracker Blue-White DPX	374	430 - 640	This UV excited ER specific stain does not stain the mitochondria and is aldehyde fixable (although with reduced fluorescence). This dye has a broad emission spectrum, which is influenced by the polarity of the environment. The dye is photo stable and non-toxic at low concentrations.
ER antibodies	conjugated to various fluorophores		Antibodies against proteins associated with the ER can be used to label the ER in fixed and permeabilised cells – but not in live cells.
Lysosomal probes			
Fluorescent macromolecules (such as labelled dextran, latex bead etc) are readily endocytosed by most cell types and will accumulate in the lysosome. Some pH indicator dyes will also strongly label the lysosomes because of their low pH.			
Fluorescent dextran	range of attached fluorophores from green to red		Fluorescently labelled high molecular weight dextran is readily endocytosed by most cell types, and will accumulate in the lysosome after several hours incubation.
Latex beads	wide range of excitation and emission spectra		A wide range of latex beads is available, ranging in size from below the resolution of light microscopy (0.014 µm) to as large as 2 µm or more. The best size for following endocytosis is approximately 0.5 µm. Several hours incubation (overnight) will result in extensive labelling of lysosomes.
LysoTracker	series of dyes with varying emission spectra (blue, green, yellow, red)		A dye developed by Molecular Probes that is specific for the lysosome.
LysoSensor	series of dyes with varying emission spectra (blue, green, yellow)		A pH indicator dye that has a high affinity for the lysosome and becomes highly fluorescent after deprotonation within the acidic organelle.
Lysosomal antibodies	conjugated to various fluorophores		Antibodies against proteins of the lysosome are difficult to obtain commercially, but a number of research laboratories do have such antibodies.

JC1, is another mitochondrial specific probe that can both be used as a marker of the membrane potential of actively respiring mitochondria and as a simple marker of non-respiring mitochondria. JC1 specifically associates with mitochondria (resulting in green fluorescence), but is greatly concentrated within the matrix of actively respiring mitochondria. The tightly packed JC1 emits red fluorescence. Dual channel fluorescence imaging can be used to monitor the ratio of active to inactive mitochondria within the cell or tissue sample. On dissipation of the membrane potential JC1 is rapidly lost from the mitochondria (changing from red to green fluorescence). However, JC1 does suffer from a number of drawbacks. The first problem is that the green/red transition is a concentration dependent phenomenon, which means that over-loading the cells can result in a significant increase in the amount of red fluorescence. Furthermore, irradiation of the tightly packed red fluorescent form of JC1 may result in a significant proportion of the dye reverting to green emission. This may be due to sudden dissipation of the mitochondrial membrane potential caused by the interaction of the laser with the dye loaded mitochondria, or it could be due to a direct effect of the laser on the packing of the JC1 dye. This can create the problem where the ratio of red/green fluorescence may change after several scans of the laser.

Lysosomal Probes

A variety of lysosomal probes are available (Table 9-8) that rely on either the low pH of the lysosome or the fact that cells will concentrate externally added macromolecules in this digestive organelle. Both of these methods are reasonably specific for the lysosome, but some additional endocytic structures and vesicles may also be labelled.

Traditionally the identification of lysosomes in living cells has relied upon the endocytosis of large fluorescently tagged macromolecules, such as dextran, and subsequent accumulation in the lysosome. This method of lysosomal detection relies on the cell being capable of endocytosis of macromolecules from the surrounding media. The endocytosis of small fluorescently labelled latex beads has the advantage that one can be sure that it is the bead that has been taken up by the cell (and not simply released dye), and that it will most likely then reside in the lysosome (the latex bead can be imaged directly within the lysosome by both fluorescence microscopy and electron microscopy).

The LysoTracker fluorescent probes developed by Molecular Probes are highly specific for acidic organelles. These relatively small fluorophores do not rely on endocytosis to gain access to the lysosome. These probes are available with a variety of emission wavelengths, making them applicable to multi-labelling applications. LysoTracker Red is aldehyde fixable, allowing one to perform antibody labelling experiments after fixation.

A number of LysoSensor fluorescent probes have been developed for tracking the acidity of cellular organelles. These acidic pH indicator probes can also be used as markers of acidic organelles, such as the lysosome.

Golgi Probes

A number of Golgi specific probes have been developed (Table 9-8). Some of these probes, such as NBD C_6-ceramide, BODIPY® FL C_5-ceramide and BODIPY® TR C_5-ceramide, are fluorescent lipid molecules that specifically associate with the Golgi apparatus in live, and in the case of NBD C_6-ceramide, also in fixed cells. These organelle specific probes can be used as both markers of the Golgi apparatus, and for the study of lipid transport and metabolism in live cells. The red emitting BODIPY® TR C_5-ceramide can be used in dual labelling experiments, whereas the green fluorescent probes emit in both the green and red, depending on the concentration and stacking of the dye.

These Golgi specific probes can be used as a marker of the Golgi apparatus within well-studied cellular systems. However, when studying unusual or disrupted cellular systems it should always be kept in mind that these probes may not necessarily associate only with the Golgi apparatus. These fluorescent lipid probes can also be metabolised to other fluorescent lipid derivatives that may associate with other subcellular organelles. For example, NBD-ceramide is metabolised to NBD-sphingomyelin and cerebroside, which returns to the plasma membrane after prolonged incubation.

> **Warning!!**
> Organelle probes may give misleading results in unusual organisms and unusual cells types, or under specific growth conditions.

Endoplasmic Reticulum Probes

$DiOC_6$ and rhodamine B hexyl esters are fluorescent probes that associate with the endoplasmic reticulum in live and fixed cells and tissue samples. As discussed above, one should verify the specificity of binding of these probes. $DiOC_6$ is a lipid soluble fluorescent dye that preferentially associates with the endoplasmic reticulum, but does associate to some extent with all cellular membranes, particularly the mitochondria.

Brefeldin A, an inhibitor of protein movement out of the endoplasmic reticulum, is the basis of the development of fluorescent derivatives that specifically associate with the endoplasmic reticulum. This "new generation" fluorescent probe is a good example of where a molecule with the required specificity (in this case binding an ER specific protein) can be made into a fluorescent marker for that particular organelle.

MEMBRANE PROBES

Fluorescent probes that label cellular membranes can be used for a wide variety of applications. This ranges from using the probe to study the lipid dynamics of the membrane, to using the probe as a marker of cellular structure. A hydrophobic membrane probe is often a great way to show cellular structure in multi-labelling applications when the other probes being used may, on their own, result in a confusing pattern of staining. Membrane probes are particularly good for staining live cell preparations, but some can be fixed in place for later immunolabelling, or even used after fixation as long as excessive organic solvent extraction of the sample has not removed most of the lipid material.

Cellular membranes are readily labelled by any fluorescent dye that is soluble in a lipid environment (Table 9-9). This includes small hydrophobic molecules, as well as larger molecules, such as fluorescently labelled phospholipids, which are modified versions of natural constituents of cell membranes (Figure 9-11). The smaller fluorescent membrane probes are often simply hydrophobic molecules that partition into the lipid domain of biological membranes (for example, DPH in Figure 9-11).

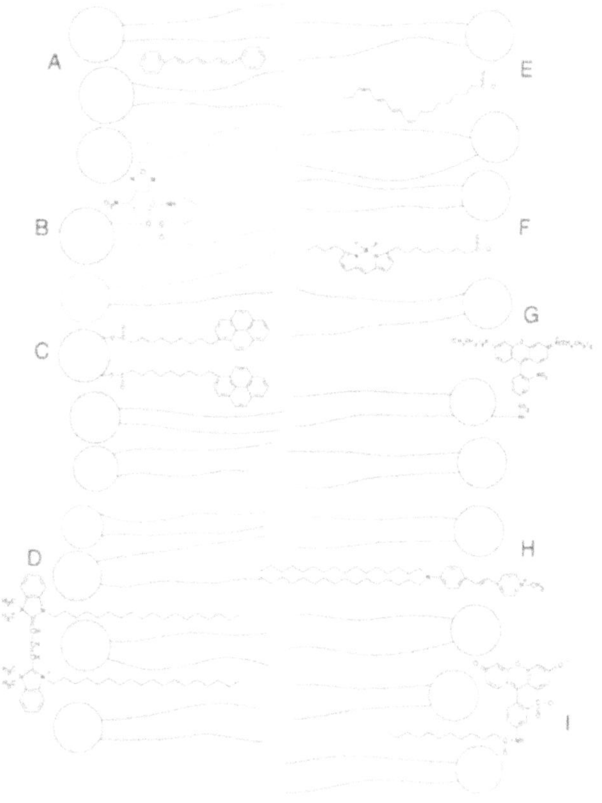

Larger hydrophobic fluorescent molecules, for example DiI (Figure 9-11), are often constructed from a fluorescent moiety attached to a hydrophobic domain (in this case a double hydrocarbon tail) that inserts into the membrane. The location of the fluorescent moiety will often determine the characteristics of the dye – particularly whether it will be influenced by ionic concentration or pH changes within the cytoplasm or organelle interior, or whether changes in the lipid environment will be more important in determining the fluorescence characteristics of the probe. For example, the fluorescent moiety of *bis*-pyrene-PC (Figure 9-11) is embedded deep within the bilayer of the membrane, whereas the fluorescent moiety of N-Rh-PE (Figure 9-11) is located on the aqueous face of the membrane. Both of these fluorescent probes are membrane lipids (phosphatidylcholine and phosphatidylethanolamine), but having the fluorescent moiety attached to different positions on the molecule will mean the behaviour of these probes will be reflecting the dynamics of very different cellular environments. Many fluorescent membrane dyes are only fluorescent when associated with the hydrophobic environment of the membrane - being virtually non-fluorescent in aqueous media. This markedly lowers any background fluorescence, which means that cells can often be imaged without the need to remove excess dye.

Figure 9-11. Membrane Probes.
A wide range of fluorescent membrane probes are available, ranging from small lipid soluble molecules to relatively large derivatives of naturally occurring phospholipids. **(A)** DPH, Diphenylhexatriene, **(B)** NBD-C_6-HPC, phospholipid, **(C)** *bis*-pyrene-PC phospholipid, **(D)** DiI, **(E)** *cis*-parinaric acid, **(F)** C_5-BODIPY®, **(G)** *N*-Rh-PE, **(H)** DiA and **(I)** C_{12}-fluorescein. This figure is kindly reproduced with permission from the Molecular Probes Handbook and is copyright by Molecular Probes.

Membrane probes are sometimes used as markers for specific subcellular organelles. For example, DiO and DiI are often considered as markers of the endoplasmic reticulum (Table 9-8), and fluorescent ceramide (BODIPY® FL – C5-ceramide) is often used as a marker for the Golgi apparatus. When you label cells with either of these types of fluorescent probes, you will notice that there is a lot more labelling in structures other than in the organelle of interest. This doesn't mean they are not any good as markers for specific cellular organelles, but it does mean that the location of the fluorescent label should be interpreted with care.

Membrane probes are often completely non-fluorescent in an aqueous environment

Membrane probes can be used to study the structure and dynamics of cellular membranes, but they are also often used to display cellular structure as a "structural background" for labelling with other fluorescent reagents. Some fluorescent membrane probes are so hydrophobic that they are virtually insoluble in an aqueous environment. Modified versions of these probes (for example DiI and DiO with varying lengths of hydrocarbon chain) are often available, but by their very nature membrane probes are generally not very soluble in culture media. One solution to this solubility problem is to use a large protein molecule, such as Bovine Serum Albumin (BSA), to assist in the delivery of the probe to the hydrophobic domain of the membrane. Delipidated BSA can be readily produced by organic solvent extraction of BSA - you can then bind your hydrophobic fluorescent probe to the carrier protein by simply mixing with a solution of the delipidated BSA. A number of hydrophobic fluorescent probes already bound to BSA can be purchased from Molecular Probes.

There are a very large number of fluorescent probes available for labelling cellular membranes; this section has only touched on a few commonly used probes. For an in-depth discussion of the probes available and their applications please refer to the latest edition of the Molecular Probes Handbook.

CELL TRACERS

Any fluorescent dye that is reasonably long lived and is relatively non-toxic can be used as a cell tracer. Of particular use as cellular markers in this category are the many GFP proteins now available (Table 9-11). Although GFP labelling is an immensely powerful tool in molecular cell biology, it does, unfortunately, require transfection of DNA into the cells under study. On the other hand, fluorescent molecules used as cell tracers can often be introduced to cell or tissue samples simply by adding a suitably diluted solution of the probe to the culture or incubation media. Fluorescent molecules used as cell tracers (Table 9-10) fall into two broad categories. The first consists of hydrophobic membrane markers that partition into cellular membranes and do not transfer between cells unless there is direct vesicle mediated lipid transfer – or transfer to the "daughter" cells by direct membrane partition between dividing cells.

Nipkow disk confocal microscopes are excellent for following embryo development using fluorescent cell tracers

The second type of cell tracer is a fluorescent molecule that can be loaded into living cells (often as the AM ester), and once inside, internal esterase activity cleaves the AM ester moiety, resulting in a charged molecule that cannot cross a biological membrane (the probe becomes "trapped" inside the cell). To further enhance the longevity of a cell tracer probe some probes become covalently attached to cell proteins so that they cannot diffuse out of the cell during cell growth. For example, the chloromethyl derivatives of fluorescein (CMFDA), see Table 9-10, also known as CellTracker Green, have a thiol-reactive chloromethyl moiety that reacts with intracellular thiols – covalently trapping the compound within the cells.

Cell tracers must be non-toxic to the cells over the period of time for which you wish to track the fate of the cells. It is important to use the lowest level of dye concentration that gives a reasonable signal – even after the minutes or hours over which you wish to follow the cells. It is also important to minimize the amount of laser light and the time used in imaging the cells (i.e. don't take images more frequently than necessary), as the combination of a fluorescent marker and laser irradiation can result in considerable molecular disruption and possibly death of the cells. It is possible to follow embryonic development by using fluorescent probes as cell tracers and, particularly the low light levels of the Nipkow disk confocal microscopes. In this way you can follow the fate of individual cells over several days, and still have embryos that can develop into perfectly normal animals.

Table 9-9. Membrane Probes

A wide range of fluorescent probes that stain biological membranes are available. These probes may consist of small highly hydrophobic molecules that preferentially dissolves into the lipid bilayer, or they may be membrane specific lipid molecules that act as markers for specific cellular membranes.

PROBE	Abs λ (nm)	Em λ (nm)	NOTES
Membrane Markers Hydrophobic fluorescent molecules that are preferentially retained by biological membranes can be used to highlight cellular membranes, and can often be used to selectively mark cells over more than one generation. The following selection of fluorescent membrane probes is only intended as an introduction to some of the more commonly used probes. The molecular structure, and the location within the lipid bilayer of a number of the probes listed in this table are shown in Figure 9-11. The Molecular Probes handbook (see Chapter 15 "Further Reading" (page 347) is a great source of information on both the fluorescent characteristics and the many different applications in which fluorescent membrane probes are used.			
$DiOC_6$	484	501 green	This hydrophobic fluorescent dye can be used as a general membrane probe, in addition to being a specific marker for the endoplasmic reticulum (see Table 9-8).
$DiIC_6$	549	565 red	A longer wavelength (red) emission fluorophore with similar characteristics to $DiOC_6$ discussed above.
$DiAC_6$	491	613 red	A hydrophobic membrane probe that partitions readily into biological membranes.
FM 1-43	479	598 red	A hydrophobic dye that is retained by cell membranes. Efficiently excited by the 488 nm line, but emits at relatively long wavelengths.
FM 4-64	506	750 far-red	A similar dye to FM 1-43 but emits in the far-red region, allowing dual labelling studies with FM 1-43 to study membrane recycling.
BODIPY® FL C_5-ceramide	505	515 green 620 red	Fluorescent derivatives of ceramide lipids are used to specifically label the Golgi apparatus in living cells (red emission), or alternatively to label a variety of cellular membranes (green emission) if the incubation is continued and the lipids are further metabolised into other lipid species that re-associate with other cellular membranes.
BODIPY® TR C_5-ceramide	568	630 red	Red emitting BODIPY® ceramide derivative used for labelling cellular membranes and the Golgi apparatus.
DPH	350	452	A relatively small hydrophobic molecule that partitions into the lipid domain of biological membranes (see Figure 9-11).
NBD-C_6-HPC	466	536	A fluorescent derivative of phosphatidylcholine with the fluorescent moiety residing within the lipid domain of the membrane.
Bis-pyrene-PC	340	473	A fluorescent derivative of phosphatidylcholine, a natural lipid found in most biological membranes. The fluorescent moiety is found at the end of the fatty acid hydrocarbon chains deep within the lipid bilayer (this is in contrast to the aqueous environment of the N-Rh-PE fluorescent moiety below).
cis-parinaric acid	303	416	A relatively small hydrophobic fluorescent molecule that partitions into the lipid domain of biological membranes.
C_4-BODIPY® C9	505	515	A hydrophobic derivative of the BODIPY® fluorescent moiety that dissolves within a lipid bilayer.
N-Rh-PE	560	581	A fluorescent derivative of phosphatidylethanolamine, a natural constituent of biological membranes. The fluorescent moiety is located in the aqueous phase, with the long fatty acid tails of the phospholipid anchoring this probe to the membrane.
C_{12}-fluorescein	495	519	A derivative of fluorescein with a long carbon chain attached to anchor it into biological membranes.

Table 9-10. Cell Tracers
There are a number of fluorescent probes that can be used as cell tracers. Most of the membrane probes described in the previous table can be used as cell tracers, but there are a number of advantages in using the cell tracer probes described below.

PROBE	Ex λ (nm)	Em λ (nm)	NOTES
Cell Tracers			
CellTracker	Various	Various	Fluorescent chloromethyl derivatives that freely diffuse across cell membranes, but once inside the cell the reactive thiol groups undergo glutathione *S*-transferase-mediated reaction to attach the dye to cellular proteins. CellTracker dyes can also be fixed in place with aldehyde fixatives.
Fluorescein diacetate	488	520 green	A useful cell tracker fluorescent probe with similar properties to the CellTracker dyes described above.
Lucifer yellow	488	500-560 yellow	Lucifer yellow (LY) has been used for a long time as an indicator of endocytosis as it does not readily cross cellular membranes. LY is also used extensively as a neuronal tracer as, not only is it retained within the neuron for an extended period, but it can also be fixed in place with aldehyde fixatives.
Membrane probes	Various	Various	Membrane probes (listed in some detail in Table 9-9) can often be used as cell tracers. Most membrane probes are readily retained within the cell membranes – only being spread from one cell to another via vesicle mediated lipid transfer or by cell division.
Latex beads	Various	Various	Fluorescently labelled latex beads of various sizes and fluorescence emission are often used as long term markers of cells. These reagents can be loaded into cells by endocytosis, bulk phase pinocytosis or microinjection. Once inside the cell fluorescent microspheres are very easily identified by their distinct and very bright punctate fluorescence. Furthermore, fluorescent latex microspheres can be readily identified by electron microscopy after suitable fixation and sectioning.
CMFDA (CellTracker Green)	488	520 green	The chloromethyl derivatives of fluorescein (CMFDA), also known as CellTracker Green, have a thiol-reactive chloromethyl moiety that reacts with intracellular thiols covalently trapping the compound within the cells.
Cell Integrity			
Calcein	494 blue	517 green	Useful for following cellular volume changes and integrity of subcellular compartments. A polyanionic fluorescein derivative that has about six negative and two positive charges at pH 7. The level of fluorescence is not unduly influenced by pH between pH 6.5 and 12. Can be purchased as the AM derivative, which will become trapped within the cell cytoplasm after cellular cleavage of the AM moiety.

CELL INTEGRITY

Calcein is a highly charged fluorescent probe that is used to determine the permeability of various cellular membranes (Table 9-9). Calcein is loaded into the cell cytoplasm by using an AM ester derivative that is cell permeable. Intracellular esterases cleave the ester group, leaving non esterified calcein within the cytoplasm. In this way the cytoplasm, for example, of living cells can be loaded with calcein, and exclusion from the mitochondria can be used as an indicator of the permeability of the mitochondria.

SOME UNUSUAL FLUORESCENT PROBES

There are fluorescent probes for an enormous array of cellular processes, many of which have been developed from dyes found to stain specific molecules or subcellular compartments. However, in more recent years there have been a number of innovative fluorescent probes and techniques developed. This includes specific fluorescent "assay" molecules such as PhiPhiLux and the exciting new technology of quantum dots.

PhiPhiLux (Apoptosis marker)

PhiPhiLux (OncoImmunin Inc) is a short hexa-peptide (DEVDGI) that contains quenched fluorochromes joined together via a bridge peptide that contains the cleavage site of the caspase-3 protease (CCP32). Cleavage of the peptide releases the fluorophores, which then become highly fluorescent.

Caspase 3 is only expressed in the early stages of apoptosis, and so in this way, early apoptotic cells can be clearly distinguished from healthy cells. PhiPhiLux is available in both a green and a red fluorescent forms, and so can be combined with other fluorochromes in dual labelling experiments. PhiPhiLux works very nicely with propidium iodide (which enters dead cells only) to mark dead cells in red and early apoptotic cells in green (PhiPhiLux). The simplicity of the PhiPhiLux assay, and the fact that it is carried out immediately on live cells, takes much of the subjective ambiguity out of scoring apoptotic cells using conventional dyes.

OncoImmunin will soon commence sales of its pro-fluorescent substrates for Caspase 1 (CaspaLux-1), Caspase 6 (CaspaLux-6), Caspase 8 (CaspaLux-8) and the Caspase 3 Processing Enzyme (CaspaLux-3PE).

Latex Beads

A very large variety of fluorescent latex beads has been developed, originally for use as standards for fluorescence activated cell sorting. Latex beads can also prove to be valuable as standards to test your confocal microscope. However, latex beads also have the possibility of being used as fluorescent probes in cell biology for both studies of endocytosis and intracellular transport, and for immunolabelling.

Use sub-resolution (0.1µm) latex beads to determine the resolution of your confocal microscope

Low molecular weight fluorescent dyes have a limited fluorescent output, particularly as there are a limited number of fluorochromes that can be conjugated to a biomolecule without unduly disrupting its function. One solution to this problem is to use highly fluorescent latex beads that have been impregnated with dyes of different emission characteristics.

Small latex beads labelled with suitable antibodies can be used in immunolabelling. This technique is particularly sensitive for locating low frequency cell surface proteins such as receptor proteins.

Latex beads should be used with caution (particularly very small beads that are not easily resolved by light microscopy), as the dye can sometimes leach out of the bead and stain subcellular structures without the bead itself entering that compartment. This type of non-specific labelling can also occur with fluorophores conjugated to protein molecules such as antibody molecules. Most commercially available latex beads and conjugated antibodies contain very low levels of free dye, but some cellular incubation conditions can result in release of free dye. If the latex bead is large enough to be imaged by light microscopy, then the correct labelling pattern can be ascertained by recognizing the bead structure under the microscope. Latex beads can, if necessary, be unambiguously identified within the cell using electron microscopy.

Quantum Dots

Quantum dots are semiconductor nanocrystals that are readily excited by blue (488 nm) light and emit fluorescence in a narrow emission band from green through to far red, depending on the composition and size (approximately 10 nm) of the quantum dot. Quantum dots can be made of a variety of materials, including europium oxide (Eu_2O_3) or a cadmium-selenium (CdSe) core with a zinc-sulphur (ZnS) coating. The semiconductor crystal core is coated with an inert polymer to which a variety of biological molecules can be attached.

Quantum dots have the important advantages of greatly reduced photobleaching and phototoxicity, high quantum yield, narrow emission peak, and excitation of various emission colours with only a single laser line. The relatively small size of the quantum dot means it can be readily taken into living cells by endocytosis – including receptor mediated endocytosis, as long as a suitable ligand molecule is attached to the surface of the quantum dot. Quantum dots can also be used as very sensitive fluorescent probes for immunolabelling of fixed cell and tissue samples. Quantum dots can also be produced with various biological molecules attached to the polymer coating. Biotin and Streptavidin coating of the quantum dots can be used to attach a range of proteins and antibodies to the quantum dots.

The stability and spectral characteristics of quantum dots make them an almost ideal fluorophore for confocal microscopy. However, these dots, although small by optical criteria (about 10 nm) are relatively large on a molecular scale when attempting to follow intracellular molecular movements. Furthermore, although quantum dots are coated with an inert polymer, the chemical composition of the internal nanocrystal is potentially highly toxic to living cells. Single quantum dots are below the resolution limit of a light microscope, but can be relatively easily imaged by confocal microscopy due to their high quantum yield and very high photo stability.

GREEN FLUORESCENT PROTEIN (GFP)

One of the most remarkable developments in recent years in the field of fluorescence imaging has been the development of the "green fluorescent protein" (GFP) as a marker of subcellular protein location and trafficking (see Table 9-11). The DNA sequence coding for GFP can be readily attached by gene fusion to any protein of interest, and the fate of the protein followed in real time in living cells.

GFP - a naturally fluorescent protein of 238 amino acids (27kD)

Green fluorescent protein is a naturally occurring fluorescent protein of 238 amino acids originally cloned from the Jellyfish *Aequorea victoria*. The role of the green fluorescent protein within the jellyfish is to transfer the blue chemiluminescence of aequorin to the green fluorescent protein and thus produce green light. The original green fluorescent protein, which is excited by light in the blue/UV region of the spectrum and emits green fluorescent light, has been modified over the past few years to produce a series of proteins of different wavelengths of excitation and emission. There are also a number of other fluorescent proteins that have been isolated from other marine organisms. Random mutagenesis has been used extensively to both enhance the fluorescence, and to alter the wavelength of excitation and fluorescence emission, of various fluorescent proteins.

This section only very briefly introduces the exciting technology of green fluorescent protein labelling of living cells. There are a large number of published articles, and several excellent books (see page 351 in Chapter 15 "Further Reading") to which you should refer if you require further information.

Structure of GFP

The fluorescent moiety of GFP is an oxidized derivative of the tripeptide Ser-Tyr-Gly, residues 65-67 of the native protein. The activation of GFP requires molecular oxygen, but no other enzymes or cofactors. This means that cloned GFP, expressed in a diverse range of organisms, displays the strong fluorescence of the native protein. This is in contrast to the highly fluorescent phycobiliproteins and peridinin-chlorophyll-a protein, which require the correct insertion of accessory pigments for the development of fluorescence.

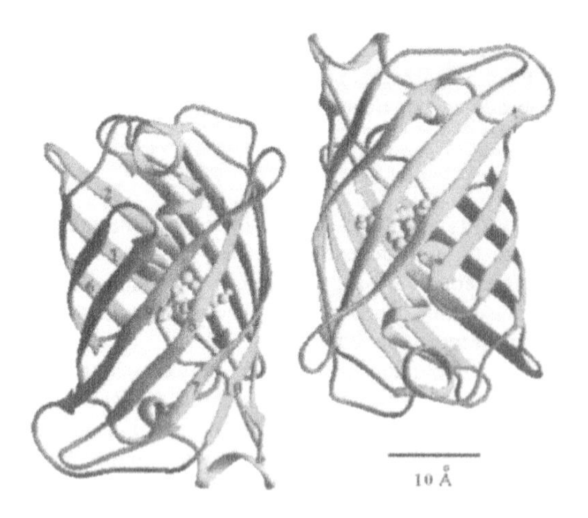

Figure 9-12. Green Fluorescent Protein Structure.
The molecular structure of Green Fluorescent Protein (GFP) clearly demonstrating the tightly constructed β-barrel threaded up the centre by an α-helix. The fluorophore is created by an oxidised derivative of the tripeptide -Ser-Tyr-Gly-. There are no other accessory molecules required for the formation of fluorescence. This figure is derived from Yang, F., Moss, L.G., and Phillips G.N. *The Molecular Structure of Green Fluorescent Protein*. Nat. Biotechnol (1996) 10:1246-1251.

The initial cyclisation between Ser^{65} and Gly^{67} to form a imidazolin-5-one intermediate is relatively fast, but this is followed by a much slower (hours) oxidation of the Tyr^{66} side chain. The formation of fluorescence in the original GFP from *Aequorea* is very dependent on temperature; at or below room temperature being optimal for correct folding of the fluorophore. Once fluorescent GFP is formed it is very stable – even up to 65°C. A number of genetic mutants of the original GFP protein are now available with improved folding activity at 37°C.

The crystal structure for GFP demonstrates that it is a tightly constructed β-barrel threaded up the centre by an α-helix (Figure 9-12), which occurs naturally as a dimer (although dimer formation in cells expressing GFP is relatively low). The residues that become the chromophore on exposure to oxygen are buried deep within the α-helix. GFP is highly resistant to denaturation, requiring 6 M guanidine hydrochloride at 90°C or very low or high pH. Renaturation occurs within minutes following reversal of the denaturing conditions.

GFP can be fixed using 4% paraformaldehyde

The compact and buried nature of the fluorophore protects it from photobleaching, and at the same time limits the environmental sensitivity of the fluorophore. In many applications (such as its function as a reporter gene) the resistance to environmental influence is of great benefit (for example, the fluorophore is not unduly affected by pH ranges found within living organisms). However, this resistance to environmental influence has created a considerable challenge to create GFP variants that can be used to determine intracellular pH or ion concentrations in living cells. The original sequence obtained from *Aequorea victoria* has been modified to produce clones with altered codon usage and improved translational initiation sequences. As will be discussed below, a number of "enhanced" variants have also been produced by genetic manipulation.

Enhanced Derivatives of GFP

The original GFP isolated is reasonably efficiently excited by short wavelength UV irradiation. Unfortunately this is not widely available when using confocal microscopy. Considerable effort has gone into developing a number of GFP proteins that have longer wavelengths of excitation and emission. In addition to changing the wavelength of excitation and emission it has also been possible to greatly enhance the quantum efficiency of the protein such that "enhanced" derivatives have up to a 35-fold increase in fluorescence (see EGFP, "Enhanced Green Fluorescent Protein", and "Emerald" GFP in Table 9-11 and Figure 9-13). Most of these mutants are due to changes in the folding of the molecule, rather than direct changes to the quantum efficiency of the fluorophore directly.

The extremely tight structure that protects the GFP fluorophore from environmental influence (resulting in good resistance to photobleaching and very long lived molecular stability), ironically, means that the original GFP protein cannot be readily used in following dynamic processes within living cells, including studies of intracellular protein turnover. Destabilised enhanced GFP, (dEGFP, see Table 9-11 and Figure 9-13) is a "destabilised" form of GFP that has been specifically constructed to allow one to follow cellular dynamics in real time. Compared to the original GFP protein, this protein has a much shorter half-life of approximately 2-hours.

Coloured Variants of GFP

The original, and enhanced versions, of the GFP protein exhibit "green" fluorescence when excited by blue or UV light. However, multi-labelling applications and FRET studies require GFP proteins with a range of emission wavelengths. A number of variant GFP proteins, that exhibit differing wavelengths of excitation and emission, have been developed by genetic manipulation. These altered GFP proteins have been constructed from minor changes to the tripeptide chromophore (sequence and substitution changes), which result in changes in the folding of the GFP fluorescent moity, rather than in changes directly affecting the fluorophore itself.

The range of colour variants of the original GFP include blue fluorescent protein (BFP) with a maximum emission in the blue region of the visible light spectrum, cyan fluorescent protein (CFP) with an emission in the longer wavelength "blue/green" region of the spectrum, and yellow fluorescent protein (YFP) with a peak emission in the yellow region of the spectrum. Together with the original green emitting protein, these derivatives of GFP result in an excellent range of wavelengths for both multi-labelling and FRET applications. Enhanced variants of these proteins, with greatly increased quantum yield, and in some cases small changes in excitation and emission wavelengths, are also available.

There are a many GFP variants that have been produced. Although, for most applications you will probably be limited to those that are commercially available, it is worth enquiring about the availability of other variants that may be more suitable for your particular experiments. Different GFP proteins from other organisms and with a range of excitation and emission wavelengths, have been isolated, and are discussed below.

GFP from the Sea Pansy (Renilla reniformis)

A GFP protein has been cloned from the sea pansy (*Renilla reniformis*). This GFP protein has a single excitation peak at 498 nm blue light (with fluorescence in the green), and interestingly, has a modified amino acid fluorophore core that is identical to that of the original GFP isolated from *Aequorea*.

Fluorescent Proteins from Coral (Discosoma sp)

A series of proteins, again with a very similar structure to the GFP isolated from *Aequorea*, but only 26-30% identical at the amino acid level, have been isolated from tropical coral (*Discosoma* sp).

Fluorescent proteins from coral – occur naturally in a variety of fluorescent colours

These naturally occurring proteins show a range of excitation and emission colours. However, the protein that has created the most interest is the red fluorescent protein (DsRed) from coral as it has a significantly longer excitation and emission wavelength compared to the original Jellyfish GFP or any of its derivatives. Red fluorescent protein has raised a lot of interest in multi-labelling and FRET studies because of the longer wavelength of emission, but a serious problem is caused by the fact that the original protein isolated from coral occurs naturally as a tetramer and takes many hours to mature into a fluorescent compound. In single labelling studies a tetramer has the disadvantage of being a significantly larger size compared to a monomer or even a dimer, but perhaps more importantly, particularly in multi-labelling and FRET applications, the formation of tetramers seriously complicates the interpretation of experimental results. It is quite possible that "false" associations could be formed, resulting in misleading conclusions as to the location, dynamics and associations of macromolecules within the cell.

DsRed – red fluorescent protein from coral

To overcome some of the limitations of DsRed a number of genetically modified variants of coral proteins have been produced. This includes versions of DsRed that have an increased rate of maturation into a fluorescent compound (DsRed2, DsRed-Express, DsRedT1 etc). Other variants of red fluorescent protein, which associate as dimers, have been derived from other coral species (HcRed1). There are also derivatives of the original red fluorescent protein that exist as monomers (mRFP1). The monomer version of DsRed may make this protein a valuable tool in not only multi-labelling applications, but also in the use of FRET to determine molecular associations. Used in association with other, shorter wavelength "donor" fluorescent proteins, red fluorescent proteins may become a valuable tool in studying the dynamics of molecular interactions in living cells.

Light Sources for GFP Excitation

The original GFP protein isolated from *Aequorea* requires relatively short wavelength (UV) light for excitation. Although there are two absorption peaks (395 and 475 nm), it is the shorter wavelength 395 nm absorption peak that dominates. This means that the original GFP protein is readily excited using the shorter (UV) wavelengths of a Hg-arc lamp used in conventional epi-fluorescence microscopy. However, the commonly used 488 nm line of the argon-ion laser used in confocal microscopy does not excite the original GFP. Modified versions of GFP (for example, enhanced GFP) with a significantly longer wavelength for optimal excitation (490 nm) have been developed. This means they can be readily excited using the argon-ion laser.

Longer wavelength emitting GFP proteins (such as YFP and DsRed) may be excited by the 514 nm line of the argon-ion laser or the 568 nm line of the krypton-argon-ion laser, making these GFP derivatives particularly well separated from EGFP in multi-labelling applications using a confocal microscope. The shorter wavelength (405 nm) violet solid state lasers that are now becoming readily available for use in confocal microscopy may become a valuable means of efficiently exciting the shorter wavelength GFP proteins, without the need to resort to UV confocal microscopy.

The GFP protein is also readily excited by two-photon excitation. Particularly, the 780 to 800 nm pulsed infrared laser used in multi-photon microscopy readily excites the shorter wavelength GFP proteins. Pulsed infrared lasers used in multi-photon microscopy may also excite longer wavelength GFP proteins (for example, CFP and YFP). With the correct combination of laser wavelengths and suitable fluorophores it may be possible to simultaneously excite a cyan or yellow excited GFP and a short wavelength UV excited GFP, with 2- and 3-photon excitation.

Photoactivation and Photoconversion GFP

Derivatives of GFP (photoactivated GPP, PA-GFP) that have greatly increased fluorescence after irradiation with shortwave blue (413 or 405 nm) light have been developed (see Patterson and Lippincott-Schwartz 2002, on page 351 of Chapter 15 "Further Reading"). These derivatives of GFP are valuable for molecular tracing studies that could be used in place of FRAP (Fluorescence Recovery After Photobleaching) experiments. Another interesting fluorescent protein derived from Coral, and called Kaede, is converted from green to red fluorescence emission by irradiation with violet 405nm light (see Ando et al. 2002, page 351). This protein can be used as a molecular tracer in living cells, with multiple markers within the one cell made possible by adjusting the ratio between green and red forms of the protein in specific regions of the cell.

GFP Applications

The various GFP proteins now available have been applied to a very wide range of biological problems using a great diversity of approaches. The most obvious applications for GFP are as a reporter for gene expression and as a means of determining the subcellular location of the gene product. However, there are many other ingenious ways in which GFP has been applied to cellular problems. A number of interesting examples of using GFP in biological systems are discussed here briefly in the enclosed box.

The discovery of GFP has made the idea of studying the fate of single molecular species within the cell a real possibility. The study of subcellular localisation and molecular associations will eventually be greatly assisted by the exciting technique of fluorescence resonance energy transfer (FRET) – (discussed in some detail on page 254 in Chapter 10 "Confocal Microscopy Techniques") – using suitable GFP derivatives as FRET pairs.

Some GFP proteins, including the original GFP isolated from *Aequorea*, have the unusual characteristic of undergoing photochemical transformations that alter the spectral characteristics of the protein over a period of time. These changes can be exploited for tracking the movement of GFP-tagged proteins within the cell by exciting a defined region of a cell or tissue sample for a short period of time, and then following the appearance of photo-converted protein over time.

GFP Applications

...a few examples....

Reporter gene – requires relatively high levels of expression. "Destabilised" GFP may be required for following dynamic processes within the cell as the original GFP protein is extremely stable in the intracellular environment.

Fusion Tag - GFP proteins have been fused with a wide variety of cellular proteins to determine their subcellular location and dynamics. Over-expression may result in misdirection of the protein into pathways that are not the normal fate of your protein of interest.

Gene Transfer – the insertion of a GFP gene into suitable vial genome transfer vectors can be used to determine the efficacy of gene transfer for the development of human gene therapy.

Cell lineage tracer – GFP expressed by specific cell lines can be used to trace cell lineage. The long lived varieties of GFP, and their resistance to photobleaching make GFP particularly valuable for studies of cell lineage over time.

pH indicator - the fluorescence of particular mutants is pH sensitive. Directing these GFP variants to particular subcellular locations allows one to determine the pH of specific subcellular organelles with considerable accuracy.

Molecular proximity - FRET pairs can be created by using GFP and a longer wavelength derivative or related protein (such as CFP or DsRed) to determine molecular associations in living cells.

FRET based protease assay - peptide linked GFP proteins of suitable wavelengths of absorption and emission have been constructed to create a FRET pair that will change in fluorescence emission on being cleaved (the FRET pair separated) by intracellular proteases.

FRAP – molecular dynamics studied by following recovery from photobleaching can be used to study the dynamics of your protein of interest. This technique is particularly suitable for determining the degree of molecular movement of your protein of interest – but be careful when interpreting your results as the GFP protein may significantly alter the dynamics of the protein under study.

Calcium concentration - calcium sensitive GFP-calcium chelator fusions have been developed. GFP based calcium indicators can be specifically directed to various subcellular compartments to determine their calcium concentrations.

Embryogenesis – cell lineage during embryogenesis can be followed using GFP fusion proteins. The low level of photobleaching makes GFP protein and its derivatives ideal for following embryonic development over several days without compromising the viability of the embryo.

Whole animal studies – whole animals can be grown with a GFP fusion present. The whole animal may become fluorescent, the best evidence that GFP expression does not unduly interfere with essential cellular metabolic processes.

Protein degradation in vivo – fusion proteins containing GFP as a reporter protein and the protein under study.

Organelle tagging – fusion of organelle retention or targeting motifs to GFP to unambiguously identify subcellular organelles.

Cellular dynamics – the ability to image live cells expressing GFP fusion proteins allows one to follow the dynamics of cellular processes in real time. The low level of photobleaching when using GFP protein and its derivatives is particularly useful for live cell studies.

......and there are many more....

The real power of GFP is in being able to study cellular processes in real time in living cells; this includes the development of embryos and whole animals, not just single transformed cells. However, if required, GFP labelled cells can be "fixed" using aldehyde fixatives such as paraformaldehyde, and the location of the fluorescence determined by using confocal microscopy at a later time. GFP fluorescence does not survive fixation with organic solvents or the use of high acidity levels during the fixation process.

GFP Limitations

The relatively large GFP moiety does not appear to unduly interfere with the expected subcellular location of many proteins – but be aware that in those experiments where there is conflict between GFP localisation results and established dogma, the established dogma will usually prevail and the GFP results will probably never be published!

Although the potential applications of GFP to the study of cell biology are extensive and most promising, there are a number of limitations due to the nature of the protein itself. One limitation is the limit of detection as there is no amplification process in the detection of the fluorescence (i.e. there is one fluorophore per GFP protein molecule). Cellular autofluorescence limits the lowest level of sensitivity available. GFP that is targeted to subcellular compartments may become significantly concentrated relative to the level of autofluorescence within the cell, making detection relatively easy. However, there is also the danger of over-expressing GFP to create a high level of labelling to make imaging easier – but is the subcellular localisation then truly representative of the original protein under study?

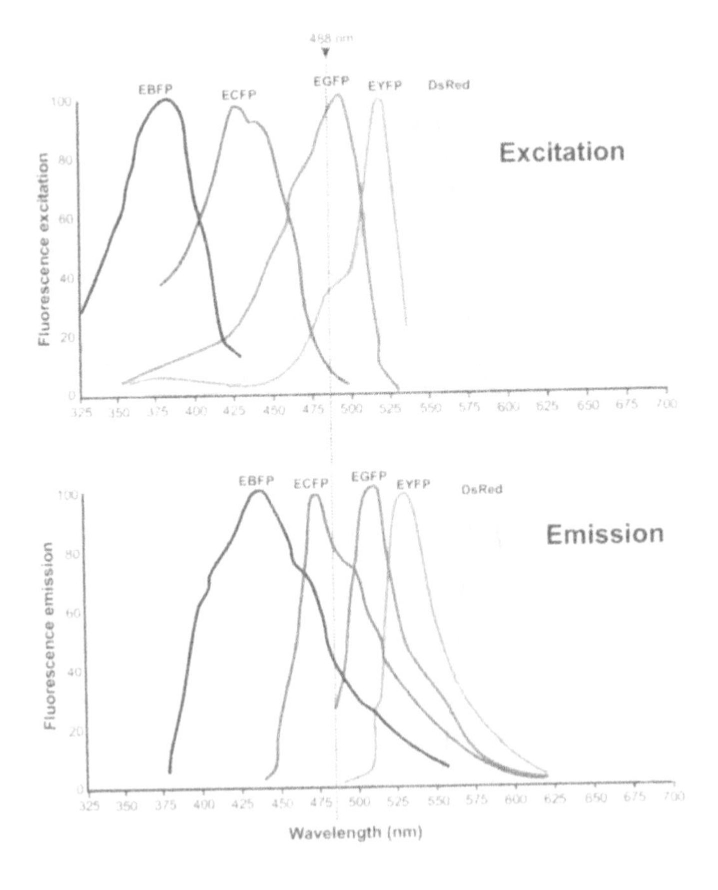

Figure 9-13. GFP Derivatives - Excitation and Emission Spectra.
The excitation (absorption) and emission spectra of several GFP variants (including the coral protein DsRed) are shown in this diagram. The 488nm line of the argon-ion laser is shown as a dotted vertical line. This diagram is derived from spectra obtained from the BD Biosciences (Clontech) web site www.clontech.com/.

Table 9-11. Green Fluorescent Protein and Derivatives

There are now many derivatives of the original green fluorescent GFP (green fluorescent protein) available. Some of these fluorescent proteins are derived from the original Jellyfish (*Aequorea victoria*), but many others are either isolated directly from, or derived from previously cloned proteins of, the sea pansy (*Renilla reniformis*) or reef coral (*Discosoma sp.*).

PROBE	Abs λ (nm)	Em λ (nm)	NOTES
***Aequorea victoria* (Jellyfish) derived fluorescent proteins**			The original GFP protein isolated from *Aequorea* has been subsequently modified to produce proteins with longer wavelengths of excitation and emission, and ironically, mutants have been constructed that are "destabilised" relative to the wild-type to allow studies of the dynamics of cellular processes. The original protein is 238 amino acids (27 kD), and is readily excited by UV/blue light, giving green fluorescence. The original GFP protein has a tendency to form dimers (particularly when targeted to membranes), but mutated versions (mCFP, mYFP etc), present within the cells predominantly as monomers are available.
wtGFP	395 UV 475 blue	509 green	**Green Fluorescent Protein:** The original wtGFP (wild-type GFP) has two excitation peaks at 395 nm (the major peak) and 475 nm (minor peak), with fluorescence emission in the green.
EGFP	490 blue	509 green	**Enhanced Green Fluorescent Protein:** This derivative of the original GFP (Ser-65 to Thr) has a red shifted excitation peak (peak excitation at 490 nm), which means it is readily excited by the commonly used 488 nm argon-ion laser used in confocal microscopy. This enhanced GFP also has a significantly increased level of fluorescence.
dEGFP	490 blue	509 green	**Destabilised Enhanced Green Fluorescent Protein:** This derivative of EGFP has been specifically designed to be more readily degraded within the cell, allowing one to follow the dynamics of gene expression in living cells. GFP has been specifically destabilised by fusing a degradation domain from ornithine decarboxylase to the C-terminus of EGFP. The resulting protein has a much shorter half-life of approximately 2-hours.
Emerald GFP	485-488 blue	508 green	**Enhanced Green Fluorescent Protein:** This derivative has a longer wavelength of excitation (485-488), making it readily excited by the blue line of the argon-ion laser.
BFP	385 UV	430 blue	**Blue Fluorescent Protein:** A "blue" fluorescent derivative of the original GFP. Relatively low quantum yield and high level of photo-bleaching.
EBFP	380 UV	440 blue	**Enhanced Blue Fluorescent Protein:** An enhanced version of BFP that has a higher quantum yield.
CFP	425 blue	475 blue	**Cyan Fluorescent Protein:** A "cyan" or blue-green fluorescent derivative of wtGFP that has a slightly longer excitation wavelength. This variant of the original wtGFP has been used successfully as the "donor" fluorophore in studies of protein-protein interactions using FRET.
ECFP	434 blue	477 blue	**Enhanced Cyan Fluorescent Protein:** An enhanced version of CFP that has enhanced folding characteristics and human codon usage.
YFP	520 green	530 yellow	**Yellow Fluorescent Protein:** A "yellow" fluorescent derivative of the original wtGFP. This fluorescent protein has been used as the "acceptor" fluorophore in studies of protein-protein interactions using FRET. This GFP derivative is also pH and Cl- sensitive.
EYFP	514 green	527 yellow	**Enhanced Yellow Fluorescent Protein:** An enhanced version of YFP that has a higher quantum yield and folds more efficiently within living cells.
Citrine GFP	514 green	524 yellow	**Citrine (yellow) Fluorescent Protein:** A derivative of YFP (Ser-65 to Gly) with greatly decreased pH and Cl- sensitivity with a fluorescent emission peak in the yellow.
Venus	514 green	524 yellow	**Venus Yellow Fluorescent Protein:** A derivative of YFP with enhanced fluorescence and less susceptibility to environmental influences.
PA-GFP			**Photoactivated GFP derivative:** GFP protein with greatly increased fluorescence after activation by irradiation with violet (405 nm) light. Used for molecular tracing studies.

Green Fluorescent Protein continued.........

PROBE	Abs λ (nm)	Em λ (nm)	NOTES
***Renilla reniformis* (sea pansy) derived fluorescent proteins**			
The GFP protein isolated from *Renilla* species, with an identical fluorophore core to that of the original GFP isolated from *Aequorea*. The original GFP from *Renilla* species forms obligate dimers.			
GFP	498 blue	509 green	**Green Fluorescent Protein:** A green fluorescent protein isolated from the Sea Pansy has a nearly identical emission spectrum to that of the original Jellyfish GFP. Occurs as an obligate dimer.
***Discosoma* (reef coral) fluorescent proteins (RCFP)**			
The red fluorescent proteins isolated from reef coral have striking structural homology to the original GFP isolated from *Aequorea*. Many of the different red fluorescent proteins listed below are derived from different genes isolated from *Discosoma*, and are not always derived from mutagenesis of the original protein. The red fluorescent proteins isolated from *Discosoma* species form obligate tetramers, although dimer and monomeric forms of this protein have now been constructed			
DsRed	563 yellow	582 red	**Red Fluorescent Protein**: longer wavelength "GFP" proteins are highly valued for use in multi-labelling applications and studies of protein-protein associations using FRET pairs. This protein does suffer from slow maturation (up to 10 hours) and for being an obligate tetramer. Slow maturation may limit the ability of the protein to be used in cellular dynamics, and the tetramer formation seriously hinders any use of this protein in studies using FRET. Modified variants of this protein have been produced.
DsRed2	563 yellow	582 red	**Red Fluorescent Protein**: modified variant of DsRed, produced by Clontech, that has some improvement in the rate of maturation of the fluorophore.
DsRed-Express	557 yellow	579 red	**Red Fluorescent Protein**: a modified version of DsRed that has a significantly increased rate of fluorophore maturation.
DsRed T1	557 yellow	579 red	**Red Fluorescent Protein**: another modified variant of DsRed that has a significantly increased rate of fluorophore maturation (less than one hour).
HcRed1	588 red	618 far red	**Red Fluorescent Protein**: dimer variant of a red fluorescent protein, derived by random mutagenesis from the reef coral *Heteractis crispa*, that emits fluorescence at a slightly longer wavelength compared to the original DsRed.
AsRed2	576 yellow	592 red	**Red Fluorescent Protein**: modified variant of DsRed with longer emission wavelength.
mRFP1	584 yellow	607 red	**Monomeric Red Fluorescent Protein:** a monomeric version of DsRed that may make this "GFP" more useful in studies of protein localisation and dynamics in vivo, including studies using the protein as the "acceptor" in a FRET pair.
AmCyan1	458 blue	489 blue	**Cyan Fluorescent Protein**: blue coral protein from Clontech.
ZsGreen1	493 blue	505 green	**Green Fluorescent Protein**: green coral protein from Clontech.
ZsYellow	529 yellow	539 yellow	**Yellow Fluorescent Protein**: yellow coral protein from Clontech.
Kaede	405 violet	520 green 600 red	**Red/green Photoconvertable Fluorescent Protein:** Irradiation of this coral protein with 405nm violet light results in photoconversion from the green to the red form. Can be used for studies involving molecular tracing.

FlAsH - HEXA-PEPTIDE FLUOROPHORE TAG

One of the potential drawbacks of using GFP fusion proteins to study the dynamics of intracellular protein movement is the large size of the GFP protein (approximately 238 amino acids, depending on the mutant).

A novel way of overcoming the problem of bulky fusion proteins is demonstrated by the development of the compound FlAsH-EDT_2 (Fluorescein Arsenical Hairpin Binder), which selectively binds a specific hexa-peptide (the tetra-cysteine, CCXXCC, where "X" is any amino acid and "C" is cysteine) where-upon it becomes fluorescent. The incorporation of the hexa-peptide unit into proteins of interest will result in far less disruption of natural cellular transport mechanisms compared to the bulky GFP proteins. The small molecular weight FlAsH -EDT2 compound can readily diffuse directly into live cells, which are subsequently fixed.

Chapter 10

Confocal Microscopy Techniques

A wide variety of novel techniques for fluorescent imaging have become available over the past few years, particularly since the introduction of the confocal and multi-photon microscopes. Many of the basic methods of collection, choice of fluorescent probes and manipulation of the sample, and a number of "techniques" (for example, immunolabelling, GFP labelling etc) have been covered in detail in earlier chapters. This chapter provides a short introduction to a number of interesting techniques in fluorescence microscopy that help to extend confocal microscopy beyond gathering simple structural information into the realm of studying molecular interactions and single cell biochemistry.

This chapter begins with a discussion on multi-labelling techniques in confocal microscopy. Although it may at first appear that multi-labelling is as simple as activating two channels on the confocal microscope and collecting images, there are many difficulties and limitations in multi-labelling that are rarely appreciated outside the specialist imaging community. This is in spite of the fact that most confocal microscopists will routinely use more than one fluorescent label!

A number of interesting molecular techniques will also be discussed, such as FRAP (Fluorescence Recovery After Photobleaching), FRET (Fluorescence Resonance Energy Transfer) and FISH (Fluorescence In Situ Hybridisation). These techniques allow one to delve into the molecular associations and environment within individual cells or organelles.

A good introduction to a number of interesting confocal microscopy techniques are found in the Bio-Rad application notes (listed at the end of this chapter), that are available from Bio-Rad on request, or can be downloaded from the Internet.

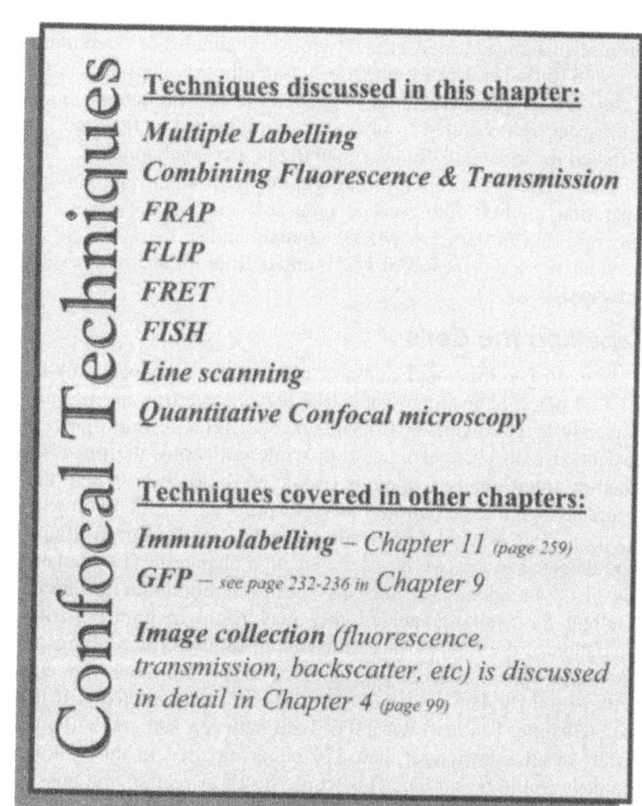

MULTIPLE LABELLING

Multiple labelling is a powerful technique used in fluorescence microscopy for determining the physical location and association between molecules or structures of interest. Confocal microscopy is particularly suited to multiple labelling techniques due to the ease with which the fluorescence emission spectra of individual fluorophores can be separated, either by directing specified regions of the light spectrum to individual detection channels, or alternatively by emission fingerprinting of the individual fluorophores.

Co-localisation by confocal microscopy – can only be determined within the limit of resolution of a light microscope

There are, however, a number of limitations when performing multiple labelling experiments using light microscopy. Of particular concern is the limit of resolution of a light microscope (resolution in a light microscope is discussed in detail on page 50 in Chapter 2 "Understanding Microscopy"), the optical aberrations inherent in any imaging device using different wavelengths of light, and the inherent problems of incorrect imaging and quantitation, possibly resulting in misleading results.

Choice of Fluorophore

Some fluorophores emit fluorescence across a broad range of the light spectrum, others emit only a narrow band of light and still others may emit fluorescence in separate regions of the light spectrum. If your interest in dual labelling is to distinguish two different proteins by dual antibody labelling then you will need to select two fluorescent probes that can be readily separated by your confocal microscope. However, dual labelling may also be used to distinguish specific areas of the cell or different subpopulations of subcellular organelles that display different characteristics – which may be variations in pH, ion concentration, membrane potential, lipid constituents etc. Irrespective of the goals of your experiment you will probably want to try and minimise the amount of fluorescent light detected in the "wrong" channel. For example, if you are attempting to separate the fluorescence emission of fluorescein from that of tetramethyl rhodamine you will want all of the fluorescein emission showing up in the "green" channel and all of the tetramethyl rhodamine fluorescence showing up in the "red" channel. Having fluorescent light emitted from fluorescein detected in the "red" channel is called bleed-through, and may be easily mistaken for true "red" fluorescence in the second channel.

Traditionally fluorophores chosen for multi-labelling applications have been chosen on the degree of physical separation in their fluorescence emission spectra. However, with the introduction of the technique of spectral fingerprinting in the Zeiss META channel and in the Leica SP confocal microscope, one can now readily separate the fluorescence emission of very similar fluorophores (for example, several "green" emitting fluorophores can be separated).

Labelling the Cells

In multi-labelling applications collecting the images is only half the experiment labelling the cells correctly in the first place is an important initial step in determining the physical association between different fluorophores. Correctly labelled cells will result in more reliable multi-labelling data, and may alleviate many of the problems associated with channel separation while collecting the images. As was discussed above, multi-probe fluorescent labelling traditionally requires a choice of fluorophore where the emission spectra of the two fluorophores is well separated. What is perhaps not as well appreciated is that you will need to "balance" the level of labelling of the two fluorophores to successfully analyse the degree of colocalisation. The "balance" that is important is the intensity of light detected in the two or more detection channels. This "balance" can be adjusted later while imaging by making changes to the gain, laser intensity etc of the individual detection channels, but a simple balancing of the amount of labelling for each channel initially may result in higher quality channel separation. In some circumstances the separation of the fluorescence emission of the two fluorophores may be improved by increasing the level of label in one particular channel. For example, in Figure 10-1 the emission spectra of FITC (fluorescein) and TMR (tetramethyl rhodamine) are shown to be of equal intensity. If the two collection channels shown in this figure are used (Channel 1, a narrow band of light between 490 and 550 nm, and Channel 2 a wide collection optical filter that results in all light longer than 570 being directed to this channel) the light intensity directed to each of the two channels would be similar. This would result in a small, but possibly significant, amount of FITC fluorescence being detected in the TMR channel. The amount of bleed-through would, in this case, be diminished by decreasing the level of labelling in Channel 1 (FITC) compared to Channel 2 (TMR). The degree of labelling for each of the

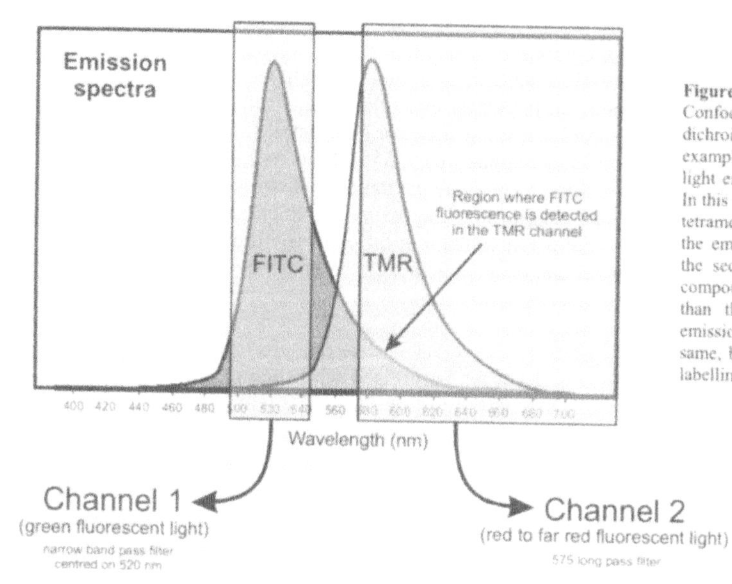

Figure 10-1. Separation of FITC and TMR Emission. Confocal microscopes use either a combination of dichroic mirrors and optical filters (as used in this example) or spectral separation to direct the fluorescent light emitted from the specimen into separate channels. In this example fluorescein (FITC) is also detected in the tetramethylrhodamine (TMR) channel due to overlap in the emission spectra. This is called bleed-through into the second channel. The problem of bleed-through is compounded if the FITC emission is significantly higher than the TMR emission. In this example the two emission spectra have been made approximately the same, but this will not necessarily be the case in a real labelling experiment.

channels can be readily adjusted by suitable dilution of the relevant dye or antibody-fluorophore complex. Alternatively, when fluorescence immunolabelling, the antibody-fluorophore combination that gives the highest level of labelling (for example the antibody with the highest binding affinity) can be used in the channel that requires the highest light level. The fluorescent probes suitable for multi-colour labelling are discussed in more detail in Chapter 9 "Fluorescent Probes" (page 201).

Multi-Channel Collection

Confocal microscopes are capable of collecting images from the fluorescence emission of several different fluorophores. The images may be collected simultaneously, or sequentially, and are often displayed as multi-coloured images (Figure 10-4 and Figure 10-5), with individual image colours (in this case red and green) representing separate collection channels.

In most confocal microscopes, multi-labelling experiments are carried out by directing the emitted fluorescent light from different regions of the light spectrum to separate photo-detectors (Figure 10-2 and Figure 10-1). This may be achieved using conventional optics (dichroic mirrors and optical filters), or by using spectral separation of the fluorescent light using either a prism or a diffraction grating. Some instruments are capable of separating the emission from multiple fluorophores by emission fingerprinting as discussed above. The confocal microscope hardware used to separate and detect multiple wavelengths is discussed in detail in Chapter 3 "Confocal Microscopy Hardware" (page 65).

The images obtained when imaging multiple labelled specimens, although often collected simultaneously, are obtained as separate grey-scale images that can be readily merged to create multi-colour composite images as required (Figure 10-2). Although the images may be displayed as coloured images on the screen during collection this colour is not part of the original data collection (a colour Look Up Table, LUT, is used to colour the image). It is advisable to display the individual channels during collection in the original grey-scale format to facilitate correctly balancing the level of fluorescence in the two channels (or use the special LUTs, such as SETCOL and GlowOver, to set the upper and lower limits of collection while at the same time using a grey-scale display to collect the image).

During collection of the images in dual or triple labelling it is important to correctly balance the light intensity of the image in the individual channels by adjusting the gain and black levels on the confocal microscope. There is usually no inherent reason why the level of labelling in the two channels should be the same biologically; although, as explained above, interpretation of multiple labelled samples is greatly facilitated if the two or three channels are approximately balanced in labelling intensity. If the intensity level in one channel is significantly different to the intensity in the other channel then you may come to the wrong conclusion in regards to colocalisation (see below).

Fluorochromes that have suitably separated emission spectra, and can be readily excited by the laser lines available, are ideal for multiple labelling applications. An important exception to this rule, as mentioned above, is when imaging with the META channel on the Zeiss confocal microscope or by spectral fingerprinting on the Leica SP confocal microscope. (see "Appendix 1: Confocal Microscopes", page 355). Fluorescent probes that may be not suitable for dual labelling experiments because they exhibit very broad excitation and emission spectra (for example the cellular dye acridine orange) can be readily used in single labelling experiments on all confocal microscopes. Multi-channel imaging to separate different regions of the cell containing different environments that influence the emission spectrum of the dye, such as acridine orange, can only be achieved by using spectral separation methods such as spectral fingerprinting.

Minimising Cross-Talk between Channels

The correct experimental controls (Figure 10-2) should always be utilised when doing multiple labelling with more than one fluorophore. The two important controls are performing the labelling without added fluorochrome (background control), and labelling the sample with each of the fluorophores separately (bleed-through controls).

The background control is used to set the limit of signal amplification that can be used before background fluorescence becomes a problem (determined for each channel). In dual or triple labelling it is most important to establish the level of background autofluorescence in each of the channels individually as there can be a significant difference in autofluorescence in each of the channels. As a general rule, autofluorescence is normally more prevalent when using shorter wavelengths for excitation (i.e. when excited by using the 488 nm blue line) rather than using wavelengths closer to the red end of the spectrum (for example, the 647 nm red line of the Kr/Ar laser).

The bleed-through controls are used to establish the amount of amplification possible in each of the channels without undue influence from the other fluorochrome. A small amount of bleed-through is readily accommodated by subtracting a percentage of one collection channel from the other. This can be done during collection by using the mixers available under LaserSharp on Bio-Rad instruments. A large amount of bleed-through cannot be readily handled by a mixer (in which case it may not be possible to know exactly how much to subtract when using the mixer). However, if the physical location of the label in the two channels is distinctly separate, then it is possible to perform very large subtractions and still get meaningful information from the dual labelling.

A much more effective method of lowering the amount of bleed-through between individual collection channels is to collect each channel separately using individual laser lines for each channel. For example, the fluorescence emission from FITC and TMR can be collected in separated collection channels by first collecting Channel 1 (Figure 10-3, panel 1) and then switching to Channel 2 to collect the second image (Figure 10-3, panel 2). Using only a single laser line each time to excite the fluorophore minimises the amount of fluorescence emitted from the second fluorescent probe. Separate channel collection may be further enhanced by altering the window of the light spectrum directed to each individual collection channel. For example, in Figure 10-3, the narrow band pass filter used to direct the FITC fluorescence to channel 1 (panel 1), to eliminate bleed-through from the second channel, could be replaced by a long-pass filter to collect a broader range of the FITC emission spectrum. If only 488 nm blue light is used to irradiate the sample there is almost no excitation of the longer wavelength TMR dye. In this way the sensitivity of the instrument for the collection of the FITC channel may be significantly increased.

Sequential collection of individual collection channels as described above is fine on fixed samples where there is unlikely to be any movement of the sample between images collected. However, when collecting multi-channel images from live cell samples a small amount of cell or organelle movement will decrease the degree of accuracy in determining the level of colocalisation. Fortunately, a number of confocal microscopes are capable of fast electronic switching between individual detection channels (multi-tracking). This includes the ability to quickly switch between individual laser lines. Automated switching between collection channels can be used to collect individual channels sequentially, but the real power of this technique is in using fast electronic switching to collect each individual channel line by line. Channel switching when using alternate line collection is so fast that the dual channel image collection speed is similar to that obtained when collecting a single channel or dual channels simultaneously without fast electronic switching. The end effect of using fast electronic line-scan switching to collect images is that the advantage of minimising bleed-through by using separate channel collection, is combined with relatively fast image collection speeds.

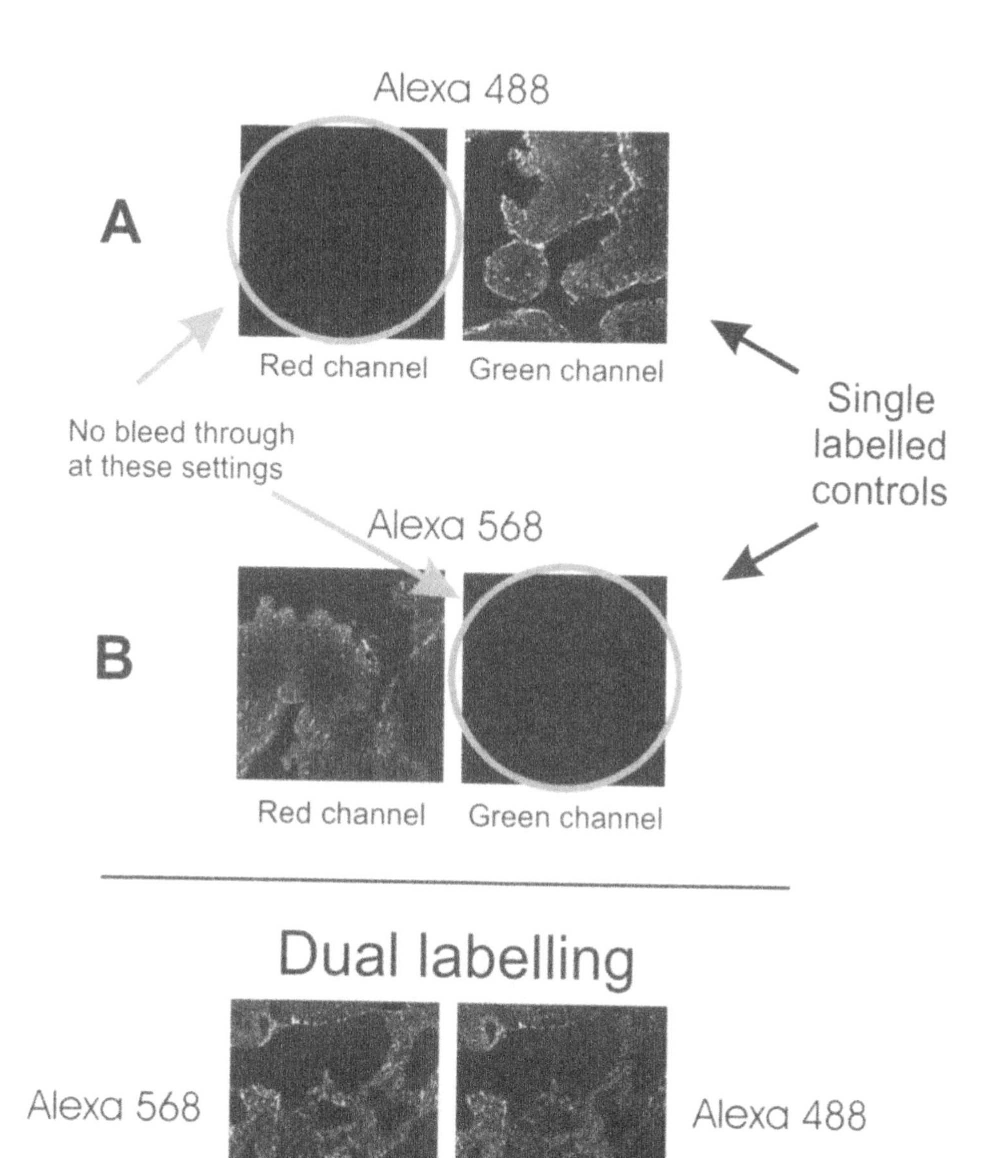

Figure 10-2. Dual Labelling.
Dual labelling normally requires the emission spectra of the two fluorophores to be sufficiently separated so that the light can be readily directed to separate light detectors. An important exception is the technique of spectral fingerprinting, using either the META channel on the Zeiss confocal microscope or the spectral imaging technique on the Leica SP and the Bio-Rad Rainbow confocal microscopes. When using separate detection channels it is important to establish whether each of the two fluorophores will be detected in the other channel (bleed-through) by the use of single labelled controls (A and B above), using the same instrument setting as used for dual labelling (lower panel). These images are cyrosections of human muscle fibres labelled with cytochrome *c* oxidase antibodies (against subunits I and IV) conjugated to Alexa Fluor dyes as indicated, and imaged using dual channel collection on a Bio-Rad MRC-1024 confocal microscope using both 488 nm blue and 568 nm yellow laser lines simultaneously.

Zeiss META, Leica SP & Bio-Rad Rainbow spectral fingerprinting can be used to separate fluorophores with highly overlapping fluorescence emission spectra

Conventional multi-channel image collection, as described above, requires that each individual fluorophore has a readily separated fluorescence emission spectrum. However, the technique of emission fingerprinting can be used to separate individual fluorophores that have highly overlapping fluorescence emission spectra. The Zeiss META confocal microscope can collect the emission fingerprint of a number of fluorophores present in the sample by using a multi-PMT array light detector to collect the complex emission spectrum from the sample, and using pre-stored data or data collected from defined regions of the sample, to calculate the degree of fluorescence emission that is due to each of the individual fluorophores. In this way up to 6 individual fluorophores with very similar emission spectra can be separated. A process called linear unmixing is used to separate out the contribution of the individual fluorophores and to display each fluorophore as a separate image (or as a separate colour in a multi-coloured merged image). For further details see page 400 in "Appendix 1: Confocal Microscopes".

The Leica SP spectral scanning confocal microscope can also be used to determine the emission fingerprint of individual fluorophores within a multi-labelled sample. In this case the emission spectrum of the sample is obtained by a technique called lambda scanning. Although the Leica instrument has individual light detectors for each channel in the instrument, the emission spectrum of the sample can be obtained by sequential collection of a series of individual images from narrowly defined regions of the visible light spectrum. The process of identifying the individual fluorophores is similar to that used on the Zeiss META instrument; although the collection time will be delayed by the necessity to sequentially collect the individual points in the lambda scan (the Zeiss META confocal microscope can collect several points in the spectrum simultaneously using the multi-PMT array).

Spectral reassignment is also available using the Bio-Rad Rainbow control microscope (see page 363 in "Appendix 1: Confocal Microscopes"). The Rainbow instrument uses a similar algorithm to the Zeiss META confocal microscope, but is designed primarily for the separation of two or three channels with overlapping emission spectra.

Spectral fingerprinting is a very powerful technique for separating the fluorescence emission of several very similar fluorophores, but this technology is not a complete replacement for more conventional image collection. Spectral fingerprinting has a significantly lower level of sensitivity compared to conventional individual channel detection, and environmental influences on the dye may result in some difficulties in clearly identifying a particular dye in various positions throughout the cell.

Determining the Degree of Co-localisation

Colocalisation is widely used to determine the relationship between various macromolecules and subcellular structures. Unfortunately, macromolecules can only be co-localised within a cell within the resolution limit of a light microscope when using visible light microscopy. The limit of resolution of a light microscope (see page 50 in Chapter 2 "Understanding Microscopy") is not sufficient to determine whether the two fluorescent labels are attached to the same macromolecular complex, or are even within the same transport vesicle. Larger subcellular structures, such as the nucleus and cell cytoplasm are clearly separated, but many subcellular organelles are close to or below the limit of resolution of a light microscope (which is theoretically 0.1 µm when using 488nm blue light, but is more realistically approximately 0.2 to 0.5 µm under most experimental conditions). However, one should always keep in mind that sometimes the result of a multi-labelling experiment is very clear indeed – when there is no colocalisation as shown in Figure 10-4. In this example immunolabelled structures (the rhoptry, shown in green), located towards one end of the small merozoite of the malarial parasite, are distinctly separate from the nucleus of the merozoite (staining strongly with the DNA stain propidium iodide and shown in red in Figure 10-4). The small amount of overlap between individual "red" and "green" labelled regions of the parasite (see right hand panel in Figure 10-4) is most likely due to close juxtaposition of the two structures in 3D, rather than any direct physical interaction. It would be considered that this labelling experiment clearly demonstrates the separation of the rhoptry organelle from the nucleus of the malarial parasite merozoite.

High-speed Channel Switching (multi-tracking) – greatly reduces cross-talk between individual channels

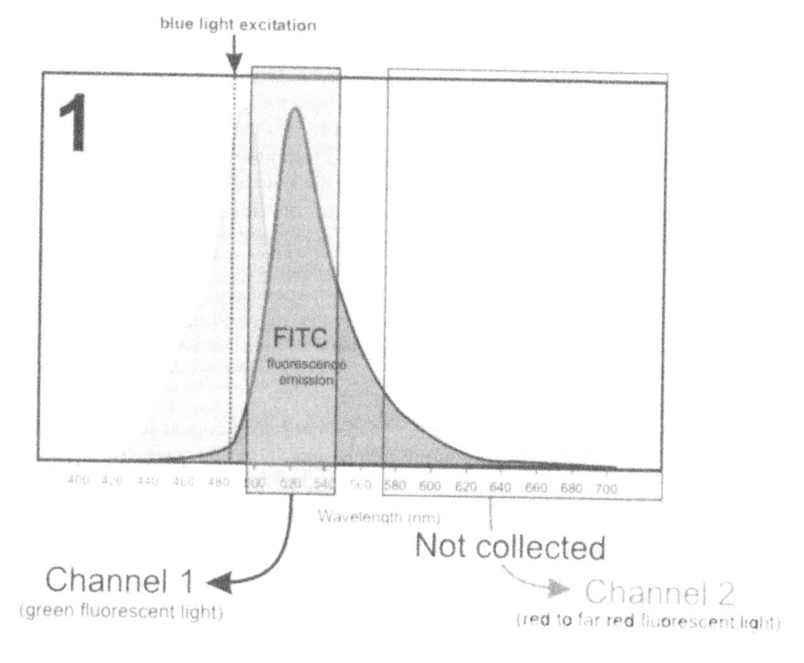

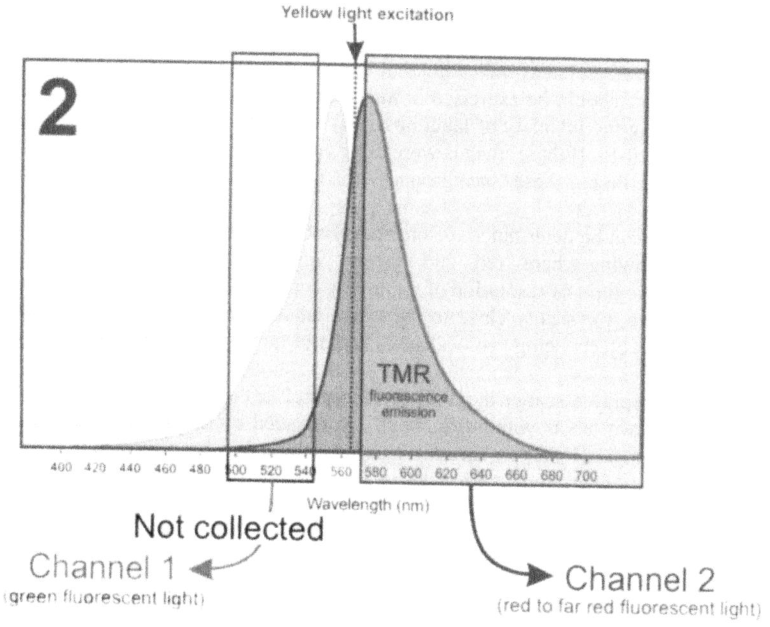

Figure 10-3. Sequential Excitation and Collection.
A simple and effective method of lowering the level of bleed-through in multi-labelling experiments is to collect images (or lines) from each channel sequentially. In this example a single laser line (488nm blue light) in "1" above is used to excite FITC, and only one channel is used for collection (Channel 1). Once an image has been collected, a second laser line (568 nm yellow light) in "2" above is used to excite TMR and Channel 2 is used to collect the fluorescent light. Some confocal microscopes are capable of very fast laser and channel switching such that each line in the image can be excited by a different wavelength and collected into a separate channel sequentially at the normal line scanning speed of the microscope.

Colocalisation of two or more fluorescent labels (Figure 10-5), in contrast to clearly physically separated structures discussed above, is much more difficult to interpret. It is fraught with a number of difficulties that may result in incorrect conclusions as to the degree of colocalisation. As mentioned above, the close proximity of two fluorescent labels can only be determined within the limit of resolution of the light microscope. However, even within this limit there are a number of other factors that can impinge on the validity of the multi-labelled data obtained. In Figure 10-5 human muscle biopsy samples have been labelled with two different antibodies against two different subunits of the mitochondrially located cytochrome *c* oxidase complex. This colocalisation experiment was carried out to determine the level of cytochrome *c* oxidase present in individual muscle fibres, the number of mitochondria present and the balance between the individual cytochrome *c* oxidase subunits within individual mitochondria. All of this information was collected to help in determining the bioenergy state of individual muscle fibres in a range of individuals of different ages. The most striking result is that at first glance there appears to be good colocalisation of the two individual cytochrome *c* oxidase subunits (the majority of mitochondria appear "yellow" in Figure 10-5). Note that the "balance" between the two channels has been adjusted to give an overall level of approximately equal staining intensity in each channel (hence the majority of the mitochondria appear "yellow", which is an equal amount of red and green in the two separate channels).

On closer inspection (see the right hand panel in Figure 10-5) there does appear to be a much greater degree of complexity to this colocalisation experiment than is initially apparent. The first thing to note is that the individual mitochondria that are labelled by both subunits, display a "shadowing" of the labelling to one side of the mitochondria. As this "shadowing" effect is predominantly in the same direction and is associated with most of the mitochondria, the most likely explanation is that the instrument was not correctly aligned when these two images were collected. As both of the images in this dual labelling experiment were collected simultaneously, sideways shift in the alignment of the mitochondria in the merged images is most likely due to an optical effect rather than movement of the sample. This small off-set between the two images will result in a significantly lower amount of colocalisation if you were to analyse these images using computer algorithms to determine the degree of colocalisation. The other compounding result in this experiment is that if you look closely at the areas of the image between the brightly dual labelled mitochondria there is a further "network" of labelling that is predominantly green. At a glance this could easily be interpreted as indicating that these particular structures are staining with only one of the two antibodies. However, care should be exercised in interpreting these finely stained structures, as the original image was collected by balancing the intensity of label observed in the strongly staining mitochondria (hence their appearance as yellow in Figure 10-5). If these images were to be collected while optimising the balance between the two channels for the level of labelling of these "background" structures you may find that in fact they are labelled by both antibodies.

The degree of colocalisation can be determined to some degree by presenting the image as a multi-colour image, and interpreting "yellow" as showing where "red" and "green" within the two channels coincides. Of course, this can only be determined within the limit of resolution of the microscope as discussed above. If the red label is located in a separate, but small structure, physically close to the green label, then the object will appear "yellow". This would be interpreted as the two objects being "colocalised", when in fact they may reside in two physically distinct compartments.

Computer algorithms that display a scatter diagram of each pixel in both images can be used to determine the degree of colocalisation of the two labels without the "bias" introduced by having one channel visually dominate when using a merged colour image. However, careful visual inspection of the images will be required, otherwise "background" low level labelling of structures within the sample as shown in Figure 10-5, may result in the analysis giving you misleading results.

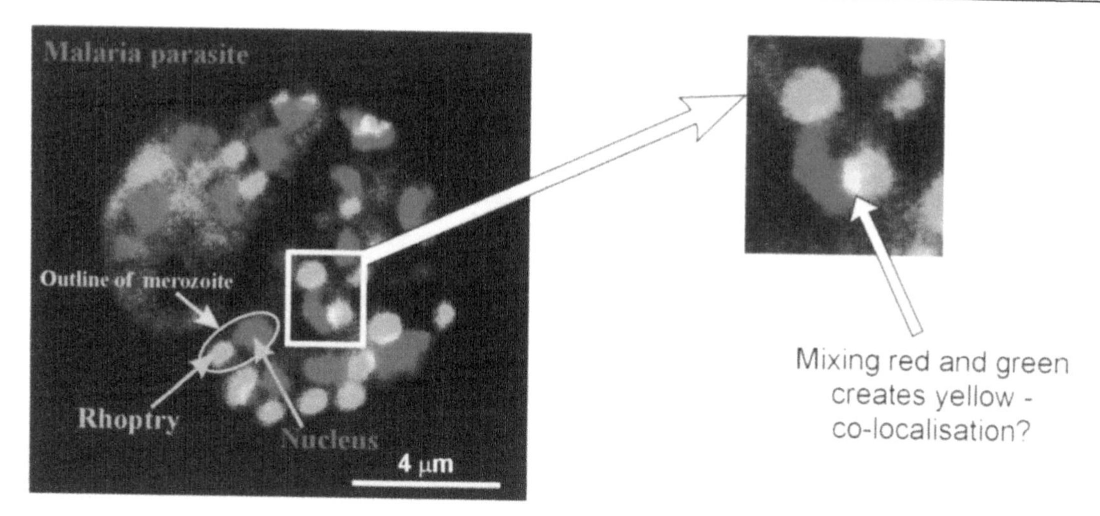

Figure 10-4. Distinct Separation of Dual Labels.
Distinct separation of two labels in a dual labelling experiment is clear evidence that the two fluorophores are not labelling the same compartment – this is self-evident, but is often forgotten in the effort to determine whether two closely associated molecules are co-localised. In this experiment human malarial parasites have been fixed and immunolabelled with an antibody against RAP1 (green) and stained with propidium iodide (red). The RAP1 antibody stains the rhoptry organelles located at one end of the malarial merozoite, and the propidium iodide predominantly stains the DNA in the nucleus of the merozoite. These objects are small, but clearly separated by confocal microscopy. The small amount of "yellow" overlap between the two labels (shown in the right hand panel) is not considered to indicate that these two labels are in overlapping compartments. This image was collected using a Bio-Rad MRC-600 confocal microscope with a Zeiss 63x 1.4 NA oil immersion objective.

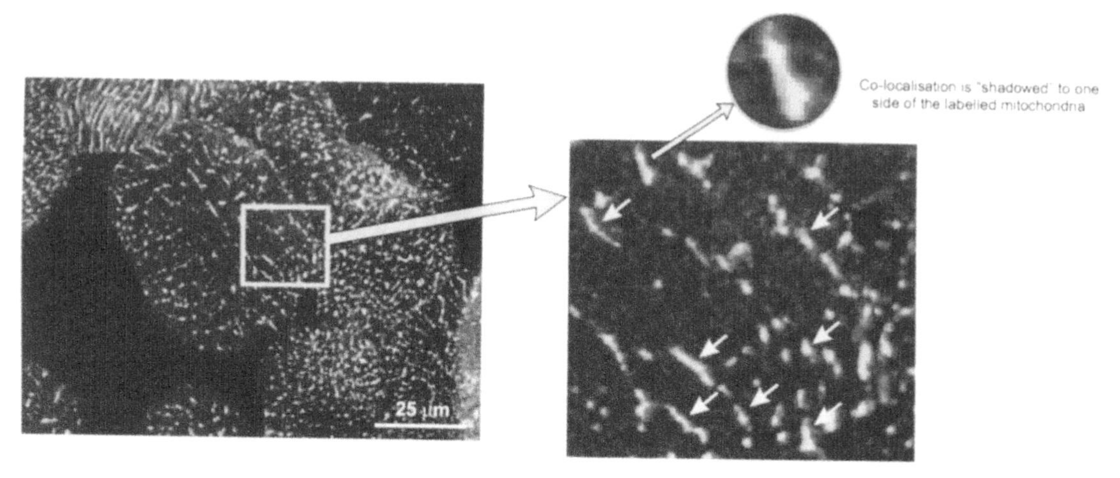

Figure 10-5. Co-localisation.
The dual labelled merged image on the left appears to have a high degree of co-localisation between two labelled subunits of cytochrome *c* oxidase (the mitochondria in the muscle fibre appear predominantly yellow), but on closer inspection (right panel) the co-localisation is quite complex. This includes both a high level of green label in the "background", and a slight horizontal shift in red verses green pixels (arrowed pixels in right hand image). This may be due to misalignment of the optics within the confocal microscope, but could also be due to optical aberrations introduced by the sample. These images are cyrosections of human muscle labelled with antibodies against cytochrome *c* oxidase subunits I and IV conjugated to Alexa Fluor 488 and 568 dyes respectively. This image was collected on a Bio-Rad MRC-1024 confocal microscope using dual excitation (488 and 568 nm light) and simultaneous dual channel collection with a Nikon 60x 1.4 NA oil immersion objective. The channels were carefully adjusted to minimize bleed-through between the two channels, and to "balance" the fluorescence intensity in each channel.

COMBINING FLUORESCENCE AND TRANSMISSION IMAGES

Grey-scale transmission images are often combined with single or multi-labelled fluorescence images to provide some orientation as to where the fluorescence is located. More accurate information may require a second fluorescent probe to identify the compartment of the cell that is labelled, but a good transmission image will certainly establish whether the fluorescence is nuclear, cytoplasmic or cell surface. The pattern of labelling in a fluorescence image may be quite confusing without the additional information provided by the transmission image. In fact it may be difficult to determine where the cells are located if rather sparsely labelled structures are all that is visible in the fluorescence image.

The software provided to control the confocal microscope will provide multi-coloured merged images (usually red and green, or red, green and blue) when performing multi-channel labelling. Some software will also provide a good grey-scale transmission image merged with the green or red fluorescence image. However, you may find it convenient to use general image processing programs such as Photoshop to create multi-coloured fluorescence and combined transmission images. This section describes two different methods of combining images using Photoshop. The first method is to paste the images into separate layers and then to adjust the colour and opacity of each layer independently before saving the image as an RGB full colour image. Alternatively, the grey-scale images from individual detection channels on the confocal microscope can be merged into excellent multi-coloured images by using "Channels".

Merging Images - Photoshop Layers

Multi-coloured merged images can be created in Photoshop using "layers". Directly pasting images into a new layer and adjusting the opacity will create a reasonable merged colour image, but to create an image with vibrant colour it will be necessary to remove all background from the image before pasting and then to adjust the colour balance. If you are attempting to merge two images to create a red/green merged image then use channels is a better way to merge the images (see following page).

Creating a Grey-scale Transmission Image with Green Fluorescence Overlay (Figure 10-6):

1. Open the grey-scale transmission image
2. Select the whole image (CTL-A) and copy (CTL-C) to the clipboard
3. Open a new file (the default size will now be the same as the copy on the clipboard)
4. Make sure this file is in RGB (on main menu select Image/Mode/ and make sure RGB is ticked)
5. Paste the transmission image into the new window.
6. Open the fluorescence image you want to be green (at this stage still a grey-scale image)
7. Select the "magic wand" tool and click on the background area of the fluorescence image (black area), with the "tolerance" set to 30 (this may need to be changed) and the "contiguous" box NOT ticked. Try with and without "Anti-aliased" and work with the best results.
8. Invert the selection (Select/Inverse, or Shift CLT-I), and copy to the clipboard (CTL-C)
9. Paste into the newly created window that contains the transmission image.
10. Open the "Layers" dialogue box to ensure the image has been pasted into a new layer.
11. Select the fluorescence image layer and adjust the colour balance (Image/Adjustments/Colour Balance, or CTL-B).
12. Adjust the colour balance of the shadows, midtones and highlights separately (preserving luminosity results in a brighter fluorescence layer). To create a green fluorescence image simply adjust the Green to 100% for the shadows, midtones and highlights, and set the other colours to 0%.
13. If required, adjust the colour balance to create any colour you would like for the fluorescence overlay.

The Result should be a vibrant green fluorescence image with a neutral grey transmission image background.

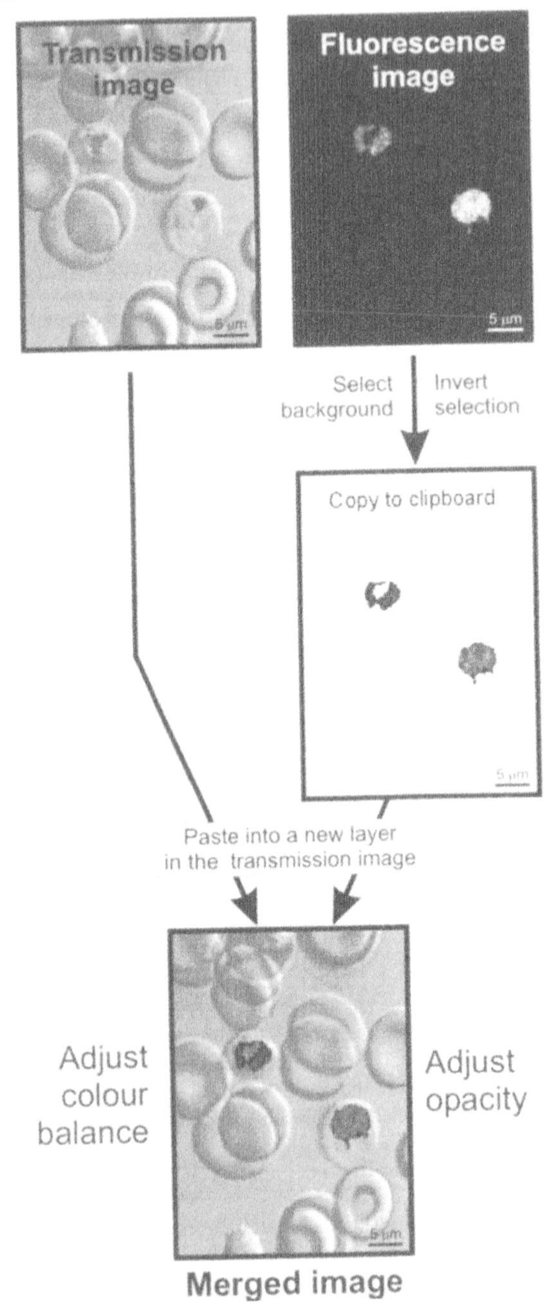

Figure 10-6. Merging Fluorescence and Transmission Images.
Information on the subcellular location of fluorescence can often be determined by merging the fluorescence image with a grey-scale transmission image (see image merging methods using Photoshop on pages 249 and 251). Both of these images were collected simultaneously on the confocal microscope, but saved as separate grey-scale images. Imaging software, such as Photoshop can be used to merge the fluorescence image with the grey-scale transmission image. The layer containing the fluorescence image can then be adjusted for opacity and colour. This image shows human red blood cells infected with malarial parasites which have accumulated Lucifer yellow. These images were collected using a Bio-Rad MRC-600 confocal microscope using a Zeiss 63x 1.4 NA oil immersion objective lens.

Merging Images - Photoshop Channels

The easiest way to create multi-coloured merged images using Photoshop is to copy and paste the images into different (red, green or blue) channels in an RGB image. If you are using only the red and green channels, then the third blue channel can be used for a grey-scale transmission image, to create a red/green dual labelled image with a grey-scale transmission image background.

Creating a dual coloured image (red / green merge):

1. Open grey-scale image one (the one you want to be green)
2. Select the whole image (CTL-A) and copy (CTL-C) to the clipboard
3. Open a new file (the default size will now be the same as the copy on the clipboard)
4. Make sure this file is in RGB (on main menu select Image/Mode/ and make sure RGB is ticked)
5. Open the "Channel" window (on main menu select Window/Show Channels) and click on the green channel in the pop up box.
6. Paste the image you want to be green into the green channel by pasting into the RGB image with the Green channel selected.
7. Open grey-scale image two (the one you want to be red)
8. Select the whole image (CTL-A) and copy (CTL-C) to the clipboard.
9. Paste this image into the red channel (this time select the red channel in the pop up box).
10. Now click on RGB in the pop up box and you should have a red/green image with a strong blue background.

Removing the blue background:

11. To remove the blue background click on the blue channel in the pop up box.
12. Go to the menu bar and choose Image/Adjust/Levels (or press CTL-L)
13. Now adjust the blue channel output to zero (output should be zero in both boxes, don't touch the input values, and don't change the sliders).
14. Alternatively select the blue channel in the pop up box, then select the whole image (CTL-A) and press delete. If the background is set to black (100% K) the blue channel will not contribute to the RGB image.
15. You can now click on the RGB channel of your image (in the layers/channel pop up box) and it should be a rich green and red colour with no blue present.
16. The image can now be saved as an RGB TIF image.

Changing the red and green colours of your image:

17. If you wish to change the red or green colours in your image the best way is to select RGB in the channels pop up box and then on the main menu select Image/Adjust/Selective colour.
18. In this way it is possible to not only change the green or red to another colour, but to also change the regions of mixed red/green colour to create an entirely different colour than yellow where the two images overlap.

Adding a grey-scale transmission image:

19. Open the grey-scale transmission image.
20. Select the whole image (CTL-A) and copy to the clipboard (CTL-C).
21. Select the merged Red/Green image.
22. Select the blue channel in the pop up box.
23. Paste the grey transmission image into the image (this should result in a red/green image with a strong blue transmission image).
24. On the main menu use Image/Adjust/Selective colour. Choose "Blue" and move the Cyan slider to –100% and the Magenta slider to –100% (leave Yellow as is).
25. You should now have a red/green image overlaying a neutral grey-scale transmission image.
26. You can change the colours in the image by selecting RGB in the channels pop up menu box and then on the main menu selecting Image/Adjust/Selective colour – as described for the red/green image above.

The Result should be excellent red / green merged images overlaying a neutral grey transmission image.

FRAP, FLUORESCENCE RECOVERY AFTER PHOTOBLEACHING

Fluorescence Recovery After Photobleaching (FRAP) and the related technique, Fluorescence Loss In Photobleaching (FLIP) are powerful tools for studying the molecular dynamics and connectivity of cellular membranes and subcellular compartments. This section briefly describes how FRAP experiments are carried out, and how they can be used to determine macromolecular movement. These techniques are described in much greater detail in the review by Klonis et al (*"Fluorescence Photobleaching Analysis of the Study of Cellular Dynamics"*, Klonis et al 2002, European Biophysics Journal, 31(1):36-51, see page 350 in Chapter 15 "Further Reading").

FRAP - for measuring molecular dynamics

Proteins attached to or embedded into biological membranes will have a mobility that is determined by the fluidity of the particular membrane, and their association with other proteins, particularly cytoskeleton proteins. Fluorescence FRAP is a technique where a defined area of a membrane or cell that has been tagged with a fluorescent probe is photo-bleached by exposure to high levels of laser light for a relatively short period of time. An image of the cell containing the area to be bleached is collected before bleaching and at suitable time periods during which the recovery of fluorescence into the bleached area is followed (Figure 10-7). Image

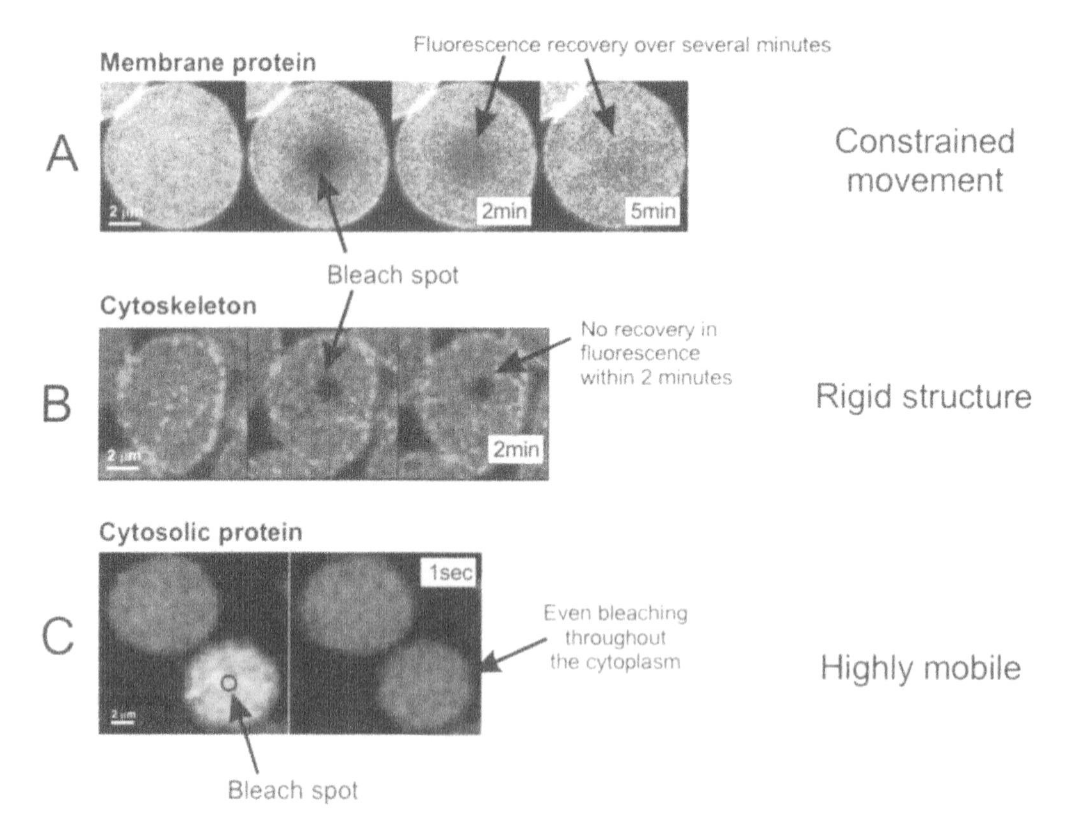

Figure 10-7. FRAP (Fluorescence Recovery after Photobleaching).
This technique allows one to study the degree of movement and thus the molecular environment of your protein or structure of interest. In the above example a membrane protein (fluorescein-glycophorin) in erythrocyte ghosts (A) is shown to diffuse back into the bleached area over a period of several minutes. The rigidly anchored cytoskeleton of the erythrocyte, stained using rhodamine-phalloidin (B) is shown to be relatively stable over a period of several minutes. Finally, a cytosolic protein (FITC-casein in resealed erythrocytes) shown in (C) is shown to relocate throughout the cytoplasm within the time frame of bleaching and collecting the first image. These images were kindly provided by Nick Klonis, La Trobe University, Australia. The images were collected using a Leica SP confocal microscope using a Leica 63x 1.4 NA objective lens.

collection is performed at a much lower level of laser irradiation, compared to the high level used for photobleaching. This minimizes any photo-damage or bleaching during the process of acquiring the images.

FRAP can be performed on all laser scanning confocal microscopes by simply focusing either the parked beam at a specific location within the cell, or scanning a small box size at high laser intensity. Bleaching may be carried out for a fraction of a second or for several seconds, depending on the mobility of the macromolecule being studied. A series of images at much lower laser intensity is then collected and the intensity changes within the bleached area can be used to calculate the mobility of the molecule under study. A number of confocal microscopes provide specialised modules within their software for FRAP studies. These dedicated modules will automatically take an initial image, bleach a defined spot or area and then collect a series of post-bleach images.

Confocal microscopes with AOTF filters for adjusting the intensity of the laser light that reaches the sample are particularly suited to studies involving FRAP photobleaching. When using an AOTF filter one can readily alter the laser intensity (between zero and 100% of full laser power), very quickly, if necessary, and yet return the laser light very accurately to the previous intensity. In this way the pre-bleach and post-bleach images can be collected with exactly the same laser intensity. Confocal microscopes that use neutral density filters can also be used for FRAP studies, but it may not be possible to change between the high intensity laser light used for photobleaching and the much lower intensity light used for collecting the images fast enough to follow molecules that are diffusing relatively quickly.

The time series of images collected after photobleaching can be used to determine the half-life of fluorescence recovery (in a period of seconds to minutes for biological membranes), and the speed of fluorescence recovery, by following the rate of movement of fluorescently tagged molecules from the surrounding membrane. Fluorescence recovery can be used to determine the fluidity of the membrane, or the movement of the macromolecule in question. In Figure 10-7 human erythrocytes have been labelled with either a membrane protein (A), a cytoskeleton dye (B) or a cytosolic protein (C). The membrane protein (fluorescein-glycophorin) is shown to be relatively mobile as it diffuses back into the bleached area over a period of a few minutes (compare the bleach spot at "0" min compared with the 2 or 5 minute image). The cytoskeleton of the erythrocyte, labelled with rhodamine-phalloidin is shown to be quite stable over a period of two minutes (B), whereas the cytosolic protein FITC-casein that has been loaded into erythrocytes is shown to be so mobile that the whole internal area of the erythrocyte is evenly bleached within the period of approximately one second required for the bleach and the collection of the first image.

FRAP can be used to study the movement of any fluorescently labelled protein or macromolecule, including GFP constructs. This technique provides valuable information on the molecular environment and thus mobility of macromolecules in living cells. However, care should be taken when interpreting the results obtained by FRAP as there is considerable cellular damage during the bleaching process. The damage is not only at the focussed "spot" that is defined in the two dimensional image shown on the screen, but also throughout the sample in three dimensions (the confocal microscope images only the focal plane, but the full depth of the sample is irradiated with the laser light). Cellular or subcellular movement will also detract from the reliability of the FRAP study. If a diffraction limited spot is used for photobleaching, even very small movements in the cell will seriously affect the level of fluorescence within the defined bleach area in subsequent images. A simultaneously collected transmission image, particularly a DIC transmission image, can be used to correct for any cellular movement before analysing the level of fluorescence intensity within the bleached area during recovery. Another potential problem area encountered when attempting to follow fluorescence recovery after photobleaching - is when photobleaching occurs while acquiring the images. A variety of approaches can be employed to minimise problems of photobleaching while collecting images during photo recovery, including using relatively photo stable probes, using a low laser intensity to collect the images and increasing the scan speed (deceasing the pixel dwell time). A correction can be made for photobleaching, if necessary, by using a region of the cell that is remote from, and unaffected by the photo bleached spot.

The related technique, called FLIP (Fluorescence Loss in Photobleaching) can be used to study the connectivity between cellular compartments. In this technique a focussed spot of laser light is directed onto a specific subcellular structure, and will eventually result in the loss of fluorescence from other regions of the cell that are physically connected to the compartment or structure that is being irradiated. In this way the connections between cellular membranes and subcellular organelles can be experimentally established.

FLIP - Fluorescence Loss In Photobleaching – for measuring connectivity of cellular compartments

FRET, FLUORESCENCE RESONANCE ENERGY TRANSFER

Fluorescence Resonance Energy Transfer (FRET) is a technique that allows you to establish molecular interactions that would otherwise be well beyond the resolution of a light microscope. FRET is due to intermolecular transfer of energy between two fluorophores that are located within relatively close molecular distances from each other. Confocal microscopy can be used to establish the presence of FRET within cell and tissue samples using a variety of FRET pairs – including GFP combinations such as CFP/YFP - to study protein interactions within living cells. Single cell biochemistry and FRET are described in more detail in the excellent review from Philippe Bastian's laboratory (*"Imaging biochemistry inside cells"*, Wouters et al 2001, TRENDS in Cell Biology 11(5):203-211, see page 350 in Chapter 15 "Further Reading").

FRET
Used to determine molecular proximity

FRET occurs when two fluorochromes, which are closely associated with each other, can interact such that the light used to excite one will be transferred as internal energy to the other fluorochrome, and then emitted as longer wavelength light characteristic of the second fluorochrome (Figure 10-9). For energy transfer to occur the fluorescence emission spectrum of the first fluorophore (donor) must overlap with the absorption spectrum of the second fluorophore (acceptor) (Figure 10-8), although the energy transfer is internal rather than the re-absorption of a donor emitted photon.

For FRET to occur the two molecules must be within approximately 2 to 6 nm. This means that the molecules have to be within molecular distances from each other for energy transfer to occur. The Förster radius is the distance at which the energy transfer is 50% efficient (i.e. 50% of molecules transfer their energy to the adjacent dye). This radius can be calculated for any two fluorochrome pairs, but the exact value observed in specific experiments will vary as environmental factors influence the efficiency of the energy transfer.

FRET can be measured using either intensity-based detection methods or fluorescence decay-kinetics based methods. Intensity-based imaging methods for detecting FRET are by detecting either the increased emission of longer wavelength light from the acceptor molecule, or by the quenching of the shorter wavelength light of the donor molecule. When using "ideal" FRET pairs in a biological assay (for example, when two fluorophores are brought into close proximity by attachment to a short peptide, which may be later cleaved within the cell and thus physically separate the two molecules – resulting in a loss of FRET) the changes in acceptor emission and donor quenching can be readily followed by fluorescence microscopy. However, using FRET in biological systems to determine molecular interactions is fraught with a number of difficulties. One of the most difficult problems is that dynamic molecular interactions may result in varying levels of FRET within the cell, and will rarely be any where near 100% energy transfer.

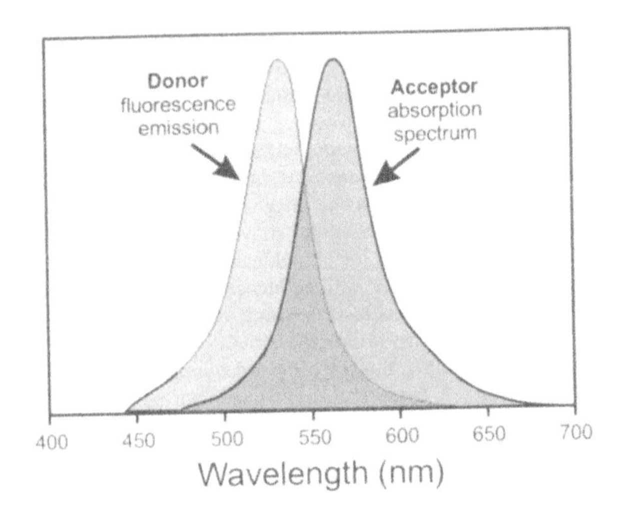

Figure 10-8. Spectral Requirements for FRET.
For FRET to occur the fluorescence spectrum of the donor molecule must overlap with the absorption spectrum of the acceptor molecule. The more the emission and excitation spectra of the two fluorophores overlap the greater the amount of FRET that is likely to occur. The light energy absorbed by the donor is transferred through intramolecular interactions to the acceptor molecule – the acceptor is not being excited by light emitted by the donor.

There are a number of methods of improving the reliability of FRET measurements in biological systems, only some of which are applicable to confocal microscopy. One of the best ways of accurately measuring the level of FRET occurring between two fluorescently labelled macromolecules is by determining the life-time of the fluorescence (FLIM – Fluorescence Life Time Microscopy). This technique takes advantage of the change in the life time of the fluorescence of the donor, depending on the amount of FRET that is occurring between the two fluorophores. However, this technique does require specialised instrumentation.

A technique for demonstrating FRET in biological systems that is readily available when using a confocal microscope is the photobleaching method. In this method a fluorescence image is taken of the sample using collection channels for both the donor and acceptor fluorescence emission (the pre-bleach image). A defined area of the sample (a small boxed area rather than a point) is then irradiated with a greatly increased laser intensity to bleach either the donor (using the shorter wavelength laser light that is optimal for exciting the donor) or the acceptor (using the longer wavelength laser light). A second dual channel image is then collected using the same zoom and laser intensity levels as were used in the original pre-bleach image collection. If acceptor bleaching has been carried out, then an increase in the donor fluorescence intensity after photobleaching the acceptor is indicative of the amount of FRET that is occurring between the two FRET pairs.

Experimentally, a number of fluorescent dye combinations work well as FRET pairs (FITC/Lissamine rhodamine, Alexa 488/Lissamine rhodamine, Cy3/Cy5, CFP/YFP etc). The Alexa dyes are ideal for the non-bleached fluorophore in FRET studies. A combination of Alexa 488 and Alexa 546 does result in a high level of FRET, but the difficulty associated with bleaching the Alexa Fluor dyes mean that one of the FRET pairs may need to be a fluorophore that is more easily photo bleached.

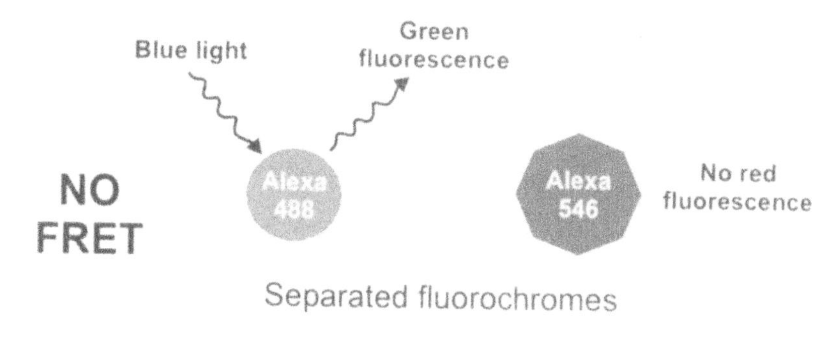

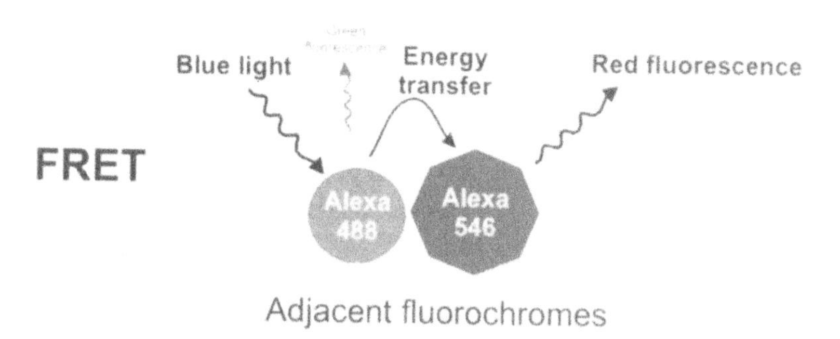

Figure 10-9. FRET (Fluorescence Resonance Energy Transfer).
For FRET to occur the two fluorochromes must be within 10 to 100 Å of each other, and have overlapping fluorescence emission and excitation spectra (see Figure 10-8) – although the energy from the absorbed light is transferred directly to the acceptor molecule and is not due to the emission of a donor photon being absorbed by the acceptor molecule. FRET occurs only when the donor and acceptor are in very close proximity (within molecular distances) and so can be used as a very sensitive measure of molecular associations at a resolution well below that which can be obtained by structural studies in light microscopy.

FISH, FLUORESCENCE IN SITU HYBRIDISATION

Fluorescence *in situ* hybridisation (FISH) is a method by which genes can be mapped using fluorescent probes. The genes can be mapped to specific chromosomes by using a specific oligonucleotide. A series of genes can also be mapped relative to each other on the one chromosome by the use of unique oligonucleotides and differently coloured fluorochromes. *In situ* fluorescence PCR can be performed to detect singly copy genes on a chromosome. Confocal microscopy can be readily used to image FISH labelled chromosomes, but as the chromosome preparation is relatively thin there is little advantage in using a confocal microscope. A conventional epi-fluorescence microscope with suitable optical filter blocks is not only sufficient for analysing FISH labelling, but perhaps significantly easier.

LINE SCANNING

The speed of image acquisition in a confocal microscope is determined by the speed of the scanning mirrors within the scan head. In turn the speed at which the mirrors can scan is determined by the amount of light that can be accumulated in each pixel within the image. In other words, if the scan rate is too fast then there may be no signal collected for most of the pixels. There is clearly an upper limit on the speed at which these scanning mirrors can operate.

One way of speeding up the rate of data acquisition without speeding up the scan speed, is to scan a single line across the image and then plot this line as a function of time. The x-x scan rate is very fast in the confocal microscope (approximately 4 msec per scan, depending on the x-scan speed of the instrument). If the confocal microscope is directed to simply scan a single line then the rate of data acquisition across that line is very fast indeed. If the lines are plotted below each other on the screen, an image is built up which consists of a very fine slice across the cell in the x-axis and information on the amount of fluorescence with time in the y-axis. In this way it is possible to image calcium pulses in live muscle cells in real time on a relatively slow scan speed laser spot scanning confocal microscope.

QUANTITATIVE CONFOCAL MICROSCOPY

Quantitation of data obtained by confocal microscopy is not a simple matter of measuring the value of the pixels within the image of interest. Quantitation of the image starts back with the preparation of the sample, continues with the method of labelling and finally relies on the way in which the image is collected. If all these parameters can be controlled then there is excellent software available that will readily give one a quantitative value on the amount of fluorescence in a particular sample or within a particular structure (discussed in detail on page 160 in Chapter 5 "Digital Images in Microscopy").

The variables associated with sample preparation and labelling are best controlled when the comparative regions of interest are within the same sample. For example, one can readily compare the amount of fluorescence associated with the plasma membrane to that associated with the endoplasmic reticulum - as long as you have suitable probes for such labelling. However, a word of caution – the optical slice capability of a confocal microscope seriously complicates accurate quantitation. A single optical slice may give a highly misleading level of label in particular structures or compartments simply due to the very fine optical slice obtained. A series of optical sections will improve the accuracy of quantitative confocal microscopy, although care needs to be taken not to unduly bleach the sample when collecting a large number of optical slices. A compromise may be to collect a relatively small number of optical slices in the region of the structure for which you want quantitative data.

Attempting to quantify the level of fluorescence between samples, and particularly on different days, is very difficult. Not only do you need to contend with samples that can be very easily labelled to a different extent (or problems associated with the degree of washing in the case of immunolabelling), but there is also the problem of laser stability and making certain that the settings on the confocal microscope are identical to those used previously.

Although quantitative confocal microscopy is very difficult, one can often obtain useful semi-quantitative data when you are mainly interested in large changes in labelling. The best way to compare samples is to label and image them all on the same day. In this way a number of the problems of laser stability and microscope settings are diminished.

INFORMATION ON CONFOCAL MICROSCOPY TECHNIQUES

Bio-Rad Application and Technical Notes

Bio-Rad Application Notes:

http://cellscience.bio-rad.com/reference/application.htm

1 Confocal microscopy in the food industry. R. Jeacocke.
2 Non-isotopic in situ hybridisation. Donna Albertson.
3 Imaging intracellular calcium using the confocal microscope. Linda McNaughton et al.
4 Visualization and localisation of DNA replication sites. Tony Mills et al.
5 The krypton-argon-ion mixed gas laser. Anna Smallcombe.
6 Ultra sensitive imaging using Fast Photon counting on the MRC-600. Andrew Dixon.
7 Three colour confocal imaging. Steve Paddock.
8 Ratiometric confocal calcium measurements. P. Lipp and E. Niggli. *(fluo-3 and Fura red combined).*
9 Localisation of mitochondrial antigens in biliary epithelial cells revealed by confocal laser scanning microscopy. Ruth Joplin and Gerald Johnson.
10 Immunocytochemistry and confocal microscopy: two complementary tools to understand the morphological diversity in ciliates. Anne Fleury et al.
11 Striated muscle: 3D structure of cellular membrane systems using confocal microscopy. AP and AD Somlyo et al.
12 Immunofluorescent localisation of cardiac gap junctions: Rob Gourdie et al.
14 Non-fading high sensitivity confocal fluorescence imaging for in situ hybridisation & immunocytochemistry. E-J Speel et al.
15 Nucleic Calcium Oscillations in Mouse Oocytes. A Pesty and B Lefevre.
16 Double Label Immunofluorescence using Cyanine Dyes and Laser Scanning Confocal Microscopy. P Sargent.
17 Using GFP to study viral invasion and spread throughout plant tissues. A Roberts, DAM Prior and A Smallcombe.
18 Triple or quadruple labelling of cultured neurones for multi-colour fluorescence imaging: Triple labelling part 1. Anna Smallcombe.
19 Triple labelling immunofluorescence and confocal microscopy: Triple labelling part 2. Alan Entwistle and Anna Smallcombe.
20 MHC Class II transport in living cells: Confocal analysis of the GFP-tagged protein. L Oomen, et al.
21 Confocal Microscopy for Diagnostic Cytology. ME Boon and LP Kok.
22 Dynamic volume measurement of porcine chondrocytes in intact cartilage explants. R Erington and N White.
23 GFP as a fluorescent tag to monitor spindle dynamics in live oocytes and embryos of Drosophila by Laser scanning confocal microscopy. Sharyn Endow and DJ Komma.
24 Tubular extensions of the nuclear envelope found deep in many mammalian interphase nuclei. Fricker et al.
25 Applications of multi-photon fluorescence microscopy in cell and developmental biology using a mode-locked, all-solid-state Nd: YLF laser. John White.
26 Confocal reflectance and fluorescence imaging of post-capillary venules. Ping He.
27 Characterisation of developmental regulation of the cytoskeleton using confocal microscopy. Nicholas Harden and Louis Lim.
28 Multi-photon microscopy at the Bio-Rad Biological Microscopy Unit. Nick White.
29 Confocal Investigation into the influence of chondroitin sulphate on axon guidance in embryonic Xenopus brain. RB Anderson, A Walz and Alan W. Decho.
30 Application of confocal and multi-photon imaging to geomicrobiology. Tomohiro Kawaguchi and Alan W. Decho.
31 Non-invasive multi-photon imaging of embryonic development. Jayne M. Squirrell.
32 CLONIS - A New Tool for Micro engineering Cellular Organization and Interactions in Tissue Culture. M. Schindler.
33 CLONIS - Micro engineering and Sorting Cells in Culture and Microdissection of Tissue. M. Schindler.
34 Multi-photon microscopy in imaging amyloid-b plaques in living mice. Brian Bacskai.
35 Applications of atomic force microscopy in conjunction with confocal microscopy.
36 FRET and FRET-FLIM microscopy imaging of localized protein interactions in living cell nucleus. Ammasi Periasamy.
37 Fluorescence recovery after photobleaching (FRAP). Justin Wudel and Nobuaki Kikyo.

Bio-Rad Technical Notes:

http://cellscience.bio-rad.com/reference/technical.htm

1 Obtaining the best performance from your Krypton/Argon laser. Andy Sowerby and Graham Hogg.
2 Key sensitivity features of the MRC-1024. Duncan McMillan.
3 Multi-photon fluorescence microscopy from Bio-Rad. Andrew Dixon.
4 The use of AOTF to achieve high quality simultaneous multiple label imaging. Dino Sharma.
5 The benefits of non-descanned (external) detectors in multi-photon microscopy. Graham Brown.
6 The use of intensity windowing to give 2 X 8 bit data acquisition in confocal microscopy.
7 The advantages of femtosecond pulsed lasers in multi-photon microscopy. Andrew Dixon.
8 Co-localisation analysis with the LaserSharp software. Anna Smallcombe and Duncan McMillan.
9 Optimum optical design characteristics for confocal and multi-photon imaging systems. Brad Amos.
10 Fluorescence Bleed-through: How to reduce or avoid it. Anna Smallcombe.
11 Co-localisation: How is it determined, and how is it analysed with the Bio-Rad LaserPix image analysis software? Nick White and Rachel Errington.
12 Setting up and managing a biological laser scanning microscope resource. W.B. Amos and S. Reichelt.
13 Bio-Rad Signal Enhancing Lens System (SELS) - turbo mode for confocal. W.B. Amos and S. Reichelt.
14 FLIM. Eric Pierce.

Chapter 11

Fluorescence Immunolabelling

Antibodies bound to cells or tissue slices can be readily visualised by using confocal microscopy if they are "tagged" with suitable fluorescent probes. This process of fluorescence immunolabelling provides a highly specific way of identifying the subcellular location of either the molecule of interest, or of the organelle or subcellular structure where the molecule is known to be located. Single, dual and triple labelling is possible using confocal microscopy, with the fluorescence images often being combined with high resolution transmission images to provide a means of locating the cells or highlighting the tissue structure. Immunofluorescence antibody labelling can be used to visualise proteins in both fixed (Figure 11-1) and live cell (Figure 11-2) preparations. In this chapter the principles of fluorescence immunolabelling are described in some detail, with many practical hints on producing highly specific labelling with excellent sensitivity and low "background" fluorescence.

Why use a confocal microscope for immunolabelling?

- Increased sensitivity
- Better resolution
- 3D information
- Accurate dual/triple labelling
- Transmission & fluorescence combined
- Great for work with live cells

WHAT IS IMMUNOLABELLING?

Antibodies provide exquisite specificity for identifying molecules (usually, but not always proteins) of interest. There are a number of ways of visualising these high specificity antibodies by microscopy. One is to precipitate an optically opaque reagent (often DAB, diaminobenzidine) at the site of antibody binding. This method is routinely used in conventional light microscopy, and is particularly suited to thin tissue sections. The DAB precipitate can also be imaged with the confocal microscope using the transmission detectors.

A far better method of detecting antibodies by confocal microscopy is to use a fluorescent probe attached to the antibody. There are two simple methods that are widely used. One method is to use a fluorescently labelled secondary antibody that is directed against the antibody class of the first antibody (for example if the first antibody is a mouse monoclonal, the secondary antibody would be an anti-mouse IgG antibody - which is often produced in goats). The second method is to directly attach a fluorescent moiety (such as FITC) directly to the first antibody by chemical means. These immunofluorescence methods are commonly known as IFA (Immunofluorescence Antibody Assay) assays.

IFA - Immunofluorescence Antibody Assay

Using a Confocal Microscope for Immunolabelling

Conventional fluorescence microscopy has been used for many years with great success in fluorescence immunolabelling. In fact, the technology was developed to suit the limitations of the conventional fluorescence microscope. In particular, fixation has traditionally been aimed at creating maximum levels of high specificity sample fluorescence with a minimum of "background" labelling or autofluorescence. Some of the methods traditionally used in fluorescence immunolabelling may result in compromised structural integrity of the sample.

The increased resolution obtainable by using confocal microscopy when imaging fluorescently labelled cells and tissue samples has meant that cell or tissue morphology has now become a more important consideration. Furthermore, the very high sensitivity of the confocal microscope means that it is no longer as important to use a procedure that maximises the level of fluorescence obtained.

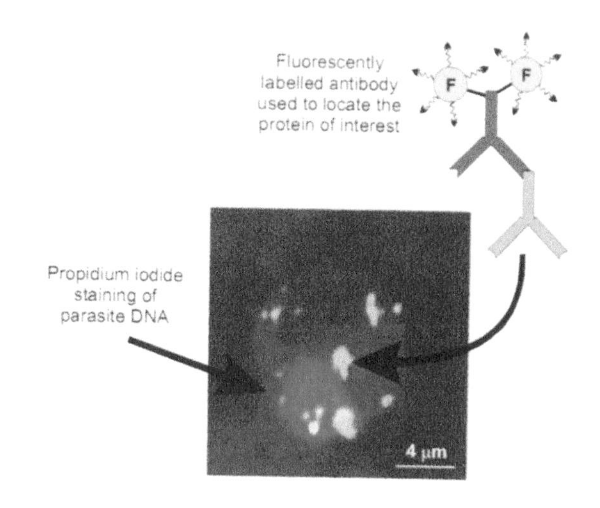

Figure 11-1. Immunofluorescence Labelling of Fixed Cells.
Human red blood cells containing malarial parasites were dried, fixed with acetone/methanol and then visualised using both a fluorescein labelled secondary antibody (indirect immunolabelling), shown in green, and propidium iodide (red) to visualise both the parasite DNA, parasite structures and to show a feint outline of the infected red blood cell (dual labelling with the added advantage of having an outline of the cell) This image was collected using a Bio-Rad MRC-600 confocal microscope with a Zeiss 63x 1.4 NA objective lens. Further information on the conditions used to collect image can be found in the publication Kun et al (1997) – see page 348 of Chapter 15 "Further Reading".

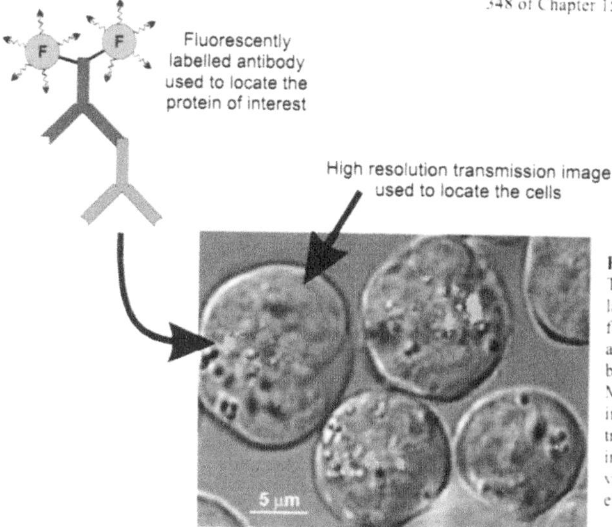

Figure 11-2. Immunofluorescence Labelling of Live Cells.
Transformed human T-lymphocytes (MOLT-4) cells were labelled with fluorescently labelled antibody against H-ferritin and then incubated in culture media for 60 minutes to allow the antibody H-ferritin receptor complex to be taken up by the cells. The live cells were imaged using a Bio-Rad MRC-600 confocal microscope with a Zeiss 63x 1.4 NA oil immersion objective lens. The simultaneously collected transmission image clearly shows the location of the cells, including considerable subcellular structure. A modified version of this image has been previously published in Moss et al (1994) – see page 349 of Chapter 15 "Further Reading".

The sensitivity of the confocal microscope is greatly increased when imaging fixed samples (most, but not all immunofluorescence labelling is done on fixed cells) due to the ability to collect a number of images as line or screen averages. This technique greatly lowers the background "noise", effectively increasing the degree of signal from fluorescently labelled structures. A large number of screen averages are possible when using fixed material as there is no movement of either the cells or subcellular structures, as long as the microscope itself is free of vibration. There is of-course a trade-off between the number of screen averages possible and the fading of the sample with the irradiating laser light, but with a number of newer dyes that show very low levels of fading, or with the use of suitable antifade reagents a large numbers of screen averages can be readily collected.

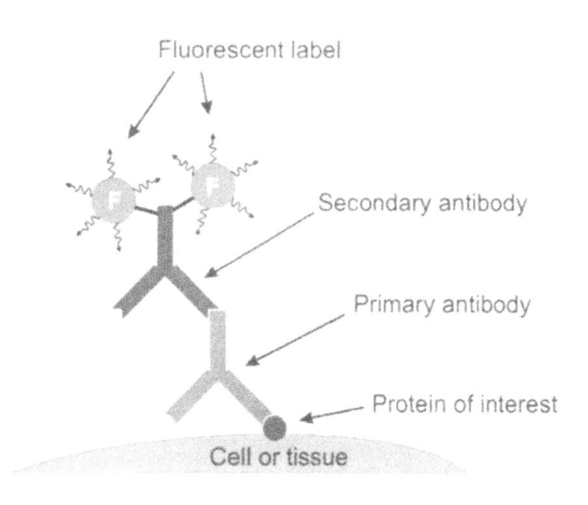

Figure 11-3. Indirect Immunolabelling.
The primary antibody (for example, a mouse monoclonal) is bound to the protein of interest. After washing to remove excess monoclonal antibody a fluorescently labelled secondary antibody directed against the primary antibody (in this example an anti-mouse antibody) is added.

Fluorescence immunolabelling can also be readily combined in dual and triple labelling experiments using a combination of several antibodies or combining with other molecular probes, for example, with propidium iodide for staining DNA (Figure 11-1), or with, for example, MitoTracker red for specifically staining mitochondria. Probes that stain specific cellular constituents after fixation (such as propidium iodide) can be used on fixed material. Other probes, such as MitoTracker red are normally used as probes for mitochondria in live cell preparations, and then the probes themselves can be fixed in place using aldehyde fixatives. Some probes, (such as CFDA, carboxyfluorescein diacetate, a cell tracer probe) are even fixed in place within the living cell, without the need to introduce any fixative. Probes that are not fixed in place either by common tissue fixatives or cellular metabolism and do not stain the organelle or molecule of interest in non-living cells, cannot be readily used on fixed tissue or cell samples.

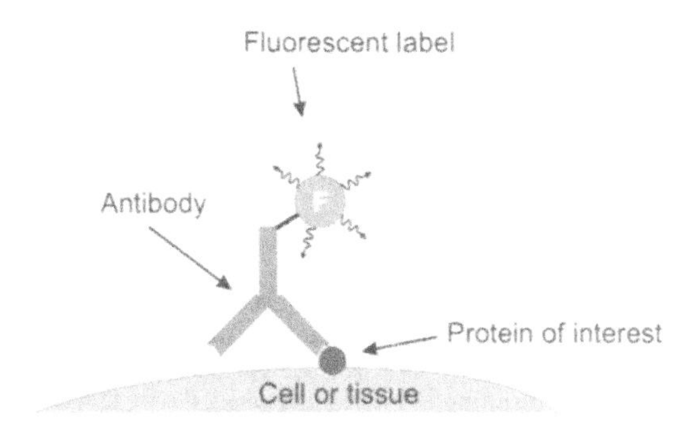

Figure 11-4. Direct Immunolabelling.
The antibody directed against your protein of interest can be directly labelled with a suitable fluorophore. This alleviates the need for a secondary antibody, but does have the disadvantage of requiring direct chemical labelling of each antibody.

LABELLING METHODS

Indirect Immunolabelling

The most common method of immunolabelling is to use a commercially available secondary antibody directed against your own antibody (Figure 11-3). This is a simple and very effective method for visualising antibodies located in cell or tissue samples. The commercially available secondary antibody can be readily tested on a sample that you know will show labelling, and then you can be sure that the secondary antibody will work on more difficult samples. Furthermore, the commercially available secondary antibodies have been produced with careful control of the number of fluorescent molecules attached, thus optimising the degree of fluorescence you would expect from immunolabelling.

Secondary antibodies are available with a variety of different fluorescent probes attached. For example,

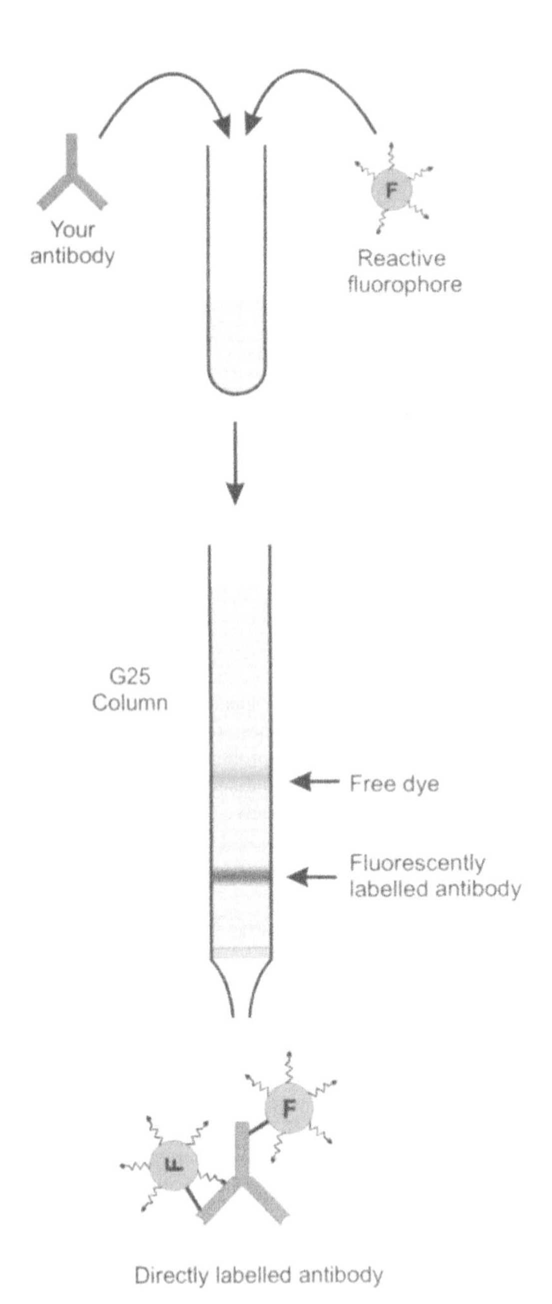

Figure 11-5. Labelling your own Antibody.
A solution of your antibody and reactive fluorophore are incubated together to fluorescently label your antibody. Free fluorophore is then separated from your labelled antibody by using a G25 size exclusion chromatography column.

FITC, TRITC, Lissamine rhodamine, Texas Red, Alexa 488 etc can all be purchased already attached to various secondary antibodies. Substituting with a different fluorescently labelled secondary antibody when a different colour of fluorescence is required is quite simple.

Secondary antibodies that are directed against primary antibodies that have been raised in a variety of organisms are readily available. For example, Goat-anti-mouse IgG (GAM) can be purchased if your primary antibody is a mouse monoclonal. Alternatively you can purchase Goat-anti-rabbit IgG (GAR) for use with a rabbit polyclonal antibody. The secondary antibody is highly specific for the IgG which the antibody is directed against, but as they are polyclonal antibodies they may contain antibodies against IgG antibodies for other species. This can be a problem if your sample contains binding sites for IgG antibodies. This problem is usually overcome by buying a more expensive antibody that has had unwanted antibodies removed. For example when working with human tissue or cells it is often advisable to use secondary antibodies that have been absorbed against human sera. Some companies, such as Molecular Probes, only sell antibodies that have been absorbed against a number of species to minimise cross-reactivity.

Direct Immunolabelling

Sometimes directly labelling your own antibody with a fluorescent molecule (Figure 11-4) may be necessary. Direct labelling is particularly important when attempting dual labelling with two different antibodies developed in the same species (for example two mouse monoclonal antibodies). If the secondary labelling antibody technique is used in this case, you will end up with some of the fluorescently labelled secondary antibody attached to the wrong antibody - resulting in misleading results!

Derivatives of fluorescein that react with proteins are readily available. Labelling your own antibody is simply a matter of following the manufacturer's instructions (see Figure 11-5). Kits are available from Molecular Probes for conjugating different Alexa Fluor dyes to your own antibody. These kits contain pre-weighed fluorescent probe and all necessary reagents (including the G25 column, see below) for successful conjugation.

Care needs to be taken when conjugating the fluorescent molecule with your antibody not to use too large an excess of fluorochrome, as this will result in too many fluorescent molecules being added to the antibody molecule. This can result in a serious drop in fluorescence, both due to the fluorochrome being attached to the antigen binding site (Figure 11-6), and to the often serious problem of fluorescence quenching (Figure 11-7).

Through a combination of fluorescence quenching and blocked antigen binding sites an incorrectly labelled fluorophore may have very little fluorescence when used in an IFA assay. For this reason directly labelling your own antibody is not a good idea, unless you have a specific requirement for direct labelling in a dual labelling experiment or you will benefit from the shorter immunolabelling protocol required.

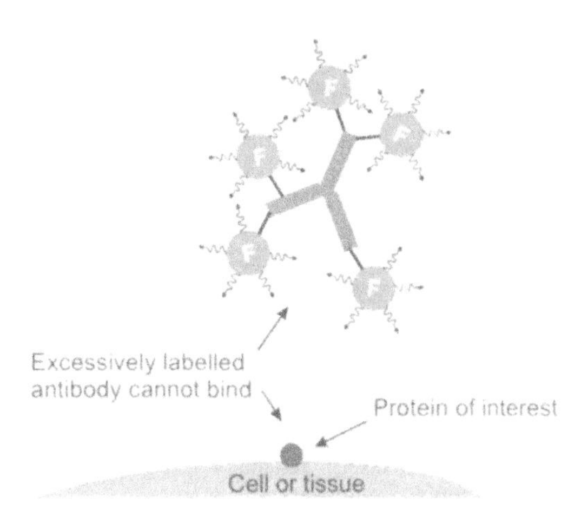

Figure 11-6. Interference with Antibody Binding.
Excessive labelling of an antibody with a fluorescent probe may result in too many fluorescent molecules being attached to the antibody. This may block the antibody binding site, resulting in diminished fluorescence.

After conjugating the fluorescent probe to your antibody the reaction mix must be passed through a G-25 size exclusion column to remove non-conjugated fluorescent probe (Figure 11-5). This step should be done with considerable care as a low level of contaminating, free (i.e. unconjugated) fluorescent probe will result in cell or tissue labelling that is not due to the binding of the antibody. Fluorescent probes themselves can often stain many cellular constituents in a very specific, but very misleading manner (i.e. the labelling may be associated with structures or proteins that are unrelated to your protein of interest).

The remarkable Alexa Fluor dyes, from Molecular Probes, show very little quenching when present in high concentrations conjugated to an antibody. This property means that the molar ratio between the reactive dye and your antibody is not as critical as with other fluorescent probes. This can be particularly useful when labelling antibodies where you only have a very small amount and have only a very rough idea (if at all) of the antibody concentration. Instead of wasting valuable antibody performing a protein assay or measuring its absorbance to estimate a protein concentration it may be advisable to simply do the conjugation with a molar excess of fluorophore. Just remember that this large excess must be removed completely in the G-25 column chromatography step! Even with the Alexa Fluor dyes, this large molar excess of fluorophore does not result in an increased level of fluorescence – in fact increasing the molar excess of fluorophore may result in a decrease in the level of fluorescence observed, even when using Alexa Fluor dyes.

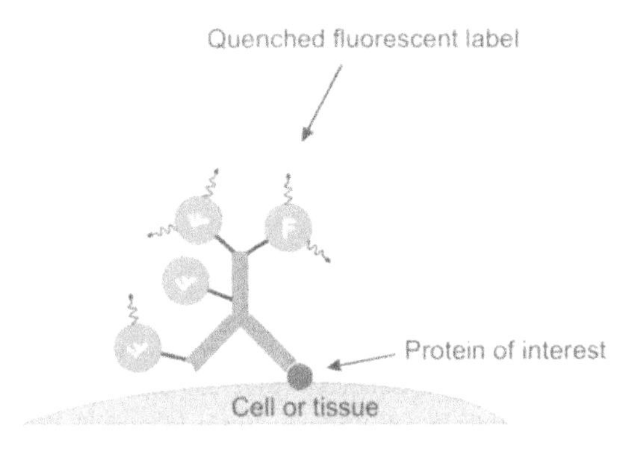

Figure 11-7. Fluorescence Quenching.
Excessive labelling of the antibody with a fluorescent probe may result in fluorescence quenching, greatly diminishing the signal available for forming an image.

Dual Immunolabelling

Labelling two different proteins within the one cell or tissue sample (Figure 11-8) is relatively simple if the two antibodies were raised in different animal species. Thus if one antibody is a mouse monoclonal and the other is a rabbit polyclonal antibody then dual labelling is carried out simply by using two different secondary antibodies, one directed against the mouse monoclonal antibody and the other directed against the rabbit polyclonal antibody. To reduce the chance of cross-reactivity, try using secondary antibodies raised in the same animal species (for example a goat or a donkey).

If both of your primary antibodies are from the same animal species (for example, both are mouse monoclonal antibodies) dual labelling is somewhat more involved as both of the antibodies will have to be directly labelled with a suitable fluorophore. When using two such closely related antibodies, simply conjugating one of the antibodies to a fluorophore and then using a secondary antibody for the second labelling is not sufficient. For example, when using two mouse monoclonal antibodies mouse secondary antibody will react with both of the primary antibodies that have been applied to the cells or tissue sample. An interesting technique, using fluorescently labelled Fab fragments (available from Jackson Immuno Research, Kirkegaard and Perry and Molecular Probes), can be used to label two different antibodies developed in the same animal species.

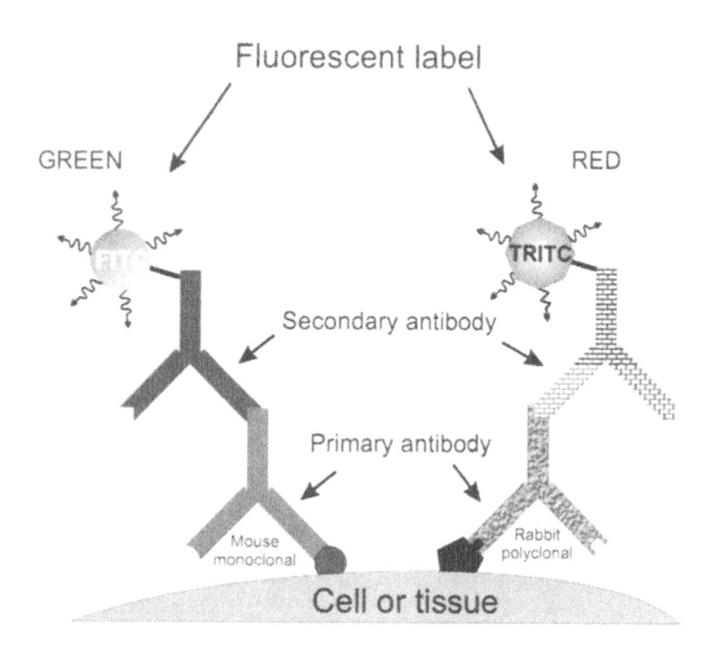

Figure 11-8. Dual Immunofluorescence Labelling.
Fluorescently immunolabelling two different proteins in the same cell or tissue sample is very simple if the two antibodies were raised in a different animal species (for example, one is a rabbit polyclonal and the other a mouse monoclonal). In this case different secondary antibodies (i.e. an anti-mouse and an anti-rabbit) are used, each conjugated to a differently coloured fluorophore.

Dual labelling does require careful choice of fluorophore combinations, as you don't want the fluorescent signal from one channel to overlap into the other channel (bleed-through). Sequential collection of the images using individual laser lines, rather than simultaneous collection with both laser lines on at the same time, is a way of significantly lowering the amount of bleed-through. Techniques to diminish the amount of cross-talk between channels is discussed in more detail in Chapter 10 "Confocal Microscopy Techniques" (page 239 and Chapter 4 "Image Collection" (page 101). These techniques include the selection of narrow-band filters (available from Chroma or Omega if they are not already installed in your confocal microscope), and the selection of suitable fluorophores with narrow emission spectra (particularly the Alexa Fluor dyes available from Molecular Probes). The spectral separation technology available from Carl Zeiss (the META system) and now Leica and Bio-Rad allows one to separate chromophores with significantly overlapping emission spectra.

Biotin/Streptavidin Immunolabelling

Biotin is a small (244 Dalton) vitamin that is very tightly and specifically bound by streptavidin (obtained from the bacterium *Streptomyces avidinii*) or avidin (a protein originally obtained from egg white). This binding has been extensively exploited in immunocytochemistry to provide a range of reagents for immunolabelling. Biotin can be readily conjugated to a variety of proteins, including antibodies, with minimal disruption of the structure of the protein. Furthermore, a number of individual biotin molecules can be bound to each antibody molecule, to which a fluorescently labelled avidin conjugate can then be bound. This results in significantly increased sensitivity when compared to using an antibody to which the fluorescent label has been directly attached, or even when using a "whole antibody" secondary fluorophore.

There are a number of derivatives of avidin and streptavidin (for example NeutrAvidin from Pierce) that have been altered to lower the level of non-specific binding, that is a problem, particularly with the original avidin molecule from egg white.

The increased versatility of using the biotin-avidin complex in immunofluorescence is of considerable benefit when performing dual labelling using a variety of antibodies. For example, FITC, TRITC, and Cy5 can be conjugated to separate preparations of streptavidin. Each molecule can then be used in turn in a dual labelling experiment with directly labelled antibodies without the need to make a new antibody-fluorophore conjugate for each experiment, and without the need to purchase several different secondary antibodies.

PREPARATION OF SAMPLES FOR IMMUNOLABELLING

Cell and tissue surface proteins can be readily labelled by incubating the antibody with the live cells, washing with culture media and then incubating the cells in the presence of the secondary antibody (discussed in more detail toward the end of this chapter). Internal antigens can be labelled in live cells by semi-permeabilising the plasma membrane with Streptolysin-O (this is explained in some detail on page 316 in Chapter 12 "Imaging Live Cells". However, for most antibody labelling experiments "fixing" and "permeabilising" the cells will be necessary.

The aim of tissue and cell fixation is to preserve the structural integrity as closely as possible to that of the live cells. Unfortunately, due to the necessity of preserving the antigenic site on the proteins of interest, a less than optimal fixation protocol may be required. Furthermore, the best fixative for preserving structural integrity, glutaraldehyde, results in a high level of autofluorescence that may cause difficulties in discerning the location of the antibody of interest. However, some background "autofluorescence" can be quite useful in outlining the position of cellular structures.

As fixation for immunolabelling requires a balance between structural integrity and the preservation of antigenic sites, there is always some compromise in choosing the best fixation procedure. In the past, fixation methods for fluorescence microscopy have been developed for producing a high level of fluorescence labelling with minimal background fluorescence. Structural integrity was not a high priority. With the greatly improved resolution of fluorescence microscopy when using a laser scanning confocal microscope, addressing the problems of structural integrity of the sample becomes important. This is particularly important when attempting to produce 3D images from a series of optical slices. Most fixation techniques involve drying the sample at some stage, and even if this is carefully controlled by progressively moving the sample through a series of alcohol steps and then drying with a critical point dryer, there is still a lot of shrinkage of the sample. The shrinkage is often much greater in the z-direction than in the x-y direction, perhaps due to adherence of the cells or tissue sample to the underlying glass slide. This high level of shrinkage may result in considerable distortion of the 3D structure of the sample when imaged by optical sectioning in a confocal microscope. Unfortunately there is no simple way of knowing how much the sample has contracted in the z-direction and so only an educated guess, rather than any exact measurement, of the amount of shrinkage can be used to help create a more "realistic" 3D reconstruction.

If the amount of shrinkage of the cells or tissue sample is particularly severe, or the cell layer was very thin in the first place, you may find that there is almost no 3D structure to your sample. The sample can still be used very successfully to locate your molecule of interest within the remaining 2D structure, but interpretation of 3D information needs to be done with care.

The greatly increased resolution of the confocal and multi-photon microscope when imaging fluorescently labelled samples (particularly in the z-direction) means that it is worthwhile considering fixation protocols that minimise sample distortion. In particular, it may be worth considering sample preparation that effectively eliminates the often highly destructive step of drying the cell or tissue samples. There are fixation methods that make use of DMSO and polyethylene glycol to preserve the hydrated state of the cell or tissue sample during the fixation process.

The following is a brief discussion of some of the fixation techniques available. For detailed protocols on using these methods see *"Confocal Microscopy, Methods and Protocols"* edited by Steve Paddock (Chapter 4 by Nadean L. Brown p75 is most useful) and *"Cell Biology, A Laboratory Handbook"* edited by Julio E. Celis (Volume 2 and 3 have useful hints on fixation and permeabilisation). Books and laboratory manuals on histochemistry (especially *"Theory and practice of histological techniques"* by Bancroft and Stevens, Butterworths, London, 1996), traditional light microscopy (for example, *"Microscopy, Immunohistochemistry, and Antigen Retrieval Methods"* by M.A. Hayat, Plenum Press, 2002) and electron microscopy (for example, *"Fixation for Electron Microscopy"* by M.A. Hayat, Academic Press, 1981) also contain a wealth of information on cell and tissue fixation. These books are listed in Chapter 15 "Further Reading" (page 347). Unfortunately published research papers often only very briefly, and often inaccurately, describe the fixation protocol used.

Fixation with Organic Solvents

A variety of organic solvents, particularly ethanol, methanol and acetone are widely used as cell and tissue fixatives. These reagents precipitate the cellular proteins, carbohydrates and nucleic acid (although mRNA is not readily fixed in place with this method). Lipid and small molecules are not fixed in place and will be lost in subsequent washing steps. Many fluorescent dyes may not only be possibly washed out during the fixation and washing of the sample, but any remaining dye may be drastically re-located within the cell or tissue sample during the fixation process. However, a number of fluorescent dyes can be fixed in place using aldehyde fixatives (see below).

Methanol Fixation: A simple, and often very effective, method of fixing cells is to immerse the cells in 100% methanol for 10 minutes. Live cells are usually first attached to a glass slide using substances such as poly-L-lysine prior to fixation, or in the case of cells that are grown attached to glass slides or coverslips no further attachment is necessary. The attached cells are immersed in methanol, either from the wet live cell condition, or after first air drying onto the glass slide. Air drying is commonly used for suspension (i.e. non-attached) cells, by either making a smear using a second glass slide or by centrifuging the cells onto the slide using a Cytospin™ centrifuge with a small plastic adaptor that fits onto the glass slide. The single cell layer created by making a smear or centrifuging is then air dried onto the glass slide. In this way suspension cells will be firmly attached to the glass slide without the need to use poly-L-lysine or other cell attachment molecules.

The fixation step can be carried out at room temperature, but fixation may be improved by immersing the slide in methanol that has been cooled to -20°C. The structural integrity of the cells is excellent when using pure methanol. However, the antigenic sites may not be readily available for antibody binding. Further permeabilisation with a non-

Figure 11-9. Solvent Fixation.
Biological tissues and cells can be readily fixed by using simple combinations of organic solvents. Acetone and methanol (often at –20°C) are common fixatives that often preserve the antigenic sites for subsequent immunolabelling. Pure acetone is highly destructive to cell or tissue structure, but is great for preserving the antigenic site and permeabilising the tissue. On the other hand, pure methanol retains the cellular structure, but is relatively poor at permeabilising the cells. A combination of acetone and methanol is often the best compromise between good retention of structure and a good level of fluorescence labelling.

ionic detergent may be necessary. Good transmission images of unstained cells can be obtained when the cells are fixed in 100% methanol, and there is reasonable retention of 3D structure of the cells.

Organic solvent fixatives: precipitate proteins, carbohydrate and DNA, but do not form cross-linkages – which means small molecules, such as non-covalently attached fluorescent probes, will leach out of the specimen.

Acetone Fixation: Pure acetone (-20°C for 10 minutes) is very effective as a fixative for immunolabelling of cells. The level of antibody binding is usually excellent (better than when using methanol as a fixative), and the amount of background fluorescence extremely low. However, the structural integrity of the cells is very poor. This is most noticeable if acetone fixed cells are imaged by electron microscopy – in which you will be alarmed to note that there is almost no fine detail remaining! However, when using light microscopy, biological structures may still be sufficiently intact after fixation with acetone for determination of the subcellular location of the protein of interest. When attempting 3D analysis of the cells by confocal microscopy, cells treated with acetone are often found to be very thin and severely lacking in 3D structure. When using 100% acetone fixation the cell outline is often almost impossible to image by transmitted light microscopy if the cells have not been stained. Acetone is a very good permeabilising agent and so treatment with detergents etc is not usually necessary.

A combination of Acetone and Methanol: A good compromise between the destructive nature of acetone and the lack of immunolabelling using methanol is to use a combination of acetone and methanol at -20°C for 10 minutes (Figure 11-9). Start with a 50:50 ratio of acetone methanol, but better results are often obtained by using 90% acetone and 10% methanol.

Fixation with Aldehyde Fixatives

Aldehyde fixatives (formaldehyde, paraformaldehyde and glutaraldehyde) are used to cross-link the protein components of the organic material within the specimen. Aldehyde fixatives offer the advantage of preserving the structure of the cell or tissue better than organic solvents. However, they can create problems of their own, such as damaging antigenic binding sites or creating unwanted fluorescence. Aldehyde fixatives are relatively slow acting, with the amount of cross-linking increasing considerably on incubating the cells in the fixative solution for a longer period (often overnight is convenient). Incubating relatively thick tissue samples for a sufficient length of time to allow the fixative to fully penetrate the tissue is particularly important. The cells can be stored for long periods of time in the fixative as long as care is taken to seal the container to avoid the sample inadvertently drying out.

If possible, the best way to allow aldehyde fixatives to penetrate a tissue sample is to perfuse the animal with fixative before removing the tissue sample. In this way the fixative becomes well dispersed throughout the tissue before the formation of cross-linkages prevents the further penetration of the aldehyde reagent.

Formaldehyde: Fixation with 2-4% formaldehyde in phosphate buffer is often very effective for immunofluorescence. However, commercially available formaldehyde (37% w/v) usually contains 10-15% methanol (obtainable from Sigma Chemical Co., Catalogue number F1635) and so may not be suitable for antigenic epitopes that are destroyed by methanol. In this case use methanol-free formaldehyde from Polysciences, Inc. (Catalogue number 04018). Formaldehyde quickly and readily penetrates cell and tissue samples, although polymerisation should be allowed to proceed for several days in relatively large tissue samples as the polymerisation process itself is relatively slow. Fixation at neutral pH make the cross-links formed reversible, whereas fixation at acidic pH may result in non-reversible cross-links. Formaldehyde solutions consist of a relatively low percentage of formaldehyde itself, with most of the solution consisting of methylene glycol. The formation of formaldehyde from methylene glycol is increased by heating (particularly microwave heating of tissue samples).

Aldehyde fixatives: cross-link macromolecules within the specimen – and will also cross-link aldehyde fixable fluorophores, such as the MitoTracker dyes, preventing them from leaching out of the specimen.

Formaldehyde and Picric Acid: A combination of 2% formaldehyde and 0.2% picric acid in phosphate buffer works well as a fixative for immunolabelling, including the fixation of both large structural proteins and small enzymes and peptides. A small amount of glutaraldehyde (0.01-0.1%) can be added to improve the structural integrity of the sample.

Paraformaldehyde: 4% paraformaldehyde is an excellent fixative for immunofluorescence microscopy (Polysciences, Inc. Catalogue number 00380). The structural integrity of the cells is not as good as when using glutaraldehyde (see below) as a fixative, but the antigenic sites are usually not greatly affected and there is very little "autofluorescence" created within the sample. Paraformaldehyde must first be made into a monomer solution by heating and slowly adding a solution of NaOH. Prepared solutions of paraformaldehyde can be readily stored frozen in small aliquots at -20°C for many months. However, paraformaldehyde does not keep well at room temperature – affecting both its solubility (polymer formation) and its ability to fix small molecules in place.

Glutaraldehyde creates strong autofluorescence of biological samples

Glutaraldehyde: 3% glutaraldehyde (electron microscopy grade) is the standard fixation procedure for electron microscopy. This results in excellent preservation of cellular structure. Unfortunately, most antibody binding sites are destroyed, or access to them is blocked, by the heavily cross-linked macromolecules within the cell. Furthermore the fixed cells will have a very high level of autofluorescence. Incubating the fixed cells in sodium borohydride solution for several minutes can lower the level of autofluorescence. A better solution may be to use a low level of glutaraldehyde (0.01 to 0.25%) in conjunction with paraformaldehyde.

Microwave Fixation

Microwave irradiation, using a normal commercially available microwave oven, or if more accuracy is required in the amount of microwave radiation used, a specially designed microwave oven for tissue and cell fixation, can be used to enhance the fixation and permeabilisation of organic tissues. Microwave fixation (1 to 2 minutes at 55°C) is particularly valuable when using relatively large tissue samples as the microwave energy readily penetrates organic tissue. Some people have found that the most suitable method for microwave fixation is to fix the tissue with a relatively low level of microwave fixation, and then to post-fix using a more traditional fixative such as an aldehyde fixative. Microwave fixation, with or without subsequent aldehyde fixation, may be particularly useful for allowing antibodies to penetrate relatively thick tissue samples.

Microwave irradiation of the sample can also be used to help aldehyde fixatives both penetrate tissue samples and speed up the process of cross-linking. Care should be taken not to heavily fix the tissue too quickly with microwave radiation as this may prevent further fixative from readily penetrating the tissue sample.

Commercially Available Fixative Preparations

Various commercially available fixatives are available. Many of these fixatives use a combination of the above solvent and aldehyde fixatives and include permeabilising reagents. Some of these fixatives have been formulated for specific applications, which means they may give superior results when used appropriately. If you want consistency in your fixation, or you have a large number of samples to process, the commercial fixatives may provide a more reproducible and reliable method of fixing your cell or tissue samples. However, the proprietary nature of many of these fixatives means that you may not be fully aware of all of the components that are used to make up the fixative preparation.

Permeabilisation

Most methods of fixation require a further step involving the permeabilisation of the tissue. This can be carried out by using a non-ionic detergent such as Ninidet p-40 (NP-40), Triton X-100 or Tween 20. However, a more recent development is to use microwave irradiation to permeabilise the tissue (see above). The heating effect of microwave irradiation penetrates deep within the tissue, unlike conventional heating methods that create a temperature gradient. Not only is the tissue very effectively fixed throughout the sample, but the tissue sample is also thoroughly permeabilised.

Solvent extraction with acetone results in the removal of most lipid, considerable destruction of cellular structure and consequently very effective permeabilisation. Methanol is not as effective as a permeabilising agent, although the addition of a small amount of acetone (10%) as discussed above, may result in an acceptable level of permeabilisation. Other solvents, such as DMSO or ethanol can be used as very effective means of permeabilising the tissue or cell sample with minimal disruption to cellular structure. These solvents are less destructive to tissue integrity compared to acetone or detergent treatment.

Lack of permeabilisation of single cell layers will usually result in greatly decreased or no fluorescence of your immunolabelled cells. A permeability problem may not be obvious if the level of labelling is sufficient for reasonable imaging, even though the best way of improving the images may be to improve the degree of permeabilisation. However, when immunolabelling tissue samples, lack of permeabilisation becomes very obvious, as the edges of the tissue will appear fluorescently labelled much more intensely than the internal areas when using the confocal microscope to obtain an internal optical slice.

Embedding Tissue Samples in Paraffin Wax

Traditional microscopy of fixed tissue samples is carried out by embedding the fixed tissue sample in paraffin wax, after suitable processing through a series of alcohol steps to dehydrate the tissue. The paraffin wax acts as a support for the fragile tissue sample for subsequent thin sectioning, staining and mounting for microscopy. Although traditional histochemical samples processed by aldehyde fixation and paraffin embedding can be imaged by confocal microscopy, there is probably very little, if any, advantage in using an expensive and often difficult to use microscope such as a laser scanning confocal microscope on such a thin sample. A much simpler method of imaging such thin tissue sections is to use a conventional light microscope (either bright-field or fluorescence microscopy) and to capture the image using a CCD camera. Previously, images were captured on photographic film, but using a sensitive CCD camera offers a number of advantages in terms of sensitivity and the ability to manipulate the digital images after collection.

Confocal microscopy offers the advantage of being able to image relatively thick tissue samples without the need to cut thin sections from fixed and embedded tissue samples. Relatively thick means you can image very large tissue samples (even whole animals!), but you can only penetrate the sample for 100 to 200 μm, and so the tissue will need to be cut into suitable pieces or even sliced for imaging deeper into the tissue. Many organs from laboratory animals can be imaged directly on the confocal microscope by mounting the whole organ in an imaging chamber. Alternatively a tissue biopsy (often several mm in size) can be directly mounted and held in place with a tissue "harp". A vibratome, which can cut relatively thick tissue sections (40 – 500 μm) from fresh tissue samples, may alleviate the necessity for tissue fixation in some labelling experiments. More detailed information on imaging live cell and tissue samples is provided in Chapter 12 "Imaging Live Cells" (page 279).

Cryo-sectioning of Tissue Samples

The technique of freezing tissue samples at very low temperatures (liquid nitrogen) by using a suitable cryogen (such as OTC) to preserve the integrity of the tissue sample is extensively used in electron microscopy, but can just as readily be utilised for samples imaged by light microscopy. A low temperature cryo-microtome is used to cut the frozen tissue block, and the resultant thin tissue sections are collected on glass slides that have been treated with a suitable reagent to allow the tissue sample to readily adhere. The samples can then be more readily fixed and permeabilised for immunolabelling than can relatively thick pieces of tissue. Care must be taken when low temperature freezing not to introduce ice-crystal freezing artefacts into the sample. In general low temperature cryo-fixation is highly effective for very quickly fixing tissue samples. Ultra fast freezing is used for electron microscopy studies to preserve fine structural integrity, but is not normally necessary for light microscopy. However, very fast freezing techniques may be useful in light microscopy for studying cellular metabolites that are changing quickly.

PRACTICAL HINTS FOR IMMUNOLABELLING

There are many ways in which immunolabelling can be performed, but the following hints may be of help if you are in a laboratory that doesn't have an established method, or the current method you are using is not working well. A detailed protocol for immunofluorescence labelling is described on page 276.

You need to first fix your sample as outlined above, which will normally result in a firm attachment of your sample to the microscope slide. However, a variety of attachment procedures can be used to make sure thicker tissue samples remain attached to the microscope slide throughout the following labelling and washing procedures.

The method outlined below uses cells or tissue samples attached to a glass slide, however, it may be advantageous to use a "free-floating" tissue sample when antibody labelling. This will allow the antibody to have access to the tissue sample from all sides. Free-floating tissue samples can be incubated for prolonged periods in the antibody solution by using a humid chamber, and sodium azide to prevent the growth of bacteria. One disadvantage of the free-floating technique is that a relatively large volume of antibody solution may be required (see details below for using very small volumes of antibody solution).

Figure 11-10. Limiting the Spread of Antibody.
A PAP water resist pen is used to mark small areas on the fixed and dried sample to limit the spread of diluted antibody solution. A very small volume (2 to 5 µl) of antibody solution is then applied to each spot.

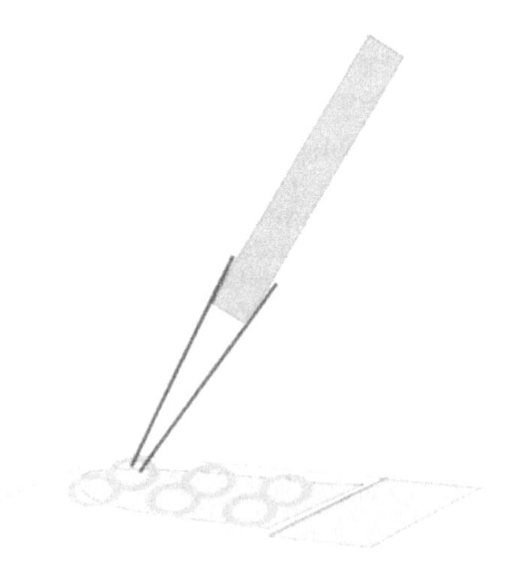

Figure 11-11. Application of your Antibody.
A suitably diluted solution of your antibody is applied to the centre of the PAP pen marked rings. The microscope slide is then incubated in a humid chamber (a Petri dish with a damp tissue) to prevent the small volume of antibody drying out.

Application of Your Antibody

An excellent way to conserve valuable antibody, and to allow several assays to be done on one microscope slide, is to use a "PAP" wax based pen (available from most microscopy supply companies) to create a hydrophobic ring around a small area of the sample (Figure 11-10). If you then carefully apply suitably diluted antibody to this area by using a micropipette (Figure 11-11) you will find that the antibody solution is contained within the PAP pen mark. The PAP pen is very stable when incubating the antibody at room temperature or 37°C for a few hours, but overnight incubation at 4°C may result in some parts of the PAP pen mark lifting. However, if this should happen all is not lost because often when the PAP pen ring has lifted the sample has remained in place. You should proceed with the labelling, although you may have trouble locating where the labelled cells actually are on the slide!

The primary antibody is normally diluted from 1:10 to 1:100, but can be diluted much more (as far as 1:10,000) for some antibodies. To improve the specificity of your antibody you should use the highest dilution you can before you start to loose significant amount of signal.

Use a humid chamber to stop the antibody solution drying out

Using a Humidity Chamber

When incubating your sample with small volumes of antibody solution, try to eliminate evaporation – otherwise you will end up with drying artefacts, often manifested as precipitation of fluorescent label over the sample. A very simple and effective way to stop evaporation of the antibody solution is to place the microscope slide in a Petri dish in which a moist tissue is used to maintain a high level of humidity in the dish. If you raise the microscope slide slightly above the bottom of the Petri dish you will stop any water in the chamber coming into direct contact with the microscope slide and possibly ruining the sample by mixing up the antibody solutions from different PAP pen rings on the slide. Take particular care when incubating the humid chamber at 4°C overnight to prevent condensation under the lid of the Petri dish dropping onto the sample.

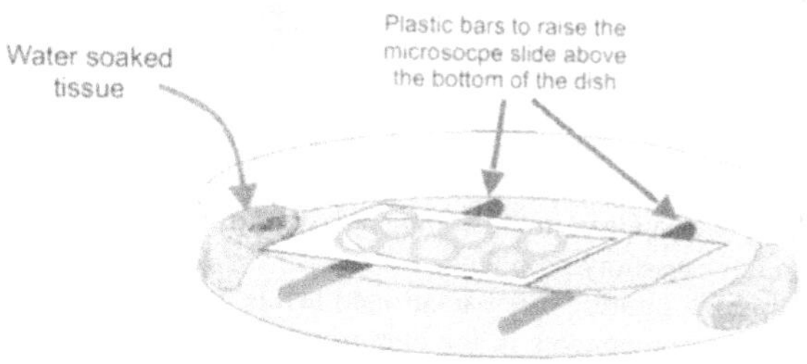

Figure 11-12. Humid Chamber for Incubating Antibodies.
A simple humid chamber can be constructed out of a plastic Petri dish to stop the small volume of antibody solution drying out. The microscope slide should be placed on plastic support bars to stop the slide coming into direct contact with the water used to maintain the humidity of the chamber.

Use the "squirt" from a wash bottle for a fast and efficient washing method

Washing to Remove Excess Antibody

A fast and efficient way of washing the sample after incubating with the primary or secondary antibody is to squirt the area containing the cells and unwanted antibody solution with a fine stream of phosphate buffered saline (PBS) from a squirt bottle (Figure 11-13). Don't be frightened of squirting hard – the best washing appears to be from a relatively short but firm squirt (i.e. move the stream around within the labelled area for 2 to 3 seconds). This method of washing, once you get the hang of it, is far superior to soaking the slides in PBS. Besides being much faster, this normally results in the background being lower, and a higher level of signal as well!

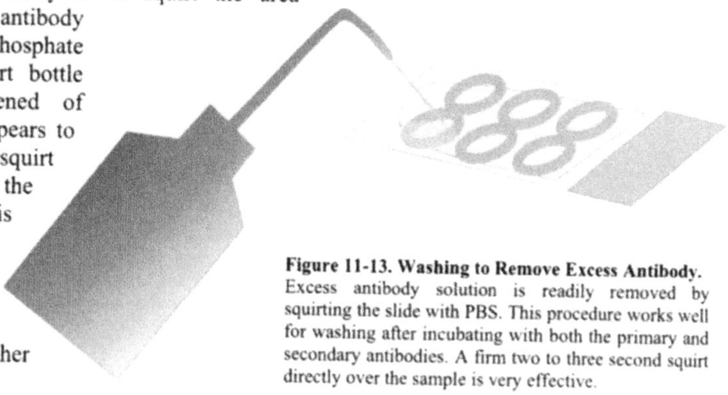

Figure 11-13. Washing to Remove Excess Antibody.
Excess antibody solution is readily removed by squirting the slide with PBS. This procedure works well for washing after incubating with both the primary and secondary antibodies. A firm two to three second squirt directly over the sample is very effective.

Application of Secondary Antibody

The secondary antibody is diluted as recommended by the manufacturer, or your own past experience, and applied as a small volume (2-5 μl) to the small circle enclosed by the PAP pen marking. Don't be tempted to use a more concentrated secondary antibody solution than that recommended in the false hope of increasing the amount of labelling – all you will succeed in doing is increasing the level of background labelling and the possibility of precipitation of concentrated fluorophore on the sample. Again, the microscope slide will need to be incubated within a humid chamber to prevent the small volume of antibody from drying out. If speckles of fluorescence are a continuing problem with your immunolabelling, try centrifuging the secondary antibody solution before using it (spin hard in a bench top Eppendorf tube centrifuge for 10 minutes), to remove any protein aggregates.

Blocking of Non-specific Background Labelling

Immunolabelling may be significantly enhanced by the addition of a "blocking" step to eliminate non-specific binding of the fluorescent antibody probe to the cells or tissue sample. A blocking step is not always necessary, and so a test slide should always be produced with and without a blocking step when a new antibody or tissue is being investigated.

Bovine serum albumin (2% BSA in PBS) placed on the fixed cells for between 5 minutes and overnight (5 minutes is ample time, but overnight is often convenient) is a very effective blocking media. Skim milk (5%) is often just as effective as a blocking agent. Another common blocking method is to simply use serum (10%) from the same species that were used to raise your secondary antibodies (often goat or donkey). The blocking of non-specific antibody binding sites appears to be complete within minutes, but long incubation times (as long as the sample does not dry out) do not cause any problems.

Occasionally high fluorescence background levels may be detected on the glass surface between individual cells. This may be due to components in the culture media that have bound to the glass slide and then subsequently the fluorescently labelled antibody. This type of background fluorescence can often be "removed" from the image obtained by confocal microscopy by simply taking an optical slice (or series of slices) that does not include the glass surface (i.e. the bulk of the cell, which is raised above the layer of the glass slide is imaged).

Preparation of Slides for Microscopy

Once the antibody labelling has been completed the sample is finally washed with PBS by using a squirt bottle, and the microscope slide is allowed to drain and then air dry. The drying process can be sped up by using a warm air blow dryer, or by simply placing the microscope slide in a high air movement area. Better structural integrity may be retained by "wet mounting", i.e. not allowing the slide to dry out before adding the glycerol based mountant. Once the microscope slide is completely dry the first thing to do is to mark, with a black marker, the outline of the PAP pen rings you originally marked on the slide, but this time on the back side of the microscope slide. At this stage the PAP pen markings are readily visible, but once the coverslip is mounted you will find that locating the PAP pen rings is very difficult.

The microscope slide is readily mounted in a simple glycerol based antifade mountant (see method for *n*-propyl gallate antifade mountant on page 198 in Chapter 8 "What is Fluorescence?"). Take care to add a small drop of mountant to the centre of each PAP pen ring (the glycerol will not spread readily past the PAP pen markings). Don't add too much glycerol mountant – there should be just enough to eventually (after several minutes) cover the area under the coverslip and to hold the coverslip firmly in place. If you add too much mountant the coverslip will move around when you are trying to move the sample under the microscope. If this should happen excess mountant should be removed with a pipette or tissue.

For longer-term preservation of your valuable immunolabelled sample the mounted sample should be sealed with nail polish. This is done by painting a line of clear nail polish along the edge of the coverslip, making sure that there are no gaps in your application. Allow the nail polish to dry for at least 1 hour in a warm place before placing under the microscope (nail polish is quite difficult to remove from expensive objectives lenses). VALAP (a mixture of paraffin, Vaseline and lanolin, see page 283 in Chapter 12 "Imaging Live Cells"), can also be used to seal the edge of the coverslip – and is more easily removed from the objective lens!

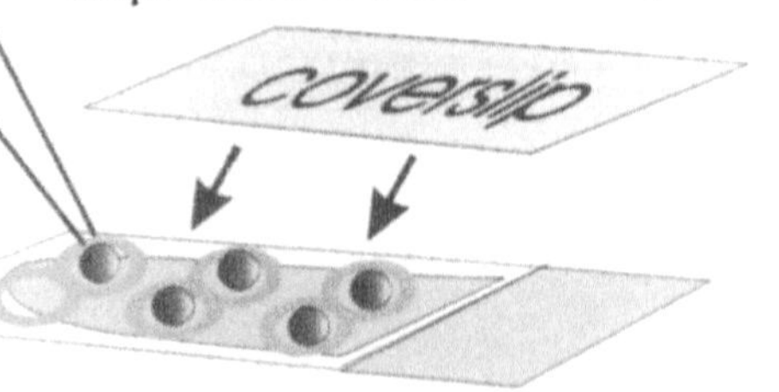

Figure 11-14. Mounting the Slide.
A small drop of glycerol mountant (often containing an antifade reagent) is placed in the centre of each PAP pen ring, and then the coverslip is placed firmly on top.

AUTOFLUORESCENCE IN IMMUNOLABELLED SAMPLES

Autofluorescence has been discussed in detail previously (see page 198 in Chapter 8 "What is Fluorescence?"). However, it is important to briefly discuss some of the difficulties encountered with autofluorescence when attempting immunolabelling.

Distinguishing between "background" fluorescence and "autofluorescence" is important. The background fluorescence is determined by the binding of the fluorescent antibody to non-specific structures (or perhaps very specific structures – but not the expected structure!) within the sample and is obviously not present when the fluorescent antibody is not added. On the other hand "autofluorescence" is present in the control sample that lacks any added fluorescent label.

The molecules present in the original sample often create autofluorescence. For example, aromatic amino acids, reduced pyridine nucleotides, Flavins, zinc-protoporphyrin, chitin, chlorophyll, lipofuscin, collagen and elastin all produce autofluorescence of biological material. Many macromolecules that show autofluorescence are more readily excited by shorter wavelengths (particularly UV light), and level of autofluorescence can often be significantly reduced simply by imaging the sample at longer (particularly red) wavelengths using a long wavelength emitting red dye such as Cy5.

Autofluorescence is often inadvertently introduced by the method of fixation used. Aldehyde fixatives, and particularly glutaraldehyde, as discussed at the beginning of this chapter, will introduce unwanted autofluorescence. This fluorescence can often be lowered to acceptable levels by washing with sodium borohydride. However, as mentioned above, a more effective method of lowering the level of autofluorescence is to either not use glutaraldehyde, or to use only a very low concentration for a limited period.

Autofluorescence is also found to be significantly greater in methanol fixed cells compared to acetone fixed cells. Lowering the methanol percentage in your fixative preparation may considerably lower the level of autofluorescence.

Not all autofluorescence is a disaster. In fact, a low level of autofluorescence can often be useful for imaging cellular detail, such that the location of the fluorescent probe can be more easily determined. A control sample, that has had no fluorescent probe added, is used to determine the location and level of the autofluorescence within the sample. This will determine whether there is any location or intensity of fluorescence that may create difficulty or confusion as to the location of your probe or antibody.

Figure 11-15. Seal Edges of Coverslip with Nail Polish. Apply a line of nail polish (or VALAP) to the edge of the coverslip to seal the glycerol mountant. This will not only stop the slide drying out, but will also limit the supply of oxygen - thus helping to limit the level of fading.

Make sure the nail polish is dry before placing the microscope slide under the microscope!

Immunofluorescence Antibody Labelling (IFA) Method

IFA method

This method uses a relatively "crude" air-drying step that may not be suitable for all samples – and is certainly not suitable for thick tissue samples. However, it is a very fast and effective method for fluorescence immunolabelling of single cell layers.

- Spread cells or tissue cryosection on a microscope slide, or remove excess liquid from adherent cells.
- Air dry sample by placing in a well ventilated position.
- Fix sample in Acetone: Methanol 1:1 for 10 minutes ← **-20°C improves fixation**
- Air dry sample – forced airflow from a moderate temperature heater.
- Block non-specific binding sites by incubation of the sample in 5% skim milk for 10 minutes
- Add suitably diluted primary antibody - ***your antibody*** ← Overnight incubation at 4°C for improved labelling

  A dilution of between 1:10 and 1:10,000 depending on your antibody.
- Incubate at room temperature for 2 hours by placing the microscope slide with the diluted antibody solution within a humidity chamber.
- Wash thoroughly with phosphate buffered saline (PBS) - "squirt" washing (see Figure 11-13) is fast and effective.
- Add diluted fluorescent secondary antibody (using the manufacturers recommended dilution).
- Incubate at room temperature for 1 hour (using a humidity chamber).
- Wash thoroughly with phosphate buffered saline (PBS) - again, "squirt" washing is very effective.
- Mount with antifade mountant (see note below) & seal with nail polish or VALAP.

Image by confocal microscopy

For high specificity immunolabelling you should dilute your antibody as much as possible – to the point at which you can only just still detect fluorescence labelling. A higher level of antibody concentration may result in non-specific cross-reactivity, giving false labelling. It is good practice to use a series of antibody dilutions on the same sample – you can then use the results from the more dilute antibody concentrations to determine antibody specificity, even though you may use images collected from a less diluted antibody concentration to obtain high quality images.

Mounting your slide in antifade reagents may result in lower overall fluorescence (this is particularly noticeable when using the Alexa Fluor dyes from Molecular Probes). Sealing the slide with VALAP or nail polish (to exclude oxygen) and using a simple glycerol mountant may help reduce fading in the absence of any antifade reagent.

For further information on antifade reagents see page 198 in Chapter 8 "What is Fluorescence?"

IMMUNOLABELLING OF LIVE CELLS

Live cells can be readily labelled with fluorescently labelled antibodies. This method of labelling is very straight forward for labelling cell surface antigens, but can also be used to label internal compartments by either allowing natural processes of endocytosis to take up the antibody (see Figure 11-2) or by using Streptolysin-O to semi-permeabilise the cells.

Cell Surface Labelling

To label the cell surface of cells all you need to do is incubate the cells in media containing the antibody. This is more easily performed using a directly labelled fluorescent antibody, but can also be done by using an unlabelled primary antibody and a suitable fluorescently labelled secondary antibody, and sequentially incubating the cells in first the primary and then the secondary antibody. If the antigenic site is available on the cell surface, you will often find that the labelling is much better and a lot faster than in fixed cell preparations. This is probably due to the antibody more readily recognising the antigenic site of live unfixed cells, and due to the fact that a readily accessible cell surface antigen is being labelled. Although a fixed cell preparation is permeabilised for immunolabelling, often the level of permeabilisation is not ideal.

If endocytosis of the cell surface receptor to which your antibody is binding is not required (or in fact is considered detrimental for surface labelling), the labelling can be carried out by keeping the incubation tubes on ice and centrifuging in a refrigerated centrifuge at 4°C. Cell "capping" and subsequent endocytosis may still occur while imaging on the microscope if the cells are not maintained at a low temperature. However, as endocytosis is a relatively slow event, excellent images of surface labelled proteins can often be obtained within a few minutes, before the cells have had a chance to internalise the label. Of course, if you are only interested in labelling specific cells and are not concerned about a surface location there is no need to maintain the cells at low temperature (which in itself may have other detrimental effects on the cells – see Chapter 12 "Imaging Live Cells" (page 279).

Live cells can be labelled for as short a period as 1 to 2 minutes

Internalisation of Cell Surface Receptors

Binding of antibodies to cell surface receptors may lead to internalisation of the fluorescently labelled antibody by receptor-mediated endocytosis (see Figure 11-2). If internalisation is required, then the cells will need to be incubated above room temperature (preferably 37°C) for mammalian cells. If internalisation is not required, maintain the cells at all times at or below 4°C as some internalisation may eventually occur at relatively low temperatures. Receptor mediated endocytosis, and thus fluorescently labelled antibody uptake, is severely limited in suspension cells that have been artificially attached to a substrate – for example, when using poly-L-lysine to attach cultured human lymphocyte cell lines to the coverslip base of a microscope imaging chamber.

To inhibit the endocytosis of surface antigens label cells at 0°C (on ice)

Streptolysin-O (SLO) Permeabilised Cells

The bacterial toxin Streptolysin-O forms proteinaceous pores within cellular membranes. These pores can be of sufficient size to allow large macromolecules, such as antibodies, to pass through the membrane without the release of subcellular organelles. In this way internal antigenic sites can be labelled with fluorescently (and also gold) labelled antibodies in "live" cells.

A series of test samples will need to be processed to determine the correct SLO concentration and incubation conditions (temperature and time) that will result in a good level of permeabilisation without excessively damaging the integrity of the cells. Once you have semi-permeabilised the cells with SLO you can add your antibody (preferably directly labelled). Relatively short incubation times (several minutes) are often very effective – as long as the cells are permeable to the antibody. Many cellular processes can still continue in the SLO semi-permeabilised cells if sufficient care is taken to supply the correct buffer and bio-energy conditions. However, some batches of SLO can severely inhibit cellular processes once the cells have been permeabilised, although this may not be a problem when structural integrity is more important than metabolic activity. For more details see page 316.

Chapter 12

Imaging Live Cells

The confocal microscope has not only resulted in a great deal of interest in studying fixed and embedded tissue by microscopy, but has also elicited enormous interest in studying the cells live, with specific probes highlighting the structure or cellular event of interest. Confocal microscopy has made this possible because it provides the ability to clearly visualise a single optical plane within the cell or tissue without the distraction of out-of-focus fluorescence.

This chapter outlines important points for maintaining healthy cells during microscopy, and in particular explains when temperature control is important and how to maintain cells at the correct temperature on the microscope.

A number of different cell and tissue culture chamber designs used in microscopy are also described in some detail in this chapter. These range from very simple chambers that can be readily made by yourself, to elaborate and expensive chambers that are commercially available. The more elaborate chambers attempt to fully control the environment of the cells, but experiments can often be performed in conventional environmental control incubators (which very accurately maintain the correct CO_2 concentration, temperature and humidity etc for growing cells) away from the microscope, and then the sample taken to the microscope for imaging.

This chapter is mainly concerned with the correct maintenance of mammalian cells during live cell microscopy. Much of the discussion is also relevant to other cell types, although, many other cell types are less demanding as neither temperature (for example insects cells, yeast cells, plant cells, etc) nor the composition of the media (yeast cells and many bacterial cells) is as critical.

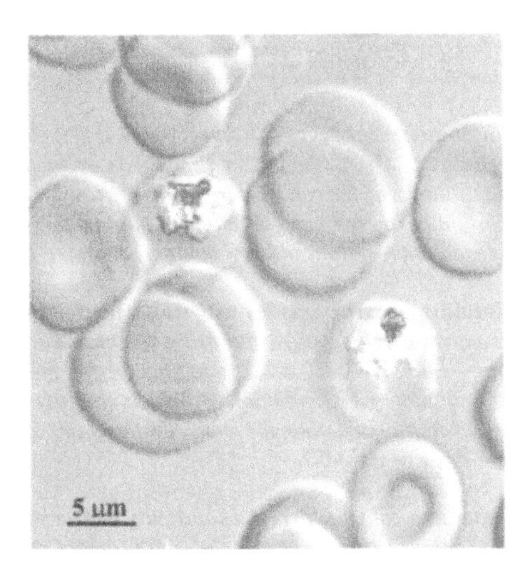

Figure 12-1. Malarial Parasites Stained with Lucifer Yellow.
Live malarial parasites were incubated in the presence of lucifer yellow – a classical marker of endocytosis in mammalian cells. The dye is clearly taken up by the parasites, but not by the uninfected red blood cells. Cells were imaged live sitting on the coverslip base of a microscope imaging-chamber. A high-resolution fluorescence image has been combined with a simultaneously collected grey-scale transmission image. These images were collected using a Bio-Rad MRC-600 confocal microscope with a Zeiss 63x 1.4 NA oil immersion objective.

MAINTAINING HEALTHY CELLS

Live cell confocal microscopy usually involves high-resolution studies on individual cells. The studies should be carried out on healthy cells that are close to the accepted normal physiological state for the particular cells being examined. Incorrect culturing and manipulation can seriously alter the metabolism and structure of cells, resulting in misleading results or even erroneous conclusions.

Keeping Cells Alive

Simply keeping the cells alive is not all there is to live cell imaging - you need to also make sure that the physiological state of the cells is not being influenced by environmental stress on the cells. Most laboratories are well equipped for growing tissue culture cells away from the microscope. What is often a routine procedure for growing cells in a temperature controlled and correctly gassed incubator can become very demanding when applied to keeping the cells alive during the imaging process.

Imaging live cells by microscopy requires careful consideration of the following:

- **Cellular structure and integrity**
  Cultured cells are quite delicate and must be treated with care when being transferred to a new flask or a microscope imaging-chamber. Physical pressure on cells can inadvertently disrupt the physiology of the cells. Some cell types can become excessively autofluorescent when stressed physically. A well-designed microscope incubation chamber, and allowing the cells to recover after being transferred to the microscopy chamber, will minimise the disruption of cellular processes due to physical trauma.

- **Temperature control**
  Full temperature control is technically difficult during microscopy. This is discussed in some detail later in this chapter, but temperature control on the microscope itself may not always be necessary. Full temperature control prior to microscopy may be sufficient for slower cellular processes such as protein trafficking, organelle structure, subcellular organization etc.

- **Maintenance of correct pH level**
  The correct pH of the culture media is critical for both long-term viability of the cultured cells and for maintaining the correct physiological state. The pH of the culture media is often maintained by using bicarbonate buffer, which requires having the correct CO_2 concentration (5%), either in the media of an enclosed microscopy culture chamber, or in the gas provided to an open chamber.

- **Maintaining correct gas levels within the media (O2, CO2)**
  The correct CO_2 level is important for maintaining the correct pH of the media as outlined in the previous point. The correct O_2 level is important for actively respiring cells. If the O_2 level is low the cells will be stressed and have significantly lowered viability in the long term. On the other hand if the O_2 level is too high (this is most important for cells that require an unusually low O_2 level), this may be detrimental to the physiology of the cells.

- **Maintenance of the correct osmolarity of the media**
  Maintaining the correct ionic strength of the media is important for many cells, but especially mammalian cells, which cannot tolerate even small variations in the osmolarity of the media. The small volumes and frequent exposure to a drying room-atmosphere can result in significant ionic changes to media used for cell culturing on the microscope. A sealed microscopy chamber or a flow through perfusion chamber, or if using an "open" microscopy chamber maintaining a fully water saturated atmosphere for the cells, is important.

Not all of the above parameters are important for all cell types or for all experiments. For example, maintenance of the correct temperature is critical for studying fast changes (such as Ca^{2+} fluxes) in live mammalian cells, but is not as critical for experiments that follow changes over a longer time frame (such as protein trafficking), where correct temperature control prior to imaging may be sufficient.

One of the most important considerations when manipulating cells for microscopy is to remember that the small volume often used in microscopy culture chambers will mean that inadvertently disrupting the environment the cells are exposed to, even during brief transfer to the microscope, is important.

Dangers of Stressing the Cells

Cells can be stressed in a variety of ways. Physical manipulation of the cell culture, changing the incubation temperature, altering the CO_2 level of the media, lowering the oxygen level in the media etc will all place considerable stress on the cells. The fluorescent dye used in your study will also induce stress in the cells. For these reasons the level of fluorescent dye used should be kept as low as possible. Furthermore, the fluorescent dye will often have increased toxicity to the cells when the cells are irradiated with the laser light during confocal imaging - therefore keep the laser intensity to the minimum necessary for collecting your data. The mercury and xenon arc lamp used for conventional epi-fluorescence imaging, and even the tungsten lamp used for bright-field imaging can be destructive to both the fluorophore and the cells (discussed in more detail on page 319)

The cellular response to this stress will vary greatly between cell types and culture conditions. The stress response in cells is not simply a matter of changes in metabolism during the period of stress, but often a complex cellular response that may involve the production of specific stress proteins. This response will then alter the observed physiological state of the cells, and it may take some time for the cells to recover after removal of the stress.

Severe environmental stress may result in the sudden and catastrophic destruction of the integrity of the cell, called necrosis (Figure 12-2). This will be readily recognised by microscopy, and thus one is unlikely to continue imaging severely damaged cells without knowing the cells are not healthy. However, programmed cell death or apoptosis, as outlined in some detail below, may result in cells that have become committed to the apoptotic pathway, and thus altered in their physiology, but are not yet visibly affected.

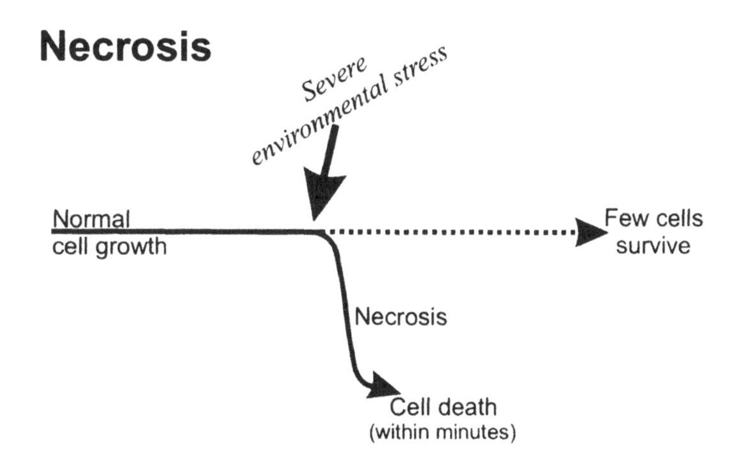

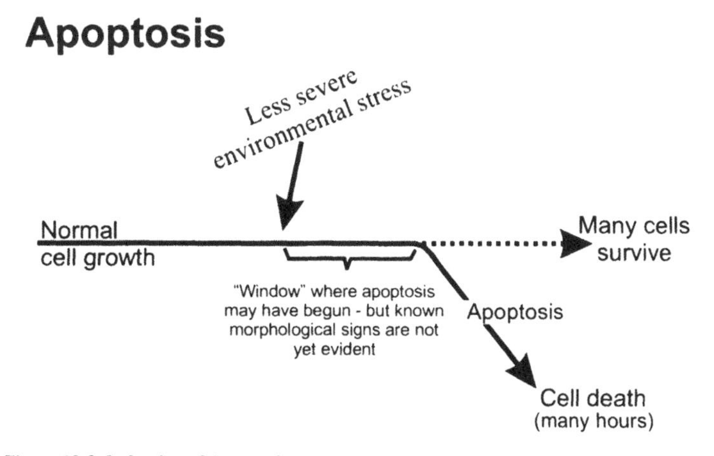

Figure 12-2. Induction of Apoptosis.
Severe environmental stress will result in immediate damage to the cells, often resulting in sudden death by necrosis. Less severe environmental stress may not appear to have an immediate effect on the cells, but after some hours it may become evident that the cells have embarked on the pathway of self-destruction called apoptosis. This "window" of apparently "normal" cells may result in a false confidence in the integrity of the cells being studied.

Apoptosis - Programmed Cell Death

Fluorescent dyes may be toxic!

A very wide variety of environmental factors that stress a cell can result in that cell going down the path of apoptosis or programmed cell death (Figure 12-2). Apoptosis is a process whereby defective cells can be removed without the release of highly toxic cellular constituents. The release of, for example, lysosomal enzymes into the surrounding tissue will not only be detrimental to other cells in the vicinity, but will also elicit an inflammatory response that can be highly destructive to the surrounding tissue. During the process of apoptosis the cell progresses through a series of developments that finally results in degradation of the nuclear DNA and eventual destruction of internal organelles. The remnant cell is finally engulfed and digested by macrophage immune cells. This process results in the removal of defective cells without eliciting an immune response.

Laser irradiation of fluorescently labelled cells may induce apoptosis!

Apoptosis also occurs naturally as a genetically determined event in embryonic development (the removal of specific cells when forming organ structures) and is occurring all the time in the continual process of maturation of immune cells.

Apoptosis is a complex cellular response that appears to be non-reversible once the process has begun. However, it may not be evident for some time from the morphology of the cells that apoptosis has begun. Although apoptosis is studied widely in many laboratories around the world, care must be taken to minimise the risk of "accidentally" studying apoptotic cells when what you want to study is the normal physiological state of the cells.

Apoptosis is a stochastic event in which the number of cells entering apoptosis will be increased when the cell culture is placed under environmental stress. Which individual cell will be tipped into apoptosis cannot be readily established. As the process of high resolution confocal microscope imaging is concerned with the subcellular distribution and concentration of specific fluorescent probes within individual cells, one may inadvertently image a cell that is committed to apoptosis, but is not yet displaying any of the characteristic morphological characteristics. Unfortunately, such a cell will have significant changes occurring to its metabolism. Thus, changes in the parameters being studied may become evident, but this may be a reflection of the apoptotic state rather than the dynamics of the healthy physiological state that you are attempting to study.

Disruption to the bioenergy of the cells is considered an important reason why cells enter the apoptotic pathway. In this way energy compromised cells will be removed from the population without the production of highly disruptive cellular constituents being released into the culture media - or tissue when this occurs within a living animal. Therefore, maintaining the correct energy supply to the cells is important, including correct O_2 levels and the correct level of nutrients and energy source. Irradiation of the cells with laser light during image collection, and particularly the interaction of the dye with the irradiating light, may induce apoptosis. Particular care should be taken when using mitochondrial dyes such as the MitoTracker probes, as disruption of the efficiency of energy production of mitochondria is a potent inducer of apoptosis.

Recognizing Signs of Cell Stress

One of the simplest ways of recognising cell stress is to image the cells by simple transmission imaging. An experienced cell biologist will be able to tell at a glance whether a particular culture is in a healthy state. However, this can be much more difficult for the beginner - particularly when the cells under study may be already in some way metabolically disrupted.

Cells that are undergoing stress will have characteristic warning signs. This may be lack of firm attachment to the coverslip for attached cells. Or perhaps the formation of protrusions and "blebs" from the plasma membrane of affected cells. The mitochondria of stressed cells may show the characteristic "swollen" mitochondria of damaged cells. This may also be an early warning sign that the cells are entering apoptosis.

Within a cell culture there will always be some cells that are not surviving well. These cells will be readily recognised as being morphologically distinct from the rest of the population. If the number of disturbed cells appears to be higher than expected for the particular strain under study, then care should be taken to improve the culture conditions used during microscopy.

Warning: Physical pressure on cells may create high levels of autofluorescence

SIMPLE IMAGING CHAMBERS

Simple chambers for microscopy can be purchased from a number of manufacturers, but they can also often easily be made from readily available materials. No attempt is made to control the temperature of these simple chambers, but they can be held at the correct temperature prior to imaging by being placed on a suitable heating block placed next to the microscope. Gas exchange (and thus pH) can be controlled during prolonged periods of incubation by placing the microscope incubation chamber in a correctly gassed and humidified tissue culture incubator. To eliminate the danger of drying out very small volumes, place the imaging chamber within its own small humidity chamber within the main tissue culture incubator.

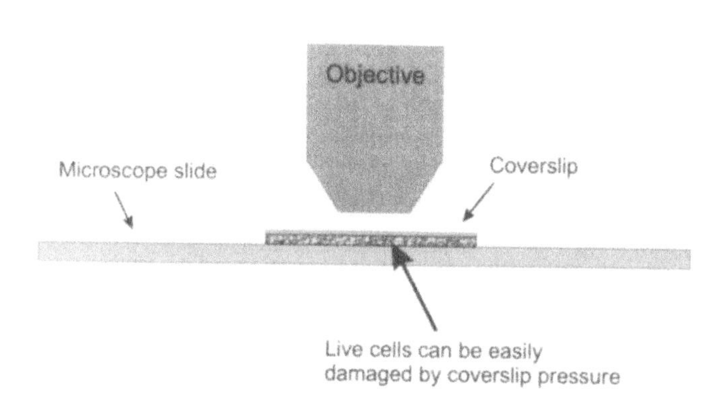

Figure 12-3. Microscope Slide for Holding Live Cells.
Simply placing a coverslip over a cell suspension can be used as a quick and easy way of imaging live cells. However, this may result in pressure damage to the cells, as well as the danger of the media drying out and the cells being disrupted by movement of the objective when using an immersion lens.

Suspension cells can be simply placed under a coverslip on a microscope slide (Figure 12-3). Attached cells that have been grown on a coverslip can be readily "floated" on media by being placed upside down on a glass microscope slide. Using a single coverslip to look at live cells may be quick and convenient if high-resolution imaging is required for the assessment of dye uptake for example. However, there is a great danger that the cells will be stressed by applying too much physical pressure to the coverslip (resulting in autofluorescence and other cellular damage). The media will also begin to evaporate around the edges of the slide (and thus alter the ionic concentration of the media), and the gas exchange with the atmosphere will quickly alter the pH balance of such a small volume of media. A larger coverslip may slow down the evaporation of the media, thus allowing a larger period of imaging before the cells are damaged by ionic concentration changes. Applying a layer of VALAP (see enclosed box), a mixture of Vaseline, Paraffin and Lanolin, around the edges of the coverslip will significantly lower the rate of evaporation and gas exchange. Do not use nail polish to seal around the edge of slides when live cell imaging.

Using a relatively small volume of liquid under the coverslip will often result in the cells being gently held in place for imaging – but take care not to unduly stress the cells or distort their structure by pressing down on the coverslip or by having such a small amount of media that the underside of the coverslip is not fully covered (resulting in undue pressure on the cells by the capillary action of the media). If the cell movement is a problem then poly-L-lysine or other cell attachment reagents can be used to hold the cells in place.

Traditional Concave Microscope Slide

The traditional concave microscope slide (Figure 12-4) can be used for imaging live cells, tissue samples and small organisms. This simple chamber has the advantage that the cells or tissue sample are not subjected to undue pressure when a coverslip is applied to the microscope slide.

However, this design of imaging chamber is not optimal for high-resolution imaging as the object under investigation may reside some distance from the coverslip. If the cells reside some distance from the coverslip this will create considerable spherical

> **VALAP recipe**
>
> **- a simple alternative to using nail polish to seal around the coverslip -**
>
> Vaseline, Lanolin, and Paraffin 1:1:1 is heated (gently) on a hotplate and then applied around the edge of the coverslip with a small pointed brush.
>
> The good thing about VALAP is that there are no solvents involved and the mixture hardens immediately on contacting the cooler microscope slide – a safer alternative to nail polish when using fixed preparations, and essential when using live cells.

aberration, particularly when imaging with an oil immersion lens. This problem is minimised when the cells are attached to the coverslip with a suitable cell attachment reagent such as poly-L-lysine.

A concave slide is simple and cheap to use and so is often ideal for initial observations. The coverslip-enclosed media will retain the correct pH and gas levels for some time. The chamber can be readily used for small organisms that would otherwise be damaged by the direct application of a coverslip to a glass slide. However, these chambers are not particularly suitable for imaging cultured cells. For this purpose the microscope slide chamber described below is more appropriate.

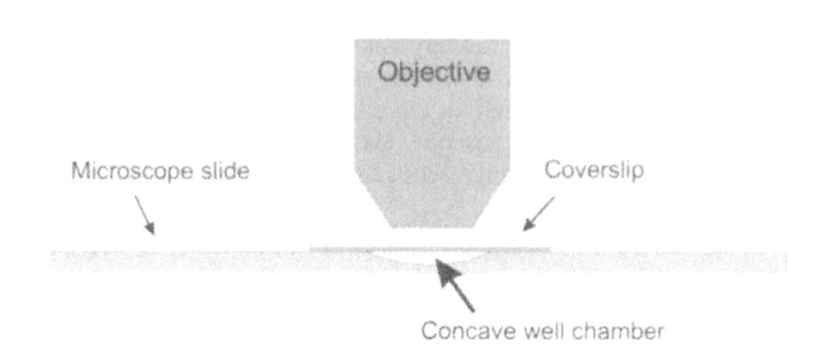

Figure 12-4. Concave Well Chambers.
A coverslip is placed over a 15 mm diameter concave well in a microscope slide. Concave well microscope slides are available from a number of microscope slide manufacturers. However, an improved version of this type of chamber slide, which can be readily made by yourself, is shown in figures Figure 12-6 and Figure 12-7.

Simple Gasket Chamber

A significant improvement on simply placing a coverslip over the cells is to enclose the cells in a small chamber formed by placing the coverslip over a silicone gasket attached to a microscope slide (Figure 12-5). The gasket will allow room for the cells or tissue sample without undue pressure being exerted from the coverslip. Furthermore, the enclosed media will be severely limited in evaporation or gas exchange with the atmosphere.

Silicone gasket enclosed chambers are available from Grace Bio-Labs. These gaskets, besides being used as simple imaging chambers, are designed to withstand chemical treatment and high temperatures used for in situ hybridisation. You can also make your own chambers by using washers made out of silicone rubber, or by using plastic reinforcement rings used for ring binders. The advantage of the reinforcement rings is that they have a sticky side and can be used stacked on top of each other to create a chamber of the desired thickness.

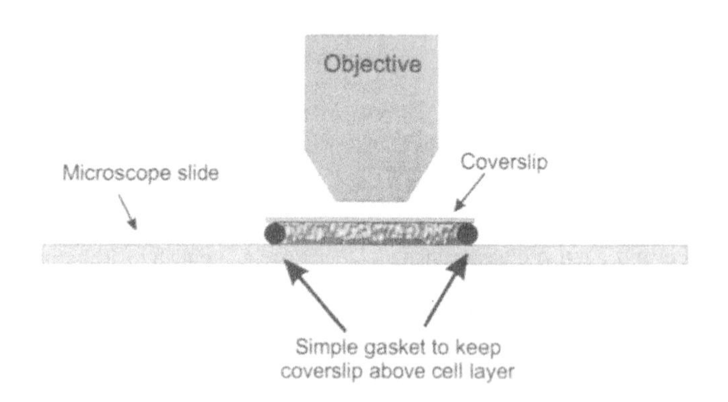

Figure 12-5. Gasket Chamber.
Placing a silicone rubber gasket between the microscope slide and coverslip can be used to form a cell chamber for imaging. This simple chamber can be constructed from readily available silicone rubber washers or plastic reinforcement rings used for ring binders. A variety of chambers based on the use of a silicone gasket are available from Grace Bio-Labs.

Microscope Slide Chamber

A simple incubation chamber can also be readily constructed by drilling a 10 mm hole through a glass microscope slide (Figure 12-6). This type of chamber can also be constructed out of suitable plastic sheeting. A coverslip is then stuck to the underside of the slide by using either nail polish (nail lacquer) for permanent chambers, silicone grease for temporary chambers or perhaps best of all, silicone rubber such as Sylgard (Dow Corning).

This type of chamber is simple to use, but does suffer from evaporation of liquid from the surface of the chamber, which may be a serious problem for maintaining correct salt concentrations. A further difficulty with the open chamber is that the meniscus around the edge of the hole will distort the transmission image. The meniscus does not distort the fluorescent image because the fluorescent image is created from light emitted from the sample and transmitted back through the objective lens - i.e. the light does not cross the surface of the liquid in the chamber. Maintaining the correct pH within such a small volume when the liquid is open to the atmosphere is quite difficult.

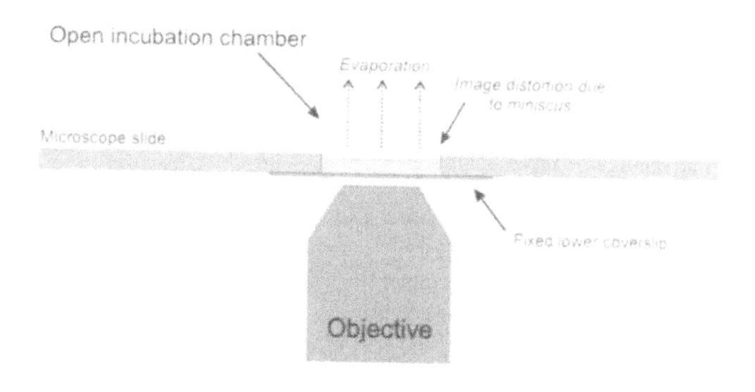

Figure 12-6. Open Microscope Slide Chamber.
A simple open incubation chamber can be constructed by fixing a coverslip over a 10 mm hole drilled in a conventional glass microscope slide, or even better, to drill a 10 mm hole in a plastic microscope slide that can be readily cut from a sheet of 2 to 3 mm thick polycarbonate or Perspex plastic.

This simple chamber can be readily loaded with live cells by pipetting in a cell suspension. The cells will slowly float down on to the fixed glass coverslip when placed on an inverted microscope. To speed up this process the incubation chamber can be centrifuged for a few minutes at a low speed (1000 rpm) in a bench top swinging bucket centrifuge. A Cytospin™ centrifuge with buckets to hold microscope slides will work best, but at a pinch the microscope slides can be gently taped to the top surface of a multi-tube swinging bucket.

Sylgard (Dow Corning)

The silicone rubber polydimethylsiloxane (Sylgard, Dow Corning) is suitable for sticking coverslips to microscope slides for forming imaging chambers, and can be autoclaved.

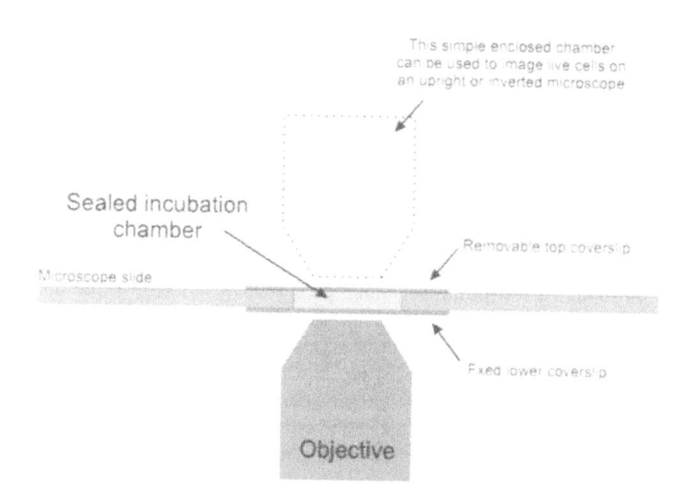

Figure 12-7. Enclosed Microscope Slide Chamber.
An improvement on the open incubation chamber depicted above in Figure 12-6 is to enclose the chamber by sliding a coverslip over the top of the incubation chamber, being careful not to trap any bubbles. The sealed incubation chamber is optically "flat" for transmission imaging, and will retain the correct pH and gas and ionic concentrations for up to 30 minutes.

A further improvement on this simple incubation chamber is to enclose the cell chamber by sliding a coverslip over the top of the chamber (Figure 12-7). If care is taken to completely fill the chamber with media, and to remove all bubbles before sliding the upper coverslip across the open chamber, a perfect bubble-free chamber can be obtained. Trapped bubbles will seriously degrade the transmission image. This enclosed chamber does not suffer from evaporation losses, and the distortion due to the meniscus is eliminated. The pH and gas levels of the media will be maintained at physiological levels for up to 30 minutes, depending on the number of cells present and the metabolic activity of the sample. Accessing the cells in the imaging chamber is a simple matter of sliding back the coverslip, changing the cells or adding a drug etc and then sliding the coverslip back in place.

An advantage of these chambers over most other simple microscopy chambers is that they can be easily used on both upright and inverted microscopes. To use on an upright microscope make sure the cells are attached to the base coverslip and then simply invert the enclosed chamber and image the "lower" coverslip now from above.

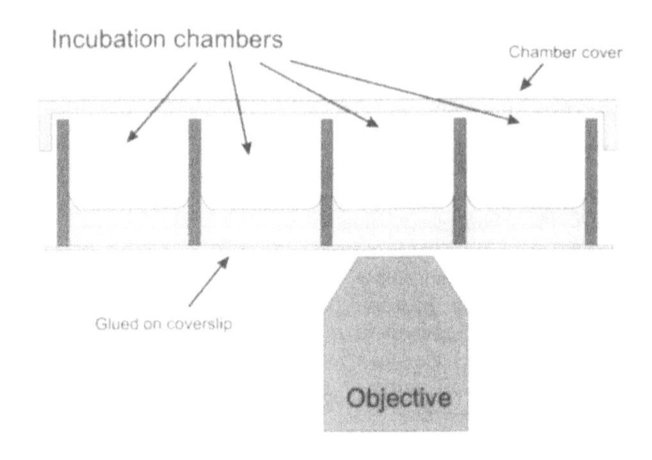

Figure 12-8. Multi-chamber Culture Dish.
These commercially available incubation chambers consist of a series of plastic chambers fixed to a coverslip or microscope slide. Lab-Tek Chambered Cover glasses are available from Nalge Nunc International.

Microscope slide chambers are very simple and convenient, but they don't provide any temperature control while the cells are being imaged. These simple incubation chambers can be readily kept at the correct temperature prior to imaging by placing on a temperature controlled dry heating block. Controlling the temperature of incubation during the experiment (dye loading, drug incubation etc) may be more easily accomplished away from the microscope. The chamber is simply placed on the room temperature microscope stage for obtaining an image. As the image can often be obtained within a few minutes or less, slower cellular processes will not be unduly affected by the lower than optimal temperature within the time frame of the experiment. Cells that can be studied at room temperature (plant cells, insect cells etc) will not need any further temperature control other than this simple chamber.

Attached cell lines can be grown in these incubation chambers by seeding 70% ethanol washed slide chambers. The nail polish does not survive autoclaving well, but is not affected by washing in ethanol. Acid cleaned incubation chambers will also be sterile if rinsed carefully in sterile water. The chamber slide can be incubated in a sterile Petri dish within a suitably gassed tissue culture incubator. Sterile water should be added to the Petri dish to stop immediate evaporation of the small volume of media within the microscope chamber slide.

These simple microscope slide chambers can be re-used by washing with detergent, rinsing well with water and air-drying. Soaking these chambers in detergent for more than a few minutes is not a good idea as this will soften the nail polish and may result in the lower coverslip coming off. If potent drugs etc are being used, removing the coverslip may be necessary for thorough cleaning (by soaking overnight in acetone if glass slides are used). Placing the slide chamber (again, glass slides only) in chromic acid for a few minutes (not longer than 5 minutes) will very rigorously clean the glass. This process will also remove any poly-L-lysine that has been used as a coating on the coverslip, and so re-coating will be necessary. Plastic chambers can be readily cleaned with acid alcohol (1% HCl in 70% ethanol) by filling the chamber with acid alcohol for a few minutes and then rinsing well with deionized water.

Sterile Culture Chambers

The above simple incubation chambers can be sterilized by washing with ethanol and drying in a sterile flow cabinet. However, there are a variety of incubation chambers available commercially (Figure 12-9 and Figure 12-8) that are supplied already sterilised with γ-irradiation.

The microscope chambers (shown in Figure 12-8) come in a variety of formats (these Lab-Tek chambers are available from Nalge Nunc International). They can be purchased with just one large incubation chamber (almost the size of a microscope slide), or several (up to 8) small incubation chambers on the one slide. The chambers have a loosely fitting plastic cover that will allow gas exchange when placed in a gassed and humidified tissue culture incubator.

These microscope chambers come in two basic forms, one has the incubation chambers attached to a microscope slide and the other has the chambers attached to a coverslip. The incubation chambers attached to a microscope slide can be "snapped off" when cell fixation and processing is required. The snapped off slide can then be mounted with a coverslip after processing. The coverslip attached culture chambers cannot be removed from the coverslip, as they are intended for high-resolution imaging of live cells on an inverted microscope.

These incubation chambers can also be purchased with a plastic microscope slide or coverslip attached. This is particularly useful for growing attached cells, where glass may not provide an agreeable surface for proper attachment.

Unfortunately, the design of the Lab-Tek incubation chambers does mean that placing a tightly fitted lid over the top of the incubation chambers is not possible, and this means that transmission imaging will be somewhat distorted by the meniscus and by the plastic cover on the chamber. These chambers are also expensive to use as they can only be washed and re-used with some difficulty.

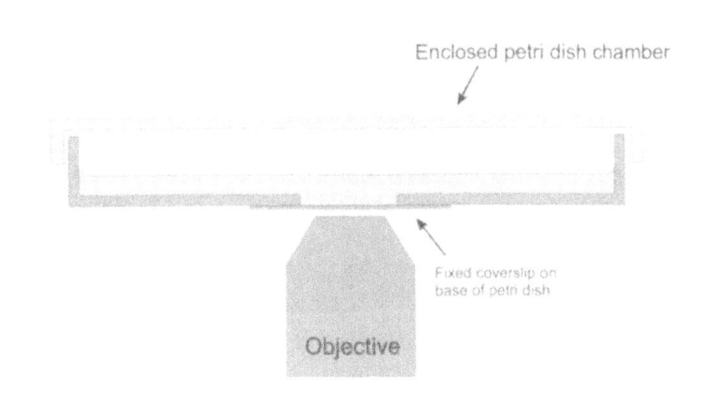

Figure 12-9. Petri Dish Culture Chamber.
A simple microscopy chamber can be made out of a plastic Petri dish. These chambers are available, with coverslip glued in place and sterile ready to use, from MatTek and Willco Wells.

Petri dish imaging chambers are available from MatTek and Willco Wells (Figure 12-9). These chambers consist of either a conventional plastic Petri dish (MatTek) which has a hole in the bottom covered with a stuck on coverslip, or a custom made Petri dish (Willco Wells). The larger volume available in the Petri dish minimises the danger of sudden changes of temperature and pH, which is always present when using very small volumes. The bottom coverslip allows high-resolution imaging of the cultured cells without the trauma to the cells caused by transferring them to an imaging chamber. These chambers come in a variety of formats, including coated cover glass for better cell adhesion, and coverslips with an engraved grid for easy relocation of the same cells.

The custom made Petri dish available from Willco Wells (similar in design to Figure 12-9) has a large round coverslip that covers the complete base of a modified Petri dish. These dishes come in a variety of sizes and designs. One design has a large coverslip base for viewing that covers the whole of a small (35 mm) Petri dish. Other designs have a small (10 mm) hole cut in the base of a plastic dish and a large round coverslip cemented over the base (with only the small 10 mm hole available for high resolution viewing). The Willco Wells dishes can also be purchased with a tight fitting plastic cover, unlike a conventional Petri dish that has a loose fitting lid for gas exchange.

Simple Flow Chamber

Many cellular reactions can be readily studied in a simple flow chamber (Figure 12-10). The advantage of the flow chamber is that, although the chamber is fully enclosed, introducing a change in the media (including the introduction of a dye or drug) while imaging, is relatively easy. Supply of media to the flow chamber is best achieved by using a motor driven syringe pump as depicted in Figure 12-10. Other types of pumps will create a pulsating effect that may be detrimental to the image under high magnification. Alternatively, a simple gravity fed siphon may provide a sufficiently steady flow through the chamber – particularly if a relatively large volume of media can be used to lower the changes in flow rate as the media level drops, or alternatively, the media supply bottle is placed on a high shelf and a needle valve is used to control the flow of media down through the chamber.

The glass tube used for this simple flow chamber can be purchased from VitroCom in a variety of optically flat rectangular and square formats. They are manufactured either from glass or quartz (for UV imaging). The flow chamber can be coated with a variety of cell adhesion compounds, such as poly-L-lysine, or specific cellular attachment molecules that you may be interested in studying.

The thickness of the glass (or quartz) wall of these rectangular tubes, and thus the "coverslip" thickness, is determined by the size of the tubing (i.e. larger diameter tubing has thicker walls), and so care must be taken when attempting high-resolution imaging to attempt to match the tubing wall thickness to close to the ideal coverslip thickness (170 μm). This will necessitate the use of a very small diameter flow tube. For lower resolution imaging the thicker walled, and hence larger diameter, tubing may not be a problem.

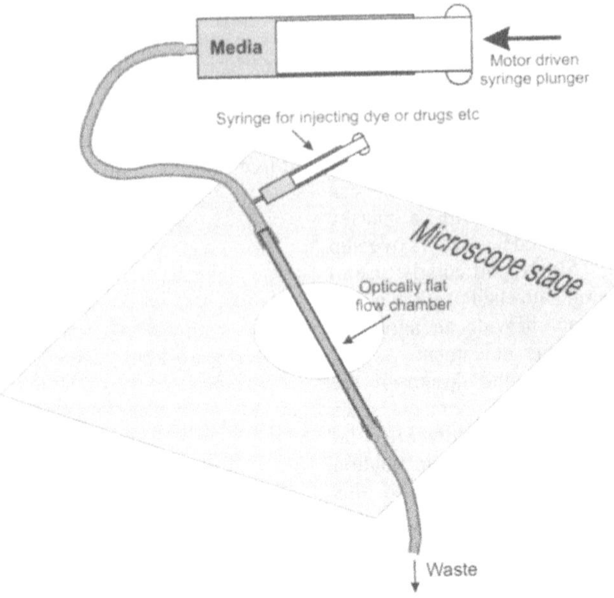

Figure 12-10. Simple Flow Chamber.
The rate of media flow is regulated by the motor driven syringe plunger in this simple flow chamber. The cells are imaged through an optically flat imaging flow chamber held in position on the microscope stage. A variety of rectangular and square glass and quartz tubing that can be used as a simple flow chamber, is available from VitroCom.

ATTACHMENT OF CELLS

Attached cells can be simply grown in the microscope chamber, but suspension cells will in most cases need to be attached to the coverslip for obtaining high quality images. This is particularly important when collecting 3D data sets or when collecting time-delayed images of the same cells. Tissue samples, on the other hand, may not need to be attached to the coverslip. However, if tissue samples do need to be attached this can be accomplished by using the cell adhesion reagents discussed below. Tissue samples can also be gently pressed against the coverslip by using a small weight (often a small wire grid device, known as a "harp") to hold the tissue in place.

Poly-L-lysine: commonly used "cell adhesive"

Attaching Suspension Cells to the Coverslip

Suspension cells can be readily attached to the bottom coverslip of a microscope incubation chamber by pre-coating the incubation chamber with a variety of compounds (see box on next page). One of the most common is poly-L-lysine, although a possibly more effective compound is polyethylene imine. By using a cell adhesion compound, the cells will be stuck firm enough such that the overlying buffer can be removed and replaced with fresh buffer containing drugs etc without disrupting the cells. The cells can still be readily removed from the chamber by more vigorous washing.

Cells attached to a glass coverslip with poly-L-lysine can be easily disrupted by high levels of serum present in the culture media (normal culture media contains up to 10% foetal calf serum). If you are having trouble attaching the cells firmly to the coverslip try attaching the cells by using serum free media. If the imaging time is only a few minutes then simply use your normal culture media, but omit the serum. The use of serum free culture media should be considered for longer-term imaging. The serum can often be added back to the cells once the cells have been attached to the cover glass. However, the replaced serum will eventually result in cells becoming detached.

Attaching suspension cells to a surface blocks endocytosis

The biochemical processes of suspension cells may be altered by making them adhere to a coverslip. This is particularly noticeable with membrane related phenomena such as "capping" and endocytosis - which is inhibited by attaching the cells to a substrate. Interference with endocytosis will also result in changes in cellular processes that are dependent on endocytosis of surface receptors for triggering intracellular events.

Cells can also be attached to microscope chambers by coating the coverslip with a variety of other reagents, such as Cell-Tak (available from Bectin Dickinson), Silanes such as 3-aminopropyltriethoxy-silane (Sigma) or Concanavalin A etc. You may also find that your suspension cells attach strongly enough to a clean glass surface, without any "cell adhesive", for you to obtain high quality images. Treating the glass in a variety of ways can often enhance this attachment. Some workers have found that autoclaving the coverslip results in increased attachment. Another "trick" is to "flame" the coverslip by passing the glass through the blue flame of a Bunsen burner two or three times. Acid washing the slides and rinsing well in distilled water can also help the cells to attach. Finally, centrifuging the cells down on to the coverslip will also help the cells to stick more firmly – whichever adhesive you use.

Sticking cells to coverslips
Poly-L-lysine
Polyethylene imine
Cell-Tak
Silanes
Concanavalin A
Autoclaving the coverslip
Acid cleaning the coverslip
"Flaming" in a Bunsen burner

Centrifuge the cells down on to the coverslip for firmer attachment

As already mentioned above no matter how the cells are made to adhere to the chamber coverslip, endocytosis is severely curtailed when suspension cells are stuck down. If your work requires functioning endocytic events in suspension cells then the experiment will have to be done with the cells either simply "sitting" on the coverslip surface, or by incubating the cells in a suitable culture chamber/tube where the cells are kept in suspension. An aliquot of the cells can then be placed in the microscope chamber for imaging.

An alternative to attaching the suspension cells to the microscope chamber coverslip is to embed the cells in low gelling temperature agarose (1%). The agarose is melted at 45°C and the temperature then lowered to 37°C (the Agarose will still remain liquid), the cells are added and the mixture is poured into a microscope-imaging chamber. The temperature of the chamber with the embedded cells is then lowered briefly to set the agarose. The agarose will now remain set when incubated at 37°C. Even highly mobile cells such as live sperm can be imaged in this way.

Growing Attached Cells for Microscopy

Cells that naturally grow attached to a surface (such as the commonly used HeLa cells) grow best when attached to specially treated plastic culture dishes. Attachment cells do not readily adhere to a glass surface. Although the cells may grow reasonably well on the glass surface, such cells are often observed to "float off" when stressed. Cells grown on a plastic surface can withstand considerably more stress before becoming detached. However, coverslips can be coated with a variety of cell attachment agents that greatly facilitate cell growth. Interestingly, most so-called attached cells may grow better when "attached" to a more "natural" molecular matrix. Growing attached cells to a bead matrix is often used for large-scale production – and may be worth investigating for confocal microscopy.

Try using low-melting agarose to immobilise active cells

All of the above simple incubation chambers can be readily used to grow attached cells. The commercially available γ-irradiated microscopy chambers are bought sterile and ready to use for cell culturing. However, sterilising the simple "home made" chambers can be somewhat difficult, as they cannot be autoclaved due to the heat intolerance of the glue used to attach the coverslip. The chambers can be sterilised by washing with ethanol or acid and then rinsing thoroughly with sterile culture media.

The commercially available incubation chambers can also be purchased with plastic coverslips attached. This is particularly useful for growing attached cells, where glass may not provide an agreeable surface for proper attachment. Plastic coverslips can also be attached to the "home made" imaging chambers. Unfortunately, although plastic coverslips provide improved cell attachment, they are not as good as glass coverslips for high-resolution imaging. Plastic coated glass coverslips may become available eventually for providing good cell adherence and high quality imaging.

Plastic coverslips;
can be used instead of regular glass ones where good cell adherence is important.

TEMPERATURE CONTROL

Mammalian cells have a very strict optimal growth temperature of 37°C. Other organisms (such as plant and insect cells) have a less stringent temperature requirement, and can often be imaged successfully at room temperature. However, most cells, including single celled organisms such as yeast and bacteria, have an optimal growth temperature, even if they are perfectly capable of surviving or even growing at a wide range of temperatures. Changes in temperature will change the physiology and growth characteristics of all cells, even if the cells do not require a particularly narrow temperature range for long-term survival. This section is mainly concerned with temperature control in mammalian cells, but the principles involved are applicable to other organisms even though their temperature requirements may not be as stringent.

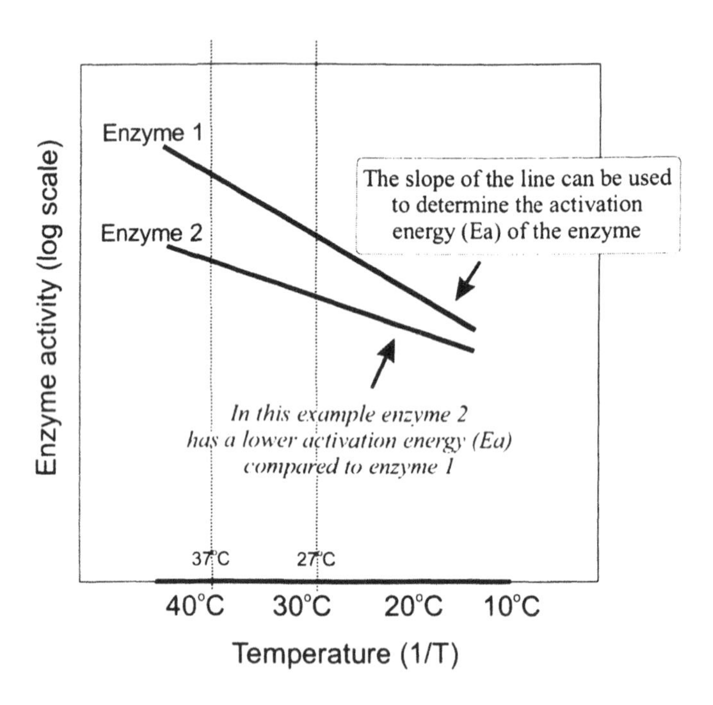

Figure 12-11. Enzyme Activity Changes with Temperature (Arrhenius plot).
The rate of reaction of all cellular enzymes is related to the temperature at which the enzyme is assayed. In a mammalian cell the thousands of individual enzymes have been optimised for operating at 37°C. Lowering the temperature, for example to 27°C, will result in not only slower enzymatic rates - but more importantly significant changes in the relative rates between individual enzymes. This temperature effect on enzymatic rates may have important consequences for imaging live cell or tissue samples on a confocal microscope.

Most cellular metabolic reactions will continue to proceed even when the cells are incubated above or below the optimal growth temperature, but some cellular functions, such as endocytosis in mammalian cells, have a very strict temperature requirement. Although many cell processes may occur at a temperature far removed from the cell's optimal growth temperature, how close will such a study be to what is happening within the living organism? Particular care should be taken with studies that involve finely balanced cellular control processes - are those Ca^{2+} changes indicative of the true physiological state of the cell? Initial pioneering experiments on cellular processes using imaging methods did yield valuable information on cellular processes even when carried out at inappropriate temperatures. However, when you are studying the intricacies of the cellular process, rather than the ability, for example, to follow Ca^{2+} changes within a cell, temperature control becomes a very important issue.

Is temperature control during imaging necessary for your sample?

Why is Temperature Control Important?

The rate of activity of cellular enzymes is well known to be dependent on the temperature at which the assay is carried out. What is often overlooked is that as the temperature is lowered the enzymatic activity of different enzymes will be altered to different extents (Figure 12-11). For example, the mitochondrial inner membrane ATPase activity may drop significantly further than for other cellular enzymes. The cell has a remarkable ability to compensate for these changes, but if you carry out the experiment away from the optimal temperature for your particular cells you may not be measuring the expected "in vivo" physiological event, but perhaps a temperature induced non-physiological change.

The sloping line in Figure 12-11 represents the change in enzymatic activity as the temperature is changed. The slope of the straight line (when graphed on a logarithmic scale) can be used to calculate the activation energy of the enzyme (Ea). The steeper the slope of the graph, the higher the activation energy of the particular enzyme being studied. However, the slope, and hence the activation energy, of each enzyme is different. Although cellular processes slow down on lowering the temperature, the fact that different enzymes have different activation energies will result in a change in the balance of enzymatic activity within the cell as you lower or raise the temperature. For example an enzyme with a high activation energy (enzyme 1 in Figure 12-11, note the steeper slope of the line compared to enzyme 2) will drop off in activity much faster with a drop in temperature than an enzyme with a lower activation energy (enzyme 2 in Figure 12-11). This will result in a different balance of enzymatic activity when the cells are grown at 27°C, compared to their optimal growth temperature of 37°C. Although cells can compensate for some of these effects of changes in temperature and continue to grow (only within a relatively narrow range for mammalian cells – see membrane phase transitions discussed below), there may still be significant changes in the metabolic or physiological state of the cells.

A small rise in temperature to 40°C can seriously disrupt cellular processes in most cells. Only cells that are specifically adapted to high temperatures, such as thermophilic bacteria, can tolerate prolonged incubation at elevated temperatures. In the case of thermophilic bacteria this can be as high as 75°C, or even much higher temperatures if the cells are also under high pressure. Once the temperature rises in most cell types the cells respond by activating a number of "heat shock" proteins. These proteins are designed to either re-fold or remove denatured enzymes. Above 40°C mammalian cells have great difficulty surviving in the long term, with most cells eventually dying by apoptosis. Even a small rise in temperature above 37° may result in many changes in the physiology of the cells due to the induction of this "heat shock" response. Care should be taken not to expose the cells to even short periods of relatively high temperature as the "heat shock" response will continue for many hours, even after the removal of the original cause of the shock response.

Most mammals closely control their body temperature, and so isolated tissue samples and cultured cells from mammals have a strict requirement for growth at 37°C. This strict temperature requirement is not simply a matter of being the optimal temperature at which their enzymes will operate, but as explained below, if the incubation temperature is lowered below 18 - 22°C there are serious changes to the structure of the cellular membranes - this has an important impact on both enzymatic activity and the viability of the cells.

Cellular processes that are associated with or dependent on the cellular membranes will have an activation energy that is influenced by the fluidity of the membrane. This will result in a "break" in the slope of the line shown in Figure 12-12 (in this example at 23°C). As the membrane lipid cools, the fluid membrane lipid will become more rigid and form a crystalline lipid phase that will alter (usually increase) the activation energy of membrane-associated enzymes. The relative enzymatic activity of membrane and non-membrane bound enzymes will become markedly different below the phase transition temperature. This will result in a significant shift in the balance of the metabolic activities within the cell. In fact mammalian cells have very little metabolic activity below their phase transition temperature. This is particularly noticeable with the production of energy by the mitochondria, where oxygen consumption almost stops as the temperature is lowered below the phase transition temperature. This phenomenon can be exploited for preserving cellular integrity in conditions where oxygen availability may be a problem, but will have serious consequences on cellular processes under study if this should happen whilst imaging.

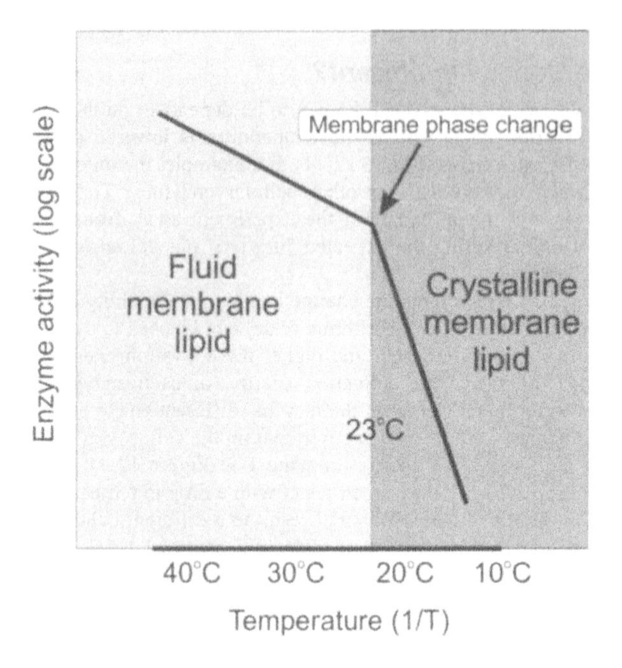

Figure 12-12. Membrane Phase Changes.
Mammalian cells undergo a lipid phase transition, often between 18° to 23°C. This phase transition has a profound effect on membrane dependent cellular functions such as ion transport, enzymatic activity and even DNA replication. Mammalian cells incubated below the membrane phase transition temperature will not survive long term culturing. Unfortunately, most confocal microscopes are maintained in a room temperature of between 22° and 24°C – right in the middle of this membrane lipid phase transition for mammalian cells!

Once the cellular plasma membrane has dropped in temperature below the lipid phase transition point, endocytosis and endocytic signalling events will also be severely curtailed. The exact temperature at which this phase change takes place will vary considerably, depending on both the origin of the cell and the incubation conditions. Most mammalian cells pass through a phase transition between 15 and 25°C. Endocytosis, although only occurring at a fraction of the rate found above the phase transition temperature, still does proceed – even down to quite low temperatures of 10°C. Therefore, if you are using low temperatures to block endocytosis (for example, to establish the level of surface labelling at the "zero" time point in studies utilising endocytosis) you must keep the temperature very low during cell manipulations and imaging (preferably at 0°C by incubating culture dishes etc containing the cells on ice).

The problem of temperature control while imaging is further exacerbated by the fact that most laboratories are maintained at the temperature at which most mammalian cells are undergoing the membrane phase transition. Other cells, such as yeast and plant cells may undergo a phase transition that is much lower than that for mammalian cells (often below 4°C). This means that these cells will not only readily tolerate a much greater range in temperature, but that changes in temperature will not have such a profound effect on the metabolic state of the cells.

Most mammals closely maintain their body temperature at 37°C. Small changes to this temperature will result in severe stress to the animal and if the change is prolonged will result in serious consequences for the viability of the animal. The exceptions to this tight temperature control are mammals that hibernate - in which case the body temperature can be lowered to just above freezing without any detrimental effects. Mammals that hibernate lower the phase transition point of their membranes before entering hibernation. In this way, although a hibernating animal may lower its body temperature to 10°C or lower, the cellular membranes never pass through the phase transition temperature. Thus hibernating animals can avoid the dangers of prolonged exposure to low cellular temperatures.

Use a miniature temperature probe to determine the temperature of the microscope chamber *where the cells will be imaged*

Another, quite unrelated, reason that temperature control is so important in microscopy is to decrease the amount of focus shift due to temperature changes when imaging over a relatively long time period. The temperature in the room where the microscope is housed is likely to fluctuate quite markedly between day and night – and sometimes there may be a steady increase in temperature due to inadequate external exhausting of the laser cooling fans. The changes in room temperature will result in changes in the temperature of the microscope and associated imaging chambers. This will result in differential expansion and contraction of various metal, plastic and glass components of the instrument – which will result in a change in the focal position! Enclosing the microscope (see page 297) is the most effective method of minimising these temperature fluctuations.

Is Temperature Control Needed while Imaging?

The above discussion demonstrates that temperature control of mammalian cells is most important. However, maintaining accurate temperature control during imaging at temperatures higher or lower than room temperature is technically difficult, as will be explained below, and so establishing whether full temperature control is necessary is most important.

Fully controlling the incubation temperature of the cells prior to placing the sample on the microscope is a relatively simple matter. This can be achieved by either placing the cells in a suitable tube in a tissue culture incubator or in a temperature-controlled heating block. An aliquot of the culture can then be removed from the incubation tube and placed in a simple microscope incubation chamber for imaging. Alternatively the whole microscope chamber, which has been maintained at the correct temperature by placing it in a tissue culture incubator, can be moved to the microscope. Images are then collected without continuing temperature control. If the cellular processes under study occur relatively slowly (for example, intracellular protein movement), excellent representative images may be obtained within a few minutes of placing the sample on the microscope – without any temperature control being necessary. Full temperature control while imaging is probably only necessary when studying fast acting cellular changes in real time, such as calcium or other ion changes in living cell or tissue samples.

As mentioned above, full temperature control during imaging is not a simple matter. Probably the only sure way of ensuring rigorous maintenance of the correct temperature is to have not only the imaging chamber, but also the whole microscope at the required temperature. This is clearly not a problem for imaging cells at room temperature, but as mammalian cells are best maintained at 37°C, placing the confocal microscope in a heated room is not a practical temperature for either the operator or the computer systems associated with the microscope! However, an enclosed incubation chamber (discussed in detail in the next section) can be built around the microscope to accurately maintain the correct temperature of the cells being imaged.

Hibernating mammals lower the phase transition temperature of their membranes – allowing their cells to survive low temperatures for many months

Temperature Control on Transfer to Microscope

In microscopy, very small samples can very easily be cooled to room temperature in the few moments taken to transfer a sample to the microscope. If temperature control is critical, take all due care to prevent the temperature dropping during transfer. This is particularly critical if you are using a small coverslip based microscope incubation chamber on which is attached a single cell layer. If you place the chamber dish for a short period on a bench top (particularly the metal bench top used for some vibration isolation tables) you may find that the cells have been exposed to a much lower temperature (or higher, if you are trying to keep them cold) than is important for your experiment.

If possible allow the sample to equilibrate at the correct temperature on the microscope stage for some time before starting the experiment. In this way you can be sure that the cells are properly maintained at the correct temperature during the course of the experiment. However, don't forget that a temperature "shock" may have lasting consequences long after the cells have been returned to the correct temperature – so take care, and always think carefully as to whether temperature control while imaging is actually important for your study.

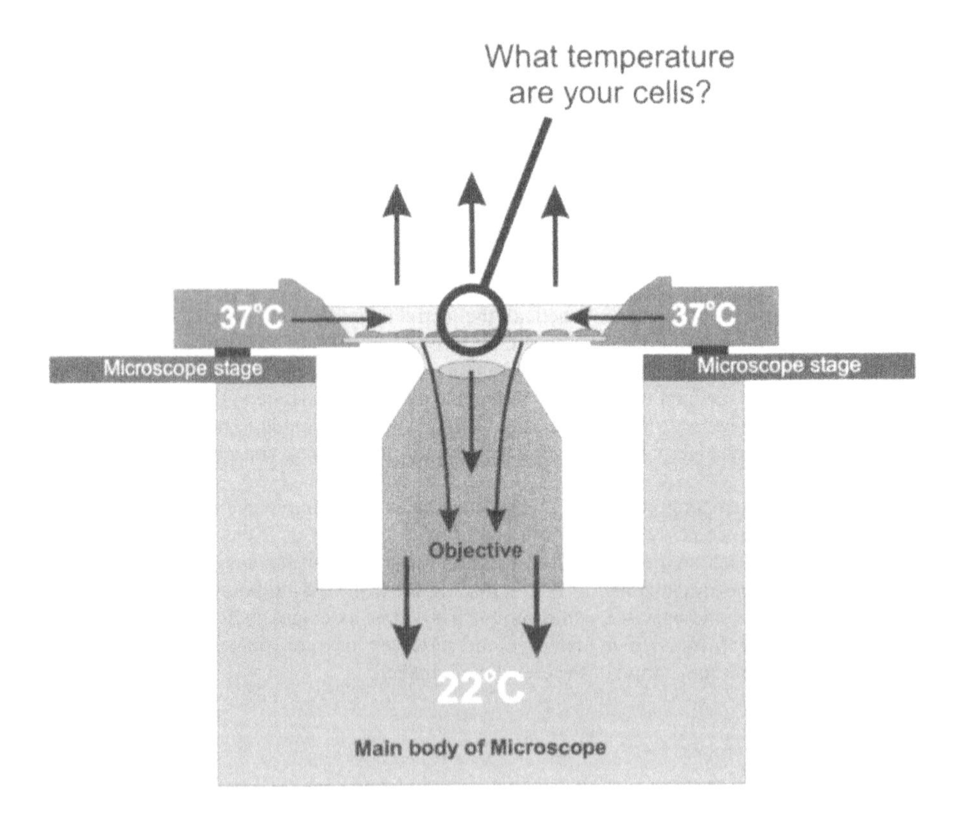

Figure 12-13. Temperature Control on the Microscope.
A heated chamber for microscopy may have very accurate temperature control - but what temperature are your cells? The surrounding atmosphere, the objective and the microscope are usually at a lower temperature to the incubation chamber. This will result in considerable heat loss from the heated imaging chamber, often resulting in the cells that are being imaged having a temperature below that of the incubation chamber.

TEMPERATURE CONTROL CHAMBERS FOR MICROSCOPY

As discussed above, accurate temperature control of live cells or tissue samples while imaging by microscopy is quite difficult. Although the incubation chamber may be very accurately maintained at the correct temperature, the surrounding atmosphere and microscope components, which are usually at a lower temperature, will constantly remove heat from the incubation chamber (see Figure 12-13). This creates a very unstable heating environment where the constant flow of heat from the sample needs to be compensated by providing more heat to the incubation chamber. This temperature imbalance creates a gradient of temperature across the imaging surface of the chamber, creating difficulties in maintaining the correct temperature of the cells that are actually being imaged.

Heated incubation chambers and microscope warming plates are often mounted on small plastic insulating blocks to isolate them from the microscope itself. This will considerably reduce the loss of heat to the microscope, but unfortunately when using an immersion lens there is also loss of heat via the immersion fluid. Heat loss via the immersion fluid is particularly serious, as the greatest heat loss will be from the area of the sample under observation. With a "dry" lens the air gap between the imaging chamber and the objective lens provides an insulating layer that helps prevent heat loss from the chamber to the microscope via the objective lens.

Further heat loss can occur across the surface of the media, particularly in an open incubation chamber. Try placing a lid over the incubation chamber to both maintain the heat within the chamber, and to lower the loss of liquid through evaporation. A chamber cover may also be necessary to maintain sterility if you are attempting long-term incubation on the microscope stage.

Maintaining the cells under observation at the correct temperature is clearly desirable, but maintaining cells in other parts of the imaging chamber at the correct temperature if they are to be imaged later in the experiment, may also be important.

As will be discussed shortly there are many different ways of tackling the problem of temperature control of live cells during microscopy. The most reliable method of providing the correct temperature to the whole of the imaging area is to in fact enclose the microscope in a heated enclosure. A fully enclosed microscope is somewhat inconvenient to use (via small covered openings in the enclosure) and as you may not have permission to enclose a multi-user microscope, using a less than ideal heated incubation chamber may be necessary.

The Bioptechs incubation chamber, which is discussed in some detail at the end of this section, is the only chamber that comes close to full temperature control of live cells on the microscope. However, many other chambers, used in conjunction with an objective heater will provide reasonably good temperature control.

In this section several temperature controlled imaging chamber designs will be described, along with some of the important advantages and disadvantages. For more detailed information on the large number of temperature control chambers available you should consult the company information available on the Internet (see Chapter 14 “Technical Supplies”, page 335 for relevant web sites).

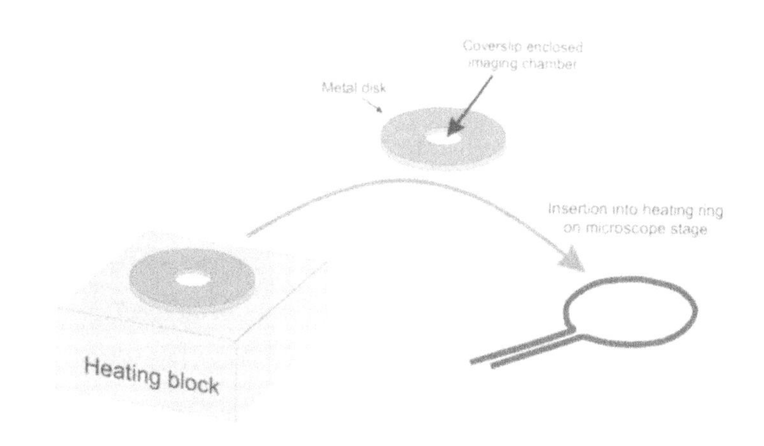

Figure 12-14. Metal Disk Chambers.
Metal disk (3mm thick) with a 10mm diameter hole in the centre that is covered on one side with a fixed coverslip and on the other with a removable coverslip. These simple chambers can be used to help maintain the correct temperature of small cultures on transfer to the microscope stage. A heating ring on the microscope can be used to maintain the chamber temperature while imaging.

An expensive temperature control chamber may not be necessary - simple heating devices that you can put together yourself may work just as well, if not better, than most commercially available chambers. These "simple" devices include both microscope enclosures and temperature controlled incubation chambers.

Simple Temperature Control Chambers

> **Metal Chambers**
>
> **Aluminium:** conducts heat fast but does not readily retain heat.
>
> **Stainless Steel:** retains heat well, but is slow to warm (but also slow to cool).
>
> Metal imaging chambers may need to be coated with a thin layer of non-toxic plastic (lacquer) if contact with the metal is toxic to your cell or tissue sample.

A simple temperature control chamber can be constructed out of a metal disk culture chamber and a heating ring mounted on the microscope stage (Figure 12-14). The heating ring on the microscope can be made from tubing through which warmed water is pumped. However, a more convenient method of heating is to make use of an electric Peltier heating ring, which removes the risk of water leaking onto the microscope. A Peltier heater, with suitable electronic controller (see page 298), can provide very accurate control of the temperature of the ring. The heat from the ring is readily transferred to the metal culture chambers. Aluminium culture chambers provide excellent heat transfer (see box on this page). In contrast stainless steel culture chambers do not provide fast heat transfer, but they do act as a very good "heat sink", resulting in less fluctuation in temperature when transferring the culture chamber from heating block to the microscope.

The disadvantage of using metal chambers is that the heating while on the microscope stage is from the edges of the media chamber – possibly resulting in your cells being at a different temperature to that of the metal surround. If full temperature control is required during imaging, then there are a number of options available. This may include heating the objective of the microscope to the required temperature, heating the microscope stage, using a heated incubation chamber or enclosing the microscope itself in a heated enclosure. Various incubator and chamber designs, that have varying success in controlling the temperature of the cells being imaged, are described in some detail on the following pages.

Heated Objectives

Probably the greatest heat loss from the sample is through the objective. Not only is the objective itself a good heat sink, but also the heat from the incubation chamber can be conducted down through the metal in the objective to the large metal body of the microscope itself.

As mentioned above, dry objectives do not result in as much heat loss, as there is an insulating layer of air between the objective and the imaging chamber. Oil and water immersion objectives have greater heat transfer properties and so controlling the incubation chamber temperature with these objectives is much more difficult. Ceramic cased objectives used for "dipping" into the culture media will not affect the temperature of the incubation chamber as much as a metal encased objective due to the insulating properties of the ceramic casing material. The electrical insulating properties of the ceramic lens make them ideal for electrophysiology.

Figure 12-15. Objective Heater. An objective heater is essential for full temperature control. The heater is usually used in conjunction with a temperature controlled microscope culture chamber. This image was kindly provided by Dan Focht, Bioptechs, USA.

Microscope objective heaters (Figure 12-15) are usually used in conjunction with a heated incubation chamber to help maintain the cells under observation at the correct temperature. A simple heating collar around the neck of the objective, as shown in Figure 12-15, is probably the simplest and most effective method of heating the objective. The heating band used in the Bioptechs objective heater shown in Figure 12-15 can be readily fitted to most objectives. A close-fitting heating band is required to ensure efficient heat transfer to the objective. Take care when heating objectives with adjustable "collars" as lubricant may leak from the moving joints and cause problems with the optics. If in doubt contact the lens manufacturer about the suitability of particular lenses to be heated above room temperature.

Although an objective heater is normally used in conjunction with other heating devices, the heater may, in fact be used alone to maintain the correct temperature of the cells under observation. However, when used in combination with other heating devices the whole of the incubation chamber can then be maintained at the required temperature (not just the few cells under observation).

There does not appear to be any problem with heating the microscope objective to 37°C, although higher temperatures are not recommended. However, continually heating and cooling an objective is not good for the integrity of the lens. This may cause undue stress on the delicate optics of the lens. If you do wish to cool the objective to low temperatures (such as 4°C and then re-heat again to 37°C) contacting the manufacturers of your objective may be advisable to determine whether the lens can withstand such temperature changes.

Heated Microscope Enclosures

Most microscope manufacturers provide a very accurate temperature controlled plexi-glass enclosure for their microscopes. Figure 12-16 shows a Leica plexi-glass enclosure mounted on a Leica inverted microscope. The PID (Proportional Integral Derivative) controller provides extremely accurate and very stable temperature control of the air within the plexi-glass enclosure. Leica manufactures a variety of plexi-glass microscope enclosures, including those that fully enclose the microscope. The microscope stage and objectives are not enclosed in the Leica plexi-glass enclosure shown in Figure 12-16. In this case additional heating devices may be required to ensure that the sample itself is maintained at the correct temperature. This may include a heated microscope stage, a heating plate or a heated incubation chamber - all of which will be discussed in some detail shortly. Finally, a heated objective, as discussed above, may be the most essential device for maintaining the actual cells being imaged at the correct temperature even when using a heated enclosure.

A simple plexi-glass enclosure can be readily constructed (Figure 12-17), although such an enclosure will normally need to be "custom made" for the particular make and model of microscope you are using. Practical experience has shown that a simple on/off thermostat controller (see boxed information) will provide sufficient temperature control for most applications. The fluctuations in temperature associated with a thermostat-controlled enclosure can be minimised by using a heating device of relatively low power (getting to the correct temperature takes longer, but there is less fluctuation in air temperature as the heater is turned on and off). The enclosed air temperature is also more easily maintained at the correct temperature if the air is recycled back through the heater rather than continually heating ambient temperature air. The large amount of metal present in the microscope, the microscope stage and the objectives will also help to dampen the air temperature fluctuations generated by the thermostat controller.

The plexi-glass enclosure shown in Figure 12-17 does enclose the microscope stage and the objectives. This results in better temperature control of the sample than the enclosure depicted in Figure 12-16 (although the temperature controller used in Figure 12-16 is a much more accurate PID controller).

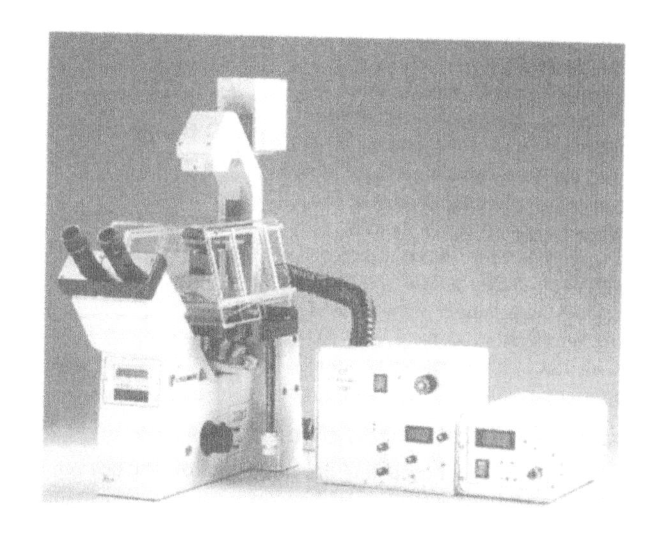

Figure 12-16. Leica Heated Microscope Stage Enclosure.
This Leica heated microscope enclosure provides accurate and stable heating of the microscope stage, using a PID electronic controller. However, the objective is not enclosed in this heated enclosure. Leica also manufactures a wide range of other enclosures, including those that fully enclose the microscope. This image was adapted from a Leica Internet published photograph.

"Home Made" Reflective Plastic Microscope Enclosure.

A simple but very effective method of providing accurate temperature control to your sample on the microscope is to totally enclose the microscope within an enclosure of reflective building insulation plastic (Figure 12-18). Reflective plastic (a reflective "bubble" plastic, which is widely used in the building industry for wall and ceiling insulation), is a relatively simple and very cheap material for constructing a microscope enclosure - and performs better than most commercially available enclosures! I would like to thank Scott Fraser (Caltech, USA) and Steve Potter (Georgia Tech, USA), for sharing with me their design for this simple microscope enclosure. Their design uses reflective bubble plastic from Reflectix Inc., USA (See page 343 for contact details), although a variety of building insulation plastics would be suitable.

A "box" of reflective plastic is formed around the microscope (excluding the Hg lamp) and held together with insulation tape. A small "window" is cut in the front of the enclosure to allow the binocular eyepieces of the microscope to be accessed. The reflective plastic should be closed reasonably tightly around the eyepieces to ensure that there is not too much loss of warm air. Small access flaps are then cut into the front and / or side of the reflective plastic enclosure to allow access to the microscope controls and the sample.

A relatively small thermostat controlled fan warmer (such as those used to warm chicken houses) is used to heat the air in the enclosure. Large diameter flexible plastic tubing is used to recycle warm air from the chamber back through the heater (this minimises temperature fluctuations within the enclosure), and to keep the fan somewhat remote from the enclosure to minimise vibration problems. The large amount of glass and metal associated with the microscope will take several hours to equilibrate to the correct temperature. In fact, good stability is best obtained by starting the heating of the microscope enclosure the day before your experiment. This simple chamber design provides surprisingly accurate and very stable temperature control of your samples over long periods of observation on the microscope.

A word about temperature controllers

Simple ON/OFF thermostat controllers:
Most household appliances that require temperature control utilise a simple on/off thermostat to regulate the temperature. The thermostat simply regulates the power supplied to the heater, with the amount of time the heater is on used to regulate the final temperature of the device. Thermostat controllers are cheap and very reliable, but do suffer from significant temperature fluctuations and sometimes instability.

P (Proportional) controllers:
Electronic temperature controller that provides less fluctuation in temperature.

PD (Proportional Derivative) controllers:
A more refined electronic temperature control device, resulting in less temperature fluctuations.

PID (Proportional Integral Derivative) controllers:
These are the most sophisticated of the electronic temperature controllers. These controllers can very accurately maintain the correct temperature with almost no fluctuation (within 0.1°C) over a very wide temperature range.

Problems with Microscope Enclosures

One of the drawbacks of enclosing the whole microscope or even just the microscope stage is that access to your sample can be difficult. A well-designed enclosure will minimise the disruption caused by many of the microscope controls being enclosed. However, a fully enclosed microscope will always create some difficulties as most of the microscope controls are within the incubation enclosure. Care needs to be taken not to leave the enclosure open more than absolutely necessary when accessing the sample or microscope controls. Leaving the enclosure open will result in large fluctuations in the air temperature within the enclosure.

A further difficulty of using an air-heated enclosure for microscopy is that the large movement of heated air within the enclosure will result in high rates of evaporation. This is particularly detrimental when using small volumes of media for microscopy. Great care needs to be taken to use either an enclosed cell incubation chamber on the microscope stage, or to use an open perfusion chamber with a sufficiently high flow rate to alleviate problems associated with evaporation. The pH of small volumes is also very easily disturbed by the movement of heated air above an open container within the enclosure. For greater long-term imaging stability, using FEP Teflon® film to cover the incubation chamber is worth considering (see the section "Long Term Growth" on page 312). This membrane blocks the transfer of water vapour, and considerably inhibits, but does not block the transfer of O_2 and CO_2 – allowing cells to be grown for extended periods on the microscope stage.

Microscope enclosures also take some time to reach the required temperature (this may be several hours for the "home made" insulating bubble plastic enclosure described above). The advantage of a completely enclosed microscope enclosure, even if the enclosure does take a long time to come up to temperature, is that the temperature is very stable once the correct temperature has been reached.

Using a heated stage and objective heater may also be beneficial, even when the microscope is fully enclosed with a temperature regulated enclosure, to further stabilise the temperature of the cells under observation.

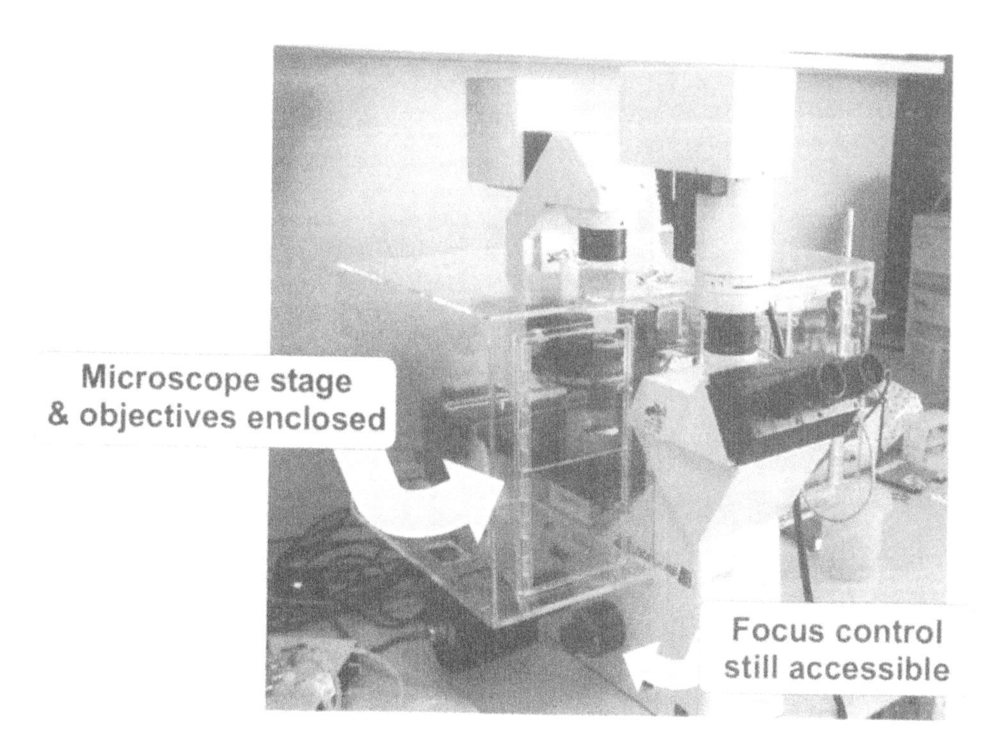

Figure 12-17. Plexi Glass Microscope Enclosure.
This heated enclosure provides surprisingly accurate and stable heating of the microscope stage, including the objectives. Heating is provided by a simple thermostat controlled fan heater (not shown). Stability is achieved by the considerable "heat sink" provided by the metal and glass of the microscope. This photograph is kindly provided by Ian Harper, Monash University, Australia.

Figure 12-18. Insulated Plastic Microscope Enclosure.
This simple "home made" microscope enclosure provides surprisingly accurate and stable heating of the microscope stage, including the objectives. Heating is provided by a simple thermostat controlled fan heater. Stability is achieved by the considerable "heat sink" provided by the metal and glass of the microscope. Access to various microscope controls can be modified as required. The magnetic strips attached to the front opening provide a convenient method of repeated access to the microscope. This photograph is kindly provided by Steve Potter, Georgia Institute of Technology, USA.

Heated Microscope Stages

Most of the major microscope manufacturers provide heated stages for a number of their microscope models. The Leica heated microscope stage (Figure 12-19) replaces the conventional stage on their inverted research microscopes. Heated microscope stages supplied by the microscope manufacturers provide very accurate temperature control of the stage itself (using PID electronics, see text box on page 298), but as is discussed throughout this chapter this may not provide sufficient temperature control of the sample under investigation. Additional temperature control devices, such as a heated objective and incubation chambers may be required.

A heated microscope stage can be readily used to warm a variety of imaging chambers. This includes the Lab-Tek microscope slide sized imaging chambers, the MatTek or Willco Wells Petri dish imaging chambers or other suitable simple chambers used for live cell microscopy.

A number of other manufacturers have developed heated stages that either fit onto the stage provided with the microscope, or are inserted as a replacement for the slide holder in the original microscope stage provided by the manufacturer. Several examples of these stage warmers are discussed here, but for a more comprehensive and very up-to-date listing see the individual web sites of the various manufacturers (these are listed on page 345 in Chapter 14 "Technical Supplies").

Figure 12-19. Heated Microscope Stage.
This Leica heated microscope stage accurately maintains the desired temperature, and will provide some help in maintaining the sample under observation at the correct temperature. A heated stage may need to be used in combination with a heated enclosure and a heated objective to provide optimal temperature control of the sample itself. This image was adapted from an image published on the Leica Internet site.

Figure 12-20. Microscope Stage Warmer.
The WS60 microscope stage warmer from Instec Inc. is placed on the existing microscope stage of an inverted microscope. This photograph was kindly provided by Instec Inc.

Petri Dish Warming Platforms

An alternative to warming the stage of the microscope itself is to place a smaller heated plate on the microscope stage. This type of heated plate can be a simple device that sits on top of a conventional microscope stage, or the heater can be a custom fitted plate that fits into the existing microscope stage.

A simple Petri dish warming plate, manufactured by Instec Inc, is shown in Figure 12-20. This heated platform can be readily placed on the stage of any inverted microscope. This warming plate can accommodate different size Petri dishes (up to 100 mm) and the cells are imaged through the hole in the centre of the warming plate.

Although the cells over the central viewing area will not be heated by the warming plate, cells maintained at the correct temperature adjacent to the viewing hole can be readily moved into place by sliding a relatively small Petri dish across the viewing area. You may have more difficulty moving a larger Petri dish into a different imaging position as the lip around the edge of the heating plate may preclude sideways movement of the Petri dish.

Figure 12-21. Heated Petri Dish Holder.
Heated Petri dish holder (TS-4LPD) from Physitemp Instruments Inc. This holder is designed to hold 60 mm or 100 mm Petri dishes. Physitemp also sells a smaller version to hold 35 mm Petri dishes. This photograph was kindly provided by Physitemp Instruments Inc.

Conventional Petri dishes can be used for low-resolution studies - the cells being imaged through the plastic base of the Petri dish. For high resolution imaging specially modified or designed Petri dishes, for example the MatTek and Willco Wells dishes (described earlier in this chapter) that have a coverslip thickness base to the Petri dish, will provide superior image quality. The full coverslip covered base of the Willco Wells microscope chambers will allow the dish to sit firmly on this type of heating plate. The attached coverslips of "home made" or MatTek dishes may result in some instability while imaging (the dish being effectively balanced on the raised coverslip).

The warming device from Physitemp Industries Inc (shown in Figure 12-21) provides excellent heating for suitably sized Petri dishes (models are available to accommodate 35mm, 60mm or 100mm dishes). Again, the coverslip bottomed Petri dishes available from MatTek and Willco Wells can be used for high-resolution imaging. The slit provided in the base of this Petri dish warming plate allows one to maintain very accurate temperature control of the cells located adjacent to the slit – which can easily be moved into the viewing area by simply turning the Petri dish. The full coverslip bottomed microscope imaging dishes from Willco Wells will, as above, provide better stability for these Petri dish warming holders.

Figure 12-22. Enclosed Microscope Slide Warmer.
The BS60 microscope slide warmer from Instec Inc. can be fitted onto most upright and inverted microscope stages. This photograph was kindly provided by Instec Inc.

Care should be taken with all warming plates that there is sufficient room for the objective to move around comfortably within the imaging "window". Many modern high NA objectives are very large and may only be accommodated by warming platforms that have a very thin metal base. Even then there may be some difficulty with very large, high NA, immersion objectives where the working distance of the lens is very small.

Heated Microscope Slide Holders

A number of manufacturers produce warming plates for conventional microscope slides. These warming plates may also be suitable for maintaining temperature control of microscope slide sized cell imaging chambers (for example, the enclosed microscope chamber shown on page 285, or the Lab-Tek chambers shown on page 287). These warming plates can be either fitted into, or placed on top of the existing microscope stage. These include the BS60 slide warmer from Instec Inc (Figure 12-22) and the TS-4 Diaphot from Physitemp Instruments Inc. (Figure 12-23). In the Physitemp microscope warming plate the microscope slide is simply clipped onto the heated plate using conventional microscope stage clips. On the other hand, the microscope slide warming plate from Instec can be used with a metal top plate to fully enclose the microscope slide within the metal temperature controlled plate. The metal top to the warming plate will result in better temperature control of the microscope slide - but unfortunately, not of your sample! The larger the viewing area in the heating plate the greater the problem of heat transfer from the heating block to the area of the sample under observation, though a small viewing area may create problems when using a physically large objective lens!

The Petri dish warmers and the slide warmers discussed above are sometimes supplied with small insulating feet to isolate the heated platform from the microscope itself to minimise heat loss to the microscope stage (if the microscope stage itself is not heated, see Figure 12-13). Alternatively the warming plate can be mounted on a relatively thin insulating sheet of plastic sheeting (1mm sheeting is sufficient). The metal stage found on most microscopes is an excellent conductor of heat – very quickly changing a chamber or warming plate placed on the stage to the temperature of the microscope and its attached stage. The heating mechanism used in these small imaging chambers and microscope slide warmers may not be sufficient to compensate for the loss of temperature to the metal "heat sink" of the microscope.

Even though all of these Petri dish and microscope slide warmers have excellent temperature control (± 0.1°C at 37°C) due to the use of highly accurate PID temperature controllers, this does not mean that your sample is at the correct temperature! As discussed above, all warming plates will create a potentially unstable heating environment as the plate is held at 37°C, while the microscope stage remains at room temperature. For these reasons, using these plate warmers in conjunction with other heating devices, such as objective heaters or heated microscope enclosures, may be necessary.

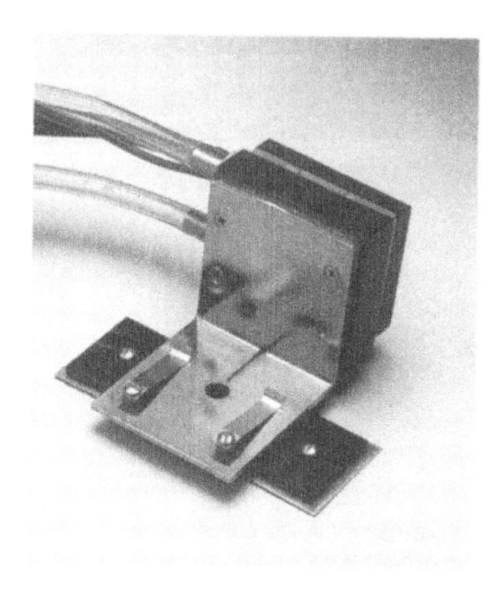

Figure 12-23. Heated Microscope Slide Holder. This heated microscope slide holder from Physitemp Instruments Inc. (TS-4Dioaphot) is designed to fit into the regular microscope slide holder on an inverted microscope. This photograph was kindly provided by Physitemp Instruments Inc.

Full temperature control may still require an objective heater

Perfusion Chambers

Perfusion chambers, in addition to providing temperature control (either directly by heating the chamber itself, or indirectly via a heating plate on which the chamber is placed) also provide a continuous supply of nutrients to the cells under study (see Figure 12-24). This has the advantage of controlling the pH and ionic concentration of the media during long-term incubation on the microscope. The added advantage of a perfusion chamber is that various reagents can be infused into the chamber without unduly disrupting the cells. This allows one to take a series of time-lapse images while, for example, introducing a drug to the system.

The supply of media to the chamber can be provided by equipment ranging from a simple siphon gravity feed to highly sophisticated pumps that minimise the effect of pump "pulsations" in the imaging area. A "syringe" pump as shown in Figure 12-10 (page 288) probably gives the smoothest flow available. A peristaltic pump that is commonly

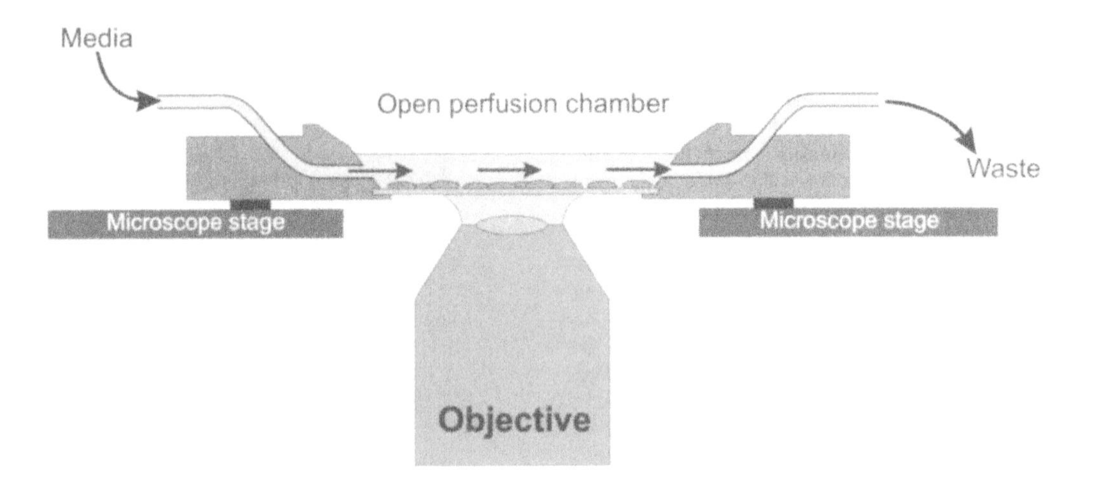

Figure 12-24. Open Perfusion Chamber.
A perfusion chamber is designed to provide a continuous flow of media across the cells being imaged. Perfusion chambers vary from simple open flow chambers to very complex environmental control chambers. The size and shape of the flow chamber is very important in determining the degree of mixing and for creating an even flow of media across the cells. A diamond flow chamber gives the best laminar flow characteristics, and round or square chambers give particularly bad flow characteristics (and very uneven mixing).

available in a biochemical/cell biology laboratory may result in some pulsation of the flow media (which may result in movement of the cover glass). If you use a gravity feed to maintain the flow of media through the perfusion chamber, as mentioned previously, using a relatively large media reservoir, or place the media bottle on a relatively high shelf and regulate the flow with a needle valve for regulating the media flow rate.

A relatively simple open flow chamber can be readily constructed from a plastic Petri dish with a pump (or gravity feed) to supply the media and a simple suction tube placed at the level you want the media to be maintained at. This type of flow chamber may not have the good laminar flow characteristics of a more elaborate chamber, but if you simply require a supply of nutrients to the cells without the need to quickly and evenly add additional reagents you may find such a simple home made chamber quite adequate.

Many of the more elaborate perfusion chambers allow for continuous flow of gas above the culture. This is particularly important when using bicarbonate buffered media that requires an elevated CO_2 level in the gas to maintain the correct pH of the media.

A further variation on the basic perfusion chamber design depicted in Figure 12-24 is to fully enclose the media within a flow chamber (i.e. to make the chamber lid directly contact the media). This has the advantage that there is no gas exchange with the overlying air and, particularly for transmission imaging, the fully enclosed chamber provides a much better optical path for high quality images. Fully enclosed perfusion chambers are made by a number of companies, including Bioptechs (discussed in some detail shortly, see Figure 12-33) and Lucas-Highland (Figure 12-30).

The shape of the perfusion chamber is very important when studying fast response events that require a laminar flow throughout the chamber. Long narrow chambers, or diamond shaped chambers are usually better at producing an even laminar flow throughout the chamber. Uneven flow may result in poor mixing of the new incoming media in different areas of the imaging chamber.

The perfusion chamber in Figure 12-25 from Intracel Inc. has a circular viewing area that can be infused with media and provided with controlled environment gasses. The gas enters the chamber via the small holes around the top of the chamber. The gas is then retained above the sample by covering it with a plastic Petri dish lid. The surrounding metal plate is accurately temperature controlled.

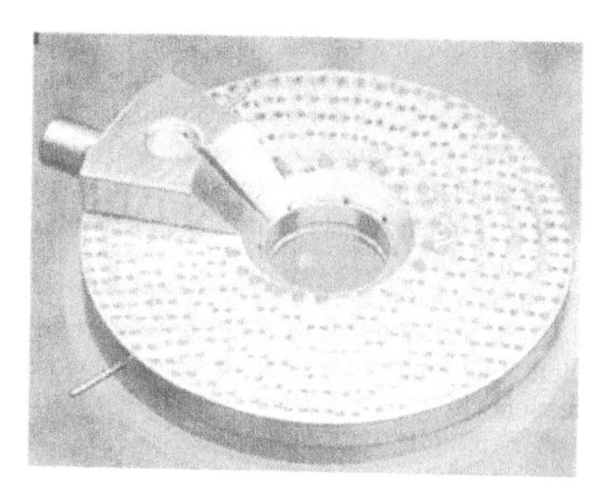

Figure 12-25. Heated Perfusion Chamber.
This open design perfusion chamber from Intracel Inc. can be mounted on the existing microscope stage of most inverted microscopes. Fresh media and surface gassing of the cells can be provided through the various ports around the perimeter of the central well. This photograph is reproduced, with permission, from the Intracel web site (www.intracel.co.uk).

Figure 12-26. Narrow Flow Chamber.
The Dagan HE200 series narrow channel perfusion chambers from Dagan Inc. can be mounted on the existing microscope stage of most inverted microscopes. This diamond shaped chamber gives good flow characteristics. This photograph is reproduced, with permission, from the Dagan web site (www.dagan.com).

The chamber depicted in Figure 12-26 (the Dagan HE200) is intended as an open top diffusion chamber, with a narrow flow chamber to improve the flow characteristics of the chamber. As mentioned above, the shape of the flow chamber is most important when attempting to image fast cellular processes under flow conditions.

Perfusion chambers are often of quite complex design (see the exploded view of the Dvorak-Stotler perfusion chamber in Figure 12-30). Although the chambers can be pulled apart for cleaning and replacement of the coverslip, their complexity means that one should not use a perfusion chamber when a simple incubation chamber, like those described in the first part of this chapter, will suffice.

Warner Instruments Inc. manufacture a very versatile range of perfusion chambers (Figure 12-28). These chambers consist of a small plastic flow chamber that is inserted into a metal temperature controlled holder. A large variety of chamber inserts are available, including a diamond flow chamber (as shown separated from the heated holder in Figure 12-28), which has excellent flow characteristics. Many of the Warner Instruments flow chambers are designed for electrophysiology. However, several designs are most useful for confocal microscopy, including special tissue sample holders with a small metal device to hold the tissue sample in place.

The highly versatile, but rather complex, Dvorak-Stotler perfusion chamber, shown as an exploded view in Figure 12-30, has interchangeable spacers and seals - allowing one to construct a chamber with the desired volume and flow characteristics.

The small Ludin perfusion chamber shown in Figure 12-27 can be placed in a rack (the size of a micro titre plate) that holds six chambers. This is particularly useful, in conjunction with a motorized x-y control on the microscope stage, for imaging multiple samples over time. The perfusion chamber itself is not heated, but the chamber can be placed within a heated plate on the microscope stage or used in conjunction with an objective heater or microscope enclosure.

The small Ludin chambers can be readily dismantled for cleaning and replacement of the "viewing" bottom coverslip. The Ludin chamber can also be autoclaved if sterility is important.

Figure 12-27. Small Ludin Perfusion Chamber. The small Ludin Perfusion chambers are available from Life Imaging Services. These chambers can be mounted in a microtitre plate sized rack of six for multiple experiments. This photograph was kindly provided by Beat Ludin from Life Imaging Services, Switzerland. (www.lis.ch/thechamber.html).

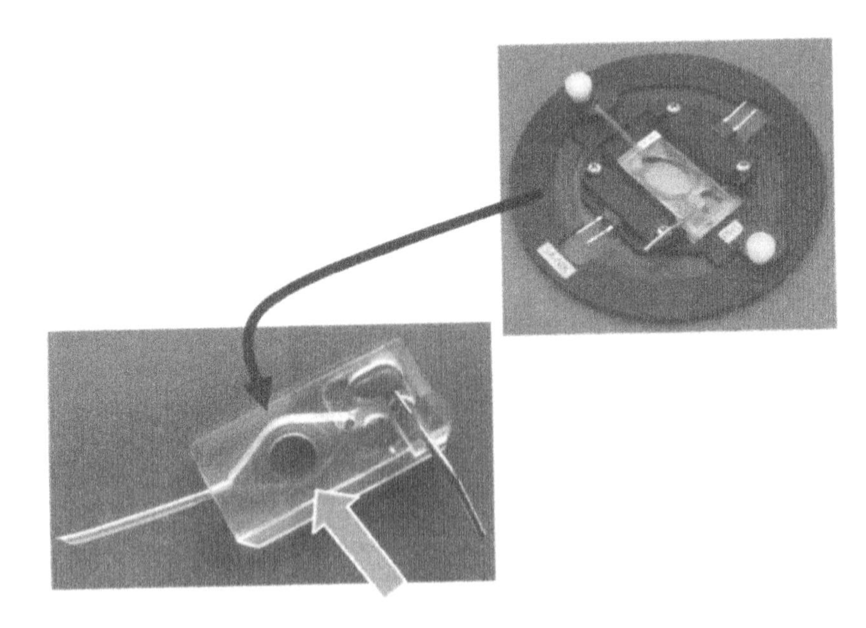

Figure 12-28. Removable Plastic Flow Chambers. Warner Instruments Inc. manufactures a range of flow chambers for both electrophysiology and microscopy. A wide variety of interchangeable plastic flow chambers is available. This image is of the Series 20 chamber fully assembled (above) and the RC-25 removable perfusion chamber (left). These images are reproduced with permission from Warner Instrument Corporation, USA, and are available on their web site (www.warnerinstruments.com).

Figure 12-29. Photograph of Dvorak-Stotler Chamber.
The Dvorak-Stotler Chamber is shown here being used on an upright light microscope, where the cells are imaged from above, through the cover glass chamber top. The DSC200 culture chamber, available from Nevtek Inc. USA is manufactured by Lucas-Highland, USA. This photograph was kindly provided by Philip Lucas, Lucas-Highland, USA.

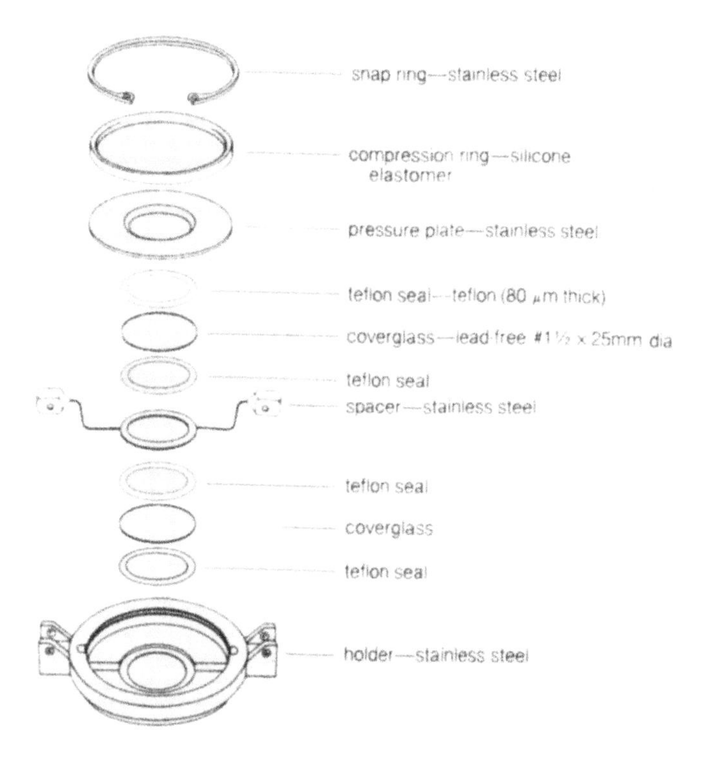

Figure 12-30. Diagram of Dvorak-Stotler Chamber.
This expanded view of the Dvorak-Stotler Chamber clearly demonstrates the complexity of many perfusion chambers. However, this chamber does have great versatility in that the size and shape of the flow chamber can be readily altered by inserting different sized and shaped gaskets or spacers. The DSC200 culture chamber, available from Nevtek Inc. USA is manufactured by Lucas-Highland, USA. This diagram was kindly provided by Philip Lucas, Lucas-Highland, USA.

Bioptechs Chamber

The Bioptechs culture chambers (Figure 12-31 to Figure 12-33) are the "Rolls Royce" of the chamber world. The Bioptechs chambers have a number of clever innovative designs that make them excellent for live cell work, but they are relatively expensive.

The Bioptechs open dish chamber system depicted in Figure 12-31 consists of a chamber holder into which are fitted the Delta T culture dishes. The chamber holder acts as a platform for holding various inlet and outlet tubes for introducing reagents (such as drugs) to the cells under observation. The chamber holder itself is not heated, but the Delta T dish has the unique characteristic of having an Indium Tin Oxide (ITO) coated coverslip that can be heated across the full base of the chamber (Figure 12-32). This is in contrast to the conventional chamber dish, where the heating of the sample is from the

Bioptechs............

Bioptechs is an innovative American optical engineering company that designs, develops and manufacturers live-cell microscopy environmental control instrumentation for qualitative and or quantitative light microscopy. Our products are dedicated to the science of live-cell microscopy and include imaging systems, micro-observation environmental control instruments, perfusion pumps, biocompatible tubing for cell culture and specialized fittings and adapters as well as numerous supplemental items. Bioptechs has developed and patented several technologies which are indispensable to the live-cell microscopist. They include:

Micro aqueduct laminar perfusion in a temperature controlled optical cavity

A unique symmetric closure mechanism for optical cavities

Objective thermal regulation devices to reduce temperature gradients when using high N.A. objectives

A unique hybrid culture dish system specifically designed for live-cell microscopy which utilizes an advanced technique of first-surface thermal transfer to quickly, accurately and uniformly maintain the thermal, optical and fluid requirements of cultured cells and tissue

Within a few short years Bioptechs products have demonstrated their ability to out-performing traditional systems and have become the preferred method of micro-environmental control. Bioptechs manufactures 32 standardized products and provides customers with the availability of a variety of related products and services

.......a note from their web site (www.bioptechs.com).

Figure 12-31. Bioptechs Open Chamber.
The Bioptechs open chamber system consists of a heating block with removable Delta T culture dishes. The dishes come in either clear plastic or black (with black lid) for extended use in experiments using light sensitive cells or compounds. The removable delta-T dish is relatively expensive, but can be re-used many times by simple washing and sterilising with alcohol. The heating block for the Delta-T dish is designed for holding a number of devices, such as injection tubing, in place during imaging. This image was kindly provided by Dan Focht, Bioptechs, USA. Further details are available on their web site (www.bioptechs.com).

surrounding chamber only. Although the Delta T dish has the advantage of heating across the entire sample area, the large heat sink of the objective may still result in some difficulty in maintaining the correct temperature in the area under observation. For this reason the Bioptechs chamber is usually operated in conjunction with an objective heater.

The small removable Delta T culture dishes have the added advantage in that they can be used to grow cells in an environmentally controlled tissue culture incubator prior to placing in the chamber holder for imaging. The Delta T dishes are supplied sterile with a loose fitting plastic lid for ready transfer of incubator gasses. The loose fitting lid does have the disadvantage that significant evaporation of small culture volumes and fast exchange of gasses can

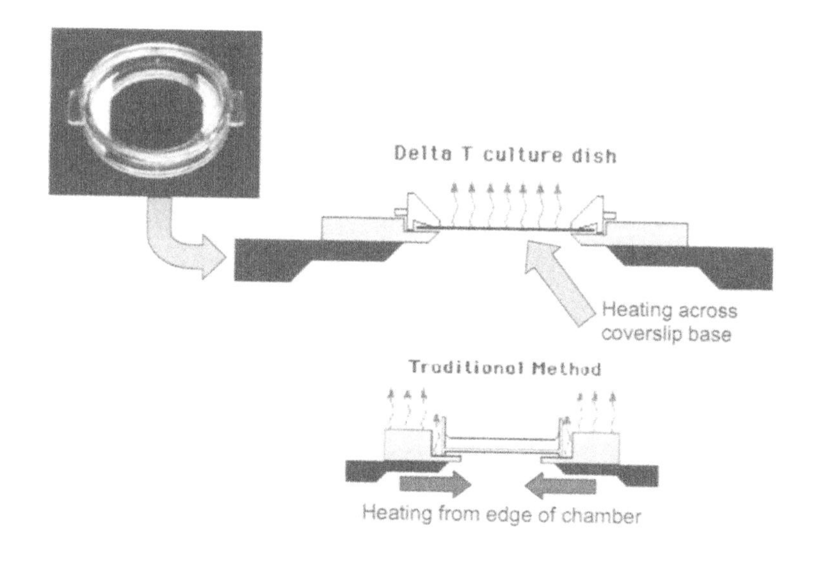

Figure 12-32. Bioptechs Delta T Dish.
The Bioptechs Delta T Dish has the unique property of being heated across the base of the coverslip, by electrical heating of an Indium Tin Oxide (ITO) coating. This results in even heating across the area of the sample being imaged, in contrast to conventional heating where the heating chamber transfers heat to the sample from the edges of the imaging chamber only. This image was kindly provided by Dan Focht, Bioptechs, USA. Further details are available on their web site (www.bioptechs.com).

occur while imaging. The Delta T dish is supplied as either clear dishes with a clear lid, or opaque black dishes and lids to minimise photo damage to the sample from room lighting. The black dishes are also useful when using conventional wide-field epi-fluorescence microscopy for preventing extraneous light from entering the optics of the microscope.

The Delta T culture dish is relatively expensive (supplied sterile), but can be easily washed clean and reused. It can be sterilised in the laboratory by rinsing with alcohol and seeded again with culture cells. If you are using suspension cells you can coat the chamber cover glass with poly-L-lysine and allow the cells to settle to the bottom of the dish. Alternatively you can gently centrifuge the suspension cells onto the poly-L-lysine coated cover glass by placing the chamber in a swinging bucket centrifuge and spinning gently (500 rpm) for a few minutes. The coverslip base of the chamber should be supported by placing the chamber bottom flat on a microtitre plate holder in the centrifuge, or alternatively placing the chamber on a small plastic support on the top of a large tube swing out bucket

– if you gently tape the chamber to the holder in the centrifuge you will find that the coverslip bottom is unlikely to break.

For longer term culturing on the microscope, using the Delta T dish with a microscope stage enclosure may be necessary for maintaining a high humidity and the correct gas concentrations for controlling the pH of the media. Covering the top of the Delta-T dish chamber with FEP Teflon® film (available from ALA Scientific, see Figure 12-34) may also be helpful for stopping evaporation and limiting, but not eliminating, gas exchange (discussed in more detail in the section on "Long-Term Growth" on page 312). Taking care, when transferring the Delta T culture chambers from an incubator to the microscope, not to inadvertently expose the cells to a large temperature change is also important. Take care not to place the chamber down on a bench surface (particularly a metal or stone surfaced vibration isolation table) as the thin layer of cells adjacent to the coverslip bottom of the chamber will very quickly equilibrate to the temperature of the bench. If you are imaging several cell preparations try having a dry heating block, at the correct temperature for your cells, sitting next to the microscope for temporary storage of the chambers prior to imaging.

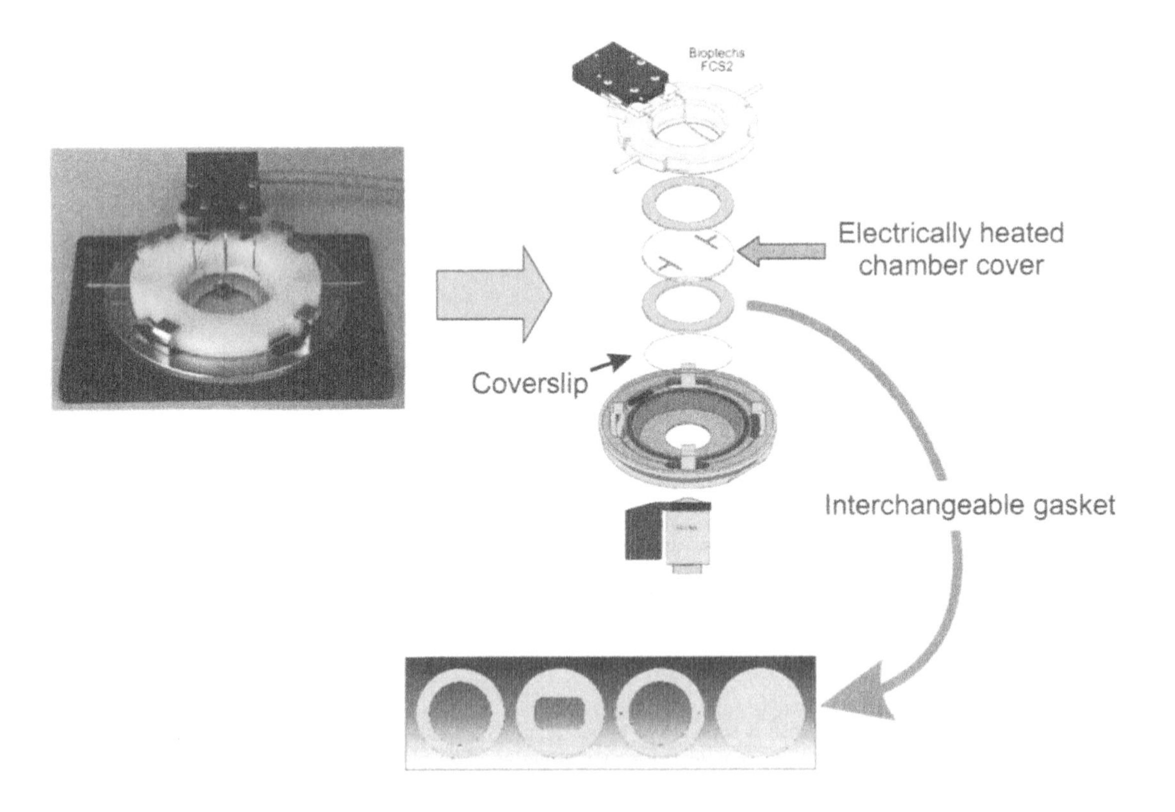

Figure 12-33. Bioptechs Enclosed Perfusion Chamber.
In addition to the heated chamber itself, the Bioptechs Perfusion chamber has an electrically heated cover that is in direct contact with the culture media. This greatly improves the temperature stability of the sample. The size and shape of the perfusion chamber can be readily altered by changing the gasket used to form the chamber. This diagram was constructed from images kindly provided by Dan Focht, Bioptechs, USA. Further details are available on their web site (www.bioptechs.com).

Bioptechs also manufactures a closed perfusion chamber (Figure 12-33). This device is much more complex than the Delta T dish system, but does have the advantage of being able to perfuse fresh media, or specific drugs through the sample while imaging. This perfusion chamber is operated as an enclosed chamber by the addition of an electrically heated glass cover to the chamber. The chamber lid is in direct contact with the culture media, both providing excellent heat transfer from the heated cover and at the same time eliminating optical distortion and unwanted gas exchange problems associated with an open chamber.

The flow characteristics of the Bioptechs perfusion chamber can be readily altered by exchanging the gasket used to form the chamber itself. This does allow great versatility in terms of chamber size and shape, but if you are growing cells under sterile conditions in the chamber you cannot change the gasket without destroying the sterility (and probably most of your cells).

Although temperature is maintained by both heating the surrounding metal casing of the chamber, and, as mentioned above, heating the glass plate that is in direct contact with the perfusion media, full temperature control of the sample under study may require the addition of a heating ring on the objective to maintain the correct temperature of the cells being imaged. The fully enclosed chamber is excellent for maintaining live cells during microscopy, but if you don't need the accurate environmental control provided by the enclosed chamber you may be better off using an open chamber as the open top design is much easier to work with, and as they are significantly cheaper you can have several samples ready for imaging.

MAINTAINING THE CORRECT PHYSIOLOGICAL STATE

Changes in the local environment of the cell or tissue sample may result in considerable stress on the cells. Cells under stress make significant changes to their cellular metabolism, including the synthesis of stress response proteins. Stress changes have been studied most extensively as a result of changes in temperature (particularly elevated temperature), but any change that is not optimal for growth may elicit a stress response within the cell.

As discussed in some detail previously, all forms of stress can potentially result in the cells entering the pathway of apoptosis (programmed cell death). It is therefore important to minimise the stress to the cells both prior to, and during the imaging process. Particular care should be taken with the following important environmental parameters.

Oxygen Level

Maintaining the correct oxygen level is critical for keeping the cells or tissue samples in a healthy state. Most cells grow best at atmospheric oxygen levels (20% oxygen). However, some specialised cells, such as malarial parasites (prefer 5% oxygen), or neurons (the oxygen level in the brain is significantly below 20%) grow better in a low oxygen environment.

Maintaining the oxygen level of cultured cells or tissue samples in a gassed tissue culture incubator is not difficult. However, when imaging live cells by microscopy greater care is needed to ensure that the correct oxygen level is maintained. This can be achieved by either having the cells maintained within an enclosed microscopy incubation chamber on the microscope stage, or by using an enclosed imaging chamber. When using an enclosed chamber, keeping the volume of cells/tissue low compared to the volume of media present to ensure sufficient oxygen is available within the relatively small volume of the enclosed chamber is most important. When using cultured cells the oxygen level within the chamber is not normally a problem for the relatively short time used for imaging if the culture media has been properly equilibrated with atmospheric oxygen before the imaging chamber has been sealed. However the oxygen level will become an important issue when the cells are imaged over a longer period of time during a time course experiment.

Maintaining the correct oxygen level when imaging tissue samples is much more difficult due to the greatly increased number of cells present within a tissue sample, and the difficulty of supplying oxygen by diffusion deep into the tissue. The maintenance of live tissue samples is discussed in more detail shortly (page 313), including the use of a semi-permeable Teflon® film to cover the microscopy chamber for controlling oxygen and CO_2 levels during long term imaging (see Figure 12-34).

pH Control

Tissue culture media often uses bicarbonate buffer to maintain the correct pH for the cells/tissue sample. The correct pH of the media is maintained by using a high level of CO_2 (usually 5%) in the incubation atmosphere. This is a very common and simple procedure when using a gassed cell culture incubator, but does become technically much more difficult when attempting to maintain the cells on a microscope stage. The very small volume often used

HEPES buffer may produce toxic products on exposure to light!

in imaging means that exposure to atmospheric CO_2 levels may very quickly change the pH of the media. Sealing gas-equilibrated incubation media within an enclosed microscope imaging chamber may provide sufficient pH control for most experiments, but long term culturing on the microscope will necessitate placing the cells within the enclosed CO_2 controlled environment chamber on the microscope. A semi-permeable Teflon® membrane (see below) can also be used to help maintain the correct CO_2 concentration within the microscope imaging chamber.

The addition of a reagent with a higher buffering capacity (such as HEPES, N-2-Hydroxyethylpiperazine-N'-2-ethanesulfonic acid) may be beneficial for maintaining the correct pH of the media. However, one should be aware that HEPES may form toxic by products when exposed to light (Lepe-Zuniga et al, *Toxicity of light-exposed Hepes media.* J Immunol Methods 103:145-150 (1987) and Spierenburg et al, *Photo-toxicity of N-2-Hydroxyethylpiperazine-N'-2-ethanesulfonic acid buffered culture media for human leukemic cell lines.* Cancer Research 44:2253-4 (1984), see page 353 in Chapter 15 "Further Reading" for more details). This is particularly noticeable when HEPES buffered media is exposed to room lighting (particularly fluorescent tubes) - resulting in poor cell growth. The high level of laser light intensity used for confocal microscopy may also result in toxic products when HEPES is used in the media. However, not all microscopists agree that HEPES is a problem when exposed to light. Perhaps a simple test to determine whether light exposure in the presence of HEPES affects your particular cells would be worthwhile. A novel alternative, discussed in some detail below, is to use a FEP Teflon® membrane that is impermeable to water vapour to cover the microscope imaging chamber.

Osmolarity of the Media

Correct osmolarity of the culture media is essential for the health and long term survival of most cultured eukaryotic cells. However, yeast, fungal, plant and of course, most prokaryotic cells can often tolerate large changes in the osmolarity of the media.

Media containing the correct osmolarity is readily prepared for tissue culturing. However, when working with very small volumes in microscopy the osmolarity of the media can be very easily and inadvertently altered by simple evaporation. Leaving small volumes of media open to the atmosphere should be avoided. Either enclose the incubation chamber, or provide a supply of fresh media. If a small volume does have to be left open to the atmosphere (this also applies to a loose fitting lid on a microscope incubation chamber) then constructing a humidity-controlled enclosure to house the incubation chamber may be advisable.

Long-Term Growth

Growing cells continuously for many months using specially designed incubation chambers is possible. Growing the cells in these special microscope chambers directly on the microscope is also possible, but particular care will be required to maintain the correct gas concentrations and temperature. Maintaining the culture for long periods of time in a conventional tissue culture incubator is considerably easier than attempting to culture the cells on the microscope stage itself. The cells can then be moved to the microscope when imaging is required.

Special care is needed during long-term incubation to make sure that there is minimal evaporation, that the O_2 and CO_2 concentrations remain constant, and that the temperature is maintained at the optimal temperature for growth. One of the simplest and possibly the best way of maintaining the correct growth conditions for the cells over many weeks or months is to use a fluorinated ethylene-propylene (FEP Teflon®, available from DuPont) film to cover the incubation chamber (Figure 12-34). This special membrane allows O_2 and CO_2 to pass through without allowing water vapour or microbes to pass. This semi-permeable Teflon® membrane can be placed on the top of a variety of microscope imaging chambers – and can be autoclaved for the preparation of sterile incubation chambers. Chambers with the FEP Teflon® membrane attached are available from ALA Scientific.

The rate of gas diffusion across this Teflon® membrane is relatively slow compared with conventional incubation chambers, where there is a small air-gap between the lid and the chamber walls. This decreased gas permeability can be most useful when imaging on a microscope as the internal CO_2 level, and hence the pH of the media, is maintained for at least 30 minutes without external maintenance of the correct CO_2 level. For longer-term incubation on the microscope providing the correct O_2 and CO_2 gas levels by using an enclosed gassed incubation chamber on the microscope stage will be necessary.

LIVE TISSUE SAMPLES

Live whole tissue samples pose a number of important difficulties when imaging by confocal microscopy. This section will briefly discuss some of these difficulties and how they may be overcome. This section is concerned with imaging live tissue samples on a conventional upright or inverted confocal microscope, but keep in mind that there are specialised confocal microscopes that are capable of imaging live tissue samples in vivo. This includes the LUCID infrared reflectance imaging confocal microscope for imaging skin, and the Optiscan Imaging miniaturised confocal microscopes that are designed for imaging skin, the colon and stomach (incorporated into an endoscope), the oral cavity and the cervix (using a miniaturised rigid confocal endomicroscope).

Figure 12-34. Long-term Growth Chambers.
A small microscope imaging chamber (left) and a multi-electrode array (right) covered with FEP Teflon® film. For long-term growth of cell cultures maintaining sterility and osmolarity, even when transferring the imaging chamber from the incubator to the microscope stage, is important. A FEP Teflon® film allows gas exchange (particularly O_2 and CO_2) without allowing any appreciable amount of water vapour or any bacteria or fungi spores to pass through. This membrane is described in "*A new approach to neural cell culture for long-term studies*" *by Potter and DeMarse (J Neuroscience Methods 110:17-24, 2001)* and is now available from ALA Scientific. This photograph is kindly provided by Steve Potter, Georgia, USA.

Tissue Chambers

Confocal microscopy is particularly valuable when attempting to image live tissue samples. In normal epi-fluorescence microscopy the tissue sample would have to be cut very thinly (extremely difficult with many fresh, unfixed tissues) for a useful image to be obtained. With confocal microscopy there is no need to cut the tissue sample thinly as the very nature of the confocal microscope will remove out of focus light (optical sectioning), resulting in superb images even when thick tissue samples (even several millimetres in thickness) are placed in a suitable imaging chamber (Figure 12-35). In fact imaging whole organs, as long as they can be suitably mounted on the microscope stage, is possible.

Obtaining high quality transmission images is not possible when examining thick tissue samples due to the absorption and scattering of the light that crosses completely through the tissue. As transmission imaging is not usually available for highlighting cellular structure when imaging thick tissue samples, a suitable secondary dye may be necessary for obtaining images of the tissue structure to physically locate the dye of interest within the cell or tissue sample.

Tissue samples can be attached to the coverslip imaging surface of the chamber by using cell adhesives like poly-L-lysine and Cell-Tak (see page289), but a more effective method, particularly for relatively "thick" tissue samples is to hold the tissue in place with a small weight (known as a "harp"). The "harp" is a small loop of metal, often with nylon threads glued across the top. The weight of the metal in combination with the threads, holds the tissue sample in place. An even more effective way is to glue a Millicel CM membrane onto the "harp" using cyanoacrylate. This membrane is in place of the fine nylon threads. The membrane is transparent when wet (images can be collected through the membrane if necessary), and the membrane will also allow nutrient and gas exchange.

When imaging live tissue samples, remember that small tissue samples will have to be treated with extreme care to avoid drying and cell damage artefacts. The tissue should be imaged as quickly as possible after removal from the animal. An exciting application of confocal microscopy is the imaging of whole live organs mounted on the microscope stage. This can be in the form of perfused isolated organs, or even the imaging of live tissues directly on an anaesthetised animal. The limitation with this type of work is that the confocal microscope can only image up to 100-200 μm into a tissue sample. This distance may be significantly greater when using multi-photon microscopy.

This is referred to as "deep" by cell biologists, but for many others (for example, pathologists and particularly surgeons) this depth would be considered to be just below the surface.

Maintenance of Live Tissue

Live tissue is difficult to maintain in a non-stressed live state during microscopy. Tissue biopsy samples may already be under considerable stress due to the fact that the tissue sample has been removed from the organism, and also now consists of a large number of cells that are surrounded by damaged and torn cells, nerves and connective tissue where the cut was made.

Furthermore, the often relatively large volume of a tissue sample means that special care needs to be taken to maintain the correct oxygen levels within the tissue. The supply of nutrients may also be compromised by the relatively large size of the tissue sample.

The level of oxygen required by the cells is greatly reduced if the temperature is lowered. This may be a good way of maintaining cell viability in a tissue sample prior to imaging. However, you will need to determine whether lowering the temperature has a detrimental effect on the parameter you are studying (see "Temperature Control", page 290). Most cells will readily recover from a period at a lower temperature. Mammalian cells have a greatly reduced oxygen requirement when the temperature is dropped to 28°C. This temperature is still high enough to avoid the problems associated with dropping below the membrane phase transition, and may result in better cell viability. Problems of maintenance of the correct oxygen level are most critical when there is some time between the time of collection of a tissue sample and the time at which imaging can begin. This is a particular problem with human biopsy samples, where considerable time may elapse after removal of the tissue sample before the tissue can be imaged in the laboratory - which is normally physically some distance from the operating theatre.

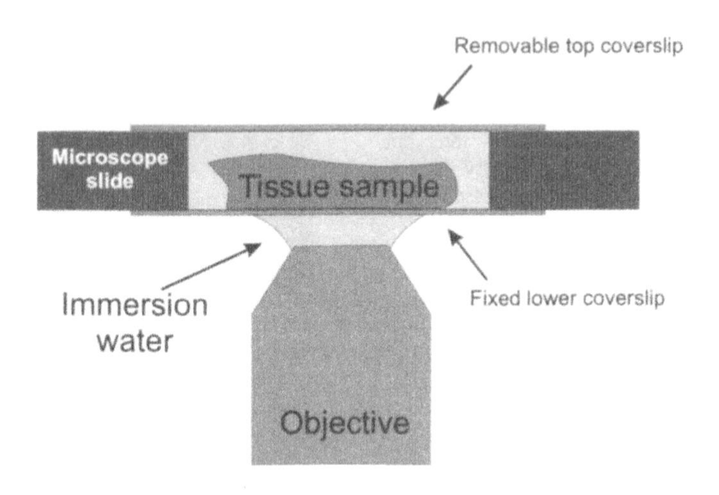

Figure 12-35. Simple Tissue Chamber.
To image deep into tissue samples the use of a water immersion objective is desirable. This is to match the immersion refractive index with that of the sample (both water). This will give the maximum resolution. However, high quality images can be obtained with an oil immersion lens as long as only the cells close to the base coverslip are imaged.

A number of specialised culture chambers are available for handling live tissue samples. Chambers used for electrophysiology can often be readily adapted for high quality confocal microscopy.

Dye Penetration

When tissue samples are being imaged, special care should be taken to establish that the dye being used does in fact fully penetrate the tissue sample. Due to the large number of cells contained within a tissue sample, the dye may have to pass through several cell layers when penetrating the tissue.

Dye penetration is most readily assessed in tissue samples by taking a single optical slice that is reasonably deep within the tissue and includes the edge of the tissue sample. A comparison of the level of label on the edge of the tissue (but viewed at a significant depth within the tissue), and the label detected deeper within the tissue sample will give a good indication of the degree of dye penetration.

Many dyes will readily diffuse into relatively large tissue samples, but dyes that are extremely hydrophobic (such as DiI / DiO) may not readily penetrate past the initial cell layer. Large molecules, such as antibodies, will also have great difficulty penetrating into the tissue. Organ perfusion, using media containing the dye solution, is an excellent way to facilitate dye access to relatively large tissue samples.

Imaging Depth

How deep can you image?
100 to 150 µm for confocal microscopy
200 to 250 µm for 2-photon microscopy

The popularity of the confocal microscope is in part due to its ability to image "deep" into a tissue sample. However, in practical terms both the absorption and scattering of the fluorescent light, and spherical aberration limit the depth of imaging possible in a tissue sample. The maximum depth that you can expect to penetrate into most tissues is about 100µm. Significantly greater depth (up to 250 µm) may be possible when imaging by multi-photon microscopy. This is mainly due to the fact that the long wavelength infrared light used in multi-photon microscopy is less scattered by biological tissue and thus more readily penetrates tissue samples. The depth of maximum penetration is also determined by the quality of image required. If establishing whether a dye is located in particular cells within the tissue is sufficient, it may be possible to obtain information from optical slices at considerable depth that may not appear particularly aesthetically pleasing.

The quality of the image when attempting to collect fluorescent light from deep within the tissue is greatly determined by the type of lens used. As discussed below, oil immersion lenses are fine for imaging aqueous samples close to the coverslip, but water immersion lenses are far superior when imaging deep within the tissue. The water immersion lens will result in less spherical aberration, and thus a better quality image.

As mentioned above, take care when discussing the idea of imaging "depth" in confocal microscopy. A cell biologist may consider 100µm a great "depth", but many others, particularly those in the medical profession interested the depth penetration of a tumour for example, will be more interested in depths of "mm" rather than "µm".

DYE LOADING INTO LIVE CELLS

There are many ways in which fluorescent dyes can be introduced into living cells. Some dyes, due to their solubility in lipid bilayers, will readily enter living cells without any further modifications. Other dyes, including many of the ion-sensitive dyes such as the calcium indicator dyes, may require chemical modification (the addition of an acetoxy-methyl-ester) to facilitate transfer across the hydrophobic domain of the cell membrane. Dyes can also be directly injected into cells, holes can be made in membranes with detergents and toxins or electroporation methods can be used to facilitate dye access. Of course, the Green Fluorescent Proteins are transfected into living cells by using DNA constructs containing the gene for the relevant GFP. Further information on getting dyes into living cells, particularly with the use of AM esters and GFP, is provided in Chapter 9 "Fluorescent Probes".

Lipid Soluble Dyes

Lipid soluble dyes will readily dissolve into the plasma membrane of cells. One problem that can be encountered with hydrophobic dyes is that they are not very soluble in water, or buffer, and will need to be introduced to the incubation media as a solution in ethanol or DMSO. These organic solvents, although they are not particularly toxic to the cells at low concentrations (1% of the culture volume or lower), they may disturb the permeability characteristics of the plasma membrane.

Lipid soluble dyes can also be introduced to the media by using a carrier molecule such as delipidated BSA (Bovine Serum Albumin in which the bound lipid has been removed). Hydrophobic dye molecules bound to the large water-soluble BSA molecule can then be readily introduced into the culture media. The hydrophobic fluorescent dye molecules will readily exchange with the plasma membrane of the cells in the culture media.

Highly hydrophobic dyes may result in extensive labelling of the plasma membrane, with very little labelling of the internal membranes of the cell. This is due to the high rate of diffusion along the plane of a membrane, but the relatively slow transfer of hydrophobic molecules from one membrane to another. This characteristic can be used to advantage to study membrane transfer within cells, or a less hydrophobic derivative of the dye can be used to facilitate faster transfer to intracellular membranes.

Methyl Ester Derivatives of the Dye

Live cells will readily take up the non-charged acetoxy-methyl-ester (AM-ester) derivatives of fluorescent dyes, even when the original dye is excluded from living cells. The methyl-ester group will then be cleaved by the abundant esterases within the cellular cytoplasm. This technique of loading a dye into a living cell often has the added advantage that the methyl-ester derivative of the dye is not fluorescent. Thus only the dye molecules that have gained entry to the cellular cytoplasm will show fluorescence. Unfortunately, methyl ester derivatives of dye molecules are not readily taken up by plant cells due to both the difficulty of crossing the cell wall and the abundance of esterase activity within the cell wall. Methyl ester derivatives of fluorescent dyes are described in more detail on page 217 in Chapter 9 "Fluorescent Probes".

Microinjection

Dyes can be readily microinjected into living cells. This alleviates the necessity to use either methyl-ester derivatives or the possible disruption to cellular physiology of having 1% ethanol or DMSO present in the culture media. However, microinjection is time consuming and requires some skill and specialised equipment. Furthermore, the process of microinjection may stress the cells. Microinjection does have the advantage that individual cells are injected in a very controlled manner, but this can easily become a disadvantage when large numbers of cells need to be studied.

Transfection

A lot of excitement is being generated in the field of fluorescent microscopy with the discovery of the Green Fluorescent Protein (GFP) and the construction of derivatives with different wavelengths of emission, and molecular derivatives that can detect changes in intracellular calcium and pH. GFP and its derivatives, being proteins, can be introduced into cells by using DNA transfection techniques. The use of GFP in confocal microscopy is discussed in more detail on page 232 in Chapter 9 "Fluorescent Probes".

Semi-permeabilisation using Streptolysin-O (SLO)

Loading fluorescent probes into live cells is a particular problem when the probes have large molecular weights (such as fluorescently labelled antibodies). Loading these large fluorochromes by the simple method of using an AM derivative is not possible. An alternative method is to semi-permeabilise the cells without unduly disrupting the cellular function under study.

Streptolysin-O (SLO) is a bacterial toxin that binds to cholesterol, forming a pore complex within the membrane, and thus creating a semi-permeabilised cell that allows direct access to cellular contents, without disruption to many cellular processes. For example, protein trafficking can be studied in SLO permeabilised "live" cells. Streptolysin-O purchased from commercial sources (for example Sigma, St Louis, USA), however, works fine for permeabilising cells to allow the penetration of large antibody molecules for labelling internal cellular structures, but can seriously inhibit many cellular functions (such as protein trafficking). A very high purity SLO, which is known not to inhibit vesicle mediated protein transport in living cells, is available from Dr. S. Bahkti, Germany.

SLO strongly binds to cholesterol within the plasma membrane even at 4°C. Elevating the temperature to 37°C is required for the SLO subunits to form a protein channel that allows macromolecules, but not cytoplasmic structures and vesicles, to diffuse across the plasma membrane. This ability to form pores only at elevated temperatures can be utilised to very specifically label only the membrane exposed to the SLO. For example, incubating cells at 4°C with SLO will result in SLO binding to the plasma membrane. Elevating the temperature to 37°C will result in pore formation and thus semi-permeabilised cells. If further SLO is added after lowering the cell temperature to 4°C, the SLO can again enter the semi-permeabilised cell and bind to intracellular membranes (organelle membranes). These "internal" membranes can then be permeabilised by warming the cells once again to 37°C.

> **Load antibodies into live cells using Streptolysin-O (SLO)**

SLO semi-permeabilisation can be readily used to introduce antibodies that are fluorescently labelled (or even gold labelled for electron microscopy studies) into the cell cytoplasm of live cells. These cells can then be imaged by confocal microscopy directly, or fixed and analysed by electron microscopy.

Introducing Reagents while Imaging

Some experiments may require the loading of reagents (including fluorescent dyes and drugs under study) into live cells while images are being collected. The main difficulty associated with loading reagents while imaging is in being able to sufficiently mix the media within the incubation chamber, without unduly disturbing the cells under observation. If the cellular process under study is relatively slow, then simply removing the top cover of a covered microscope chamber and then replacing the media will be sufficient for collecting high quality images. However, when the cellular process under examination is fast, for example changing Ca^{2+} fluxes, then a more sophisticated method of reagent addition and mixing will be required. This is usually achieved by injecting the reagent into the media which is being pumped into a perfusion chamber. A narrow chamber with good laminar flow is necessary for following fast cellular using time lapse imaging as the "wave" of drug or metabolite passes across the cells under observation.

THE CORRECT LENS FOR LIVE CELL IMAGING

A wide variety of objective lenses can be used for live cell imaging. However, some knowledge of the advantages and disadvantages of the various lens designs will help in choosing the correct lens for the application at hand. For an introduction to the characteristics of objectives lenses see Chapter 2 "Understanding Microscopy" (page 31).

Low Magnification "Dry" Lenses

Long working distance "dry" lenses are routinely used for microdissection. However, a dry lens when used with a live cell imaging-chamber will require careful adjustment of the collar (often designated as a coverslip thickness correction collar) on the objective. This ring is designed to correct for spherical aberration while adjusting the focus control of the microscope. If the correction collar ring is not correctly adjusted, a very poor quality image (low contrast and out of focus) will result when using bright-field illumination or collecting transmission images on the confocal microscope. Confocal microscopy images (fluorescence or backscatter images) will not appear "out-of-focus", but the sensitivity of the instrument will appear to be very poor. Careful adjustment of the correction ring while adjusting the focus control should be carried out to maximise the fluorescence intensity of the image. Unfortunately the necessity to adjust both the correction collar and the focus control does mean that this adjustment cannot be carried out while collecting a z-series.

Relatively high magnification (40x) dry long working distance lenses are available. However, these lenses are very expensive and have a relatively low NA (0.65). This will result in less light being gathered and a lower resolving power compared to an oil or water immersion lens of the same magnification. Therefore, only use an expensive long working distance lens when the extra working distance is actually required.

The front lens element of a "dry" lens must be kept scrupulously clean – any residue of glycerol, oil, media or even water will seriously degrade the image.

Oil Immersion High NA Lenses

An oil immersion lens, although not normally optimal for imaging within an aqueous environment due to serious spherical aberration problems, may often be used successfully for imaging live samples by collecting images close to the coverslip (within 5µm). In this way the oil/water refractive index mismatch is minimised. An oil immersion lens is excellent for imaging a single cell layer, but if you want to image deep into a tissue sample, a water immersion lens will give a superior image.

Water Immersion Lenses

Live cells are grown in an aqueous environment (and are themselves close to the refractive index of water) and so image quality will often benefit from the use of an objective lens compatible with the refractive index of water. These lenses are called water immersion lenses and will normally be denoted "water" or "w" on the barrel of the lens. A water immersion lens uses water as the continuum between the coverslip (or chamber base) and the lens. Some lenses can be used with oil, glycerol and water by turning a ring on the neck of the lens. When using water as the immersion media make sure that the adjustment ring on the lens is actually turned to the water setting.

A high NA water immersion lens will give superior images to an oil immersion lens when attempting to obtain images deep (deeper than 10 μm) within an aqueous environment. This is particularly noticeable when imaging tissue samples, where the plane of focus is often many tens of microns into the aqueous environment of the tissue.

Take care when using a water immersion lens to ensure that immersion oil has not been accidentally smeared on to the front lens element – as this will seriously degrade the image.

Dipping Lenses

Another way of imaging live cells is to use an upright microscope and to use a special "dipping" lens (Figure 12-36). This lens is a specialised water immersion lens that is designed to be used without a coverslip. This type of lens is often used in micromanipulation and electrophysiology, and can be used very successfully with confocal microscopy. The disadvantage of the dipping lens is that the movement of the lens may disturb the sample. This may be particularly relevant if attempting to image a single cell over a period of time.

Originally long working distance "dipping" objective lenses had a relatively low numerical aperture (low NA), which means these lenses had a lower resolution and would gather less light compared to a shorter working distance lens. In more recent years the major objective lens manufacturers have developed a number of water immersion "dipping" objective lenses with high NA (up to 0.95 for a Zeiss 63 x W LWD lens) and some with an amazing working distance measured in mm rather than fractions of a mm (up to 3.61 mm, depending on the design of the lens). The major lens manufacturers, Olympus, Nikon, Zeiss and Leica produce a range of long working distance water immersion lenses from 20x up to 63x, with an NA range from 0.8 up to 0.95.

One advantage of a "dipping" lens is that images can often be collected deep within the tissue sample due to the long effective working distance of this type of lens. With a dipping lens it is possible to approach the surface of the cell or tissue sample very closely due to the fact that a coverslip is not required. The depth of imaging therefore limited by the loss of light due to spherical aberration, rather than the physical constraint imposed by the coverslip. The "dipping" lens is designed to be used on an upright microscope, but can be used successfully on an inverted microscope by incorporating a flexible rubber seal between the objective and a hole in the base of an imaging bath.

To minimise evaporation and gas exchange problems when using a dipping lens on an upright microscope a latex tube can be used to connect the objective lens with the perimeter of the imaging chamber (described in Potter, S.M. "*Two-Photon Microscopy for 4D Imaging of Living Neurons*", *In:* Yuste, R. (editor) "*Imaging Neurons, A Laboratory Manual*". Cold Spring Harbour Press, 2000 - see Chapter 15 "Further Reading" for more details).

A further advantage of a "dipping" lens is that the casing material of the objective is often constructed from a ceramic material with low conductivity. This will mean that the lens will have less effect on the temperature of the sample compared to a traditional metal cased lens.

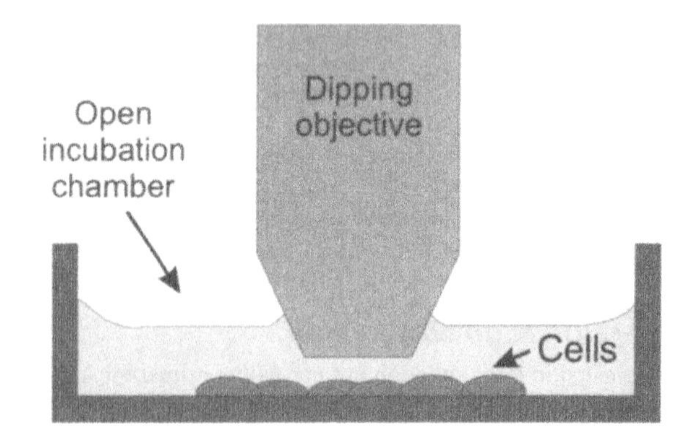

Figure 12-36. Dipping Objective.
Live cells can be imaged from above in an upright microscope by using an objective that dips into the incubation media. This objective must be a special aqueous media lens - do not use a regular water immersion lens.

MINIMISING LASER DAMAGE

The lasers used in confocal microscopy are sufficiently powerful to cause considerable molecular damage to both the tissue sample and the fluorescent dye being used. The laser is usually used significantly below maximum power, and should be increased only with considerable care when imaging live tissue samples. Laser power and minimising laser photo damage is discussed in detail on page 88 in Chapter 3 "Confocal Microscopy Hardware" and on page 110 Chapter 4 "Image Collection". In this section a brief overview will be given in regard to photo damage caused by the laser that is particularly applicable to the imaging of live cell and tissue samples.

What is Photo Damage?

Damage to live cells and tissue samples occurs in confocal microscopy due to the high intensity of the light used to scan the sample. Photo damage will occur at the focal point, as well as throughout the sample - both above and below the point of focus - as the laser light passes through the specimen. In multi-photon microscopy the pulsed infrared laser will cause very little photo damage above and below the focal plane, however the photo damage at the point of focus will possibly be higher than when using single photon excitation. Photo damage is caused by the laser light being absorbed by either the fluorescent dye being used or by normal cellular constituents. When this absorbed energy creates chemical changes within the sample, rather than the re-emission of fluorescent light of a longer wavelength, photo damage will occur.

Although high intensity laser light may result in some localised heating, the formation of free radicals within the sample is the main way in which the laser light damages the tissue and bleaches the fluorophore. Free radical formation is discussed in more detail on 196 in Chapter 8 "What is Fluorescence?". Free radicals are highly reactive chemical intermediates that will react with any suitable organic molecules in their immediate vicinity. The reactivity may result in degradation of the fluorescent dye being used as a molecular tag, or in the destruction of other cellular constituents and therefore affect the cellular phenomena under study.

Photo Damage:
Minimise laser power
Minimise laser on time

Using a minimum of laser power (at the sample) to obtain a suitable image will significantly lower the incidence of free radical formation. Keeping to a minimum the amount of time the laser is used to irradiate the sample is also important. Be aware that utilisation of the zoom function while scanning causes the total laser light to be scanned over a smaller area. This will result in a significant increase in the average intensity of laser within the reduced area - even though in the later model confocal microscopes the intensity of the on screen image may remain the same.

The mercury and xenon arc lamps used for conventional epi-fluorescence imaging, and often used to locate suitable cells for confocal microscopy, are particularly destructive to both the fluorophore and the cells. Even the tungsten lamp, used for conventional bright-field illumination, can cause photo-damage to the cells. When extended bright-field imaging is required, photo damage can be minimised by using a red-coloured glass filter to illuminate the cells with longer wavelength red light.

Free radical formation is greatly reduced by lowering the oxygen content of the sample. Unfortunately, this may be deleterious to the physiology of the cells under study. However, lowering the oxygen level may be effective when imaging cellular events that do not change rapidly.

The fading of many fluorophores seriously limits their applicability in live tissues. There has been a great deal of effort in recent years to develop dyes that are less affected by photobleaching. One such series of dyes that has recently been developed is the Alexa Fluor range of dyes from Molecular Probes. These probes show almost no fading when used in buffered aqueous media, even at relatively high levels of laser power. Some of the older but still popular dyes, such as fluorescein, may fade so fast when imaging live cells that only one or two full screen scans may be possible.

Antifade Reagents

Most antifade reagents cannot be used with live cells, simply because the compounds themselves are toxic to the cells. Vitamin C (Na ascorbate) can be used as a free radical scavenger (antifade reagent) on live cells as this compound is a natural component of biological tissues and is relatively non-toxic. However, the high concentration of Vitamin C required (2mg /ml) to be effective as an antifade reagent may result in physiological changes to the cells. A derivative of Vitamin E (Trolux) is also effective in reducing free radical damage in live cells. (see page 198, for a table of suitable antifade reagents in live and fixed cell preparations).

Laser Power

The power of the laser has an important influence on the amount of fading caused by free radical damage. The total laser power directed at the tissue is determined both by the power of the laser light used for irradiation and the amount of time the laser irradiates the sample. The laser power that is important in confocal microscopy is the power level at the sample. A combination of both the laser power setting and the neutral density or level of AOTF filter employed will determine the intensity of the laser at the sample.

As mentioned above laser light in confocal microscopy is of sufficient intensity throughout the depth of the sample for free radical formation to occur. For example, when collecting the 10th optical slice of a 3D image stack, all the other "optical slices", both above and below the 10th optical slice, will be irradiated by the laser. This will mean that free radical formation will occur through the full depth of the scanned area. The effect of this high level of irradiation throughout the sample may become particularly noticeable when attempting to collect a stack of optical slices for the purpose of creating a 3D reconstruction of the object. The optical slices collected last may have a greatly diminished level of fluorescence. Multi-photon microscopy, as discussed above, does not have sufficient power outside the focal plane to cause appreciable photo damage.

A large amount of fading can occur during preliminary observation of the sample. This can often be minimised by lowering the amount of laser light reaching the sample, opening the pinhole, and turning the gain to its highest setting. This will result in a poor quality image, but at least you will be able to locate the cells of interest without destroying them in the process! Another trick to minimising photo damage is to use a very low level of laser light and to take advantage of transmission imaging (or bright-field illumination if observing the cells directly through the microscope) to first locate the cells or area of interest.

AUTOFLUORESCENCE IN LIVE CELLS

Some cells and tissues naturally fluoresce when irradiated with visible light. The problem of autofluorescence is greater at shorter wavelengths, with UV light giving the highest levels of autofluorescence and far red light showing very little autofluorescence. The level of interfering autofluorescence can often be lowered by exciting a suitable dye with the somewhat longer wavelength 568 nm yellow light of the krypton-argon-ion laser, rather than the shorter wavelength 488 nm blue line.

As a Probe for Cellular Structure and Function

Autofluorescence is not all bad, as sometimes this phenomenon can be put to good use in studying the cells of interest. For example NADH is strongly fluorescent at UV wavelengths (and is also readily excited by 2-photon excitation) and will show up strongly in active mitochondria. You can use this "autofluorescence" to study the bio-energy state of the mitochondria – or you can treat this fluorescence as "background" to provide further structural information when imaging live tissue samples. Another strongly fluorescent natural constituent of many cells is Lipofuscin, a fluorescent lipid that accumulates in cells as they age. These lipid granules are particularly prevalent in muscle cells, and can be used as a marker of the age of the muscle tissue.

Dying cells are often fluorescent

Autofluorescence can also be utilised to produce excellent images in plants (chlorophyll), insects and crustaceans (chitin) and many other cell types. In fact most cells will show some level of autofluorescence if the gain on the confocal microscope is increased sufficiently. Autofluorescence may be used to create a "background" image of the specimen so that the fluorochrome of interest can be more easily localised within the cell or tissue sample. Many dying or sick sells often show greatly increased levels of autofluorescence. This may create some confusion when attempting to establish levels of apoptosis or necrosis by the use of fluorescence probes, such as PhiPhiLux and propidium iodide, which are specific for apoptotic cells and dead cells respectively.

Autofluorescence is often observed in both channels when dual labelling (both the red and green channel). Although this doesn't solve the problem, the spread of fluorescence across more than one channel may at least identify when there is a problem that may result in misleading conclusions in your experiments. Autofluorescence may sometimes be at least partially removed from your images by subtracting one of the other imaging channels that is showing the autofluorescence in a similar "pattern" to the channel you are using for your fluorophore.

Controlling Unwanted Autofluorescence

Autofluorescence may be influenced by the metabolic state of the cells and by available nutrients in the culture media. If you suspect that the culture media is creating autofluorescence then first try leaving out the phenol red from the media. However, a number of other nutrients, particularly vitamins, may also increase the level of autofluorescence of the cells. Not only is there a problem of the media being fluorescent, but the live cells may concentrate relatively low levels of potentially fluorescent compounds, particularly within subcellular structures. As has already been discussed above, the infrared laser used in multi-photon microscopy may result in significantly less autofluorescence than the blue and yellow lasers routinely used in confocal microscopy.

Some cell lines will have a much higher level of autofluorescence compared to other cell lines. Changing to a different cell line, if this is not detrimental to your experiments, may be a simple way of illuminating unwanted autofluorescence.

Autofluorescence is quite a different problem when using fixed tissue or cell preparations. In this case the method of fixation is often the cause of the autofluorescence. In particular, aldehyde fixatives can result in high levels of autofluorescence (autofluorescence induced by fixation is discussed in more detail on page 275 in Chapter 11 “Fluorescence Immunolabelling”).

IMAGE COLLECTION TIPS FOR LIVE CELLS

The successful collection of high quality and informative images when using live cells may require some modification of the approach used in imaging fixed tissue or cell samples. Fixed tissue and cell preparations are best imaged by averaging several scans per optical slice. In this way the high noise level of a low light level sample can be greatly reduced. Imaging live cells is not as straight forward. In this section a number of ideas are put forward for helping you to improve the quality of your live cell images.

High Speed Collection

A number of cellular processes occur at a rate that requires high-speed image collection. The laser spot scanning confocal microscope is limited in the speed of image collection by the speed of the scanning mirrors in the confocal microscope scan head. However, there are a number of tricks that can allow one to obtain significantly faster collection rates than are normally possible. Specialised high-speed confocal microscopes are discussed below.

The simplest way to increase the image collection rate is to collect a smaller image. This is achieved by scanning a smaller box size (Figure 12-37). In this example a box size of 150 x 150 pixels, rather than the normal 512 x 512 pixel box size, has been used to collect the image. The zoom function has been utilised in order to allow a single cell to fill the smaller box size. This results in the same pixel resolution in the smaller image as compared to the large sized image, but with only one cell instead of several cells within the image.

Care should be taken however, when increasing the “scan rate” in laser scanning confocal microscopy. One problem of increasing the scan speed without decreasing the pixel “box” size is that the dwell time per pixel is significantly reduced. This may result in a large increase in the “noise” of the image. The technique described above, where the box size is decreased, does not alter the dwell time even though the scan rate has been increased significantly.

Increasing the "scan rate" on some instruments will result in faster image collection, but with a subsequent loss of detail due to there being less pixels within the image. The fast scan mode on the Bio-Rad instruments, for example, results in an image with larger sized pixels. This higher speed scan rate is normally used for quickly moving around the sample to locate cells of interest, although this fast scan rate can be used for following fast cellular changes if high resolution images are not required.

Specialised High Speed Confocal Microscopes

The Nipkow disk based confocal microscopes (Yokogawa, Perkin-Elmer Life Sciences, VisiTech International and Atto Bioscience) allow for greatly increased speed in confocal microscopy imaging (discussed in detail on page 91 in Chapter 3 “Confocal Microscopy Hardware”, and on pages 431-443 in “Appendix 1: Confocal Microscopes”).

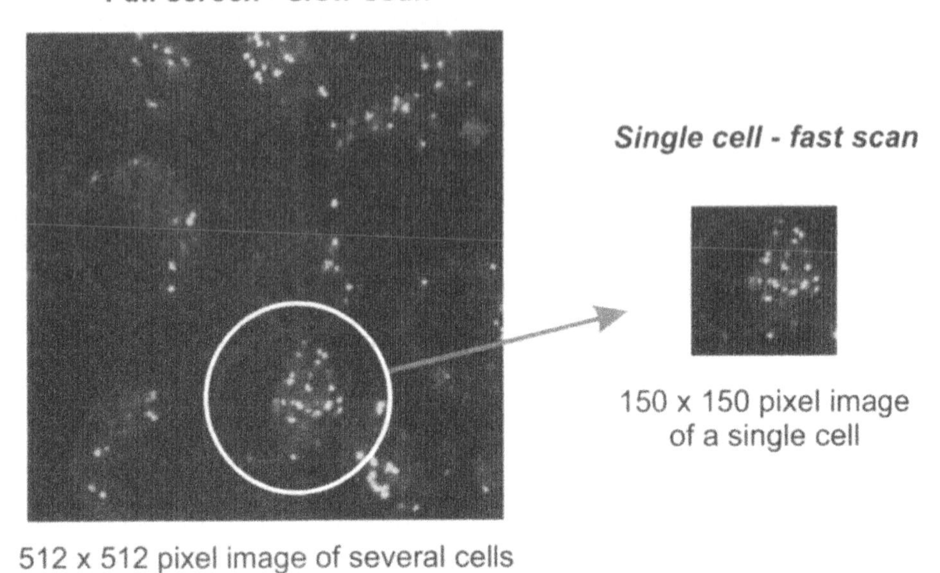

Figure 12-37. Faster Scanning.
The rate of scanning can be significantly increased by imaging a single cell. This is achieved by decreasing the scan box size (in this case from 512 x 512 down to 150 x 150 pixels) and then increasing the scan zoom such that the single cell fills the new box size. Human T-lymphocytes (MOLT-4 cells) were labelled with endocytosed fluorescent latex beads (microspheres). The latex beads clearly mark the position of endocytic vesicles and lysosomal structures. Further information on the experimental conditions used to obtain this image can be found in Hibbs et al (1997) see page 349 in Chapter 15 "Further Reading".

The Nipkow disk based confocal microscope can be routinely used to collect images at 8 to 14 frames per second (using a box size of 512 x 512 pixels) and can reach speeds of over 100 frames per second if a suitable high-speed cooled CCD camera is used.

The spinning Nipkow disk contains a large number of small "pinholes" that allow the laser light to be scanned across the object as many thousands of points at the one time. The Perkin-Elmer and VisiTech instruments utilize the Yokogawa scan head, which has the useful feature of having an additional disk containing thousands of micro lenses. This has the effect of increasing the laser intensity at each "spot". These instruments can be used for conventional fixed sample confocal microscopy imaging, but their real advantage is in the high speed at which high quality live cell images can be collected. An additional advantage when imaging live cells is that the simultaneous irradiation of many points at a relatively low laser intensity (compared to a "spot" scanning confocal microscope) means that there is less photo damage to the sample and less photo-bleaching of the fluorophore. The unique design of these instruments does result in some loss of resolution, which is particularly noticeable when attempting to image small subcellular organelles – and this is compounded by the fact that the instruments do not have an "optical zoom" function like the point scanning confocal microscopes. However, the somewhat reduced resolution is a small price to pay when imaging live cells – where fluorophore fading and speed of image acquisition are critical issues.

The relatively low light intensity, and thus the minimal photo damage caused when using these Nipkow disk based confocal microscopes, results in the capability of being able to use very long time course collections on live cell or tissue samples without affecting the ability of the cells to continue to grow and divide. Particularly spectacular images of cellular changes during embryo development have been obtained with these confocal microscopes.

Multi-photon Imaging of Live Cells

Multi-photon microscopy (discussed in detail on page 94 in Chapter 3 "Confocal Microscopy Hardware") is particularly suited to live cell imaging of samples that are relatively "thick". This includes not only tissue samples, but also cultured cells grown under conditions that favour the formation of larger cellular structures. The term "thickness" in microscopy can be somewhat misleading as we are talking about the difference between confocal microscopy (100 to 150 μm depth) and multi-photon microscopy (200 to 250 μm depth) being only a matter of 100 μm or less. This is a large distance in terms of individual cells (i.e. you will be able to image several cells deeper into the tissue sample), but in terms of the tissue itself (for example, when imaging a tumour) the increased depth is often not sufficient to allow a great deal more structural information to be obtained.

The reason multi-photon microscopy is an advantage when imaging live tissue/cell preparations is that the infrared laser used can more readily penetrate biological tissue, due to less scattering and little or no absorption. There is also less autofluorescence in the infrared region of the light spectrum, and the unique characteristics of multi-photon microscopy means that only the focal plane is exposed to sufficient light intensity to cause photo damage. However, the infrared laser used in multi-photon microscopy can create a high level of localised heating and cause considerable damage to biological tissue samples. In fact, you can easily make single cells "explode" with a multi-photon laser! Cells that are pigmented are much more likely to "explode" compared to cells that lack pigmentation. Avoiding pigmented cells or tissues may not always be possible, but if they can be avoided better imaging results may be obtained. Using scan or line averaging to create low noise images, rather than increasing the laser power, will help retain the integrity of the cells.

"Slow Scan" Collection

Most confocal microscopes have a "slow scan" mode that results in a significantly longer dwell time for each pixel in the image. The speed of scanning can often be adjusted by either altering the speed of the scanning mirrors, or by line averaging. Slow scanning will result in a much higher quality image (as more photons can be collected), and has the advantage that small movements within the image will not result in a "blurred" image, as would happen if a screen averaging algorithm were used. The image will be slightly distorted if cellular movement has occurred, but depending on the speed of movement within the image, this may not be at all noticeable. "Slow scanning" is particularly useful when collecting images of live suspension cells that have simply settled onto the coverslip base of the chamber. These cells, not being attached firmly to the coverslip, will still move somewhat during the imaging process. Slow scan imaging can result in excellent high-resolution images, as this cellular movement is relatively slow compared to the scan speed.

Some laser scanning confocal microscopes can also collect images by line averaging. This has the effect of further slowing the rate of collection, and greatly increases the signal to noise ratio. Again, the image will turn out slightly distorted if there is slow movement of the cells - but at least the image won't be unduly blurred.

Single-Line Collection

If very fast collection times are required (often used in calcium imaging) then using a more innovative approach to increasing collection speed may be necessary. One way this can be done is to collect a single line across the image (the horizontal line scan speed may be as high as 500 Hz, which is much faster than the relatively slow maximum vertical scan rate of 4 frames per second). The single-line collection can be collected over time, and then displayed as an image, in which the vertical axis is time and the horizontal axis is a slice across the cell. In this way highly transient calcium spikes within a cell can be readily studied.

Chapter 13

The Internet

The Internet contains a wealth of information on confocal microscopy. There are a number of well-maintained University based sites that have many links to other useful confocal microscopy and related sites. A number of companies involved in confocal microscopy, microscopy supplies, imaging systems and fluorescent probes also have excellent sites containing information on microscopy related technologies. Each of the major confocal microscope manufacturers maintain web sites, some of which have extensive information on the technology.

In addition to web sites relevant to confocal microscopy, there is a well-utilised Listserver for confocal microscopy. On the following pages there is a brief introduction to a number of important sites on the Internet which are relevant to confocal microscopy, and information how to subscribe to the confocal microscopy List.

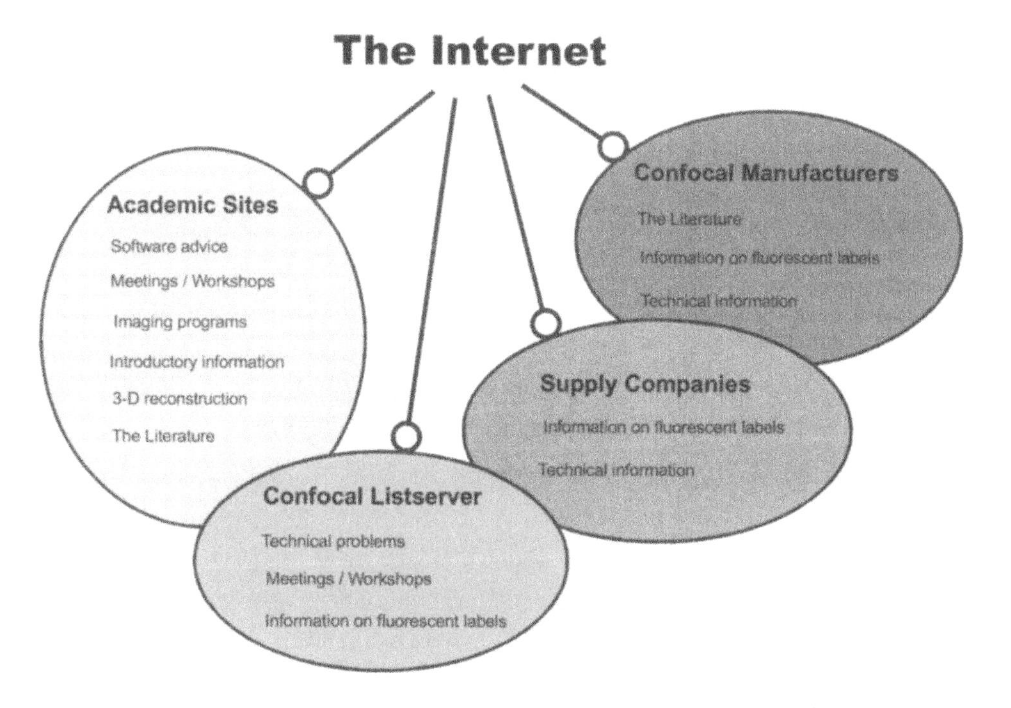

The Internet sites listed in this chapter can be accessed from the authors own web site at **www.BIOCON.com.au**

CONFOCAL MICROSCOPY LISTSERVER

The confocal microscopy Listserver is an automatic mailing list where you can ask questions concerning confocal microscopy. Your message to the List will be sent out to the many subscribers around the world. Members of the list include many research personnel, students, confocal microscopy experts as well as the major confocal microscopy manufacturers. Due to the large number of highly knowledgeable subscribers your query will often be answered in considerable detail very quickly (within 24 hours). However, the down side to a Listserver is that you do need to frame your question carefully, for if you don't give enough detail you'll end up with the right answer to the wrong question! Use a concise and clear subject line in your email, otherwise your message may be deleted before it's even read! Please note that accidentally sending messages to this list will result in your messages being read by over 1500 people around the world - so be careful!

An archive of previous email exchanges on the confocal microscopy Listserver is available on the Internet. The archive is grouped into topics discussed, and one can search the archive for topics of interest. A lot of common problems are frequently discussed on the Listserver and so the archives will often be a good starting point for solving technical problems.

To send messages to the list:

Ask a precise question in the body of the message and use a clear short note in the subject line. Send the message to all members of the list,

confocal@listserve.acsu.buffalo.edu

The Confocal List

Warning:

take care when using the "reply" button that you do not accidentally send a personal email message out to everyone on the list!

To subscribe:
Send the message 'subscribe confocal'

To get help:
Send the message 'help confocal'

To unsubscribe:
Send the message 'signoff confocal'

These commands should be sent in the body of the message (with nothing in the subject line), and without the quotation marks.

Subscription requests should be sent to;

listserv@ubvm.cc.buffalo.edu

DO NOT send these requests to the list (which will send them out to everyone on the list).

Subscribe on-line at:
http://listservacsu.buffalo.edu/archives/confocal.html

Archives of the List are available from:http://listserv.acsu.buffalo.edu/archives/confocal.html

WEB SITES RELEVANT TO CONFOCAL MICROSCOPY

A number of interesting Internet sites, sometimes with addresses for specific pages within the site, are listed below. This is not intended to be a comprehensive listing – just a few sites to get you started. Due to the highly dynamic nature of the Internet you are advised to consult the author's web site (www.biocon.com.au) for an updated list of the Internet sites listed below.

International Mailing Lists

Confocal Microscopy Listserver http://listserv.acsu.buffalo.edu/cgi-bin/wa?SUBED1=confocal&A=1	This international mailing list is intended as a forum for people using confocal microscopy in research. You need to send a message to the Listserver to subscribe (see previous page for details).
http://listserv.acsu.buffalo.edu/archives/confocal.html	**Archives of the List:** The archive can be searched for topics of interest, or you can browse though the monthly accumulations.
http://listserv.acsu.buffalo.edu/cgi-bin/wa?SUBED1=confocal&A=1	**Subscribing to the list:** a simple form for subscribing.
Histonet List www.histosearch.com/listserver.html	Histology techniques Listserver.
Microscopy Listserver www.msa.microscopy.com/MicroscopyListserver/MLInstructions.html	Discussion group for all aspects of microscopy and microanalysis.
Multi-Photon Microscopy Listserver http://groups.yahoo.com/group/mplsm-users/	This list is intended as a means of communication for people interested in multi-photon (2-photon) microscopy. **Archives of the List** are available for members only.

Useful Research and Core Facility Sites

3D Confocal - Lance Ladic, Vancouver www.cs.ubc.ca/spider/ladic/confocal.html	A good, but somewhat dated site, for technical information on software and confocal microscopy imaging problems.
3D reconstruction, Stanford http://biocomp.stanford.edu/3dreconstruction/index.html	This page is dedicated to providing information, pointers, references, and results of 3D reconstruction work for medical/life-science research (MRI, CT, PET, microscopy).
Albert Einstein College of Medicine www.aecom.yu.edu/aif/welcome.htm	Core facility web page with information on microscopy, confocal microscopy and electron microscopy.
BIOCON, specialists in confocal microscopy www.biocon.com.au	Web site maintained by the Alan R Hibbs, the author of this book, listing the Internet sites in this chapter. This site also contains information on confocal and multi-photon microscopy and courses provided by the author.
Biomedical Structural Research http://ourworld.compuserve.com/homepages/pvosta/pcrimg.htm	Comprehensive listing of biomedical imaging sites on the Internet.
Caltech Biological Imaging Center http://bioimaging.caltech.edu/	Core facility web page with information on light microscopy.
Cell Sciences Imaging Facility, Stanford http://taltos.stanford.edu/	Core facility web page with information on fluorescence microscopy and electron microscopy.
http://taltos.stanford.edu/fmc.html	Direct link to **fluorescence microscopy** web page, including an excellent question and answer section.
Confocal Club, Karolinska Institute, Sweden www.ki.se/cns/confocal/uselinks.html	Information on confocal microscopy and interesting microscopy and confocal microscopy links and courses on confocal microscopy.

Dynamic Cell Imaging (DCI) www.univ-reims.fr/Labos/INSERM514/DCI/	A site dedicated to live cell imaging. The site provides links, a bibliographic database, discussion forum and educational material related to live cell imaging. Registration (free) is required to access this site.
Forest Research – New Zealand www.forestresearch.co.nz/ www.forestresearch.co.nz/topic.asp?topic=Microscopy%20Suite&title=Services%20%26%20Research	Imaging facilities at the Forest Research include confocal and electron microscopy. The site also includes information and links on imaging wood structure. Go to "Search Tips" on this web site and search for "microscopy" to access information on light, electron and confocal microscopy.
IHC (Immunohistochemistry) World www.ihcworld.com/	Web site for information and protocols on immunohistochemistry and immunolabelling.
www.ihcworld.com/protocol_database.htm	**Protocol Database**.
www.ihcworld.com/forum/	**Discussion forum for immunohistochemistry.**
Jagiellonian University, Poland	Laboratory of Confocal Microscopy and Image Analysis
http://helios.mol.uj.edu.pl	Department of Biophysics, Faculty of Biotechnology.
http://helios.mol.uj.edu.pl/conf_c/main.htm	**Introductory course** on confocal microscopy by Jurek Dobrucki.
Key Centre, Sydney http://web.emu.usyd.edu.au/kc/index.html	The Australian Key Centre for Microscopy and Microanalysis at the University of Sydney is a major source of confocal microscopy expertise in Australia. A number of microscopy and confocal microscopy courses are run by the Key Centre (see "Courses on Fluorescence Imaging & Confocal Microscopy" below).
LRSM – The Laboratory for Research on the Structure of Matter www.lrsm.upenn.edu/	Center for materials research at the University of Pennsylvania, USA.
http://glinda.lrsm.upenn.edu/	Confocal microscopy home page
www.physics.emory.edu/~weeks/confocal/	Short outline of "**How does a confocal microscope work?**".
Micro-Scope, Kiel, Germany. www.micro-scope.de/toc.html	An **on-line reference guide** to microscopy (in English).
Microscopy at University of Wisconsin - Madison www.microscopy.wisc.edu/	**Web portal** to research and service microscopy at the University of Wisconsin – Madison, USA.
Microscopy Info www.mwrn.com/default.asp	A wealth of **contact sites** for commercial, academic, research, professional societies and more.
Multi-Photon excitation of common fluorophores www.uga.edu/caur/dyechart.pdf	This document provides a data sheet of the single-photon and 2-photon excitation of a range of commonly used fluorescent dyes.
Princeton Confocal & Electron Microscopy www.molbio.princeton.edu/facility/confocal/index.html	Core facility web page with information on confocal and electron microscopy.
Purdue University Cytometry Laboratories www.cyto.purdue.edu/index.htm	Extensive coverage of both flow cytometry and microscopy, including confocal microscopy.
www.cyto.purdue.edu/flowcyt/confocal/confocal.htm	**Confocal microscopy** web site direct link.
www.cyto.purdue.edu/hmarchiv/index.htm	**Flow cytometry** discussion group.
www.cyto.purdue.edu/flowcyt/books/refsconf.htm	A list of **books on confocal microscopy**.
www.cyto.purdue.edu/flowcyt/cdseries.htm	**CD ROM series**, including 1998 and 2004 on Microscopy.
Tsien Laboratory University of California, San Diego (UCSD), USA http://tsienlab.ucsd.edu/Default.htm	The Tsien laboratory are pioneers in the development of fluorescent probes for studying living cells.

Uni of Michigan Microscopy and Image Analysis www.med.umich.edu/cdb/sub_pages/programs_facilities/microscopy.htm	Core facility laboratory web site with educational material on confocal microscopy.
www.med.umich.edu/cdb/mial/ZeissMETAInstructions_files/frame.htm	**Zeiss LSM 510 PowerPoint tutorial**.
University of Arizona	
Southwest Environmental Health Sciences Center http://swehsc.pharmacy.arizona.edu/exppath/index.html	Extensive **links to other useful sites** relevant to confocal microscopy. Articles on confocal microscopy, fluorescence imaging, fixation etc.
Molecular and Cellular Biology Imaging Facility www.mcb.arizona.edu/IPC/	Core facility site with **information on confocal microscopy**.
www.mcb.arizona.edu/IPC/spectra_page.htm	**Fluorescence database:** overlay excitation and emission spectra.
University of Edinburgh www.dns.ed.ac.uk/Confocal_web/set_Home.htm	Core facility web page with a range of links related to confocal microscopy.
University of Florida, Microscopy Primer http://microscopy.fsu.edu/primer/index.html	Good **introductory tutorials** on various aspects of microscopy, including confocal microscopy (including animated Java tutorials demonstrating many of the principles of microscopy).
University of Georgia Centre for Ultrastructure www.uga.edu/caur/	Core facility web page with information on light microscopy.
University of Manchester Confocal and Multi-Photon Microscopy www.biomed2.man.ac.uk/ireland/homepage.html	Core facility web page with information on fluorescent probes, confocal and multi-photon microscopy.
University of Otago Department of Anatomy and Structural Biology http://anatomy.otago.ac.nz/occm/	Core facility web page with links, information and references on microscopy, including confocal microscopy.
University of Pennsylvania Biomedical Imaging www.uphs.upenn.edu/morphlab/light.html	Core facility web page with information on light microscopy.
University of Victoria Advanced Imaging Lab. http://web.uvic.ca/ail/	Core facility web site with extensive information on the instruments available, including a Zeiss LSM 410 confocal microscope.
http://web.uvic.ca/ail/lsm_index.html	Direct link to Zeiss LSM 410 information.
http://web.uvic.ca/ail/fluorochromes.html	List of commonly used **fluorescent probes.**
Virtual Library: Microscopy www.ou.edu/research/electron/www-vl/	Wealth of information on microscopy and related techniques.
W.M. Keck Laboratory for Biological Imaging www.keck.bioimaging.wisc.edu/	University of Wisconsin, Madison, USA, core facility site.
www.keck.bioimaging.wisc.edu/keck/linx.htm	Links to various Internet sites related to confocal microscopy.
Watt Webb Laboratory – Cornell. www.drbio.cornell.edu/	The Webb laboratory are pioneers in the area of multi-photon microscopy. This site contains valuable information on the multi-photon cross-section data of a number of fluorescent probes, including a range of GFP proteins.
www.drbio.cornell.edu/MPE/mpe2.html	Direct access to page containing **multi-photon cross-section data**.
Center for Biofilm Engineering – Montana. www.erc.montana.edu/CBEssentials-SW/research/microscope%20facilities/	The Montana State University core facility for biofilm engineering has created an internet site with valuable information on various aspects of microscopy and fluorescence imaging, which can be accessed directly, or via images of various biofilms utilising the different imaging approaches.

Courses on Fluorescence Imaging & Confocal Microscopy

Advanced Techniques in Microscopy Temple University, Philadelphia, USA. http://astro.temple.edu/~jbs/courses/333/Schedule-2003.html	**Semester-long course on microscopy** designed as an undergraduate/graduate level course, but can be taken as "continuing education" by people outside the university.
ACS Microscopy Course Pittsburgh, USA. www.microscopyeducation.com/acs_crse.htm	American Chemical Society **3-day course on microscopy** is intended for chemists, biologists, geneticists and forensics and materials scientists. Organised annually to precede the PITTCON American Chemical Society meeting.
Advanced Course in Laser Scanning Microscopy Carl Zeiss – Center for Microscopy, Jena, Germany www.zeiss.de/C12567BE0045ACF1?Open	A range of courses on **microscopy, including confocal microscopy** are held in Jena, Germany (in English) at the Carl Zeiss Center for Microscopy. This includes basic and advanced courses in laser scanning confocal microscopy and fluorescence correlation spectroscopy.
BIOCON Courses in Confocal Microscopy Melbourne, Australia. www.biocon.com.au/	The author of this book conducts **2 and 3-day in-house courses** in fluorescence imaging, confocal microscopy and multi-photon microscopy. These courses are designed around the type of instrument and interests within your department/institute or company.
Cellular Imaging and Confocal Techniques Karolinska Institute, Stockholm, Sweden. www.ki.se/cns/confocal/course.html	**Annual 5-day course** on cellular imaging, including confocal microscopy held at the Karolinska Institute, Sweden.
Chicago Microscopy Course The Chicago Medical School North Chicago, Illinois, USA. www.microbrightfield.com/microscopycourse2003.html	**Qualitative and quantitative microscopy for biomedical research**. Covering fluorescence, confocal microscopy, stereology, specimen preparation, histology etc.
Confocal Light Microscopy: Fundamentals and Biological Applications Swammerdam Institute for Life Sciences, The Netherlands. wwwmc.bio.uva.nl/PhDconfocal/PhDconfocal.html	**Annual 5-day course** covering fluorescence and confocal microscopy as well as digital imaging, sample preparation and live cell microscopy.
EMBO Live Specimen Light Microscopy Course European Molecular Biology Laboratory, Heidelberg, Germany. www.embl-heidelberg.de/ExternalInfo/stelzer/embocourse_courseoutline.html	**Annual 2-week EMBO practical course run on light microscopy** held at the European Molecular Biology Laboratory (EMBL), Heidelberg, Germany. Organised by Ernst Stelzer and Rainer Pepperkok. Variations of this course are run in a number of countries, including Singapore, Brazil and Japan.
International 3D Microscopy of Living Cells Vancouver, Canada www.3dcourse.ubc.ca	**Annual 12-day live cell imaging course** organised at the University of British Columbia, by James B. Pawley. A 12 day live-in course with a wide variety of instruments (confocal, multi-photon, wide-field etc). Followed by a 3-day "3D **Image Processing Workshop**".
International Cytometry School Jagiellonian University, Krakow, Poland. http://helios.mol.uj.edu.pl/	**Annual course in flow cytometry and confocal microscopy** organised by the Laboratory of Confocal Microscopy, Jagiellonian University in collaboration with the Polish Cytometry Society.
Leica Advanced Microscopy Course Heidelberg, Germany and Washington DC, USA. www.confocal-microscopy.com/website/sc_llt.nsf	**Advanced microscopy courses** are organised by Leica Microsystems are held annually in Heidelberg, Germany, and the National Institute of Health (NIH) in Washington DC, USA.

MBL - Analytical and Quantitative Light Microscopy Marine Biological Laboratory (MBL), Woods Hole, USA. www.mbl.edu/education/courses/special_topics/aqlm.html	**Annual 8 day course** on light microscopy for researchers in biology, medicine and materials sciences held at the Marine Biological Laboratory (MBL), Woods Hole, USA. An in-depth examination of the theory of image formation and application of video imaging. Designed for professional researchers, postdoctoral fellows and advanced graduate students who wish to expand their knowledge of microscopy.
MBL - Optical Microscopy and Imaging in the Biomedical Sciences Marine Biological Laboratory (MBL), Woods Hole, USA. www.mbl.edu/education/courses/special_topics/om.html	**Annual 8 day course** on microscopy in the animal, plant and material sciences held at the Marine Biological Laboratory (MBL), Woods Hole, USA. An introductory course for research scientists, physicians, postdoctoral trainees and advanced graduate students.
Microscopy/Microscopy Education (MME) USA www.microscopyeducation.com/	**Annual courses in fluorescence imaging and microscopy**, including in-house training tailored to specific laboratory or institute requirements are organised by Barbara Fostèr of MME, USA.
Monash Micro Imaging Courses Monash University, Melbourne, Australia. http://microimaging.monash.org/train.html	**Basic light microscopy, fluorescence and confocal microscopy** 3-day courses are organised by Ian S. Harper. Specialised live cell imaging workshops are also held throughout the year.
Pittsburgh Quantitative Fluorescence Microscopy Center for Biologic Imaging, University of Pittsburgh, USA. www.cbi.pitt.edu/qfm/index.html	**6-day course** on Quantitative Fluorescence Microscopy at the Center for Biologic Imaging at the University of Pittsburgh organised by Simon Watkins. Covering microscopy, CCD cameras, confocal & multi-photon microscopy and fluorescent probes and fluorescent imaging in biology.
Sydney Key Centre Courses in Microscopy. University of Sydney, Australia. http://web.emu.usyd.edu.au/kc/educational_products/courses.html	The Australian Key Centre for Microscopy and Microanalysis at the University of Sydney conducts **regular training courses;** including light microscopy, confocal microscopy, specimen preparation, TEM, SEM, Stereology, image analysis etc.
Live Cell Imaging. Instituto Gulbenkian de Ciência, Portugal. http://uic.igc.gulbenkian.pt/micro_course	**Annual 4 to 5 day** course on live cell imaging held in Portugal.

Confocal Microscope Manufacturers

Each of the major confocal microscope manufacturers maintains a web site that details the range of products they sell. Contact details for each of these companies is listed in the following chapter (Chapter 14 "Technical Supplies", page 335). A number of confocal microscope manufacturers that provide information on the principles of confocal microscopy and useful information on fluorescence and other confocal microscopy related matters are listed below.

Bio-Rad Cell Sciences Division	www.microscopy.bio-rad.com/	The Bio-Rad confocal microscopy site has information on their own confocal microscopes, the Bio-Rad bibliographic listing of research articles that utilise confocal microscopy and a series of application notes that can be downloaded or ordered on the web.
	http://fluorescence.bio-rad.com/	**Fluorescence database**, an excellent database on fluorescent probes. You can choose your fluorochrome of interest and graph the excitation and emission spectra.
	http://microscopy.bio-rad.com/customer/customers.htm	**Technical support page**, search their information database, or email a technical problem.
	ftp://ftp.genetics.bio-rad.com/Public/confocal/	**FTP site** for downloading software (for example, Photoshop plugins for reading "PIC" files and Confocal Assistant).
Carl Zeiss	www.zeiss.de	Carl Zeiss main web page.
	www.zeiss.de/C12567BE0045ACF1?Open	**Carl Zeiss Microscopy** web page, including confocal microscopy.
	www.zeiss.de/C12567BE0045ACF1?Open	**Courses** in confocal microscopy.

Leica Microsystems	www.leica.com	Information on the Leica confocal microscopes and technical information sheets on various aspects of imaging.
	www.leica-microsystems.com/website/sc_llt.nsf	**Leica confocal microscopy home page**.
	www.leica-microsystems.com/website/download.nsf/(ALLIDs)/8214E366D24DF836C1256ABD0048AC51	**Scientific and Technical information notes.** A series of detailed articles on various confocal microscopy and microscopy related subjects.
Nikon Microscopy		An extensive site with lots of information on microscopy. A number of **Java applets on various aspects of microscopy** are particularly informative.
	www.microscopyu.com/articles/confocal/index.html	A direct link to the **confocal microscopy site**, including Java applets.
Olympus	www.olympus.com	The Olympus Internet site has extensive information on their broad range of microscopes and related products.
	www.olympus.com/product.asp?c=26&p=18&s=11&product=133&x=25&y=5	**Fluoview** the Olympus confocal microscope.
	www.olympusmicro.com/primer/	Excellent **tutorial on microscopy**, including interactive Java applets.

Supply Companies

Companies that supply products for microscopy and fluorescence reagents often have a wealth of information relating to their products. The better sites have extensive technical information that is highly informative without being too "pushy" as regards sales. Many more companies are listed in Chapter 14 "Technical Supplies". The following companies have particularly useful pages on technical information related to their products.

BD Biosciences	www.bdbiosciences.com/	Supplier of reagents for biomedical research.
	www.bdbiosciences.com/spectra/	**Fluorescence database:** overlay excitation and emission spectra.
Chroma Technology	www.chroma.com	Manufacturers of optical filters.
	www.chroma.com/resources/application_notes.cfm	**GFP application notes** for downloading.
Clontech	www.clontech.com	Extensive range of **GFP clones** and related products.
	www.clontech.com/gfp/index.shtml	BD Living Colors™ Fluorescent Proteins.
Molecular Probes	www.probes.com	This site contains a wealth of information on thousands of fluorescent probes. This includes an extensive literature database, technical information and information on most of their products (including **excitation/emission spectra** and **molecular structure**).
Omega Filters	www.omegafilters.com	Major supplier of optical filters for confocal microscopy.
	www.omegafilters.com/curvomatic/spectra.php	**Fluorescence database:** interactive database of fluorophore excitation and emission spectra and optical filter spectra.
Universal Imaging	www.image1.com/	Suppliers of imaging and microscopy equipment and software. Their site is quite extensive, with downloadable articles on a number of aspects of microscopy, fluorescence and digital imaging.
	www.image1.com/products/apps/fluor.cfm	Information on **GFP proteins**,
	www.image1.com/products/apps/3d.cfm	Excellent article on **3D reconstruction,**
	www.image1.com/products/apps/fret.cfm	Information on **FRET**.

Imaging Software

The URL for a number of internet sites that provide information and downloads of imaging software are provided on pages 165 - 167 in Chapter 6 "Imaging Software", with company information and contact details provided on page 337 of Chapter 14 "Technical Supplies".

SVI Group University of Utrecht, The Netherlands http://sidb.sourceforge.net/	**Web-driven database for images.** Free program, with source code available, for archiving and retrieving images.

Microscope Imaging Chambers

The large number of microscope chambers available can make purchasing somewhat confusing. Most of the following companies have extensive information about their chambers on the web, including good diagrams and photos that allow one to get a feel for their instruments. For more information on the types of chambers available see page 343 in Chapter 14 "Technical Supplies".

20/20 Technologies	www.20-20tech.com	Warming plates for microscopy.
ALA Scientific Inst	www.alascience.com/productindex2.html	FEP Teflon membrane covered microscopy chambers.
Bellco Glass	www.bellcoglass.com	Flow chamber for microscopy.
Bioptechs	www.bioptechs.com	Open and closed temp controlled chambers for microscopy.
Dagan	www.dagan.com	Heated stages for microscopy.
Grace Bio-Labs	www.gracebio.com	Gasket (Lab-Tek) chambers for microscopy.
Hanna	www.hannainst.co.uk	Extensive range of digital thermometers.
Harvard Instruments	www.haicellbiology.com	Perfusion and culture chambers for electrophysiology.
Immunetics	www.immunetics.com	Cell adhesion flow chambers.
Instec	www.instec.com/	Stage warmers for microscopy.
Intracell	www.intracel.co.uk/cellperfid.htm	Temperature controlled perfusion chambers.
Kentscientific	www.kentscientific.com	Organ baths and live tissue chambers.
Live Imaging Services	www.lis.ch	Ludin open perfusion chambers and heated enclosures.
MatTek	www.mattek.com	Petri dish microscopy chambers.
Molecular Probes	www.probes.com/lit/bioprobes23/atto.html	Suppliers of imaging chambers for microscopy.
Nalge Nunc	http://www.nalgenunc.com/	Lab-Tek chambers for microscopy.
Neuroprobe	www.neuroprobe.com/	Specialist chambers for studying live cells.
Nevtek (Lucas-Highland)	www.nevtek.com/	Dvorak-Stotler controlled environment microscopy chamber.
Physitemp	www.physitemp.com/	Temp control systems for microscopy (-20°C to 100°C).
Solent Scientific	http://www.solentsci.com/	Microscope incubator enclosures
Vitrocom	www.vitrocom.com/	Glass & quartz square / rectangular tubing for flow chambers.
Warner Instruments	www.warnerinstruments.com/	A large range of perfusion chambers frames.
WillCo Wells	www.willcowells.com/	Petri dish imaging chambers for microscopy.

Chapter 14

Technical Supplies

CONFOCAL MICROSCOPES

Atto Bioscience

15010 Braschart Road
Rockville, MD 20850
USA

Atto Bioscience manufacture a Nipkow spinning disk confocal microscope for high-speed live cell imaging. The Nipkow disk used by Atto Bioscience instrument does not incorporate a micro-lens array disk like that used in the Yokogawa scan head.

Phone: +1 (301) 340 7320 **FAX:** +1 (301) 340 9775 **Web:** www.atto.com/

Bio-Rad Cell Sciences Ltd

Bio-Rad House
Maylands avenue, Hemel Hempstead
Hertfordshire, HP 7TD
UK

Bio-Rad is the manufacturer of the MRC and Radiance series of confocal microscopes, and is also a major supplier of reagents for cell biology. Bio-Rad is not a microscope manufacturer, but the Bio-Rad confocal microscope scan heads can be attached to microscopes from most microscope manufacturers.

Phone: +44 (0) 208 328 2141 **FAX:** +44 (0) 208 223 2500 **Web:** http://cellscience.bio-rad.com/

Carl Zeiss Pty Ltd

Carl Zeiss, Mikroskopie
D-07740 Jena
Germany

Zeiss is a major manufacturer of both confocal and conventional light microscopes. This includes both the electronics and optics of the scan head as well as the lens components of the microscope itself.

Phone: +49 (364) 164 1616 **FAX:** +49 (641) 64 3144 **Web:** www.zeiss.de/

Chromaphor Analysen-Technik GmbH

Keniastrasse. 12
47259 Duisburg
Germany

Chromaphor Analysen-Technik are the German distributors (with some modifications) of the VisiTech Nipkow spinning disk confocal microscope based on the Yokogawa scan head.

Phone: +49 (203) 99 86 28 **FAX:** +49 (203) 99 86 30 **Web:** www.Chromaphor.de/

Leica Instruments Pty Ltd

Leica Microsystems Heidelberg
Am Friedensplatz 3, 68165 Mannheim
Germany

Leica is a major confocal microscope and conventional light microscope manufacturer. This includes both the electronics and optics of the scan head as well as the lens components of the microscope itself.

Phone: +49 (621) 70 28 0 **FAX:** +49 (217) 28 1028 **Web:** www.leica-microsystems.com/

McBain Instruments

9601 Variel Avenue
Chatsworth, CA 91311-4914
USA

McBain Instruments are distributors (with some modifications) of the VisiTech Nipkow spinning disk confocal microscope based on the Yokogawa scan head.

Phone: +1 (818) 998 2702 **FAX:** +1 (818) 718 0363 **Web:** www.mcbaininstruments.com/

Nikon Pty Ltd

Fuji Building
2-3 Marunouchi 3-chome
Chiyoda-ku, Tokyo 100
Japan

Nikon is a major optics manufacturer that produces a large range of optics based instrument. In addition to producing a laser scanning confocal microscope (the C1) Nikon optical microscopes are highly respected in the scientific research community. The Nikon microscopy site contains extensive information (including Java applets) on light microscopy, including confocal and multi-photon microscopy. Also see Chapter 13 "The Internet", page 332.

Phone: +81 3 3216 1039 Japan, +1 (631) 547 8540 USA **FAX:** +81 3 3201 5856 Japan, +1 (631) 547 4033 USA **Web:** www.nikon.com/

Olympus Pty Ltd

San-Ei Building
22-2 Nishi Shinjuku 1-chrome
Shinjuku-ku, Tokyo 163-0914
Japan

Olympus is an important manufacturer of confocal microscopes as well as conventional light microscopes. The Olympus Internet site contains a wealth of information on microscopy, including confocal and multi-photon microscopy. Also see Chapter 13 "The Internet", page 332.

Phone: +81 (03) 3340 2111, Japan, 1 800 645 8160 USA **FAX:** **Web:** www.olympus.com/

Optiscan Imaging Limited

PO Box 1066
Mt. Waverley MDC, VIC 3149
Australia

Highly miniaturised scan heads based on fibreoptics technology have allowed Optiscan to produce a range of medical instruments that are capable of imaging in a variety of relatively inaccessible regions in vivo on human patients and animal models. This includes a confocal endomicroscope, a hand held skin confocal microscope probe and a rigid endoscope.

Phone: +61 3 9538 3333 **FAX:** +61 3 9562 7742 **Web:** www.optiscan.com.au/

PerkinElmer Life Sciences

549 Albany Street, Boston, MA
USA

PerkinElmer Life Sciences assembles and sells a Nipkow disk confocal microscope based on the Yokogawa scan head.

Phone: +1 (617) 350 9263 **FAX:** +1 (617) 482 1380 **Web:** www.perkinelmer.com/lifesciences/

Quorum Technologies Inc.

1 Hunters Lane
Guelph, Ontario N1C 1B1
Canada

Quorum Technologies are distributors (with some modifications) of the VisiTech Nipkow spinning disk confocal microscope based on the Yokogawa scan head.

Phone: +1 (519) 824 0854 **FAX:** +1 (519) 824 5845 **Web:** www.quorumtechnologies.com/

Solamere Technology Group

1427 Perry Ave.,
Salt Lake City UT 84103
USA

Solamere are distributors (with some modifications) of the VisiTech Nipkow spinning disk confocal microscope based on the Yokogawa scan head.

Phone: +1 (801) 322 2645 **FAX:** +1 (801) 322 2645 **Web:** http://www.solameretech.com/Confocal.htm

VisiTech International Ltd.

Unit 92 Silverbriar
Sunderland Enterprise Park (East)
Sunderland, SR5 2TQ
UK

VisiTech International assemble and sell a Nipkow spinning disk confocal microscope based on the Yokogawa scan head. VisiTech also produce a high-speed spot scanning confocal microscope based on using an acoustic-optical reflector (AOD) to scan the x-axis at high-speed.

Phone: +44 (191) 5166 255 **FAX:** +44 (191) 5166 258 **Web:** www.visitech.co.uk/

Visitron Systems GmbH

Gutenbergstr. 9
D-82178 Puchheim
Germany

Visitron are distributors (with some modifications) of the VisiTech Nipkow spinning disk confocal microscope based on the Yokogawa scan head.

Phone: +49 (089) 890 2450 **FAX:** +49 (089) 890 24518 **Web:** www.visitron.de/

Yokogawa Electric Corporation

Corporate Headquarters
9-32, Nakacho 2-chome, Musashino-shi,
Tokyo 180,
Japan

Yokogawa are the manufacturers of a Nipkow spinning disk confocal microscope scan head that utilizes an array of micro lenses to enhance the sensitivity of the instrument. The Yokogawa scan head can be purchased alone, or as an integrated confocal microscope package from a number of companies.

Phone: +81 422 52 5535, Japan; +1 (770) 253 7000, USA **FAX:** +81 422 52 6985, Japan; +1 (770) 251 2088, USA **Web:** www.yokogawa.com/

IMAGING SOFTWARE

A list of imaging software (including free programs) is provided on page 166 of Chapter 6 "Imaging Software".

AutoQuant Imaging

AutoQuant Imaging, Inc.
877 25th Street
Watervliet, New York 12189
USA

Producers of **AutoDeblur**(deconvolution, including its use in confocal microscopy) and **AutoVisualize**(3D rendering, image processing and analysis) imaging software packages.

Phone: +1 (518) 276 2138 **FAX:** +1 (518) 276 3069 **Web:** www.aqi.com/

Bitplane AG

Badenerstrasse 682
CH-8048 Zurich
Switzerland

Producers of **Imaris** 3D image manipulation and analysis program.

Phone: +41 1 430 11 00 **FAX:** +41 1 430 11 01 **Web:** www.bitplane.com/

Applied Precision (API)

1040 12th Avenue Northwest
Issaquah, Washington 98027
USA

Deconvolution software, **SoftWoRx**, used in the DeltaVision wide-field live cell imaging platform can be purchased separately.

Phone: +1 (425) 557 1000 **FAX:** +1 (425) 557 1055 **Web:** www.api.com/

Iatia Ltd.

46 Rutland Road
Box Hill VIC 3128
Australia

Produces of QPm software for creating phase and DIC images using conventional bright-field images collected on either a confocal microscope or on a conventional wide-field light microscope.

Phone: +61 3 9898 6388 **FAX:** +61 3 9899 6388 **Web:** www.iatia.com.au

Indeed - Visual Concepts GmbH

Ihnestr. 23
D-14195 Berlin-Dahlem
Germany

Producers of **Amira** 3D visualisation and volume modelling software.

Phone: +49 30 84185 221 **FAX:** +49 30 82701 747 **Web:** www.indeed3d.com

Intelligent Imaging Innovations Inc

5124 North Washington Street
Denver, CO 80216
USA

Producers of **Slidebook** image processing and analysis software. Intelligent Imaging Innovations also put together a number of imaging platforms for wide field, spinning Nipkow disk and FRAP applications.

Phone: +1 (303) 607 9429 **FAX:** +1 (303) 607 9430 **Web:** www.intelligent-imaging.com/

Improvision Inc

1 Cranberry Hill
Lexington, MA 02421
USA

Producers of **Openlab** image acquisition and analysis and **Volocity** 3D visualisation and measurement programs. Improvision also produce **Phylum**, a digital image management program.

Phone: +1 (781) 402 0134 **FAX:** +1 (781) 402 0251 **Web:** www.improvision.com/

Media Cybernetics

8484 Georgia Avenue, Suite 200
Silver Spring, MD 20910-5611
USA

Producers of **ImageProPlus** image capture and manipulation program and **IQbase**, an image archive and management program.

Phone: +1 (301) 495 3305 **FAX:** +1 (301) 495 5964 **Web:** www.mediacy.com/

Research Systems Inc (RSI)

An Eastman Kodak Company, see web site for contact addresses

Producers of **IDL** (Interactive Data Language) programming software for data analysis and visualisation.

Phone: +1 (303) 786 9900 **FAX:** **Web:** www.rsinc.com/

Scanalytics

8550 Lee Highway, Suite 400
Fairfax, VA 22031-1515
USA

Producers of **IPLab** image acquisition and analysis program.

Phone: +1 (703) 208 2230 **FAX:** +1 (703) 208 1960 **Web:** www.scanalytics.com/

Universal Imaging Corporation

Universal Imaging CorporationTM
402 Boot Road
Downingtown PA 19335
USA

Producers of **Metamorph** image acquisition and analysis program.

Phone: +1 (610) 873 5610 **FAX:** +1 (610) 873 5499 **Web:** www.universal-imaging.com/

VayTek Inc.

305 West Lowe Avenue, Suite 109
Fairfield, IA 52556
USA

Producers of **VoxBlast** 3D rendering and image manipulation software and **VayTek Image** software for image acquisition and processing.

Phone: +1 (641) 472 2227 **FAX:** +1 (641) 472 8131 **Web:** www.vaytek.com/

Vital Images Inc.

3300 Fernbrook Lnae North
Suite 200, Plymouth, MN 55447
USA

Producers of **Vitrea 2** imaging software for 3D rendering and analysis.

Phone: +1 (763) 852 4100 **FAX:** +1 (763) 852 4110 **Web:** www.vitalimages.com/

FLUORESCENT PROBES AND ANTIBODIES

Amersham Pharmacia Biotech Pty Ltd

Amersham Place, Little Chalfont,
Buckinghamshire HP7 9NA
UK

Supplier of antibodies and fluorescent secondary antibodies, including the Cy probes (Cy2, Cy3, Cy5 etc).

Phone: +44 (0) 870 606 1921 **FAX:** +44 (0) 1494 544350 **Web:** www4.amershambiosciences.com/

Biostatus Limited

56 Charnwood Road
Shepshed, Leicestershire LE12 9NP
UK

Producers of DRAQ5™ (live cell DNA stain) and APOPTRAK™ (apoptosis indicator stain).

Phone: +44 (0) 1509 558163 **FAX:** +44 (0) 1509 651061 **Web:** www.biostatus.co.uk/

Biotium Inc.

3423 Investment Blvd. Suite 8
Hayward, CA 94545
USA

Manufacturer and supplier of fluorescent reagents for biological research.

Phone: +1 (510) 265 1027 **FAX:** +1 (510) 265 1352 **Web:** www.biotium.com/

Helix Research

1940 Don Street, Suite 280,
Springfield, OR 97477
USA

Manufacturer and supplier of fluorescent probes for biological research.

Phone: +1 (541) 988 0464 **FAX:** +1 (541) 744 9179 **Web:** www.helixresearch.com/

ICN Pharmaceuticals Inc.

3300 Hyland Avenue
Costa Mesa, CA 92626
USA

ICN is a major supplier of reagents and laboratory ware for cell biology.

Phone: +1 (714) 545 0100 **FAX:** +1 (714) 641 7223 **Web:** www.icnpharm.com/

Jackson Immuno Research

872 West Baltimore Pike
West Grove PA 19390
USA

Supplier of antibodies and fluorescent secondary antibodies, including the Cy probes.

Australian agent: Medical Dynamics

Phone: +1 (610) 869 4024 **FAX:** +1 (610) 869 0171 **Web:** www.jacksonimmuno.com/

Kirkegaard and Perry Laboratories Inc.

2 Cessna Court
Gaithersburg, Maryland 20879-4145
USA

Producers of affinity purified antibodies, including fluorescently conjugated anti-IgG of many species.

Australian agent: CSL

Phone: +1 (301) 948 7755 **FAX:** +1 (301) 948 0169 **Web:** www.kpl.com/

Molecular Probes Inc.

PO Box 22010
Eugene, OR 97402-0469,
USA

Molecular probes is the premier supplier of fluorescent probes for research. Their catalogue and web site contain a wealth of information on thousands of fluorescent probes.

Australian agent: BioScientific.

Phone: +1 (503) 465 8353 **FAX:** +1 (503) 465 4593 **Web:** www.probes.com/

OncoImmunin, Inc.

207A Perry Parkway, Suite 6
Gaithersburg, MD 20877,
USA

Supplier of *PhiPhiLux*, the fluorescent assay for apoptosis.

Phone: +1 (301) 987 7881 **FAX:** +1 (301) 987 7882 **Web:** www.phiphilux.com/

Pierce Biotechnology Inc.

3747 N. Meridian Rd
P.O. Box 117, Rockford, IL 61105
USA

Supplier of biochemical reagents.

Australian agent: LabSupply.

Phone: +1 (815) 968 0747 **FAX:** +1 (815) 968 8148 **Web:** www.piercenet.com/

QuantumDot Corp.

26118 Research Road
Hayward, CA 94545
USA

Manufacturer and supplier of Qdot™ nanocrystals for biological research. Quantum dots with a range of fluorescence emission and surface coatings are available. This includes Avidin and Biotin coated Quantum Dots that can be used for the attachment of a variety of proteins and antibodies.

Australian agents: Australian Biosearch.

Phone: +1 (510) 887 8775 **FAX:** +1 (510) 783 9729 **Web:** www.qdots.com/

Roche Applied Science

Roche Diagnostics (Schweiz) AG
Industriestraße 7, 6343 Rotkreuz
Switzerland

Roche Applied Science is a major manufacturer and supplier of biological reagents, including an extensive range of fluorescently labelled secondary antibodies.

Phone: Please see country specific contact information on their web page. **Web:** www.roche-applied-science.com/

Research Diagnostics Inc.

Pleasant Hill Road
Flanders, NJ 07836,
USA

Distributor of immunolabelling reagents.

Phone: +1 (973) 584 7093 **FAX:** +1 (973) 584 0210 **Web:** www.researchd.com/

Sigma-Aldrich Inc.

Sigma-Aldrich Corp.
St. Louis, MO,
USA

Sigma is a major supplier of chemicals and biological reagents, including secondary antibodies for immunolabelling. Sigma-Aldrich provides chemicals and reagents from Sigma, Aldrich and Fluka. Be careful when ordering – the different divisions sometimes sell reagents by the same name that are not identical (for example, N-propyl gallate from Sigma is more soluble in glycerol than the product from Aldrich!)

Australian agents: BioScientific and Sigma Australia.

Phone: +1 (314) 771 5765 **FAX:** +1 (314) 771 5757 **Web:** www.sigmaaldrich.com/

GREEN FLUORESCENT PROTEIN AND DERIVATIVES

Bio-World Inc.

P. O. BOX 888
Dublin, OH 43017
USA

Supplier of Green Fluorescent Protein (GFP) Blue Fluorescent Protein clones and related products.

Phone: +1 (614) 792 8680 **FAX:** +1 (614) 792 8685 **Web:** www.bio-world.com/

Clontech Laboratories (BD Biosciences)

1020 East Meadow Circle,
Palo Alto, CA 94303
USA

Supplier of Green Fluorescent Protein (GFP) clones and related products. The "Living Colour™" fluorescent proteins are a series of proteins derived from either the original Jellyfish or Coral Reef organisms. There is a wealth of information on GFP and its derivatives available on their internet site.

Australian agent: LabSupply

Phone: +1 (650) 424 8222 **FAX:** +1 (650) 424 1352 **Web:** www.clontech.com/

CPG Inc.

3 Borinski Road, Lincoln Park, NJ 07035
USA

Supplier of Green Fluorescent Protein (GFP) and Blue Florescent Protein (BFP) clones and related products.

Phone: +1 (973) 305 8181 **FAX:** +1 (973) 305 0884 **Web:** http://www.cpg-biotech.com/

Invitrogen Corp.

1600 Faraday Avenue
PO Box 6482
Carlsbad, California 92008
USA

Supplier of Green Fluorescent Protein (GFP) and Blue Florescent Protein (BFP) clones and related products.

Phone: +1 (760) 603 7200 **FAX:** +1 (760) 602-6500 **Web:** www.invitrogen.com/

MoBiTec GmbH

Lotzestrasse 22a
37083 Göttingen
Germany

Supplier of molecular biology reagents, including yeast 1 and 2 hybrid systems incorporating Green Fluorescent Protein (GFP). They also supply a range of modified Calcium indicator fluorescent probes.

Phone: +49 (551) 707 220 **FAX:** +49 (551) 707 2222 **Web:** www.mobitec.de/

Qbiogene Inc.

2251 Rutherford Road
Carlsbad, CA 92008
USA

Supplier of Green Fluorescent Protein (GFP) and Blue Florescent Protein (BFP) clones and related products.

Phone: +1 800 424 6101 **FAX:** +1 (760) 918 9313 **Web:** http://www.qbiogene.com/

MICROSCOPY EQUIPMENT AND REAGENTS

Dow Corning

see web page for contact details.

Manufacturers of a range of silicone rubber products that are suitable for sealing microscope imaging chambers. For example, Sylgard 184 is clear, inert once cured, and can be autoclaved.

Phone: **FAX:** **Web:** www.dowcorning.com/

Electron Microscopy Sciences

Box 251,
Fort Washington, PA 19034,
USA

Extensive catalogue of equipment and reagents for EM and light microscopy

Phone: +1 (215) 646 1566 **FAX:** +1 (215) 646 8931 **Web:** www.emsdiasum.com/ems/

ProSciTech Pty. Ltd.

PO Box 111 Thuringowa
Queensland 4817,
Australia

ProSciTech has an extensive range of microscopy reagents and equipment. This mail order business usually delivers within one to two days (within Australia).

Phone: +61 7 4773 9444 **FAX:** +61 7 4773 2244 **Web:** www.proscitech.com.au/

Sutter Instrument Company

51 Digital Drive.
Novato, CA 94949
USA

Manufacturers of laboratory instruments, including optical filter wheels and high-speed wavelength switchers for microscopy.

Phone: +1 (415) 883 0128 **FAX:** +1 (415) 883 0572 **Web:** www.sutter.com/

OPTICAL FILTERS

Barr Associates Inc.

2 Lyberty Way, Westford, MA 01886,
USA

Extensive range of glass and interference filters for microscopy.

Phone: +1 (978) 692 7513 **FAX:** +1 (978) 692 7443 **Web:** www.barrassociates.com/

Bioptechs Inc.

3560 Beck Road,
Butler, PA 16002,
USA

Bioptechs (biological optical technologies) manufacture heated stages and objective warmers for most microscopes.

Phone: +1 (412) 282 7145 **FAX:** +1 (412) 282 0745 **Web:** www.bioptechs.com/

BioScientific Pty. Ltd.

P.O. Box 78
Gymea NSW 2227,
Australia

BioScientific are agents for a large number of biochemical and equipment supply companies. They are the agents for a large number of companies, including Molecular Probes, Polysciences.

Phone: +61 2 9521 2177 **FAX:** +61 2 9542 3100 **Web:** www.biosci.com.au/

Chroma Technology Corp.

74 Cotton Mill Hill,
Brattleboro, VT 05301
USA

Specialize in the design and manufacturer of optical filters.

Phone: +1 (802) 257 1800 **FAX:** +1 (802) 257 9400 **Web:** www.chroma.com/

DuPont High Performance Films

P.O. Box 89
Route 23 and DuPont Road
Circleville, OH 43113
USA

Manufacturers of the **FEP Teflon hydrocarbon film** for microscope imaging chamber covers. This film is semi-permeable to gasses (O_2 and CO_2) but not water vapour. See Figure 12-34 in Chapter 12 "Imaging Live Cells" for a description of using this film for long term cell culturing.

Phone +1 800 967 5607 **FAX** +1 800 879 4481 **Web** www.dupont.com/teflon/films/H-55007-2.html/

Intor, Inc.

1445 Frontage Rd. NW
Socorro, NM 87801
USA

Extensive range of glass and interference filters for microscopy.

Phone: +1 (505) 835 2200 **FAX:** +1 (505) 835 2004 **Web:** http://www.intor.com/

Omega Optical

210 Main Street, Brattleboro, VT 05301
USA

Omega Optical has a very extensive stock of optical filters and dichroic mirrors for confocal microscopy and conventional fluorescence microscopy.

Phone: +1 (802) 254 2690 **FAX:** +1 (802) 254-3937 **Web:** www.omegafilters.com/

Polysciences Inc

400 Valley Road
Warrington, PA 18976,
USA

Polysciences has an extensive range of chemicals, biological reagents and equipment for microscopy.

Australian agent: BioScientific.

Phone: +1 (215) 343 6484 **FAX:** +1 (215) 343 0214 **Web:** www.polysciences.com/

Reflectix Inc.

#1 School Street
P.O. Box 108, Markleville, IN 46056
USA

Producers of the **insulating reflective bubble plastic** used for constructing a simple microscope enclosure for keeping your cells at the correct temperature while imaging. The use of the this plastic sheeting to make your own enclosure is discussed in Chapter 12 "Imaging Live Cells" (see Figure 12-34).

Phone +1 (765) 533 4332 **FAX** +1 (765) 533 2327 **Web** www.reflectixinc.com/

Schott Glas

Business Segment Optics for Devices
Hattenbergstr. 10
55122 Mainz
Germany

Extensive range of glass and interference filters for microscopy.

Phone: +49 (0) 6131/66 38 35 **FAX:** +49 (0) 6131/66 19 98 **Web:** www.schott.com/optik/english/products/optical_filters/

SIMPLE MICROSCOPY CHAMBERS

Bellcoglass

340 Edrude Rd.
Vineland New Jersey, 08360-3493,
USA

Manufacturers of the 1943-Sykes-Moore Culture Chamber. This chamber can be used as a simple imaging chamber with the introduction of dyes by injection through peripheral holes around the chamber. Alternatively the holes can be used to create a perfusion chamber.

Phone +1 (856) 691 1075 **FAX** +1 (856) 691 3247 **Web** www.bellcoglass.com/

Grace Bio-Labs

PO Box 228
Bend OR 97709,
USA

Manufacturers of a variety of microscopy chambers using a **silicone gasket** to retain the media. A variety of simple gasket chambers are available, including imaging chambers, perfusion chambers, hybridisation (able to withstand heating), etc.

Australian agent: ASTRAL.

Phone: +1 (541) 318 1208 **FAX:** +1 (541) 318 0242 **Web** www.gracebio.com/

MatTek Corporation

Ashland Technology Centre
200 Homer Avenue
Ashland, MA 01721,
USA

Manufacturers of **coverslip bottomed plastic Petri dish** microscopy chambers. A small (10 mm) hole has been cut in the base of a plastic Petri dish and then covered with a coverslip. A variety of configurations are available, including uncoated, coated (poly-D-lysine) and gridded coverslips. No. 1.5 and No. 0 coverslips are available. Supplied sterile in packs of 10. The "glass-bottom-dishes" web site also has a listing of microscopy positions and second-hand confocal microscopes for sale directly from the owners.

Phone +1 (508) 881 6771 **FAX** +1 (508) 879 1532 **Web** www.glass-bottom-dishes.com/

Nalge Nunc International

2000 North Aurora Road
Naperville, IL 60563,
USA

Manufacturers of the Lab-Tek **microscope chambers**. These plastic chambers come attached to either a coverslip (glass or plastic) for high resolution imaging of live cells, or attached to a microscope slide (glass or plastic) for conventional fixation and staining.

Australian agent: Medos.

Phone +1 (630) 983 5700 **FAX** +1 (630) 416 2519 **Web** www.nalgenunc.com/

VitroCom

8 Morris Avenue
Mountain Lakes, NJ 07046-0125,
USA

Manufacturers of **precision square and rectangular glass and quartz tubing**. A variety of sizes are available, with custom sizes available on request. These tubes can be used to construct a very simple flow chamber for microscopy.

Phone +1 (973) 402 1443 **FAX** +1 (973) 402 1445 **Web** www.vitrocom.com/

Willco Wells

WG Plein 287
1054 SE Amsterdam
The Netherlands

Manufacturers of the **Willco-dish microscope incubation chamber**. These dishes are available as either 35mm or 50mm diameter dishes with a coverslip imaging window on the base of from 12 to 40 mm. Coverslips are available made out of Quartz, glass or plastic film and can be purchased with an ITO-coating.

Phone: +31 0 20 285 0171 **FAX:** +31 0 20 685 0333 **Web** www.willcowells.com/

MICROSCOPY PERFUSION CHAMBERS

ALA Scientific Instruments

1100 Shames Drive
Westbury, NY 11590
USA

Suppliers of a wide range of instruments for physiology. Including a number of tissue and cell perfusion chambers for microscopy. The also sell the FEP Teflon covered chambers discussed in Chapter 12 "Imaging Live Cells" (see Figure 12-34).

Phone +1 (516) 997-5780 **FAX** +1 (516) 997-0528 **Web** http://www.alascience.com/

Harvard Apparatus

84 October Hill Road
Holliston, Massachusetts 01746,
USA

Manufacturers and suppliers of a wide range of instruments for physiology. This includes a number of tissue and cell perfusion chambers, **mainly for electrophysiology** - but some instruments are for microscopy.

Phone +1 (508) 893 9888 **FAX** +1 (508) 429 5732 **Web** www.haicellbiology.com/

Intracel

Unit 4, Station Road
SheprethRoyston, Herts SG8 6PZ,
UK

The Controlled Cell Perfusion Chamber provides accurate temperature control from 4°C to 45°C. The perfusion chamber consists of removable imaging chambers made out of either Perspex or stainless steel that can be dismantled for removal of the coverslip. Also supply the Bioptechs incubation chambers and WillCo glass bottomed imaging chambers.

Phone +44 (763) 26 2680 **FAX** +44 (763) 26 2676 **Web** www.intracel.co.uk/

Kent Scientific

457 Bantam Road#16
Litchfield, CT 06750,
USA

Suppliers of a wide range of instruments for physiology. Particularly organ baths and live tissue chambers. This also includes a number of tissue and cell perfusion chambers for microscopy. Also supply the *Bioptechs* incubation chambers.

Phone +1 (860) 567 5496 **FAX** +1 (860) 567 4201 **Web** www.kentscientific.com/

Ludin Chamber

Live Imaging Services
Muehletalweg 22
CH-4600 Olten,
Switzerland

The **Ludin Chamber** is a small stainless steel perfusion chamber with removable cover glass. Six chambers can be fitted into a frame the size of a micro titre plate. The chamber can be used as either an open or closed perfusion chamber.

Phone +41 79 235 7154 **FAX** +41 860 622 963 160 **Web** www.lis.ch/

Lucas-Highland

(see Nevtek)

HCR03 Box 99
Burnsville VA 24487,
USA

DSC200 Dvorak-Stotler Controlled-Environment Culture Chamber: made of Teflon and stainless steel, can be dismantled for cleaning. The chamber can be autoclaved as a fully enclosed chamber both before and after use. The chamber itself is not temperature controlled, but can be readily fitted to a heated stage or Nevtek air stream incubator.

Phone +1 (540) 925 2322 **FAX** +1 (540) 925 2322 **Web** www.nevtek.com/chamber.html/

Molecular Probes Inc

PO Box 22010
Eugene, OR 97402-0469,
USA

Molecular probes supplies the **Attofluor Cell Chamber** (A-7816). This simple chamber is constructed from stainless steel, and can be readily screwed apart to replace the coverslip. The chamber itself does not have any temperature control, but can be readily used on a heated microscope stage.

Phone +1 (503) 465 8353 **FAX** +1 (503) 465 4593 **Web** www.probes.com/

Warner Instruments

1141 Dixwell Avenue
Hamden, CT 06514,
USA

A large range of imaging and recording chambers are available. The **Series 20 Perfusion Chambers** consist of a polycarbonate insert (with a coverslip base) and an aluminium platform. Heated platforms are available. The diamond-shape of the perfusion chamber provides good linear flow characteristics.

Phone +1 (203) 776 0664 **FAX** +1 (203) 776 1278 **Web** http://www.warneronline.com/

MICROSCOPE HEATERS

In addition to the specialist microscope heater manufacturers listed below, all of the major microscope manufacturers provide heated microscope stages specifically designed for their microscopes.

Nevtek

HCR03 Box 99
Burnsville VA 24487,
USA

ASI 400 Air Stream Incubator: a precision control air stream incubator used for heating a live cell incubation chamber while on the microscope. Rapid and accurate temperature control of the specimen can be obtained using a continuous flow (non-cyclical) of accurately heated air.

Phone +1 (540) 925 2322 **FAX** +1 (540) 925 2322 **Web** www.nevtek.com/

TEMPERATURE CONTROLLED MICROSCOPE STAGES

20/20 Technologies

Building 2, Unit A
311 Judges Road
Wilmington, NC 28405,
USA

Biostage 600 warming plates for microscopy (25°C to 55°C).

Bionomic System for heating, cooling and atmospheric control (-10°C to 60°C). Accommodates a variety of Petri dish sizes and a number of open and closed perfusion chambers.

Phone +1 (910) 791 9226 **FAX** +1 (910) 791 4126 **Web** www.20-20tech.com/

Dagan Corporation

2855 Park Avenue
Minneapolis, Minnesota 55407,
USA

HE-100 and **HE-200 Thermal Stages:** A number of Peltier heated microscope warming stages are available. Models are available for fitting either Petri dishes or the Dagan thermally conductive imaging chamber. Design is for both microscopy and electrophysiology.

Phone +1 (612) 827 5959 **FAX** +1 (612) 827 6535 **Web** www.dagan.com/

Instec Temperature Control

5589 Arapahoe Avenue
Boulder, CO 80303,
USA

A variety of **microscope stage warmers** that are designed to hold either Petri dish chambers or a microscope slide. The Peltier heated stages can heat from ambient to 60°C.

Phone +1 (303) 444 4608 **FAX** +1 (303) 444 4607 **Web** www.instec.com/

Linkam Scientific Instruments

8 Epsom Downs Metro Centre
Waterfield, Tadworth,
Surrey KT20 5HT,
UK

A variety of **Temperature Control Stages** with a very wide range (-196°C to 1500°C). A number of highly specialised stages are also available, such as the FDCS 196 Freeze Drying Vacuum Stage. A series of warm stages (ambient to 100°C) to fit Nikon, Leica, Olympus and Zeiss upright and inverted microscopes are available.

Phone +44 0 1737 363476 **FAX** +44 0 1737 363480 **Web** www.linkam.co.uk/

Physitemp Instruments Inc.

154 Huron Avenue
Clifton, New Jersey 07013,
USA

Temperature controlled microscope stages are available (-20°C to 100°C. Models are available to fit 35 mm, 60 mm or 100 mm Petri dishes. Thermoelectric (Peltier) temperature control with water cooling is used to very accurately control the stage temperature.

Phone +1 (973) 779 5577 **FAX** +1 (973) 779 5954 **Web** www.physitemp.com/

TEMPERATURE CONTROLLED IMAGING CHAMBERS

Bioptechs Inc.

3560 Beck Road,
Butler, PA 16002,
USA

Bioptechs manufacture heated chambers and objective warmers. The unique design of metal coated self-heating coverslips results in very even heating across the bottom of the chamber. Used in conjunction with an objective heater these chambers give excellent temperature control. **ΔT Open Culture Dish and FCS2 closed perfusion chambers are available.**

Phone +1 (412) 282 7145 **FAX** +1 (412) 282 0745 **Web** www.bioptechs.com/

FLOW CHAMBERS

Immunetics, Inc.

63 Rogers Street
Cambridge, MA 02142,
USA

Cell Adhesion Flow Chamber for performing dynamic cell adhesion assays in a laminar flow environment. The chamber is not temperature controlled.

Phone +1 (617) 492 5416 **FAX** +1 (617) 868 7879 **Web** www.immunetics.com/

Chapter 15

Further Reading

The literature for confocal and multi-photon microscopy relevant for biological scientists is unfortunately scattered amongst a large number of specialist publications. There are many journal articles in the biological research literature that make, often extensive, use of various types of fluorescent and confocal microscopy techniques, but unfortunately both the indexing of research publications, and the fact that they are written primarily from the perspective of the problem at hand (microscopy was simply a "tool" employed), does make accessing this important literature exceedingly difficult.

The following books and occasional journal articles have been used extensively as a reference source not just during the writing of this book, but also over the past few years as my own work became more involved in confocal and multi-photon microscopy. This list is by no means comprehensive – just my own biased view of a number of useful publications.

OPTICS

An understanding of optics is not essential, but is helpful in making the most out of your work involving microscopy. Light does have unique characteristics that create a wealth of microscopy techniques, but also severely limit how far you can push light microscope.

Hecht, E. (1998). *Optics*, 3rd edition. Addison-Wesley Publishers. A comprehensive textbook covering most aspects of optics. This text is aimed at students of optics, but is useful as a reference source for microscopists who are looking for more comprehensive information on optics than is available in general microscopy texts.

MICROSCOPY

Light microscopy is an old and established science, with many books written on the more technical aspects of light and electron microscopy. Unfortunately there are few books on microscopy that cover light microscopy from a biologist's point of view. Most books on the subject are either very slender rudimentary volumes, or they go into way too much detail – and will never be read by most biologists! The following books are a select few that do cover aspects of light microscopy that are helpful for biologists interested in getting the most out of using a confocal or multi-photon microscope. The book "Fundamentals of Light Microscopy and Electronic Imaging" by Douglas Murphy is highly recommended as a well written introductory book on microscopy for biologists using various aspects of microscopy, including confocal microscopy, in their work.

Bradbury, S. (1976). *The Optical Microscope in Biology*. Edward Arnold Publishers. A good introduction to light microscopy for biologists. The coverage of the basics of light microscopy is still relevant today, but this book pre-dates the introduction of the confocal microscope in biology!

Foster, B. (1997). ***Optimising Light Microscopy for Biological and Clinical Laboratories.*** **American Society for Clinical Laboratory Science.** An introductory book on basic aspects of light microscopy. The book contains a wealth of practical information and some theoretical explanations of the basics of optics and microscopy.

Goldstein, D. J. (1999). ***Understanding the light microscope.*** **Academic Press.** A book designed to assist in using and understanding a series of computer programs (on accompanying CD) showing ray diagrams of various microscope illumination methods and lens configurations. The book and CD has a wealth of information, but is aimed at a somewhat advanced user.

Kiernan, J. A. and Mason, I., editors. (2002). ***Microscopy & Histology for Biologists. A Users Guide.*** **Portland Press.**

Lacey, A. J., editor. (1989). ***Light microscopy in biology. A practical approach.*** **IRL Press.** Lots of practical information on microscopy and its application to biology.

Murphy, D. B. (2001). ***Fundamentals of Light Microscopy and Electronic Imaging.*** **Wiley-Liss Inc.** An excellent introductory text covering basic aspects of microscopy, the properties of light, confocal microscopy, CCD cameras and image processing.

Oldfield, R. (1994). ***Light Microscopy. An Illustrated Guide.*** **Wolfe Publishing.** A comprehensive introduction to light microscopy written for a biology course unit in microscopy.

Rawlins, D. J. (1992). ***Light Microscopy.*** **BIOS Scientific Publishers Limited.** Small but comprehensive introduction to light microscopy.

Rost, F. W. D. (1995). ***Fluorescence Microscopy.*** **Cambridge University Press, Volume I and II.** Extensive single author work on most aspects of fluorescence microscopy. Very comprehensive coverage of epi-fluorescence microscopy and a large number of fluorescent probes. Limited coverage of confocal microscopy.

Slayter, E. M. and Slayter, H. S. (1992). ***Light and Electron Microscopy.*** **Cambridge University Press.** A relatively advanced outline of the fundamental principles of light and electron microscopy.

Zeiss., (1997). ***Microscopy from the very beginning.*** **Carl Zeiss, Jena GmbH.** A good introduction to light microscopy. Also includes a double page spread on the founders of modern light microscopy (Zeiss, Abbe, Schott and Köhler – all involved in the original Carl Zeiss works in Jena, Germany).

CONFOCAL MICROSCOPY

Considering the great upsurge in interest in confocal and multi-photon microscopy, somewhat surprising is that there are very few books on confocal microscopy that are really suitable for the person first starting out in this exciting area of science. "The book Handbook of Biological Confocal Microscopy" edited by James B. Pawley is considered the "classic" reference text for users of confocal microscopes – but this book is a daunting book for those just starting out. The other books listed below are compiled multi-author books covering many aspects of confocal microscopy. They all contain a wealth of information on various aspects of fluorescence, and confocal and multi-photon microscopy. Don't forget however, that some of the articles on biological topics, of which you have no direct interest, may contain just the information you want on using these wonderful microscopes.

Bio-Rad. (1998). ***Confocal Microscopy Bibliography.*** **Bio-Rad.** Extensive bibliography of the research literature, covering both advances and applications of confocal microscopy. Divided into broad subject areas. The fluorescent probe used is noted for most papers. A paper and electronic version is available free of charge from Bio-Rad.

Celis, J. E., editor. (1999). ***Cell Biology A Laboratory Handbook,*** **(four-volumes). Academic Press.** Large multi-author four-volume manual for cell biology. Written by a large selection of world leaders in cell biology. Volume 3 contains detailed information on light microscopy, video microscopy, fluorescence microscopy, confocal microscopy, image processing and histology.

Cheng, P. C. (1996). ***Multidimensional Microscopy.*** **Singapore. River Edge, NJ: World Scientific.**

Conn, P. M., editor. (1999). Confocal Microscopy. *Methods in Enzymology* Volume 307, Academic Press. Authoritative coverage of many molecular cell biology applications of confocal microscopy.

Corle, T. R. and Kino, G. S. (1996). *Confocal Scanning Optical Microscopy and Related Imaging Systems.* Academic Press. An extensive coverage of the theory of operation of a many different confocal microscope designs.

Diaspro, A., editor. (2002). *Confocal and Two-Photon Microscopy: Foundations, Applications and Advances.* Wiley. Multi-author book covering many aspects of confocal and multi-photon microscopy.

Hibbs, A. R. and Saul, A. J. (1994). Plasmodium falciparum: highly mobile small vesicles in the malaria infected red blood cell cytoplasm. *Experimental Parasitology* 79:260-269. Acridine orange was used to follow the intracellular movement of small vesicles in live malarial parasites using confocal microscopy. Figure 1 and 2 in this publication are reprinted in this book with permission from Elsevier (Figure 1-7, Figure 6-9, Figure 7-1 and Figure 7-2).

Hibbs, A. R., Stenzel, D. J., and Saul, A. J. (1997). Macromolecular Transport in Malaria - Does the Duct Exist? *European Journal of Cell Biology*, 72:182-188. Small fluorescently labelled latex beads were used to demonstrate the lack of existence of the malarial parasite transport duct using live cell confocal microscopy. Figure 3 in this publication is reproduced with modification in Figure 4-12 in this book with permission from Wissenschaftliche Verlagsgesellschaft. Further information on the experimental conditions used to obtain Figure 12-37 in this book can be found in the methods section of this publication.

Kun, J. F. J., Hibbs, A. R., Saul, A., McColl, D. J., Coppel, R. L. and Anders R. F. (1997). A putative Plasmodium falciparum exported serine/threonine protein kinase. *Molecular and Biochemical Parasitology* 85:41-51. Multi-labelling confocal microscopy was used to determine the subcellular location of the malarial protein FEST1. Further information on the immunolabelling protocols used to obtain the images shown in Figure 1-11 and Figure 11-1 can be found in this publication.

Mason, W. T., editor. (1999). *Fluorescent & Luminescent Probes for Biological Activity.* Academic Press. Comprehensive multi-author book covering many aspects of confocal microscopy, with a particular emphasis on fluorescent probes and their use.

Masters, B. R., editor. (1996). *Selected Papers on Confocal Microscopy.* SPIE Milestone Series Volume MS 131. SPIE Optical Engineering Press. An excellent collection of important papers on laser scanning confocal microscopy, including a survey of early history and the memoir paper by Marvin Minsky listed below.

Matsumoto, B., editor. (2003). Cell Biological Applications of Confocal Microscopy. *Methods in Cell Biology* Volume 70. Academic Press. An updated version of the 1993 edition of Methods in Cell Biology Vol. 38. Extensive information on a range of topics in confocal microscopy in the biological sciences.

Min Gu. (1996). *Principles of Three-dimensional Imaging in Confocal Microscopes.* Singapore; River Edge, NJ: World Scientific. Advanced coverage of 3D imaging in confocal microscopes.

Minsky, M. (1957). U.S. Patent #3013467, *Microscopy Apparatus.* The original patent for confocal microscopy by Marvin Minsky.

Minsky, M. (1988). Memoir on inventing the confocal laser-scanning microscope. *Scanning* 10:128-138. An article by the inventor of the confocal microscope on the history of its invention.

Moss, D., Hibbs, A. R., Stenzel, D., Powell, L. W. and Halliday, J. W. (1994). The endocytic pathway for H-ferritin established in live MOLT-4 cells by laser scanning confocal microscopy. *British Journal of Haematology* 88:746-753. Live MOLT-4 cells are shown to recycle transferrin over a 2 hour period, whereas H-ferritin is directed to the lysosome using confocal microscopy. A modified version of Figure 11-2 in this book was originally published in this journal article. Further information on the experimental procedures used to obtain the images shown in Figure 1-12, Figure 1-16 and Figure 11-2 in this book can be found in this publication.

Paddock, S. W., editor. (1999). ***Confocal Microscopy Methods and Protocols.*** **Totowa, Humana Press, NJ.** Multi-author work covering many aspects of confocal microscopy. Excellent technique based information, does assume considerable knowledge about confocal microscopy and fluorescence labelling.

Pawley, J. B., editor. (1995). ***Handbook of Biological Confocal Microscopy.*** **Plenum Press, NY.** Very comprehensive coverage of most aspects of confocal microscopy in 1995. Somewhat out of date now, but still full of lots of valuable help and information. However, most chapters assume a considerable knowledge of confocal microscopy. An updated edition of this important text is currently in preparation (2004).

Pawley, J. B. (2002). Limitations on Optical Sectioning in Live-Cell Confocal Microscopy. ***SCANNING*** **24:241-246.** An article that describes serious problems of spherical aberration in live cell imaging that is created by the sample itself.

Reichelt, S. and Amos, W. B. (2001). SELS: A New Method for Laser Scanning Microscopy of Live Cells. ***Microscopy and Analysis.*** **November 2001, pages 9-11.** A brief article that describes the function of the Bio-Rad developed Scanning Enhancement Lens (SELS) and its use in live cell microscopy.

Sheppard, C. and Shotton D. M. (1997). ***Confocal laser scanning microscopy.*** **Oxford: BIOS Scientific.** Small concise introductory texts on laser scanning confocal microscopy.

Shotton, D. M. (1993). ***Electronic Light Microscopy.*** **Wiley-Liss.** Very comprehensive, but highly technical, coverage of the physical aspects of video and confocal microscopy.

Sluder, G. and Wolf D. E., editors. (1998). Video Microscopy. ***Methods in Cell Biology.*** **Volume 56. Academic Press.**

Spector, D. L., Goldman, R. D., Leinwand, L. A., editors. (1998). ***Cells, A Laboratory Manual.*** **Cold Spring Harbor Laboratory Press.** A large and very comprehensive three-volume manual on working with cells. Includes sections on wide-field epi-fluorescence microscopy, confocal microscopy, multi-photon microscopy and deconvolution. Background information and laboratory protocols are included.

Wilson, T., editor. (1990). ***Confocal Microscopy.*** **Academic Press.** An extensive multi-author treatise on confocal microscopy and instrumentation.

Yuste, R., Lanni, F. and Konnerth A., editors. (2000). ***Imaging Neurons, A Laboratory Manual.*** **Cold Spring Harbour Press.** A comprehensive coverage of imaging methods that applies to living cells. Covers many aspects of imaging cells based on neurobiology, but applicable to many other cells.

Zeiss., (2004). ***Confocal Laser Scanning Microscopy.*** **Carl Zeiss, Jena GmbH.** A good introduction to the principles of confocal microscopy, particularly image formation and signal processing.

Zimmermann, T., Rietdorf, J. and Pepperkok, R. (2003). Spectral imaging and its applications in live cell microscopy. ***FEBS Letters*** **546:87-92.** Excellent discussion on using spectral imaging and linear unmixing in microscopy in the biological sciences.

Zucker, R.M. and Price, O.T. (2001). Evaluation of Confocal Microscopy System Performance. Cytometry 44:273-294. Detailed article on determining the performance of your confocal microscope using a variety of readily available specimens.

Zucker, R.M. and Price, O.T. (2001). Statistical Evaluation of Confocal Microscopy Images. Cytometry 44:295-308. A comprehensive article on evaluating the signal to noise ratio in images collected from a confocal microscope.

CONFOCAL MICROSCOPY APPLICATIONS

Bio-Rad, ***Application notes for Confocal Microscopy.***
An extensive range of application notes available from Bio-Rad. These notes cover many aspects of confocal microscopy, from elementary coverage of basic imaging to advanced multi-photon imaging. A list of application notes available is printed on page 257 of this book.

Klonis, N., Rug, M., Harper, I., Wickham, M., Cowman, A. and Tilley L. (2002). Fluorescence Photobleaching Analysis for the Study of Cellular Dynamics. *European Biophysics Journal*, 31:36-51. Review article on using FRAP for studying the dynamics of biological membranes.

Leitch, A. R., Schwarzacher, T., Jackson, D., and Leitch I. J. (1994). *In Situ Hybridisation.* BIOS Scientific Publishers. Short and succinct coverage of most aspects of in situ hybridisation. Mainly concerned with epi-fluorescence microscopy, but can be readily adapted to confocal microscopy.

Nuccitelli, R., editor. (1994). A Practical Guide to the Study of Calcium in Living Cells. *Methods in Cell Biology.* Volume 40. Academic Press.

Rizzuto, R. and Fasolato, C., editors. (1999). *Imaging Living Cells.* Springer Lab Manual. A practical based book covering instrumentation, fluorescent dyes and recombinant proteins (aequorin, luciferase and GFP).

Wouters, F. S., Verveer, P. J. and Bastiaens, P. I. H. (2001). Imaging biochemistry inside cells. *TRENDS in Cell Biology* 11(5):203-211. Excellent review of using imaging to perform single cell biochemistry, including information on FRET and how it can be measured by fluorescence microscopy.

FLUORESCENCE

Fluorescence is a relatively established science, with the following books giving excellent coverage for biologists interesting in understanding the choice and limitations of various fluorescent probes.

Haugland, R. P. (2002). *Handbook of Fluorescent Probes and Research Chemicals.* Molecular Probes, 9th edition 2002 (print edition), available from distributors of Molecular Probes products. A CD-ROM edition is published regularly, and is available by mail order or by directly accessing the Molecular Probes web site. The molecular probes handbook contains a wealth of information on their very extensive range of fluorescent probes. Molecular Probes (see page 332) maintain an extensive bibliographic database that covers a large number of fluorescent probes used in confocal microscopy. The database can be accessed via their web page, by sending a request to their technical department, or as a listing at the end of most sections of their manual.

Lakowicz, J. R. (1999). *Principles of Fluorescence Spectroscopy.* Kluwer Academic / Plenum Press, NY. Comprehensive single author textbook on all aspects of the chemistry of fluorescence. Does not cover fluorescence imaging directly, but contains a wealth of information on all aspects of fluorescence, including multi-photon fluorescence.

Mason, W. T., editor. (1999). *Fluorescent and Luminescent Probes for Biological Activity.* Academic Press. Comprehensive multi-author book covering many aspects of confocal microscopy, with a particular emphasis on fluorescent probes and their use. The chapters are written by highly respected authorities in their field.

GREEN FLUORESCENT PROTEIN (GFP)

A large number of journal articles in the bio-medical sciences that make, often extensive, use of green fluorescent protein (GFP) and its multi-coloured derivatives are currently being published. The technology for producing fusion proteins of GFP and your protein of interest are well established and available either in the literature or from the commercial suppliers of GFP and related fluorescent proteins. What is more difficult to come by are critical accounts of using GFP in a pioneering manner (i.e. not simply to confirm what is already known). The following compilations are a good source of information on GFP and its use in a wide variety of applications.

Ando, R., Hama, H., Yamamoto-Hino, M., Mizuno, H., Miyawaki, A. (2002). An optical marker based on the UV-induced green-to-red photoconversion of a fluorescent protein. *Proc. Natl. Acad. Sci.* USA 9(20):12651-6. Short wavelength violet light (405 nm) coverts the Kaede fluorescent protein from green to red emission. This protein can be used as a molecular tracer in living cells.

Campbell, R. E., Tour, O., Palmer, A. E., Steinbach, P. A., Baird, G. S., Zacharias, D. A. and Tsien R. Y. (2002). A monomeric red fluorescent protein. *Proc. Nat. Acad. Sci.* USA 99: 7877-7882. An article describing the genetic manipulation of the original red fluorescent protein from coral (*Discosoma*) to make this protein into a monomeric molecule that is more easily used in studies of protein dynamics in living cells.

Chalfie, M. and Kain, S., editors. (1998). *Green Fluorescent Protein: Properties, Applications and Protocols.* Wiley Liss.

Conn, P. M., editor. (1999). Green Fluorescent Protein. *Methods in Enzymology.* Volume 302, Academic Press. Authoritative coverage of many aspects of green fluorescent protein.

Mills, C. E. (1999-2002). *Bioluminescence of Aequorea, a hydromedusa.* Electronic internet document available at http://faculty.washington.edu/cemills/Aequorea.html. An interesting account of the extent of use and abuse of Jellyfish photos in research publications. Don't ever use a Jellyfish photograph without first reading this article! Don't be caught using a "flash" photograph of a Jellyfish (the most common one around) as a fluorescence photograph!

Patterson, G. H. and Lippincott-Schwartz, J. (2002). A photoactivatable GFP for selective photolabeling of proteins and cells. *Science* 297(5588):1873-7. A novel green fluorescent protein derivative that is highly fluorescent only after activation with relatively short wavelength blue light. Can be used as a molecular tracer in living cells.

Sullivan, K. F. and Kay, S. A., editors. (1999). Green Fluorescent Proteins. *Methods in Cell Biology.* Volume 58. Academic Press. Extensive coverage of most aspects of green fluorescent protein.

Tsien, R. (1998). The Green Fluorescent Protein. *Ann. Rev. Biochem.* 67:509-44. Excellent overview of green fluorescent protein and its derivatives.

Yang, F., Moss, L. G. and Phillips, G. N. (1996). The Molecular Structure of Green Fluorescent Protein. *Nat. Biotechnol.* 10:1246-1251. The X-ray derived 3D molecular structure of the original green fluorescent protein isolated from the Jellyfish *Aequorea*.

Zhang, J, Campbell, R. E., Ting, A. Y and Tsien, R. (2002). Creating New Fluorescent Probes for Cell Biology. *Nature Reviews* 3:906:906-918. Excellent review of green fluorescent protein and genetically expressed small molecular probes.

HISTOLOGY AND SAMPLE PROCESSING

Some of the best books on cell and tissue fixation can be found in the "older" literature, particularly the technical literature concerning electron microscopy. The preservation of the structure of biological material is critical for high quality imaging when using an electron microscope, and therefore the techniques developed are often better for confocal and multi-photon microscopy than the techniques traditionally used for light microscopy.

Bancroft, J. D. and Stevens, A. (1996). *Theory and practice of histological techniques.* Butterworths, London. A very comprehensive book on laboratory histology techniques.

Gao, K., editor. (1993). *Polyethylene Glycol as an Embedment for Microscopy and Histochemistry.* CRC Press. A range of articles on using polyethylene glycol to preserve tissue and cell structure for preparing samples for microscopy.

Griffiths, G. (1993). *Fine Structure Immunocytochemistry.* Springer-Verlag, Berlin.

Hayat, M. A. (1981). *Fixation for Electron Microscopy.* Academic Press, NY. Comprehensive guide to fixation for electron microscopy; highly applicable to confocal microscopy where the cells cannot be imaged live (particularly immunolabelling).

Hayat, M. A. (2002). ***Microscopy, Immunohistochemistry, and Antigen Retrieval Methods for Light and Electron Microscopy.*** **Kluwer Academic / Plenum Publishers, NY.** A comprehensive guide to sample preparation for light and electron microscopy.

Hayat, M. A. (2000). ***Principles and Techniques of Electron Microscopy; Biological Applications,*** **4th ed. Cambridge Uni Press.**

CELL BIOLOGY TECHNIQUES

The following comprehensive texts (often multi-volume) contain a wealth of information on cell biology that is often of great importance for successful confocal or multi-photon microscopy. Some of the volumes also contain specific chapters on confocal and multi-photon microscopy itself.

Celis, J. E., editor. (1999). ***Cell Biology A Laboratory Handbook.*** **Academic Press.** Four-volume work containing extensive information on cell culturing, immunolabelling, microscopy, gene expression etc.

Lepe-Zuniga, J. L., Zigler, J. S. and Gery, I. (1987). Toxicity of light-exposed Hepes media. ***J Immunol Methods*** **103:145-150.** An article describing the toxicity of the common buffer HEPES when exposed to light – including room lighting (particularly fluorescent room lighting).

Pierce Chemical Company (1994-). ***Pierce Catalogue and Handbook.*** **Pierce Chemical Company.** The Pierce catalogue contains very extensive background information on the reagents sold. Very good for information on immunolabelling methods and reagents.

Potter, S. M. and DeMarse, T. B. (2001). A new approach to neural cell culture for long-term studies. ***Journal of Neuroscience Methods.*** **110:17-24.** An article describing a method of keeping cells alive in a gas tight chamber for many months. The membrane is permeable to O_2 and CO_2 for gas exchange in a conventional incubator, but is impermeable to H_2O and bacteria and fungus spores. This chamber is designed for long-term studies involving electrophysiology or microscopy.

Spector, D. L., Goldman, R. D., Leinwand, L. A., editors. (1998). ***Cells, A Laboratory Manual.*** **Cold Spring Harbor Laboratory Press.** A large and very comprehensive three-volume manual on working with cells. Includes sections on wide-field epi-fluorescence microscopy, confocal microscopy, multi-photon microscopy and deconvolution. Background information and laboratory protocols are included.

Spierenburg, G. T., Oerlemans, F., Vanlaarhoven, J. and Debruyn, C. (1984). Photo-toxicity of N-2-Hydroxyethylpiperazine-N'-2-ethanesulfonic acid buffered culture media for human leukemic cell lines. ***Cancer Research*** **44:2253-4.** An article describing photo-toxicity effects of HEPES buffer on cultured cell lines.

IMAGE PROCESSING

The extensive literature on image processing is often of little direct value to most people using confocal or multi-photon microscopy. However, if you do wish to go beyond the very basic level of image processing provided by most of the confocal microscope manufacturers the very comprehensive reference book "The Image Processing Handbook" by J.C. Russ is a good starting point.

Russ, J. C. (1999). ***The Image Processing Handbook,*** **3rd edition. CRC Press.** A comprehensive book on image processing and the principles involved, with many practical examples. A CD-ROM with worked examples is also available.

Sedgewick, J. (2002). ***Quick Photoshop for Research.*** **Kluwer Academic / Plenum Publishers, NY.** An excellent "hands on" guide to Photoshop for biologists. Aimed at the research scientist, rather than the general photographic market.

Weinmann, E. and Lourekas, P. (1999). ***Photoshop for Windows and Macintosh.*** **Peachpit Press.** A very readable text covering most aspects of Photoshop.

Appendix 1: Confocal Microscopes

A range of confocal microscopes is available – from the highly versatile but very expensive "top end" instruments, to specialised high-speed imaging and even miniaturised medical instruments. Confocal microscopes are at the forefront of the upsurge in interest in light microscopy in the past 15 years. The technology is still evolving fast – so you should expect major changes and innovations in the coming years.

This appendix gives an overview of the hardware for most of the laser scanning confocal microscopes currently available. Although not all manufacturers are described in as much detail as others, this does not in any way reflect on the instruments – this imbalance simply reflects the difficulties in compiling the necessary information for each of the instruments in the time available. I hope that this appendix is an evolving project that will include not only updated, but also expanded information on each of the instruments in subsequent editions of this book.

Laser Spot Scanning Confocal Microscopes

These instruments are based on the use of a finely focussed laser "spot" that is scanned across the sample. The returning fluorescent or backscattered light is directed through a confocal iris or pinhole that eliminates out-of-focus light.

Nipkow Spinning Disk Confocal Microscopes

The Nipkow disk contains a large number of fine "pinholes" arranged in a spiral array such that a confocal image is created when the disk is spinning. These instruments operate at higher speeds than a laser single spot scanning system. They also result in less intense light irradiation at each spot on the sample, significantly reducing fluorophore fading and cellular damage.

Nipkow Disk / Micro-lens Array Confocal Microscopes

A further modification of the basic Nipkow disk confocal microscope is the incorporation of a second disk, containing an array of micro-lenses, that focuses the laser light through the pinholes on the Nipkow disk. Compact and robust micro-lens array Nipkow disk scan heads (CSU10 and CSU21) are produced by Yokogawa Electric Corporation, and are assembled into confocal microscopes by a number of manufacturers.

Confocal Microscopes using the Yokogawa Scan Head

Bio-Rad Cell Science Division

The Bio-Rad laser scanning confocal microscopes are capable of superb resolution and have enormous versatility for multiple fluorescence and transmission imaging. This appendix is aimed at giving an overview of the hardware involved in the Bio-Rad confocal and multi-photon microscopes. Although the latest confocal microscopes from Bio-Rad are almost fully controlled from the computer, a basic understanding of the physical design of a confocal microscope scanning system is important for both getting the most out of your instrument and minimising your frustrations when learning to use the software. This appendix explains in some detail the various hardware components of the Bio-Rad Radiance and MRC confocal microscopes. For more detailed information please refer to the Bio-Rad instrument technical and user manuals.

BIO-RAD INSTRUMENTS

The latest Bio-Rad confocal microscopes (the Radiance series) have a relatively small scan head that can be readily transferred from one microscope to another (Figure A-1). The small and compact scan head is factory aligned, and so no maintenance or alignment is required once the instrument is installed. The scan head itself will need to be re-aligned into the microscope if you do move the scan head from one microscope to another. This is not a difficult task and can be readily accomplished in a few minutes.

Bio-Rad is a specialist manufacturer and innovative developer of the laser scanning confocal microscope scan head, associated optics, electronic components and computer software for controlling the microscope. Bio-Rad laser scanning confocal microscopes are manufactured in the UK, with a major sales distribution and technical support division in the USA.

Bio-Rad, unlike a number of the other laser scanning confocal microscope manufacturers, is not an optics or microscope manufacturer. The Bio-Rad scan head can be fitted on microscopes from a number of manufacturers, but in more recent years Bio-Rad has favoured selling a complete package that includes a research grade light microscope from Nikon.

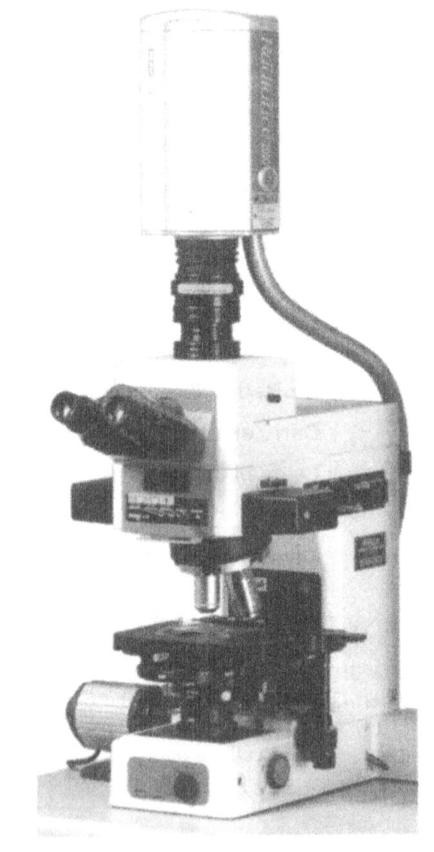

Figure A-1. Bio-Rad Radiance Confocal Microscope. The small Radiance scan head is shown attached to the video port of a Nikon upright microscope. The associated controller box, laser and computer are not shown in this photo. (This figure is adapted from a photograph kindly provided by Bio-Rad Cell Science, UK).

Bio-Rad Cell Science Division

Web: http://cellscience.biorad.com/

Address:	Bio-Rad House Maylands Avenue, Hemel Hempstead Hertfordshire, HP 7TD **UK**	Life Sciences Group. 2000 Alfred Nobel Drive Hercules, CA 94547 **USA**	Unit 1 block Y Regents Park Industrial Estate 391 Park Rd, Regents Park NSW 2143 **Australia**
Phone:	+44 (0) 208 328 2141	+1 (510) 741 6868	+61 2 9914 2800
FAX:	+44 (0) 208 223 2500	+1 (510) 741 5811	+61 2 9914 2888

Bio-Rad Confocal Microscopes

Radiance Confocal Microscopes

Fully computer controlled small and robust scan head, with associated ICU (Instrument Control Unit), which is located under the work bench. The scan head can be readily moved between an upright and an inverted microscope, and can be mounted on a number of different light microscopes by using the correct adaptor.

Radiance 2100™ Rainbow	Spectral imaging version of the Radiance 2100 confocal microscope (can also be installed as an upgrade to an existing Radiance 2100 instrument). See Figure 3-14 (page 85).
Radiance 2100™	Single, dual or triple channel, fully computer controlled confocal microscope using Windows NT/2000 LaserSharp software. Can also be purchased multi-photon ready.
Radiance 2100™ MP	Multi-photon combined with a confocal, or multi-photon dedicated (MPD). See Figure 3-19 (page 94).
CellMap™ ID	"Personal" confocal microscope with 405nm and 488 nm solid state laser confocal microscope for DAPI/CFP and GFP/FITC.
CellMap™ IC	"Personal" confocal microscope with 488nm & 532 nm solid state lasers designed for GFP/FITC/TRITC/Cy3.
Micro Radiance™	Same optical design as the Radiance, but with more limited versatility.
RTS 2000	Fully computerised 3 channel high-speed video-rate imaging using a resonant galvanometer.
RTS 2000 MP	Multi-photon version video-rate instrument.

Earlier MRC Confocal Microscopes

The large scan head of the MRC series houses all of the optics and the PMT tubes. These instruments are no longer manufactured, but continue to be supported technically by Bio-Rad where possible.

MRC-1024	Most scan head optical components controlled via the computer using LaserSharp software, originally operating under OS/2, but can be upgraded to run under Windows NT. Triple fluorescence and triple transmission imaging.
MRC-1024 MP	Multi-photon version of the MRC-1024 confocal microscope.
MRC-1000	Multi-channel confocal microscope with improved sensitivity compared to MRC-600.
MRC-600	Significantly more sensitive fluorescence detection than the MRC-500. Improved COMOS software running under DOS.
MRC-500	Original commercially available laser scanning confocal microscope from Bio-Rad, with settings on the scan head being manually controlled. Images collected using SOM software under DOS.

Bio-Rad Cell Science Division is a specialist manufacturer of confocal microscopes. This includes their unique scan head and controller box design, which can be readily attached to a number of microscopes. Bio-Rad is not a manufacturer of microscopes or lenses.

Bio-Rad first introduced biologists to confocal microscopy in the late 1980's with the MRC-500. This was later upgraded to the MRC-600, MRC-1000 and subsequently the MRC-1024. These upgrades provided greatly increased sensitivity, further computer control of the microscope and a much more versatile instrument. The MRC-1024 has the capability of triple fluorescence and triple channel transmission imaging simultaneously. A completely redesigned confocal microscope, (the Radiance series of confocal microscopes) has now replaced the MRC series.

The Bio-Rad confocal microscopes have the unique characteristic of using infinity optics within the scan head (parallel light beams) that allows the use of a large (mm size) variable iris to function as the pinhole. This has the advantage that alignment of the laser and scan head is very robust and greatly simplified. Due to this simplicity of scan head alignment, the Bio-Rad confocal microscope scan heads can be installed on many different conventional light microscopes.

A significant change to the Bio-Rad confocal microscopes in more recent years, as far as the use is concerned, has come about due to the reliability and cheapness of quite powerful computers. This has resulted in very significant change in the microscope user interface, where the instrument is now completely controlled by the computer.

The development of the Bio-Rad laser scanning confocal microscopes has traditionally been closely associated with the academic research environment, and many innovative changes have come through feedback from customers all over the world.

RADIANCE SERIES OF CONFOCAL MICROSCOPES

The Radiance series of confocal microscopes from Bio-Rad are based on a relatively small factory aligned scan head, which is attached by fibreoptic cable to a controller box, which houses the lasers, optical filters and PMT detectors (Figure A-2). This simple modular design means that various components can be readily upgraded or further optical filters or lasers added without the need to replace the scan head. This section describes in some detail the design of the Radiance series of confocal microscopes, with particular emphasis on the Radiance 2100 instrument.

Radiance 2100: is a high-end confocal microscope based on the use of dichroic mirrors and optical filters to separate the fluorescent light into various regions of the light spectrum. The instrument is capable of simultaneous triple labelling, transmission and backscatter (reflectance) imaging. Instrument settings are all computer controlled via the LaserSharp software running under Windows NT/2000. The **Radiance 2100 Rainbow** confocal microscope is capable of separating fluorophores with highly overlapping emission spectra by spectral analysis using a series of computer controlled optical filters.

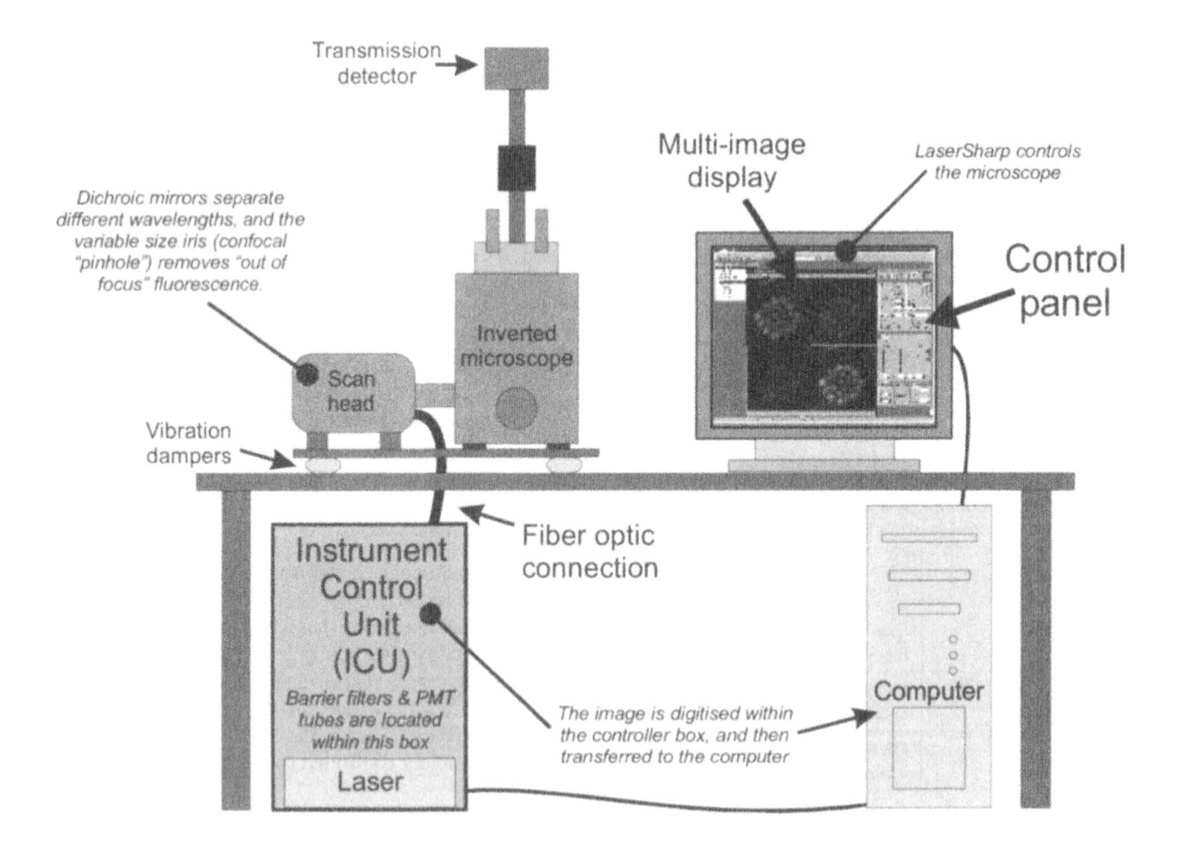

Figure A-2. Radiance Confocal Microscope.
The Radiance confocal microscopes from Bio-Rad have a relatively small scan head that can be readily moved from one light microscope to another. Furthermore, the scan head is factory aligned and all that is required from the user is for the scan head alignment into the microscope to be adjusted. Most of the optical components, PMTs, lasers etc are located in the Instrument Control Unit (ICU) that is located below the microscope bench.

Radiance 2100

Components of the Radiance Confocal Microscopes

The Radiance laser spot scanning confocal microscopes can be configured for visible light and multi-photon microscopy on either upright or inverted light microscopes.

Design

Fluorescent light separation is based on dichroic mirrors and optical filters. The small scan head is directly attached to a light microscope, with the detection optics and lasers housed in a remote unit. The instrument is a high-end confocal microscope that has great versatility.

Scan Head

Permanently factory aligned small and robust scan head can be easily moved between microscopes.

Polarisation beamsplitter: The primary beam splitter (for separating irradiating laser light from fluorescent light) is based on polarisation – a very efficient means of collecting the fluorescent light.

Dichroic mirrors: Mounted inside the scan head is a computer-controlled wheel that contains a number of dichroic mirrors that are used to provide an initial separation of the fluorescence emission.

Variable size confocal iris: A computer controlled variable size "pinhole" or iris is mounted in front of each fibreoptic pickup (one for each channel) within the scan head. The scan head optics is designed such that the confocal "pinhole" is relatively large (mm size), which greatly simplifies alignment.

Fibreoptic pickup: The fluorescent light, after passing or being deflected by suitable dichroic mirrors, is transmitted to the controller box by multi-mode fibreoptic cable.

Scanning mirrors: X & Y fast galvo scan mirrors with two associated concave mirrors to ensure the back aperture of the objective is fully filled throughout the scan. Zoom, bi-directional & scan rotation.

SEL (Signal Enhancing Lens): A removable lens system that is used to direct more out-of-focus light to the detectors (greatly increasing instrument sensitivity) with only a small loss in x-y resolution.

Controller Box

Modular Instrument Control Unit (ICU) contains most of the optical components. Built in a modular way such that individual components can be easily replaced or upgraded. The scan head is connected to the controller box via a large diameter multi-mode fibreoptic connection.

Barrier filters: A computer-controlled wheel of removable barrier filters is mounted in front of each of the PMT tubes. Spectral imaging is available on the Rainbow model by using a series of computer controlled Long Pass and Short Pass interference filters.

Light detectors: Prism enhanced photomultiplier tube for each channel.

Lasers

Lasers: A wide range of lasers can be attached via fibreoptic connection, including the Kr/Ar, Ar-ion, HeNe, 405 nm violet, red laser diode and 561 yellow laser diode, HeCd and single line Kr-ion. Ti:sapphire multi-photon lasers can be attached to suitably configured instruments.

Laser attenuation: Computer selected ND or AOTF filters located within the Instrument Control Unit.

Laser line selection: Laser line selection is controlled by computer controlled optical filters or AOTF filters located in front of each laser, prior to fibreoptic cable transfer to the scan head.

Microscope

The relatively small scan head can be easily transferred between an upright or inverted microscope. The Bio-Rad scan head can be mounted onto a variety of research grade light microscopes from different manufacturers. Bio-Rad is not a manufacturer of light microscopes or lenses.

Imaging Modes

Multi-channel fluorescence (up to 3 channels), photon counting, backscatter (reflectance), transmission. Box sizes user selectable up to 1280 x 1024 pixels. Line, ROI, zoom, panning and rotated scanning. Simultaneous and sequential multi-channel scanning. Emission ratio and mixer combinations. 12-bit data acquisition and processing, saved as 8-bit or 12-bit image files. Time lapse imaging.

Transmission

Single channel photodiode detector or PMT detector (with additional filters). Capable of bright-field, phase and DIC imaging, depending on the optical settings available on the light microscope.

Manual Control

Programmable manual control unit can be used to control all of the essential acquisition parameters.

Computer

High-end Windows computer controls the confocal microscope and collects the images. A single large computer screen or a dual display configuration is possible.

Software

LaserSharp acquisition and analysis running under Windows NT/2000 controls all scanning associated functions. Also used for some image manipulation, colocalisation and 3D reconstruction. Bio-Rad also sells a separate 2D image analysis program (LaserPix) and 3D visualisation and measurement software (LaserVox). Confocal Assistant (free software available from Bio-Rad; see page 168) is a handy program for basic image viewing and manipulation.

Models

Multi channel (up to 3 fluorescence + 1 transmission), single channel system can be upgraded to multi-channel after purchase. **Multi-photon** (page 94), with optional external detectors. **High-speed** video-rate resonant scanning system. **Spectral imaging** is available on the **Rainbow** instrument (page 83).

Radiance Scan Head

The Radiance scan head (Figure A-3) contains the scanning mirrors, the polarisation beamsplitter for separating the irradiating laser light from the returning fluorescent light, the dichroic mirrors for separating the different wavelengths of fluorescent light, the variable iris and a fibreoptic pickup port for transmitting the light to the Instrument Control Unit. All other optical components are housed within the Instrument Control Unit that is usually located under the microscope bench (Figure A-4). The operation of the Radiance scan head is fully controlled by the Bio-Rad produced LaserSharp software.

The x-y galvometric mirrors located within the Radiance scan head, are factory aligned and cannot be adjusted by the user. The relatively small and compact nature of the Radiance scan head makes this scan head particularly easy to transfer from one computer to another. There are only minor alignments to be done when the scan head is moved to another microscope. This includes both lateral adjustment of the scan head and beam focus adjustments.

The scan head is connected to the associated Instrument Control Unit via a large diameter flexible metal tubing that houses the control cables and fibreoptic cables for each channel.

The fast galvo scanning mirrors in the Radiance 2100 confocal microscope are designed such that the scan can be accomplished in any direction. This means that a narrow scan can be executed at an angle across the screen. This is useful, for example, when scanning an individual neuron presented at an angle across the sample. The scanning mirrors in the earlier Micro Radiance instruments cannot be "angled" under OS/2, but angled scanning is possible if these instruments are upgraded to Windows 2000. Two concave mirrors, located on either side of the x and y galvo scan mirrors (shown in Figure A-3), provide a scanning geometry that ensures that the back aperture of the objective is filled throughout the scan cycle (this is important for maximising the optical resolution of the instrument).

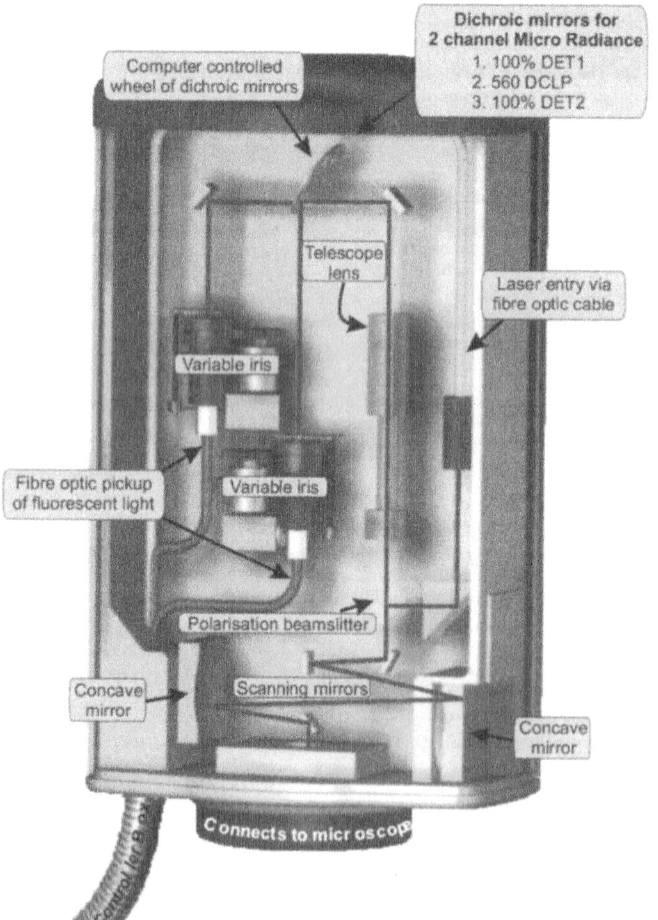

Fluorescent and reflected light returning from the sample is de-scanned by passing back through the scanning mirrors. The light is then transmitted through a polarisation beam splitter to a

Figure A-3. Radiance Scan Head.
The Radiance series of confocal microscopes has a relatively small and robust scan head. Many optical components, previously housed in the scan head, have now been transferred to the Instrument Control Unit (ICU), see Figure A-4. This figure, adapted from a figure kindly provided by Bio-Rad, UK, is of a dual channel Micro Radiance confocal microscope.

telescope lens arrangement (allowing this scan head to be much smaller than the previous MRC models). This lens arrangement is necessary to obtain the correct optical configuration when aligning the light through the variable pinholes. The polarisation beam splitter is used to separate the highly polarised irradiating laser beam from the essentially randomly polarised returning fluorescent light. This type of polarisation beam splitter, prevents any fluorescent light that is polarised in the same direction as the irradiating laser from entering the detectors, but does allow the full range of the light spectrum to be detected. Unlike a dichroic mirror, there are no regions of the spectrum where light cannot penetrate the beamsplitter.

A computerised wheel containing a number of dichroic mirrors is used to split the fluorescent light into the relevant component wavelengths. In the two channel Micro Radiance scan head (shown in Figure A-3) there is a

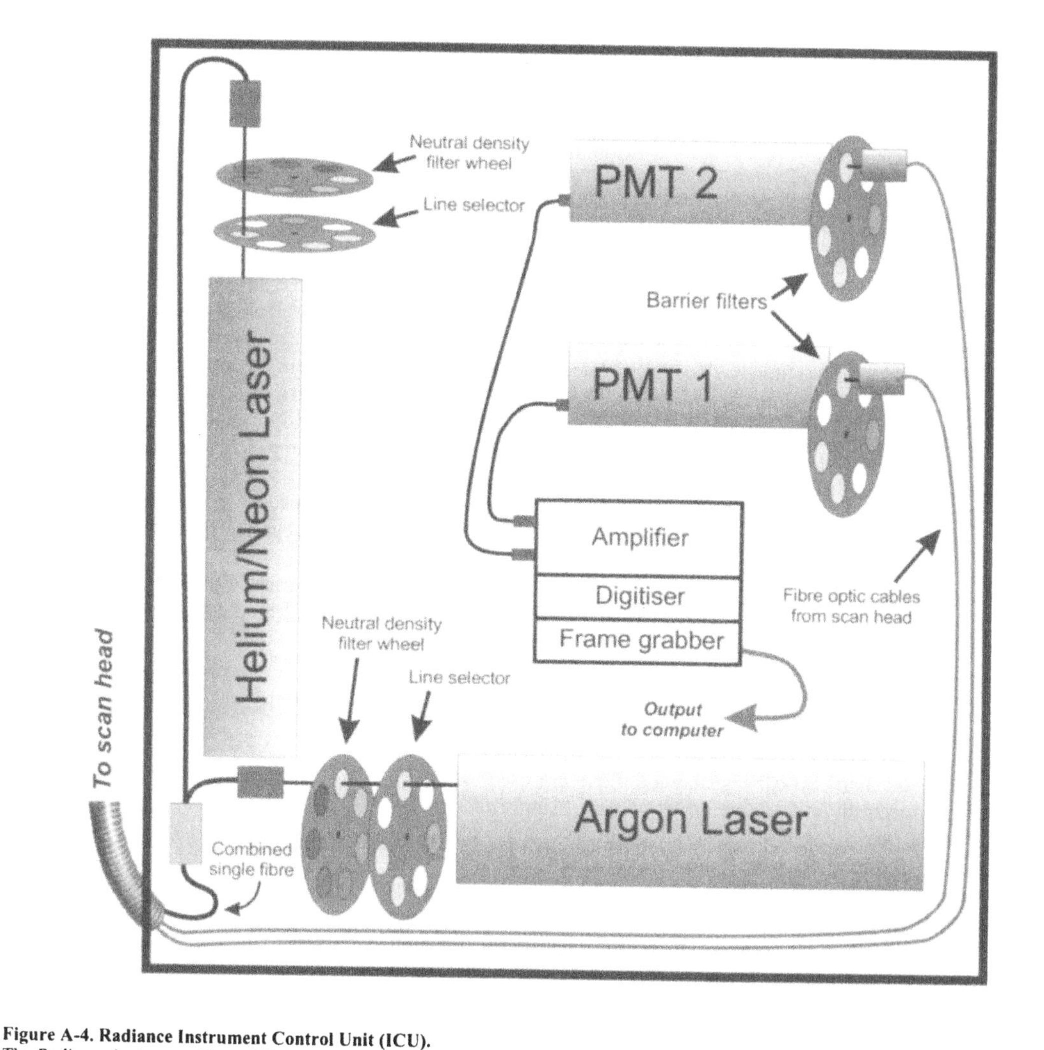

Figure A-4. Radiance Instrument Control Unit (ICU).
The Radiance Instrument Control Unit now houses the lasers (including the laser attenuation neutral density or AOTF filters and laser line selector optical filter wheels), the PMT tubes for detecting fluorescent light (including the barrier filter wheel), the amplifier, digitiser and frame grabber. The digitised signal is then transferred to the computer for processing.

single dichroic mirror wheel. In the Radiance 2100, which is capable of simultaneous triple labelling, there are two wheels of computer controlled dichroic mirrors.

After suitable reflection or transmission via the dichroic mirror the light is aligned into the variable iris mounted in front of the fibreoptic cable pickup ports. Due to the infinity optics design of the scan head the "pinhole" is relatively large, being mm in diameter and circular in shape. The fluorescent light is transferred to the Instrument Control Unit via a large diameter fibreoptic cable to minimise loss of light. The large diameter multi-mode fibre also means that the optic fibre does not act as a "pinhole", and hence the requirement for the variable iris or confocal "pinholes" mounted within the scan head.

Radiance Instrument Control Unit (ICU)

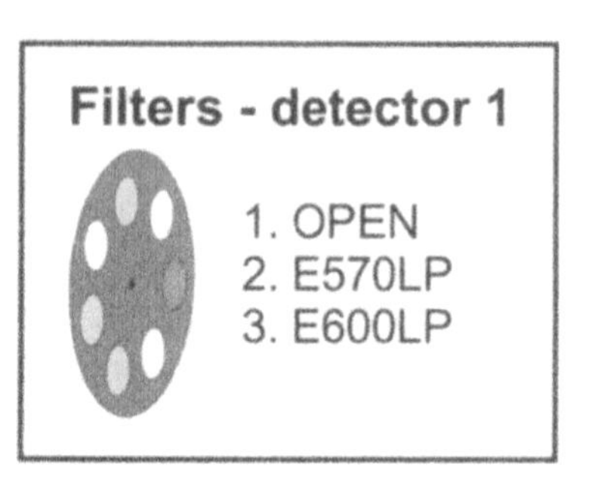

Many of the optical components previously housed in the scan head have been transferred to the Instrument Control Unit in the Radiance series of confocal microscopes (Figure A-4). This has been made possible by using large diameter multi-mode fibreoptic cables between the scan head and the Instrument Control Unit as discussed above.

Mounted within the Instrument Control Unit are the lasers and their power supplies. The argon-ion or krypton-argon-ion lasers require a cooling fan that is mounted outside the Instrument Control Unit, but the helium-neon lasers do not generate excessive heat and so do not need an external-cooling fan. The argon-ion / krypton-argon-ion laser cooling fan is shut off automatically when the laser has cooled sufficiently (approximately 5 minutes) after the laser unit has been turned off. The power supply to the Instrument Control Unit should be switched to "standby" when an imaging session has been completed, to ensure the fan will remain on until the laser has cooled.

Computer controlled neutral density filters and line selection filters (Figure A-4), or computer controlled AOTF filters are used to attenuate the amount of laser light that enters the microscope and to select individual laser lines. These filters are mounted in front of each of the installed lasers. The laser light, after passing through the attenuation and line selection filters is then picked up by fibreoptic cable, combined into a single fibreoptic cable, and transferred to the scan head. Large diameter multi-mode fibreoptic cables are then used to transfer the fluorescent or reflected light from the scan head back into the Instrument Control Unit (with a separate fibreoptic cable for each detector channel).

There are separate computer controlled barrier filter wheels mounted in front of each of the PMT tubes within the Instrument Control Unit to further define the region of the light spectrum that will be detected by each individual channel. For specialised applications the barrier filters loaded in the filter wheels can be changed.

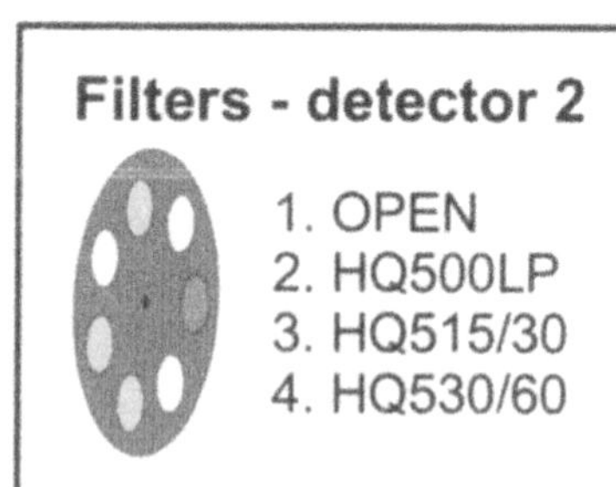

Additional optical filter wheels are mounted inside the Instrument Control Unit of the Radiance 2100 Rainbow confocal microscope for spectral imaging (see Figure 3-14 on page 85). A combination of Long Pass and Short Pass sharp cut-off interference filters are used to dissect the visible light spectrum into narrow bands of light. Automated optical filter selection is used to generate a series of images that make up a spectral scan of the sample. Using these optical filter wheels for both spectral imaging and spectral reassignment in multi-labelling imaging applications is discussed in detail below.

A prism enhanced PMT tube (a more efficient light collector compared to a conventional PMT tube) is used for each individual detection channel. The output from the PMT tubes is amplified and then digitised within the Instrument Control Unit, before transfer to the computer for display and further processing. The modular nature and large size of the Instrument Control Unit now makes the replacement of defective components or the upgrading of individual components a relatively simple task.

Changing Optical Filters and Dichroic mirrors

A number of dichroic mirrors (in the scan head) and barrier filters (in the Instrument Control Unit) can be mounted within the computer controlled filter wheel and so there should be little need to consider changing the filters. However, these optical filters and dichroic mirrors can be changed if necessary. The filter is designed to slip out by holding the loop at the top of the filter holder. Another filter can then be placed in the empty position (see the instrument manual). You will then need to update the information in the LaserSharp software to reflect the new filters that have been installed.

Radiance 2100 Rainbow

The Rainbow version of the Radiance confocal microscope uses two computer controlled interference filter wheels (see page 85) to separate fluorescent light into separate regions of the light spectrum. One optical filter wheel has a set of short-pass interference filters and the other a set of long-pass filters. In combination these two optical filter wheels can be used to select a defined region of the light spectrum. Each of the fluorescence imaging channels (up to 3) available on the Radiance 2100 Rainbow confocal microscope has a separate dual filter wheel assembly, pinhole and PMT detector.

Radiance 2100 Rainbow
Spectral separation of fluorophores using computer controlled interference filters (see page 85).

The dual optical filter wheel assembly in the Rainbow is a very convenient method of separating regions of the light spectrum, with the flexibility of allowing one to choose individual regions to be directed to each detector. The filter wheels not only provide considerable flexibility when choosing spectral regions, but they also provide an intuitive method of choosing the correct optical filter. The on-screen filter display is listed as a series of optical filters on a drop down menu (the blocking filter set and the emission filter set – see page 380). Each optical filter in the series has a 10 nm difference in the spectral cut-off point, and as the filters are arranged in sequential order it is relatively easy to choose the correct filter. Furthermore, Bio-Rad now provide a visual on-screen spectral diagram that includes a coloured outline of the part of the light spectrum directed to each channel.

The dual filter wheel assembly in the Rainbow instrument can also be used to obtain spectral information from the sample. A spectral image series or lambda stack is generated by collecting a series of images, each from a narrow region of the light spectrum (as narrow as 10 nm if required). The spectral image series can be used to study spectral changes within the cell or tissue sample, or for spectral separation of individual fluorophores.

Spectral information can also be used, in a process called spectral reassignment, to separate probes with highly overlapping emission spectra. If fluorescent probes with overlapping emission spectra are imaged using two or more collection channels then there will be bleed-through between the channels. On Bio-Rad instruments a small amount of bleed-through can be corrected while imaging by subtracting a small percentage of one collection channel from the other collection channel using on-screen mixers. However, when the bleed-through is more substantial, and is occurring in more than one channel then the process of spectral reassignment is more appropriate. This process is similar to the Zeiss META spectral unmixing algorithm, but is performed on only two or possibly three spectrally separated images rather than a lambda stack. This has the advantage of very fast simultaneous collection of the spectral information. The degree of bleed-through between individual channels can be calculated from images collected using single labelled control samples.

In the Bio-Rad spectral reassignment process the light from defined regions of the light spectrum is directed to a individual (single PMT) detection channels. This greatly increases the sensitivity of the instrument, and also allows one to easily adjust the relative intensity of the signal for each of the channels. For spectral reassignment to be successful both fluorophores must show similar intensity in the image. Adjusting the individual collection channels is readily accomplished by changing the gain for each of the detection channels. The Bio-Rad method of spectral separation does have the advantage of increased sensitivity and speed of acquisition and analysis, but under conditions of high levels of fluorescence the Zeiss META linear unmixing algorithm used on a full lambda stack may be able to separate a larger number of fluorescent probes.

SELS (Signal Enhancing Lens)

The Signal Enhancing Lens (SELS) is a small removable optical device that greatly increases the amount of out-of-focus light that reaches the detectors without greatly reducing the x-y resolution of the instrument. Inserting the SELS lens reduces the effective magnification of the light from the Radiance telescope lens, resulting in more light passing through the confocal pinhole or iris without the need to greatly increase the size of the iris. The greatly increased signal provided by the SELS lens is ideal for live cell imaging where photobleaching and photo damage often seriously limit the time allowed for following cellular events in live cells (see "SELS: A New Method for Laser Scanning Microscopy of Live Cells by Reichelt and Amos, *Microscopy and Analysis,* November 2001 – further details in Chapter 15 "Further Reading", page 350).

Photon Counting

Bio-Rad confocal microscopes are capable of "photon counting", where electronic circuitry is used to "count" each photon as it is collected by the photomultiplier tube (see page 37 and 110). Photon counting is only applicable when using samples with low levels of fluorescence. Excellent images, with low noise levels, can be collected with photon counting if the "accumulate" mode is used when collecting several images. When using the photon counting mode an image may not be visible with a single scan – a number of scans may need to be accumulated before the image becomes visible.

Photon counting can also be utilised on the Bio-Rad Radiance Rainbow instrument when collecting multi-channel images for the separation of highly overlapping emission spectra using spectral reassignment.

Computer and Software

A DELL IBM compatible computer is used to operate the Radiance series of confocal microscopes. Only a relatively modest computer is required for operating the confocal microscope and collecting images (sufficient to run Windows XP or NT), however, 3D image processing will be smoother and faster with a computer with a large amount of memory (RAM) a large hard disk and a fast video card.

The Micro Radiance series of confocal microscopes are operated by using either the older version of LaserSharp, which runs under OS/2, or utilising the new version of LaserSharp that runs under Windows NT. Confocal microscopes still operating under the OS/2 operating system can be upgraded to the latest Windows operating system. However, as this requires the installation of both the operating system and new software for controlling the microscope, don't attempt an upgrade unless you are familiar with making such major changes to a computer!

The LaserSharp software is designed to control the confocal microscope, collect images and to perform most elementary image processing. This includes 3D reconstruction and colocalisation algorithms. The LaserSharp software is discussed in more detail later in this appendix (page 377). The spectral reassignment algorithm is available on the Radiance 2100 Rainbow confocal microscope using SpectraSharp.

MRC SERIES OF CONFOCAL MICROSCOPES

The MRC series of confocal microscopes, from the MRC-500, which was one of the earliest confocal microscopes used extensively in the biological sciences (first introduced in the early 1980's) to the highly versatile MRC-1024 (produced up to 1998) are used in a large number of Universities and Research Centres throughout the world.

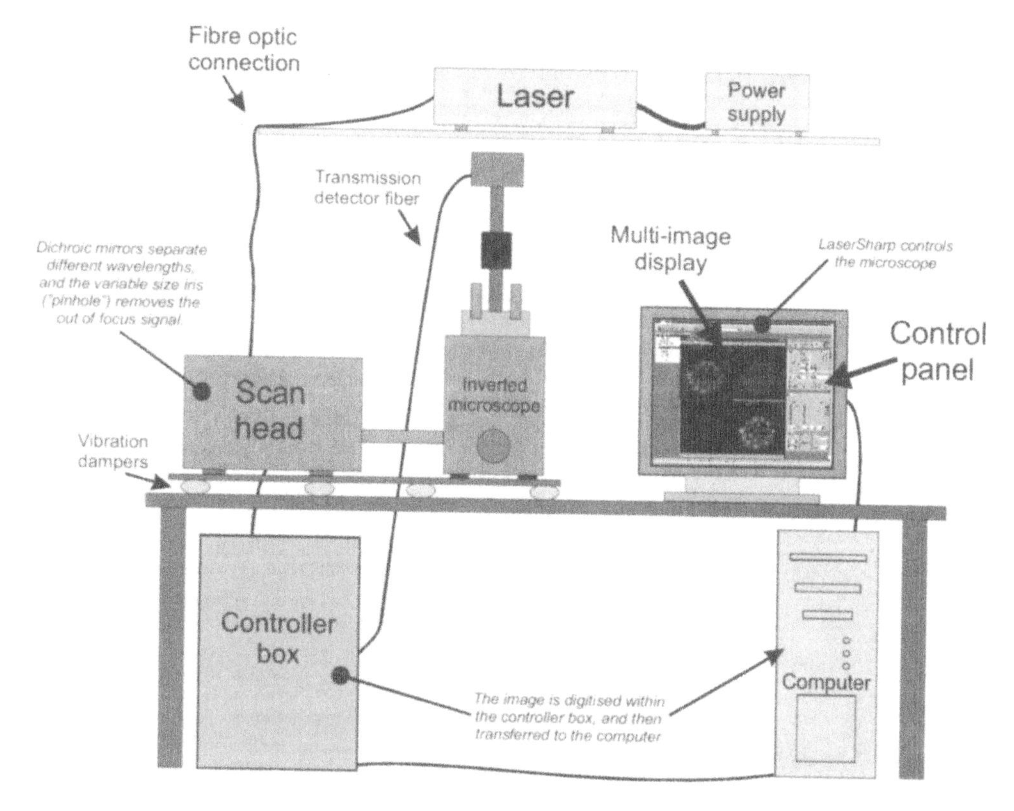

Figure A-5. MRC-1024.
The MRC series of confocal microscopes have significantly larger scan heads compared to the Radiance confocal microscopes. This is due to most of the optical components, and the PMT tubes being located in the scan head. The MRC scan head can be mounted on both upright and inverted research grade microscopes from a variety of microscope manufacturers.

Although the MRC instruments are now superseded by the Radiance confocal microscope design (see previous section), they are still excellent instruments with many still in operation – and continue to be technically supported by Bio-Rad. In this section a detailed description is given of the latest in the MRC series, the MRC-1024 confocal microscope. The previous members of the MRC confocal microscopes (MRC-500, MRC-600 and MRC-1000) are based on the same fundamental design used in the MRC-1024. The major difference between these earlier model MRC instruments and the MRC-1024, as far as the user is concerned, is the extensive use of computer software in the latter to control the settings of the confocal microscope scan head. There are also many important technical advances utilized in the MRC-1024 instrument that make this confocal microscope significantly more sensitive and more versatile than the earlier MRC instruments. The LaserSharp software, used for controlling the new Radiance confocal microscopes, can also be implemented on the MRC-1024 and MRC-1000, but the earlier versions of the MRC confocal microscope (MRC-500/600) are controlled by the DOS based COMOS software.

MRC-1024

Components of the MRC-1024 Confocal Microscope

The MRC-1024 laser spot scanning confocal microscope can be configured for visible light, UV light and multi-photon microscopy on either upright or inverted light microscopes.

Design

A large scan head, which houses all of the necessary optics, pinholes and detectors is directly attached to an upright or inverted microscope. The instrument is computer controlled, with the exception of the changeover of specific optical filter blocks in the scan head.

Scan Head

The relatively large and heavy scan head of the MRC series of confocal microscopes means that mounting the scan head on another microscope is difficult – even though Bio-Rad advertised this instrument as being readily moved from one microscope to another.

The MRC-1024 scan head contains all of the necessary optics and PMT tubes for multiple channel labelling. As a consequence the scan head is not only rather large, but also a delicate instrument.

Filter blocks: Removable optical filter and dichroic mirror blocks are housed within the scan head. Changing the laser may require a different combination of optical filter blocks.

Dichroic mirrors: Dichroic mirrors for specific applications are changed by substituting different optical filter blocks.

Variable size confocal iris: A variable size iris, or "pinhole", is mounted in front of each PMT tube and its size is under computer control.

Light detectors: Three photomultiplier tubes (PMT), allowing for 3 colour simultaneous fluorescence imaging, are housed within the scan head.

Scanning mirrors: X & Y fast galvo scan mirrors. Zoom and bi-directional capability.

Controller Box

The image is digitised within the controller box and then transferred to the computer. The transmission image (3 channels in the MRC-1024) is also detected within the controller box, after collection on the microscope and transfer to the controller box by a fibreoptic cable.

Laser

Lasers: A wide range of lasers can be attached via fibreoptic connection, including the Kr/Ar, Ar-ion, HeNe, red laser diode, HeCd and single line Kr-ion. UV and Ti:sapphire multi-photon lasers can be attached to suitably configured instruments.

Laser attenuation: Computer selected ND or AOTF filters located within the laser box.

Laser line selection: Computer controlled filters for laser line selection are located in front of each laser, prior to fibreoptic cable transfer to the scan head.

Microscope

The scan head can be attached to either an upright or inverted microscope, but moving the scan head to another microscope is somewhat difficult due to the relatively large size and delicate nature of the scan head. Bio-Rad has designed adaptors for the MRC-1024 scan head to be attached to a research grade light microscope from each of the major microscope manufacturers.

Imaging Modes

Multi-channel fluorescence, photon counting, backscatter (reflectance), transmission. Box sizes user selectable up to 1280 x 1024 pixels, zoom and panning. Simultaneous and sequential multi-channel scanning. Mixer combinations. 12-bit data acquisition and processing, saved as 8-bit or 12-bit image files. Time lapse imaging.

Transmission

A mirror located after the condenser (above the condenser on an inverted microscope) is used to direct transmitted light to up to three detectors located within the controller box. The operation of the transmission detectors is via the LaserSharp software. However, a mirror on the microscope must be manually changed to direct the laser light into the fibreoptic pickup cable.

Manual Control

Programmable manual control unit is not available for the MRC instruments.

Computer

High-end PC computer with a single large computer screen to display the images as they are collected and to display the controls for scanning and changing microscope settings. More complex image analysis is usually done using software installed on a separate computer.

Software

LaserSharp acquisition and analysis, running originally under OS/2 but can be upgraded to run under Windows NT, controls all scanning associated functions. Also used for some image manipulation and 3D reconstruction. Bio-Rad also sells a separate 2D image analysis program (LaserPix) and 3D visualisation and measurement software (LaserVox). Confocal Assistant (free software available from Bio-Rad; see page 168) is a handy program for basic image viewing and manipulation.

Models

The MRC confocal microscopes are no longer manufactured, but **multi channel** (3 fluorescence + 3 transmission), **multi-photon** and **UV** systems are still in use around the world.

MRC-1024 Scan Head

The Bio-Rad MRC-1024 scan head (Figure A-6) contains an extended light path that is folded back several times (to gain sufficient length for optical requirements without using a long optical box). The scan head contains all of the optical components, a variable iris for each channel and the photomultiplier (PMT) tubes. This results in a rather large and delicate scan head that, although it can be attached to both upright and inverted microscopes, is not easily transferred from one microscope to another. The new Radiance series of confocal microscopes from Bio-Rad has a radically "down sized" scan head, where many optical components have been transferred to the Instrument Control Unit (discussed in detail in the previous section), and the light path has been significantly shortened by using a telescope lens.

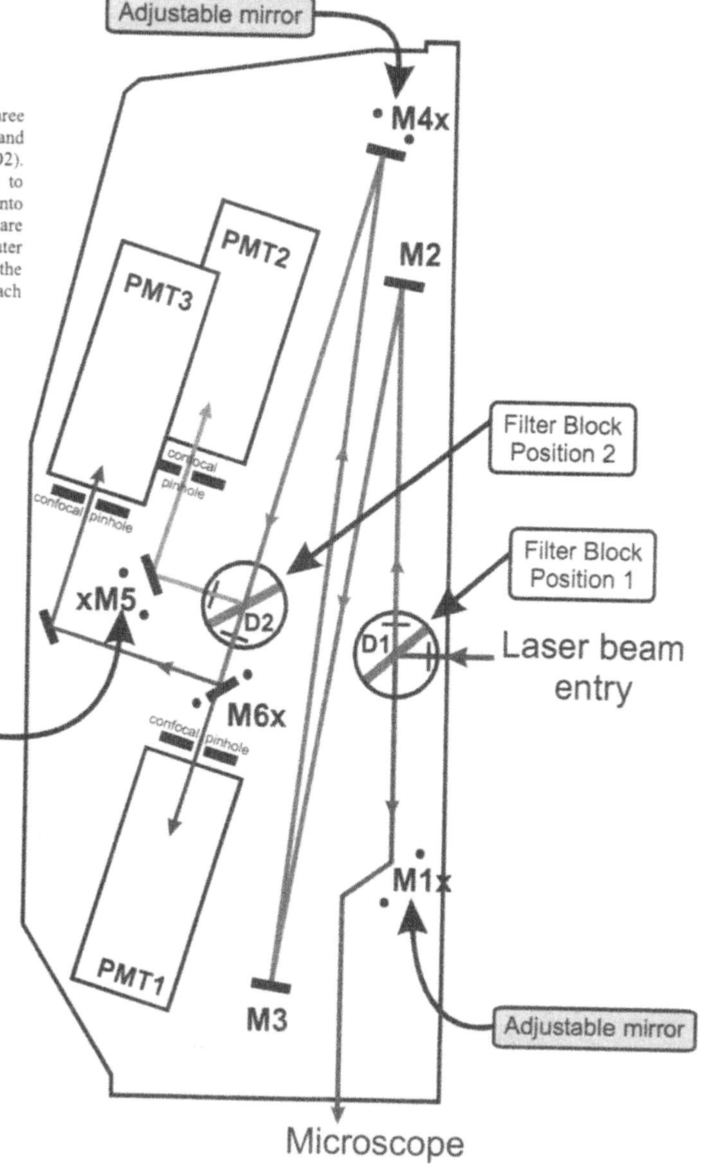

Figure A-6. MRC-1024 Scan Head.
The MRC-1024 scan head contains three photomultiplier tubes (PMT1, PMT2, PMT3), and two removable dichroic mirror sets (D1 and D2). M1, M4 and M5 are user adjustable mirrors to allow one to correctly align the laser beam into the PMT tubes. Mirrors M2, M3 and M6 are factory set. The scan head also contains computer controlled optical filter wheels for refining the wavelengths of light that will be detected by each of the PMT tubes.

There are a number of mirrors positioned within the MRC-1024 scan head for the purpose of both aligning the laser light into the objective on the microscope, as well as aligning the returning fluorescence light into the iris and PMT tubes (denoted M1 to M6 in Figure A-6). A number of these mirrors are factory aligned (M2, M3 and M6) but some of these mirrors are user adjustable (M1, M4 and M5). Extreme care should always be taken when attempting to adjust these mirrors, as incorrect alignment of the mirrors can cause considerable loss of sensitivity and uneven detection across the field of view.

The Bio-Rad confocal microscope scan head can be fitted to a variety of light microscopes from a number of manufacturers. This includes both upright and inverted microscopes. The scan head is best connected to the microscope as close to the objective as possible (to minimise problems with light loss from the many optical components within the microscope).

Laser Alignment

Less sensitivity than usual?

Try re-aligning the scan head mirrors:

M1, aligns the laser beam into the scan head. Use the "bulls eye" prism, provided by Bio-Rad.

M4, careful adjustment of M4 will align the fluorescent light into PMT1. Poor alignment results in low signal and uneven lighting.

M5, this mirror allows you to align the fluorescent light into PMT2.

See Figure A-6 *for location of alignment mirrors.*

When the scan head is mounted on an inverted microscope the coupling is via an attachment that allows the laser to be delivered directly to the objective. This means that the confocal microscope image quality obtained on an inverted microscope is exactly the same as that obtained on an upright microscope when using the same objectives. However, the scan head can be mounted in a variety of places, including the video mount on the top front of an inverted microscope, with only a small loss in image quality.

Scan Head Alignment

The alignment of the scan head with the microscope is most important for high quality imaging. Although the scan head should be correctly aligned by the Bio-Rad technical personnel during installation of the instrument, the scan head should be checked periodically to make sure there has been no movement between the scan head and the microscope.

If the scan head is moved to another microscope, re-alignment will be essential. Aligning the scan head is a reasonably simple operation. The Scan Head should be positioned in approximately the correct position (determined by the length of the connecting tube). The microscope and scan head should be connected with the enclosed connecting tube. There are two alignment procedures required; the first is to align the laser scan centrally, and the second is to focus the scan onto the back focal plane of the objective.

To align the laser scan centrally, first make sure the plastic prism "bulls eye" provided by Bio-Rad is screwed into one of the objective sockets on the microscope. While the laser is scanning, gently move the scan head to align the scan to the centre of the "bulls eye". Once the scan is central, and while continuing to scan, gently turn the threaded turret on the scan head (microscope-coupling tube) until the scanning spot in the "bulls eye" stops flickering (pulsing). If this cannot be achieved, try turning the turret in the other direction. The laser should now be a spot of light in the centre of the "bulls eye" with only a hint of "flickering" as the laser scans. Once the scan has been aligned, both the microscope and the scan head should be bolted tightly to the table to prevent accidental movement. The scan head should not need re-aligning into the microscope again unless either the microscope or scan head is moved.

Low/High power on a Kr/Ar laser

For critical applications (such as real colour transmission imaging) switch the laser to high power. The switch is located on the side of the laser.

Don't forget - switch back to low power!

Optical Filter Blocks

Don't forget - you must physically change the filter blocks in the scan head!

The Bio-Rad MRC-1024 scan head contains two removable optical filter blocks that house both dichroic mirrors and sometimes, suitable barrier filters (further barrier filters are provided on computer controlled rotating wheels in front of each PMT tube within the scan head). Several different optical filter blocks are available and can be used in various combinations for different labelling requirements. Below is a brief description of the characteristics of the filter blocks available for the Bio-Rad MRC-1024 scan head, with the most common filter blocks described in more detail shortly (Figure A-7, Figure A-8 and Figure A-9). Optical filter block combinations used for specific imaging methods are also described in some detail (Figure A-10, Figure A-11, Figure A-12).

A requested change of filter block on the menu screen of the LaserSharp software requires a physical change of filter block within the scan head. Changing the filter block is a simple process of depressing the button on the top side of the filter block to release the catch and lifting the filter block out of the scan head. The replacement filter block is then simply pressed into place in the scan head such that the catch "clicks" to indicate the filter block is correctly seated.

Filter Blocks for the Bio-Rad MRC-1024 Confocal Microscope

Krypton-argon-ion laser

T1: (triple dichroic mirror) for use with the krypton-argon-ion laser – reflects 488, 568 and 647 nm.

T2A: (560 DRLP) dichroic mirror for splitting the green (PMT2) and red (PMT1) fluorescent light.

Live cell imaging with any laser

B1: beamsplitter optical filter block where 80% of all light is transmitted and 20% reflected. Used in position 1 for live cell studies, and in combination with the T1 triple dichroic mirror in position 2 for backscatter imaging.

Argon-ion laser

A1: (527 DRLP) for use with the argon-ion laser for dual labelling - reflects 488 and 514 nm.

A2: (565 DRLP) for splitting the green (PMT2) & red (PMT1) fluorescent light.

VHS: (510 DRLP) Reflects 457 & 488 nm blue light – for use with Ar-lasers for fluorophores such as Lucifer yellow.

Green HeNe laser

EG1: reflects 488 & 543 nm for dual labelling with argon-ion/green HeNe lasers.

E2: (560 DRLP ext R) splits fluorescent light into green/blue (PMT2) and red in PMT1.

Specialised optical filter blocks

OPEN: contains no filters or dichroic mirrors for single colour imaging.

SA2R: (610 DRSP) contains emission filters 580/32 and 640/40 for SNARF @ 514 nm excitation.

SK2R: (605 DRSP) contains emission filters 570/40 and 640/40 for SNARF @ 488 nm excitation.

FF2R: contains emission filters 600LP and 530/40 for Fura red plus Fluo-3 imaging.

UV laser

UBHS: reflects 351, 363 and 488 nm light.

INI 1: (380 DCLP) reflects UV lines.

INI 2: (440 DCLP) splits violet (for example, Indo-1) into PMT2 and blue light into PMT1

E2: (560 DRLP) splits green and blue light into PMT2 and red light into PMT1.

Note: DRLP = Dichroic Long Pass Filter & DRSP = Dichroic Short Pass Filter

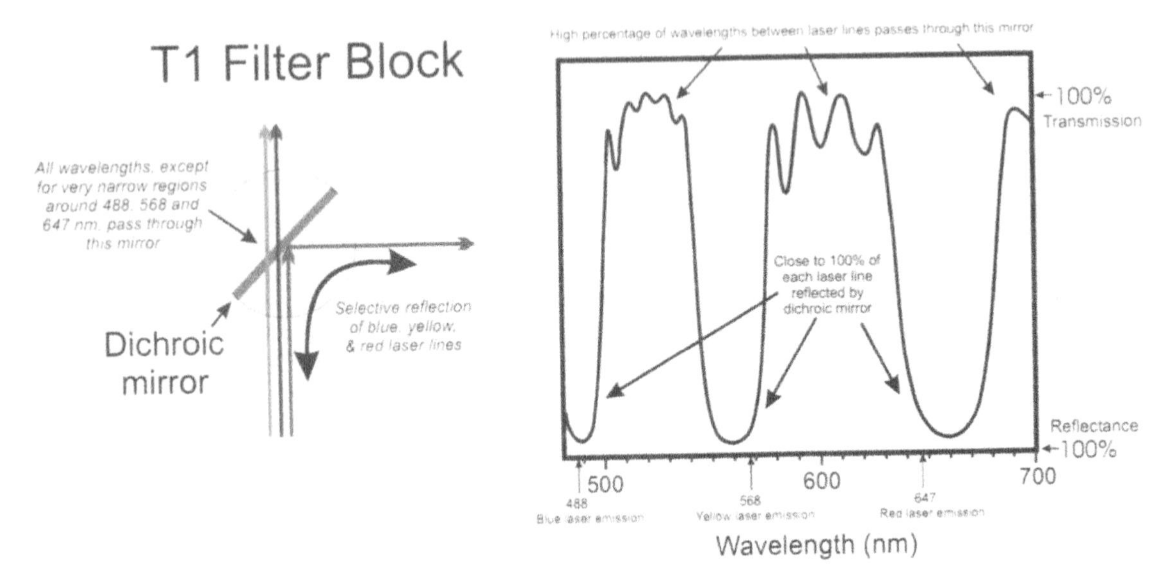

Figure A-7. MRC T1 Optical Filter Block.
The T1 optical filter block is designed to reflect each of the 488 nm (blue), 568 nm (yellow) and 647 nm (red) laser lines of the krypton-argon-ion mixed gas laser. Wavelengths outside these three regions can pass through the dichroic mirror. This optical filter block is used in the position closest to the laser intake when doing standard dual and triple labelling (Figure A-10. The T1 filter block can also be placed in the 2nd filter block position (away from the laser input) when doing backscatter (reflectance) imaging (Figure A-12).

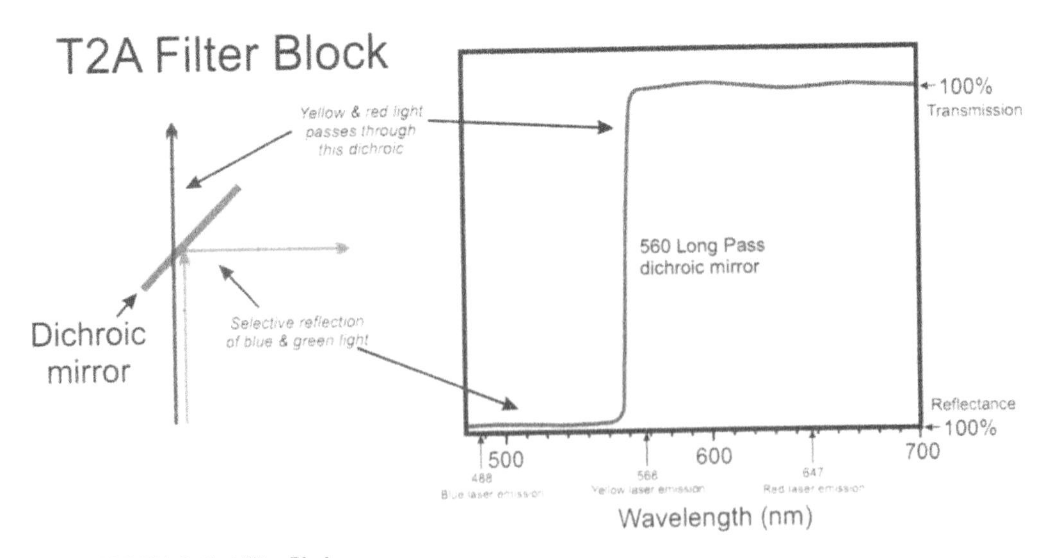

Figure A-8. MRC T2A Optical Filter Block.
The T2A filter block is designed to split red and green wavelengths of light. Short wavelengths (blue and green) are efficiently reflected by the dichroic mirror, whereas longer wavelengths (yellow to far-red) readily pass through the mirror. This filter block is used in the 2nd position (away from the laser intake) in combination with the T1 triple dichroic mirror (in the first position, near the laser intake) for dual and triple labelling.

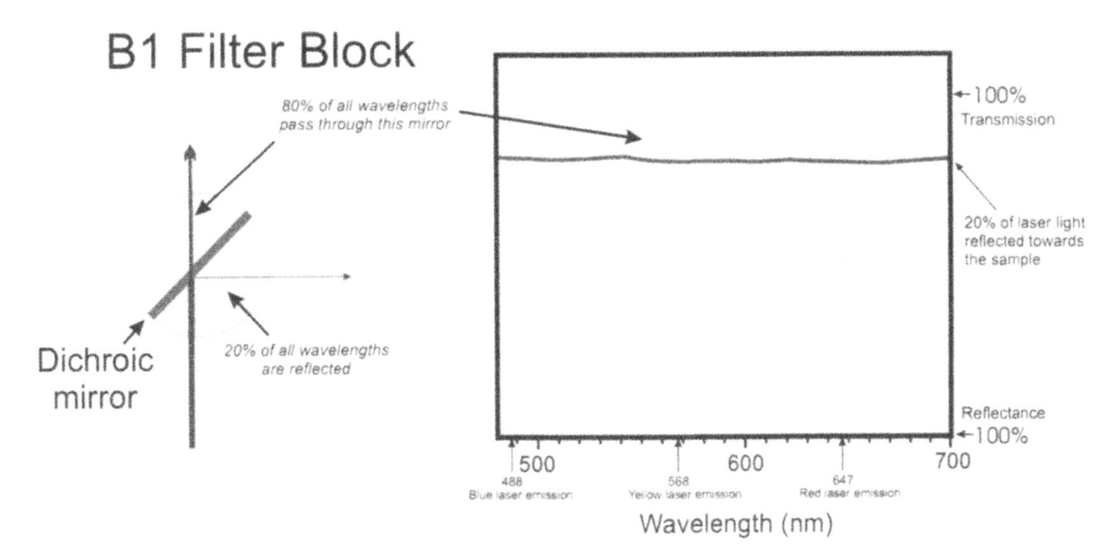

Figure A-9. MRC B1 Optical Filter Block.
The B1 filter block is designed to reflect 20% of all wavelengths of light, and to transmit the remaining 80% of all wavelengths. The filter block is particularly suitable for live cell imaging where maximum sensitivity of fluorescence detection is required (Figure A-11). The B1 filter block can also be used in combination with the T1 triple dichroic mirror for backscatter (reflectance) imaging (Figure A-12).

Open Filter Block, *contains no dichroic mirrors or filters, resulting in all light being directed to PMT1. Placed in the position in the scan head furthest away from the microscope. Don't forget that when "OPEN BLOCK" is selected on the computer the Open Block must be physically inserted in the scan head.*

640SP (Short Pass)
Permanently installed dichroic mirror (in 3-PMT systems) for directing far-red light into PMT3.

For special labelling applications it is possible to install your own dichroic mirrors and filters.

Emission Filter Wheels

Within the MRC-1024 scan head are emission filter wheels that are controlled by the LaserSharp software. These optical filters are used to prevent unwanted excitation laser light entering the PMT tube, and to more closely define the region of the light spectrum that will be detected by each channel on the confocal microscope (for dual and triple labelling applications). The combination of optical filters available will be dependent on the lasers you have installed on your system.

Emission Filters

When using the krypton-argon-ion laser:
PMT1: OPEN, 585LP, 605/32, OG515,680/32
PMT2: OPEN, 522/35, Blue Reflection.
PMT3: OPEN, 680/32

When using the argon-ion laser:
PMT1: OPEN OG515, 580/32, 585LP,680/32
PMT2: OPEN, 540/30, Blue Reflection
PMT3: OPEN, 680/32

Photomultiplier Tubes (PMT)

The photomultiplier tubes are very sensitive light detectors that are mounted within the scan head. The MRC-1024 contains three separate photomultiplier tubes, which allows simultaneous 3-channel fluorescence imaging. There is a separate "pinhole" (confocal iris) and an individual excitation filter wheel associated with each PMT.

Controller Box

The controller box for the MRC-1024 houses the electronic components for operating the confocal microscope. The controller box houses the "frame grabber" card, which is where the image is initially collected before transfer to the computer. The transmission detectors are located within the controller box, with a fibreoptic cable transferring the transmitted light from the microscope down to the detectors.

The controller box should be turned on fully (not just "stand-by" mode) before the computer is turned on. If the computer attempts to complete the initialisation routine without the controller box being turned on, an error will occur when the computer attempts to down-load control programs to components within the controller box. The controller box should be turned to "stand-by" mode when not in use.

Computer Requirements

The MRC-1024 is usually installed with an IBM compatible DELL computer, although earlier versions of this microscope were operated with a Compaq computer. The more computer memory (RAM) and the faster the video card then the faster the 3D reconstruction will be and the smoother the "rocking" motion when displaying 3D images. Neither a high-end computer, nor a large amount of computer memory is required for the actual collection of confocal microscope images.

A large hard disk is essential for collecting confocal microscope images. This is particularly important when collecting optical slices for 3D reconstruction or a series of time-lapse images. Imaging on a confocal microscope can very quickly result in the collection of a very large amount of image data. Regular removal of the collected images to a more permanent form of storage, such as a CD, is most important. The software for operating the MRC-1024 confocal microscope was originally written for OS/2, an operating system that is now obsolete. A Windows NT version of LaserSharp is available for operating the MRC-1024 confocal microscope, but obtain help from Bio-Rad or other users before upgrading so as to avoid the many problems that can be caused. People who have upgraded successfully are convinced the effort is worthwhile as you then have "in house" support for a Windows operating environment, which greatly facilitates such things as linking to your network and installing additional drives etc.

Hardware set up for specific imaging methods

There are a number of "standard" optical filter block combinations that are routinely used for a variety of fluorophores. These are described in some detail on the following pages.

Dual and Triple labelling - T1 and T2A Filter Blocks

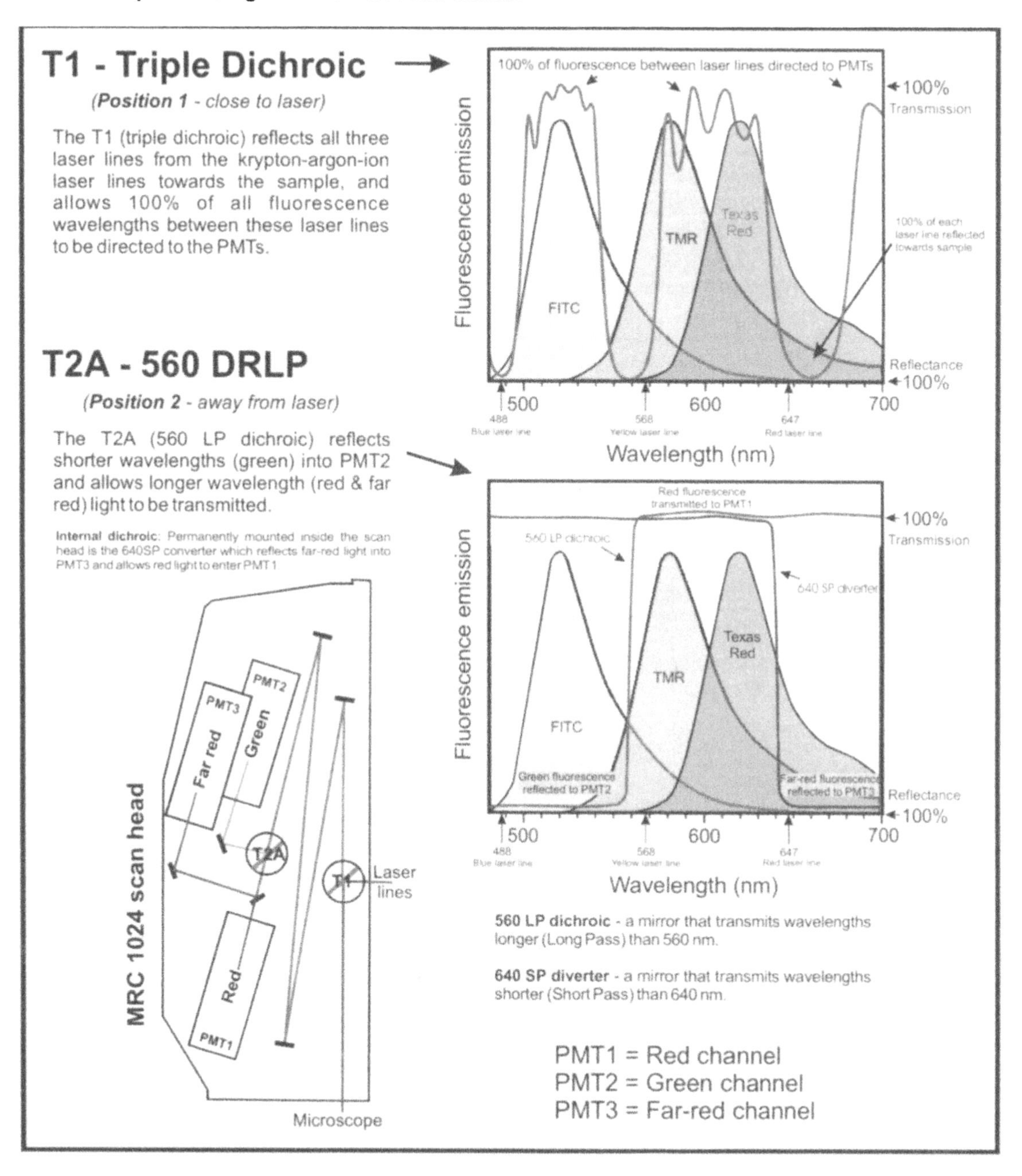

Figure A-10. MRC T1 and T2A Filter Block Combination.
The T1 and T2A optical filter block combination is ideal for dual and triple labelling using fluorophores such as FITC, Texas Red and Cy5. The T1 block directs all laser lines (blue, yellow and red) to the sample, and the fluorescent green, red and far-red light is transmitted to the photomultiplier tubes. The T2A block contains the 560LP dichroic mirror, which splits the emission into green and longer (red) wavelengths. The 640SP diverter (permanently installed in 3-PMT systems) results in far-red light being directed into PMT3.

Live Cell Work - B1 and OPEN BLOCK Filter Blocks

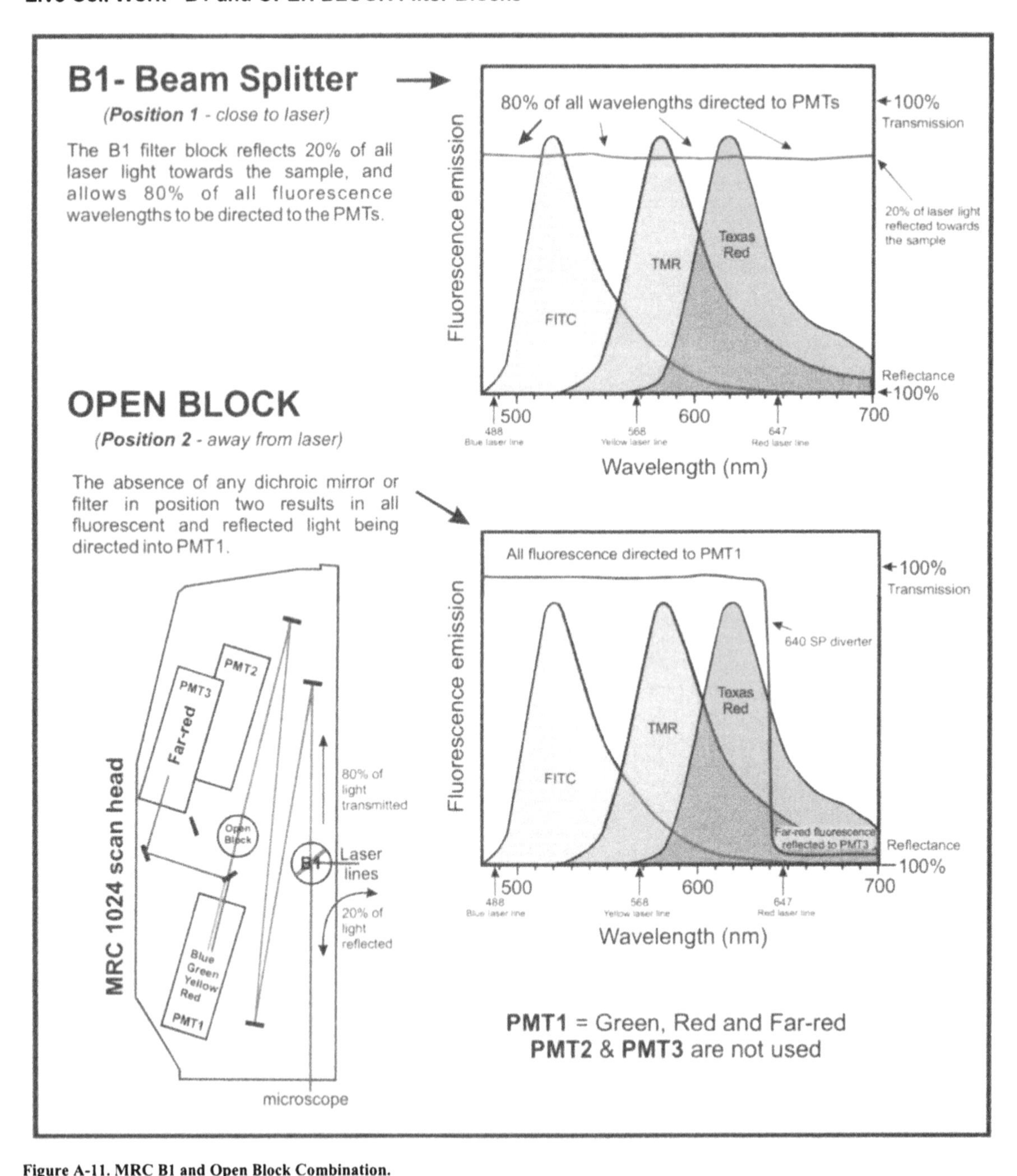

Figure A-11. MRC B1 and Open Block Combination.
The B1 and OPEN filter block combination is ideal for live cell work as 20% of all laser light is directed to the sample, and all fluorescent light is directed into PMT1 with approximately 80% efficiency (longer wavelength red light is directed to PMT3 by the 640SP diverter). This results in less laser light reaching the sample, but excellent efficiency in the collection of fluorescent light.

Back Scatter Imaging - B1 and T1 Filter Blocks

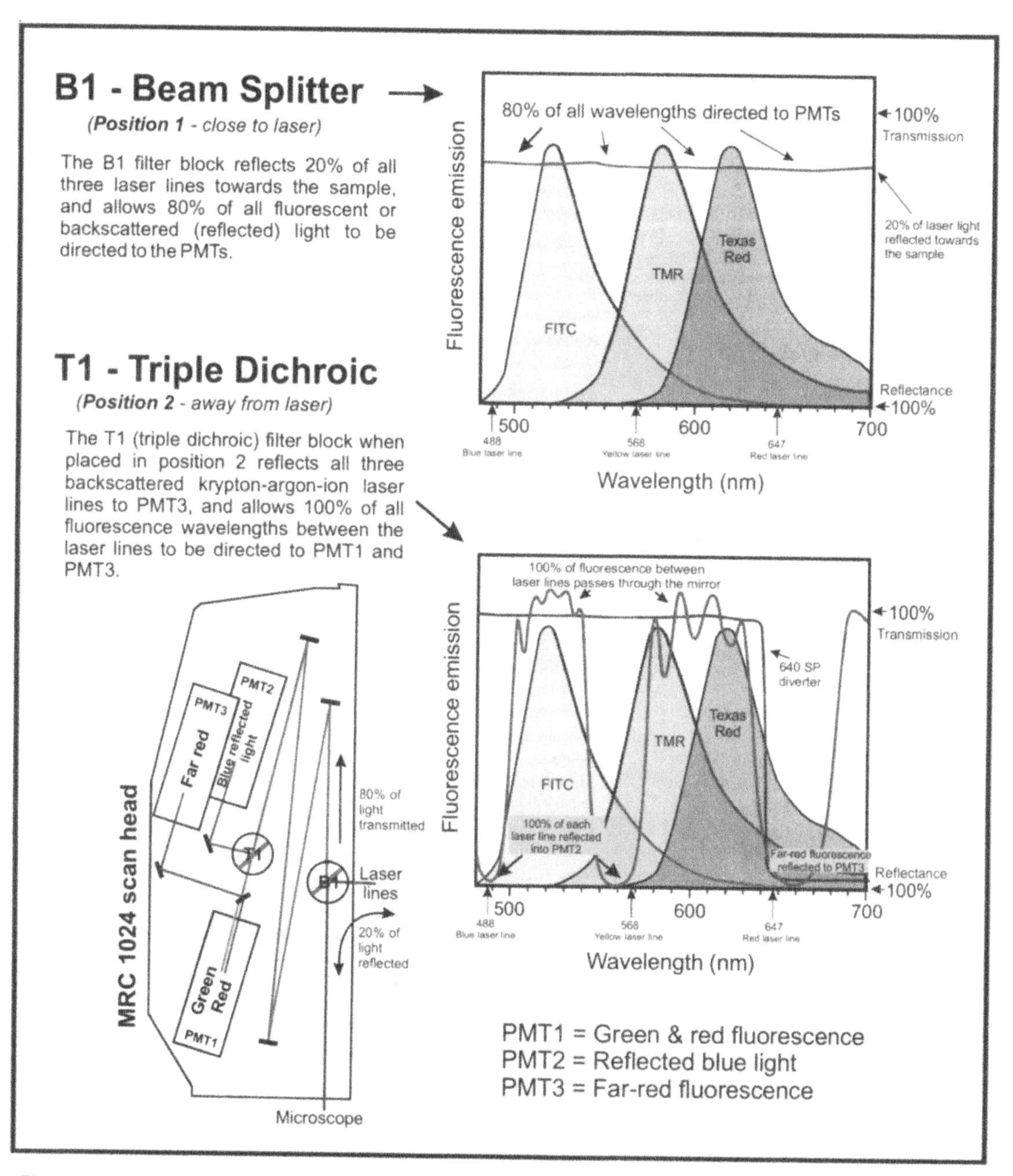

Figure A-12. MRC Back Scatter (Reflectance) Imaging.
Backscatter imaging, also known as reflectance imaging, is a highly sensitive method of imaging, which gives excellent resolution. Some biological material is naturally highly reflective, or specific structures can be imaged by using silver enhanced gold immunolabelling. In this filter block combination all backscattered (reflected) light is directed to PMT2, and all fluorescent light is directed to either PMT1 (red and green light) or PMT 3 (far-red light).

TROUBLE SHOOTING (MRC AND RADIANCE)

Problem	Cause	Remedy
Diminished sensitivity	Laser out of alignment	See page 6-11 of the MRC-1024 manual. **Warning** - do not attempt to align the laser unless you have been instructed how to do so.
	Scan head not correctly aligned into microscope	**MRC:** Adjust mirror M1 (see Figure A-6 on page 367). Take care to make only small adjustments while aligning the laser within the "bulls eye" of the prism on the microscope.
		Radiance: Refer to section 7.2.2 of the Micro Radiance operating manual. Two adjustments can be made within the "neck" of the scan head where the scan head attaches to the microscope. One adjusts the lateral alignment of the laser; the other adjusts the focus of the laser.
	Laser not correctly aligned into PMT tubes	**MRC:** Adjust mirror M4 to align laser into PMT1 (the red channel when using the T1/T2A filter block combination for dual/triple labelling). Adjust mirror M5 to align mirror into PMT2 (the green channel when dual/triple labelling).
		Radiance: Alignment into the PMT tubes (now housed in the controller box) is factory set and should not require adjustment by the user.
Uneven lighting of image	Laser alignment	See "diminished sensitivity" above.
	Objective not planer	An objective that does not have a completely flat field of view (a Planar lens) will result in dark areas of the image towards the edges. This problem is often more obvious on a low magnification lens. A lower quality lens will also display considerable unevenness in the field of view. Considerable improvement in the lack of planar field of view can often be accomplished by re-scanning the image at a higher zoom.
No signal in 3rd channel	Loss of 647 nm red line (Kr/Ar laser)	Check to see if the red line is present by setting the scan to 647 nm line only (using LaserSharp) and looking at the sample (from the side, not through the eyepieces of the microscope) while imaging. First wait for a further 15 minutes to make sure the laser is fully warmed up.
No transmission image	**Dark screen:** Pickup mirror incorrectly set	Swing the bar on the transmission mirror towards the back of the microscope. The transmission mirror is located above the condenser on an inverted microscope. The mirror position should be automatically set by the software when attempting to change the gain on the transmission setting when using the Radiance series of confocal microscopes.
	Bright screen: set gain and offset controls.	Use the SETCOL LUT to set the gain and black (offset) levels to their optimum settings.
Software problems	Software frequently "freezes"	Make sure that there is sufficient room on the hard disk for the system swap file (try to have at least 500 MB free).
Poor quality image	Immersion oil on a dry lens	Clean objective with ethanol and lens tissue to remove any residual oil that may have accidentally stuck to the lens.
	Incorrect setting on lens	Some lenses have a correction collar for oil, glycerol or water immersion, make sure that the correct setting is selected for the immersion media in use.
Jagged edges to lines	Vibration during scan	Mount the microscope and scan head on a vibration-damping platform. Placing the scan head and microscope on a metal plate that is placed on the top of two small bike tubes can create a simple vibration platform. The air cushion created will damp out most vibration.
Blurring of image when using Kalman collection	Vibration during scan	Vibration may not be obvious with a single scan, but may manifest itself as a blurred image after averaging several scans when using the Kalman collection mode. Do not test suspected vibration problems using live cells as slight movement of the cells during imaging can cause "blurring" effects.
	Cell movement when imaging lives cells	Try "SLOW" scan or line averaging instead of "Kalman" screen averaging for improving image quality.
Horizontal "bars" on images	Interfering electrical equipment	Electrical equipment, particularly computer screens can cause banding patterns on the collected image. Place any additional equipment or screens further away from sensitive electrical components.
	Laser unstable	Have laser serviced by service engineer. Press "Ctrl-L", or run the lasercheck script on Radiance systems, to obtain a listing of the anode current, laser power and hours used for the laser. This information will be useful for a suitable technician to diagnose the problem.

BIO-RAD LASERSHARP SOFTWARE

The latest version of LaserSharp software produced by Bio-Rad (LaserSharp2000, LS2K) is designed to run under Windows NT/2000 (currently used for the Radiance 2000/2100 and Micro Radiance confocal microscopes). A Windows NT version of LaserSharp is also available for the MRC-1024 confocal microscope. The spectral imaging Radiance 2100 Rainbow confocal microscope is operated with the Bio-Rad SpectraSharp software.

> **LaserSharp Analysis**
>
> The analysis component of the LaserSharp program can be loaded onto other computers for opening Bio-Rad *"PIC"* files with all microscope information intact

The original software used for controlling the MRC-1024 confocal microscope was called LaserSharp, and was developed to run under the OS/2 operating system. The OS/2 operating system was an operating system developed by IBM but is no longer being produced or supported. OS/2 did have significant advantages over the earlier versions of Windows and the DOS operating system. This was particularly important in the way the OS/2 operating system easily handled very large file sizes. However, with the advent of more advanced versions of the Windows operating system the OS/2 operating system is no longer advantageous. In fact there is now often considerable difficulty obtaining expert advice on the OS/2 operating system. If you have an older Bio-Rad MRC confocal microscope using the older OS/2 software you should contact Bio-Rad about upgrading to the latest Windows version. An emulation / data processing version of LaserSharp2000 is available for people still using an OS/2 based instrument. This software can be used to manipulate the "PIC" format files and give you some idea of the format of the new LaserSharp interface.

The LaserSharp software is divided into two components - acquisition and analysis. The acquisition component can only be run on the confocal microscope computer and is used to collect the images from the microscope and to control the scan head settings. The analysis part of LaserSharp can be used on the confocal microscope computer, or loaded onto another computer, which can free up valuable time on the confocal microscope.

Bio-Rad has also developed a separate program called LaserPix, which is designed as a "stand alone" image analysis program for confocal microscope images. This program is based on the popular image analysis program "ImageProPlus" with added features for confocal microscopy. LaserPix will run under Windows NT/2000 and can be loaded on a separate computer to the one attached to the confocal microscope. LaserPix does require a separate license and associated "dongle" before installation can proceed.

The following information is intended to give you an introductory overview of the LaserSharp software. If you do require more detailed information please refer to the Bio-Rad user manual, or probably a much better way – ask someone who is already using LaserSharp!

The proprietary Bio-Rad "PIC" file format can be read by using a Photoshop plugin (available for download from Bio-Rad, ftp://ftp.genetics.bio-rad.com/Public/confocal/). Confocal Assistant, a small free imaging processing program designed to handle Bio-Rad "PIC" format images, is also available from the Bio-Rad FTP site.

> **LaserSharp Login Procedure**
>
> LaserSharp software set up for a multi-user system should be installed with a password for each user to log on to the instrument. This password is not a security system, but simply a means of directing all of your collected images to your own subdirectory on the computer's hard disk. Anyone who has logged on under their own password has full access to your files! If you use the LaserSharp login procedure you will also have in your subdirectory a copy of all your "methods" and colour Look Up Tables (LUT) . These files are copied from the "Default" user directory when you are registered as a new user. "Methods" that you save will be retained in your own subdirectories. "Methods" developed by other users of the microscope will not be directly available to you – unless they are copied into your subdirectory.
>
> ***The Login password is NOT a security system!***

Image Collection Panel

The Bio-Rad LaserSharp software can be operated using a single screen, in which case the control panel and image being collected will be displayed as shown in Figure A-13. Alternatively, a double computer screen can be used with one monitor displaying the main menu and associated control panels and the other screen displaying the images as they are collected.

The images displayed on the Bio-Rad display screen are the output of one of 3 "mixers", rather than the output of individual collection channels. In many applications the output for mixer 1 would be simply 100% PMT1 and the output for mixer 2, 100% PMT2. However, one can readily mix different combinations of the channels. For example, when multi-channel labelling you can subtract, say, 10% of PMT2 from PMT1 by making Mixer 1 a combination of 100% PMT 2 and -10% of PMT1. This is a simple and dynamic method of eliminating unwanted bleed through in multi-labelling applications.

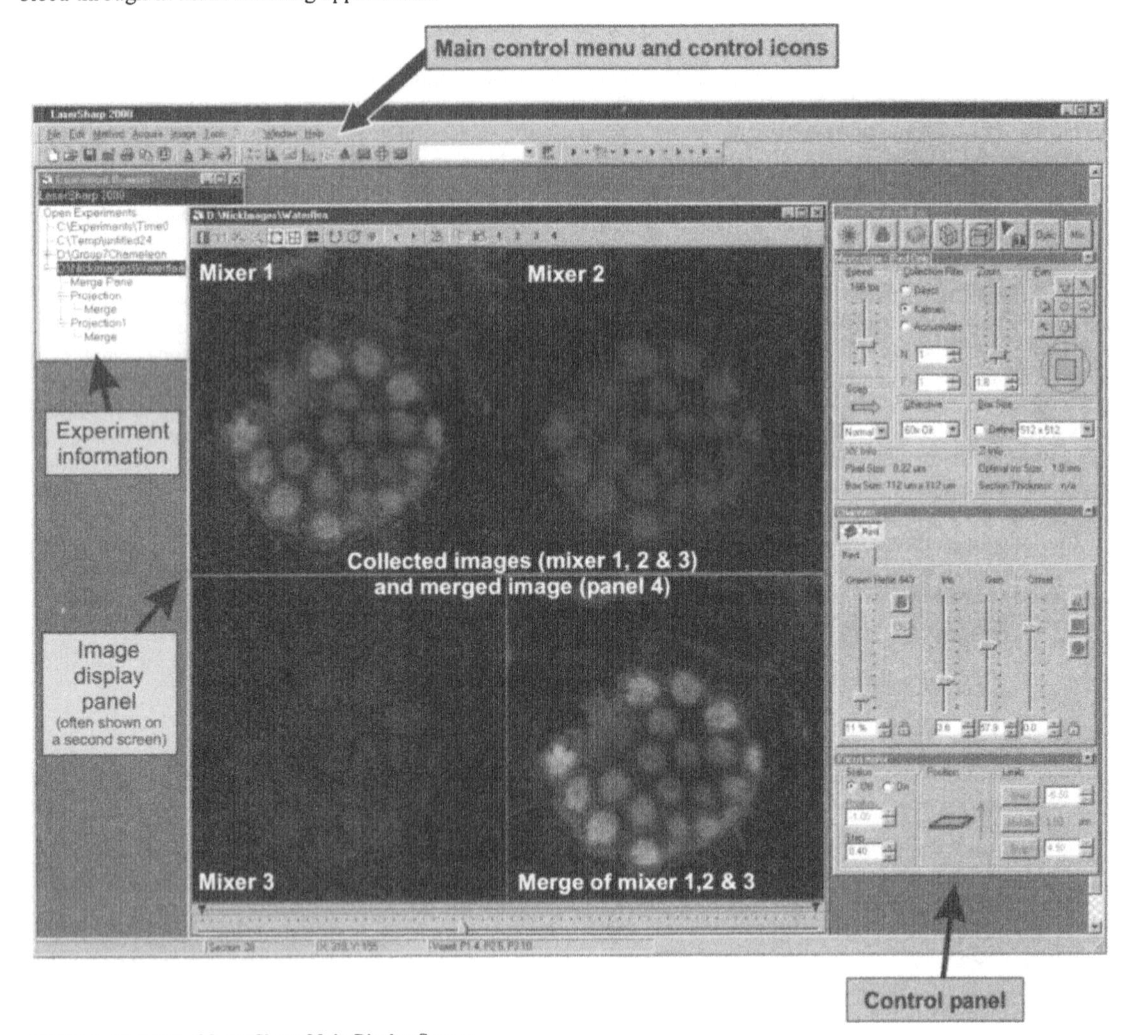

Figure A-13. Bio-Rad LaserSharp Main Display Screen
The Bio-Rad LaserSharp software can be used on a single computer monitor, in which case the control panel and collected images would be displayed as shown above. However, the software does support the use of two monitors, where the control panel and other menu options can be displayed on one monitor and the collected images displayed on a second monitor. On-screen image display is the output of individual "mixers" as described above, and is not necessarily the output from a single detection channel.

Main Control Panel

The main control panel (Figure A-14) is the central part of the LaserSharp software for controlling the Bio-Rad confocal microscopes, and displays in one simple panel all of the main control functions you will need to operate the instrument. Scanning is initiated by pressing the "Start scanning" button or simply depressing "F12" on the keyboard.

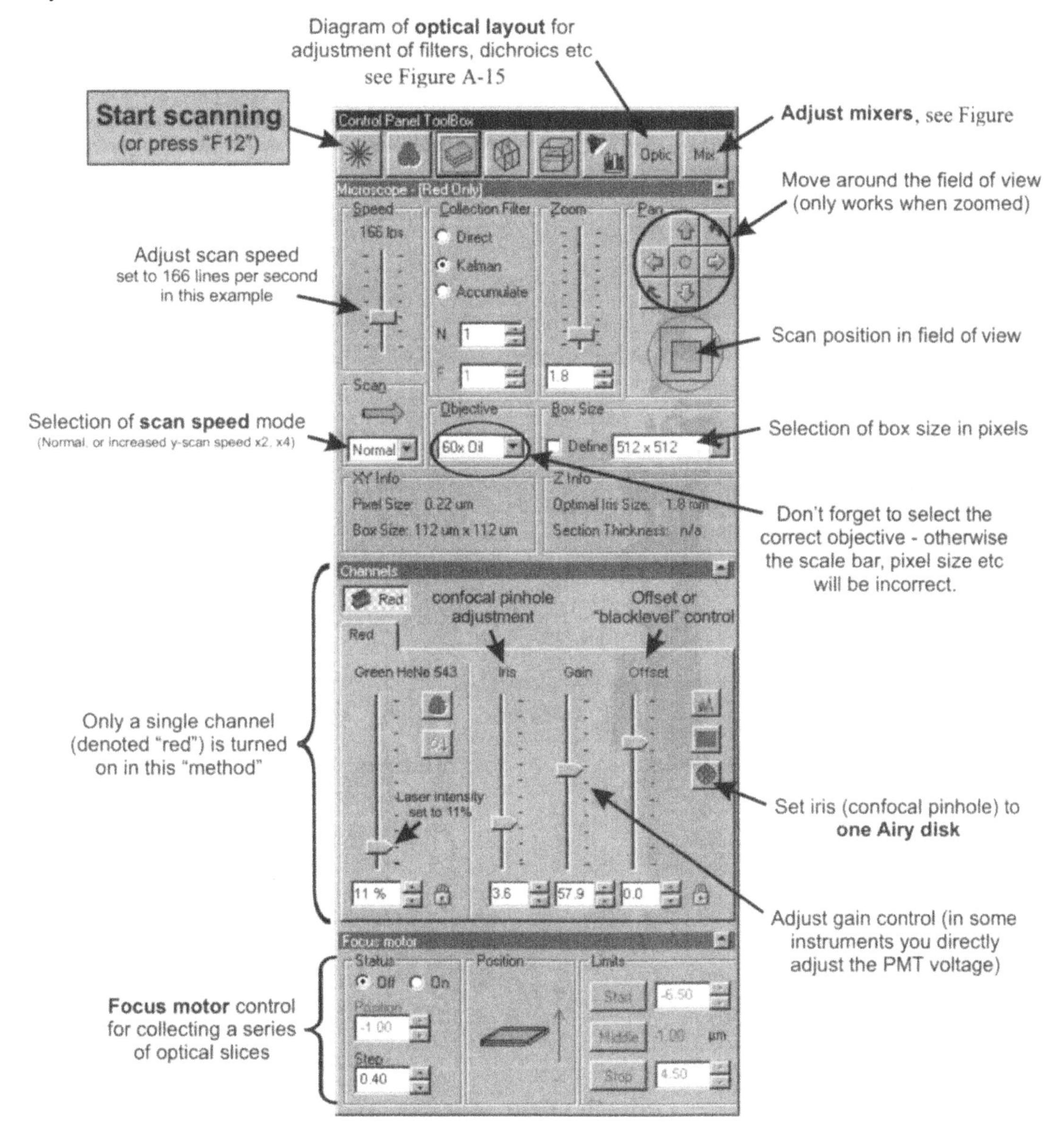

Figure A-14. Bio-Rad LaserSharp Main Control Panel
The Bio-Rad LaserSharp main control panel is the central window through which all of the other microscope controls are accessed. Pre-saved "methods" can be used to set up the controls on the instrument. You can also save your own instrument settings for later recall.

Optics Control Panel

The "Optic" control panel (see Figure A-15) displays graphically the optical settings for the confocal microscope. The settings shown in this diagram are "pre-set" when you select a specific "method". However, check these settings even when using a known "method" as one can quite easily inadvertently alter a method and still save the method under the same name. In fact, in multi-user facilities take great care in regard to filter settings when using specific methods if you don't have your own login and personal file and method allocations.

In this example the 405 nm violet laser diode (also known as a solid state violet laser) and the 543 nm green HeNe lasers are active. A long pass dichroic mirror (560DCLP) is used to split the fluorescent light into red (wavelengths longer than 560 nm) and green (wavelengths shorter than 560 nm). Various "blocking" and "emission" filters are used to further define the wavelengths allowed to enter each detection channel (only two, PMT2 and PMT3, are active in this example). The optical layout shown in the "Optic" panel is a combination of components located in both the scan head and the controller box.

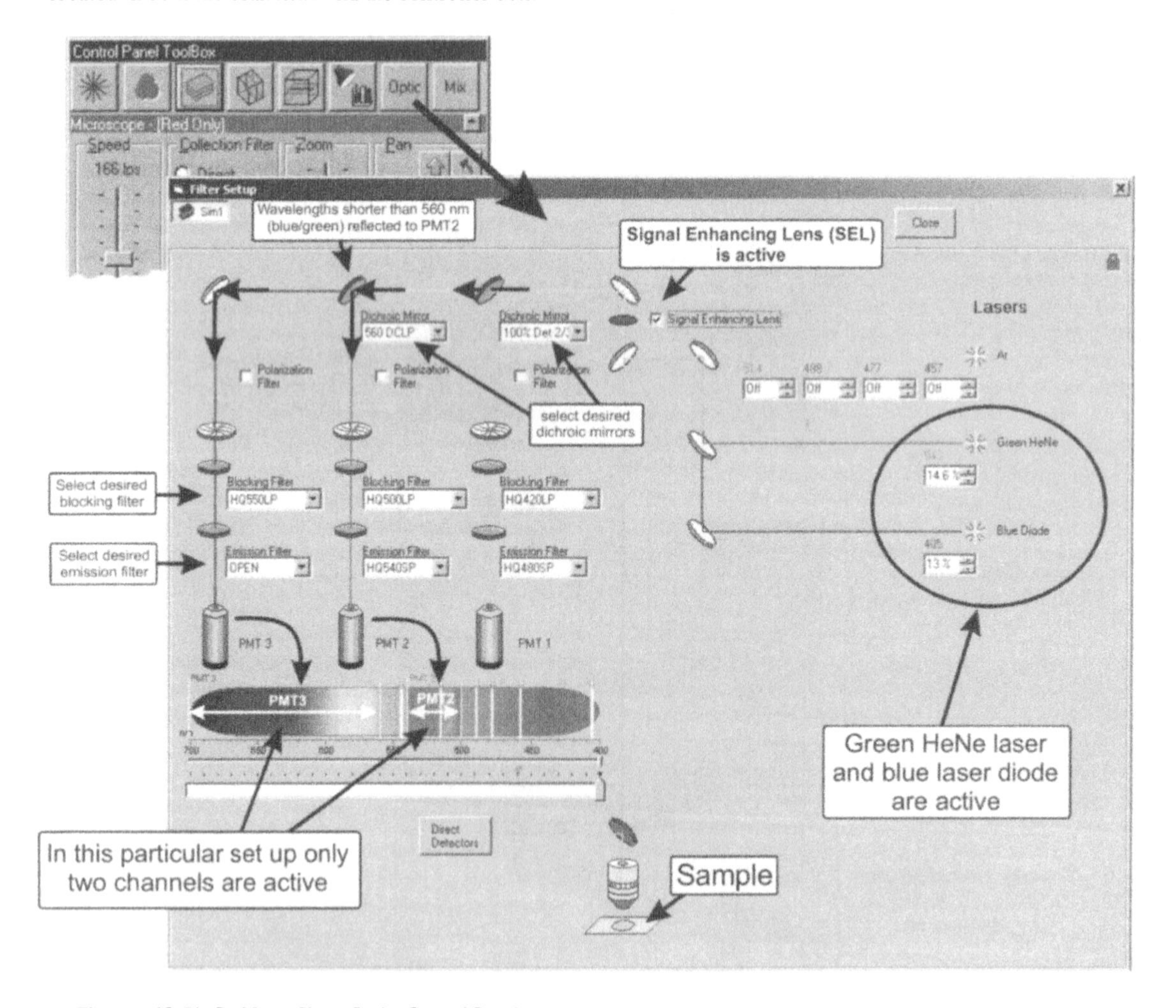

Figure A-15. Bio-Rad LaserSharp Optics Control Panel
Selection of the "Optic" control panel allows one to change the optical filters and dichroic mirrors using a graphic display of the optics layout of the instrument. In this example the blue laser diode (13% power) and the green HeNe laser (14.6% power) are activated. The "red" (PMT3) and "green" (PMT2) image collection channels are active.

Mixer Control Panel

The LaserSharp software has the unique characteristic that each of the collected images is created by the output of various detectors as determined by the user (or method used). Common practice is to assign 100% PMT 1 to mixer one, as shown below. However, other combinations of detectors can be assigned to mixer 1. For example, you could assign a level of minus 10% of PMT 2 to mixer one (in addition to assigning 100% of PMT 1 to mixer 1) to correct for bleed through when dual channel labelling. This ability to display the output as a combination of different detection channels is a powerful tool for real-time dynamic bleed-through correction.

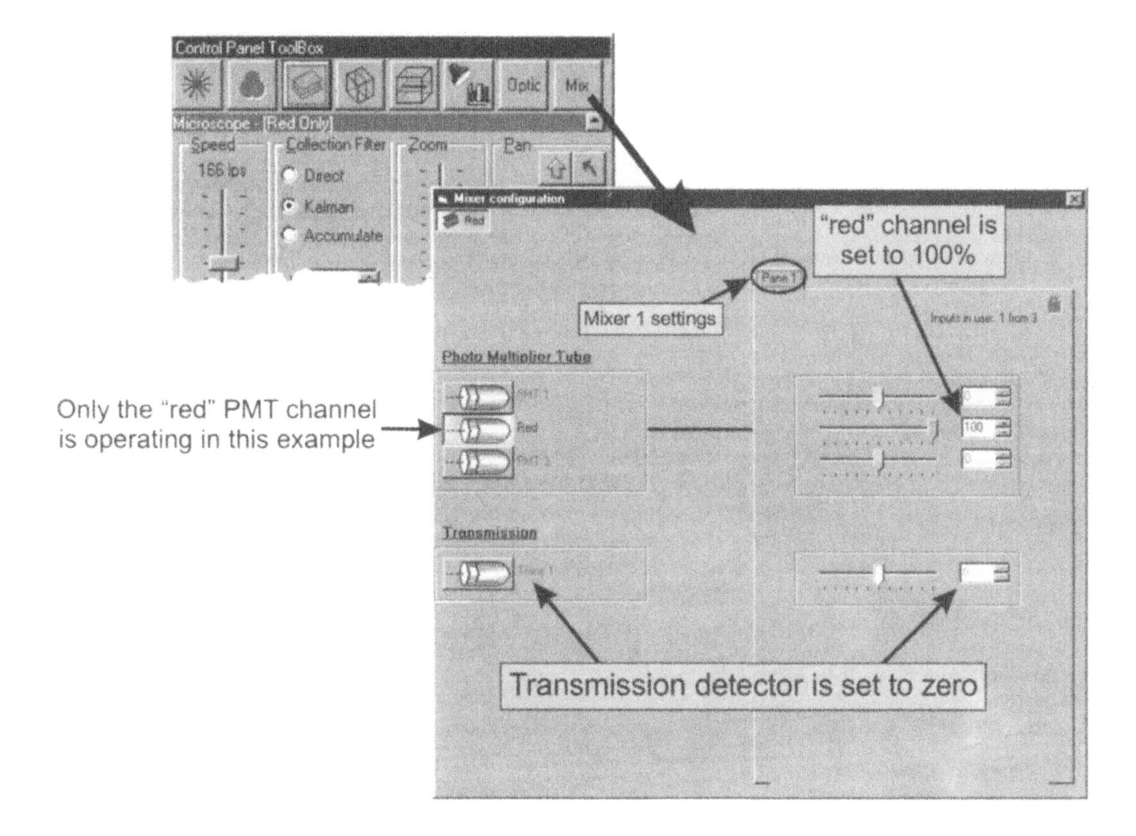

Figure A-16. Bio-Rad LaserSharp Mixer Control Panel
Each of the image display panels in the LaserSharp software is the output of various PMT tubes or transmission detectors (called Mixers). In the above example Mixer 1 is comprised of 100% PMT2, without any input from any other channel. However, one can assign set percentages of other detectors to any of the mixers, including subtracting a percentage of one PMT output from another PMT output (used in dual and triple labelling applications to subtract bleed-through).

Press "F12" to start scanning

Carl Zeiss Microscopy

The Carl Zeiss optics company has been producing superb optical instruments since the very beginning of the commercial development of optical instruments. The coming together of the Carl Zeiss lens manufacturing with the development of specialised optical glass from Otto Schott, and the theoretical input on optical design from Ernst Abbe resulted in a highly innovative company dedicated to high quality optics. Although the company has had a traumatic history, including the splitting of the company into an East and West German entity – today the company once again operates as a single unit. The production and development of the laser scanning confocal microscopes is located in Jena, the town where Carl Zeiss made his first microscopes.

ZEISS CONFOCAL MICROSCOPES

The latest laser-scanning instrument from Carl Zeiss is the LSM 510 META confocal microscope (Figure A-17). This instrument is capable of multiple channel fluorescence imaging using both conventional beam splitting optics and the innovative spectral imaging META system where multiple overlapping fluorophores can be imaged simultaneously.

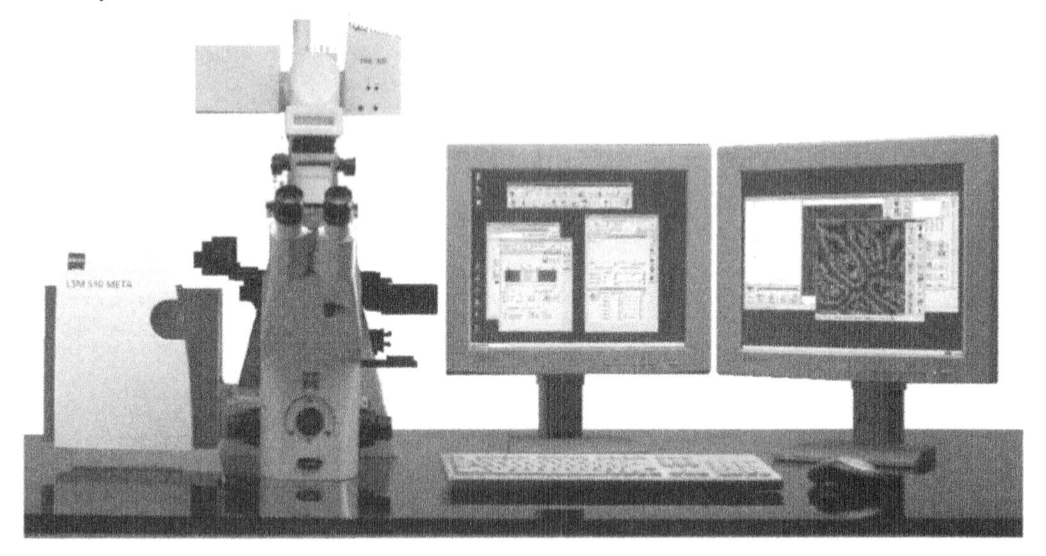

Figure A-17. Zeiss LSM 510 META Confocal Microscope.
The Zeiss LSM 510 META confocal microscope is designed for multi-channel fluorescence imaging in the biological sciences. The META scan head and associated optics is shown here attached to a Zeiss inverted research grade light microscope. The lasers, and the computer used to control the instrument are located under the microscope bench. This photograph was kindly provided by Gavin Symonds, Carl Zeiss, Australia.

Carl Zeiss **Web:** http://www.zeiss.de/lsm or www.zeiss.com/micro

Address:	Carl Zeiss, Mikroskopie D-07740 Jena **Germany**	Carl Zeiss Micro imaging, Inc. One Zeiss Drive Thornwood, NY 10594 **USA**	Carl Zeiss Pty. Ltd. 114 Pyrmont Bridge Road Camperdown, NSW 2050 **Australia**
Phone:	+49 3641 64 1616	+1 800 233 2343	1800 112 401
FAX:	+49 3641 64 3144	+1 (914) 681 7446	+61 (02) 9519-5642

Zeiss Confocal Microscopes

Laser Scanning Confocal Microscopy

Zeiss has continued to further develop and automate their confocal microscopes, with a number of innovative developments in recent years, including the development of the META – multi-array detector for spectral imaging.

LSM 510 META	Fully computer controlled confocal microscope with both "conventional" detection channels and the multi-PMT detector array (META channel), allowing the simultaneous detection of several fluorescent labels by analysing the fluorescence emission spectrum at each pixel within the image.
LSM 510	A fully computer controlled confocal microscope with up to 4 fluorescence detection channels.
LSM 5 PASCAL	Fully computer controlled "personal" confocal microscope available in a variety of configurations, including a choice of lasers and detection channels.
LSM 510 NLO LSM 510 META NLO	Multi-photon version of the above LSM 510 and LSM 510 META. Fully automated microscopes, with a mirror connected Ti-Sapphire laser providing ultra short pulsed infrared light for multi-photon microscopy.

Earlier Zeiss Confocal Microscopes

The earlier Zeiss confocal microscopes were high quality instruments with superb resolution, but with fewer detection channels, a lower degree of automation and somewhat lacking in sensitivity, and therefore of limited use in the biological sciences.

LSM 410/310	3- or 2-channel confocal microscopes, for use with inverted (410) or upright (310) microscope stands. Windows control of scanning modes and advanced graphics boards with mega pixel-resolution made these instruments suitable for both industry and biology.
LSM 10/44	2- or 1-channel confocal microscopes. Built in the pioneering 1980's, these confocal microscopes were primarily designed for use in industry.

Fluorescence Correlation Spectroscopy (FCS)

Fluorescence correlation spectroscopy is used to study molecular interactions by following random molecular movement within a defined microscopic volume.

ConfoCor 2	An FCS microscope for use in the biological sciences. This instrument can be combined with a variety of configurations of the Zeiss LSM 510 and LSM 510 META confocal microscopes.

Carl Zeiss has been producing microscopes since the very beginning of light microscopy in the mid-1840's. By the early 1900's Zeiss was a major manufacturer of high quality compound microscopes for both industry and the research community.

The company was split into two at the end of WWII, with West German Zeiss being established in Oberkochen (near Stuttgart, West Germany) with the transfer of a large number of technical and scientific staff from the Jena factory by the USA occupying forces. Zeiss continued in Jena as a state enterprise (VEB Carl Zeiss JENA) of the German Democratic Republic. On re-unification of Germany, Zeiss was again brought back together, with the production of microscopes continuing in both towns, but the research and development and continued innovative changes to the confocal microscope being established in Jena.

Zeiss first commercially produced a laser scanning confocal microscope (LSM 44) in 1982. This was upgraded with the LSM 10 in 1988. The early Carl Zeiss confocal microscopes were designed for use in materials science (particularly microchip manufacturing) and were not sufficiently sensitive to be of general use in biology. With the introduction of the windows controlled 2-channel LSM 310 (1991) and 3-channel LSM 410 (1992) with individual pinholes, and greatly increased sensitivity, the Zeiss confocal microscope became highly regarded in the biological research community.

The LSM 510 in 1997 has improved on the original excellent design, particularly with the addition of flexibility, speed and greater functionality of the software. This instrument is a highly versatile and very sensitive multi channel confocal microscope.

The LSM 510 META is capable of spectral imaging using a multi-PMT array. This innovative design allows the simultaneous imaging of several fluorophores without the necessity for optical filters etc to collect specific "windows" of light. Several fluorophores can be separated into different "channels" by automated analysis of the spectrum obtained. As many as 6 fluorochromes can be readily separated in this manner, in addition to two "conventional" single PMT detection channels.

Carl Zeiss also produces a high-quality but less versatile instrument called the LSM 5 PASCAL confocal microscope. This instrument is of similar design to the Zeiss 510 META, but without the META multi-PMT array channel. The LSM 5 PASCAL is significantly less expensive compared to the LSM 510 META confocal microscope, being designed for use in individual laboratories as a "personal" confocal microscope, rather than being based in large University or Research Institute facilities.

The ConfoCor2 fluorescence correlation spectroscopy instrument is a specialised addition to the conventional confocal microscope for studies on molecular interactions and movement of macromolecules.

ZEISS LSM 510 META CONFOCAL MICROSCOPE

The Zeiss LSM 510 META confocal microscope, as mentioned above, is a conventional confocal microscope based on the use of dichroic mirrors and optical filters with an additional META spectral imaging channel. The META channel can be used as a multiple channel imaging device, but the real power of the META channel is the ability to separate the fluorescence emission from several very similar fluorochromes by a process called linear unmixing – allowing one to perform multiple labelling experiments using remarkably similar probes.

A wide range of visible light and UV lasers can be connected to the Zeiss instruments. Modified versions of the LSM 510 META confocal microscope are also available for use with pulsed infra-red lasers for multi-photon imaging.

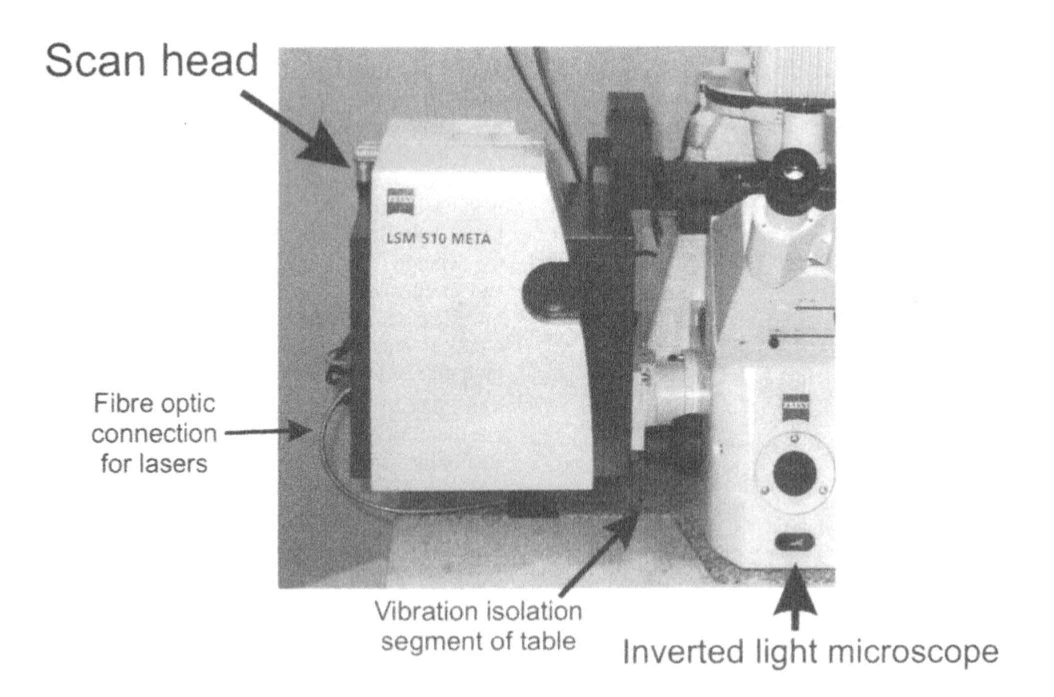

Figure A-18. Zeiss LSM 510 META Scan Head.
The LSM 510 META confocal microscope scan head is shown attached to a Zeiss research grade fully automated inverted microscope. The scan head contains the scanning mirrors, dichroic mirrors, optical filters, detectors (including the META 32-PMT array detector) and the confocal pinholes. The lasers are located remote from the scan head, with the laser light being directed into the scan head via fibreoptic connections. In this photograph the microscope and attached confocal microscope scan head are shown mounted on a vibration isolation segment of the main instrument work desk. This photograph was taken at the Baker Heart Research Institute, Melbourne, Australia.

LSM 510 META

Components of the Zeiss 510 META

The Zeiss LSM 510 META laser scanning confocal microscope can be configured for visible light, UV light or multi-photon microscopy on either upright or inverted microscopes.

Design

The Zeiss LSM 510 META instrument is based on both a multi-channel META detector and two dichroic mirror / optical filter based channels. A very high-end confocal microscope that has great versatility, but due to the relatively high cost is often located in central facilities rather than individual laboratories.

Scan Head

The Zeiss LSM 510 META scan head is mounted directly onto a Zeiss research grade inverted or upright microscope. The scan head houses all of the optics, pinholes and detectors necessary for confocal imaging. The scan head can be moved from inverted to upright microscope if required.

META detection channel: The META detection channel consists of a spectral separation diffraction grating and a 32-PMT array that can be used as additional channels for fluorescence imaging, as a multi-channel detector – or as a spectral imaging channel for multi-colour labelling. One of the unique advantages of the META detector is that a computer algorithm called "linear unmixing" can be used to separate fluorophores with very similar emission spectra.

Conventional detection channels: Two "conventional" detection channels that employ dichroic mirrors and optical filters to separate out different wavelengths of fluorescent light. Independent photomultiplier tubes for each channel are used to detect the fluorescent light.

Dichroic mirrors: Computer selected dichroic mirrors to both separate the fluorescence emission from the excitation light (the primary dichroic mirror for conventional and META channels) and to further separate the fluorescent light into various regions of the light spectrum (only used on the conventional channels).

Barrier filters: A computer-controlled wheel of removable barrier filters is mounted in front of each of the conventional photomultiplier tubes. The META detection channel does not require barrier filters as the light is separated into the light spectrum by a diffraction grating.

Variable size confocal iris: A computer-controlled variable size "pinhole" is mounted in front of each detector channel, including a single variable size confocal iris for the META channel. For optimal resolution and sensitivity, the META pinhole can be aligned in x, y & z and the other pinholes in x & y.

Light detectors: Individual photomultiplier tubes are used for each of the "conventional" detection channels, and an additional 32-PMT array is used for the META detector. Up to 4 optional external non-descanned PMT detectors are available for detection of fluorescence generated by multi-photon excitation.

Scanning mirrors: X & Y fast galvo scan mirrors with zoom, bi-directional & scan rotation capabilities.

Lasers

Lasers: A wide range of lasers can be attached via fibre optic connection, including the Kr/Ar, Ar-ion, HeNe, 405 nm violet diode laser and UV-Ar-ion lasers. Ti:sapphire multi-photon lasers can be attached to suitably configured instruments via fibre optic connection or direct coupling.

Line selection and laser light intensity attenuation: Laser lines are selected and may be rapidly switched using acousto-optical tuneable filters (AOTF) under software control. AOTF filters also control laser intensity (giving accurate levels of between 0 and 100% of the current laser power setting).

Microscope

The scan head is attached to either an upright or inverted Zeiss automated research microscope. High quality Zeiss objectives are most important for producing a high-resolution confocal image. The scanning laser light enters the microscope just below the objective in an inverted microscope, resulting in an image quality at least as good as that on an upright microscope.

Imaging Modes

Multi-channel fluorescence, backscatter, transmission. Image collection box sizes user selectable from 1 x 4 up to 2048 x 2048 pixels. Scanning modes include spot, line, free-hand line, ROI, z-stack, zoom, panning and rotated scanning. These may be combined with time lapse imaging and spectrally resolved imaging (Lambda Stacks). Simultaneous and sequential multi-channel scanning optionally combined with localized bleaching. Online calculations such as emission ratio and linear unmixing (Online Fingerprinting). 12-bit data acquisition and processing, saved as 8-bit or 12-bit image files.

Transmission

Single channel transmission detection using a photo multiplier detector. DIC, Phase and VAREL images can be obtained using the transmission detection channel and suitable microscope optics.

Manual Control

The instrument is fully computer controlled with no programmable manual control device.

Computer

High-end Windows based computer used to control the confocal microscope and collect the images. Dual computer screens may be used to display the control panel and images on separate monitors.

Software

The Zeiss proprietary software for controlling the microscope, image collection, 2D to 4D image manipulation, 3D- and 4D reconstruction, visualization, animation and image file management system is operated under Windows NT/2000. The software can be readily loaded onto other instruments not associated with the microscope (such as a laptop) and the basic image database management utilised to access both the images and the information on microscope settings that are stored with the images.

Models

LSM 510 META - with META and "conventional" channels. **LSM 510** – various configurations with up to 4 single-channel PMT detectors. **LSM 510 NLO** and **LSM 510 META NLO Multi-photon** – having both confocal and multi-photon capabilities; optionally equipped with up to 4 additional external PMT detectors. **LSM 5 PASCAL** – "personal" confocal microscope for single-user applications.

Zeiss LSM 510 META Scan Head

The relatively large LSM 510 META scan head (Figure A-18) contains all of the optical components required for scanning the laser beam across the sample and separating the fluorescence emission into various regions of the light spectrum. This includes the dichroic mirrors, optical filters, diffraction grating, META detector and photomultiplier tubes. The Zeiss LSM 510 scan head is directly connected to a conventional Zeiss research grade light microscope, both of which are mounted on a vibration isolation segment of the microscope work bench. The scan head can be readily moved between various microscopes if required. The scan head can be connected to either manually operated or fully automated microscopes. When the scan head is connected to a fully automated microscope, changing between bright-field imaging and confocal imaging is a simple matter of pressing a "button" on the Zeiss LSM controlling software. The scan head is fully motorised, allowing full computer control.

The lasers used for irradiating the sample are connected to the scan head via fibreoptic connections, except for the pulsed infrared laser used in multi-photon imaging, which is directly optically coupled to the scan head.

A schematic diagram of the optical components of the Zeiss LSM 510 META scan head is shown in Figure A-19, with the on-screen diagram used for changing various scan head settings shown in Figure A-21. A number of different primary dichroic mirrors can be software selected (for example, the triple dichroic mirror that reflects 488, 543 and 633 nm light as shown in Figure A-21) for the separation of the irradiating laser light from the fluorescent light emanating from the sample, shown as optical filter wheel (1) in Figure A-19 and Figure A-21. A second removable mirror, shown as optical filter wheel (2) in Figure A-19 and Figure A-21 is used to direct the fluorescent light to channel 2 and 3 (Ch2 and Ch3). The mirror shown in position "2" in Figure A-21 can be replaced by a dichroic mirror to direct specified regions of the emission spectrum to either of these two channels. Further specified wavelengths can be directed to the META channel, denoted ChS in Figure A-19 and Figure A-21 by choice of the correct dichroic mirror at position "2" in Figure A-21. Alternatively, the mirror at position "2" can be removed altogether to allow all fluorescent light to be directed to the META channel.

A third dichroic mirror filter wheel, denoted optical filter wheel (3) in Figure A-19 and Figure A-21 is used to separate the emitted fluorescent light into different regions of the light spectrum for detection in either channel 2 (Ch2) or channel 3 (Ch3). The META channel utilises a diffraction grating, denoted (7) in Figure A-19, to separate the emitted fluorescent light into its constituent wavelengths, which are then detected using the 32-PMT array META detector (ChS in Figure A-19 and Figure A-21). There are independently controlled confocal pinholes for each of the "conventional" detection channels, but only one confocal pinhole for the META channel (Figure A-19). Alignment of the confocal pinholes associated with each of the single PMT detection channels can be adjusted in x and y using software control, and the pinhole use for the META detector can be adjusted in x, y and z directions. Pinhole alignment is important for gaining maximum sensitivity from each of the channels -- and should be performed each time the dichroic mirror/optical filter settings are changed.

Lasers

A wide variety of visible light lasers can be readily connected via fibreoptic coupling to the Zeiss 510 META scan head. This includes the commonly used lasers such as the argon-ion and krypton-argon-ion lasers and helium-neon lasers. A wide range of lasers and laser lines can then be software selected, including high-speed switching for line by line Multi Track imaging.

UV lasers can also be connected by fibreoptic coupling to the scan head. These lasers are usually relatively high-powered argon-ion lasers in which the shorter wavelength UV lines are selected. There is considerable difficulty (and danger) associated with using UV lasers for confocal microscopy. The relatively short wavelength solid-state lasers, such as the 405 nm violet laser diode, may eventually replace most uses of the UV laser.

A pulsed infrared laser for multi-photon microscopy can also be connected to the scan head. Zeiss has experimented with both fibre optical coupling and direct connection of the laser, although the simpler direct coupling of the laser is favoured for most installations.

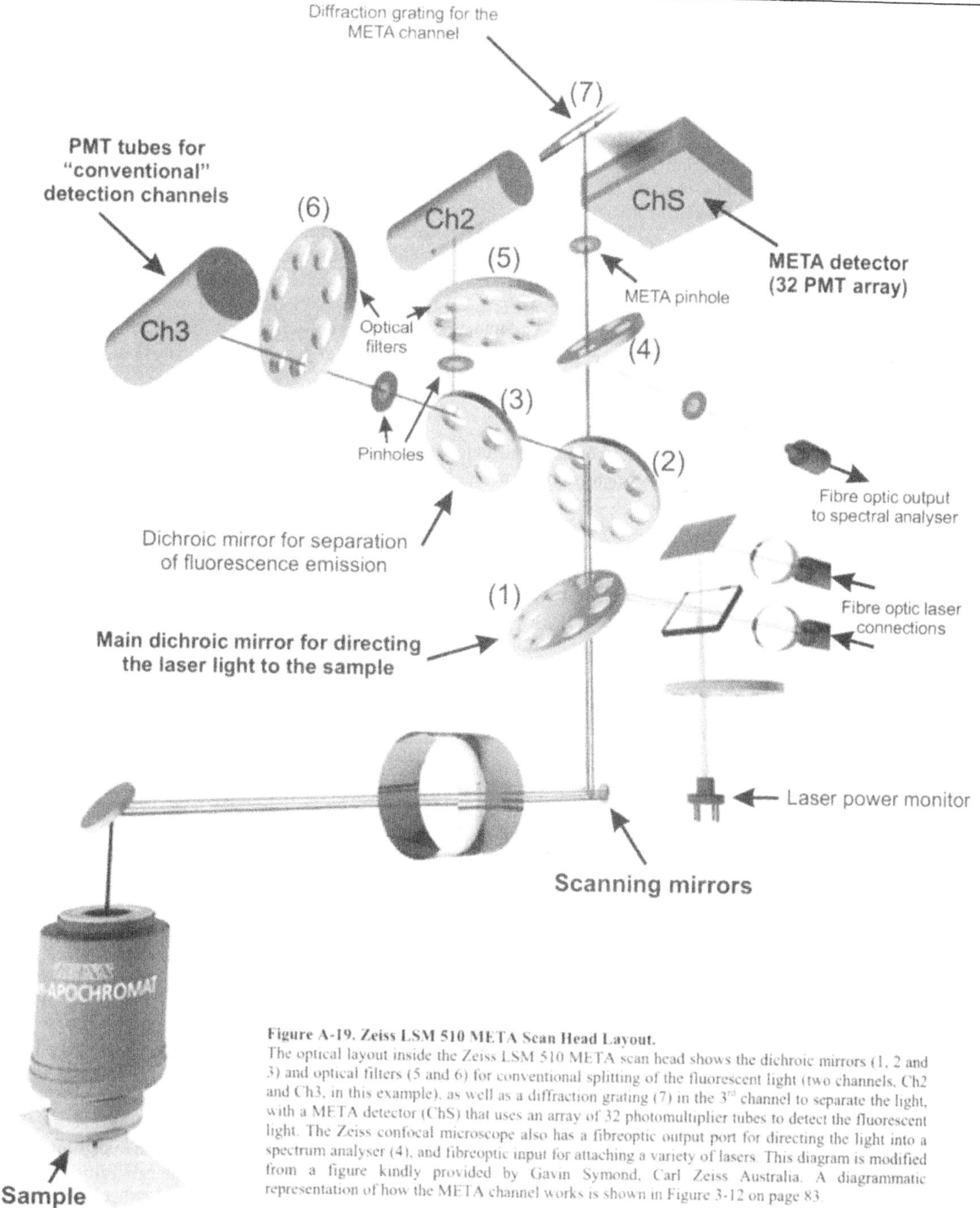

Figure A-19. Zeiss LSM 510 META Scan Head Layout.
The optical layout inside the Zeiss LSM 510 META scan head shows the dichroic mirrors (1, 2 and 3) and optical filters (5 and 6) for conventional splitting of the fluorescent light (two channels, Ch2 and Ch3, in this example), as well as a diffraction grating (7) in the 3rd channel to separate the light, with a META detector (ChS) that uses an array of 32 photomultiplier tubes to detect the fluorescent light. The Zeiss confocal microscope also has a fibreoptic output port for directing the light into a spectrum analyser (4), and fibreoptic input for attaching a variety of lasers. This diagram is modified from a figure kindly provided by Gavin Symond, Carl Zeiss Australia. A diagrammatic representation of how the META channel works is shown in Figure 3-12 on page 83.

Single Track and Multi Track Image Collection

Image acquisition can be performed using the single tack mode (shown in Figure A-21), where a single scan of the sample is made using either single or multiple channel acquisition. One can also collect images in the "Multi Track" mode (the "Multi Track" tab is shown in the Configuration Control panel in Figure A-21), where individual screens or lines are collected using specified collection parameters. For example, when dual labelling using Cy2 and Cy3 it is possible to collect both channels simultaneously using the "Single Track" mode. However, if bleed through is a problem then each individual channel can be excited separately and the fluorescent light collected separately using the "Multi Track" mode. This can be done by collecting, for example, the Cy2 fluorescence using all necessary settings for this dye (including the optimal laser line for excitation), and then collecting the subsequent image using settings (and appropriate laser line) for the Cy3 dye. Sequential frame collection is fine when collecting images from fixed material, but when imaging live cells any slight movement between collection frames will detract from accurate image registration when comparing areas of colocalisation. A much better way of sequentially collecting dual labelled samples is to use the "Multi Track" mode using line collection. In this case each individual scan line is collected using a different laser line, with only one channel active at each line scan. The resultant images will have minimal bleed through between individual collection channels.

Conventional Detection Channels

In addition to the META channel, described in detail below, the Zeiss LSM 510 META confocal microscope scan head also contains two conventional detection channels. Each conventional channel contains a single highly sensitive photomultiplier tube that is designed to detect any light that is directed to it by means of dichroic mirrors and optical filters. Each conventional detection channel has a confocal pinhole that can be aligned (in x and y) and varied in diameter. The gain and offset can also be adjusted independently for each photomultiplier tube. These single channel detectors can also be combined with the META channel to create a highly versatile instrument with a large number of imaging channels.

META Channel

The META channel consists of a diffraction grating, which is used to split the fluorescent light into its component colours, and a multi-PMT array (32 individual PMT tubes arranged in an array) to detect the light. The META channel can be used as up to eight "conventional" channels (in addition to the single PMT channels already available on the instrument), or a method called "linear unmixing" can be employed to derive the contributions from the individual fluorophores by means of spectral information.

The META channel, when used as a multi-band pass optical filter (Figure A-25), has on-screen sliders to delineate the region of the light spectrum that will be collected into a set number of individual photomultiplier tubes in the multi-PMT array. In this way the META detector can be used as additional detection channels (up to 8 channels). The conventional single channel detectors (described above) can be used in combination with these META imaging channels. However, all these additional channels are limited by bleed-through between individual channels. This will depend on the fluorophore combinations used, but as a general rule up to 3 and possibly 4 different fluorophores can be separated in this manner. Once you attempt to separate more fluorophores the narrow spectral band allocated to each channel no longer contains fluorescent light unique to that particular fluorophore.

The META channel can also be used in a very different mode, called emission fingerprinting, to separate several overlapping fluorescence emission spectra. In this mode a series of images is collected, which is called a "lambda stack". Each image in the stack corresponds to a single PMT within the 32-PMT array of the META detector. A full stack would consist of 32 images (collected as four scans using eight detectors in each pass) covering a defined range of the visible light spectrum. In this way spectral information is collected for each pixel in the image. A narrower region of the light spectrum can be collected by using fewer individual detectors in the multi-PMT array, or the collection step size can be increased by binning the output of individual PMTs on the array.

A computer process called linear unmixing is used to separate out the contribution made to the spectrum from several fluorophores with highly overlapping fluorescence emission spectra. The instrument can utilise stored spectral data to determine the contributions from each of the individual fluorophores, or you can collect individual spectra from different regions of the sample or from single labelled control specimens. The remarkable separation achieved by the META channel means that not only can many more fluorophores be imaged simultaneously, but that the fluorophores can all be green! As long as there is a recognisable spectral difference between the fluorophores they can be separated by this method. The final output from the META channel (after linear unmixing) is not the spectral information, but an individual image from each of the designated linearly unmixed channels.

If the META channel is as remarkable as it sounds then why bother at all with conventional channels? The answer is that there are limitations to the use of the META channel. These include issues of detector sensitivity, the limitations of spectral separation, and the constraints of a single pinhole and only one PMT gain and off-set control for the multi-channel META detector.

The Zeiss META PMT array detector and associated electronics does produce a detection channel with high sensitivity and excellent signal to noise ratio. However, there is often a discernable difference between the sensitivity of the conventional single PMT detection channels used on the Zeiss LSM 510 instrument compared to the META channel. An apparent lower sensitivity in the META channel may be due in part to differences between the two types of PMT tube, but perhaps more importantly there are other design differences between the two channels that could account for the observed differences.

One constraint associated with a PMT array is that the light for each channel is now spread over several PMT tubes instead of being collected by a single PMT tube. When using conventional dichroic mirrors and optical filters, light from a specified region of the light spectrum (which in some cases may be relatively broad) is directed to a single PMT tube. In the case of the META channel the light is collected by up to 32 PMT tubes. When there is an abundance of light the META channel will perform superbly, but as the light becomes much less, the limit of sensitivity of the META channel will be reached before the limit of sensitivity for a conventional channel is reached.

Another important difference between the META channel and the conventional channels is that a diffraction grating (with associated loss of light) is used to separate the various wavelengths in the META channel. There will also be some loss of light when using dichroic mirrors and optical filters in the conventional channels, but the amount of light lost will depend greatly on the type of optical filter and the number of filters or dichroic mirrors used.

Another important consideration when using the META channel is that there is only a single variable size pinhole and a single detector gain and offset control for the whole PMT array. This means that care needs to be taken to "balance" the fluorescence emission of each fluorophore when using several fluorescent probes. Conventionally this "balance" is achieved by adjusting the individual detector gain levels for each imaging channel. When using the META channel you will need to adjust the fluorescence emission of each fluorophore by adjusting the individual laser line intensities or by adjusting the relative concentrations of the individual fluorophores. If there are significant differences in the fluorescence intensity between individual fluorophores when using the META channel, then using 12-bit instead of 8-bit data acquisition will result in a significant improvement in the image from the lower intensity fluorophore.

A very powerful feature of the META channel is the ability to quickly collect the individual reference spectra from a real live sample, for use in the linear unmixing algorithm. This would at first appear to eliminate any discrepancies between the stored spectral data and your sample. However, many dyes show spectral changes depending on the environment of the dye. This characteristic is exploited in a number of dyes (for example, the changes from green to red fluorescence, depending on the membrane potential of the mitochondria when using a number of mitochondrial dyes). Cellular ions, including pH changes also often influence not only the level of fluorescence (fluorescein, for example, has a much higher level of fluorescence at alkaline pH) but may also alter the wavelength of emission (acridine orange ranges from green through to red fluorescence, depending on pH and interaction with other molecules). This variability in the emission spectrum of the dye being used could result in misleading results if you use an emission spectrum derived from an inappropriate control for use in the process of linear unmixing. However, you can also exploit these differences in florescence emission to obtain valuable information about the local environment of the dye.

Image Collection Modes on the Zeiss LSM 510 META

The Zeiss LSM 510 META confocal microscope is capable of imaging a wide range of fluorescent molecules using a variety of spectral separation techniques.

Single detector channels – *light is directed to individual photomultiplier tubes using conventional dichroic mirrors and optical filters. Two individual channels are available.*

Single channels + META detector– *the META detector can be used as an additional detection channel.*

META detector as a band pass filter – *the META channel can be used to collect up to 8 different regions of the light spectrum.*

META emission fingerprinting – *the META channel can be used to collect the emission spectrum of the sample and to then use a process called linear unmixing to derive the components of the light that can be attributed to individual fluorophores.*

META + 2-photon excitation fingerprinting – *emission spectra are collected using a series of 2-photon excitation wavelengths. Linear unmixing is used to derive individual fluorophore components.*

The unique fluorescence separation method used in the META channel is a very powerful tool in confocal microscopy that will allow you to utilise fluorophores that could not be separated by more conventional optical means. The META channel, in combination with the conventional single PMT channels available on the Zeiss 510 META, results in a highly versatile instrument for biological imaging.

Transmission Imaging

The Zeiss LSM 510 META confocal microscope is capable of simultaneous transmission imaging, including Phase and DIC imaging if the correct optical elements are installed on the attached light microscope. The Wollaston prism used for DIC imaging in the Zeiss microscope does not appear to affect the quality of simultaneously collected fluorescence images. The light used to create the transmission image does not pass through the confocal pinhole and therefore the transmission image is not confocal.

ZEISS CONFOCAL MICROSCOPE CONTROL SOFTWARE

The software used to control the Zeiss confocal microscope has a graphical based user interface that displays all of the imaging and microscope parameters in a number of display panels. The software is reasonably easy to use, but the complexity of the instrument does mean that a great deal of information often needs to be presented on-screen – creating some confusion for people not familiar with the instrument. The program is designed to operate on dual screens, with the control panels displayed on monitor 1 and the images displayed on monitor 2. The software can be used in two different configurations, the "expert" mode, where all of the controls are available to the user, and the "routine" mode, where only a select group of relevant controls for the particular method selected are displayed. The examples shown on the following pages are all taken from an instrument in which the "expert" mode was selected.

The image manipulation and database component of the Zeiss image collection software (Zeiss Image Examiner can be loaded onto a separate computer (available from the Zeiss web site free of charge) to allow easy access to images collected on the Zeiss confocal microscope.

Main Control Panel

When using the Zeiss software a small concise main control panel is visible as a box on your desktop (Figure A-20). The three buttons on the bottom right of the control panel (data collection mode) allow one to readily switch between conventional bright-field or epi-fluorescence microscopy (VIS) and laser scanning confocal microscopy (LSM). If you are using a fully automated light microscope these buttons make all the necessary changes to allow you to either look down the eyepieces of the microscope or to use the laser scanning microscope, without having to manually change optical filters and settings as you move between the two imaging systems.

The Zeiss software utilises an excellent filing system for managing your images (accessible via the "file" button, see Figure A-20). This database provides information on microscope and scan head settings when the image was collected, a good thumbnail display of your images, and the opportunity to add notes etc to annotate the images.

The collection of images on the Zeiss 510 confocal microscope is achieved by activating the "Acquire" button (Figure A-20) on the main control panel. Activating the acquire button will display 8 simple buttons that are used to operate the microscope, the lasers, and the scan head for image acquisition. Detailed information on the menu options available for the "Micro", "Config" and "Scan" buttons are shown in Figure A-21, Figure A-22 and Figure A-23 respectively.

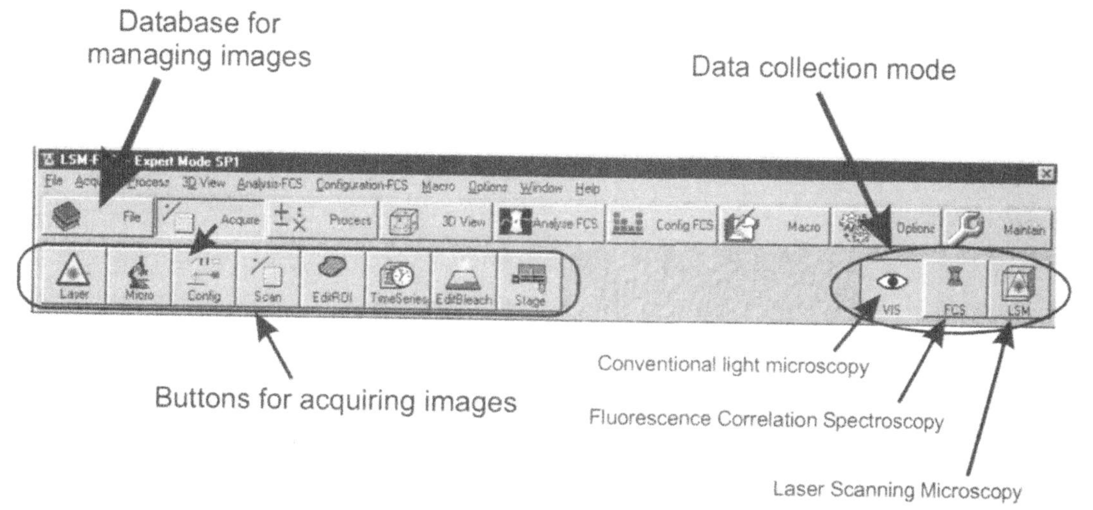

Figure A-20. Zeiss LSM 510 Main Control Panel.
The Zeiss LSM main control panel when using the "expert" mode is a small "box" that sits directly on your computer desktop. This panel gives you access to all the controls, including microscope set up, image acquisition, programming macros, etc. The three buttons on the lower right of the control panel allow one to switch between bright-field conventional microscopy and laser scanning microscopy without the need to adjust sliders etc on the microscope. The software used to control the Zeiss confocal microscope can be configured (during installation) in the "expert" mode, where all parameters are available for the user to adjust, or in the "routine" mode, where only a select group of parameters relevant to the particular labelling being performed are available.

Setting up the Microscope and Imaging Channels

Single track imaging is the standard method for single or multi-channel imaging (Figure A-21). However, if bleed through is a problem then multi-track imaging (see page 242), where each frame or alternate line is collected using different settings, may provide much better separation between channels.

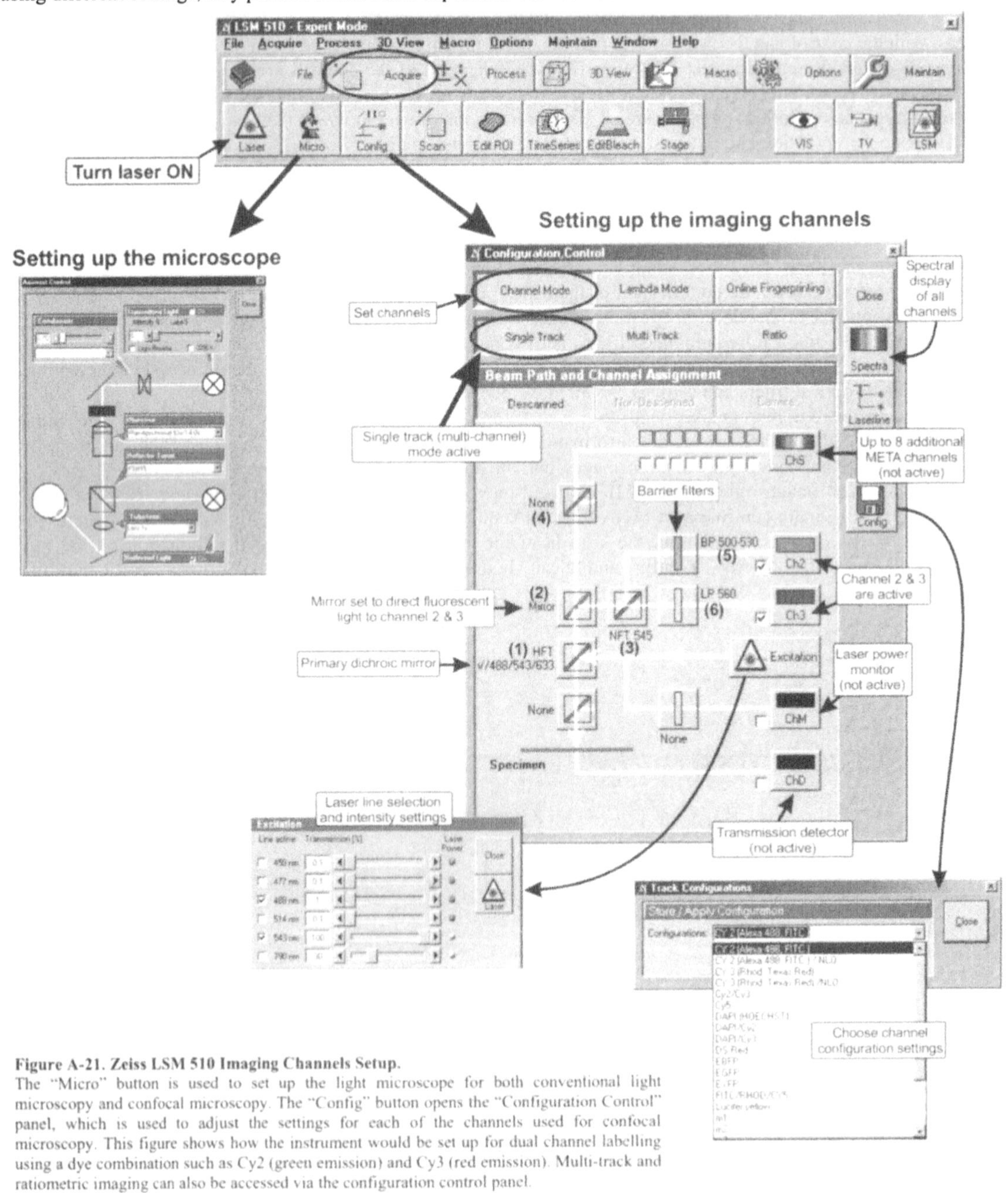

Figure A-21. Zeiss LSM 510 Imaging Channels Setup.
The "Micro" button is used to set up the light microscope for both conventional light microscopy and confocal microscopy. The "Config" button opens the "Configuration Control" panel, which is used to adjust the settings for each of the channels used for confocal microscopy. This figure shows how the instrument would be set up for dual channel labelling using a dye combination such as Cy2 (green emission) and Cy3 (red emission). Multi-track and ratiometric imaging can also be accessed via the configuration control panel.

Setting the Scan Control Parameters

Various scanning parameters, such as the zoom factor, collection box size, scan speed, collection filter etc., all need to be continually adjusted to optimise image collection (Figure A-22). Although these parameters can be obtained from a previously collected image (using the "reuse" button), it is necessary to be always aware of the settings and to continually make adjustments to improve the quality and resolution of the image collected.

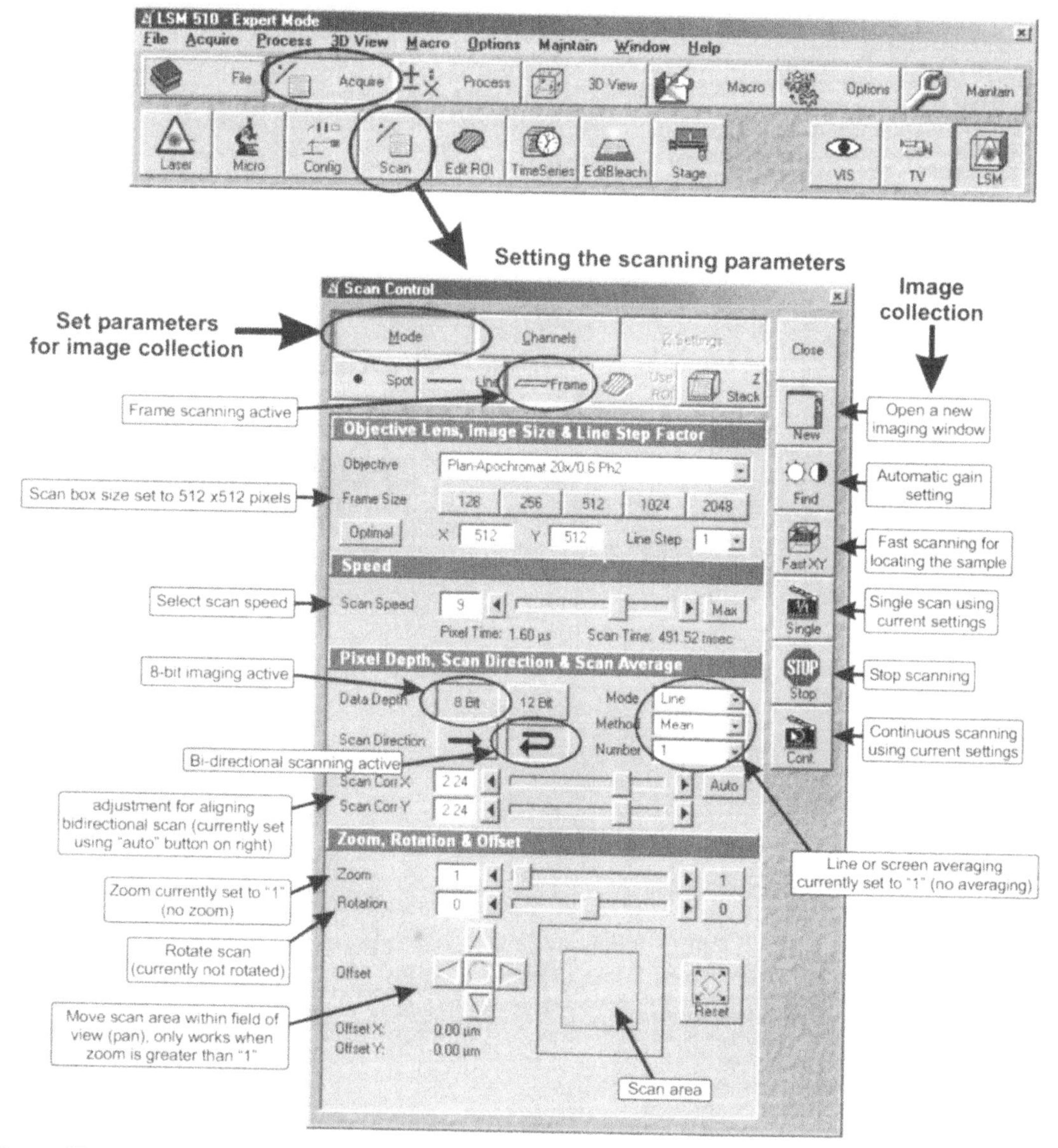

Figure A-22. Zeiss LSM 510 - Setting the Scan Parameters.
The "Scan" button is used to set the scan parameters for confocal microscopy. On the right hand side of the Scan Control panel are a number of buttons used for initiating image acquisition. The "find" button is useful for initially setting the gain for "locating" your sample. When this button is depressed the instrument collects a number of images in a relatively fast acquisition mode while varying the detector gain, and if necessary adjusting the amplifier gain (adjusting the sensitivity of the instrument). Once a suitable gain level has been found, a conventional single optical slice image will be displayed - often an excellent starting point for further adjustments.

Adjusting Imaging Channel Settings

Each channel in the confocal microscope has a number of parameters that need to be continually adjusted (Figure A-23). These include the pinhole size, detector gain, amplifier offset (black level) and the laser intensity. These parameters need to be continually adjusted during image collection to maximise the resolution and signal to noise ratio that can be obtained. Each channel (denoted Ch2 and Ch3 in Figure A-23) has an independent set of controls.

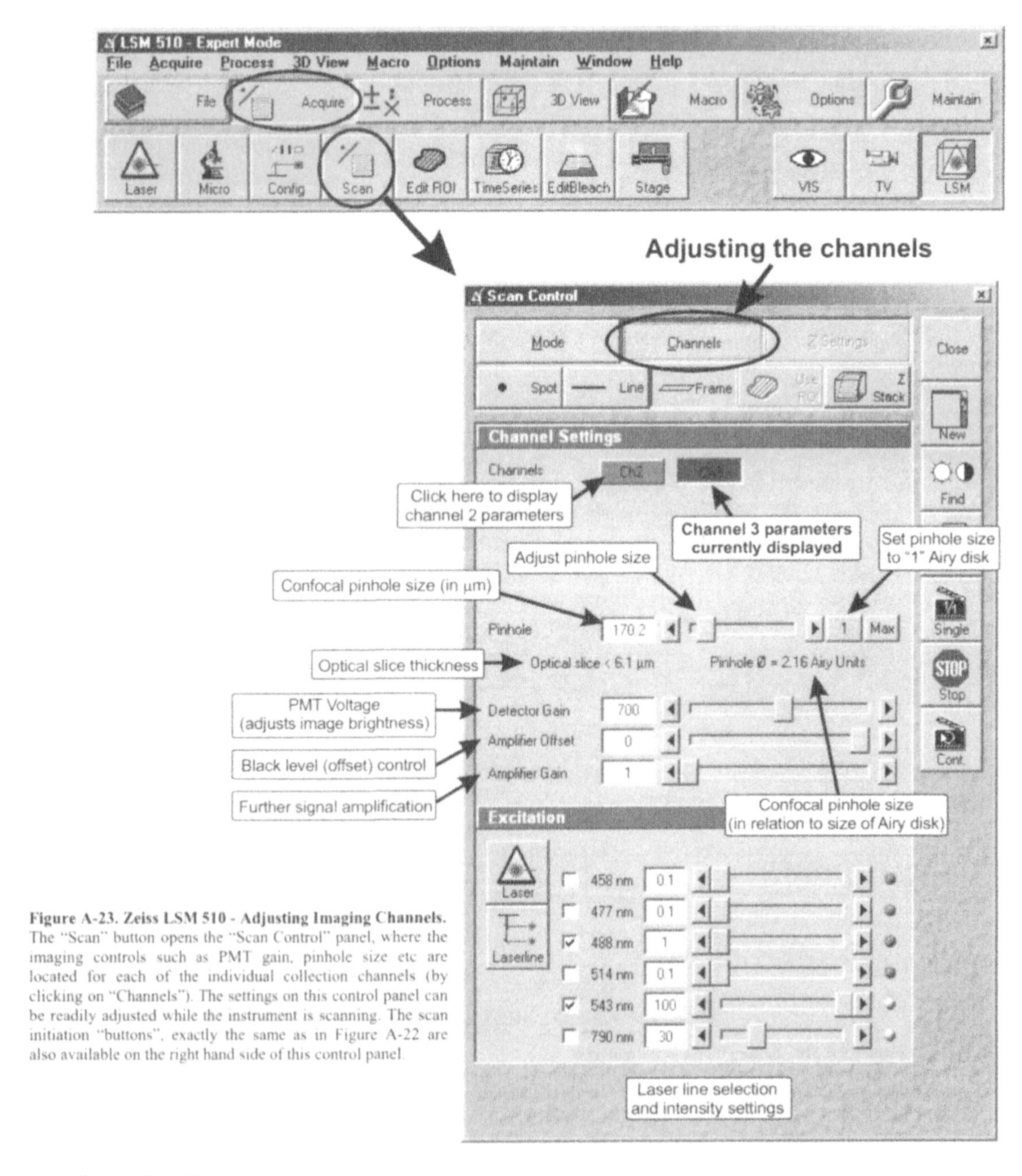

Figure A-23. Zeiss LSM 510 - Adjusting Imaging Channels. The "Scan" button opens the "Scan Control" panel, where the imaging controls such as PMT gain, pinhole size etc are located for each of the individual collection channels (by clicking on "Channels"). The settings on this control panel can be readily adjusted while the instrument is scanning. The scan initiation "buttons", exactly the same as in Figure A-22 are also available on the right hand side of this control panel.

Image Collection Panel

The image collection panel is where the images are displayed during collection, and is usually located on the second screen in a dual screen system (Figure A-24). Individual channels, or a split screen showing all active channels as shown in Figure A-24, are used to assess the quality of the image during collection. The images should be shown as a 1:1 (100% size) ratio when resolution of fine structures is important. A merged image (image 3 in Figure A-24) is used to display the individual channels as a multi-coloured overlay. If the "new" button is pressed on the image scan control menu (Figure A-22) a new window similar to that shown below will be created. Use the "reuse" button to select all of the microscope settings from a previously saved image to start imaging with the same conditions as used previously.

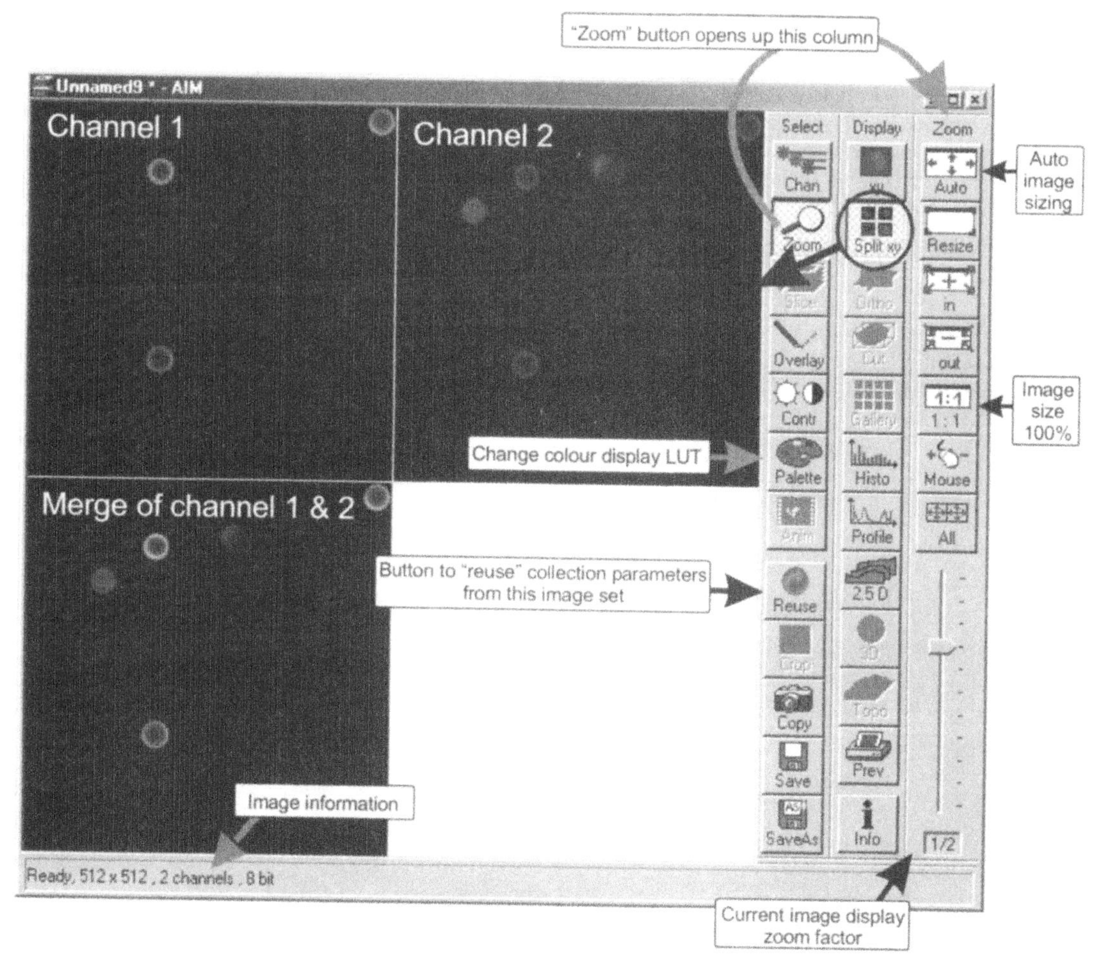

Figure A-24. Zeiss LSM 510 - Image Collection Panel.
A new image collection panel is opened by pressing the "new" button on the "scan control" screen (see Figure A-21, Figure A-22 and Figure A-23, top right of "Scan Control" panel). The images are displayed "live" while scanning in a variety of formats, including the "split xy" format shown above, where each individual channel, as well as a merged image, are all displayed on the screen. Individual channels ("xy Display") can also be displayed separately. The image display box size (this does not affect the image acquisition frame size, see Figure A-22) is adjusted by first selecting the "zoom" button and then selecting the appropriate image display size.

A very useful item found on the image display panel is the "reuse" button. A simple click on this button on any image (even images collected in a previous session, or by another person) will set up all of the image collection parameters to those used to collect the original image on display. In this way complex settings from previous experiments can be very easily replicated at a later date.

Using the META Channel

The META channel is a powerful and versatile tool for collecting images on the Zeiss LSM 510 META confocal microscope. As discussed above, the META channel can be used in a number of different imaging modes to collect images. These methods are discussed below in relation to how they are implemented in the Zeiss imaging software.

META Channel as a multi-Band Pass Optical Filter

The Zeiss META detector can be readily used as a multi-channel band pass optical filter by setting the region of the light spectrum that is directed to each of up to eight separate channels (Figure A-25). In this example only three channels are active (one collecting the green, one the yellow and the third the red region of the light spectrum). Another five channels are available, but have not been activated. In this mode the fluorescent light emanating from the sample is being separated by a diffraction grating and then the various spectral regions are detected by defined regions of the META multi-PMT array. Using the META as a multi-band pass optical filter does cause potential bleed-through problems, and so this type of multi-channel detection is normally used to separate two or more fluorophores that have well separated emission spectra. Although the sliders used to define the spectral range for individual imaging channels can be positioned such that each channel collects light from overlapping regions of the light spectrum, overlapping settings should be used with caution as they will result in serious bleed-through

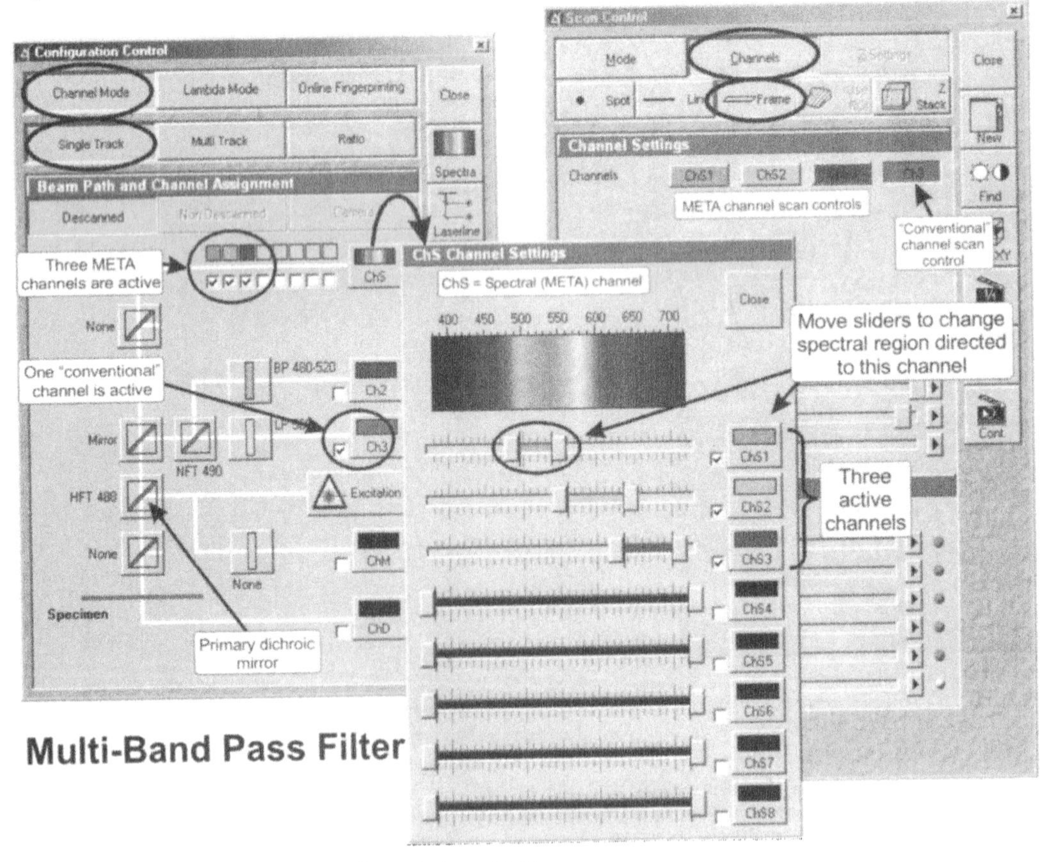

Figure A-25. Zeiss LSM 510 META Channel – Spectral Channel Settings.
The META channel can be used to collect up to eight different regions of the light spectrum as independent imaging channels. In the above example the "green" region of the visible light spectrum (500 to 550 nm) is directed to channel 1 (ChS1), the yellow region to ChS2 and the red to Chs3. The spectral region directed to each channel needs to be selected carefully to minimise bleed-through between channels. These META spectral channels can also be combined with "conventional" imaging channels. The position of each of the imaging channels can also be set by using the "Apply to Hardware" button (Figure A-28) on the Lambda display panel (Figure A-27).

problems for most fluorophore combinations. These META collection channels can also be combined with the two conventional channels available on the Zeiss LSM 510 META confocal microscope (Figure A-21).

Separating the light into separate channels by using the META channel as a band-pass filter should not be confused with linear unmixing (discussed shortly), where the fluorescence emission from fluorophores, even with highly-overlapping emission spectra, can be separated into individual imaging channels by computer analysis of the collected spectral image stack.

META Channel - Lambda Mode (Spectral Imaging)

The META detector can also be used to collect spectral information from defined regions of the light spectrum (Figure A-26). In the example shown in Figure A-26 a region of the light spectrum, from 497 nm ("Start") to 657 nm ("End"), is allocated (by using the on screen sliders) as the spectral region to be collected. The "step" size denotes the width of the spectral region (in this case 10.7 nm) that will be allocated to each individual image. A larger step size of 21.4 nm can be formed by binning two adjacent PMT cells in the META multi-PMT array, with even larger step sizes being created by binning more than two adjacent PMT cells. Eight individual spectral steps can be collected in a single pass of the scanning laser. Further spectral steps can be collected by using multiple passes of the laser (a total of 4 passes of 8 steps each is possible).

The images collected with the settings shown in Figure A-26 are displayed using the wavelength or lambda (λ) display mode (Figure A-26, lower panel). In this example each individual image displayed is the light collected from a 10.7 nm wide spectral region. The first eight images displayed were collected in a single pass of the laser, with the subsequent 8 images (only three of which are shown) being collected during a second pass of the laser.

> **The META Channel**
>
> **As a Multi-Band Pass Optical Filter:**
> up to 8 individual image channels can be specified.
>
> **For Spectral Imaging:**
> create a wavelength (λ) stack, where each image is derived from a narrow (usually 10.7 nm) region of the light spectrum.
>
> → ***Extracting Channels:***
> up to 8 individual image channels, from defined regions of the light spectrum, can be extracted from the wavelength stack.
>
> → ***Emission Fingerprinting:***
> unmix individual fluorophores (often with highly overlapping fluorescence emission spectra) by separating mixed colour spectra using reference spectra.

The spectral information collected for each pixel within the image can be used to provide valuable information on the spectral properties of fluorescent dyes within defined subcellular locations. Furthermore, individual image channels from defined regions of the light spectrum can be extracted from the wavelength or lambda (λ) stack. Spectral information can also be used in a process called "linear unmixing" to separate fluorophores - even those with highly overlapping emission spectra.

Extracting Image Channels from a Spectral Stack

The spectral information collected when using the META channel (Figure A-26 and Figure A-27, upper panel) can be separated into defined regions of the light spectrum and displayed as individual images (Figure A-27, lower panel). A graphical display of the spectral information within the wavelength or lambda (λ) stack is used to help define the region of the light spectrum to be allocated to individual channels. The defined spectral regions are displayed as individual images (Figure A-27, lower panel), including a merged multi-coloured display if required. This type of spectral separation does require that the fluorophores have relatively non-overlapping spectra, or at least regions of their spectra that are distinctly separate from one another. Overlapping spectra will result in problems of bleed-through as happens when using "conventional" channels defined by optical filters.

The smooth graphical display of the spectral data for each ROI (Region of Interest) shown in Figure A-27 is simply a "curve of best fit", and should not be interpreted as indicating that the spectral data within the lambda stack produces a "perfect" spectrum. In fact, the "real" data may often be relatively "noisy". At some point the "noise" level is such that a reliable spectrum can no longer be obtained – although a "smooth" curve will still be displayed!

Spectral Imaging

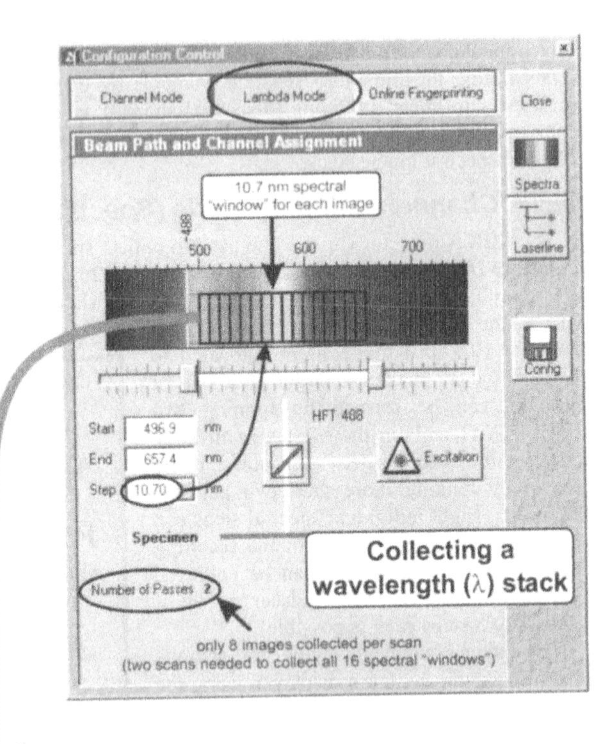

Figure A-26. Zeiss LSM 510 META Channel – Lambda Mode. The META channel can be used to collect a wavelength (lambda) image stack. In the "Lambda" mode (see panel to the right) a region of the light spectrum is delineated by moving the on screen sliders. This region is then divided into a number of "steps" or spectral "windows" for collecting the images. Up to 32 spectral "windows" of 10.7 nm can be collected (a total range of 320 nm). However only 8 steps can be collected per scan (hence the requirement for two scans in this example). The spectral "window" size can be doubled, to 21.4 nm, or greater by binning adjacent PMTs in the multi-PMT array.

Each spectral step collected using the META channel is displayed below as an individual image (λ display). Each image represents a 10.7 nm "window" of the light spectrum – with only 11 out of a total of 16 images collected being displayed in this example.

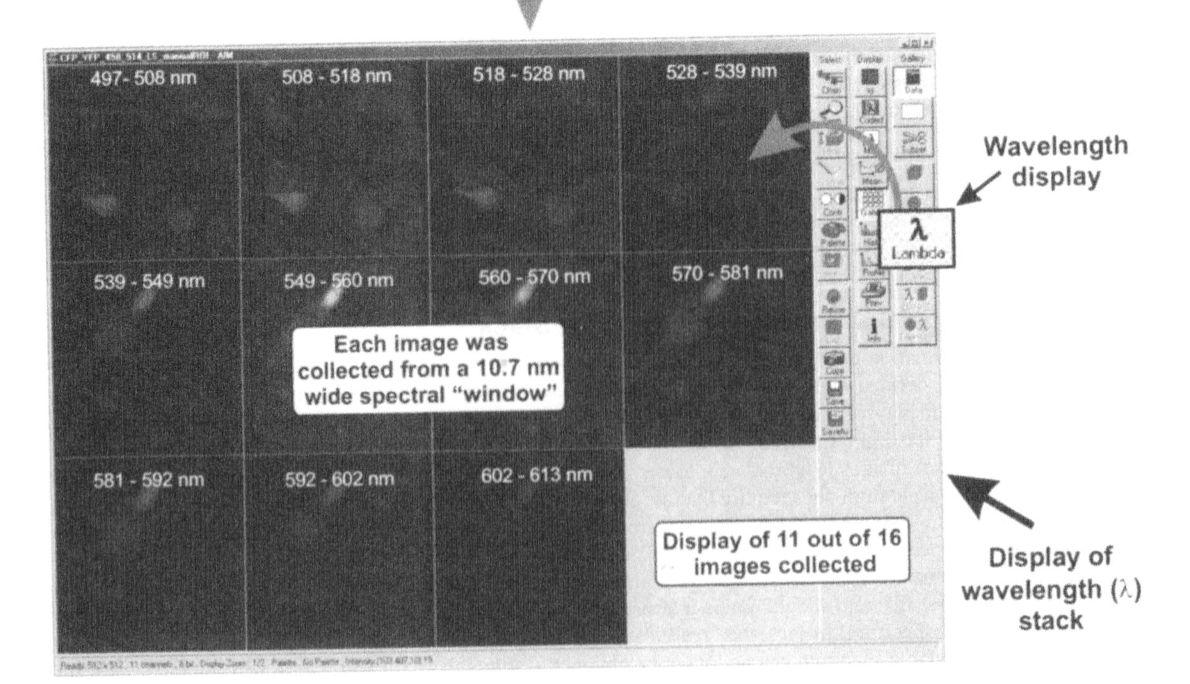

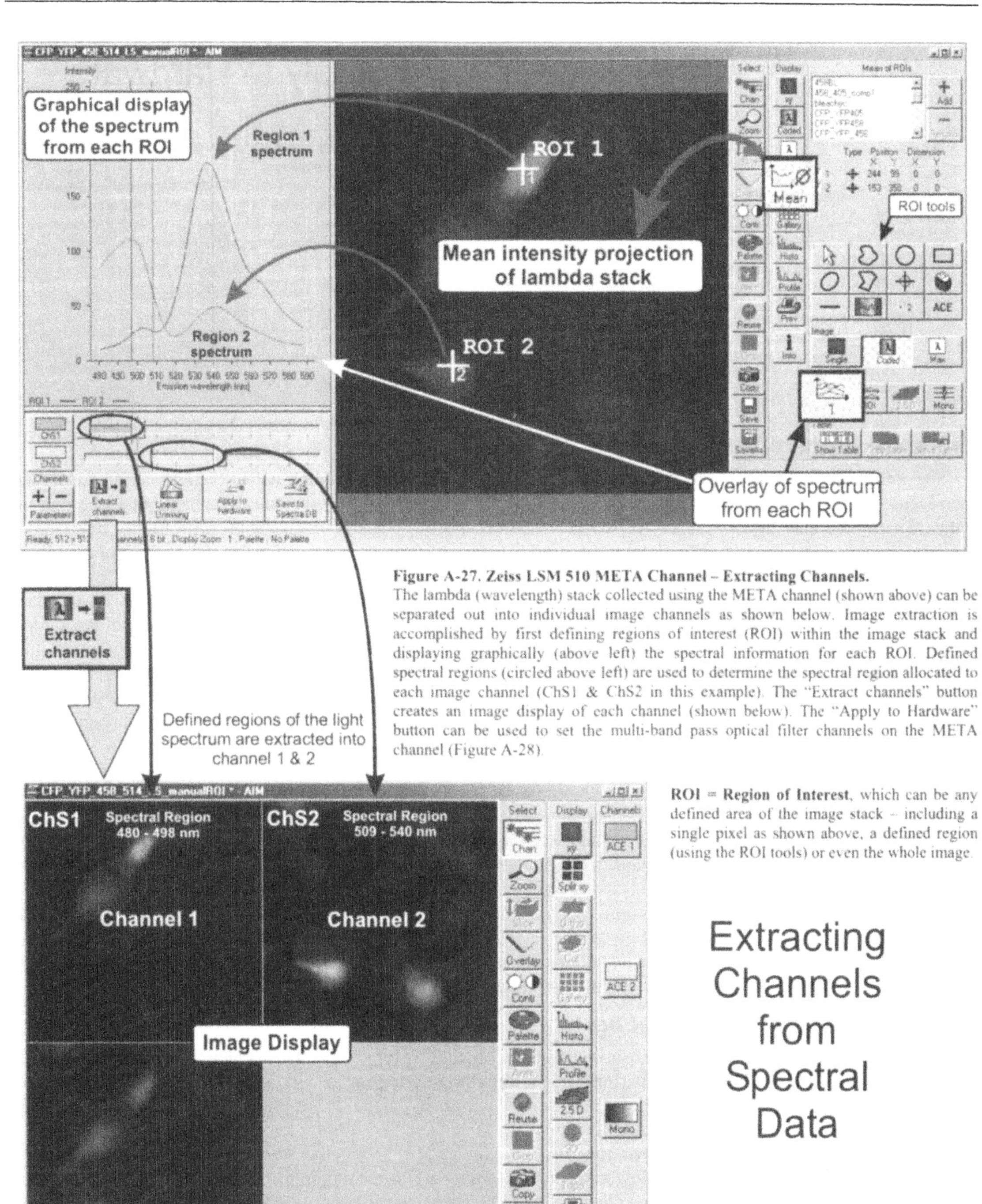

Figure A-27. Zeiss LSM 510 META Channel – Extracting Channels. The lambda (wavelength) stack collected using the META channel (shown above) can be separated out into individual image channels as shown below. Image extraction is accomplished by first defining regions of interest (ROI) within the image stack and displaying graphically (above left) the spectral information for each ROI. Defined spectral regions (circled above left) are used to determine the spectral region allocated to each image channel (ChS1 & ChS2 in this example). The "Extract channels" button creates an image display of each channel (shown below). The "Apply to Hardware" button can be used to set the multi-band pass optical filter channels on the META channel (Figure A-28).

ROI = Region of Interest, which can be any defined area of the image stack – including a single pixel as shown above, a defined region (using the ROI tools) or even the whole image.

Extracting Channels from Spectral Data

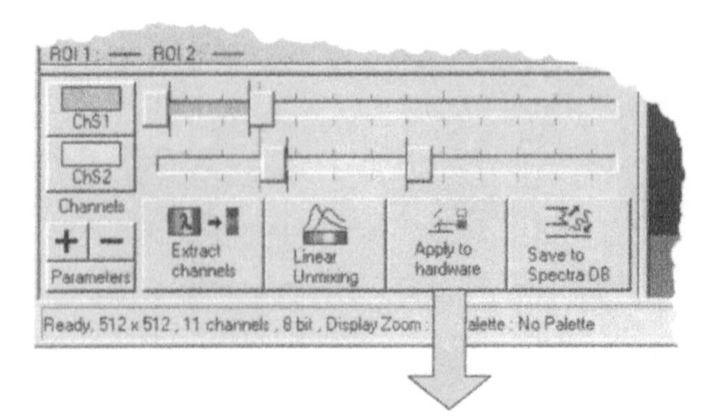

Apply to Hardware

Figure A-28. Zeiss LSM 510 META Channel – Apply to Hardware.
The "apply to hardware" button (found on the lower left of the display panel in Figure A-27 and Figure A-29 is used to transfer channel settings to the META hardware for multi-band pass imaging (Figure A-25).

Emission Fingerprinting

Spectral image acquisition and linear unmixing are combined in a procedure called "emission fingerprinting" (see Figure A-29) that may be performed both, on- and offline. Emission fingerprinting is capable of separating not only fluorophores that have physically separated emission spectra, but also fluorophores with highly overlapping emission spectra. For example, several green emitting fluorophores can be separated as long as each fluorophore has a sufficiently distinctive spectral characteristic.

The process of emission fingerprinting first requires one to collect a fluorescent emission spectrum image stack of the sample (Figure A-29, upper panel). The spectral characteristics of the fluorophores used to label the sample are then determined by choosing a region of interest (ROI) and displaying the emission spectrum as a graph (Figure A-29, upper panel where four regions of interest have been marked). The emission spectra of the individual fluorophores can also be determined from a single labelled control, or taken from a stored spectral database. Linear unmixing is the process by which the spectral data is used to "unmix" the various fluorophores present in the sample, and to present them as images in separate channels (Figure A-29, lower panel).

Emission fingerprinting can also be used to distinguish between different regions of a cell or tissue sample that show different emission spectra for the one fluorophore. For example, propidium iodide, which has an emission spectrum that shows distinct differences depending on the local environment of the probe, can be used with emission fingerprinting to distinguish between DNA and RNA bound dye. Dye bound to DNA has a distinctly different emission spectrum compared to dye bound to RNA, even though the fluorescence emissions of both forms of the dye are highly overlapping.

A further powerful feature of emission fingerprinting is that unwanted "background" fluorescence can be spectrally characterised and then removed from the final composite image. Samples with relatively high background fluorescence, particularly naturally occurring autofluorescence that cannot be easily removed by other means, can be very successfully imaged by using emission fingerprinting.

On-Line Emission Fingerprinting

Linear unmixing can be used on-line in real time to display several imaging channels. Pre-recorded spectral information from the sample is used to perform linear unmixing during image acquisition. In this mode images are displayed in separate channels consisting of the emission from individual fluorophores. The spectral information used for linear unmixing is not displayed or saved to disk.

The ability to do linear unmixing in real time on live samples is of great value when imaging dynamic processes. In this way cellular processes can be followed on-screen as they are happening in the same way as performing time-lapse imaging using conventional imaging channels. On-line emission fingerprinting also has the advantage of being able to display images in which the background has been removed by linear unmixing in real time. One can use the dynamic removal of background from the image to optimise the image collection of samples that may be difficult to image using conventional single or multiple channel imaging.

Emission Fingerprinting

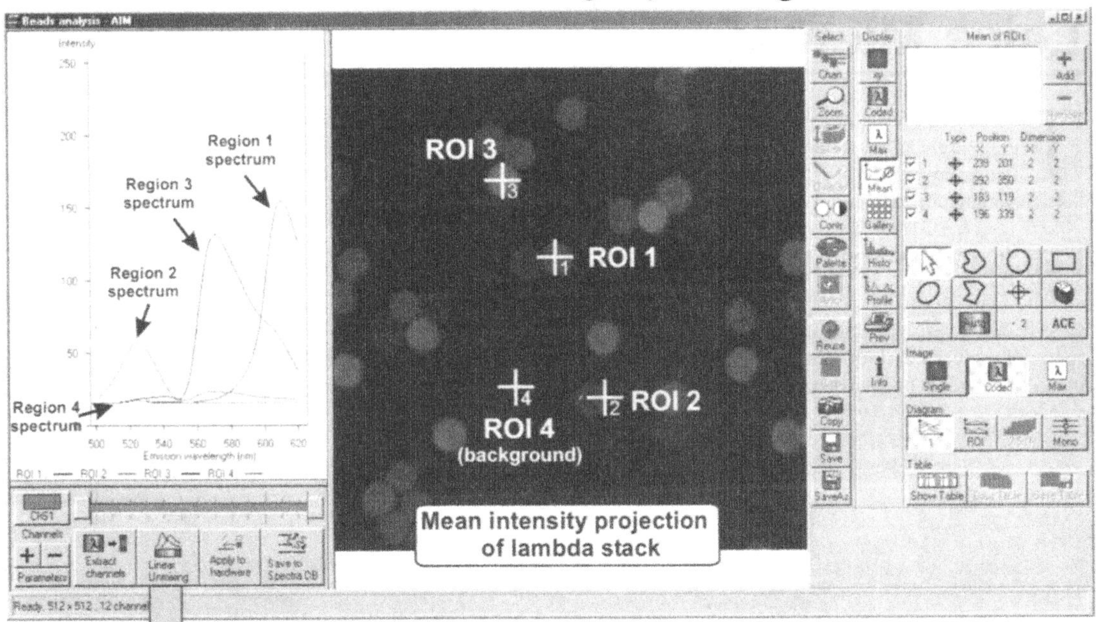

Figure A-29. Zeiss LSM 510 META Channel – Emission Fingerprinting.
Emission fingerprinting is a method by which spectral acquisition and linear unmixing can be used to separate fluorophores based on their spectral characteristics, rather than just their degree of spectral separation. Even fluorophores with highly overlapping emission spectra can be separated in this way. Spectral information can be obtained from the wavelength (lambda) image stack of the sample (as shown above), from a spectral database, or from a lambda image stack of a single labelled control. The process of Linear Unmixing separates each defined spectra, as shown above, into separate image channels (shown below).

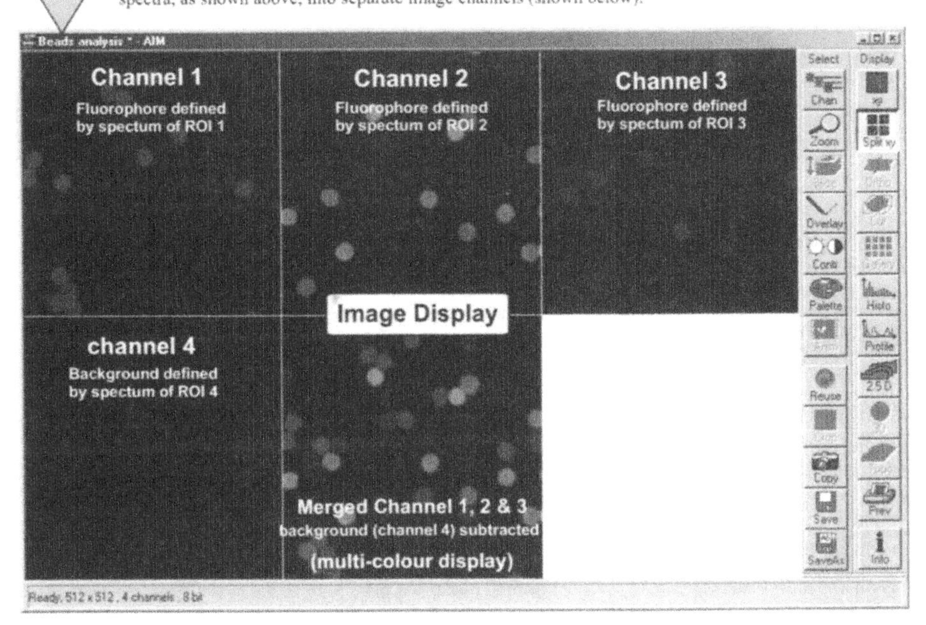

Leica Microsystems

The Leica optics company produces a wide range of high quality optical instruments from cameras to research grade light microscopes. Leica is particularly renowned for producing superb quality lenses. A subsidiary of Leica, Leica Microsystems Heidelberg GmbH, is responsible for the development and production of the Leica laser scanning confocal microscopes.

LEICA CONFOCAL MICROSCOPES

The Leica confocal microscopes have been developed over many years into sophisticated instruments that have superb optics, are capable of multi-channel labelling and posses a variety of innovations, such as the ability to spectrally select the fluorophore of interest. All this makes these instruments capable of not only creating excellent images, but also makes them highly versatile for using confocal microscopy as a major tool in the biological sciences.

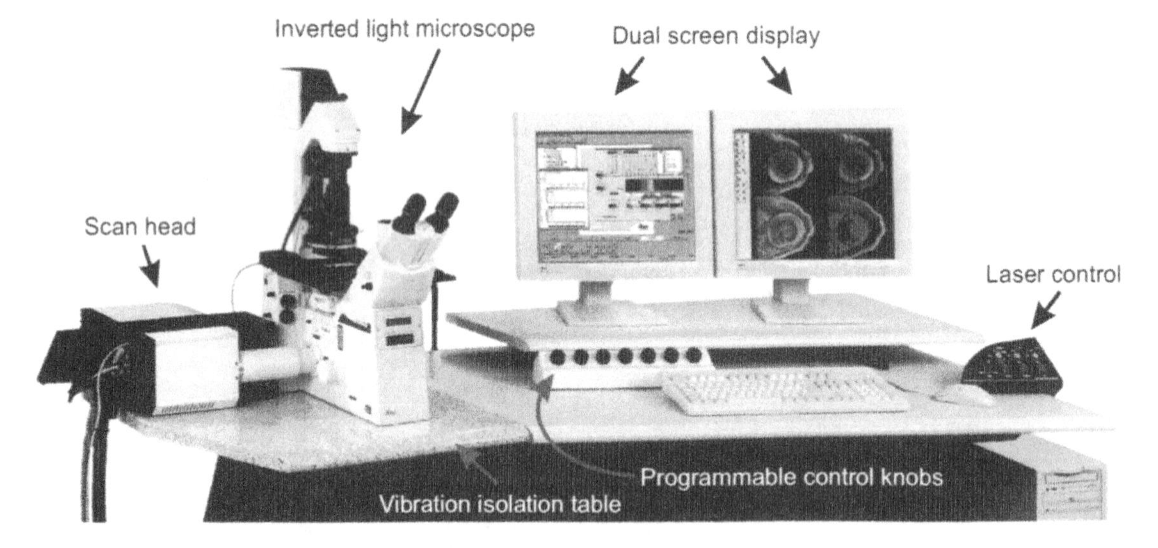

Figure A-30. Leica SP2 RS High Speed Confocal Microscope.
The Leica confocal microscopes are arranged with a two monitor display, a programmable panel box (normally used to control PMT gain, z-position, zoom etc) and laser control mounted on a workbench next to the vibration isolation table on which the microscope and scan head is mounted. The lasers are mounted under the main table, with a fibreoptic connection to the scan head. This photograph is kindly provided by Werner Knebel, Leica Microsystems, Mannheim Germany.

Leica Microsystems Heidelberg GmbH

Web: http://www.llt.de/ or http://www.leica-microsystems.com/website/lms.nsf

Address:	Leica Microsystems Heidelberg GmbH Am Friedensplatz 3 68165 Mannheim **Germany**	Leica Microsystems Inc. 410 Eagleview Blvd. Exton, PA 19341 **USA**	Leica Microsystems Pty Ltd Level 2, Building B, Glade View Business Par 482 Victoria Rd., Gladesville, NSW 2111 **Australia**
Phone:	+49 (0) 621 70 28 0	+1 (610) 321 0460	+61 2 9879 9700
FAX:	+49 (0) 621 70 28 10 28	+1 (610) 321 0426	+61 2 9817 8358

Leica Confocal Microscopes

Leica Confocal Microscopes

The latest confocal microscope from Leica uses a spectral imaging device that allows you to choose the "window" of wavelength (colour) for each of up to 4 channels. Furthermore, extremely narrow band separation of laser lines from fluorescence is possible using the AOBS filter, as well as continuous attenuation of the laser beam using AOTF filers. The instruments are built to customer specifications, which may include the more traditional dichroic mirror / optical filter separation of fluorescent light instead of the spectral separation technique.

Leica TCS SP2 AOBS	Full featured visible/UV confocal microscope (depending on lasers attached) with spectral (SP) imaging, and replacement of the primary dichroic mirror beam splitter with the AOBS filter.
Leica TCS SP2 RS	High-speed confocal/multi-photon microscope for live cell imaging. Utilises a very high-speed resonant scanner that allows up to 100 frames/sec at 512 x 512 pixels box size.
Leica TCS SP2 MP	Multi-photon version of the SP2 confocal microscope.
Leica TCS SP1	First introduction of the spectral separation technology in 1998.
Leica TCS SL	"Personal" confocal microscope capable of spectral imaging in 2 channels.
Leica ICM 1000	A confocal microscope designed for surface measurements in the materials science and industry.

Earlier Confocal Microscopes

The earlier confocal microscope models from Leica used dichroic mirrors and optical filters to separate the various wavelengths of fluorescent light. They also used neutral density filters to control the amount of laser light reaching the sample.

Leica TCS NT	Introduction of NT software.
Wild Leitz CLSM Aristoplan	Compact CLSM scanning head for upright microscopes.
Wild Leitz CLSM Fluovert	Inverted confocal microscope for biomedical research.
Leica TCS 4D	Introduction of a smaller scan head in 1992.
Wild Leitz CLSM	First introduced for biological imaging in 1989. High-resolution confocal microscope with powerful, but difficult to use software running under OS9.

Leica has a long tradition of production of high quality optics. The company has evolved over many years from the amalgamation of a number of optical and instrumentation companies, dating back to the middle of the 1800's with Spence Lens and American Optical Instruments and the "Optical Institute" in Wetzlar, Germany. In 1990 Wild Leitz and Cambridge Instruments merged to become Leica. The headquarters for Leica are located in Wetzlar, Germany, with manufacturing facilities and sales offices in North America, Europe and the Asia / Pacific region.

The Leica confocal microscope is produced by a wholly owned subsidiary of Leica, Leica Microsystems in Mannheim, Germany, with research laboratories still maintained in the original town, Heidelberg, in which the microscope was developed.

Although the basic optical components of a light microscope have not changed substantially over the past 10 years, Leica, like the other confocal microscope manufacturers, has introduced a number of innovative technologies that have made the confocal microscope a very powerful research tool in the biological sciences. Early models of the Leica confocal microscope were difficult to use due to what was relatively powerful, but not at all "user friendly", software. The most recent software from Leica is based on using a very intuitive graphical interface, where, although the scan head controls are all computer operated, the on screen "dial" often resembles the dial or slider and layout of the manually operated earlier instruments.

Early in the development of confocal microscopes, Leica introduced a number of innovative developments. This includes the AOTF (Acoustic Optical Tuneable Filter) for attenuating the amount of laser light reaching the sample and the AOBS (Acoustic Optical Beam Splitter) for separating the irradiating laser light from the fluorescent light. Another innovative development in the late 1990's was the introduction of the Spectral (SP) confocal microscope detection channels. This eliminated the need for specific optical filters for each fluorophore, with the user being able to simply move a slider to define the "window" of light that could be directed to each channel.

Leica TCS SP2

Components of the Leica Spectral Imaging Instrument

The Leica SP2 laser spot scanning confocal microscope can be configured for visible light, UV light and/or multi-photon microscopy on either upright or inverted microscopes (or interchangeable).

Design
Fluorescent separation using a prism with up to four separate detection channels. The scan head contains all of the necessary optics, pinholes and detectors. Remote lasers are fibreoptic connected. The instrument is a very high-end confocal microscope that has great versatility.

Scan Head
The relatively large scan head is build onto either an inverted or upright Leica light microscope.
AOBS: In the AOBS model the selection of the laser line used for excitation is performed by the "Acoustic-Optical Beam Splitter" filter - which allows for very narrow separation of laser lines from emitted fluorescence.

Dichroic mirrors: In instruments without the AOBS beam splitter, various dichroic mirrors are available.

Spectral prism: Returning fluorescent light in the "SP" model is spectrally separated using a prism. Carefully positioned computer operated "slits" are then used to direct a user-defined region of the spectrum to each of up to four photomultiplier tubes.

Variable collection slits: Light entering each of the photomultiplier tubes must first pass through a variable slit, which controls the region of the spectrum that is detected by that particular channel.

Variable size confocal iris: Single computer controlled variable size "pinhole".

Barrier filters: Computer-controlled wheel of removable barrier filters is available for selection of suitable wavelengths in models that are not equipped with spectral selection of wavelengths.

Light detectors: Photomultiplier tubes (the number depends on the number of detection channels available - up to 4, depending on the model), are housed within the scan head.

Scan mirrors: X & Y fast galvo scan mirrors with zoom, bi-directional & scan rotation capabilities, assembled into a single unit called the K-scanner.

Controller Box
The controller box contains the electronics associated with controlling the scan head and the collection and digitisation of the signal from the photomultiplier tubes.

Lasers
Lasers: Wide range of lasers can be attached via fibreoptic connection, including the Kr/Ar, Ar-ion, HeNe, 405 nm violet and red laser diode, HeCd and single line Kr-ion. UV and Ti:sapphire multi-photon lasers can be attached to suitably configured instruments.

Laser power: Adjusting the laser power knob sets the maximum power available - the actual power level at the sample is determined by the AOTF setting.

AOTF laser attenuation: Computer controlled AOTF filters control laser intensity at the sample.

Microscope
The relatively large scan head and associated optics can be attached to either an upright or inverted Leica microscope, but this is usually carried out by technical experts from Leica. The scan head cannot be readily attached to microscopes from other manufacturers. Leica is a manufacturer of light microscopes, and lenses as well as the confocal microscope scan head.

Imaging Modes
Multi-channel fluorescence, backscatter (reflectance) and transmission. Box sizes up to 4096 x 4096 pixels. Line, ROI, zoom, panning and rotated scanning. Time lapse imaging. Simultaneous and sequential frame or line multi-channel scanning. 12-bit data acquisition and processing saved as 8-bit or 12-bit image files.

Transmission
Single channel photomultiplier detector. Capable of bright-field, phase and DIC imaging, depending on the optical settings available on the light microscope.

Manual Control
Programmable manual control knobs (panel box) used to control all essential acquisition parameters.

Computer
High-end PC computer to control the confocal microscope, the settings on the light microscope if a fully automated microscope is installed, and to collect the images. Dual computer screens are used to display both the images as they are collected and to display the controls for scanning and changing microscope settings.

Software
The Leica software is used to control the scan head, collect the image, and some image manipulation and 3D reconstruction. A limited version of the Leica software (LCS Lite) provides basic image manipulation, and can be loaded onto another computer without incurring any further costs.

Models
Multi-channel spectral separation (SP) - detailed specifications (number of channels, spectral verses conventional optics, AOTF, AOBS etc) are required for assembly to customer specifications. **Multi-photon** system available. **High-speed** video rate resonant scanning system available.

Leica TCS SP2 Confocal Microscope

The Leica SP2 spectral separation confocal microscope is a highly versatile instrument based on the use of a prism to separate the fluorescent light into various regions of the light spectrum. This instrument can be used as a multi-channel confocal microscope by directing different spectral regions to up to four independent detection channels. The instrument can also be used to collect spectral information for each pixel within the image, which can then be used to separate highly overlapping fluorophores by a process called linear unmixing.

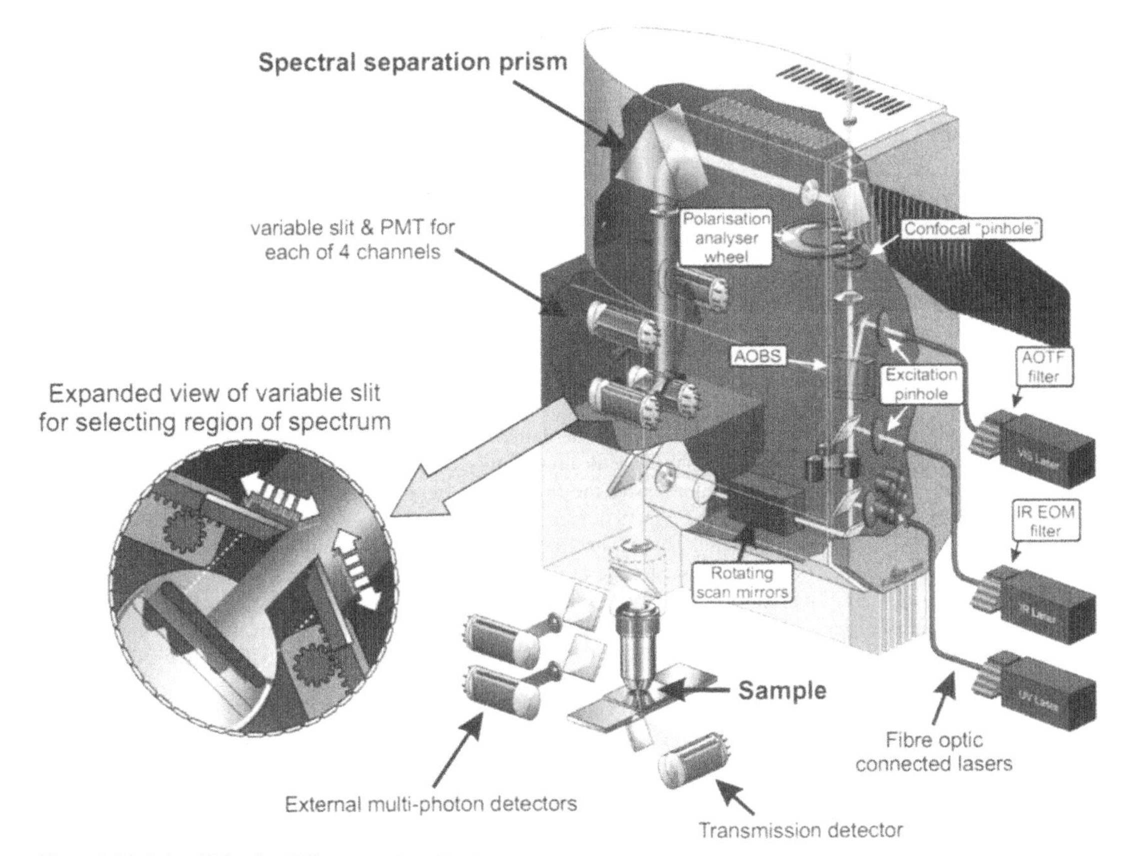

Figure A-31. Leica SP Confocal Microscope Scan Head.
Leica spectral selection (SP) confocal microscope scan head uses a prism to separate the returning fluorescent light into separate colours. Computer controlled variable slits are used to select the region of the light spectrum directed to each of up to four detection channels. In this example the highly efficient AOBS device is used to separate the laser excitation light from the returning fluorescent light. This figure is adapted from a diagram kindly supplied by Werner Knebel, Leica Microsystems, Mannheim Germany. A diagrammatic representation of how the Leica SP spectral separation scan head works is shown in Figure 3-11 on page 82.

Leica TCS SP2 Scan Head

The latest confocal microscope scan head from Leica Microsystems contains a number of highly innovative technologies (Figure A-31). These includes the Acoustic Optical Beam Selection (AOBS) device to very efficiently separate the irradiating laser light from the returning fluorescent light, a prism to spectrally separate the fluorescent light, and computer controlled slits to direct specific regions of the light spectrum into up to four independent channels.

The relatively large scan head contains all of the necessary optical components for confocal microscopy. Only the lasers, which are connected by fibreoptics, and the external detectors used in multi-photon microscopy, are located remote from the scan head. Various lasers can be readily connected via three main laser entry ports. One port is for connecting visible light lasers (a wide range of lasers can be connected to this port). In the example shown in Figure A-31 the AOBS beam splitter (described in more detail below) is used to split the incoming visible light from returning fluorescent light from the sample. However, the IR laser used for multi-photon microscopy, and UV lasers are connected via separate connections that utilise dichroic mirrors to separate the irradiating light from the fluorescent light.

Each laser port contains an excitation pinhole, but there is only a single very small (micron sized) v-shaped variable sized confocal pinhole in the Leica SP scan head. This pinhole is located before the fluorescent light is separated into different spectral regions, which greatly simplifies pinhole alignment, but does mean that the pinhole size cannot be matched to the wavelength of light being examined in multi-labelling applications.

The Leica confocal microscope scan head assembly is based on customer specifications at the time of purchase, which means that Leica confocal microscopes will be found to contain a range of features and innovations that vary between individual instruments.

Acoustic Optical Beam Splitter (AOBS)

The Acoustic Optical Beam Splitter device (described in more detail on page 78 in Chapter 3 "Confocal Microscopy Hardware") on the latest Leica confocal microscopes allows for highly efficient separation of the irradiating laser light from fluorescent light returning from the sample. The AOBS element is an optional replacement for the primary dichroic mirror used in the Leica SP confocal microscope. The advantage of the AOBS element is that a very narrow band of light (essentially a single wavelength, for example 488 nm) can be selectively directed to the sample - with the full spectrum of returning fluorescent light (except for the very narrow band of 488 nm light) being allowed to pass through the device to the detectors in the scan head. Up to 8 individual laser lines can be high-speed switched and separated from the returning fluorescent light.

The AOBS element results in significantly less of the fluorescent light being rejected, resulting in an increase in sensitivity of the instrument. Furthermore the AOBS device, being capable of very high-speed line switching, means that images can be collected using different excitation lines for each alternate scan line in the image. Laser line switching can also be achieved by electronic switching of the AOTF filters used to attenuate the intensity of the laser light, but the AOBS element can both select out specific narrow bands of light and switch between multiple bands at very high speeds in the one optical device. The AOBS filter has the additional advantage of being readily programmable to select out the wavelengths of choice if a new laser is installed without the need to install additional optical filters or dichroic mirrors.

Spectral Separation

The innovative method of spectral separation of various regions of the light spectrum in the Leica confocal microscope results in a highly versatile instrument that has a simple and intuitive means of separating different wavelengths of light in multi-labelling applications. This includes on screen software that mimics the physical settings of the variable width slits (see Figure A-34).

Spectral separation of the light spectrum using a prism is a very well known phenomenon, but where the Leica instrument excels is the way in which up to four detection channels are designed to collect specific regions of the light spectrum without any "dead" area between the channels. The way this is achieved is by using computer controlled variable width slits that have mirrored surfaces. The position and width of a slit determines the region of the light spectrum that is directed to that particular channel. Light that is excluded from the slit is reflected from the mirror surface of the slit to a second mirrored slit where a further region of the light spectrum can be separated – up to four such channels may be available on the Leica SP confocal microscope, depending on how many channels were specified by the customer when the instrument was assembled (see also page 81 for further discussion on how this spectral separation technique works).

Spectral separation can be used for routine multi-labelling applications, although a simple dichroic mirror system may be just as efficient at separating the fluorescence emission of well separated fluorophores. The Leica software does provide a number of pre-programmed and programmable menu selections that will set the slit position and width for a wide range of common fluorophores. However, one of the more powerful features of the flexibility of the Leica system of spectral separation is the ability to readily change the position and width of each individual collection channel while imaging. This allows one to adjust each channel for optimised collection of multiple labels even when using unusual fluorophores or when attempting to separate specific spectral regions for a single fluorophore.

The spectral separation technology of the Leica confocal microscope scan head can also be used to determine the spectral characteristics of the image (lambda scanning). In this technique an image is collected by using a slit width covering a narrowly defined portion of the light spectrum, with subsequent images being colleted by using narrow windows of shorter or longer wavelengths of light. In this way a series of images that specify the spectral characteristics of each pixel within the image can be collected. Spectral information of individual fluorophores carries valuable information on the local environment of the dye, but perhaps more importantly spectral imaging can be used to separate fluorophores with distinct but highly overlapping spectra. This process, called linear unmixing, which was pioneered by Carl Zeiss using the META detector (see page 388), is now available on the Leica SP confocal microscope.

UV Excitation

An Ultra Violet (UV) emitting laser can be readily connected to the Leica confocal microscope scan head. However, the UV laser is a difficult laser to use in confocal microscopy, both due to the destructive power of the relatively high-power UV light and the optical difficulties of dealing with wavelengths that are not ideally suited to the design of the scan head or light microscope being used. With the aid of special UV correction optical elements the UV (and also 405 nm violet laser line) focus is shifted into the position for visible light. This ensures that the emission light passes through the confocal detection pinhole.

Multi-Photon Excitation

The Leica confocal microscope is capable of multi-photon excitation by the attachment of an ultra-short pulsed infrared laser. Multi-photon excitation is particularly suited to deep tissue imaging as the infrared light can penetrate much deeper into living tissue than can visible light lasers. The multi-photon laser is also capable of exciting many commonly used UV dyes – making the multi-photon microscope an alternative to the UV confocal microscope.

Multi-photon imaging does not require a confocal pinhole (see page 94 in Chapter 3 "Confocal Microscopy Hardware") and so a more efficient means of collecting the fluorescent light is to attach external detectors (see Figure A-31). These detectors are housed outside the scan head, and collect fluorescent light that has not passed through the confocal pinhole or back through the scanning mirrors. External detectors result in a significant increase in sensitivity of the instrument when multi-photon imaging.

Transmission (DIC) Imaging

The Leica confocal microscope is capable of collecting excellent transmission images, which can be combined with single or multi-channel fluorescence images. The transmitted light detector is located after the condenser in the light microscope, in which case care should be taken to correctly adjust the condenser as this is now part of the optical system of the microscope (see 42 in Chapter 2 "Understanding Microscopy").

Excellent DIC images can be collected by using properly installed and adjusted DIC optics on the light microscope. However, if high-resolution DIC and fluorescence images are required, collect the two images sequentially by removing the Wollaston prism from the optics path when collecting the fluorescence image. Otherwise you will find that the fluorescence image will consist of a slightly offset "double" image – this is only obvious when you image at high zoom with a high NA lens, but will detract from the quality of the fluorescence image even at lower zoom levels. The position of the slight off-set is not affected by making adjustments to the Wollaston prism or by removing the polariser or analyser – removing the Wollaston prism is the only effective way of eliminating this artefact. DIC can be implemented in a number of related, but technically different ways, not all of which have an effect on the quality of the fluorescence image.

LEICA CONFOCAL MICROSCOPE CONTROL SOFTWARE

The software used to control the Leica confocal microscope has a particularly easy to use interface – although, as with all confocal microscopes, the complexity of the instrument does mean that the software can at first appear daunting to the user. The highly graphical interface, and the programmable desktop knobs (programmable panel box), mean that most controls are within easy reach. All of the "buttons" available on the control screen can be "hidden" or brought out by using the settings in the "Tools" menu – so if a "button" has gone missing, try looking in the customising tools menu.

The following pages provide a brief introduction to the software used to control the Leica confocal microscopes. For a more detailed explanation please consult the Leica confocal microscope manual, the on-line help, or perhaps best of all ask somebody who is already familiar with the software.

The Leica confocal microscope software runs under Windows NT/XP. If you have been set-up as a user on the system, the password log-in is designed to allocate you a specific folder for images and to allow you to recall previously saved microscope settings (imaging methods), but is not intended as a security system for your files.

Main Control Panel – Monitor 1

The main control panel (Figure A-32) is used to control the settings on the confocal microscope, and to save images under a specified folder name. The lower part of the control panel contains a series of "buttons" that are used to control or access submenus that are used to operate the instrument.

The "Continuous" scan button is used to provide continuous scanning while locating the sample and adjusting the instrument. This button does not result in any of the images being saved to a file. To save images you will need to press the "Single Scan" or "Series" buttons, which will scan and save the image as a TIF file (saving it to a temporary folder, until you have specified a specific folder name).

"Single Scan" will collect an image using the settings specified on the control panel. These settings include the pinhole size, speed, direction of scan etc, but also whether frame or line averaging has been specified. For example, if frame averaging is set to 8 then "Single Scan" will collect a single image, consisting of 8 scans that have been averaged. The "Series" button also collects an image using the pre-set parameters and saves the images to a file. The "Series" button is used to collect a z-series or a time-series image set. As the images are saved as single "TIF" image files (one file for each image in the series, or a single image file when not collecting a series), the Windows thumbnail display will show each of the images in the folder. These individual "TIF" files can also be readily accessed through most image processing programs, including Photoshop, if you save the images as 8-bit files rather than 16-bit files.

Information on instrument settings is all conveniently displayed in a "window" as shown in Figure A-32. This information is saved with the image file, but can only be readily accessed when using the Leica software to look at the saved images. However, the information is saved as a text file in the folder containing the images and so can be accessed with a simple text editing program if necessary.

You can specify your own file folder name for the images by saving the file – or you will be prompted for a name when you attempt to close the file or shutdown the computer.

Leica Lite

A free, cut-down version of the Leica software that can be used to open Leica image files

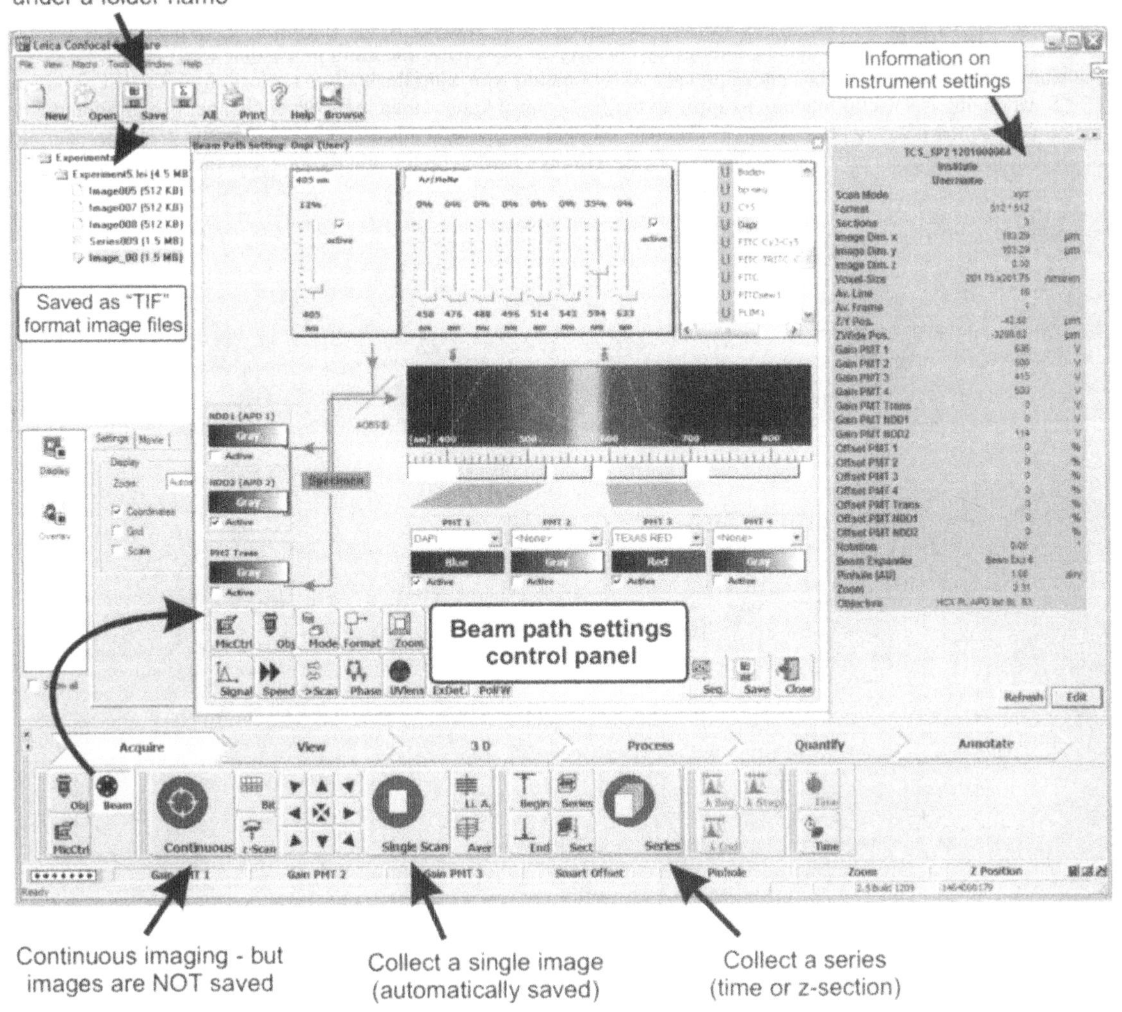

Figure A-32. Leica SP2 Screen One - Main Control Panel
Leica uses two computer screens; screen one (above) being used to display the control panel for operating the confocal microscope. This particular control panel is from a Leica confocal microscope that has spectral imaging and the AOBS beam splitter for separating the laser line from the emitted fluorescence. In this control panel a number of images (some with multiple channels) have all been saved under the folder name "Experiment5". The images are saved as individual "TIF" format files that can be readily accessed by most image processing software.

Changing the Desktop Knob Functions.

The physical desktop knobs (programmable panel box) provided with the Leica confocal microscopes can be programmed to control a variety of useful functions. The default settings are for the first knob (from the left) to control the gain on PMT1, and second the gain on PMT2 etc as outlined in the control panel below (Figure A-33). You can readily change these default settings by clicking on any one of the knob labels displayed on the bottom of the main control panel – and then choosing the control function you want the knob to perform from the list provided.

"Smart Gain" is a useful function to apply to the first control knob – then, whichever channel is highlighted with the mouse (on the imaging screen) will be controlled by this knob. This allows one to control the gain, for example, of all four imaging channels by using the one physical knob – freeing up the other knobs for other functions.

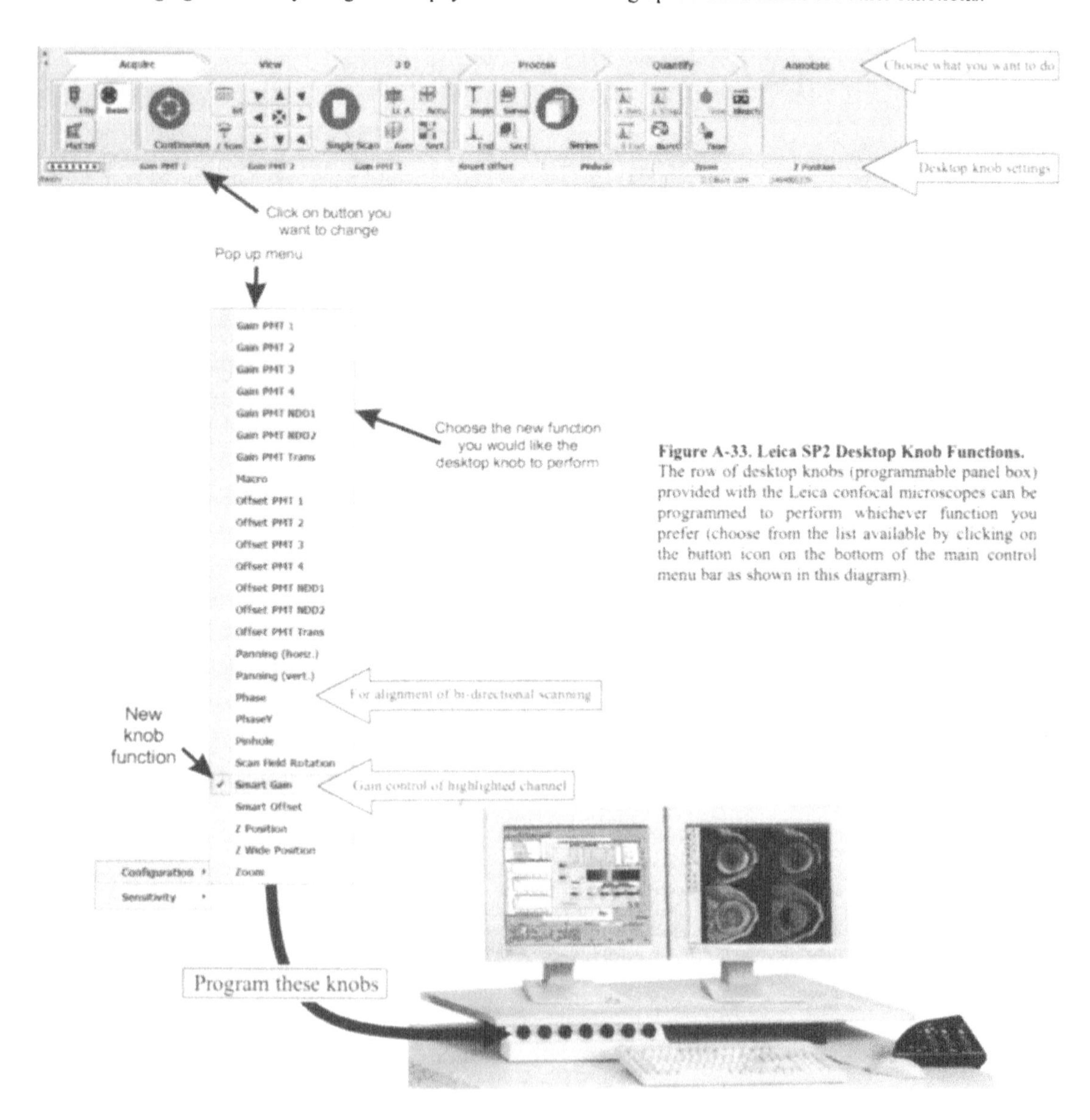

Figure A-33. Leica SP2 Desktop Knob Functions. The row of desktop knobs (programmable panel box) provided with the Leica confocal microscopes can be programmed to perform whichever function you prefer (choose from the list available by clicking on the button icon on the bottom of the main control menu bar as shown in this diagram).

"Beam Path Setting" Control Panel

One of the best features of the Leica confocal microscope controlling software is the "Beam Path Setting" control panel shown below (Figure A-34). The great feature of this panel is the graphic display of the colour spectrum, including an overlay of the emission spectrum of the fluorophore you are using. Simply "grabbing" the ends of the slider windows and expanding or narrowing the collection window is a very simple way of modifying the regions of the light spectrum that are directed to the individual channels.

A series of pre-set parameters can be obtained by clicking on the fluorophore setting displayed in the top right panel (in this example the "U" or "user" settings are displayed – these are the settings that you can modify and save under your own file name. There is also a set of "L", or "Leica" settings that are installed with your instrument and can be called up as required but cannot be modified by the user).

The LUT (colour display) used for each channel can be easily changed (including the transmission channel) by clicking on the graded colour display below the PMT channel label and choosing the colour gradient of your choice. Don't forget to "tick" the "active" box when you want data collected from a particular channel (only PMT1 and 3 channels are active in this example).

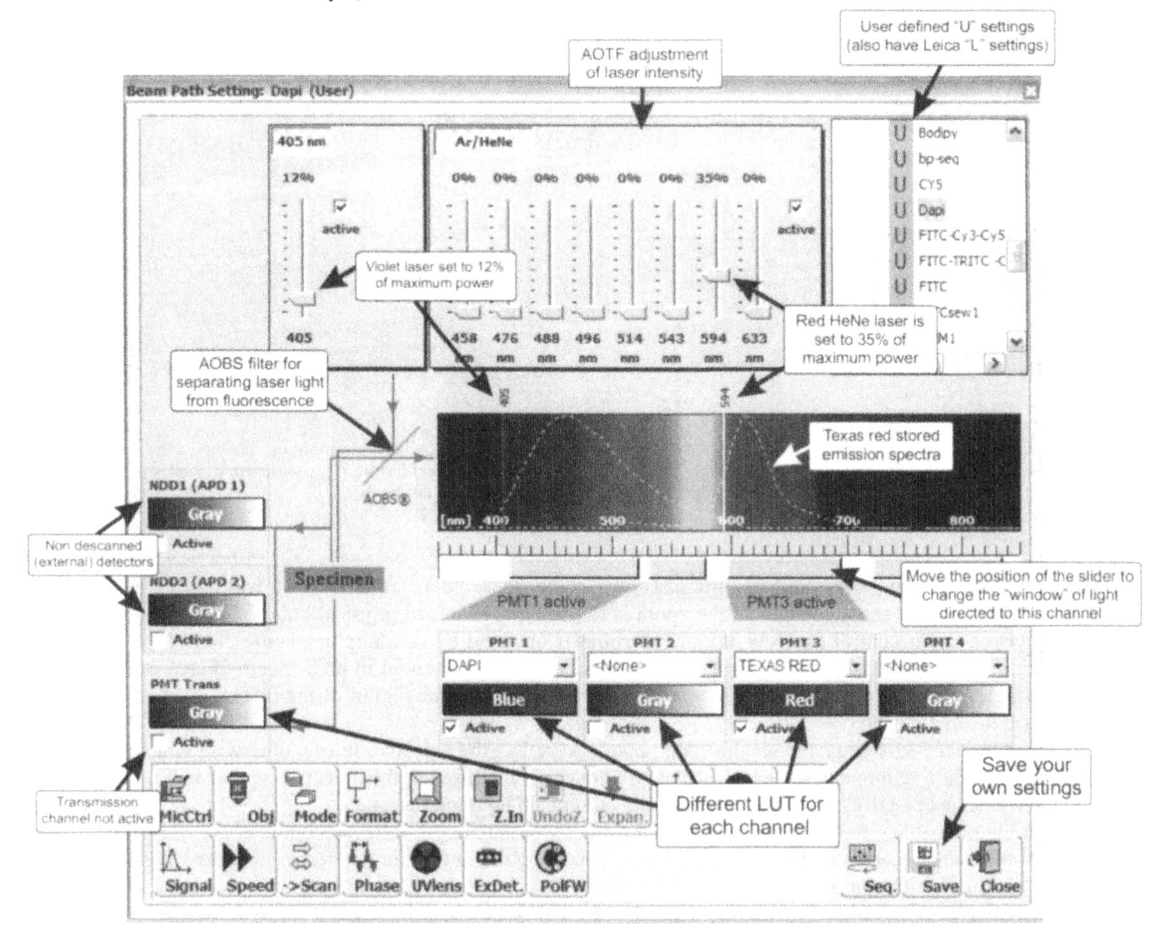

Figure A-34. Leica SP2 Beam Path Settings.
Monitor 1 - beam path settings screen is used to set the "window" of light directed to each channel. To display this panel click on the "beam" icon in the lower bottom left of the main control panel (see Figure A-32). **Please note:** The fluorophore spectra overlaid on the colour spectrum in this panel is taken from stored spectral data – and are not spectra derived from your sample. Use these spectra only as an approximate guide to the spectral emission of the fluorophores used to label your sample.

Control "Buttons" for Image Collection

The main control "buttons" for image collection (Figure A-35 and Figure A-36) are found in a group, usually at the bottom of the control panel screen or within the "Beam Path Settings" panel (on monitor 1). The buttons displayed and the layout used can be readily altered by the user (from the "Tools" options on the top of the control panel) - so you may find the instrument you are using has a somewhat different layout. The most common buttons are explained below in Figure A-35, and on the next page in Figure A-36.

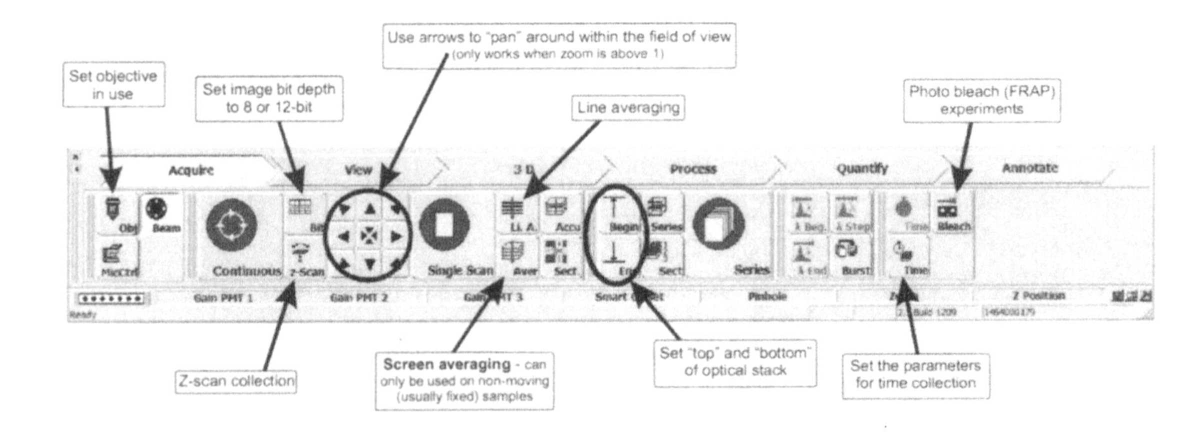

Figure A-35. Leica SP2 Main Control Panel.
The main control panel is normally located on the lower part of the screen, and contains a variety of "buttons". The common "Acquire" buttons used are shown above. However, you may find that your instrument has been programmed to display a slightly different set of buttons. See Figure A-36 for a description of further buttons that can be displayed.

Zoom control on the Leica confocal microscope can be achieved in a number of ways (Figure A-36). A convenient way to make small changes in the zoom is to use a programmed physical knob on the desk top (Figure A-33). However, better control over the amount of zoom is obtained by selecting a specific zoom level using the "Zoom" button on the control panel (Figure A-36). An excellent way to zoom in on a specified area of the image, such as a single cell, is to use the "Z.In" button to draw a box around the region of the image of which you wish enlarge to fill the imaging box.

Take care to make sure you have set the "Obj" button to reflect the objective in use, otherwise you will find the one Airy disk pinhole setting and scale bar display are not correct. However, the objective setting will be taken care of when you change objectives if you are using a fully automated light microscope.

Another button that has an important impact on the quality of the images collected, but is often overlooked, is the "Expan." button, which controls a beam expander lens. Changing the setting for the beam expander will have a large impact on both the resolution and sensitivity of the instrument. The beam expander must be correctly set for each objective – simply try changing the beam expander to find which setting is best for your particular lens and application.

The "Scan" button is used to switch between bi-directional and single direction scanning. However, when switching to bi-directional scanning take care to adjust the "Phase" by using a programmed desktop knob to bring the alternate scan lines back into alignment.

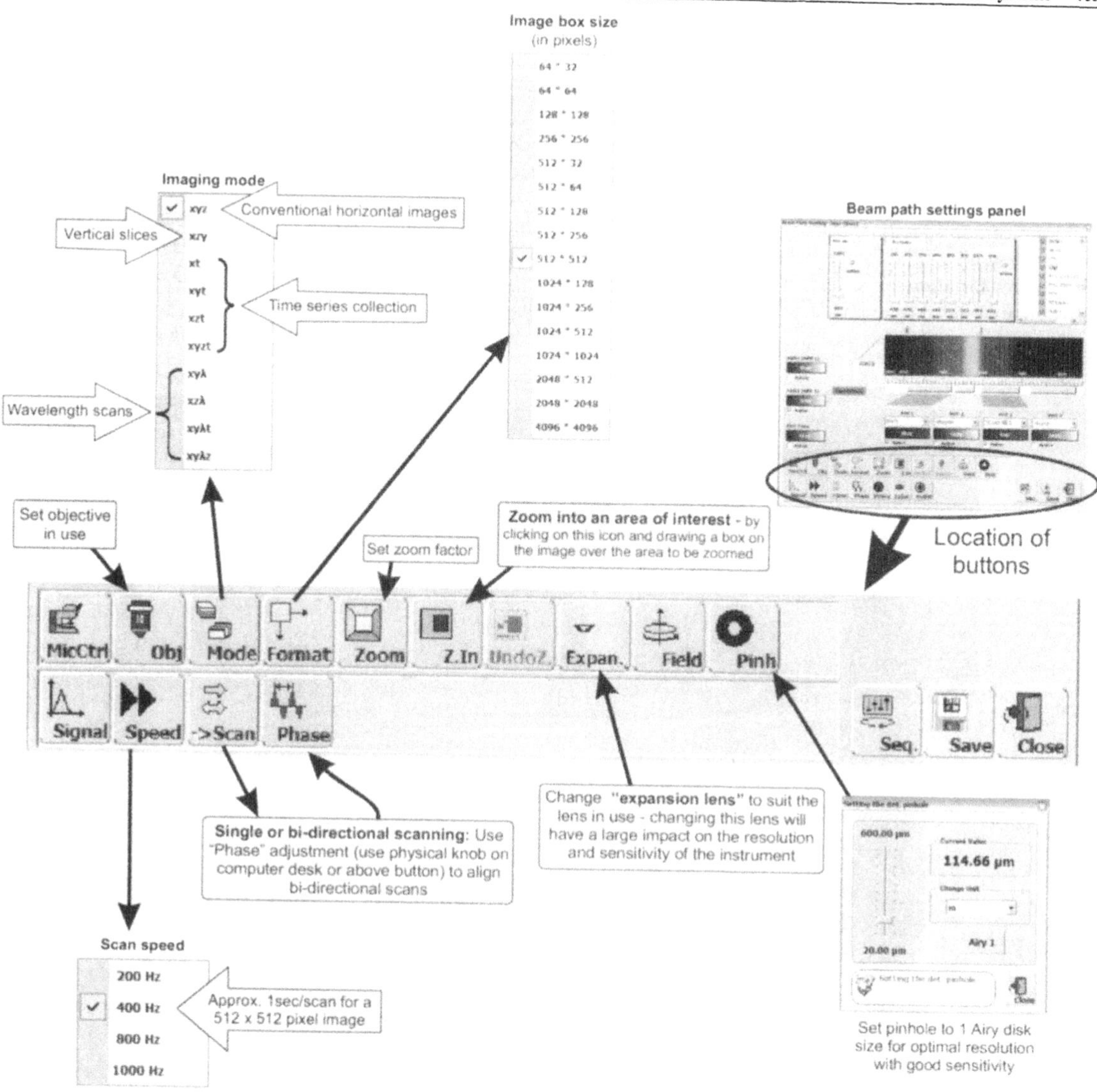

Figure A-36. Leica SP2 Control Buttons
This figure is a continuation of the description of some of the common "acquire" buttons located on the main control panel (see also Figure A-35). These buttons can be displayed in the main control menu bar, or alternatively, they can be displayed in the "Beam Path Settings" panel (as shown in Figure A-34). Which buttons are displayed and their layout on the screen can be changed by the user which means you will find the instrument you are using will have a somewhat different display to the above.

> **Warning: -** be patient when using the Leica software. Make sure each operation is completed before you click on the next button – or you may "crash" the program!

Image Collection Panel – Monitor 2

The image collection display panel is normally located on monitor 2 (Figure A-37), but this display panel can be placed on monitor 1 if required. Each experiment is displayed as a separate window on this screen. To close and save a particular experiment simply click on the normal windows close function (the top right hand cross) and you will be prompted for a file name (if you have not already specified one) and whether you wish to save the image – before the window is closed.

Take care to specify 1:1 image display (Figure A-37) using the "Display" button when high-resolution imaging. In this way all of the image data collected will be displayed on the screen. Displaying the images as "Auto" is convenient for allowing several images to be displayed on the one screen in multi-labelling applications, but don't forget that in this display mode not all of the image pixels collected will necessarily be displayed.

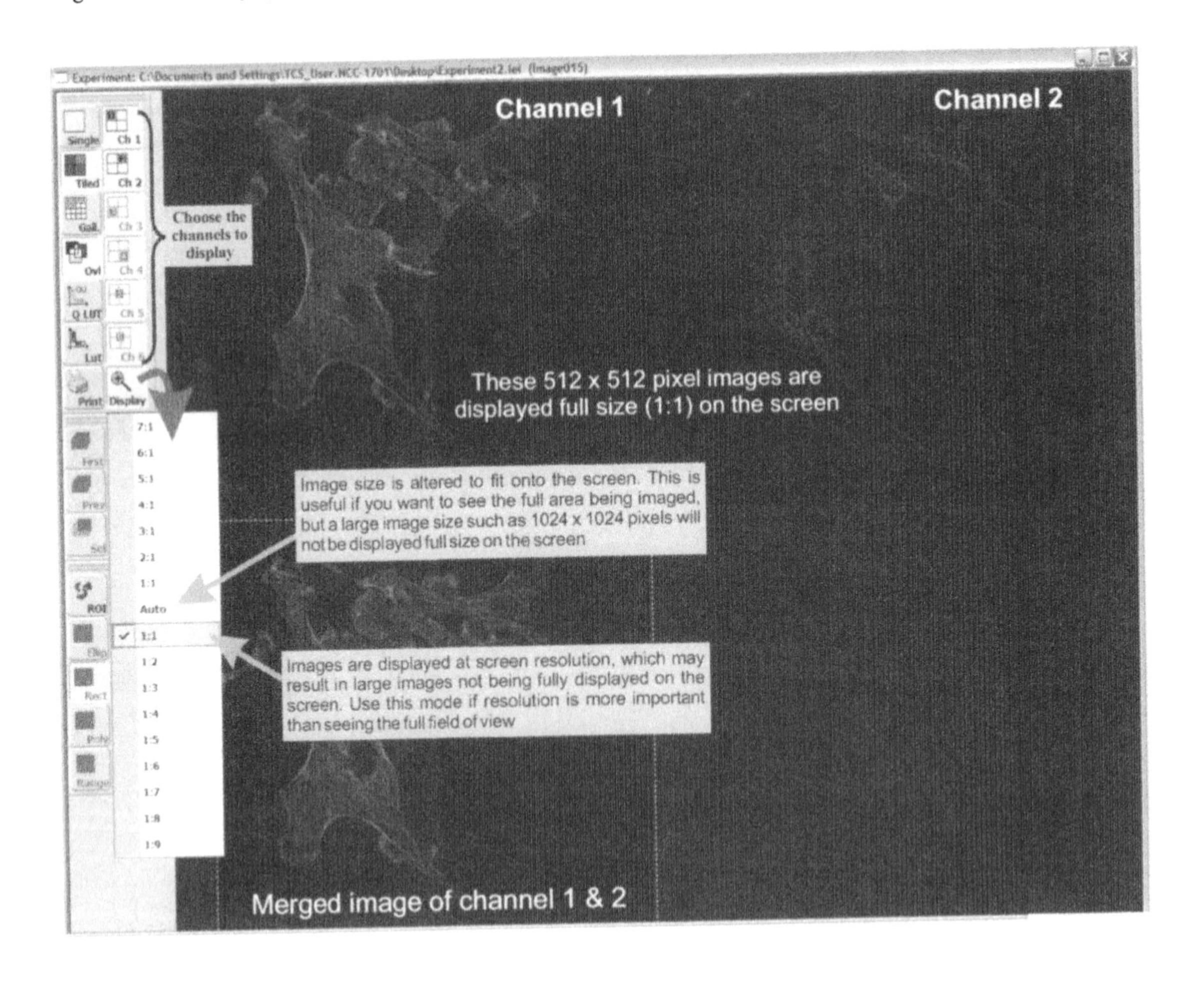

Figure A-37. Leica SP2 Screen Two - Image Display.
The second computer screen is used to display the images as they are collected. The images can be displayed in a variety of sizes, depending on whether seeing all of the pixels (the 1:1 view) or the entire image is more important. You can set the LUT colour display used for the images during collection by changing the settings on the individual channels on the "beam path settings" panel (Figure A-34).

Nikon Instruments

The Nikon optics company produces a wide range of high quality optical instruments, including a variety of research grade light microscopes and superb lenses. The Nikon research grade light microscopes are used for both the Nikon confocal microscopes, and as the main microscope used by Bio-Rad for attachment to the Radiance or MRC scan heads.

NIKON CONFOCAL MICROSCOPES

The Nikon confocal microscopes use the established technology of dichroic mirrors and optical filters to separate the fluorescent light into individual imaging channels. The Nikon instruments are aimed at the "personal" user, where high quality optics for specific applications is perhaps more important than the greater versatility of the more complex (and more expensive) models available from other manufacturers.

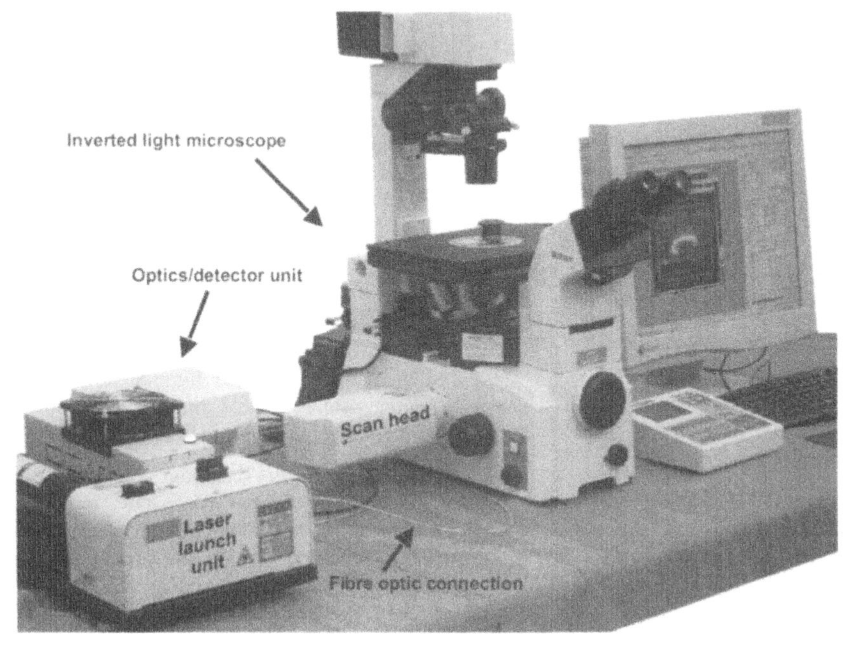

Figure A-38. Nikon C1 Confocal Microscope.
The Nikon C1 confocal microscope consists of a small scan head (shown here directly attached to the side port of a Nikon inverted research grade light microscope), and a fibreoptic connected optics/detector unit and a fibreoptic connected laser launch unit. The Nikon C1 confocal microscope is designed on a modular basis for ease of instrument upgrading and transfer between microscopes.

Nikon Instruments **Web:** www.ave.nikon.co.jp/inst/Biomedical/c1/index.htm

Address:	Nikon Instech Co., Ltd. Parale Mitsui Bldg., 8, Higashida-cho, Kawasaki-ku, Kawasaki, Kanagawa 210-0005 **Japan**	Nikon Instruments Inc. 1300 Walt Whitman Road Melville, New York 11747-3064 **USA**
Phone:	+81-44-223-2167	+1 (631) 547 8500
FAX:	+81-44-223-2182	+1 (631) 547 0299

Nikon Confocal Microscopes

Nikon "Personal" Confocal Microscopes

The Nikon confocal microscopes have been designed with the "single" laboratory user in mind rather than the large multi-user facilities where more versatile, but significantly higher cost, confocal microscopes are often installed.

C1 Digital Eclipse	Three fluorescence channel "modular" confocal microscope, plus transmission channel, that can be readily upgraded after purchase and installation. The confocal microscope is designed on a modular basis, which includes interchangeable optical filter blocks for use with various dyes. The small and lightweight fibreoptic connected scan head can be readily mounted on a variety of microscopes.

PCM 2000	Manually operated "personal" or laboratory based confocal microscope with two fluorescence channels and a single transmission channel, or three fluorescence channels without transmission imaging. The small, easily moved, scan head can be attached to either an upright or inverted microscope. Manually controlled pinhole size selection and dichroic mirror and optical filter changes.

Nikon High-Speed Confocal Microscope

This laser scanning confocal microscope can achieve image acquisition rates as high as 30 frames per second. This instrument is designed as a high-speed confocal microscope for studies involving fast cellular processes such as intracellular calcium ion changes.

RCM 8000	High-speed dual fluorescence channel confocal microscope capable of imaging 30 frames per second with an image size of 512 x 483 pixels. Specially designed for ratiometric imaging using visible or UV laser lines. Real time ratiometric display of image data is possible at high-speed acquisition rates.

Nikon is a major manufacturer of a wide range of optical based instruments, ranging from conventional cameras to digital cameras and a variety of high quality light microscopes.

Nikon also manufactures optical lenses for use in a variety of instruments, from cameras to research microscopes. Nikon produces a wide range of very high quality microscope objectives, including non-immersion or "dry" objectives as well as oil, water, glycerol and multi-immersion objectives. Nikon produces a superb "air" objective with relatively high NA and a very long working distance – with an adjustable collar to allow for various differences in coverslip thickness or to correct for spherical aberration within the sample.

Nikon manufactures excellent "research grade" upright and inverted microscopes that are used in many laboratories around the world. The Nikon confocal microscope scan heads can be attached to a variety of microscopes, including microscopes from other manufacturers. The small scan head of the Nikon confocal microscope makes moving the scan head from, for example, an upright to an inverted microscope, a relatively simple matter.

Nikon research grade light microscopes are often the microscopes of choice for attachment of the Bio-Rad confocal microscope scan head when purchasing a Bio-Rad confocal or multi-photon microscope.

The Nikon laser scanning confocal microscopes are designed as a number of separate "modules" that can be readily upgraded or replaced. The scan head is relatively small, with most of the optics and the detectors being located in a fibreoptic connected "optics box" that is placed on the bench near the microscope.

The Nikon confocal microscopes are intended as a lower cost alternative to the more expensive confocal microscopes available – without compromise to the optics of the instrument. The cost savings are due to the more limited versatility of the instrument – which is often fine for a "personal" confocal microscope where a laboratory often has a relatively closely defined use for the instrument.

C1 – digital eclipse

Components of the Nikon C1 Confocal Microscope

The Nikon C1 is a visible light laser spot scanning confocal microscope that can be attached to either an upright or inverted light microscope.

Design

The C1 confocal microscope uses dichroic mirrors and optical filters in the form of "filter blocks" that are used in conventional epi-fluorescence microscopy. The instrument is intended as a "personal" confocal microscope that can be purchased by individual laboratories.

Scan Head

The Nikon C1 scan head is a small and compact design that is connected to the controller box and lasers via fibreoptic coupling. The compact design means that the Nikon C1 scan head can be readily transferred from an upright to an inverted microscope. The scan head optics are factory aligned, although minor adjustment of the alignment of the scan head into the microscope will be required.

Dichroic mirror: Mounted inside the C1 scan head is a dichroic mirror filter slider, that can be manually changed, for initial separation of irradiating laser light from returning fluorescent light.

Changing the confocal iris size: A computer-controlled wheel containing three different pinhole sizes and a 4th "open" pinhole setting is mounted inside the scan head.

Fibreoptic pickup: The fluorescent light, after passing or being deflected by the primary dichroic mirror mounted within the scan head, is transmitted to the optics box by fibreoptic cable.

Scan mirrors: X & Y fast galvo scan mirrors with zoom capabilities.

Optics box

The returning fluorescent light is transferred from the scan head to the controller box via a fibreoptic cable. The light then passes through suitable dichroic mirrors and barrier filters, and into the PMT tube. The digitised image is then transferred to the computer.

Barrier filters: Filter cubes for a variety of fluorophore combinations can readily be changed (manually) to match the particular fluorophore combination used. These filter cubes are physically the same as those used in the Nikon epi-fluorescence microscope, but often with special dichroic and optical filter combinations for labelling applications particularly applicable to confocal microscopy.

Light detectors: The small compact optics box can be readily swapped over from the standard two photomultiplier tube detector unit to a three detector unit when such an upgrade is required.

Lasers

Lasers: A wide range of lasers can be attached via fibreoptic connection, including the Kr/Ar, Ar-ion, HeNe, 405 nm violet and red laser diode and HeCd (up to three lasers at a time).

Neutral density or AOTF filters: A series of neutral density filters (or optional AOTF filters) are located within the controller box. The appropriate filter is manually selected (ND filters) or computer adjusted (AOTF filters) to adjust the amount of laser light that reaches the sample.

Laser line selection: Suitable filters can be placed in front of the laser beam before the laser light enters the fibreoptic cable for transfer to the scan head. A computer controlled shutter is used to select the appropriate laser.

Microscope

The scan head can be attached to either an upright or inverted microscope. The scan head is designed to be fitted to Nikon research grade microscopes, but the scan head can be readily connected to other microscopes perhaps an existing research grade microscope you already have.

Imaging Modes

Multi-channel fluorescence, backscatter (reflectance), transmission. Box size user selectable up to 2048 x 2048 pixels. Zoom and panning. Time lapse imaging. 12-bit data acquisition and processing.

Transmission

A separately supplied transmitted light detector with bright-field, Phase and DIC imaging capability if the microscope is fitted with suitable optics.

Manual Control

The Nikon C1 scan head is partially manually operated (scan head dichroic mirror sliders are changed by hand), with computer control of image acquisition parameters.

Computer

High-end PC computer to collect images, adjust PMT voltages and to move the z-position when collecting z-sections. However, many functions within the scan head, and the optics box are operated manually. A single large computer screen is used to both display the images as they are collected, and to display the controls for scanning and changing microscope settings.

Software

Nikon acquisition and analysis software (including VBA macro language) under Windows 2000/XP.

Models

A modular design allows this instrument to be upgraded to incorporate more channels (up to 3) or to separate specified regions of the light spectrum by purchasing different optical filter blocks.

Nikon C1 Scan Head

The Nikon C1 confocal microscope manufactured by Nikon is based on the well-established technology of dichroic mirrors and optical filters, with a manually controlled scan head that offers excellent optics for a very affordable price. The Nikon C1 scan head (Figure A-39, upper photo) is small and robust, allowing for easy transfer from upright or inverted microscope. The scan head houses the scanning mirrors, confocal pinhole wheel, primary dichroic mirror, and associated optics. All other optical components, including the dichroic mirrors and optical filters used to separate various regions of the light spectrum are housed in the fibreoptic connected Optics/Detector Unit. The confocal pinhole size is altered by changing the position of the pinhole wheel, which contains a range of pinhole sizes, including an "open" no-pinhole position.

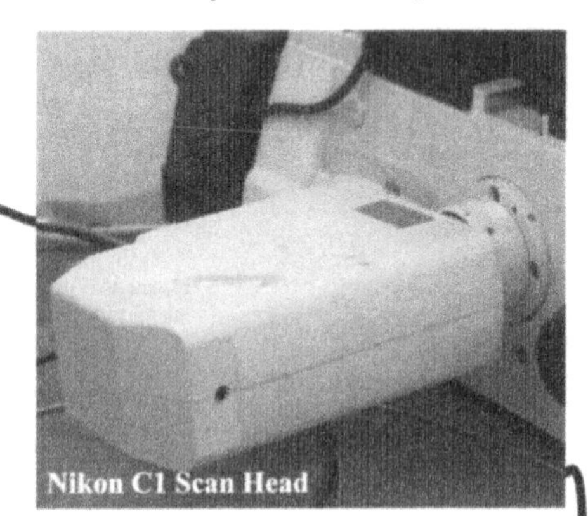

Optics/Detector Unit

Various regions of the light spectrum are separated and directed to individual detection channels within the optics/detection unit (Figure A-39, photograph on the left). This unit can have up to three separate detection channels (3 photomultiplier tubes and 3 filter blocks). The optical filter blocks are physically the same as those used in the Nikon epi-fluorescence wide-field microscope – although you may require specialised optical filters and dichroic mirrors, depending on the fluorophores being used. The filter bocks can be readily changed for optimal detection of various fluorescent dyes.

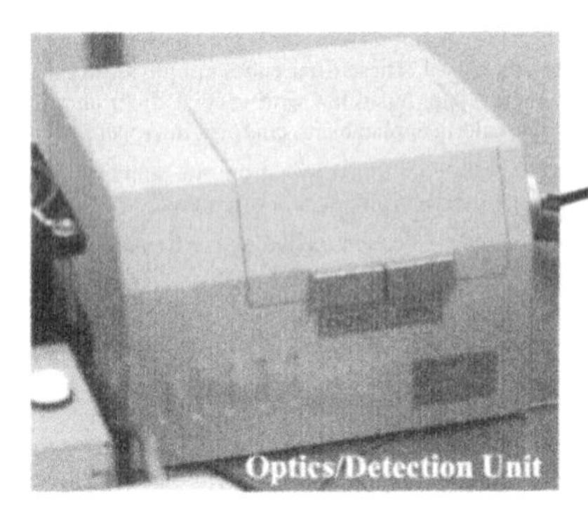

Figure A-39. Nikon C1 Confocal Microscope.
The small robust Nikon C1 confocal microscope scan head is shown above attached to the side port of a Nikon inverted microscope. The C1 optics/detection unit (shown on the left) houses the necessary optics for splitting the light spectrum into various regions using Nikon fluorescence microscope optical filter cubes. The laser launch unit (shown below) can house up to three independent lasers, with the light being transferred to the scan head via fibreoptic connection.

Laser Launch Unit

The laser launch unit (Figure A-39, lower photo) is an integrated laser platform on which several different lasers can be mounted. The laser launch unit shown in Figure A-39 has an air-cooled argon-ion single-line laser and a green HeNe laser installed. Laser line selection and intensity control are provided by the unit. Laser lines from the different lasers can be combined if required and then transferred to the scan head by fibreoptic connection.

NIKON C1 USER INTERFACE

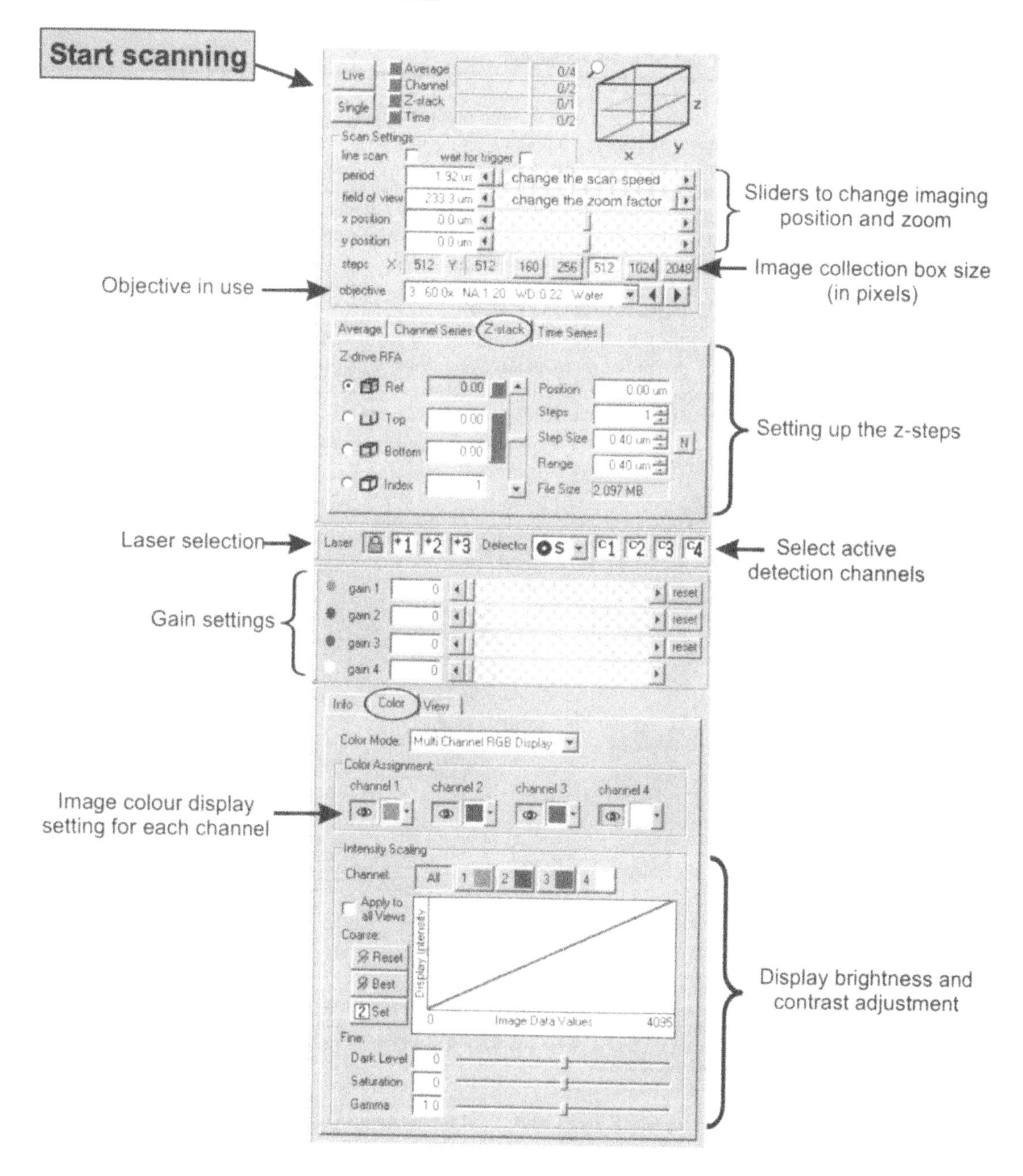

Figure A-40. Nikon C1 Main Control Panel.
The Nikon C1 confocal microscope is controlled from the above panel. All scan head and laser controls are available from this one control panel (which is usually displayed alongside the imaging panes). The Nikon C1 computer interface is usually operated on a single large computer screen that displays both the above menu and the images as they are acquired.

Olympus Corporation

The Olympus Corporation produces a large range of high quality optical instruments, from digital cameras to research grade light microscopes, including a number of objective lenses specifically designed for use in the biological sciences. The Olympus laser scanning confocal microscope scan heads, which can be attached to a wide range of upright and inverted microscopes, are capable of excellent resolution and offer great versatility. The latest instrument from Olympus utilises spectral separation as well as the established technology of dichroic mirrors and optical filters for separating the fluorescent light into various regions of the light spectrum.

OLYMPUS CONFOCAL MICROSCOPES

Olympus has recently introduced the fully computer controlled spectral separation dual scan Fluoview 1000 (FV1000) confocal microscope, but the previous manually controlled scan head of the Fluoview 300 is still manufactured for laboratories that require high quality optics, but do not necessarily require the added versatility (and greater expense) of the computer controlled scan head.

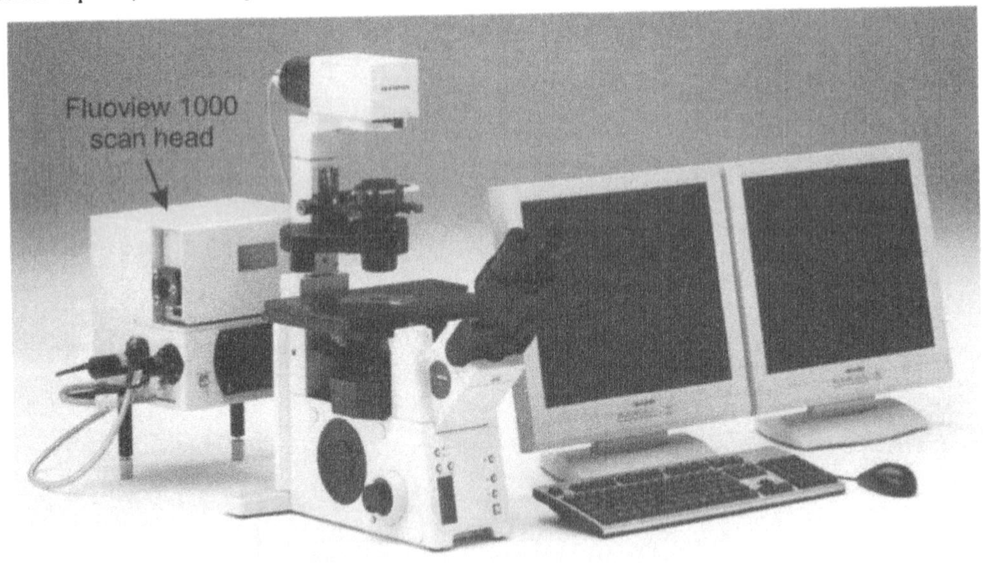

Figure A-41. Olympus Fluoview 1000 Confocal Microscope.
Olympus manufactures two laser scanning confocal microscopes for use in the biological sciences, the fully computer controlled spectral separation Fluoview 1000 (above) and the manually adjusted Fluoview 300 (Figure A-44). The scan head can be attached directly to either an upright or inverted research grade light microscope, and in this case is attached to the rear access port of an Olympus inverted microscope. This photograph was kindly provided by Olympus Australia.

	Olympus Corporation	Olympus Europe GmbH	Olympus America Inc.	Olympus Australia Pty Ltd
Address:	Shinjuku Monolith, 3-1 Nishi Shinjuku, 2-chrome, Shinjuku-ku, Tokyo 163-0914, **Japan**	Wendenstrasse 14-18 D-20097 Hamburg **Germany**	2 Corporate Center Drive Melville NY 11747-3157 **USA**	31 Gilby Rd Mt Waverley VIC 3149 **Australia**
Phone:	+81 3 3340 2111	+49 40 23773-0	+1 631 844 5000	+61 3 9265 5400
Web:	www.olympus-global.com/en/global/	http://cf.olympus-europa.com	www.olympusamerica.com	www.olympus.co.jp/en/lineup/

Olympus Confocal Microscopes

Olympus Confocal Microscopes

Olympus currently manufacture two laser spot scanning confocal microscopes (Fluoview 1000 and Fluoview 300) that are used extensively in the biological sciences. Olympus also manufacture a Nipkow disk scan head (MZX50-CF) that is mainly used as a semiconductor inspection microscope.

Laser Scanning Confocal Microscopes

Fluoview 1000	A dual scan confocal microscope with spectral separation capabilities. Two spectral separation channels and up to three optical filter separation channels, in addition to a single transmission detection channel. Fully computer controlled with a single adjustable confocal pinhole. Multiple lasers can be attached via fibreoptic connections.
Fluoview 500	The Fluoview 500 laser scanning confocal microscope is sold as a confocal microscope suitable for multi-user facilities that offers considerable versatility with tried and true dichroic mirror and optical filter technology. The Fluoview 500 scan head is fully computer controlled, with up to four fluorescence channels and one transmission channel. Multiple lasers can be attached via fibreoptic connections. This instrument has now been replaced by the Fluoview 1000.
Fluoview 300	Designed as a "personal" confocal microscope, with a manually operated scan unit, with excellent optics using conventional dichroic mirrors and optical filters. Two fluorescence channels and one transmission channel are available. A single laser can be attached directly to the scan head via a fibreoptic connection, or multiple lasers can be used by incorporating the additional lasers into the Olympus laser combiner.

Nipkow Disk Confocal Microscope

MX50-CF	Real time confocal microscope based on a Nipkow disk scan head and mercury or xenon lamp illumination. Mainly used as a semiconductor inspection microscope.

Earlier Confocal Microscope

FVX	Early model confocal microscope from Olympus, with either two fluorescent channels or one fluorescent and one transmission channel.

Olympus Corporation is a major manufacturer of optically based instruments, including scanners, cameras, medical instruments and microscopes.

Olympus manufactured the first Japanese made microscope in 1919 under the name Takachiho Seisakusho. The company, known as Olympus since 1921, now designs and manufactures a wide range of "student" level and "research grade" microscopes, including stereomicroscopes, upright microscopes and inverted microscopes.

Although Olympus was originally known for their production of good quality "student" microscopes, over the past few years the quality of the research grade microscopes, and the range of high quality objective lenses that Olympus produces, have increased greatly.

The "research" grade objective lenses manufactured by Olympus include oil and water immersion objectives, dipping objectives and non-immersion objectives, many of which are specifically designed for the biological sciences, including specialised lenses for confocal microscopy.

Olympus produces two confocal microscopes for the biological sciences, the Fluoview 1000, which is fully computer controlled and capable of spectral separation, and the Fluoview 300, which has a manually operated scan head and utilises dichroic mirrors and optical filters for light separation. The well established technology of using dichroic mirrors and optical filters to both select the laser lines used, and to separate the fluorescence signal, means that the Fluoview 300 can be produced at considerably lower cost compared with the "high end" confocal microscope instruments from Olympus and other manufacturers. The manually operated Fluoview 300 model is still an excellent choice where high quality optics are required but a lower purchase price and in some cases the ability to readily modify the various optical components are important.

Olympus also produces a Nipkow disk based confocal microscope specifically for use in the materials sciences.

Components of the Olympus Fluoview 1000

The Olympus Fluoview 1000 laser spot scanning confocal microscope is sold as a visible light instrument, but can be readily configured by the user to incorporate UV or multi-photon microscopy.

Design

The Olympus Fluoview 1000 confocal microscope is a fully computer controlled dual scan instrument with two spectral separation channels and an additional two detection channels that utilises dichroic mirrors and optical filters.

Scan Head

The Fluoview 1000 scan head is attached to the rear light path of either an inverted or an upright Olympus research microscope.

Fully automated scan head (computer controlled) and is relatively large, containing all of the optics, as well as the scanning mirrors, pinhole and photomultiplier tubes.

Spectral channels: Two independent spectral separation channels with separate diffraction gratings.

Dichroic mirrors: Mounted inside the scan head are computer-controlled wheels that contain a number of dichroic mirrors used to separate the laser excitation light from the emitted fluorescence, and also to separate different fluorescence emissions in multi-labelling applications. These mirrors are used for the initial separation of different wavelengths for both the spectral separation channels and the optical filter based separation channels.

Barrier filters: Computer-controlled wheels of barrier filters are mounted in front of each of the photomultiplier tubes used for channel 3 and 4 (the optical filter separation channels). These barrier filters, in combination with the dichroic mirrors, provide the means by which various fluorescent colours can be separated.

Variable size confocal iris: A single common computer controlled variable size confocal iris.

Light detectors: Up to four photomultiplier tubes.

Scanning mirrors: X & Y fast galvo scan mirrors, zoom, bi-directional & scan rotation capabilities.

SIM (Simultaneous scanner): the instrument is capable of simultaneous but independent scanning with two different lasers or laser lines.

Lasers

Lasers: A variety of lasers can be mounted directly onto the scan head via fibre optics. However, a more versatile option is to use the Olympus laser combiner, which can be used to combine the output from a number of different lasers. This includes the Kr/Ar, Ar-ion, HeNe, 405 nm violet and 440 nm blue diode, single line Kr-ion, UV and IR lasers.

Laser attenuation: AOTF filters located within the laser combiner unit are used to attenuate the amount of laser light reaching the sample (in the range of 0.1 to 100%).

Laser line selection: Suitable computer controlled optical filters for selecting the appropriate laser line can be placed in front of the laser beam before the laser light enters the fibreoptic cable for transfer to the scan head.

Microscope

Olympus research grade inverted or upright microscopes.

Imaging Modes

Multi-channel fluorescence, spectral imaging, backscatter (reflectance), transmission. Box size user selectable up to 4096 x 4096 pixels, zoom and panning. Dual scanning, simultaneous and sequential multi-channel scanning. Time lapse imaging. 12-bit data acquisition and processing, saved as 8-bit or 12-bit files. Region of interest scan (clip scanning), including two independent sets of regions of interest when dual scanning. Rotated scan, line scanning and free-line scanning.

Transmission

A single photomultiplier tube transmitted light detector. Bright-field, Phase and DIC imaging capabilities, depending on the optics installed in the light microscope.

Manual Control

The Fluoview 1000 confocal microscope is fully computer controlled with no provision for manual adjustment of acquisition parameters.

Computer

High-end PC computer to control and acquire images. Dual screen computer monitors are used for displaying the control panel and the images as they are collected.

Software

Fluoview software running under Windows XP operates the Fluoview 1000 scan head.

Options

Dual or single scanner, optional 4th fluorescence detection channel, fibre optic output and non-confocal detectors.

Models

Fluoview 1000 (spectral + filter system, or filter system only), the manually operated **Fluoview 300** (see page 427) and the previously manufactured **Fluoview 500** (see page 424).

FLUOVIEW 1000 CONFOCAL MICROSCOPE

The Fluoview 1000 dual scan spectral separation confocal microscope is designed for high resolution imaging in the biological sciences. The scan head is directly attached to a variety of ports on either an upright or inverted research grade light microscope. The Fluoview 1000 scan head (Figure A-42) contains all of the necessary components for scanning the laser beam across the sample, the separation of various wavelengths of fluorescent or reflected light and their detection by up to four independent channels. A range of lasers can be connected to the scan head either directly by fiber optic connection, or by using the Olympus laser combiner platform.

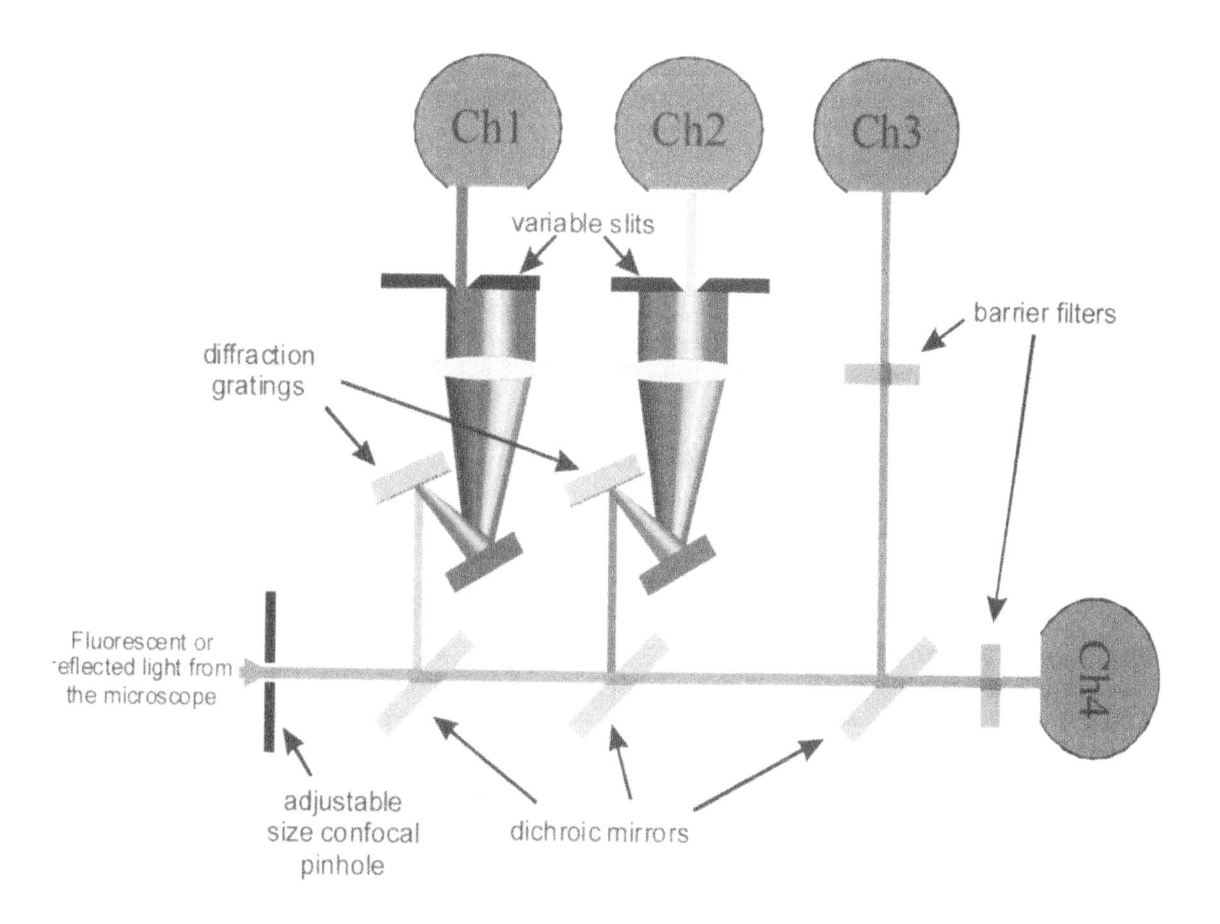

Figure A-42. Olympus Fluoview 1000 Scan Head Layout.
The Olympus Fluoview 1000 scan head contains up to four detection channels, two of which are based on spectral separation using a diffraction grating and two use conventional dichroic mirrors and barrier filters (a more detailed description of the method of light separation used in the Olympus Fluoview 1000 scan head is given in Figure 3-13 on page 84). A single variable size confocal pinhole or iris is used for all detection channels. The lasers are mounted on a separate laser combiner and attached to the scan head via separate fibreoptic connections. The scan head is attached to the rear light entry port on an inverted microscope (as shown in Figure A-41) or the top illumination port on an upright microscope. This figure is adapted from a diagram kindly provided by Olympus Australia.

The scan head contains up to four independent detection channels (Figure A-42). The two spectral separation detection channels contain diffraction gratings and variable width slits that can be moved to select specified regions of the light spectrum. The slits can be used to specify specific wavelength ranges that are to be directed to each channel, or they can be used to collect a very narrow band of light (down to 1 nm) for spectral imaging. The various spectral separation techniques used by different confocal microscope manufacturers are described in detail on page 81 – 85 in Chapter 3 "Confocal Microscopy Hardware". The scan head also contains one, or if required, two "conventional" detection channels that utilize dichroic mirrors and optical filters to separate various regions of the light spectrum. The spectral separation and the optical filter based channels can be used to simultaneously image up to four different fluorophores. Alternatively, the spectral imaging channels can be used to collect spectral information from fluorophores with highly overlapping fluorescence emission spectra. Linear unmixing is then used to separate the contribution from the individual fluorophores, which is then displayed as separate images.

A single computer controlled variable size confocal iris or pinhole is used to select only the focal plane within the specimen. Non-confocal detectors (positioned within the scan head prior to the confocal pinhole) for multi-photon microscopy are also available as a optional accessory.

Feedback control of laser intensity, high-sensitivity photomultiplier tubes with photon counting capability and the ability to collect spectral information over a broad range (400-790 nm) at high resolution make this instrument a highly versatile microscope for biology, and particularly for live cell imaging.

SIM (Simultaneous) Scanner

The Fluoview 1000 scan head is capable of simultaneous and independent dual scanning with two different laser lines or lasers. This feature allows one to perform real-time fluorescence photobleaching (FRAP) and uncaging experiments by using one laser to uncage or bleach a specified region of the specimen while a second laser is used to collect the image. This feature also allows the simultaneous collection of two independent sets of regions of interest (ROI) within the sample.

FLUOVIEW 500 CONFOCAL MICROSCOPE

The Fluoview 500 confocal microscope has a fully automated scan head with a fibreoptic connected computer controlled laser platform. The instrument is capable of imaging up to four fluorescence channels and a single transmission channel simultaneously. Although the instrument is fully computer controlled the relatively simple optics do mean that the scan head can be readily customised by more experienced users. This includes changing the dichroic mirrors and optical filters and the attachment of pulsed infrared multi-photon lasers.

Fluoview 500 Scan Head

The Fluoview 500 scan head shown in Figure A-43 is mounted directly onto the side of an Olympus inverted microscope, but can also be mounted in a number of other positions, including to the rear of the microscope as shown in Figure A-41. The separation of the fluorescent light into defined regions of the light spectrum is achieved by using the established technology of optical filters and dichroic mirrors (Figure A-43). The scan head contains computer controlled variable size pinholes for each of up to four detection channels. This is in contrast to the single pinhole (with a choice of sizes) used for both channels in the Fluoview 300 scan head, and the single variable size pinhole used in the Fluoview 1000 scan head.

Custom designed optical filters and dichroic mirrors can be readily fitted into the filter wheels used in the Fluoview 500 scan head.

A number of different lasers can be directly fibreoptic connected to the scan head, or they can be mounted on a separate laser platform which contains laser line selection and attenuation filters. Pulsed infrared lasers for multi-photon microscopy can be readily attached directly to the Fluoview 500 scan head.

Galvanometric scan mirrors can be rotated to scan any defined area and box size within the microscope field of view. Although several defined regions of interest (ROI) can be scanned simultaneously, this instrument is not capable of dual scanning (where two completely independent ROI can be scanned simultaneously).

Fluoview 500 / 300

Components of the Olympus Fluoview 300 and 500

Design

The Olympus Fluoview 300 and 500 laser spot scanning confocal microscopes are sold as visible light instruments, but can be readily configured by the user to incorporate UV or multi-photon microscopy.

The Olympus Fluoview 300 confocal microscope is a manually operated instrument that is affordable for individual laboratories. The previously manufactured Fluoview 500 is fully computer controlled, but has now been replaced by the Fluoview 1000. The Fluoview 300 and 500 instruments are based on the use of dichroic mirrors and optical filters.

Scan Head

The Fluoview scan head is attached to either an inverted or an upright Olympus research microscope. The attachment locations include the familiar top position for an upright microscope and the side position for an inverted microscope, or the rear light path of either an upright or inverted microscope.

Fluoview 500

Fully automated scan head (computer controlled) and is relatively large, containing all of the optics, as well as the scanning mirrors, pinholes and photomultiplier tubes.

Dichroic mirrors: Mounted inside the scan head are computer-controlled wheels that contain a number of dichroic mirrors used to separate the laser excitation light from the emitted fluorescence, and also to separate different fluorescence emissions in multi-labelling applications.

Barrier filters: Computer-controlled wheels of barrier filters are mounted in front of each of the photomultiplier tubes. These barrier filters, in combination with the dichroic mirrors, provide the means by which various fluorescent colours can be separated.

Variable size confocal iris: A computer controlled variable size "pinhole" is mounted in front of each photomultiplier tube within the scan head.

Light detectors: Up to four photomultiplier tubes.

Scanning mirrors: X & Y fast galvo scan mirrors, zoom, bi-directional & scan rotation capabilities.

Fluoview 300

Manually operated dichroic mirrors, optical filters and pinhole size adjustment within the scan head.

Dichroic mirrors: Various dichroic mirrors are mounted on manual sliders within the scan head.

Barrier filters: Manually adjusted sliders are used to change the barrier filters within the scan head. These barrier filters, in combination with the dichroic mirrors, provide the means by which various fluorescent colours can be separated.

Confocal iris: Fluoview 300 scan head has a manually operated five-position pinhole unit.

Light detectors: Two photomultiplier tubes.

Scanning mirrors: X & Y fast galvo scan mirrors, zoom, bi-directional & scan rotation capabilities.

Lasers

Lasers: A variety of lasers can be mounted directly onto the scan head (three attachment ports for the Fluoview 500 and one for the Fluoview 300). However, the Olympus laser combiner, can be used to combine the output from a number of different lasers. This includes the Kr/Ar, Ar-ion, HeNe, 405 nm violet and 440 nm blue diode, single line Kr-ion, UV and IR lasers.

Laser attenuation: Neutral density filters or AOTF filters located within the scan head or laser combiner unit, are used to attenuate the amount of laser light reaching the sample.

Laser line selection: Suitable optical filters for selecting the appropriate laser line can be placed in front of the laser beam before the laser light enters the fibreoptic cable for transfer to the scan head. These filter wheels are controlled manually in the Fluoview 300 and by computer in the Fluoview 500

Microscope

Olympus research grade inverted or upright microscopes.

Imaging Modes

Multi-channel fluorescence, backscatter (reflectance), transmission. Box size user selectable up to 1024 x 1024 pixels, zoom and panning. Simultaneous and sequential multi-channel scanning. Time lapse imaging. 12-bit data acquisition and processing, saved as 8-bit or 12-bit files. Single region of interest scan (clip scanning), rotated scan, line scanning and free-line scanning.

Transmission

A single transmission photo diode transmission detector. Bright-field, Phase and DIC imaging capabilities, depending on the optics installed in the light microscope.

Manual Control

The Fluoview 500 confocal microscope is fully computer controlled with no provision for manual adjustment of acquisition parameters. The Fluoview 300 is a manually operated confocal microscope.

Computer

High-end PC computer to control and acquire images in the Fluoview 500 confocal microscope. The Fluoview 300 scan head is a manually operated instrument, but uses a computer for image acquisition. A single large computer screen is used for both instruments.

Software

Fluoview software running under Windows NT/2000 operates both the Fluoview 500 and Fluoview 300 scan heads (with the appropriate components loaded during installation).

Models

Fluoview 300 (manually controlled), the previously manufactured **Fluoview 500** (computer controlled) and the spectral separation Fluoview 1000 (see page 423).

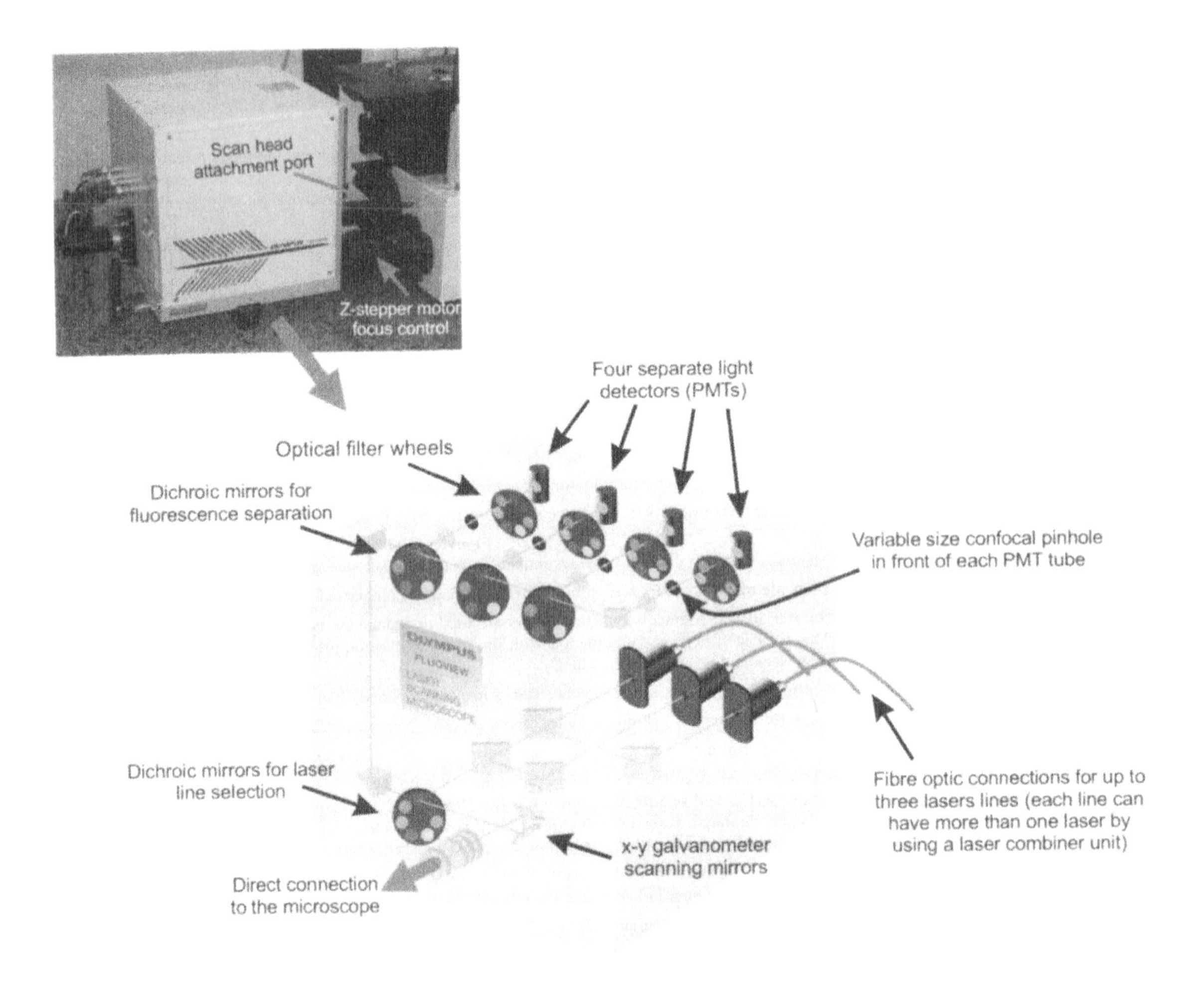

Figure A-43. Olympus Fluoview 500 Scan Head.
The Olympus Fluoview 500 scan head contains most of the optical components necessary for confocal microscopy. This includes the dichroic mirrors, optical filter wheels, scanning mirrors, variable size pinholes, photomultiplier tubes (up to four) and laser line selector. However, the lasers are mounted on a separate laser combiner and attached to the scan head via three separate fibreoptic connections. The scan head is shown here attached to the side port of an Olympus inverted microscope, but can also be attached to the rear light entry port on an inverted microscope (Figure A-41) or the top video port on an upright microscope. This figure is adapted from a diagram kindly provided by Olympus Australia.

FLUOVIEW 300 CONFOCAL MICROSCOPE

The Fluoview 300 confocal microscope is a manually operated confocal microscope with somewhat limited versatility but with excellent optics based on dichroic mirrors and optical filters for separating the fluorescent light. Several lasers can be connected via fibreoptic connections.

Fluoview 300 Scan Head

The Fluoview 300 scan head, shown here attached to an unusual upright microscope that has a rigidly fixed microscope stage (Figure A-44), is manually operated. Simple sliders are used to position the correct dichroic mirror and optical filter for imaging various fluorophores. The manual settings are very simple to adjust, with a graphical computer display of user settings for specific dye combinations. There is a choice of 5 different sizes for the single confocal pinhole (in contrast to the variable size pinhole for each channel in the Fluoview 500 instrument).

The Fluoview 300 confocal microscope scan head is significantly less expensive compared to many other confocal microscopes, but with excellent optics, and the added advantage that the scan head can be readily customised for individual use. This includes using customised dichroic mirrors and optical filters, as well as the direct attachment of pulsed infrared multi-photon lasers.

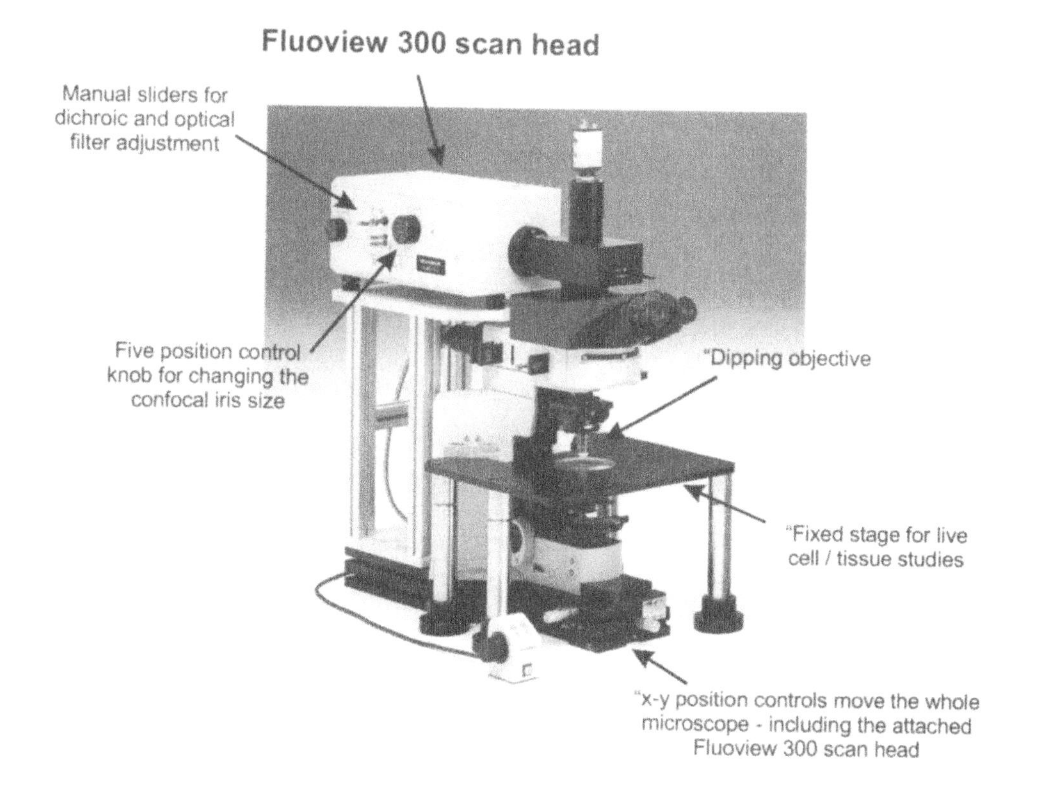

Figure A-44. Olympus Fluoview 300 Scan Head Attached to a BX61W1 Fixed Stage Microscope.
The Fluoview 300 manually operated confocal microscope scan head is shown here attached to the unusual Olympus designed fixed stage upright microscope. The scan head has been mounted on the top camera port on the microscope, but has been conveniently positioned to the rear of the microscope using a simple mirror junction box (which also allows for the attachment of a CCD camera in this example). The stage on this microscope, as the name suggests, is fixed to the bench – with all movement and focus controls being achieved by moving the microscope. This photograph was kindly provided by Olympus Australia.

Fluoview Software

The Fluoview software provided by Olympus is designed to operate the Fluoview 300, Fluoview 500 and the Fluoview 1000 scan heads. During installation of the software the various components necessary for each of the scan heads is installed. The following figures are designed to give you a brief overview of the main menus for operating the Olympus confocal microscope, based on the display used for the Fluoview 500 instrument. For a more detailed discussion you should refer to the Olympus confocal microscope operating manual.

Image Collection Panel

The main display screen (Figure A-45) contains the main control panel (described in more detail in Figure A-46) on the left hand side of the active "live" image display. A series of "tabs" on the top of the display screen can be used to access a number of recently imaged or previously opened image files.

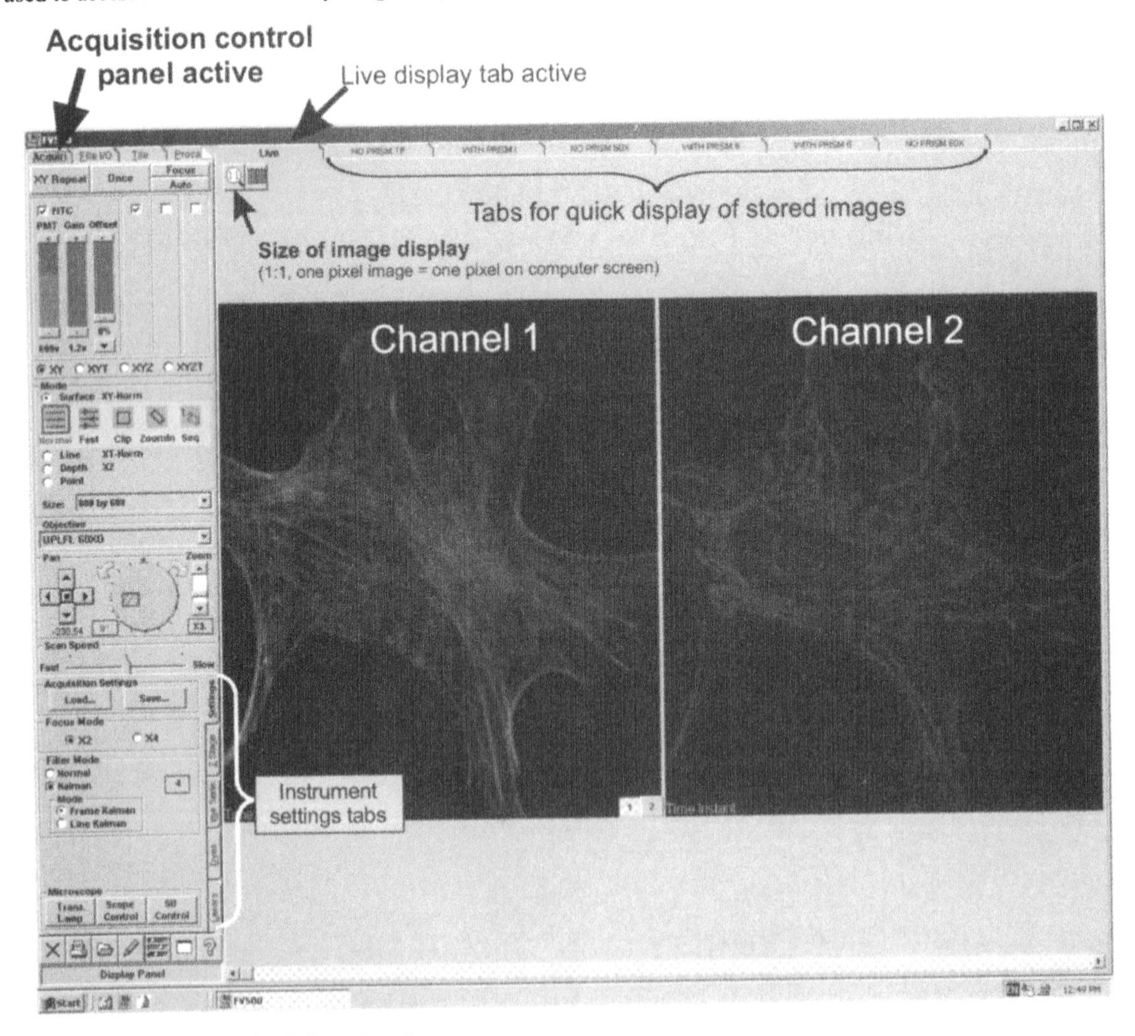

Figure A-45. Olympus Fluoview Software Interface.
The Fluoview 500 and 300 confocal microscopes are both operated by the Fluoview software, which is customised for installation on each of these instruments. All of the main control features are shown in the panel on the left. Two images are shown in this particular display, with the top tab on the screen allowing you to display at the click of a mouse a number of previous experiments without the need to go looking for the file. Detailed information on the control panel on the left hand side of this screen display is given in Figure A-46.

Main Control Panel

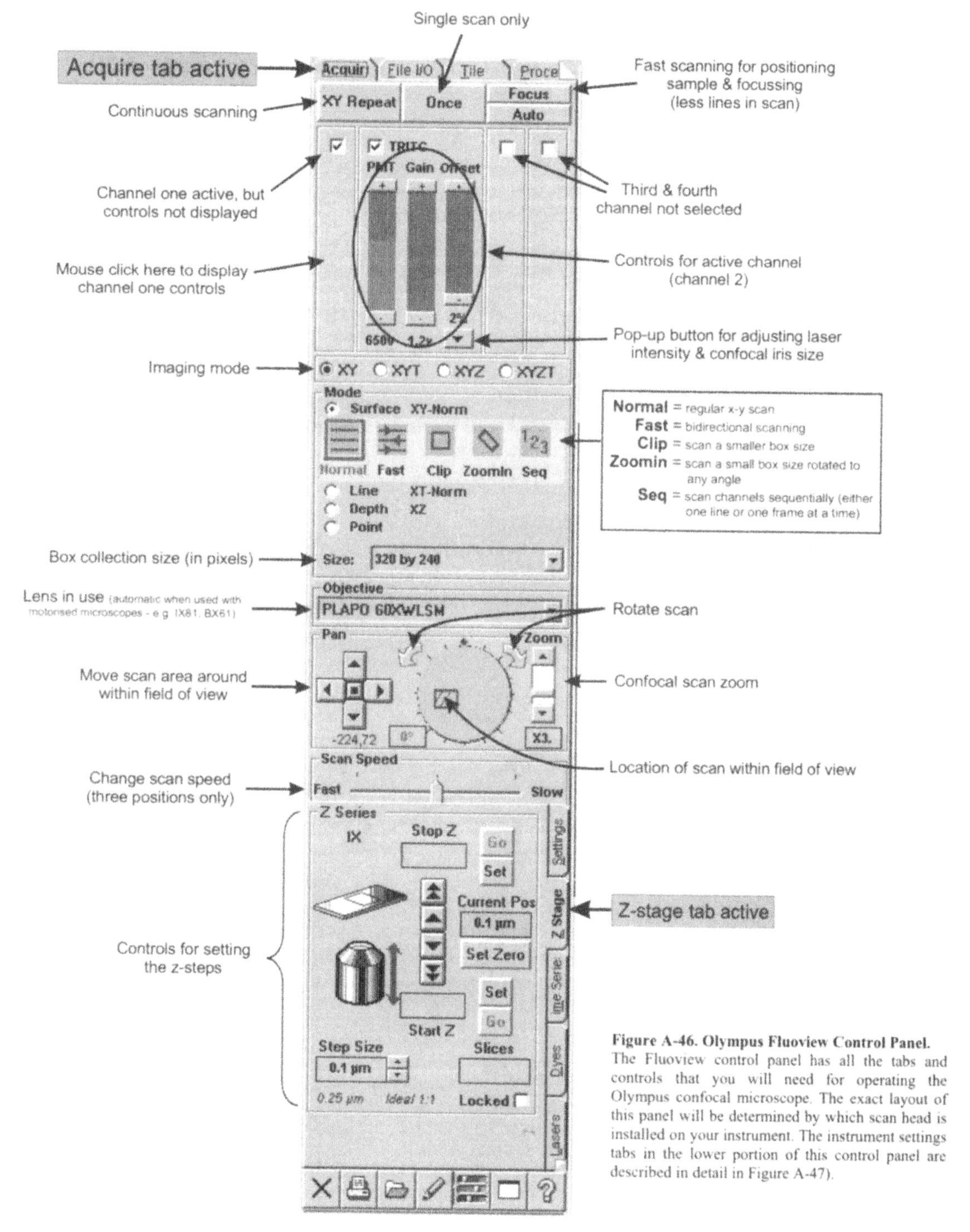

Figure A-46. Olympus Fluoview Control Panel. The Fluoview control panel has all the tabs and controls that you will need for operating the Olympus confocal microscope. The exact layout of this panel will be determined by which scan head is installed on your instrument. The instrument settings tabs in the lower portion of this control panel are described in detail in Figure A-47).

Instrument Settings Tabs

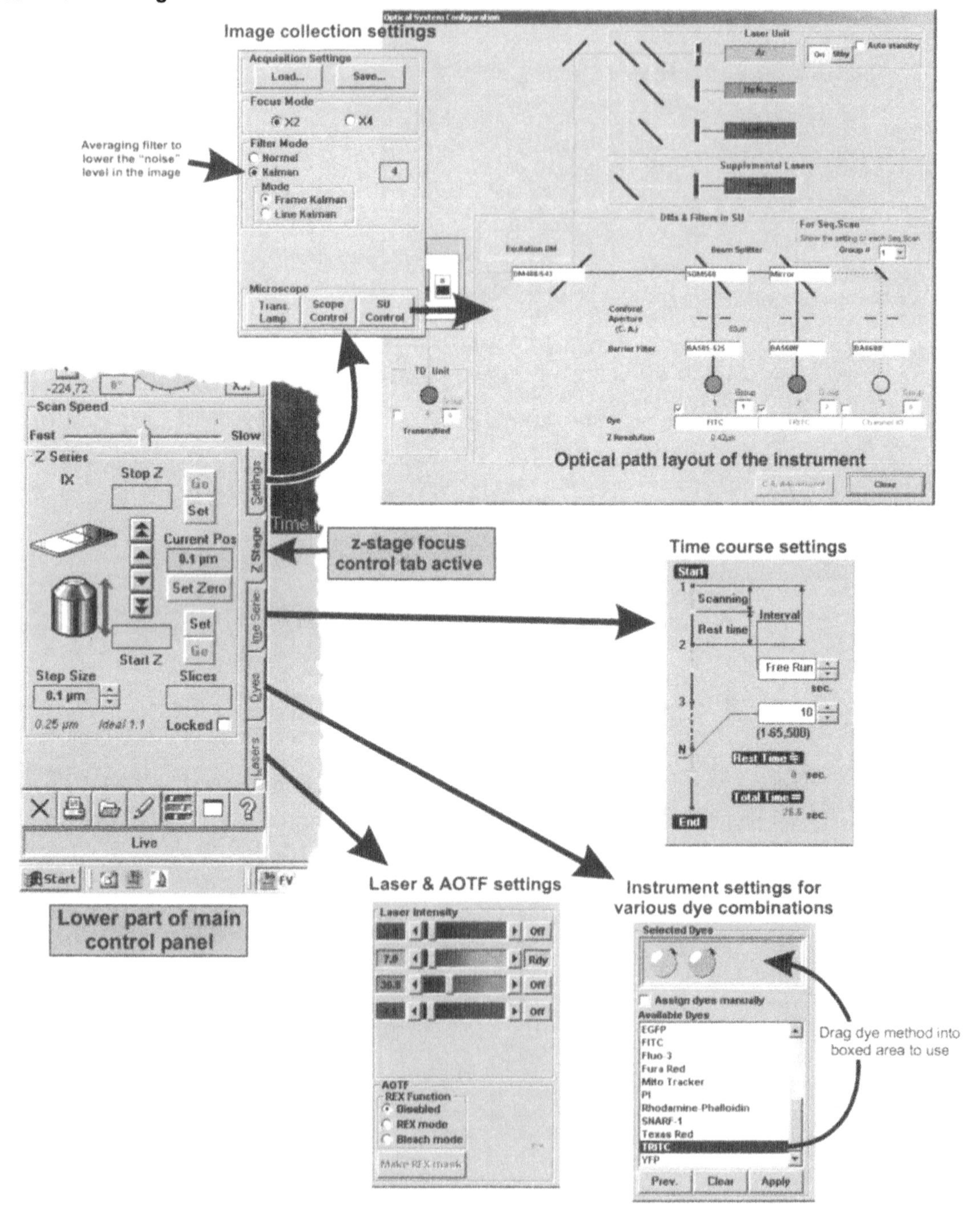

Figure A-47. Olympus Fluoview Instrument Settings Tabs.
On the lower portion of the main control panel (shown in Figure A-46) are a series of tabs for displaying submenus for changing various settings on the microscope.

Atto Bioscience

Atto Bioscience produces a number of confocal microscopes based on the spinning Nipkow disk technology. These instruments are particularly suited to live cell imaging due to the high acquisition speed and relatively low excitation light intensity required. The Nipkow spinning disk scan head produced by Atto Bioscience, called CARV, is sold as an independent unit that can be attached to a new or existing laboratory microscope (Figure A-48), or as an integrated unit, including lamp source, CCD camera and associated software from both Atto Bioscience and a number of other vendors.

ATTO BIOSCIENCE CONFOCAL MICROSCOPES

Atto Bioscience produces the CARV confocal microscope scan head (Figure A-48) as an independent unit that can be attached to a variety of upright or inverted microscopes. Atto Bioscience also produces the Pathway HT confocal microscope (based on the technology used in the CARV scan head), that is a fully integrated instrument designed for high-throughput screening of living cells.

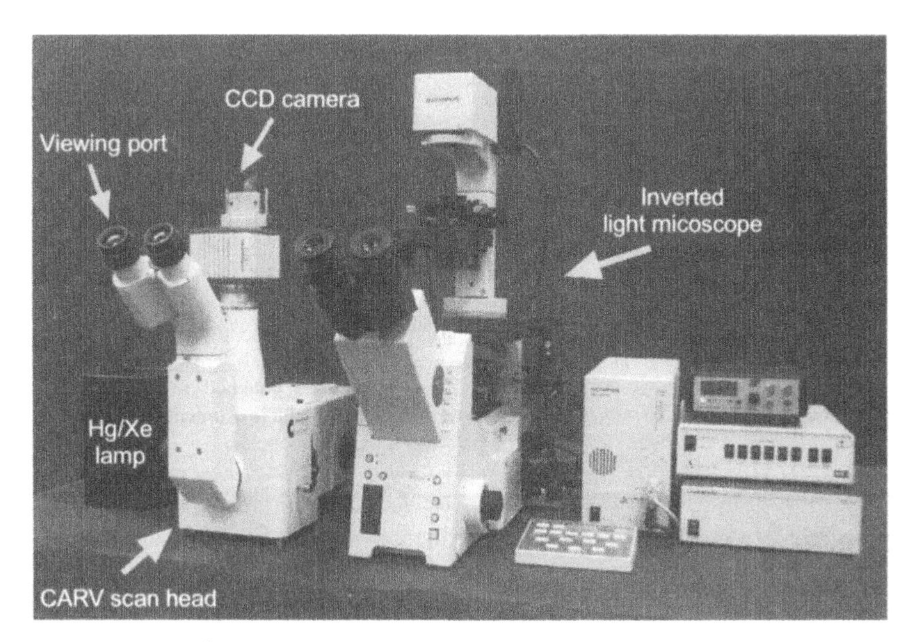

Figure A-48. Atto Bioscience CARV Nipkow Disk Confocal Microscope.
The CARV scan head (shown on the left) can be attached to either an inverted or upright microscope from a variety of manufacturers. The image can be viewed by looking directly down the eyepieces on the scan head, or a CCD camera (shown here attached to the top c-mount port on the scan head) can be used to capture the image. This instrument is capable of high-speed imaging (up to 200 frames per second, depending on the speed and sensitivity of the CCD camera attached). This photograph was kindly provided by Baggi Somasundaram, Atto Bioscience, USA.

Atto Bioscience web: www.atto.com/

Address:	15010 Broschart Road Rockville, MD 20850, **USA**		
Phone:	+1 (301) 340 7320	**FAX:**	+1 (301) 340 9775

Atto Bioscience Confocal Microscopes

Nipkow Disk Confocal Microscopes

The confocal microscopes produced by Atto Bioscience utilise a Nipkow disk with a an array of 20,000 pinholes. The image can be viewed directly by looking down the viewing port on the scan head, or a CCD camera can be attached to capture the image.

The instruments are designed for live cell imaging, and are particularly suited to either fast cellular fluxes or relatively long time series.

CARV	Single channel Nipkow disk confocal microscope scan head that attaches to a range of commercially available light microscopes. This includes upright and inverted microscopes. The image can be viewed directly through the viewing port on the scan head, or a high-sensitivity CCD camera can be used to capture the images. The instrument is capable of very high speed imaging (the Nipkow disk is scanning at 200 frames per second, but the actual speed of imaging is determined by the sensitivity and speed of the attached CCD camera). The relatively inexpensive CARV scan head is sold as an independent unit, but you will need to purchase a CCD camera and suitable software to capture images.
Pathway HT	High-throughput automated confocal imaging instrument based on the CARV Nipkow disk technology. Designed to handle 96 or 384 well culture plates, or conventional microscope slides. Automated software provides for the collection of images from a large number of samples, and the ability to analyse each individual cell within a large number of images for multi-parameter response to various stimuli. This instrument is sold as a fully integrated unit and is thus significantly more expensive than the CARV scan head alone.

The Atto Bioscience CARV Nipkow disk confocal microscope scan head was originally marketed by Atto Instruments, an optical instrumentation company based in the USA, through an exclusive relationship with Carl Zeiss, Germany. The CARV instrument, and more recently the Pathway HT confocal microscope, are now produced and marketed directly by Atto Bioscience. The CARV confocal microscope is marketed by Atto Biosciences and through a number of world-wide distributors.

The Atto Bioscience confocal microscopes, being based on the Nipkow disk, are particularly suited to live cell imaging. These instruments are significantly faster than the laser spot scanning confocal microscopes, and due to the large number of fine points of light that are scanned across the sample the light intensity at any one point is relatively low.

The Atto Bioscience confocal microscopes utilise conventional mercury or xenon lamp irradiation to excite the fluorophores. This has the advantage that one can choose wavelengths of excitation from UV through to infrared, and the existing lamp using for conventional epi-florescence microscopy can be attached to the scan head when fitting the instrument to an existing research microscope.

CARV CONFOCAL MICROSCOPES

The CARV confocal microscopes are instruments based on a scan head that incorporates a single Nipkow disk spinning at high-speed. These instruments are capable of scanning at significantly higher speeds compared to a laser spot scanning confocal microscope. Full frame image collection can routinely be obtained at several frames per second (up to 100 fps, depending on the CCD and light levels), compared to approximately 1 fps for a laser spot scanning system.

The CARV scan head can be purchased as a separate independent unit that can be attached to a light microscope. In this case you will then need to attach a light source, a CCD camera, and install suitable software for image acquisition. However, a number of companies put together an integrated unit that includes the microscope, a CCD camera and image collection software.

A fully integrated Nipkow disk confocal microscope, the Pathway HT, is designed as a high-throughput instrument for screening large numbers of cells. This instrument is based on the CARV single Nipkow disk scan head, with the added versatility of dedicated software to collect and analyse images from a large number of samples.

Atto Bioscience CARV

Components of the CARV Confocal Microscope

A real time Nipkow scanning disk confocal microscope manufactured by Atto Bioscience. The relatively small and compact CARV scan head can be purchased as a separate unit that can be attached to an existing research microscope. Alternatively you can purchase a fully functional confocal microscope, which includes the scan head, the microscope, the CCD camera and operating software.

Design

High-speed Nipkow spinning disk confocal microscope designed around the CARV scan head. Illumination by using a mercury or xenon arc lamp. The image can be directly viewed through the scan head eyepieces, or the image can be acquired using a CCD camera.

The scan head is relatively inexpensive compared to other confocal microscopes, but the total cost of the system will be determined by the price of the attached CCD camera (high-speed cooled CCD cameras are very expensive), and the type of software required (no software is provided).

A simple sliding lever allows one to readily switch between confocal and non-confocal viewing.

Scan Head

CARV Nipkow disk unit contains a single Nipkow disk (without the micro-lens array disk), dichroic mirrors and optical filters for separation of various regions of the light spectrum. The relatively small robust scan head is factory aligned. A mercury or xenon arc lamp is attached to the side of the scan head.

Fixed "pinhole" Nipkow disk: The pinhole disk contains 20,000 pinholes (at any one time about 1000 pinholes scan the sample) that reject the out-of-focus light. Due to the design of this instrument the size of the pinholes cannot be adjusted.

Dichroic mirrors and optical filters: Mounted within the CARV scan head are conventional (the same as those used in the Zeiss wide-field epi-fluorescence microscope) dichroic mirror / optical filter blocks, which can be changed depending on the light separation required. A three position turret for choice of optical filters is provided.

CCD Camera

The disk scan rate is at 200 frames per second, although the speed at which the CARV confocal microscope can operate is determined by the sensitivity and speed of the CCD camera that is attached to the scan head. A highly sensitive and high-speed CCD camera will be required for low light fluorescence imaging applications and can readily achieve speeds of 50 fps at 512 x 512 pixels.

Light Source

Mercury or xenon arc lamp: A wide range of excitation wavelengths, from 340nm UV through to near IR, are available, depending on the filter block used. The mercury or xenon lamp housing on a conventional wide-field epi-fluorescence microscope can be transferred to the CARV scan head when retrofitting the scan head to an existing microscope. Excitation wavelengths are selected by the use of specific optical filters.

Microscope

The CARV scan head can be attached to a number of upright or inverted microscopes, including easy attachment to existing laboratory microscopes.

Imaging Modes

High-speed single channel fluorescence confocal or backscatter (reflectance) confocal imaging. Easily switched to bright-field / DIC imaging and wide-field epi-fluorescence imaging. Direct viewing of the confocal image is possible.

Very high-speed imaging (up to 100 fps, depending on the speed of the attached camera).

Multi-fluorescence imaging can be achieved by sequential scanning and adjustment of optical filter positions. Zoom, ROI, panning etc are not possible due to the way in which the image is collected via a CCD camera. Excitation / emission filters can be fitted.

Transmission

Transmission images can be obtained by moving a simple slider in the scan head.

Manual Control

The scan head is manually operated, with the computer being used to capture images from the CCD.

Computer

The instrument is manually operated and the image can be viewed directly by looking through the eyepieces on the scan head. Image capture is via a CCD camera, using commercially available software.

Software

Commercially available software for capturing images from the CCD camera is used to collect digital images.

Models

CARV: The CARV scan head can be readily adapted to a variety of microscopes.

Pathway HT: A high-throughput automated confocal microscope based on the CARV scan head.

CARV Scan Head

The CARV confocal microscope scan head is a finely engineered, robust, factory aligned independent instrument that can be readily attached to a conventional light microscope (see Figure A-48). The scan head contains the single Nipkow disk (Figure A-49), which is spinning at adjustable speeds of up to 200 frames per second, although the actual image collection speed will normally be significantly lower than this (15 to 25 frames per second), and is highly dependent on the speed and sensitivity of the camera and the amount of fluorescence in the sample.

A conventional mercury or xenon arc lamp is attached to the scan head, providing excitation light from UV through to IR – depending on the particular dichroic mirrors and optical filters used.

A series of conventional wide-field epi-fluorescence microscopy optical filter blocks are used to separate out the various regions of the light spectrum and to direct this light to either the binocular viewing port or to an attached high-resolution CCD camera. A wheel containing up to three user selectable optical filter blocks is mounted in the scan head. Further optical filter blocks can be swapped for those already in the filter wheel.

A z-stepper, when installed on the microscope can be used to collect optical sections for 3D reconstruction. The ability to collect images at relatively high-speed means that a full 3D optical stack can be collected in seconds or less – allowing one to collect almost real time 4D data sets (3D + time).

Sophisticated 4D or time lapse imaging is dependent on having suitable software installed to control the CCD camera acquisition time and z-stepper.

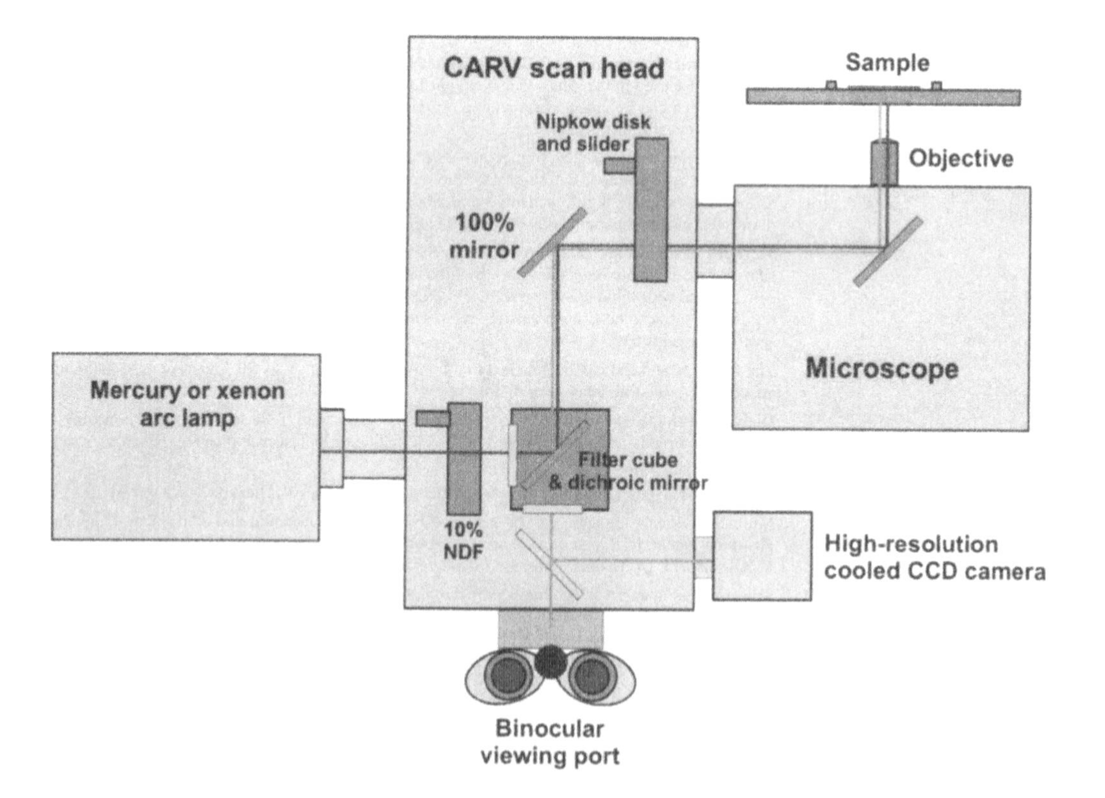

Figure A-49. Atto Bioscience CARV Nipkow Disk Scan Head.
The light path for the CARV scan head is shown diagrammatically. The light is focussed through the large array of pinholes on the Nipkow disk (see page 91), resulting in a large number of finely focussed spots scanning the sample at a rate of 200 frames per second. Illumination is provided by a conventional mercury or xenon arc lamp attached to the scan head. The image can be viewed directly by looking through the binocular viewing port on the scan head. A CCD camera can be attached to the scan head to capture the image. Excitation and emission filters can be changed by using a different filter cube (the conventional wide-field epi-fluorescence filter blocks used in Zeiss research grade light microscopes)..

Yokogawa Electric Corporation

The Yokogawa CSU10 and CSU21 spinning Nipkow disk scan heads are produced in Japan by the Yokogawa Electric Corporation. The design of these scan heads has resulted in an instrument that is unusually fast for a laser-scanning microscope (routinely 10 to 20 frames per second, but can be as high as 100 or more frames per second). Furthermore, the multi-point scanning technique used in this instrument, means that the light intensity at each individual point of light is much less intense compared to a conventional single point scanning confocal microscope – resulting in greatly decreased photo bleaching of the sample. The Yokogawa scan heads are assembled into confocal microscopes by a number of manufacturers, including PerkinElmer, and VisiTech International, which are discussed in some detail on the following pages.

The principles involved in confocal image acquisition in the Yokogawa Nipkow disk / micro-lens array scan head are described in more detail in on page 91.

Further details on the Yokogawa scan head can be found in Methods in Cell Biology Vol. 70 edited by Brian Matsumoto, chapter 5 "*Direct-View High-Speed Confocal Scanner – the CSU10*" by Shinya Inoue and Ted Inoue (see page 349 in Chapter 15 "Further Reading" for more details).

Figure A-50. Yokogawa CSU10 Scan Head
The CSU10 micro-lens array Nipkow disk scan head is produced in Japan by Yokogawa, and is assembled into complete confocal microscope systems by a number of companies. The scan head can be attached to either upright or inverted microscopes from a variety of manufacturers. The instrument is capable of very high-speed confocal imaging – the final image acquisition speed achieved depends on the speed and sensitivity of the attached CCD camera. This image was kindly provided by George L. Kumar (PerkinElmer, USA).

CSU10 SCAN HEAD

The compact and precisely engineered Nipkow disk CSU10 scan head (Figure A-50) is designed and manufactured as an independent unit. The scan head contains a dual spinning disk – where the first disk is an array of micro-lenses that are accurately aligned with a corresponding array of "pinholes" on a second disk (see diagram in Figure A-51, and Figure 3-18 on page 92).

Light passing through each micro-lens is directed through a dichroic mirror and

Yokogawa Electric Corporation

Web: www.yokogawa.com/
Email: csu@csv.yokogawa.co.jp

Address:	Confocal Scanner Business Group BIO Centre, ATE Business Division Yokogawa Electric Corporation 2-9-31 Nakacho, Musashinoshi, Tokyo, 180-8750, **Japan**
Phone:	+81 442 52 5550 — FAX: +81 442 52 7300

focussed onto the "pinhole" of the second Nipkow disk. The technique of focusing the light through each individual pinhole greatly increases the light efficiency of this instrument compared to those using a single Nipkow disk. The light from each of the individual pinholes then passes through the objective lens of the microscope – resulting in thousands of points of light being projected onto the sample. The returning fluorescent light is directed back through the pinholes on the first disk (rejecting any out-of-focus light) and reflected by the dichroic mirror (a "reverse" dichroic mirror compared to the point scanning confocal microscopes) onto a CCD camera.

The micro-lens array disk and the Nipkow disk are physically attached to each other to eliminate any problems of

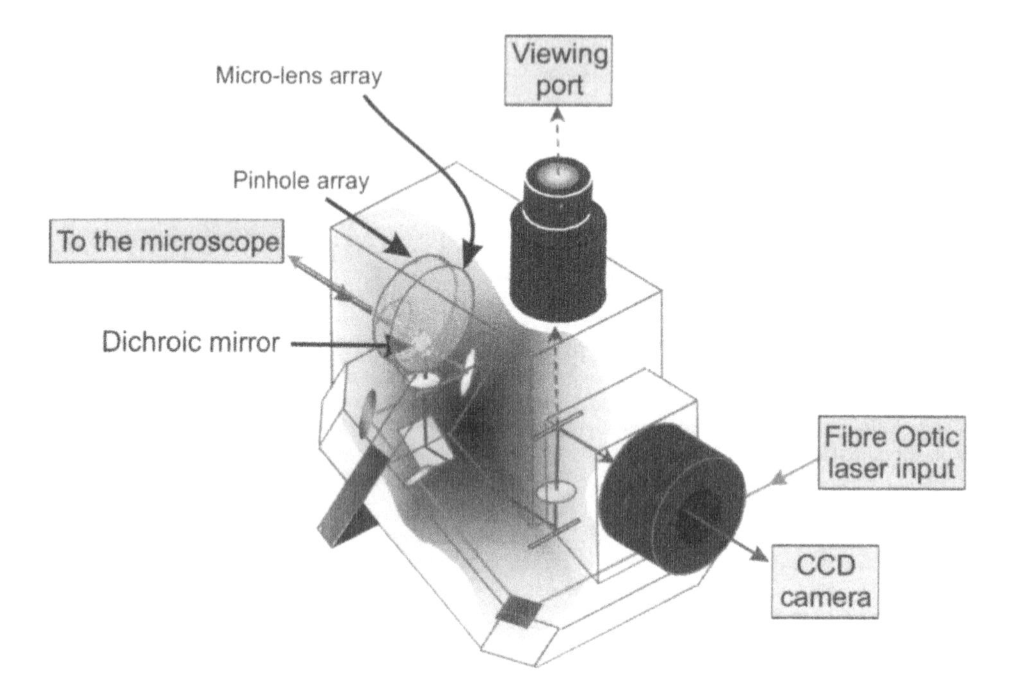

Figure A-51. Optical Path of Yokogawa CSU10 Scan Head.
The Yokogawa CSU10 scan head contains a Nipkow disk and a micro-lens array disk (described in detail on page 93) that are spun at high-speed together to illuminate the sample with a large number of light points. The confocal image can be viewed live by looking down the viewing port on the scan head, or images can be captured by using an attached CCD camera. Some instruments provide two CCD cameras for simultaneous dual channel imaging. Irradiation is provided by a variety of lasers. This diagram is modified from a diagram kindly provided by George L. Kumar (PerkinElmer, USA). The layout of the Nipkow disk / micro lens array disk in the Yokogawa scan head is shown diagrammatically in Figure 3-18 on page 92.

alignment. The dual disk system is spun at 1,800 rpm (30 rps), which results in 1200 pinholes scanning the sample at any particular instant. The disks are spinning at 12 times the video rate of 30 fps (i.e. 360 fps), which gives 12 frame averages for each video frame. Further enhancement can be achieved by on chip averaging. The 20,000 pinholes in the Nipkow disk have a diameter of 50µm (250 µm apart) to give a light throughput of only 4%. However, by focusing the light through the pinholes using the micro lenses the light throughput of the disk is significantly improved to approximately 40%. The holes are arranged in a "constant-pitch helical pattern" to create even illumination across the field without creating scanning artefacts.

The CSU10 scan head is connected to the camera port of a conventional light microscope (Figure A-51). The confocal image can be observed by direct viewing through the eyepiece (viewing port) on the scan head, or by simply moving a slider so that the light can be directed to an attached CCD camera.

The CSU10 scan head is made up into fully integrated confocal microscopes by a number of manufacturers (see the following pages). This includes providing a computer with appropriate software for acquiring images and controlling laser intensity and line selection. A grey-scale or three-colour CCD camera, or sometimes two grey-scale CCD cameras are attached for simultaneous dual channel image capturing.

Nipkow disk based confocal microscopes do have a number of important advantages compared to the laser spot scanning instruments, but there are also a number of limitations inherent in the technology (discussed in detail on page 91-93 in Chapter 3 "Confocal Microscopy Hardware").

An important advantage of the Nipkow disk based instruments is the ability to do relatively high-speed imaging. Although the CSU10 scan head Nipkow disk is spinning at 360 frames per second, the actual speed of image acquisition achieved is considerably less and is highly dependent on the type of CCD camera attached. However, even with a relatively modest CCD camera speeds of up to 20 frames per second can be achieved. This is in contrast to a typical acquisition speed of only one frame per second for a spot scanning system.

CSU21 SCAN HEAD

The CSU21 scan head (see Table A-1 below), also produced by Yokogawa, is similar in design to the CSU10 scan head but does have a number of features that make it particularly suited to very high-speed imaging. The CSU21 scan head has a variable scan speed from 1,800 rpm to as high as 5,000 rpm, which results in a frames scan rate as high as 1,000 frames per second. However, the actual image capture rate is still dependent on the sensitivity and speed of the attached CCD camera. The CSU21 scan head is exclusively distributed outside of Japan by PerkinElmer.

Table A-1. Comparison of the Yokogawa CSU10 and CSU21 Scan Head.

Yokogawa produces two Nipkow disk / micro-lens array disk scan heads, the CSU10 and the CSU21. The CSU21 scan head is significantly faster than the CSU10, has motorised optical filter changes, can have two CCD cameras attached and has reduced noise levels. A number of manufacturers (discussed on the following pages) produce instruments with modified versions of the CSU10 scan head that may have some of the features of the CSU21 scan head.

	CSU10 scan head	CSU21 scan head
Nipkow disk	20,000 holes (50μm)	20,000 holes (50μm)
Micro-lens array disk	20,000 lenses	20,000 lenses
Disk rotation speed	Fixed 1,800 rpm	Variable 1,800 5,000 rpm
Eye viewing	Yes	Yes
CCD camera	Single grey-scale or three-colour CCD camera, with modification two cameras can be attached.	Single camera or three-colour CCD, with modification two cameras can be attached.
ND and excitation filters	Manual change	Motorized
Lasers	Various: Fibreoptic coupled	Various: Fibreoptic coupled
Frame scan speed	360 frames per second	As high as 1,000 frames per second
	(image capture speed depends on CCD camera and light level)	
Installation	C-mount on microscope	C-mount on microscope
Emission filters	Manual filter change	Motorised filter change PC or front panel control.
Synchronisation	NTSC & PAL or slower speeds	NTSC, PAL and synchronisation with various exposure times of CCD cameras.

PerkinElmer Life Sciences

The Nipkow spinning disk confocal microscope produced by PerkinElmer is based on the CSU10 micro-lens array Nipkow disk scanning head produced by Yokogawa (page 435). The design of this confocal microscope has resulted in an instrument that is unusually fast for a laser-scanning confocal microscope (routinely 10 to 20 frames per second, but can be higher than 100 frames per second). The multi-point scanning technique used in this microscope means that the light intensity at each individual point of light is much less than with a conventional single point scanning confocal microscope – resulting in greatly decreased photo bleaching of the sample.

PERKINELMER CONFOCAL MICROSCOPES

The principles involved in the Nipkow disk confocal microscope are described in more detail on pages 91-93 in Chapter 3 "Confocal Microscopy Hardware", and the Yokogawa CSU10 scan head used in the PerkinElmer instruments is described on page 435 of this appendix.

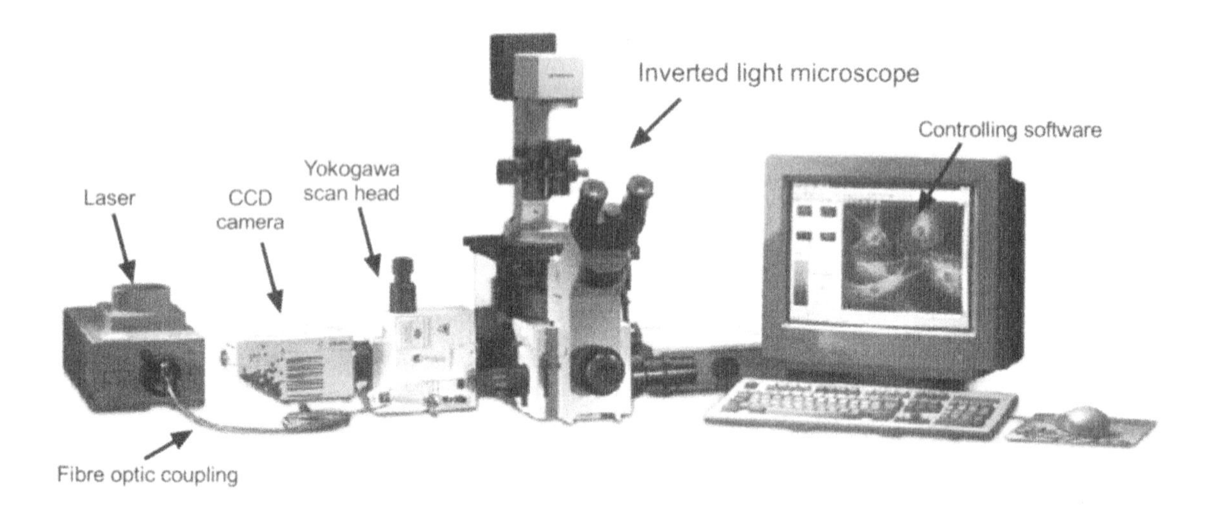

Figure A-52. PerkinElmer Ultra VIEW Confocal Microscope.
The PerkinElmer (Wallac) Nipkow disk confocal microscope is designed around the Yokogawa micro-lens Nipkow disk scan head (CSU10). Unlike the point-scanning confocal microscopes, this instrument creates an image by using a CCD camera and you can directly view the image by using the eyepiece directly mounted on the scan head. The instrument is capable of very high-speed confocal imaging but the speed achieved does depend on the type of CCD camera attached. This image was kindly provided by George L. Kumar, PerkinElmer, USA.

PerkinElmer Life Sciences **Web:** www.perkinelmer.com/lifesciences

Address:	PerkinElmer Life Sciences 549 Albany Street Boston, MA **USA**	PerkinElmer Life Sciences Imperiastraat 8, B-1930 Zaventem **Belgium**	PerkinElmer Life Sciences P.O. Box 600 Knoxfield MDC VIC 3178 **Australia**
Phone:	1 800 551 2121	+32 2 717 7911	+61 3 9212 8500
FAX:			+61 3 9212 8595

Nipkow disk based confocal microscopes are particularly suited to live cells imaging where either high speed imaging of cellular events is required (for example, when imaging fast changes in calcium fluxes) or minimal laser damage is required during long term time lapse imaging (for example, when imaging developing embryos). The instrument can still be used for more conventional imaging of fixed samples such as immunolabelling, although there are some limitations when compared to a single point scanning confocal microscope (see pages 91- 93 in Chapter 3 "Confocal Microscopy Hardware").

Ultra VIEW LC1 layout

The Ultra VIEW LC1 confocal microscope is sold as an integrated unit, consisting of the Yokogawa scan head, a single CCD camera, a microscope, and with computer and proprietary software to control laser line selection and attenuation, scan head, and image acquisition. The Yokogawa scan head is a small and compact instrument that takes up a relatively small area on the microscope bench (Figure A-53). The scan head can be attached to any camera port on an upright or inverted microscope, although the preference would be to install this type of scan head on an inverted microscope due to the significant advantages for using this instrument for imaging live cell or tissue preparations.

The PerkinElmer Nipkow disk confocal microscopes are provided with z-focus control using a stepper motor, or for more accurate and significantly faster z-sectioning high-speed piezoelectric focusing is used. In this way fast 3D data sets can be readily collected to follow cellular functions in 3D over time (4D imaging).

Various lasers are readily accommodated by the Ultra VIEW LC1 confocal microscope by attachment to the scan head via a fibreoptic connection. The laser does have the advantage of having very precisely defined wavelengths that can be used for excitation, but there is the disadvantage that you will be limited to the excitation wavelength of the lasers you have on hand. Nipkow disk confocal microscopes that utilise the much cheaper and the broad excitation wavelengths of a mercury or xenon arc lamp, are available.

Automated optical filter control is available on the Ultra VIEW LC1 to allow fast switching of excitation and emission wavelengths for multiple labelling and ratiometric labelling applications.

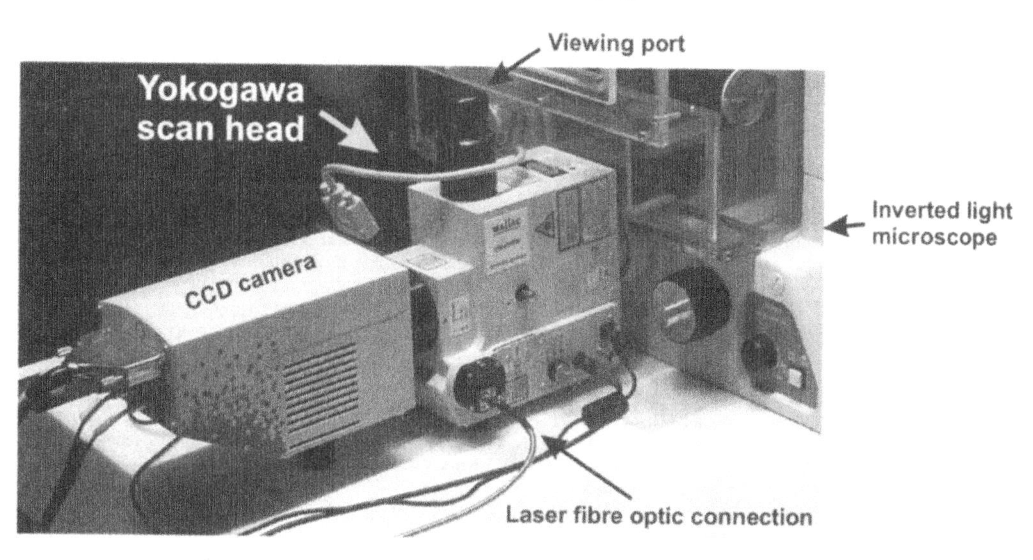

Figure A-53. PerkinElmer Ultra VIEW Scan Head.
An enlarged view of the Yokogawa scan head and attached CCD camera as assembled by PerkinElmer Life Sciences. The image can be directly viewed by looking down the "viewing port" while the scanner is operating. The CCD Camera can be of a variety of designs; a high-speed cooled CCD camera is required for low light level high-speed imaging as found in live cell confocal microscopy. A dichroic mirror (located between the micro-lens array disk and the pinhole array disk), is used to direct the returning fluorescent light to the CCD camera. A variety of lasers can be attached via the fibreoptic connection. This photograph was taken at the European Molecular Biology Laboratories (EMBL), Heidelberg, Germany.

Ultra VIEW LCI

Components of the PerkinElmer Ultra VIEW

The PerkinElmer Ultra VIEW Nipkow scanning disk confocal microscope is assembled using the Yokogawa micro-lens array CSU10 scan head.

Design

A Nipkow spinning disk confocal microscope that uses a single CCD camera to capture the image.

Scan Head

The Yokogawa Nipkow disk-scanning unit (see page 435) is the "heart" of the PerkinElmer confocal microscope. This compact and carefully engineered scan head is attached directly to the microscope. The image can be viewed in real time through the eyepiece on the top of the Yokogawa scan head, or the image can be captured electronically using a CCD camera attached to the scan head.

Micro-lens array Nipkow disk: The Yokogawa scan head contains two finely engineered spinning disks, the micro-lens array disk and the Nipkow (pinhole) disk. The micro-lens array disk contains a large number of very small lenses that focus the laser irradiating light into a corresponding "pinhole" that is spun in tandem with the micro-lens array disk.

Fixed "pinhole" Nipkow disk: The Nipkow disk contains a large number of pinholes that reject out-of-focus light. Due to the design of this instrument the size of the pinholes cannot be adjusted.

Dichroic mirrors: Mounted within the Yokogawa scan head is a dichroic mirror (between the micro-lens array disk and the pinhole disk) for directing the fluorescent light towards the CCD camera.

CCD Camera

The sensitivity and speed of the CCD camera attached to the Yokogawa scan head determines the number of frames per second that can be collected in low light fluorescence imaging applications.

The faster Ultra VIEW LC1 FES 250 uses a CCD with frame transfer digital technology, resulting in high speed imaging of over 50 frames per second.

Other cameras used by PerkinElmer include interline CCD cameras for high quality, but somewhat lower speed image acquisition (up to 20 frames per second).

These cameras provide 12-bit digitisation to create a wide dynamic range that is particularly useful for fluorescence imaging. The PerkinElmer assembled confocal microscope using the Yokogawa CSU10 scan head utilises a single CCD camera with multiple imaging being performed by sequential scanning.

Lasers

Lasers: Argon-ion Krypton-argon-ion, argon-ion / helium-neon combinations or other lasers are mounted separately from the scan head with the laser light being transferred using a fibreoptic connection.

Laser power: The laser power can be adjusted by rotating the knob on the laser casing. This controls the laser output not necessarily the amount of laser light reaching the sample.

AOTF laser attenuation: The amount of laser light is controlled by using software controlled AOTF filters (giving accurate levels of between 0 and 100% of the current laser power setting).

Microscope

The Yokogawa scan head can be attached to most upright or inverted light microscopes.

Imaging Modes

High-speed single channel fluorescence. Dual channel possible by attaching a second CCD camera. Simultaneous transmission imaging also possible if a bright-field CCD camera is attached to the microscope directly. Direct viewing of the confocal image is possible. Zoom, ROI, panning etc are not possible due to the way in which the image is collected via a CCD camera.

Transmission

Transmission images can be overlaid with fluorescence images if a CCD camera is attached to the microscope directly.

Manual Control

The Yokogawa scan head controls are operated manually.

Computer

The microscope is operated via a standard PC compatible computer. A single large computer screen is used to display the images as they are collected and the controls for scanning and changing microscope settings. The "basic" Ultra VIEW confocal microscope requires manual operation of the Yokogawa scan head controls.

Software

Proprietary Perkin-Elmer software for controlling the microscope and CCD camera. The software can also perform basic image manipulation. **Temporal Module:** This software is for high-speed time-lapse imaging (including 4-D and multi-labelling). **Spatial Module:** Software for high quality imaging of samples with minimal movement or fluorescence changes.

Models

Ultra VIEW LC1 is a fully computer controlled instrument with a single CCD camera.

Ultra VIEW is a basic system with manual filter changes.

A variety of configurations of the basic instrument are available, including a range of CCD cameras.

VisiTech International

VisiTech International supplies and manufactures a range of imaging related instruments for use in the life sciences. VisiTech manufactures two high-speed confocal microscopes that are particularly suited to live cell imaging. One instrument is based on the spinning Nipkow disk scan head manufactured by Yokogawa, and the other is a high-speed spot scanning system using an acoustic-optical deflector (AOD) to scan the laser across the sample. VisiTech has a number of distributors, some of which make substantial contributions to the design and layout of their confocal microscopes, particularly the instruments based on the Yokogawa scan head.

VISITECH CONFOCAL MICROSCOPES

The VT-Eye confocal microscope (Figure A-54) is based on a scan head manufactured by VisiTech International that uses a high-speed acousto-optical deflector (AOD) to scan a focussed laser beam across the sample at very high speed. This instrument is particularly suited to live cell imaging where high-speed image acquisition rates and the imaging advantages of a confocal microscope are required.

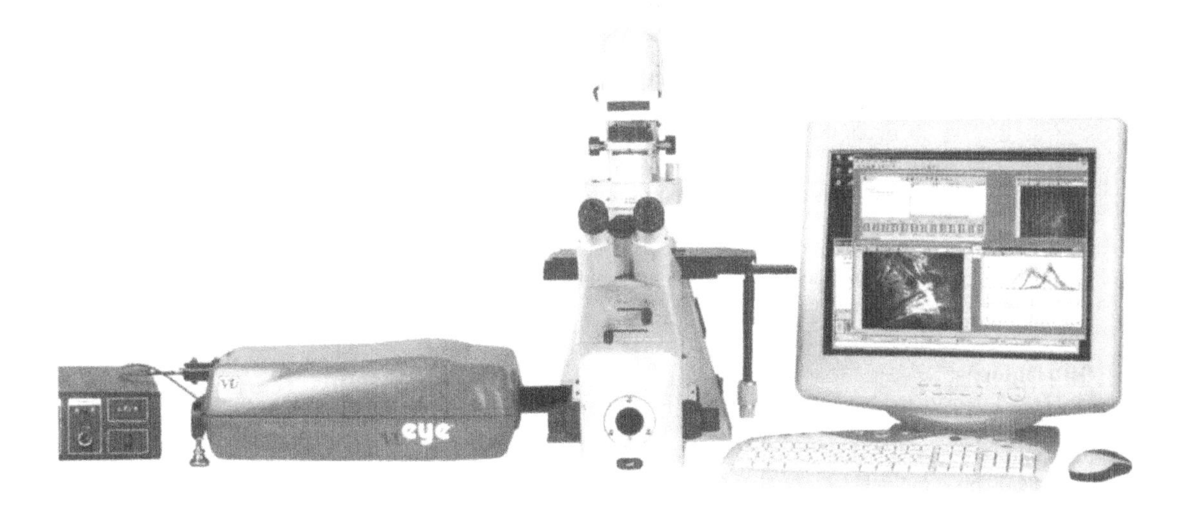

Figure A-54. VisiTech International VT-Eye Confocal Microscope.
The VisiTech International VT-Eye confocal microscope is a high-speed laser scanning confocal microscope ideally suited to live cell imaging. Very high-speed scanning is achieved by using an acousto-optical deflector (AOD) device for scanning the laser in the x-direction. A conventional galvometric mirror is used for the y-scan direction. The VT-Eye scan head is shown above attached to the side port of a Zeiss inverted research grade light microscope. This photograph is reproduced with permission from VisiTech International, UK.

VisiTech International Ltd. **Web:** www.visitech.co.uk/

Address:	Unit 92 Silverbriar, Sunderland Enterprise Park (East) Sunderland, SR5 2TQ, **UK**		
Phone:	+44 (0) 191 5166 255	FAX:	+44 (0) 191 5166 258

VisiTech International Confocal Microscopes

VisiTech International produces two high-speed laser scanning confocal microscopes. The VoxCell Scan confocal microscope, assembled using the spinning Nipkow disk scan head manufactured by Yokogawa, and the VT-Eye, which utilises a very high-speed acousto-optical deflector (AOD) to create a spot scanning confocal microscope capable of very high-speed scanning.

VisiTech Instruments

VT-Eye (Figure A-54)	An acousto-optical deflector (AOD) based scanning mechanism resulting in a confocal microscope with unusually high scan speeds (30 frames per second for a 1024 x 1024 image), even though a single laser spot is used to scan the sample. Up to 4 detection channels for multi-labelling applications. AOTF filters for laser line selection and operated using software running under Windows XP. The VT-Eye confocal microscope is based on the same technology as used in the Noran OZ confocal microscope, but with significant modifications made by VisiTech to the original design.
VoxCell Scan (Figure A-55)	Visitech International, and a number of distributors (listed below) put together a complete confocal microscope based on the Yokogawa CSU10 scan head (see page 435). The instrument, being based on the same CSU10 scan head as the PerkinElmer Ultra VIEW LCI confocal microscope, has many of the same specifications as the instrument assembled by PerkinElmer (see page 440 for details). However, due to the modular design of this instrument there are often important differences, not just between the VoxCell Scan and the Ultra VIEW, but also between each VoxCell Scan assembled. Various lasers are mounted on a separate platform, with the attenuated laser lines of choice being transferred to the scan head by fibreoptic connection. Motorised optical filter wheels are often inserted between the scan head and the CCD camera. A dual camera attachment for the instrument is also manufactured by VisiTech to allow for simultaneous dual labelling.

Distributors:

VisiTech International has established a world wide distribution network by utilising a range of relatively small companies that have considerable technical expertise of their own. VisiTech, and a number of the distributors, often assemble a confocal microscope with specialised components as requested by the customer. This may include specialised CCD cameras, AOTF filters for laser line selection, optical filter wheels for multi-labelling applications and a nanofocus device for highly accurate positioning of the microscope objective. The following distributors use licensed components from VisiTech International, as well as items sourced from other specialist manufacturers.

Chromaphor Analysen-Technik GmbH (Germany)	www.Chromaphor.de
McBain Instruments (California, USA)	www.mcbaininstruments.com
Quorum Technologies Inc. (Canada)	www.quorumtechnologies.com
Solamere Technology Group (USA)	solameretg@aol.com
Visitron Systems GmbH (Austria/Switzerland)	www.visitron.de

The VoxCell Scan confocal microscope (Figure A-55, with a short technical description on the previous page) is a spinning Nipkow disk instrument that utilises the QLC100 Nipkow disk scan head, which is based on the Yokogawa CSU10 scan head (see page 435). The VoxCell Scan is capable of high-speed imaging due to the multi-pinhole design of the spinning Nipkow disk (see page 91). High-speed image acquisition is particularly suited to the imaging of dynamic processes in living cells.

The micro-lens array disk incorporated into the design of the Yokogawa scan head does mean that more light is focussed onto the sample than when using the Nipkow disk alone (increasing the sensitivity of the instrument). However, compared to a conventional spot scanning confocal microscope, the amount of light per spot in a Nipkow disk based instrument is relatively low. The low excitation light level at the sample results in significantly less photo-damage, both to the fluorophore (manifested as less photo fading) and to the integrity of the cells (allowing one to image live cells or embryos over many hours). The principles involved in the Nipkow spinning disk technology are described in more detail on pages 91 to 93 in Chapter 3 "Confocal Microscopy Hardware" and in the sections on the Yokogawa scan head (page 435) and the Ultra VIEW confocal microscope produced by PerkinElmer (page 438) in this appendix.

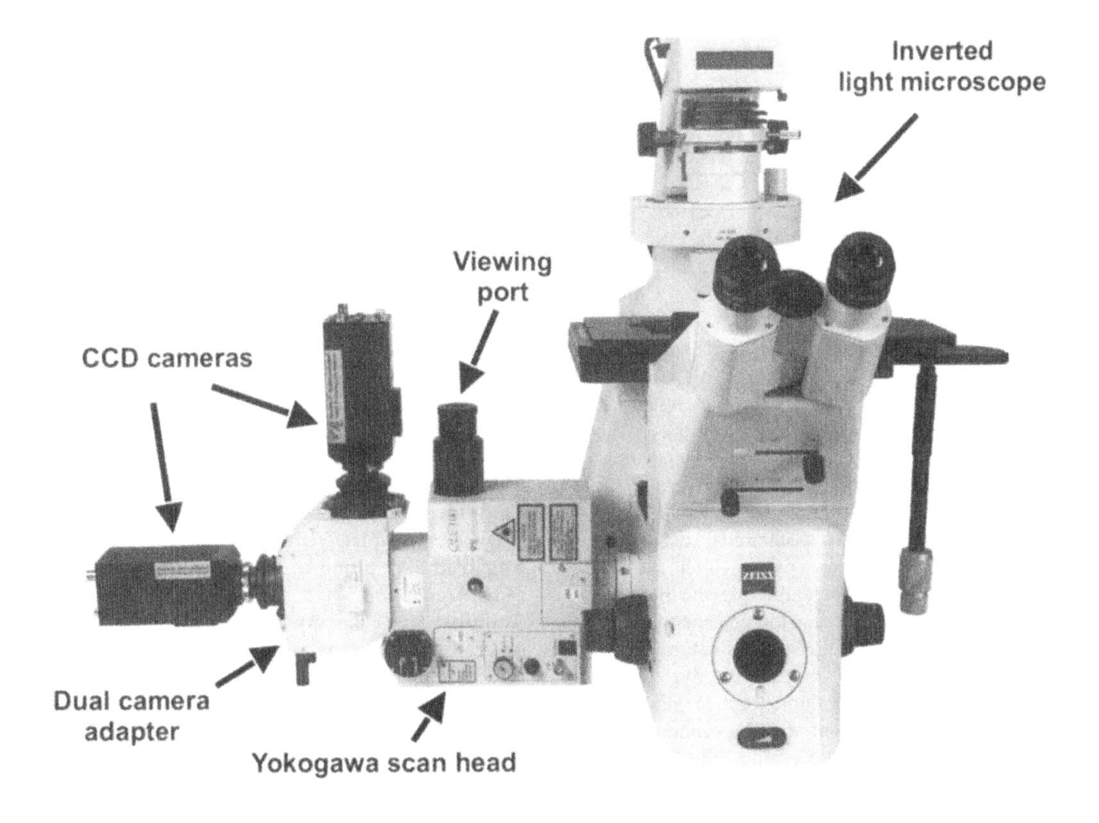

Figure A-55. VisiTech VoxCell Scan Confocal Microscope.
The VisiTech VoxCell Scan confocal microscope is a high-speed spinning Nipkow disk confocal microscope built using the QLC100 Nipkow disk / micro-lens array scan head manufactured by Yokogawa (see page 435). VisiTech also manufactures a dual camera adaptor (shown in the above photograph) for simultaneous dual channel imaging. This photograph is reproduced with permission from VisiTech International, UK.

Glossary

8-bit: each pixel in an 8-bit (1 byte) digital image contains 256 (2^8) grey levels, usually from zero (black) to 255 (white). Some image processing programs consider zero = white and 256 = black (i.e. the image will appear to be grey-scale inverted). All confocal microscopes can collect 8-bit images, and many can also collect 12-bi images if required.

12-bit: each pixel in a 12-bit digital image contains 4096 (2^{12}) grey levels. Most confocal microscopes have the option of collecting images in 12-bit.

16-bit: each pixel in a 16-bit digital image contains 65,536 (2^{16}) grey levels.

3D stack: a series of images collected on a confocal microscope by moving the focal plane in discrete z-steps with either a fine focus motor drive or a piezoelectric z-stepper that moves either the microscope stage or the objective.

A/D converter (ADC): an electrical device used to covert analogue signals to digital signals.

Aberration: an optical error caused by an imperfect optical system or specimen. See also "Chromatic aberration" and "Spherical aberration".

Absorption curve / spectrum: the absorption of specific wavelengths (colours) of light, normally plotted as intensity verses wavelength. Each fluorophore has a characteristic absorption profile. In single photon excitation the absorption spectrum is the same as the excitation spectrum for a given fluorophore.

Acetoxymethyl ester (AM): a chemical derivative of many common fluorescent dyes which makes the molecule non-charged, thus allowing the fluorophore to more readily diffuse across the cell membrane. The methyl ester is removed by esterases naturally present inside a living cell, thus trapping the dye within the cell. An added bonus for fluorescent labelling is that the AM-ester derivative of the fluorophore is often non-fluorescent, only becoming fluorescent on cleavage of the AM ester group within the cell.

Achromat: (also achromatic) a lens designed to reduce chromatic aberration of the red and blue wavelengths and spherical aberration of green wavelengths.

Acoustic-Optical Tuneable Filter: see "AOTF".

Acousto-Optical Deflector (AOD): high-speed laser scanning can be achieved in a confocal microscope by using an AOTF type active crystal for scanning the laser beam at very high-speed in the x-direction. A conventional galvometric mirror is normally used to scan more slowly in the y-direction.

Airy Disk: diffraction pattern created by an object under the microscope. Rings of bright/dark pattern may be seen around small objects at high magnification. The Rayleigh Criterion is one way to describe the limit of resolution of the microscope. This criterion describes the amount of overlap of neighbouring Airy disks - when the two Airy disks merge the object can no longer be resolved.

Aliasing: the formation of artefactual objects or edges within a digital image that did not exist within the original sample, a result of digital sampling errors. Aliasing is avoided by sufficient sampling (see also Nyquist criteria).

Alignment: the laser must be aligned properly into the objective, and the returning fluorescent light must also be aligned into the pinhole(s). In some confocal microscopes these alignments are factory set and then tested when the machine is installed. Other confocal microscopes require the user to periodically check the alignment. If the sensitivity of the machine appears lower than expected then check the alignment (particularly the alignment of the laser into the confocal pinhole).

AM: see "Acetoxymethyl ester".

Analogue: a continuously variable electrical signal.

Analogue-to-Digital converter: see "A/D converter".

Analyser: a light polariser filter used in DIC (Differential Interference Contrast) microscopy.

Angstrom (Å): a unit of length, 1/10th of a nanometre (10^{-10} meters).

Annulus: see "Phase Annulus".

Antifade: compounds such as *n*-propyl-gallate and *p*-phenylenediamine help slow down the rate of fading (or photobleaching) of the sample. Antifade reagents are antioxidants, as photobleaching is due to light induced oxidising reactions. Lowering the amount of laser light reaching the sample or limiting the amount of oxygen available also helps to reduce fading.

Antioxidant: molecules that lower the level of oxidative damage to the cell or tissue by reacting with free radicals that are created by the illuminating light. Antioxidants will lower the amount of fading of the fluorophore.

Anti-reflection coating: a special coating on air-glass surfaces of optical components (lens, optical filters etc) that minimises the amount of light reflected at the surface interface.

AOTF (Acoustic-Optical Tuneable Filter): is an active crystal device that acts as a variable diffraction grating. Varying the radio-frequency acoustical vibrations in the crystal allows one to diffract out a desired wavelength of light. An AOTF filter can be used to select one or multiple wavelengths from a multi-line laser, and to adjust the power level transmitted (a continuously variable "neutral density" filter).

Aperture diaphragm: an adjustable diaphragm located in the condenser optics of the microscope. Used to adjust the effective NA of the condenser.

Apochromat: (also Apochromatic) a lens in which chromatic aberration has been corrected for three or more colours (red, green, blue and UV). Also corrected for spherical aberration for two or more wavelengths (green and red).

Astigmatism: an optical aberration that results in the horizontal and vertical focus not coinciding along the optical axis (resulting in a circular object appearing ellipsoid).

Autofluorescence: the inherent ability of a specimen to fluoresce. In mammalian cells the principle cause of autofluorescence are flavin coenzymes (FAD, FMN and NADH). In plant cells lignin (green fluorescence) and chlorophyll (red fluorescence) are the major sources of autofluorescence. Aldehyde fixatives induce fluorescence in biological material, which is often incorrectly called autofluorescence.

Autumn LUT (Bio-Rad): red colour "Look Up Table" used by Bio-Rad (dark red - light red - yellow - white colour gradient) for colouring digital images.

Averaging: multiple images are averaged to reduce Poisson noise, such that each pixel intensity in the final image is an average of intensities from the corresponding pixels in the constituent images.

Avidin: a 66KDa glycoprotein derived from egg white that specifically binds biotin. Biotinylated probes are detected with fluorescently labelled avidin. The bacterial equivalent is called Streptavidin.

Axial illumination: occurs in conventional bright-field imaging when the condenser aperture is closed down. Increases the contrast and depth of field. Not recommended for high resolution, but is useful for "finding" the specimen.

Axial resolution: resolution along the z-axis, i.e. resolution perpendicular to the plane of focus. In confocal fluorescence microscopy the axial resolution increases inversely proportional to the square of the NA of the objective - in contrast to lateral resolution, which is to the first power of the NA of the objective (i.e. high NA objectives have correspondingly higher axial resolution).

Back focal plane (BFP): the focal plane of a lens located on the side of the lens away from the object. Phase rings are inserted at the BFP. The BFP can be viewed using a Bertrand lens or phase telescope.

Backscatter: a term used to describe the light that is scattered by the specimen back towards the objective lens. The confocal microscope can create an excellent high-resolution image using "backscattered" light (sometimes referred to as "reflectance" imaging).

Band Pass filter (BP): an optical filter that transmits a band of colour, the centre of which is the centre wavelength (CWL). The width of the band is indicated by the full width at half maximum transmission (FWHM), also known as the half bandwidth (HBW).

Barrier filter: see "Emission filter".

Beam splitter: a partial mirror, polarisation filter, prism or diffraction grating that diverts part or all of the light from one direction to another. A beam splitter is often used to divert some or all of the light from the eyepieces to the camera, to select specific wavelengths for irradiating the sample in a confocal microscope, or for directing specific wavelengths of light to individual channels (PMTs) in a confocal microscope. The primary beam splitter in a confocal microscope splits the irradiating laser light from the returning fluorescent light.

Bertrand lens: a lens that allows the back focal plane of the objective to be viewed through the eyepieces.

Binary image: an image comprised of only 2 intensity values – usually represented by black and white.

Bit: the basic unit of digital information ("**bi**nary uni**t**"), conveys 2 possible states denoted on/off or 1/0.

Bit depth: see "Dynamic range".

Black level: the amount of signal offset (alters the blackness level of the image).

Bleaching: see "Photobleaching".

Bleed-through (bleed-over): occurs when the emission of one fluorophore (e.g. ethidium bromide) is also detected in the second channel (e.g. the fluorescein channel). There is always some bleed-through (minimised by careful setting of the gain control), but unwanted bleed-through can be lowered considerably by suitable choice of optical filter blocks. The Bio-Rad software allows on screen "mixing" of each of the detection channels, thus allowing one to subtract a predetermined percentage of one channel from the other to minimise the effects of bleed-through.

BMP: Windows Bit Map, a standard image format for Windows-compatible computers. You can also specify RLE compression.

Box size: size of the image collection box in pixels. Commonly used sizes for LSCM are 512 x 512, 768 x 512 or 1024 x 1024 pixels.

Bright-field: this is the "normal" wide-field illumination method in microscopy when viewing a specimen down through the eyepieces (see Köhler illumination). In the absence of a specimen, the background appears "bright". See also, "Dark field" and "Epi-illumination".

Byte: a unit of information storage employed in a computer, comprised of 8-bits.

Caged probes: molecular probes that are converted to their active derivative upon irradiation with UV light. In this way highly reactive species can be targeted to subcellular destinations before being activated.

CCD and CID: electronic chips as the detector in an electronic camera. CCD = charge-coupled device, CID = charge-induced device. Both cameras use an array of photodiodes on a chip which record an image as a mosaic of charges.

CD "burner": a device for writing information to CD-R disks, also known as "CD-ROM disks". The "burned" disk can then be read in a regular CD-ROM drive. A disk can be written as a single session or several times as a multi-session disk.

CD-R: Recordable CD (optically written disk) of 650 or 700 MB capacity. Also known as CD-ROM (Compact Disk - Read Only Memory). Once data has been written to the disk, the data can be read and copied, but cannot be erased or modified. However, re-writing to a multi-session disk may result in previous files of the same name no longer being accessible (the directory information is changed).

CD-RW: Re-writable CD laser disk, with the same capacity as the CD-ROM (650 or 700 MB), but can be written/erased many times. Although the CD-RW disk can be read using the CD-ROM drive present in most computers, reading the disk on other computers from which it was originally written may require the installation of additional software.

Centre wavelength (CWL): the arithmetic centre of the window of transmission of a Band Pass filter. This value is not necessarily the same as the peak wavelength.

Channels: different wavelengths (colours) of light are collected in separate "channels" in a confocal microscope. Each channel will have its own set of discriminating filters and light detector.

Charged-Coupled-Device: see "CCD".

Chromatic aberration: glass refracts different colours (wavelengths) of light to different extents. Most microscope lenses use combinations of elements to minimise chromatic aberration. Monochromatic light avoids the problem of chromatic aberration, but multi-channel imaging on a confocal microscopy may result in misregistration between images made with fluorescent dyes that emit at different wavelengths.

CLSM: see "Confocal Laser Scanning Microscope".

C-mount: standard screw-in lens mount for the attachment of a camera to a microscope (or other device).

Coherent light: light consisting of waves vibrating in the same phase, but not necessarily the same plane. Laser light is monochromatic, linearly polarised and highly coherent.

Collector lens: a lens used to collect light from an emitting light source and to pass the light through the field diaphragm and the aperture diaphragm to the specimen.

Colour Look Up Table: see "Look Up Tables".

Colour saturation: colour saturation refers to the degree of colour present (i.e. intensity of colour).

Colour temperature: a measure of the bluish (high temperature) or reddish (low temperature) hue of the "white" light expressed as the absolute temperature (degrees Kelvin). This nomenclature is used to express both the sensitivity of an electronic light detector and the colour display of a computer monitor.

Compound microscope: produces magnification using an objective and an eyepiece - resulting in an upright image.

Compression: see "File compression".

Condenser: lens system usually under the stage, but in an inverted microscope the condenser resides above the stage. This lens is designed to focus and project illuminating light onto the specimen. In epi-illumination, as used in both wide-field epi-fluorescence microscopy and laser scanning confocal microscopy, the objective itself acts as both condenser and objective. With the confocal microscope a transmission image can be obtained by collecting the laser light that passes through the condenser.

Condenser aperture: a variable aperture (iris) in the condenser used in Köhler illumination to vary the numerical aperture of the illumination.

Confocal: an image of only the in-focus plane, obtained by passing the returning light through a pinhole.

Confocal aperture (pinhole or iris): defining feature of a confocal microscope. To reach the detector, the light returning from the sample is focussed through an aperture that removes any out-of-focus light rays, thereby producing an "optical section". In Bio-Rad instruments the confocal iris is millimetres in size, compared to a very fine micron sized iris for most other confocal microscopes. The Nipkow disk based confocal microscopes have a large array of pinholes within the disk that allow only in-focus light to penetrate the disk.

Confocal Laser Scanning Microscope (CLSM): uses a scanning system consisting of rotating mirrors or opto-acoustic deflectors to scan a point of laser light over the specimen.

Conjugate planes: planes in a complex optical system at which a reference plane is mutually in focus.

Constructive interference: phenomenon of increased light intensity that results when light waves are in step, travelling and vibrating in the same direction and are at the same location at the same time.

Contrast: the degree of visibility between two objects or of an object against its background. A low-contrast feature blends into the background, and a high-contrast feature stands out distinctly. Contrast can be manipulated during acquisition by altering the gain and blackness level on the instrument, as well as changed after collection by subsequent image processing.

Contrast range: see "Dynamic range".

Contrast stretch: a process applied to digital images that remaps intensities in the collected image to spread them over the maximum possible range of values (increases contrast within the image).

Cooled CCD: a CCD camera that operates below ambient temperature to reduce or eliminate dark current "noise".

Cooling fan: argon-ion and krypton-argon-ion lasers are air cooled by a fan. The fan must be left on for several minutes after turning the laser off to ensure the laser has cooled sufficiently.

Correction collar: an adjustable ring on an objective. There are two types, one controlling a diaphragm in the back focal plane which allows the NA to be adjusted when setting darkfield illumination, and the other allowing compensation for different immersion media or coverslip thickness (this collar can also be used to correct for spherical aberration created by the sample).

Coverslip thickness: the correct coverslip thickness (usually 0.17 μm, number 1.5 coverslip) is important for optimal resolution, and in fluorescence confocal microscopy also for optimal sensitivity. Coverslip thickness is especially critical when using a "dry" lens. The required coverslip thickness is engraved on the side of the objective.

Critical illumination: a type of illumination in which the image of the light source is focussed onto the specimen. This type of illumination requires light scramblers to avoid having a pattern of the illumination source on the image plane.

Cryo-fixation: rapid cooling of a specimen to preserve its structure (usually to liquid nitrogen temperatures).

Cryostat: a trade name for a refrigerated microtome used for cutting frozen sections.

Curvature of field: when the focal plane of a lens is curved instead of flat.

Cyan: a light blue colour formed when red light is subtracted from white light (leaving green + blue). CMYK (Cyan, Magenta, Yellow and Black) inks are used in most printers.

Dark current: the background current produced by a photodetector. Cooling a CCD or PMT detector significantly reduces the dark current noise.

Dark field illumination: illumination that passes through the specimen at such an angle that light rays cannot enter the objective lens directly. Dark field images have a black background on which structural features appear lighter. Useful for observing live unstained cells.

Deconvolution: mathematical process used to improve the clarity of images by re-allocating out-of-focus light back to its plane of origin. Deconvolution can be performed on a z-stack obtained from a wide-field fluorescence, confocal or multi-photon microscope.

Depth of field: the thickness of the optical slice (the distance along the z-axis of the specimen that is in-focus). High numerical aperture (NA) objectives have a very small depth of field resulting in a very thin optical slice when using a confocal microscope. Not to be confused with "depth of focus", which is the z-distance in the image plane of the microscope that is in focus.

Depth of focus: the thickness of the image plane that is in focus. In a high NA lens the depth of focus is large (the position of the eyepiece is not critical), but the "depth of field" (the physical z-distance within the specimen in focus) is very small.

Destructive interference: is where light waves are out of step - resulting in decreased light intensity. If the two waves are out of step by 1/2 wavelength then the peak of one will exactly co-inside with the trough of the second and result in zero intensity if the two initial light rays have the same intensity.

DF (Discriminating Filter): an optical filter with very steep-sided transmission windows, with especially deep attenuation of energy close to the band. For example, 488DF10, describes a Discriminating Filter (DF) that allows a narrow band of 10 nm light, in the vicinity of 488 nm, to pass through.

Diascopic illumination: light transmitted through the specimen, using a condenser to focus the light.

DIC microscopy: see "Differential Interference Contrast".

DIC prism: see "Wollaston prism".

Dichroic mirror: (also dichroic beamsplitter) a mirror that reflects certain wavelengths and allows others through unhindered. They can be made with one, two or three reflective ranges (singe, double and triple dichroic mirrors). For example, the triple dichroic mirror in the T1 filter block of the Bio-Rad MRC-1024 reflects a region of light around 488, 568 and 647 nm – light not reflected passes through this mirror.

Differential Interference Contrast (DIC): (Nomarski optics) is an illumination technique in microscopy where polarised light is used to produce an apparent 3D effect by creating light and dark shadows at opposing edges of features in the specimen. Excellent DIC images can be collected using the transmission channel on a confocal microscope equipped with DIC optics. The "shadows" in the DIC image are created by small changes in refractive index, and are not a true 3D representation of the sample.

Diffraction: a change in the direction of light caused by the interaction of the light with an object. Diffraction is greatly increased when light passes objects that are close to or smaller than the wavelength of light.

Diffraction grating: a series of parallel lines, forming alternating grooves and ridges with spacings close to the wavelength of light, scribed on a reflecting or transparent substrate. A mixture of various wavelengths, including white light, can be readily split into the component colours using a diffraction grating. The Meta channel on the Zeiss 510 LSM confocal microscope uses a diffraction grating to separate the fluorescent light into its component colours.

Digital: the collection and/or storage of information as a series of numbers. Images from confocal microscopes are stored as digital image files.

Digital zoom: zooming (enlarging) the image by spreading the available pixels over a larger area - this may result in a pixilated image showing relatively large pixels, or a smooth image in which the enlarged pixels have been "smoothed" using a computer algorithm. Digital zoom does not result in increased resolution. Also see "Scanning zoom".

Diopter setting: the facility for focusing one eyepiece separately from the other to compensate for differences in focus between the microscopist's eyes.

Dispersion: the property of light whereby it splits into separate component colours when undergoing diffraction, refraction etc. (i.e. the amount of diffraction, refraction etc. varies with wavelength).

DRLP: "Dichroic Long Pass" optical filters that transmit a broad range of longer wavelengths of light, while efficiently reflecting shorter wavelengths.

Dry objective: an objective that is used without immersion media (i.e. only an air gap) between the lens and the specimen. Highly susceptible to spherical aberration, and must be used with the correct coverslip thickness (some objectives have an adjustment collar for differences in the thickness of the coverslip).

Dual imaging: the ability to collect two images from two separate light detecting channels on a confocal microscope.

Dual labelling: the labelling of a sample with two differently coloured fluorophores so that the co-localisation of the two structures/molecules of interest can be determined.

DVD disk: very high capacity laser disk (4-15 x the capacity of a CD-R disk). Re-writable DVD disks are available, but there are currently several formats and standards for DVD disks.

Dwell time: the amount of time the excitation light illuminates a spot in the sample, and hence the collection time for each pixel in the sample. A longer dwell time may result in more photo damage to the sample. Typical dwell times for a laser scanning confocal microscope are 0.1 to 1.0 μs.

Dye saturation: refers to the highest level of light that can be used to produce fluorescence - further increase in the laser light intensity will not increase the level of fluorescence.

Dynamic range (or Bit Depth): the number of possible shades of grey between black and white in an image. An 8-bit image has 256 (2^8) possible grey levels, a 12-bit image has 4096 (2^{12}) possible grey levels and a 16-bit image has 65,536 (2^{16}) possible grey levels.

Edge enhancement: an algorithm for sharpening the edges on features within digital images.

Edge Filter: another term for a short pass or long pass optical filter with a very sharp cut-on or cut-off.

EFLP (Long Pass Edge Filter): an optical filter that reflects more than 99.999% of shorter wavelength light, with a very sharp cut-off wavelength.

ELF (Enzyme Labelled Fluorescence): substrates that yield fluorescent precipitates at the site of enzymatic activity.

Emission: release of light from a fluorophore when an excited electron returns to the ground state.

Emission filter: an optical filter that ensures that only light of the desired wavelengths reaches the eyepieces or CCD camera of a wide-field epi-fluorescence microscope, or the light detectors of a confocal microscope.

Emission fingerprinting: the process of collecting spectral information from the sample and applying linear unmixing to separate fluorophores with highly overlapping emission spectra.

Emission spectrum: the spread of wavelengths over which the fluorophore emits fluorescent light. The radiation results from electrons returning from the first singlet excited states to ground level. The fluorescence emission wavelengths in single photon excitation are always of longer wavelength (towards the red) compared to the wavelength used for excitation.

Emission wavelength: usually the wavelength of maximum emission (519 nm green light for fluorescein). See also "Peak emission wavelength".

Empty magnification: greater magnification than is useful. No increase in resolution, and sharpness and contrast will decrease. This concept once depended on the acuity of the eye, but now depends on the size of the image pixels relative to the resolution limit of the microscope/lens being used (>2.5 pixels/resel).

Epi-illumination (epi-fluorescence): excitation light is reflected off a 45 degree dichroic beamsplitter and through the objective to the specimen. Fluorescence or reflected light is detected after passing back through the objective lens, and directly through the dichroic beam splitter. The objective therefore also acts as a condenser. Conventional fluorescence microscopy, confocal fluorescence microscopy and back scatter imaging are all performed using epi-illumination.

Esterase derivatives: see "Acetoxymethyl (AM) ester".

Excitation: absorption of energy in the form of light causes an electron to jump from the ground state to a higher energy level (the excited state). The subsequent drop back to the ground state results in the emission of a photon of longer wavelength (lower energy) than that of the original photon that excited the electron.

Excitation filter: an optical filter (usually of the interference type) which limits the incident light used to excite the fluorochrome to specific wavelengths.

Excitation spectrum: the spread of wavelengths of light that can be used to excite the fluorophore.

Excitation wavelength: the wavelength of peak excitation for a fluorophore (for example, close to 488 nm for fluorescein).

Exit pupil: the point just above an eyepiece where an image is formed that can be viewed by placing the surface of the cornea of the eye at that position.

Extended focus image: a series of optical slices can be collected and displayed as a single composite image (simply added together or projected in a variety of different ways - see "Projection"). An extended focus image is usually only useful for samples with a relatively small number of fluorescent structures.

Extinction coefficient (ε): The amount of light absorbed by a particular molecule. The molar extinction coefficient is the optical density of a one molar solution of the compound through a one cm light path (usually quoted at the wavelength of maximum absorption).

Eyepiece graticule: a transparent disk marked with a scale or other measuring device that fits into the eyepiece at the level of the intermediate image, and is therefore superimposed on the image of the focus plane. Used for approximate measurements in microscopy. Different scales are needed for different magnification objectives.

Eyepieces (oculars): are lenses that project the intermediate image through the optics of the eye to form an image on the retina. A typical eyepiece has a magnification of 10x, but lower (5x) and higher (20x) oculars are common. The oculars are not used when scanning the laser in a confocal microscope to generate an image.

FACS (Fluorescence Activated Cell Sorting): is a machine that sorts (or counts) individual fluorescently tagged cells (normally using fluorescently conjugated cell specific antibodies) using a laser. Similar fluorochromes to those used in confocal microscopy (such as FITC, TRITC etc) can be used in a FACS instrument.

Fading: see "Photobleaching".

False colour: all colour displayed in digital images collected on a confocal microscope is "false" colour - i.e. the colour is added by the computer, and does not directly reflect the range of wavelengths of light directed to that particular channel.

Fast X-Y: fast scan speed on the Zeiss LSM 510 for locating the sample (increased y-scan speed, and decreased lines, resulting in larger pixels).

Fibreoptic coupling: the use of fibreoptic cable to connect various lasers to the confocal microscope scan head, and on some instruments to direct the fluorescent light from the scan head to an optics box where the light is separated into individual channels.

Field aperture: see "Field diaphragm".

Field curvature: an optical distortion where the centre of the image is in focus and the edges remain out of focus. A high quality "plan" objective will have a "flat" field of view, but at a penalty of more elements and less throughput (transparency). Out of focus areas of an image will not be "seen" by confocal microscopy - resulting in an image truncated around the edges when field curvature is present. A lens with field curvature can often be used successfully in confocal microscopy by using the centre of the field of view only (i.e. zooming the image).

Field diaphragm: an iris controlling the size of the illuminated field in the sample. Usually located close to the light port (between the condenser and the light bulb). In Köhler illumination the field diaphragm should be focussed at the same plane as the specimen.

Field iris: see "Field diaphragm".

File compression: used to reduce the storage space required by an image data file. Lossless techniques compress image data without removing detail (but may lose colour information); lossy techniques compress images by removing detail.

Filter (optical): an optical device that may change the intensity of light, or block specific ranges of wavelengths of light from passing through the filter. Optical filters may be simple coloured glasses, interference filters or tuneable acoustic-optical devices.

DF= Discriminating Filter LP= Long Pass
SP= Short Pass KP= Kurz (short)Pass
DRLP= Dichroic Long Pass EFLP= Long Pass Edge Filter
NB= Narrow Band OG= Orange Glass
RG= Red Glass WB= Wide Band
ND= Neutral Density
AOTF= Acoustic-Optical Tuneable Filter
AOBS= Acoustic-Optical Band Selection
AOD= Acoustic-Optical Deflector

FISH (Fluorescence In Situ Hybridisation): a technique used to locate genes on specific chromosomes, and to map genes relative to known genetic markers.

Fixation: a process of cross-linking (mainly aldehyde fixatives) or precipitating (organic solvents) the protein constituents of the cell to retain the structural integrity of cells. Care should be taken to retain not only 2D structural integrity, but also the 3D volume of the cell if 3D confocal microscopy is to be attempted. Glutaraldehyde, although resulting in excellent structural integrity, does create a high level of autofluorescence.

FLIM (Fluorescence Lifetime Imaging): is a powerful technique for the analysis of molecular interactions using FRET. The lifetime of the fluorescence is measured directly, which means FRET interactions can be monitored independent of the concentration of the individual molecules under study.

FLIP (Fluorescence Loss in Photobleaching): is used to study the molecular dynamics and connectivity of cellular membranes.

Flow cytometry: see "FACS, Fluorescence Activated Cell Sorting".

Fluorescence: the emission of light of specific wavelengths from a molecule that is irradiated with light of a shorter wavelength (single photon excitation). Multi-photon excitation results in shorter wavelength fluorescence emission after irradiation with high-energy pulsed longer wavelength infrared light. Fluorescence emission continues (with a half-life of 10^{-8} seconds) only as long as the irradiating light is present.

Fluorescence microscopy: a form of light microscopy using fluorescent dyes to highlight the area of the specimen of interest. Conventional (wide-field) epi-fluorescence microscopy is carried out using a mercury or xenon arc lamp. Laser scanning confocal microscopy usually, but not always, uses fluorescence imaging to detect the region of interest.

Fluorescence Recovery After Photobleaching: see "FRAP".

Fluorescence Resonance Energy Transfer: see "FRET".

Fluorescent probe: a fluorochrome, i.e. a molecule that exhibits fluorescence.

Fluorite: (CaF_2) a type of "glass" used in some microscope objectives, particularly for use in UV fluorescence microscopy. Fluorite lenses are only partially colour corrected. Not suitable for polarisation microscopy.

Fluorochrome: a molecule that exhibits fluorescence. Also known as a "fluorescent probe".

Fluorophore: the part of the molecule responsible for fluorescence. Fluorophores, such as fluorescein, rhodamine etc are often attached to other molecules such as antibodies to create a fluorescent probe, of which the fluorescein or rhodamine moity is the fluorophore, and the whole molecule is known as a fluorochrome.

Focal length: the distance between the optical centre of the lens and its focal point. Each lens has a focal point on each side of the lens.

Focal plane: an imaginary 2-dimensional plane at right angles to the optical axis and passing through the focal point. The focal plane can be thought of as the imaginary "screen" on which the image is formed. See also "back focal plane" and "front focal plane".

Focal point: the point on the optical axis at which light entering the lens parallel to the optical axis comes to a focus.

Focus: the ability of a lens to converge light rays to a single point. A high quality colour corrected lens will result in all wavelengths being focussed to a single point. The size of the "point", known as the "Point Spread Function", is determined by the NA of the lens and the wavelength of light (diffraction limited microscopy).

Focus motor: computer controlled motor to move the fine focus (or sometimes the course focus) on the microscope, used for collecting a series of optical slices.

Fourier space: a space containing the Fourier transform information of the object. The back focal plane of the objective is a Fourier space.

Fourier transform: the separation of an image into its spatial or frequency components. The Fourier transform image can be filtered and manipulated to alter or reduce periodic signals within the image, and then an inverse transform function performed to regain the original image (with selected frequencies accentuated or removed).

Frame grabber: (frame store) a computer card that allows images from video cameras, PMT tubes etc (analogue outputs) to be digitised and stored as a computer file.

FRAP (Fluorescence Recovery After Photobleaching): als o known as Fluorescence Photobleaching Recovery (FPR). This is a quantitative fluorescence optical technique used to measure the dynamics of 2D molecular mobility.

FRET (Fluorescence Resonance Energy Transfer): is a technique for studying molecular associations that are beyond the limit of light microscopy. The energy absorbed by a fluorophore (donor) is transferred to an absorber (acceptor) molecule which then emits light at a longer wavelength. The fluorophore (donor) and the absorber (acceptor) molecules must be within a few nanometres of each other, allowing one to study molecular associations using light microscopy.

Front Focal Plane (FFP): is the focal plane of a lens located on the side of the lens towards the object.

Gain: the level of signal amplification (alters the whiteness level of the image). Used in conjunction with the offset control (or black level) to produce an image with the best contrast.

Gamma: used to describe the relationship between the original signal and the display image. A gamma of 1 indicates that the relationship is linear. Changing the gamma can highlight areas of the image that show particularly low fluorescence intensities.

GEOG LUT: multi-coloured "Look Up Table" that has series of bands of colour ranging from cool colours (blue) to hot colours (yellow and white). Often used to highlight areas of interest within the image.

GFP: see "Green Fluorescent Protein".

GIF (Graphics Interchange Format) is a file format commonly used to display indexed-colour graphics and images in hypertext mark-up language (HTML) documents over the World Wide Web and other online services. GIF is an LZW-compressed format designed to minimize file size and electronic transfer time.

Gigabyte (GB or Gbyte): a unit of storage capacity, 1,073,741,824 (2^{30}) bytes.

GlowUnder and GlowOver (Leica): Look Up Table algorithm used by Leica confocal microscope software to highlight areas of the image that are at the maximum and minimum intensity level - used to accurately set the upper and lower light levels (gain and offset control).

Glycerol immersion objective: an objective designed for use with glycerol to form a continuum between the lens and the coverslip. Often used for glycerol mounted specimens (particularly immunolabelling) to reduce spherical aberration. Multi-variable (oil, glycerol, water) immersion lenses are also available.

Graticule: see "Eyepiece graticule".

Green Fluorescent Protein (GFP): is a very powerful method of "tagging" cellular proteins with a naturally fluorescent protein derived from the Jellyfish *Aequorea victoria*. The DNA sequence coding for GFP can be readily attached by gene fusion to any protein of interest, and the fate of the protein followed in real time in living cells. A number of derivatives of the original protein that have different wavelengths of emission, have been developed. These include RFP (Red Fluorescent Protein), YFP (Yellow Fluorescent Protein), CFP (Cyan Fluorescent Protein) etc. Also see "Reef Coral Fluorescent Protein" (RCFP).

Green LUT: green "Look Up Table" used to colour images with a gradient from dark green through yellow to white (often used to display green fluorophores such as fluorescein).

Grey levels: number of actual intensity values present in an image. This is the product of the number of sampling events (photons absorbed), Poisson noise and electronic noise. The maximum number of grey-levels may be 256 (8-bit), 4096 (12-bit), 65,536 (16-bit) or even higher.

Heat filter: an optical filter that blocks infrared radiation, but transmits visible light. They can be either absorption or interference (reflection) filters.

Histogram: see "Image histogram".

IFA (Immunofluorescence Antibody Assay): see "Immunofluorescence".

Image: An array of intensity values representing light intensity at spatial or temporal locations in the sample.

Image analysis: performing measurements on features in a stored image without changing the image itself.

Image histogram: a bar graph depicting the number of pixels at each grey level within the total image, or within a selected region of the image. The image histogram can be used to both analyse the distribution of fluorescence within the image and to manipulate the Look Up Table (LUT) levels within the image.

Image mathematics: image processing where whole images are added, subtracted, multiplied or divided into one another.

Image processing: altering the image to improve contrast, sharpness, feature extraction etc.

Image segmentation: the partitioning of a digital image into non-overlapping regions according to grey levels, texture etc.

Immersion media: the material used between the objective and the sample (air, water, glycerol, oil etc). The type of immersion media used is marked on the objective (no marking indicates air). Some objectives have a moveable collar that allows for adjustment for different immersion media.

Immersion oil: an oil with a refractive index of 1.515 that is used to provide a continuum of the same refractive index between the objective front lens and the coverslip. Different immersion oils are not usually miscible - if you change immersion oil make sure the lens is cleaned well before applying the new oil. The stated refractive index of the oil is at a specific temperature (usually 23°C).

Immunofluorescence (IFA): "Immunofluorescence Antibody Assay" (also known as immunolabelling) is where antibodies are bound to cells or tissue slices and are visualised by using fluorescently labelled secondary antibodies, or by directly labelling the antibody with a fluorescent probe. Immunolabelling provides a highly specific way of identifying the subcellular location of either the molecule of interest, or of the organelle or subcellular structure where the molecule is known to be located.

Indium Tin Oxide (ITO): an electrically conductive, optically clear coating used on coverslips for heating across the surface of the coverslip. ITO coated coverslips are used in the Bioptechs heated chambers.

Infinity optics: most modern microscopes now use infinity corrected objective lenses (parallel light rays) rather than the older method of having a fixed focal length. The Bio-Rad confocal microscope scan-head also uses infinity optics, allowing the "pinhole" to be located physically distant from the objective, and to be of relatively large, and variable, size (in millimetres).

Infinity-corrected objectives: have a tube-length of infinity and so require a separate lens to form an image. Their advantage is that the light from the image is kept as parallel rays until just before the eyepieces which means that optical elements can be introduced any-where along the light path without affecting the focus or magnification of the final image. Most modern microscopes now use infinity corrected objectives.

Infrared (IR): light from the region of the spectrum with wavelengths between 750 nm (red) and 0.1 mm (microwave).

Intensity of light: the flow of energy per unit area. Intensity is a function of the number of photons per unit area and their energy (the shorter the wavelength the higher the energy of the photon).

Interference: the interaction between one wave and another, resulting in the addition or subtraction of light in the overlapping area. Constructive interference is when two waves are in step, travelling and vibrating in the same direction, at the same time in the same location - resulting in greater intensity. Destructive interference is where the waves are out of step - resulting in less intensity. If the two waves are out of step by 1/2 wavelength then the peak of one will exactly co-inside with the trough of the second and result in zero intensity.

Interference filter: an optical filter that contains a series of optical coatings or layers that filter light by causing reflectance or destructive interference of specific wavelengths.

Interpolation: the addition or subtraction of pixels in an image to reduce aliasing effects when changing the image size. Generally preserves shapes by altering intensities in replicated or remaining pixels.

Inverted microscope: used for examining specimens in Petri dishes or incubation chambers. The condenser is mounted above the stage with the objectives underneath. Modern inverted microscopes produce images as good as those of upright microscopes. Using a confocal scan head on an inverted microscope is optically the same as using an upright microscope.

Isobestic point: a wavelength point in an absorption or fluorescence emission spectrum at which the indicator fluorescence is insensitive to ion binding (at this point the amount of absorbance or fluorescence is directly related to the dye concentration and is not influenced by the binding of specific ions).

Jaz disk: a relatively high-speed, reasonably high capacity (1 to 2 Gbyte) removable magnetic disk. Convenient for daily use, but expensive for long term storage. Available for SCSI, parallel and USB ports.

JPG (JPEG): the Joint Photographic Experts Group (JPEG) format is commonly used to display photographs and other continuous-tone images in hypertext mark-up language (HTML) documents over the World Wide Web and other online services. Unlike the GIF format, JPEG retains all colour information in an RGB image but compresses file size by selectively discarding data. A JPEG image is automatically decompressed when opened. A higher level of compression results in lower image quality.

Kalman averaging: an averaging algorithm which displays a "running average" as images are collected. Each image of the averaged set contributes a weight to the final result that is proportional to its position in the scan order. Kalman averaging greatly improves the signal to noise ratio in the image, and can also be used to follow movement – where the latest images collected create a "ghosting" effect where objects have moved.

KG: a short pass colour absorption glass that transmits visible light while attenuating both longer and shorter wavelength energy.

Kilobyte (KB or Kbytes): 1024 (2^{10}) bytes, 1000 Kb equal to 1 Mb (megabyte).

Köhler illumination: the most common type of illumination used in transmitted light microscopy, resulting in an evenly illuminated back focal plane of the objective for maximum resolution and an evenly illuminated background. Developed by the German biologist named Köhler (1866-1948).

Kompenzatione (K): notation located on the eyepieces of some microscopes, indicating that the eyepiece is responsible for some of the colour correction of the objective lens. German for "compensating".

Kurz pass (KP): (German for Short Pass) an optical filter that allows short wavelengths to pass. For example KP 490 would indicate that all wavelengths shorter than 490 (violet) pass through, whereas longer wavelengths are blocked.

Laser emission: all lasers emit light at discreet wavelengths (laser lines). Different lasers emit at different wavelengths, often emitting several different discrete wavelengths. The argon-ion laser has two main emissions, 488 nm (blue) and 514 nm. The Kr/Ar laser has 3 main emissions at 488 nm (blue), 568 nm (yellow) and 630 nm (red).

Lateral resolution: resolution in the plane of focus (x-y resolution). Lateral resolution is proportional to the inverse of the NA of the objective (i.e. high NA objectives have significantly higher lateral resolution).

Latex beads: small latex (plastic) spheres (also called latex microspheres or FluoSpheres) from 0.02 µm up to 15 µm in diameter and stained with a range of fluorescent dyes. Used to study endocytosis and intracellular movement. Also used as a very sensitive method of antigen detection by surface coating with a suitable antibody. Sub-resolution latex beads can also be used to measure the PFS (resolution) of the microscope.

Line averaging: each line is scanned several times, and the average is displayed. This will increase both the sensitivity and the signal to noise ratio of the image. The advantage over screen averaging is that small movements (for example, when imaging live cells) will not detract unduly from the quality of the image.

Line scanning: continual scanning of a single, fixed-line, can be used for rapidly monitoring the level of fluorescence. The image can be collected as a line-time image in which the horizontal (x) axis represents a physical line across the sample, but the vertical (y) axis represents time – i.e. each subsequent line collected is displayed one below the other.

Line selection: some lasers have several specific wavelengths of light available (lines). A laser line selection filter wheel, AOTF filter or AOBS filter allows one to select individual lines or combinations of lines for multi-labelling applications.

Linear unmixing: a mathematical process whereby fluorescent light can be re-allocated back to the correct channel in multi-labelling applications.

Local area contrast: computer algorithm that increases image contrast over a small area. Can be useful for enhancing very small differences in intensity, but may dramatically increase any spotty or patchy areas of the image.

Local contrast enhancement: image-processing algorithms that increases image contrast and sharpness over a selected area.

Long Pass filter: see "LP"

Long Working Distance (LWD): an objective that has an unusually long distance between the objective lens front element and the object (coverslip). This type of lens is especially useful for tissue culture and microinjection work. A long working distance objective is usually significantly more expensive and may have a lower NA (lower resolution) compared to a more conventional working distance lens.

Look Up Tables (LUT): computer algorithm for adding colour to your images. Each grey level in the image is denoted a specific colour or shade of colour for creating colour gradients. A dark green - green - yellow - white gradient LUT can be used for displaying green fluorochromes such as FITC. Other LUTs can be used to delineate particular areas of interest, for example a LUT may be used to set the correct dynamic range during collection by applying discrete colour to pixels that are under- or over-saturated. A LUT is applied to grey-scale images without the intensities being altered. The LUT coloured image can be converted to an RGB (Red, Green, Blue) colour image.

Low contrast image: an image comprised of midrange grey tones with little or no black or white in the image.

LP (Long Pass) optical filter: allows longer wavelength light to pass through and blocks the transmission of shorter wavelength light.

LSCM: Laser Scanning Confocal microscopy.

LUT: see "Look Up Tables".

Magenta: a light purple colour formed when green light is subtracted from white light (leaving red + blue). CMYK (Cyan, Magenta, Yellow and Black) inks are used in most printers.

Magnification: an image that is the result of an enlarged view of the object. An increase in image size beyond the resolution of the microscope will result in a larger image, without any increase in resolution (see "Empty magnification"). Historically the "magnification", denoted for example as 1000X was displayed along with the image. However, a more accurate method of displaying the magnification is to use a scale bar within the image. In this way changes in "magnification" due to the size at which the image is printed or displayed do not affect the denoted scale of the image.

Median filter: image processing method that has the property of smoothing the image (removing noise) while maintaining edges. This algorithm has a tendency to create "false" edges in the image and so should be used with care.

Megabyte (MB or Mbyte): 1,048,576 (2^{20}) bytes.

Mercury lamp: powerful arc lamp used as a light source in normal fluorescence microscopy. The emission spectrum is from the UV to the far red. The Hg arc lamp has a limited life span of approximately 200 hours, and should be changed at the time recommended by the manufacturers.

Micro-lens array: the Yokogawa Nipkow spinning disk confocal microscope scan head also incorporates a micro-lens array disk that focuses the irradiating light through the pinholes in the Nipkow disk - thus increasing the amount of light available at the sample for exciting fluorescent dyes.

Micron (μ or μm): 1μm = 1000 nm (10^{-6} meters). The limit of resolution of light microscopy is 0.1 to 0.2μm (100 to 200 nm), depending on the wavelength of light and the NA of the objective.

Multi-photon: excitation of a fluorophore using long wavelength infrared light to produce short wavelength blue or green light. Also known as 2-photon and 3-photon excitation. Multi-photon microscopes use very short pulses of high-intensity infrared light to excite fluorescent probes. Only the focus point has sufficient light intensity for multi-photon fluorescence to occur - which results in only the focal plane being visible, i.e. optical sectioning is achieved by selectively exciting the focus plane rather than eliminating the out-of-focus light later as in a confocal microscope. A multi-photon microscope is particularly suited to relatively deep penetration of whole tissue samples, and can be used to excite "UV" dyes such as DAPI as well as visible wavelength dyes.

Multi-tracking: image collection using laser line switching to collect alternate channels with pre-set laser line excitation and collection optics (for eliminating bleed-through between channels). High-speed electronic switching of laser lines can be used to collect alternate image lines with different laser settings.

NA: see "Numerical Aperture".

Nanometre (nm): unit of length used to measure wavelengths of light. 10^{-9} meters. The Kr/Ar laser produces 488 nm (blue), 568 nm (yellow) and 647 nm (red) lines.

NB (Narrow Band) filter: allows a narrow range of wavelengths to pass through. For example a 605DF32 (discriminating filter) is an optical filter that allows a narrow band of light (32 nm) to pass, centred around a wavelength of 605 nm.

ND (Neutral Density) filter: optical filter that blocks light of all wavelengths equally. Used to attenuate the amount of laser light reaching the sample.

Near Infrared (NIR): the light spectrum from approximately 750 to 2500 nm.

Nipkow disk: a disk with a spiral pattern of holes arranged so that, as the disk spins, light shining through the disk scans every part of the specimen with a series of small points of light. A number of confocal microscopes are based on the use of a Nipkow disk, either alone or in combination with a micro-lens array disk.

Noise: in a confocal microscope noise is due to both optical or shot (Poisson) noise (due to the statistical nature of light, and of particular concern when the fluorescence intensity of the sample is relatively low), and electronic noise of the instrument (this is usually quite low, unless the gain level is set too close to maximum). Noise is manifested as "speckle" in the image.

Nomarski optics: see "Differential Interference Contrast (DIC)" optics.

Normal incidence: an angle of incidence of zero degrees.

Numerical Aperture (NA): is the sine of the angle under which light enters the objective. This number determines the amount of light an objective lens lets through (the light gathering power of the lens) and is also related to the resolving power of the objective (the higher the NA the better the resolution). The highest resolution lens generally available is an oil immersion 60x or 63x NA 1.4.

Nyquist criteria: (also Nyquist sampling) pixel resolution should be set at half the separation of the optical resolution (this means an optical resolution of 0.2μm would require a pixel size of no large than 0.1μm). In practical terms you should aim for approximately four pixels across the object you wish to resolve. This means 16 pixels for 2D images (4 x 4 pixels) and 64 pixels for 3D images (4 x 4 x 4 pixels).

Objective: a complex system of lens components that produces most of the magnification, and its numerical aperture (NA) limits the resolution achievable. High NA lenses are very expensive, delicate and are critically important for a good quality image.

Ocular: see "Eyepieces"

OG (Orange Glass): long pass colour absorption glasses that absorb more than 99.999% of light of shorter wavelength energy (blue light).

Oil immersion objective: an objective lens designed for use with oils that have a refractive index equal to that of glass (1.515). Effectively this means that the lens is in continuum with the coverslip. Resolution and contrast is not affected by small movements of the objective (as when focusing on different parts of the specimen).

Optical density: a logarithmic unit of transmission. OD = -log (T).

Optical sectioning (optical slices): the process of recording images at different focus planes through the specimen using a confocal or multi-photon microscope. A stack of optical sections can then be reconstructed into a 3D representation of the object by using suitable software.

Parfocal: is where two or more objective lenses are in focus at the same focus control position – particularly useful when changing lenses.

PDF (Portable Document Format): Adobe's electronic publishing document format software for Windows, Mac OS, UNIX®, and DOS. You can view PDF files using the free Acrobat Reader® software. PDF files can represent both vector and bitmap graphics, and can contain electronic document search and navigation features such as electronic links. PDF documents are generated using the Adobe Acrobat software.

Peak emission or excitation wavelength: wavelength of maximum emission/excitation of a fluorophore. This is the emission/excitation wavelength usually give in tables of fluorophores.

Permeabilisation: cells must be permeabilised to allow large molecules such as antibodies to penetrate to the cellular interior. Detergents, organic solvents or microwave heating are often used to permeabilise fixed cell and tissue samples. Streptolysin O or some detergents can be used to permeabilise live cells.

Phase annulus: a ring shaped aperture placed in the condenser to produce illumination for phase contrast microscopy. A specific phase annulus is required to match the phase plate in the objective.

Phase contrast: uses the retardation of light by the specimen to produce phase differences, which are converted into contrast. This technique is used extensively to image unstained live cells. However, phase contrast imaging is rarely used on the confocal microscope as contrast can be readily increased by altering the gain and blackness levels on the instrument, or alternatively very high quality transmission images can be obtained using DIC optics.

Phase plate: the glass plate placed at the back focal plane of a phase objective that contains the phase ring.

Phase ring: a darkened ring on the phase plate which (normally) retards light less than the phase plate. This ring is used in phase contrast microscopy together with the phase annulus to produce contrast from phase differences between light rays from different parts of the specimen and the background.

Photobleaching (fading): loss of fluorescence from the area of the sample that has been intensely irradiated with light. Caused predominantly by the production of reactive free radicals. The rate of fading can be controlled by addition of antifade compounds (anti-oxidants) such as *n*-propyl-gallate or *p*-phenylenediamine for fixed samples, or Vitamin C (sodium ascorbate) for live cells. The amount of bleaching is dependent on the fluorescent molecules used, the amount of light used to illuminate the object, and the length of time the sample is irradiated.

Photodynamic therapy (PDT): a cancer treatment based on using photosensitising chemicals (specific fluorophores) to kill cells that are exposed to light of a specific wavelength. Also called photoradiation therapy, phototherapy or photochemotherapy.

Photomultiplier tube (PMT): the photoelectric device used to detect light by converting light into an electric current that can be amplified. Each separate channel in a confocal microscope has its own PMT tube, except in the case of the META channel where a multi-PMT array is used to detect the light.

Photon counting: a pulse counting mode whereby the PMT circuitry is set to only count pulses over a preset threshold. One pulse correlates with a single photon. This method of image acquisition is only useful in relatively low light level images, and results in a significant reduction in noise levels.

Photon of light: a quantum of light, based on Planck's quantum theory of light.

PIC (Bio-Rad): file format for Bio-Rad confocal images, not to be confused with other PIC formats. Can be readily converted to many other formats using Confocal Assistant or LaserSharp. Many image processing programs can import Bio-Rad PIC files directly.

Pincushion distortion: a geometrical distortion of the image on the computer screen that makes a square appear to bow inwards. On better quality monitors this pincushion distortion can be altered by using controls on the monitor. Pincushion effects can be quite severe when photographing the screen directly - in which case avoid rectangular or square line objects in the image.

Pinhole: see "Confocal aperture".

Pinholes: small breaks in the coating of an interference filter.

Pixel: "Picture Element", is the smallest unit of a digital image. Pixels may be described by their spatial or temporal coordinates and intensity.

Plan: or planar, refers to a lens that has been corrected so that the resultant image is in-focus across the whole field of view.

Plan Apo: plan apochromatic objective lenses are highly corrected for colour (red, green, blue and UV), spherical aberrations (for green and red or more wavelengths) and flatness of field.

Plane polarised light: light oscillating in one plane only.

PMT: see "Photomultiplier tube".

Point Spread Function (PSF): the three-dimensional diffraction limited shape formed by an objective lens in a light microscope. The PFS is determined both by the lens and the media in which the sample is mounted.

Polar: anything that transmits light in one plane only. They are often made of a stretched plastic film sandwiched between two layers of glass, which transmits light oscillating in a plane parallel to the stretch.

Polarisation beam splitter: an optical filter that deflects the polarised laser light to the sample while at the same time allowing the randomly polarised fluorescent light emanating from the sample to pass through to the detectors.

Polariser: a polar placed in the incident light (usually below the condenser) to produce plane polarised light.

Polarisation: restriction of the orientation of the vibration of the electromagnetic waves in light. When vibrations are restricted to one particular angle, the light is plane-polarised.

Polychroic: a dichroic beamsplitter that has multiple reflection bands and transmission regions.

Primary colour: the colours red, green and blue (RGB). These colours are denoted by specific wavelengths within the electromagnetic spectrum and are perceived as separate colours due to the method of light detection in the human eye (the red, green and blue colour cones). Computer screens generate the large range of colours visible by simply mixing the amount of red, green and blue light present at each pixel.

Primary image plane: the image plane where an image of the specimen is first formed.

Projection: the process of displaying a series of optical sections as a single composite image. The composite image can be "projected" in different ways, including "average view", "maximum intensity", or "topographical".

Pseudocolour: the colouring of a grey-scale image by assigning specific grey levels to specific colours or gradients of colour using a Colour Look Up Table (LUT). For example the grey-scale image collected from FITC fluorescence can be coloured with a colour gradient of dark green, green, yellow and white to denote the degree of fluorescence intensity – but the image could just as easily be coloured blue through to white!

PSF: see "Point Spread Function".

Quantum Dots: are semiconductor nanocrystals that can be excited by blue light (for example, 488 nm) and emit fluorescence in a narrow emission from green through to far red, depending on the size (approximately 10nm) and composition of the quantum dot. Quantum dots can be coated with a protective outer layer, and then surface labelled with a variety of biological molecules, including Biotin and Streptavidin.

Quantum efficiency: the efficiency with which a fluorophore converts absorbed light into emitted fluorescent light.

Quantum yield: the fraction of excited fluorophore molecules that emit a photon of light.

Quarter wave plate: retards light by a quarter of a wavelength, producing circularly polarised light.

Quenching: the loss of fluorescence caused by a variety of phenomena, particularly the loss of energy by production of heat. Anti-fluorophore antibodies can be used to quench fluorescence (useful for studying accessibility to externally added ligands).

Raster: the sequential scanning pattern used in confocal microscopy and video imaging.

Ratiometric dye: a fluorescent dye with either two different excitation or two different emission wavelengths that change in fluorescence intensity, but in opposite directions, on binding specific ions. Calculating the ratio of the fluorescent light emitted at the two wavelengths, or at one wavelength in the case where two excitation wavelengths are used, can be used to accurately establish the amount of fluorescent change that is due to binding specific ions (for example calcium) independent of the concentration of the dye.

Rayleigh criteria: two objects are said to be resolved if their separation is such that their diffraction patterns (Airy disks) show a detectable drop in intensity between them. This somewhat arbitrary criterion was decided to be approximately 20% of the peak intensity, which corresponds to the first dark ring of the Airy disk.

Real image: an image that is located on the opposite side of the lens from the object and can be projected onto a camera film or the retina of your eye.

Real time collection: fast collection of images showing molecular fluxes (e.g. Ca^{2+}) or movement of subcellular components within cells. The collected images can be displayed as a time lapse "movie".

Red, Green and Blue: see "RGB".

Reef Coral Fluorescent Protein: (RCFP) is a series of fluorescent proteins cloned from various reef coral species (*Discosoma* sp) that can be readily attached by gene fusion to any protein of interest, allowing one to follow the fate of the protein in real time in living cells. Also see "Green Fluorescent Protein" (GFP).

Reflectance imaging: see "Backscatter".

Refraction: the change in direction of oblique light passing through a material of a different refractive index. See "Snell's law".

Refractive index (□ or RI): the ratio of the speed of propagation of light through a vacuum to that through the specimen. Vacuum = 1.0, glass = 1.5, water = 1.333, immersion oil = 1.515. The direction of travel of light is altered (refracted) on moving from media with one refractive index to another.

Registration shift: a shift in the apparent position of the specimen when an optical element is inserted or removed.

Resel: diffraction limited resolution limit of light microscopy. This depends primarily on the NA of the lens. The resel limit for a 1.4 NA oil immersion lens is 0.1 to 0.2 □m.

Resolution: the ability of an optical system to distinguish fine detail in a specimen (i.e. the ability to image separately two neighbouring points of information). The resolution limit of an objective is determined by the Numerical Aperture (NA) of the lens (the higher the NA the better the resolution) and the wavelength of light used (shorter wavelengths results in higher resolution).

RG (Red Glass): long pass colour absorption glass that absorbs more than 99.999% of shorter wavelength energy. RG glasses absorb blue and green light (thus appearing red).

RGB (Red / Green / Blue): colours used on the computer screen to generate all other colours. Also denotes an image file consisting of 3 images representing the red, green and blue components of a multi-colour image.

RLE compression: Run Length Encoded image compression algorithm (used by Windows BMP file format).

ROI (Region Of Interest): a defined area (or several defined areas) of an image may be used to scan only that region of the sample. The defined area may be a circle, square, line or a randomly drawn shape.

Saturation: see "Colour saturation" and "Dye saturation".

Scan head: the confocal microscope scan head contains the scanning mirrors, and for most manufacturers the dichroic mirrors and sometimes the barrier filters. The scan head is mounted directly on to a conventional light microscope.

Scan speed (scan rate): the speed of scanning the image. The final image scan rate is determined by the scan speed of the scanning mirrors, or in the case of a Nipkow disk instrument the speed of the spinning disk, and whether line averaging or binning has been implemented.

Scanning mirrors: mirrors in the scan head that scan the laser fast in the "x" direction and more slowly in the "y" direction.

Scanning zoom: a real increase in resolution (and magnification) can be obtained by zooming the scanned image (scanning a smaller area) in laser scanning confocal microscopy. When in zoom mode the scan area can be moved by using the panning keys.

Screen averaging: screen averaging is used to collect multiple screens and display them as an average (thus lowering the noise level in the image).

Secondary colours: colours derived by mixing the three primary colours. Red + blue = magenta (light purple), red + green = yellow, blue + green = cyan (light blue). The secondary colours can also be thought of as the "inverse" of the primary colours, for example, cyan is the full visible light spectrum without green. Most printers use the ink colours CMYK (Cyan, Magenta, Yellow and Black).

Segmentation: see "Image segmentation".

Semi-silvered mirror: a mirror that reflects half the incident light and transmits the rest.

Serial sections: adjacent physical or optical sections of tissue or cells.

SETCOL (Bio-Rad): Bio-Rad "Set Colour" Look Up Table (LUT) used to accurately set the upper and lower light levels (gain and black level) in the image. The lowest light levels are displayed green and the highest light levels red, with a grey scale gradient in between. An ideal image will contain a touch of green and a touch of red with most of the image displayed as shades of grey.

Sharpening: image-processing method used to enhance edges or other transitions in the image. This type of image manipulation is often useful when producing a slide for presentation, as "blurred" images are particularly distracting for the audience.

Short Pass optical filter: see "SP"

Signal-to-noise ratio (S/N): is the ratio of the signal coming from the sample and the unwanted signal caused by various optical and electronic components within the microscope and associated electronics. An image with a higher signal to noise ratio is a better quality image.

Slide maker: an instrument for making photographic slides from digital files. The instrument consists of a high-resolution grey-scale monitor that is photographed using colour filter wheels to create a colour image. Image quality is better than that obtained by the best colour monitors. Also known as a "film recorder".

Smoothing: an image-processing method to remove high frequency noise, or widely variant pixel values. This often has the effect of blurring the image, but is particularly useful for making noisy, or "grainy" images look much better.

Snell's law: light bends towards the normal (an imaginary reference line drawn perpendicular to the surface) as light passes from lower to higher refractive index material.

SP (Short Pass) optical filter: allows shorter wavelength light to pass through and blocks the transmission of longer wavelength light.

Spectrofluorometer: an instrument for measuring the excitation (absorption) and emission spectra of fluorescent molecules.

Spectrum: the visible light spectrum ranges from violet light (400 nm) through green (500 nm) to red (700 nm). Ultraviolet (less than 400 nm) and infrared (greater than 700 nm) are wavelengths of light often used in microscopy that are beyond the limits of the human eye to detect.

Spherical aberration: the inability of a lens to focus axial and marginal light rays to the same point. Spherical aberration can be caused by the objective itself, the improper use of the objective, incorrect immersion media or the sample. In confocal microscopy spherical aberration is manifested as a serious loss of light in the image or allocation of fluorescence to the wrong image plane, often in particular areas of the image or in the periphery of the field of view.

Spring LUT (Bio-Rad): Bio-Rad green Look Up Table (LUT). The image is coloured with a gradient of dark green - light green - yellow - white.

Standard tube-length optics: a fixed focal length of the objective (usually 160 mm). This is in contrast to infinity-corrected objectives, where the light rays are focussed to infinity (parallel light rays). Most modern optical microscopes now use infinity corrected optics.

Stereology: a type of image analysis based on the statistical analysis of numbers of objects, their size, and orientation.

Stokes shift: the difference between the wavelength of excitation and the wavelength of emission of a fluorophore.

Three dimensional (3D) reconstruction: computer software rendition of a 3D volume of interest contained within a z-series of digital images.

Three-photon: see "Multi-photon".

Thresholding: selection of regions of interest within an image based upon pixel grey level value. Thresholding usually selects pixels above or below a single specified level, sometimes called "binary thresholding". Greyscale thresholding, or "density slicing", selects pixels lying within the limits of a range of intensities. The specified grey levels are usually replaced with a solid area of colour.

TIFF (Tagged-Image File Format): is used to exchange files between applications and computer platforms. TIFF is a flexible bitmap image format widely supported by virtually all paint, image editing, and page-layout applications and does not alter image intensities or colour table. Can be either compressed or uncompressed.

Transmission fluorescence: fluorescence microscopy where the illuminating light is transmitted through the condenser (as in conventional Bright-field microscopy) and the fluorescent light is collected by the objective. This type of fluorescence microscopy is quite dangerous for the user, as a correctly placed barrier filter is critical for stopping the powerful illuminating light (often UV) from entering the observers eyes (or the PMT tube used in confocal microscopy). Transmission fluorescence is now only used in very specialised instruments.

Transmission image: a transmission image of the object in a confocal microscope can be obtained by collecting the laser light that passes through the condenser. A high quality transmission image of living cells can be obtained by collecting a very evenly lit low contrast image (Kalman average if possible) and then contrast stretching before saving. The transmission image is NOT confocal (i.e. the transmission image contains light from other focal planes – it is not an optical slice). DIC and Phase contrast images can also be collected by using the necessary optical components on the microscope.

Tube length: the physical distance between the objective and the eyepiece.

Tungsten lamp: the lamp traditionally used for conventional transmitted light microscopy.

Two-photon: see "Multi-photon".

Ultraviolet (UV): light from the region of the electromagnetic spectrum with wavelengths between 100 and 400 nm (not visible to the human eye). UV light is highly destructive to living tissue, and so must be used with care on living cells. However, there are many excellent dyes available that require UV irradiation to generate fluorescence. Confocal microscopes that use a UV laser are available, but they are difficult and expensive to run. UVA = 320 to 380 nm, and UVB = 280 to 320 nm.

UV laser: UV lasers are very expensive, but do allow one to use UV excited fluorophores such as Hoechst dyes and various calcium indicators. Many UV dyes can also be excited by using a multi-photon microscope or the violet (405 nm) solid state laser now available.

Virtual image: a magnified image located on the same side of the lens as the object. The image cannot be projected onto film, but can be seen when looking through the lens (an example of a virtual image is that seen when looking through a hand held magnifying glass that is held relatively close to the specimen).

Visible light spectrum: light from the region of the electro magnetic spectrum with wavelengths between 400 nm (blue) and 750 nm (red).

Vital stain: a dye that is tolerated by living material. Many fluorescent stains can be used as vital dyes as long as a sufficiently low concentration of the dye is used. Some dyes are excluded from living cells, but will readily stain dead cells (the basis of the commercially available Live/Dead staining kits).

Voxel: a three dimensional pixel (Volume Element) in a digital image stack.

Water dipping objective: an objective designed to be immersed into the culture dish from above (used for microinjection and electrophysiology). This type of lens is usually made of a ceramic material and is used without a coverslip.

Water immersion objective: an objective lens designed for use with water to form a continuum between the lens and the coverslip. Water immersion objectives may result in better resolution when imaging water based biological samples. Some immersion objectives have an adjustable collar to allow them to be used with either water, glycerol or oil as the immersion media. A water immersion objective is designed to be used with the correct thickness coverslip, unlike the water "dipping" objective where no coverslip is required.

Wavelength of light (λ): The distance in nanometres between nodes in the wavelength of light. Shorter wavelength light has higher energy. The colour of the light is determined by the wavelength.

WB: "Wide Band" optical filters combine rectangular band shapes with broad regions of transmission.

White light: the combination of all three primary colours, red, green and blue. Illuminating all three primary colours together produces white on a computer screen. However, white on a printed page is produced by the absence of printing ink (displaying white paper).

Wide-field: an epi-fluorescence microscope in which the full field of view is illuminated.

Wollaston prism: prism used in interference imaging (see "Differential Interference Contrast (DIC)" microscopy), consisting of a beam splitter made of two wedges of a birefringent crystal such as quartz. The position of the Wollaston prism can be adjusted to create a lighter or darker "shadowing" effect, which is particularly valuable in live cell imaging. The Wollaston prism in some instruments may interfere with the collection of high-resolution fluorescence images.

Working distance: the distance between the front lens of the objective and the coverslip. A long working distance objective is very useful for microinjection, but usually has a lower NA value (lower resolution and lower light gathering capacity) and is more expensive compared to a normal short working distance objective.

X-Z scanning: the ability to collect an image in the X-Z plane by scanning a single line at a series of focal positions.

Yellow: a colour that looks exactly the same whether produced as a "pure" wavelength of light (around 560 nm), or a colour produced by the combination of red and green light (600 and 500 nm), which is often used to denote regions of co-localisation in a digital image. Yellow is an important colour used in printing (CMYK; Cyan, Magenta, Yellow and Black) inks are used in most printers.

Z-axis: the vertical axis in a microscope. Z-sections refer to a series of images collected by stepping the microscope fine focus with a focus motor drive, or by moving the microscope stage or objective lens with a piezo electric stepper.

Zip disk: a reasonably fast and relatively cheap disk with 100 or 250 MB capacity. The drive can be readily moved to different computers by attachment though the printer parallel port or USB port.

Zoom: method of increasing the resolution (and magnification) of the confocal microscope (scanning zoom) or of making a digital image larger without an increase in resolution (digital zoom).

Z-resolution: the resolution in the "z" direction (at right angles to the scan direction). The z-resolution on a confocal microscope is less than the x-y resolution, and is about 0.5 to 0.8 μm when using a high resolution 1.4 NA 63x oil immersion objective (the x-y resolution in this case would be approximately 0.1 to 0.2 μm).

Z-series: a series of 2D images encompassing a sample volume collected at intervals in the focal axis (z-stepping) by stepping the focus motor drive between acquisitions.

Index

Lightning Source UK Ltd.
Milton Keynes UK
UKOW07n2155040316

269619UK00003B/33/P